Student Solutions Manual

College Algebra

ELEVENTH EDITION

R. David Gustafson
Rock Valley College

Jeffrey D. Hughes
Hinds Community College

Prepared by

Michael G. Welden

BROOKS/COLE
CENGAGE Learning™

Australia • Brazil • Japan • Korea • Mexico • Singapore • Spain • United Kingdom • United States

BROOKS/COLE
CENGAGE Learning™

ISBN-13: 978-1-133-10347-9

ISBN-10: 1-133-10347-2

Brooks/Cole
20 Davis Drive
Belmont, CA 94002
USA

Cengage Learning is a leading provider of customized learning solutions with office locations around the globe, including Singapore, the United Kingdom, Australia, Mexico, Brazil, and Japan. Locate your local office at **www.cengage.com/global**

Cengage Learning products are represented in Canada by Nelson Education, Ltd.

To learn more about Brooks/Cole, visit **www.cengage.com/brookscole**

Purchase any of our products at your local college store or at our preferred online store **www.cengagebrain.com**

Printed in the United States of America
2 3 4 5 6 20 19 18 17 16

Contents

Preface

This manual contains detailed solutions to all of the odd exercises of the text *College Algebra*, eleventh edition, by R. David Gustafson and Jeff Hughes. It also contains solutions to all chapter review, chapter test, and cumulative review exercises found in the text.

Many of the exercises in the text may be solved using more than one method, but it is not feasible to list all possible solutions in this manual. Also, some of the exercises may have been solved in this manual using a method that differs slightly from that presented in the text. There are a few exercises in the text whose solutions may vary from person to person. Some of these solutions may not have been included in this manual. For the solution to an exercise like this, the notation "answers may vary" has been included.

Please remember that only reading a solution does not teach you how to solve a problem. To repeat a commonly used phrase, mathematics is not a spectator sport. You MUST make an honest attempt to solve each exercise in the text without using this manual first. This manual should be viewed more or less as a last resort. Above all, DO NOT simply copy the solution from this manual onto your own paper. Doing so will not help you learn how to do the exercise, nor will it help you to do better on quizzes or tests.

I would like to thank Paul McCombs from Rock Valley College and Cynthia Ashton of Brooks/Cole Publishing Company for their help and support. This solutions manual was prepared using EXP 5.1.

This book is dedicated to John, who helps me to realize that mathematics cannot describe everything in life.

May your study of this material be successful and rewarding.

Michael G. Welden

Exercises 0.1 (page 13)

1. set

3. union

5. decimal

7. 2

9. composite

11. decimals

13. negative

15. $x + (y + z)$

17. $5m + 5 \cdot 2$

19. interval

21. two

23. positive

25. Every natural number is a whole number, so $\mathbf{N} \subset \mathbf{W}$. **TRUE**

27. The rational number $\frac{1}{2}$ is **not** a natural number, so $\mathbf{Q} \not\subset \mathbf{N}$. **FALSE**

29. Every whole number is an integer, so $\mathbf{W} \subset \mathbf{Z}$. **TRUE**

31. $A \cup B = \{$a, b, c, d, e, f, g$\}$

33. $A \cap C = \{$a, c, e$\}$

35. $\frac{9}{16} = 0.5625$; terminates

37. $\frac{3}{11} = 0.272727...$; repeats

39. natural: $1, 2, 6, 7$

41. integers: $-5, -4, 0, 1, 2, 6, 7$

43. irrational: $\sqrt{2}$

45. composite: 6

47. odd: $-5, 1, 7$

49.

51.

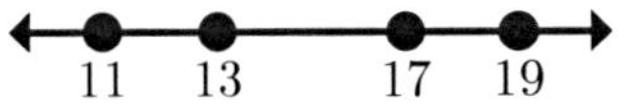

53.

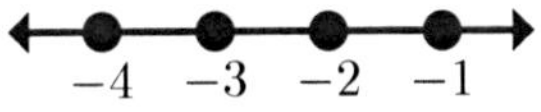

55.

57. $x > 2 \rightarrow (2, \infty)$

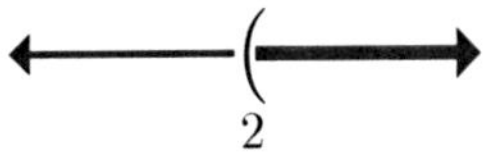

59. $0 < x < 5 \rightarrow (0, 5)$

61. $x > -4 \rightarrow (-4, \infty)$

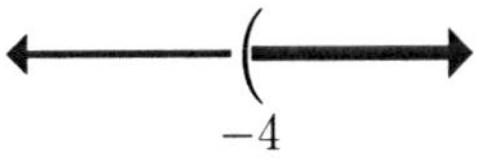

63. $-2 \le x < 2 \rightarrow [-2, 2)$

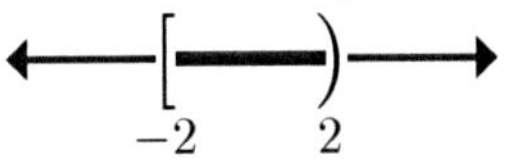

65. $x \le 5 \rightarrow (-\infty, 5]$

67. $-5 < x \le 0 \rightarrow (-5, 0]$

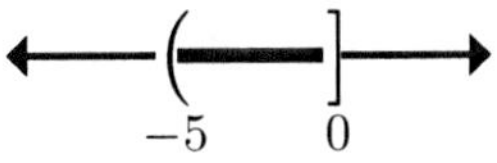

69. $-2 \le x \le 3 \rightarrow [-2, 3]$

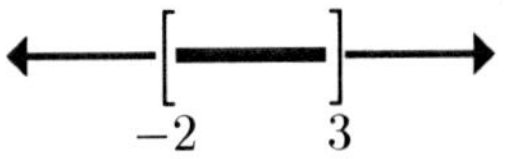

71. $6 \ge x \ge 2 \rightarrow 2 \le x \le 6 \rightarrow [2, 6]$

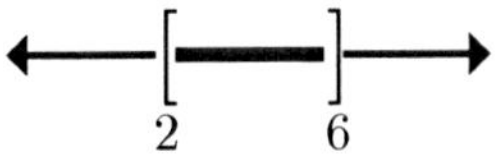

73. $x > -5$ and $x < 4 \rightarrow (-5, \infty) \cap (-\infty, 4)$

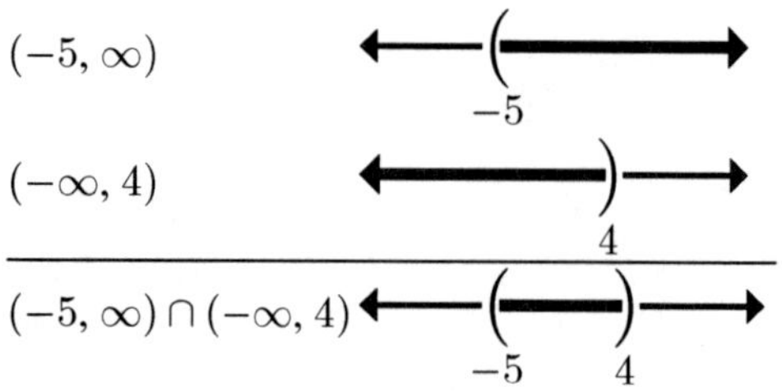

75. $x \geq -8$ and $x \leq -3 \rightarrow [-8, \infty) \cap (-\infty, -3]$

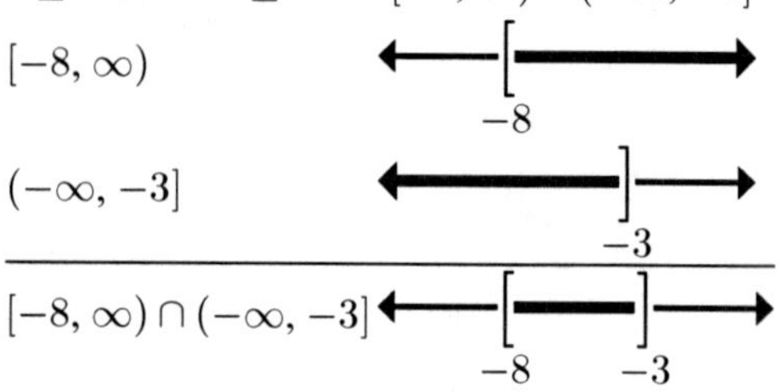

77. $x < -2$ or $x > 2 \rightarrow (-\infty, -2) \cup (2, \infty)$

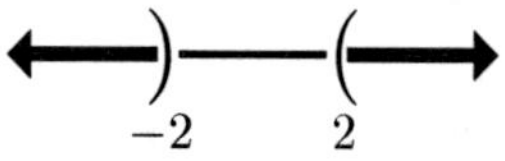

79. $x \leq -1$ or $x \geq 3 \rightarrow (-\infty, -1] \cup [3, \infty)$

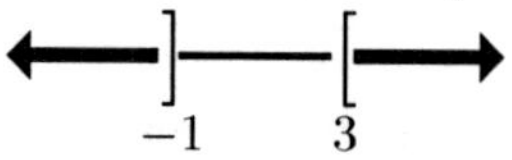

81. Since $13 \geq 0$, $|13| = 13$.

83. Since $0 \geq 0$, $|0| = 0$.

85. Since $-8 < 0$, $|-8| = -(-8) = 8$.
$-|-8| = -(8) = -8$

87. Since $32 \geq 0$, $|32| = 32$.
$-|32| = -(32) = -32$

89. Since $\pi - 5 < 0$,
$|\pi - 5| = -(\pi - 5) = -\pi + 5 = 5 - \pi$.

91. $|\pi - \pi| = |0| = 0$

93. If $x \geq 2$, then $x + 1 \geq 0$. Then
$|x + 1| = x + 1$.

95. If $x < 0$, then $x - 4 < 0$. Then
$|x - 4| = -(x - 4)$.

97. distance $= |8 - 3| = |5| = 5$

99. distance $= |-3 - (-8)| = |5| = 5$

101. Since population must be positive and never has a fractional part, the set of **natural numbers** should be used.

103. Since temperatures are usually reported without fractional parts and may be either positive or negative (or zero), the set of **integers** should be used.

105. $-x$ will represent a positive number if x itself is negative. For instance, if $x = -3$, then $-x = -(-3) = 3$, which is a positive number.

107. The statement is always true.

109. The statement is not always true. (For example, let $a = 5$ and $b = -2$.)

111. The statement $a < b > c$ could be interpreted to mean that $a > c$, when this is not necessarily true.

Exercises 0.2 (page 24)

1. factor

3. $3, 2x$

5. scientific, integer

7. $x^m x^n = x^{m+n}$

SECTION 0.2

9. $(xy)^n = x^n y^n$

11. $x^0 = 1$

13. $13^2 = 13 \cdot 13 = 169$

15. $-5^2 = -1 \cdot 5 \cdot 5 = -25$

17. $4x^3 = 4 \cdot x \cdot x \cdot x$

19. $(-5x)^4 = (-5x)(-5x)(-5x)(-5x)$

21. $-8x^4 = -8 \cdot x \cdot x \cdot x \cdot x$

23. $7xxx = 7x^3$

25. $(-x)(-x) = (-1)(-1)x^2 = x^2$

27. $(3t)(3t)(-3t) = (3)(3)(-3)t^3 = -27t^3$

29. $xxxyy = x^3y^2$

31. $2.2^3 = 10.648$

33. $-0.5^4 = -0.0625$

35. $x^2x^3 = x^{2+3} = x^5$

37. $(z^2)^3 = z^{2 \cdot 3} = z^6$

39. $(y^5y^2)^3 = (y^7)^3 = y^{21}$

41. $(z^2)^3(z^4)^5 = z^6z^{20} = z^{26}$

43. $(a^2)^3(a^4)^2 = a^6a^8 = a^{14}$

45. $(3x)^3 = 3^3x^3 = 27x^3$

47. $(x^2y)^3 = (x^2)^3y^3 = x^6y^3$

49. $\left(\frac{a^2}{b}\right)^3 = \frac{(a^2)^3}{b^3} = \frac{a^6}{b^3}$

51. $(-x)^0 = 1$

53. $(4x)^0 = 1$

55. $z^{-4} = \frac{1}{z^4}$

57. $y^{-2}y^{-3} = y^{-5} = \frac{1}{y^5}$

59. $(x^3x^{-4})^{-2} = (x^{-1})^{-2} = x^2$

61. $\frac{x^7}{x^3} = x^{7-3} = x^4$

63. $\frac{a^{21}}{a^{17}} = a^{21-17} = a^4$

65. $\frac{(x^2)^2}{x^2x} = \frac{x^4}{x^3} = x^{4-3} = x^1 = x$

67. $\left(\frac{m^3}{n^2}\right)^3 = \frac{(m^3)^3}{(n^2)^3} = \frac{m^9}{n^6}$

69. $\frac{(a^3)^{-2}}{aa^2} = \frac{a^{-6}}{a^3} = a^{-6-3} = a^{-9} = \frac{1}{a^9}$

71. $\left(\frac{a^{-3}}{b^{-1}}\right)^{-4} = \frac{(a^{-3})^{-4}}{(b^{-1})^{-4}} = \frac{a^{12}}{b^4}$

73. $\left(\frac{r^4r^{-6}}{r^3r^{-3}}\right)^2 = \left(\frac{r^{-2}}{r^0}\right)^2 = (r^{-2})^2 = r^{-4}$

$= \frac{1}{r^4}$

75. $\left(\frac{x^5y^{-2}}{x^{-3}y^2}\right)^4 = \left(\frac{x^5x^3}{y^2y^2}\right)^4 = \left(\frac{x^8}{y^4}\right)^4 = \frac{x^{32}}{y^{16}}$

77. $\left(\frac{5x^{-3}y^{-2}}{3x^2y^{-3}}\right)^{-2} = \left(\frac{3x^2y^{-3}}{5x^{-3}y^{-2}}\right)^2 = \left(\frac{3x^2x^3y^2}{5y^3}\right)^2 = \left(\frac{3x^5}{5y}\right)^2 = \frac{9x^{10}}{25y^2}$

79. $\left(\frac{3x^5y^{-3}}{6x^{-5}y^3}\right)^{-2} = \left(\frac{6x^{-5}y^3}{3x^5y^{-3}}\right)^2 = \left(\frac{2y^3y^3}{1x^5x^5}\right)^2 = \left(\frac{2y^6}{x^{10}}\right)^2 = \frac{4y^{12}}{x^{20}}$

81. $\frac{(8^{-2}z^{-3}y)^{-1}}{(5y^2z^{-2})^3(5yz^{-2})^{-1}} = \frac{8^2z^3y^{-1}}{5^3y^6z^{-6} \cdot 5^{-1}y^{-1}z^2} = \frac{64z^3y^{-1}}{5^2y^5z^{-4}} = \frac{64z^3z^4}{25y^5y^1} = \frac{64z^7}{25y^6}$

SECTION 0.2

83. $-\dfrac{5[6^2+(9-5)]}{4(2-3)^2} = -\dfrac{5[36+4]}{4(-1)^2} = -\dfrac{5[40]}{4(1)} = -\dfrac{200}{4} = -50$

85. $x^2 = (-2)^2 = 4$

87. $x^3 = (-2)^3 = -8$

89. $(-xz)^3 = [-1\cdot(-2)\cdot 3]^3 = 6^3 = 216$

91. $\dfrac{-(x^2z^3)}{z^2-y^2} = \dfrac{-[(-2)^2\cdot 3^3]}{3^2-0^2} = \dfrac{-[4\cdot 27]}{9-0} = \dfrac{-108}{9} = -12$

93. $5x^2-3y^3z = 5(-2)^2-3(0)^3(3) = 5(4)-3(0)(3) = 20-0 = 20$

95. $\dfrac{-3x^{-3}z^{-2}}{6x^2z^{-3}} = \dfrac{-1z^3}{2x^2x^3z^2} = \dfrac{-z}{2x^5} = \dfrac{-3}{2(-2)^5} = \dfrac{-3}{2(-32)} = \dfrac{-3}{-64} = \dfrac{3}{64}$

97. $372{,}000 = 3.72\times 10^5$

99. $-177{,}000{,}000 = -1.77\times 10^8$

101. $0.007 = 7\times 10^{-3}$

103. $-0.000000693 = -6.93\times 10^{-7}$

105. $1{,}000{,}000{,}000{,}000 = 1\times 10^{12}$

107. $9.37\times 10^5 = 937{,}000$

109. $2.21\times 10^{-5} = 0.0000221$

111. $0.00032\times 10^4 = 3.2$

113. $-3.2\times 10^{-3} = -0.0032$

115.
$$\begin{aligned}\frac{(65{,}000)(45{,}000)}{250{,}000} = \frac{(6.5\times 10^4)(4.5\times 10^4)}{2.5\times 10^5} = \frac{(6.5)(4.5)}{2.5}\times 10^{4+4-5} &= 11.7\times 10^3\\ &= 1.17\times 10^1\times 10^3\\ &= 1.17\times 10^4\end{aligned}$$

117.
$$\begin{aligned}\frac{(0.00000035)(170{,}000)}{0.00000085} = \frac{(3.5\times 10^{-7})(1.7\times 10^5)}{8.5\times 10^{-7}} &= \frac{(3.5)(1.7)}{8.5}\times 10^{(-7)+5-(-7)}\\ &= 0.7\times 10^5\\ &= 7\times 10^{-1}\times 10^5 = 7\times 10^4\end{aligned}$$

119.
$$\begin{aligned}\frac{(45{,}000{,}000{,}000)(212{,}000)}{0.00018} = \frac{(4.5\times 10^{10})(2.12\times 10^5)}{1.8\times 10^{-4}} &= \frac{(4.5)(2.12)}{1.8}\times 10^{10+5-(-4)}\\ &= 5.3\times 10^{19}\end{aligned}$$

121.
$$\begin{aligned}3.31\times 10^4\text{ cm/sec} = \frac{3.31\times 10^4\text{ cm}}{1\text{ sec}}\cdot\frac{1\text{ m}}{100\text{ cm}}\cdot\frac{60\text{ sec}}{1\text{ min}} &= \frac{(3.31\times 10^4)(6\times 10^1)}{1\times 10^2}\text{ m/min}\\ &= \frac{(3.31)(6)}{1}\times 10^{4+1-2}\text{ m/min}\\ &= 19.86\times 10^3\text{ m/min}\\ &= 1.986\times 10^4\text{ m/min}\end{aligned}$$

123. mass $= 1{,}000{,}000{,}000(0.00000000000000000000000167248\text{ g})$
$= \left(1 \times 10^{9}\right)\left(1.67248 \times 10^{-24}\text{ g}\right) = 1.67248 \times 10^{-15}\text{ g}$

125. Earth: $n = 3$
$d = 9{,}275{,}200\left[3\left(2^{n-2}\right)+4\right]$
$= 9{,}275{,}200\left[3\left(2^{3-2}\right)+4\right]$
$= 9{,}275{,}200\left[3\left(2^{1}\right)+4\right]$
$= 9{,}275{,}200[10]$
$= 92{,}752{,}000 \approx 93{,}000{,}000$
9.3×10^{7} mi

Mars: $n = 4$
$d = 9{,}275{,}200\left[3\left(2^{n-2}\right)+4\right]$
$= 9{,}275{,}200\left[3\left(2^{4-2}\right)+4\right]$
$= 9{,}275{,}200\left[3\left(2^{2}\right)+4\right]$
$= 9{,}275{,}200[16]$
$= 148{,}403{,}200 \approx 148{,}000{,}000$
1.48×10^{8} mi

127. polar radius $= 6.3567505 \times 10^{3}$ km
equatorial radius $= 6.378135 \times 10^{3}$ km

129. $x^{n}x^{2} = x^{n+2}$

131. $\dfrac{x^{m}x^{2}}{x^{3}} = \dfrac{x^{m+2}}{x^{3}} = x^{m+2-3} = x^{m-1}$

133. $x^{m+1}x^{3} = x^{m+1+3} = x^{m+4}$

135. In the expression $-x^{4}$, the base of the exponent is x, while in the expression $(-x)^{4}$, the base of the exponent is $-x$.

137. $(-2, 4) \Rightarrow$ 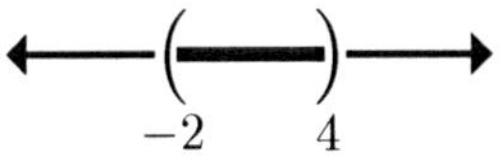

139. Since $\pi - 5 < 0$, $|\pi - 5| = -(\pi - 5) = -\pi + 5 = 5 - \pi$.

Exercises 0.3 (page 38)

1. 0

3. not

5. $a^{1/n}$

7. $\sqrt[n]{ab}$

9. $\neq$

11. $9^{1/2} = \left(3^{2}\right)^{1/2} = 3$

13. $\left(\dfrac{1}{25}\right)^{1/2} = \left[\left(\dfrac{1}{5}\right)^{2}\right]^{1/2} = \dfrac{1}{5}$

15. $-81^{1/4} = -\left(3^{4}\right)^{1/4} = -3$

17. $(10{,}000)^{1/4} = \left(10^{4}\right)^{1/4} = 10$

19. $\left(-\dfrac{27}{8}\right)^{1/3} = \left[\left(-\dfrac{3}{2}\right)^{3}\right]^{1/3} = -\dfrac{3}{2}$

21. $(-64)^{1/2} \Rightarrow$ not a real number

23. $\left(16a^{2}\right)^{1/2} = \left[(4a)^{2}\right]^{1/2} = 4|a|$

25. $\left(16a^{4}\right)^{1/4} = \left[(2a)^{4}\right]^{1/4} = 2|a|$

27. $\left(-32a^{5}\right)^{1/5} = \left[(-2a)^{5}\right]^{1/5} = -2a$

29. $\left(-216b^{6}\right)^{1/3} = \left[\left(-6b^{2}\right)^{3}\right]^{1/3} = -6b^{2}$

SECTION 0.3

31. $\left(\dfrac{16a^4}{25b^2}\right)^{1/2} = \left[\left(\dfrac{4a^2}{5b}\right)^2\right]^{1/2} = \left|\dfrac{4a^2}{5b}\right| = \dfrac{4a^2}{5|b|}$

33. $\left(-\dfrac{1{,}000x^6}{27y^3}\right)^{1/3} = \left[\left(-\dfrac{10x^2}{3y}\right)^3\right]^{1/3} = -\dfrac{10x^2}{3y}$

35. $4^{3/2} = \left(4^{1/2}\right)^3 = 2^3 = 8$

37. $-16^{3/2} = -\left(16^{1/2}\right)^3 = -(4)^3 = -64$

39. $-1{,}000^{2/3} = -\left(1{,}000^{1/3}\right)^2 = -(10)^2 = -100$

41. $64^{-1/2} = \dfrac{1}{64^{1/2}} = \dfrac{1}{8}$

43. $64^{-3/2} = \dfrac{1}{64^{3/2}} = \dfrac{1}{\left(64^{1/2}\right)^3} = \dfrac{1}{8^3} = \dfrac{1}{512}$

45. $-9^{-3/2} = -\dfrac{1}{9^{3/2}} = -\dfrac{1}{\left(9^{1/2}\right)^3} = -\dfrac{1}{3^3} = -\dfrac{1}{27}$

47. $\left(\dfrac{4}{9}\right)^{5/2} = \left[\left(\dfrac{4}{9}\right)^{1/2}\right]^5 = \left(\dfrac{2}{3}\right)^5 = \dfrac{32}{243}$

49. $\left(-\dfrac{27}{64}\right)^{-2/3} = \left(-\dfrac{64}{27}\right)^{2/3} = \left[\left(-\dfrac{64}{27}\right)^{1/3}\right]^2 = \left(-\dfrac{4}{3}\right)^2 = \dfrac{16}{9}$

51. $\left(100s^4\right)^{1/2} = 100^{1/2}\left(s^4\right)^{1/2} = 10s^2$

53. $\left(32y^{10}z^5\right)^{-1/5} = \dfrac{1}{\left(32y^{10}z^5\right)^{1/5}} = \dfrac{1}{32^{1/5}\left(y^{10}\right)^{1/5}\left(z^5\right)^{1/5}} = \dfrac{1}{2y^2z}$

55. $\left(x^{10}y^5\right)^{3/5} = x^{30/5}y^{15/5} = x^6y^3$

57. $\left(r^8s^{16}\right)^{-3/4} = r^{-24/4}s^{-48/4} = r^{-6}s^{-12} = \dfrac{1}{r^6s^{12}}$

59. $\left(-\dfrac{8a^6}{125b^9}\right)^{2/3} = \dfrac{(-8)^{2/3}a^{12/3}}{125^{2/3}b^{18/3}} = \dfrac{(-2)^2a^4}{5^2b^6} = \dfrac{4a^4}{25b^6}$

61. $\left(\dfrac{27r^6}{1{,}000s^{12}}\right)^{-2/3} = \left(\dfrac{1{,}000s^{12}}{27r^6}\right)^{2/3} = \dfrac{1{,}000^{2/3}s^{24/3}}{27^{2/3}r^{12/3}} = \dfrac{10^2s^8}{3^2r^4} = \dfrac{100s^8}{9r^4}$

63. $\dfrac{a^{2/5}a^{4/5}}{a^{1/5}} = \dfrac{a^{6/5}}{a^{1/5}} = a^{5/5} = a$

65. $\sqrt{49} = \sqrt{7^2} = 7$

67. $\sqrt[3]{125} = \sqrt[3]{5^3} = 5$

69. $\sqrt[3]{-125} = \sqrt[3]{(-5)^3} = -5$

71. $\sqrt[5]{-\dfrac{32}{100{,}000}} = \sqrt[5]{\left(-\dfrac{2}{10}\right)^5} = -\dfrac{2}{10} = -\dfrac{1}{5}$

73. $\sqrt{36x^2} = \sqrt{(6x)^2} = |6x| = 6|x|$

75. $\sqrt{9y^4} = \sqrt{(3y^2)^2} = |3y^2| = 3y^2$

77. $\sqrt[3]{8y^3} = \sqrt[3]{(2y)^3} = 2y$

79. $\sqrt[4]{\dfrac{x^4y^8}{z^{12}}} = \sqrt[4]{\left(\dfrac{xy^2}{z^3}\right)^4} = \left|\dfrac{xy^2}{z^3}\right| = \dfrac{|x|y^2}{|z^3|}$

81. $\sqrt{8} - \sqrt{2} = \sqrt{4}\sqrt{2} - \sqrt{2} = 2\sqrt{2} - \sqrt{2}$
$= \sqrt{2}$

83. $\sqrt{200x^2} + \sqrt{98x^2} = \sqrt{100x^2}\sqrt{2} + \sqrt{49x^2}\sqrt{2} = 10x\sqrt{2} + 7x\sqrt{2} = 17x\sqrt{2}$

85. $2\sqrt{48y^5} - 3y\sqrt{12y^3} = 2\sqrt{16y^4}\sqrt{3y} - 3y\sqrt{4y^2}\sqrt{3y} = 2(4y^2)\sqrt{3y} - 3y(2y)\sqrt{3y}$
$= 8y^2\sqrt{3y} - 6y^2\sqrt{3y} = 2y^2\sqrt{3y}$

87. $2\sqrt[3]{81} + 3\sqrt[3]{24} = 2\sqrt[3]{27}\sqrt[3]{3} + 3\sqrt[3]{8}\sqrt[3]{3} = 2(3)\sqrt[3]{3} + 3(2)\sqrt[3]{3} = 6\sqrt[3]{3} + 6\sqrt[3]{3} = 12\sqrt[3]{3}$

89. $\sqrt[4]{768z^5} + \sqrt[4]{48z^5} = \sqrt[4]{256z^4}\sqrt[4]{3z} + \sqrt[4]{16z^4}\sqrt[4]{3z} = 4z\sqrt[4]{3z} + 2z\sqrt[4]{3z} = 6z\sqrt[4]{3z}$

91. $\sqrt{8x^2y} - x\sqrt{2y} + \sqrt{50x^2y} = \sqrt{4x^2}\sqrt{2y} - x\sqrt{2y} + \sqrt{25x^2}\sqrt{2y}$
$= 2x\sqrt{2y} - x\sqrt{2y} + 5x\sqrt{2y} = 6x\sqrt{2y}$

93. $\sqrt[3]{16xy^4} + y\sqrt[3]{2xy} - \sqrt[3]{54xy^4} = \sqrt[3]{8y^3}\sqrt[3]{2xy} + y\sqrt[3]{2xy} - \sqrt[3]{27y^3}\sqrt[3]{2xy}$
$= 2y\sqrt[3]{2xy} + y\sqrt[3]{2xy} - 3y\sqrt[3]{2xy} = 0$

95. $\dfrac{3}{\sqrt{3}} = \dfrac{3}{\sqrt{3}} \cdot \dfrac{\sqrt{3}}{\sqrt{3}} = \dfrac{3\sqrt{3}}{3} = \sqrt{3}$

97. $\dfrac{2}{\sqrt{x}} = \dfrac{2}{\sqrt{x}} \cdot \dfrac{\sqrt{x}}{\sqrt{x}} = \dfrac{2\sqrt{x}}{x}$

99. $\dfrac{2}{\sqrt[3]{2}} = \dfrac{2}{\sqrt[3]{2}} \cdot \dfrac{\sqrt[3]{4}}{\sqrt[3]{4}} = \dfrac{2\sqrt[3]{4}}{\sqrt[3]{8}} = \dfrac{2\sqrt[3]{4}}{2} = \sqrt[3]{4}$

101. $\dfrac{5a}{\sqrt[3]{25a}} = \dfrac{5a}{\sqrt[3]{25a}} \cdot \dfrac{\sqrt[3]{5a^2}}{\sqrt[3]{5a^2}} = \dfrac{5a\sqrt[3]{5a^2}}{\sqrt[3]{125a^3}} = \dfrac{5a\sqrt[3]{5a^2}}{5a} = \sqrt[3]{5a^2}$

103. $\dfrac{2b}{\sqrt[4]{3a^2}} = \dfrac{2b}{\sqrt[4]{3a^2}} \cdot \dfrac{\sqrt[4]{27a^2}}{\sqrt[4]{27a^2}} = \dfrac{2b\sqrt[4]{27a^2}}{\sqrt[4]{81a^4}} = \dfrac{2b\sqrt[4]{27a^2}}{3a}$

105. $\sqrt[3]{\dfrac{2u^4}{9v}} = \dfrac{\sqrt[3]{2u^4}}{\sqrt[3]{9v}} = \dfrac{\sqrt[3]{u^3}\sqrt[3]{2u}}{\sqrt[3]{9v}} \cdot \dfrac{\sqrt[3]{3v^2}}{\sqrt[3]{3v^2}} = \dfrac{u\sqrt[3]{6uv^2}}{\sqrt[3]{27v^3}} = \dfrac{u\sqrt[3]{6uv^2}}{3v}$

107. $\dfrac{\sqrt{5}}{10} = \dfrac{\sqrt{5}}{10} \cdot \dfrac{\sqrt{5}}{\sqrt{5}} = \dfrac{5}{10\sqrt{5}} = \dfrac{1}{2\sqrt{5}}$

109. $\dfrac{\sqrt[3]{9}}{3} = \dfrac{\sqrt[3]{9}}{3} \cdot \dfrac{\sqrt[3]{3}}{\sqrt[3]{3}} = \dfrac{\sqrt[3]{27}}{3\sqrt[3]{3}} = \dfrac{3}{3\sqrt[3]{3}} = \dfrac{1}{\sqrt[3]{3}}$

111. $\dfrac{\sqrt[5]{16b^3}}{64a} = \dfrac{\sqrt[5]{16b^3}}{64a} \cdot \dfrac{\sqrt[5]{2b^2}}{\sqrt[5]{2b^2}} = \dfrac{\sqrt[5]{32b^5}}{64a\sqrt[5]{2b^2}} = \dfrac{2b}{64a\sqrt[5]{2b^2}} = \dfrac{b}{32a\sqrt[5]{2b^2}}$

113. $\sqrt{\frac{1}{3}}-\sqrt{\frac{1}{27}}=\frac{\sqrt{1}}{\sqrt{3}}-\frac{\sqrt{1}}{\sqrt{27}}=\frac{1}{\sqrt{3}}\cdot\frac{\sqrt{3}}{\sqrt{3}}-\frac{1}{\sqrt{27}}\cdot\frac{\sqrt{3}}{\sqrt{3}}=\frac{\sqrt{3}}{3}-\frac{\sqrt{3}}{\sqrt{81}}=\frac{\sqrt{3}}{3}-\frac{\sqrt{3}}{9}$

$=\frac{3\sqrt{3}}{9}-\frac{\sqrt{3}}{9}=\frac{2\sqrt{3}}{9}$

115. $\sqrt{\frac{x}{8}}-\sqrt{\frac{x}{2}}+\sqrt{\frac{x}{32}}=\frac{\sqrt{x}}{\sqrt{8}}-\frac{\sqrt{x}}{\sqrt{2}}+\frac{\sqrt{x}}{\sqrt{32}}=\frac{\sqrt{x}}{\sqrt{8}}\cdot\frac{\sqrt{2}}{\sqrt{2}}-\frac{\sqrt{x}}{\sqrt{2}}\cdot\frac{\sqrt{2}}{\sqrt{2}}+\frac{\sqrt{x}}{\sqrt{32}}\cdot\frac{\sqrt{2}}{\sqrt{2}}$

$=\frac{\sqrt{2x}}{\sqrt{16}}-\frac{\sqrt{2x}}{\sqrt{4}}+\frac{\sqrt{2x}}{\sqrt{64}}$

$=\frac{\sqrt{2x}}{4}-\frac{\sqrt{2x}}{2}+\frac{\sqrt{2x}}{8}$

$=\frac{2\sqrt{2x}}{8}-\frac{4\sqrt{2x}}{8}+\frac{\sqrt{2x}}{8}=-\frac{\sqrt{2x}}{8}$

117. $\sqrt[4]{9}=9^{1/4}=\left(3^2\right)^{1/4}=3^{2/4}=3^{1/2}=\sqrt{3}$

119. $\sqrt[10]{16x^6}=\left(16x^6\right)^{1/10}=\left(2^4x^6\right)^{1/10}=2^{4/10}x^{6/10}=2^{2/5}x^{3/5}=\left(2^2x^3\right)^{1/5}=\sqrt[5]{4x^3}$

121. $\sqrt{2}\sqrt[3]{2}=2^{1/2}\cdot 2^{1/3}=2^{3/6}\cdot 2^{2/6}=\sqrt[6]{2^3}\sqrt[6]{2^2}=\sqrt[6]{8}\sqrt[6]{4}=\sqrt[6]{32}$

123. $\frac{\sqrt[4]{3}}{\sqrt{2}}=\frac{3^{1/4}}{2^{1/2}}=\frac{3^{1/4}}{2^{2/4}}=\frac{\sqrt[4]{3}}{\sqrt[4]{2^2}}=\frac{\sqrt[4]{3}}{\sqrt[4]{4}}=\frac{\sqrt[4]{3}}{\sqrt[4]{4}}\cdot\frac{\sqrt[4]{4}}{\sqrt[4]{4}}=\frac{\sqrt[4]{12}}{\sqrt[4]{16}}=\frac{\sqrt[4]{12}}{2}$

125. $\sqrt[4]{x^4}=|x|$. Since $|x|=x$ if $x\geq 0$, then $\sqrt[4]{x^4}=x$ if $x\geq 0$.

127. $\left(\frac{x}{y}\right)^{-m/n}=\frac{x^{-m/n}}{y^{-m/n}}=\frac{x^{-m/n}}{y^{-m/n}}\cdot\frac{x^{m/n}y^{m/n}}{x^{m/n}y^{m/n}}=\frac{y^{m/n}}{x^{m/n}}=\frac{\left(y^m\right)^{1/n}}{\left(x^m\right)^{1/n}}=\left(\frac{y^m}{x^m}\right)^{1/n}=\sqrt[n]{\frac{y^m}{x^m}}$

129. $-2<x\leq 5\Rightarrow(-2,5]$

131. $x^2-y^2=(-2)^2-3^2=4-9=-5$

133. $617{,}000{,}000=6.17\times 10^8$

Exercises 0.4 (page 50)

1. monomial, variables

3. trinomial

5. one

7. like

9. coefficients, variables

11. yes, trinomial, 2nd degree

13. no

15. yes, binomial, 3rd degree

17. yes, monomial, 0th degree

19. yes, monomial, no degree

21. $\left(x^3-3x^2\right)+\left(5x^3-8x\right)=x^3-3x^2+5x^3-8x=x^3+5x^3-3x^2-8x=6x^3-3x^2-8x$

SECTION 0.4

23. $(y^5 + 2y^3 + 7) - (y^5 - 2y^3 - 7) = y^5 + 2y^3 + 7 - y^5 + 2y^3 + 7$
$= y^5 - y^5 + 2y^3 + 2y^3 + 7 + 7 = 4y^3 + 14$

25. $2(x^2 + 3x - 1) - 3(x^2 + 2x - 4) + 4 = 2(x^2) + 2(3x) + 2(-1) - 3(x^2) - 3(2x) - 3(-4) + 4$
$= 2x^2 + 6x - 2 - 3x^2 - 6x + 12 + 4$
$= 2x^2 - 3x^2 + 6x - 6x - 2 + 12 + 4 = -x^2 + 14$

27. $8(t^2 - 2t + 5) + 4(t^2 - 3t + 2) - 6(2t^2 - 8)$
$= 8(t^2) + 8(-2t) + 8(5) + 4(t^2) + 4(-3t) + 4(2) - 6(2t^2) - 6(-8)$
$= 8t^2 - 16t + 40 + 4t^2 - 12t + 8 - 12t^2 + 48$
$= 8t^2 + 4t^2 - 12t^2 - 16t - 12t + 40 + 8 + 48 = -28t + 96$

29. $y(y^2 - 1) - y^2(y + 2) - y(2y - 2) = y(y^2) + y(-1) - y^2(y) - y^2(2) - y(2y) - y(-2)$
$= y^3 - y - y^3 - 2y^2 - 2y^2 + 2y$
$= y^3 - y^3 - 2y^2 - 2y^2 - y + 2y = -4y^2 + y$

31. $xy(x - 4y) - y(x^2 + 3xy) + xy(2x + 3y)$
$= xy(x) + xy(-4y) - y(x^2) - y(3xy) + xy(2x) + xy(3y)$
$= x^2y - 4xy^2 - x^2y - 3xy^2 + 2x^2y + 3xy^2$
$= x^2y - x^2y + 2x^2y - 4xy^2 - 3xy^2 + 3xy^2 = 2x^2y - 4xy^2$

33. $2x^2y^3(4xy^4) = 2(4)x^2xy^3y^4 = 8x^3y^7$

35. $-3m^2n(2mn^2)\left(-\frac{mn}{12}\right) = (-3)(2)\left(-\frac{1}{12}\right)m^2mmnn^2n = \frac{6}{12}m^4n^4 = \frac{m^4n^4}{2}$

37. $-4rs(r^2 + s^2) = -4rs(r^2) - 4rs(s^2) = -4r^3s - 4rs^3$

39. $6ab^2c(2ac + 3bc^2 - 4ab^2c) = 6ab^2c(2ac) + 6ab^2c(3bc^2) + 6ab^2c(-4ab^2c)$
$= 12a^2b^2c^2 + 18ab^3c^3 - 24a^2b^4c^2$

41. $(a + 2)(a + 2) = a^2 + 2a + 2a + 4$
$= a^2 + 4a + 4$

43. $(a - 6)^2 = (a - 6)(a - 6)$
$= a^2 - 6a - 6a + 36$
$= a^2 - 12a + 36$

45. $(x + 4)(x - 4) = x^2 - 4x + 4x - 16$
$= x^2 - 16$

47. $(x - 3)(x + 5) = x^2 + 5x - 3x - 15$
$= x^2 + 2x - 15$

49. $(u + 2)(3u - 2) = 3u^2 - 2u + 6u - 4$
$= 3u^2 + 4u - 4$

51. $(5x - 1)(2x + 3) = 10x^2 + 15x - 2x - 3$
$= 10x^2 + 13x - 3$

53. $(3a - 2b)^2 = (3a - 2b)(3a - 2b) = 9a^2 - 6ab - 6ab + 4b^2 = 9a^2 - 12ab + 4b^2$

SECTION 0.4

55. $(3m+4n)(3m-4n) = 9m^2 - 12mn + 12mn - 16n^2 = 9m^2 - 16n^2$

57. $(2y-4x)(3y-2x) = 6y^2 - 4xy - 12xy + 8x^2 = 6y^2 - 16xy + 8x^2$

59. $(9x-y)(x^2-3y) = 9x^3 - 27xy - x^2y + 3y^2 = 9x^3 - x^2y - 27xy + 3y^2$

61. $(5z+2t)(z^2-t) = 5z^3 - 5tz + 2tz^2 - 2t^2 = 5z^3 + 2tz^2 - 5tz - 2t^2$

63. $\left(\sqrt{5}+3x\right)\left(2-\sqrt{5}x\right) = 2\sqrt{5} - 5x + 6x - 3\sqrt{5}x^2 = -3\sqrt{5}\,x^2 + x + 2\sqrt{5}$

65.
$$\begin{aligned}(3x-1)^3 &= (3x-1)(3x-1)(3x-1)\\ &= \left(9x^2-3x-3x+1\right)(3x-1)\\ &= \left(9x^2-6x+1\right)(3x-1)\\ &= 9x^2(3x)+9x^2(-1)-6x(3x)-6x(-1)+1(3x)+1(-1)\\ &= 27x^3-9x^2-18x^2+6x+3x-1 = 27x^3-27x^2+9x-1\end{aligned}$$

67.
$$\begin{aligned}(3x+1)\left(2x^2+4x-3\right) &= 3x\left(2x^2\right)+3x(4x)+3x(-3)+1\left(2x^2\right)+1(4x)+1(-3)\\ &= 6x^3+12x^2-9x+2x^2+4x-3 = 6x^3+14x^2-5x-3\end{aligned}$$

69.
$$\begin{aligned}&(3x+2y)\left(2x^2-3xy+4y^2\right)\\ &\quad= 3x\left(2x^2\right)+3x(-3xy)+3x\left(4y^2\right)+2y\left(2x^2\right)+2y(-3xy)+2y\left(4y^2\right)\\ &\quad= 6x^3-9x^2y+12xy^2+4x^2y-6xy^2+8y^3 = 6x^3-5x^2y+6xy^2+8y^3\end{aligned}$$

71. $2y^n\left(3y^n+y^{-n}\right) = 2y^n\left(3y^n\right)+2y^n\left(y^{-n}\right) = 6y^{n+n}+2y^{n+(-n)} = 6y^{2n}+2y^0 = 6y^{2n}+2$

73.
$$\begin{aligned}-5x^{2n}y^n\left(2x^{2n}y^{-n}+3x^{-2n}y^n\right) &= -5x^{2n}y^n\left(2x^{2n}y^{-n}\right)-5x^{2n}y^n\left(3x^{-2n}y^n\right)\\ &= -10x^{2n+2n}y^{n+(-n)}-15x^{2n+(-2n)}y^{n+n}\\ &= -10x^{4n}y^0-15x^0y^{2n} = -10x^{4n}-15y^{2n}\end{aligned}$$

75. $(x^n+3)(x^n-4) = x^nx^n - 4x^n + 3x^n - 12 = x^{2n} - x^n - 12$

77.
$$\begin{aligned}(2r^n-7)(3r^n-2) &= 2r^n(3r^n)-2r^n(2)-7(3r^n)+14\\ &= 6r^{2n}-4r^n-21r^n+14 = 6r^{2n}-25r^n+14\end{aligned}$$

79. $x^{1/2}\left(x^{1/2}y+xy^{1/2}\right) = x^{1/2}x^{1/2}y+x^{1/2}xy^{1/2} = x^{2/2}y+x^{3/2}y^{1/2} = xy+x^{3/2}y^{1/2}$

81.
$$\begin{aligned}\left(a^{1/2}+b^{1/2}\right)\left(a^{1/2}-b^{1/2}\right) &= a^{1/2}a^{1/2}-a^{1/2}b^{1/2}+a^{1/2}b^{1/2}-b^{1/2}b^{1/2}\\ &= a^{2/2}-b^{2/2} = a-b\end{aligned}$$

83. $\dfrac{2}{\sqrt{3}-1} = \dfrac{2}{\sqrt{3}-1}\cdot\dfrac{\sqrt{3}+1}{\sqrt{3}+1} = \dfrac{2\left(\sqrt{3}+1\right)}{\left(\sqrt{3}\right)^2-1^2} = \dfrac{2\left(\sqrt{3}+1\right)}{3-1} = \dfrac{2\left(\sqrt{3}+1\right)}{2} = \sqrt{3}+1$

SECTION 0.4

85. $$\frac{3x}{\sqrt{7}+2}=\frac{3x}{\sqrt{7}+2}\cdot\frac{\sqrt{7}-2}{\sqrt{7}-2}=\frac{3x\left(\sqrt{7}-2\right)}{\left(\sqrt{7}\right)^2-2^2}=\frac{3x\left(\sqrt{7}-2\right)}{7-4}=\frac{3x\left(\sqrt{7}-2\right)}{3}=x\left(\sqrt{7}-2\right)$$

87. $$\frac{x}{x-\sqrt{3}}=\frac{x}{x-\sqrt{3}}\cdot\frac{x+\sqrt{3}}{x+\sqrt{3}}=\frac{x\left(x+\sqrt{3}\right)}{x^2-\left(\sqrt{3}\right)^2}=\frac{x\left(x+\sqrt{3}\right)}{x^2-3}$$

89. $$\frac{y+\sqrt{2}}{y-\sqrt{2}}=\frac{y+\sqrt{2}}{y-\sqrt{2}}\cdot\frac{y+\sqrt{2}}{y+\sqrt{2}}=\frac{\left(y+\sqrt{2}\right)\left(y+\sqrt{2}\right)}{y^2-\left(\sqrt{2}\right)^2}=\frac{y^2+2y\sqrt{2}+2}{y^2-2}$$

91. $$\frac{\sqrt{2}-\sqrt{3}}{1-\sqrt{3}}=\frac{\sqrt{2}-\sqrt{3}}{1-\sqrt{3}}\cdot\frac{1+\sqrt{3}}{1+\sqrt{3}}=\frac{\sqrt{2}+\sqrt{6}-\sqrt{3}-\left(\sqrt{3}\right)^2}{1^2-\left(\sqrt{3}\right)^2}=\frac{\sqrt{2}+\sqrt{6}-\sqrt{3}-3}{1-3}$$
$$=\frac{\sqrt{2}+\sqrt{6}-\sqrt{3}-3}{-2}$$
$$=\frac{-\left(\sqrt{2}+\sqrt{6}-\sqrt{3}-3\right)}{2}$$
$$=\frac{\sqrt{3}+3-\sqrt{2}-\sqrt{6}}{2}$$

93. $$\frac{\sqrt{x}-\sqrt{y}}{\sqrt{x}+\sqrt{y}}=\frac{\sqrt{x}-\sqrt{y}}{\sqrt{x}+\sqrt{y}}\cdot\frac{\sqrt{x}-\sqrt{y}}{\sqrt{x}-\sqrt{y}}=\frac{\sqrt{x^2}-\sqrt{xy}-\sqrt{xy}+\sqrt{y^2}}{\left(\sqrt{x}\right)^2-\left(\sqrt{y}\right)^2}=\frac{x-2\sqrt{xy}+y}{x-y}$$

95. $$\frac{\sqrt{2}+1}{2}=\frac{\sqrt{2}+1}{2}\cdot\frac{\sqrt{2}-1}{\sqrt{2}-1}=\frac{\left(\sqrt{2}\right)^2-1^2}{2\left(\sqrt{2}-1\right)}=\frac{2-1}{2\left(\sqrt{2}-1\right)}=\frac{1}{2\left(\sqrt{2}-1\right)}$$

97. $$\frac{y-\sqrt{3}}{y+\sqrt{3}}=\frac{y-\sqrt{3}}{y+\sqrt{3}}\cdot\frac{y+\sqrt{3}}{y+\sqrt{3}}=\frac{y^2-\left(\sqrt{3}\right)^2}{y^2+y\sqrt{3}+y\sqrt{3}+\sqrt{9}}=\frac{y^2-3}{y^2+2y\sqrt{3}+3}$$

99. $$\frac{\sqrt{x+3}-\sqrt{x}}{3}=\frac{\sqrt{x+3}-\sqrt{x}}{3}\cdot\frac{\sqrt{x+3}+\sqrt{x}}{\sqrt{x+3}+\sqrt{x}}=\frac{\left(\sqrt{x+3}\right)^2-\left(\sqrt{x}\right)^2}{3\left(\sqrt{x+3}+\sqrt{x}\right)}$$
$$=\frac{x+3-x}{3\left(\sqrt{x+3}+\sqrt{x}\right)}$$
$$=\frac{3}{3\left(\sqrt{x+3}+\sqrt{x}\right)}=\frac{1}{\sqrt{x+3}+\sqrt{x}}$$

SECTION 0.4

101. $\dfrac{36a^2b^3}{18ab^6} = 2a^{2-1}b^{3-6} = 2a^1b^{-3} = \dfrac{2a}{b^3}$

103. $\dfrac{16x^6y^4z^9}{-24x^9y^6z^0} = -\dfrac{2}{3}x^{6-9}y^{4-6}z^{9-0} = -\dfrac{2}{3}x^{-3}y^{-2}z^9 = -\dfrac{2z^9}{3x^3y^2}$

105. $\dfrac{5x^3y^2 + 15x^3y^4}{10x^2y^3} = \dfrac{5x^3y^2}{10x^2y^3} + \dfrac{15x^3y^4}{10x^2y^3} = \dfrac{x}{2y} + \dfrac{3xy}{2}$

107. $\dfrac{24x^5y^7 - 36x^2y^5 + 12xy}{60x^5y^4} = \dfrac{24x^5y^7}{60x^5y^4} - \dfrac{36x^2y^5}{60x^5y^4} + \dfrac{12xy}{60x^5y^4} = \dfrac{2y^3}{5} - \dfrac{3y}{5x^3} + \dfrac{1}{5x^4y^3}$

109.

$$\begin{array}{r|l} & 3x + 2 \\ \hline x+3 & 3x^2 + 11x + 6 \\ & \underline{3x^2 + 9x} \\ & 2x + 6 \\ & \underline{2x + 6} \\ & 0 \end{array}$$

111.

$$\begin{array}{r|l} & x - 7 + \frac{2}{2x-5} \\ \hline 2x-5 & 2x^2 - 19x + 37 \\ & \underline{2x^2 - 5x} \\ & -14x + 37 \\ & \underline{-14x + 35} \\ & 2 \end{array}$$

113.

$$\begin{array}{r|l} & 2x^2 + 2x + 2 + \frac{3}{x-1} \\ \hline x-1 & 2x^3 + 0x^2 + 0x + 1 \\ & \underline{2x^3 - 2x^2} \\ & 2x^2 + 0x + 1 \\ & \underline{2x^2 - 2x} \\ & 2x + 1 \\ & \underline{2x - 2} \\ & 3 \end{array}$$

115.

$$\begin{array}{r|l} & x - 3 \\ \hline x^2 + x - 1 & x^3 - 2x^2 - 4x + 3 \\ & \underline{x^3 + x^2 - x} \\ & -3x^2 - 3x + 3 \\ & \underline{-3x^2 - 3x + 3} \\ & 0 \end{array}$$

117.

$$\begin{array}{r|l} & x^2 - 2 + \frac{-x^2+5}{x^3-2} \\ \hline x^3 - 2 & x^5 + 0x^4 - 2x^3 - 3x^2 + 0x + 9 \\ & \underline{x^5 \qquad\qquad - 2x^2} \\ & -2x^3 - x^2 + 0x + 9 \\ & \underline{-2x^3 \qquad\qquad + 4} \\ & -x^2 \qquad + 5 \end{array}$$

119.

$$\begin{array}{r|l} & x^4 + 2x^3 + 4x^2 + 8x + 16 \\ \hline x-2 & x^5 + 0x^4 + 0x^3 + 0x^2 + 0x - 32 \\ & \underline{x^5 - 2x^4} \\ & 2x^4 + 0x^3 \\ & \underline{2x^4 - 4x^3} \\ & 4x^3 + 0x^2 \\ & \underline{4x^3 - 8x^2} \\ & 8x^2 + 0x \\ & \underline{8x^2 - 16x} \\ & 16x - 32 \\ & \underline{16x - 32} \\ & 0 \end{array}$$

121.

$$\begin{array}{r|l} & 6x^2 + x - 12 \\ \hline 6x^2 + 11x - 10 & 36x^4 + 72x^3 - 121x^2 - 142x + 120 \\ & \underline{36x^4 + 66x^3 - 60x^2} \\ & 6x^3 - 61x^2 - 142x \\ & \underline{6x^3 + 11x^2 - 10x} \\ & -72x^2 - 132x + 120 \\ & \underline{-72x^2 - 132x + 120} \\ & 0 \end{array}$$

SECTION 0.4

123. Area = length · width = $(x+5)(x-2)$ ft^2 = $(x^2-2x+5x-10)$ ft^2 = $(x^2+3x-10)$ ft^2

125. Volume = $l \cdot w \cdot h = (12-2x)(12-2x)x$ in.^{3} = $\left(144-48x+4x^2\right)x$ in.^{3}
$= \left(144x-48x^2+4x^3\right)$ in.^{3}

127.
$$(a+b+c)^2 = (a+b+c)(a+b+c) = a(a+b+c)+b(a+b+c)+c(a+b+c)$$
$$= a^2+ab+ac+ab+b^2+bc+ac+bc+c^2$$
$$= a^2+b^2+c^2+2ab+2bc+2ac$$

129. Answers may vary.

131. Check the formula with $a=1$ and $b=2$.

133. $9^{3/2} = \left(9^{1/2}\right)^3 = 3^3 = 27$

135. $\left(\dfrac{625x^4}{16y^8}\right)^{3/4} = \dfrac{625^{3/4}(x^4)^{3/4}}{16^{3/4}(y^8)^{3/4}} = \dfrac{125x^3}{8y^6}$

137.
$$\sqrt[3]{16ab^4} - b\sqrt[3]{54ab} = \sqrt[3]{8b^3}\sqrt[3]{2ab} - b\sqrt[3]{27}\sqrt[3]{2ab} = 2b\sqrt[3]{2ab} - b(3)\sqrt[3]{2ab} = 2b\sqrt[3]{2ab} - 3b\sqrt[3]{2ab}$$
$$= -b\sqrt[3]{2ab}$$

Exercises 0.5 (page 61)

1. factor

3. $ax+bx = x(a+b)$

5. $x^2+2xy+y^2 = (x+y)(x+y) = (x+y)^2$

7. $x^3+y^3 = (x+y)(x^2-xy+y^2)$

9. $3x-6 = 3(x-2)$

11. $8x^2+4x^3 = 4x^2(2+x)$

13. $7x^2y^2+14x^3y^2 = 7x^2y^2(1+2x)$

15. $a(x+y)+b(x+y) = (x+y)(a+b)$

17. $4a+b-12a^2-3ab = 4a+b-3a(4a+b) = 1(4a+b)-3a(4a+b) = (4a+b)(1-3a)$

19. $4x^2-9 = (2x)^2-3^2 = (2x+3)(2x-3)$

21. $4-9r^2 = 2^2-(3r)^2 = (2+3r)(2-3r)$

23. $81x^4-1 = (9x^2)^2-1^2 = (9x^2+1)(9x^2-1) = (9x^2+1)(3x+1)(3x-1)$

25. $(x+z)^2-25 = (x+z)^2-5^2 = (x+z+5)(x+z-5)$

27. $x^2+8x+16 = (x+4)(x+4) = (x+4)^2$

29. $b^2-10b+25 = (b-5)(b-5) = (b-5)^2$

31. $m^2+4mn+4n^2 = (m+2n)(m+2n)$
$= (m+2n)^2$

33. $12x^2-xy-6y^2 = (4x-3y)(3x+2y)$

35. $x^2+10x+21$: $a=1$, $b=10$, $c=21$; key number $= ac = 1(21) = 21$
$x^2+10x+21 = x^2+7x+3x+21 = x(x+7)+3(x+7) = (x+7)(x+3)$

SECTION 0.5

37. $x^2 - 4x - 12$: $a = 1, b = -4, c = -12$
key number $= ac = 1(-12) = -12$

$$\begin{aligned} x^2 - 4x - 12 &= x^2 - 6x + 2x - 12 \\ &= x(x-6) + 2(x-6) \\ &= (x-6)(x+2) \end{aligned}$$

39. $6p^2 + 7p - 3$: $a = 6, b = 7, c = -3$
key number $= ac = 6(-3) = -18$

$$\begin{aligned} 6p^2 + 7p - 3 &= 6p^2 + 9p - 2p - 3 \\ &= 3p(2p+3) - (2p+3) \\ &= (2p+3)(3p-1) \end{aligned}$$

41. $t^3 + 343 = t^3 + 7^3 = (t+7)[t^2 - (t)(7) + 7^2] = (t+7)(t^2 - 7t + 49)$

43. $8z^3 - 27 = (2z)^3 - 3^3 = (2z-3)\left[(2z)^2 + (2z)(3) + 3^2\right] = (2z-3)(4z^2 + 6z + 9)$

45. $3a^2bc + 6ab^2c + 9abc^2 = 3abc(a + 2b + 3c)$

47. $3x^3 + 3x^2 - x - 1 = 3x^2(x+1) - 1(x+1) = (x+1)(3x^2 - 1)$

49. $2txy + 2ctx - 3ty - 3ct = t(2xy + 2cx - 3y - 3c) = t[2x(y+c) - 3(y+c)] = t(y+c)(2x-3)$

51. $ax + bx + ay + by + az + bz = x(a+b) + y(a+b) + z(a+b) = (a+b)(x+y+z)$

53. $x^2 - (y-z)^2 = [x + (y-z)][x - (y-z)] = (x+y-z)(x-y+z)$

55.
$$\begin{aligned} (x-y)^2 - (x+y)^2 &= [(x-y) + (x+y)][(x-y) - (x+y)] = (x - y + x + y)(x - y - x - y) \\ &= (2x)(-2y) = -4xy \end{aligned}$$

57. $x^4 - y^4 = (x^2)^2 - (y^2)^2 = (x^2 + y^2)(x^2 - y^2) = (x^2 + y^2)(x+y)(x-y)$

59. $3x^2 - 12 = 3(x^2 - 4) = 3(x+2)(x-2)$

61.
$$\begin{aligned} 18xy^2 - 8x &= 2x(9y^2 - 4) \\ &= 2x(3y+2)(3y-2) \end{aligned}$$

63. $x^2 - 2x + 15 \Rightarrow$ prime

65.
$$\begin{aligned} -15 + 2a + 24a^2 &= 24a^2 + 2a - 15 \\ &= (6a+5)(4a-3) \end{aligned}$$

67. $6x^2 + 29xy + 35y^2 = (3x + 7y)(2x + 5y)$

69. $12p^2 - 58pq - 70q^2 = 2(6p^2 - 29pq - 35q^2) = 2(6p - 35q)(p + q)$

71. $-6m^2 + 47mn - 35n^2 = -(6m^2 - 47mn + 35n^2) = -(6m - 5n)(m - 7n)$

73. $-6x^3 + 23x^2 + 35x = -x(6x^2 - 23x - 35) = -x(6x+7)(x-5)$

75. $6x^4 - 11x^3 - 35x^2 = x^2(6x^2 - 11x - 35) = x^2(2x-7)(3x+5)$

77. $x^4 + 2x^2 - 15 = (x^2 + 5)(x^2 - 3)$

79. $a^{2n} - 2a^n - 3 = (a^n - 3)(a^n + 1)$

81. $6x^{2n} - 7x^n + 2 = (3x^n - 2)(2x^n - 1)$

83.
$$\begin{aligned} 4x^{2n} - 9y^{2n} &= (2x^n)^2 - (3y^n)^2 \\ &= (2x^n + 3y^n)(2x^n - 3y^n) \end{aligned}$$

SECTION 0.5

85. $10y^{2n} - 11y^n - 6 = (5y^n + 2)(2y^n - 3)$

87. $2x^3 + 2{,}000 = 2(x^3 + 1{,}000) = 2(x^3 + 10^3) = 2(x + 10)(x^2 - 10x + 100)$

89. $$\begin{aligned}(x+y)^3 - 64 = (x+y)^3 - 4^3 &= [(x+y) - 4]\left[(x+y)^2 + 4(x+y) + 4^2\right] \\ &= (x + y - 4)\left(x^2 + 2xy + y^2 + 4x + 4y + 16\right)\end{aligned}$$

91. $$\begin{aligned}64a^6 - y^6 = \left(8a^3\right)^2 - \left(y^3\right)^2 &= \left(8a^3 + y^3\right)\left(8a^3 - y^3\right) \\ &= (2a + y)\left(4a^2 - 2ay + y^2\right)(2a - y)\left(4a^2 + 2ay + y^2\right) \\ &= (2a + y)(2a - y)\left(4a^2 - 2ay + y^2\right)\left(4a^2 + 2ay + y^2\right)\end{aligned}$$

93. $a^3 - b^3 + a - b = (a - b)(a^2 + ab + b^2) + (a - b)1 = (a - b)(a^2 + ab + b^2 + 1)$

95. $$\begin{aligned}64x^6 + y^6 = \left(4x^2\right)^3 + \left(y^2\right)^3 &= \left(4x^2 + y^2\right)\left(\left(4x^2\right)^2 - 4x^2y^2 + \left(y^2\right)^2\right) \\ &= \left(4x^2 + y^2\right)\left(16x^4 - 4x^2y^2 + y^4\right)\end{aligned}$$

97. $$\begin{aligned}x^2 - 6x + 9 - 144y^2 = (x-3)(x-3) - 144y^2 &= (x-3)^2 - (12y)^2 \\ &= (x - 3 + 12y)(x - 3 - 12y)\end{aligned}$$

99. $(a+b)^2 - 3(a+b) - 10 = [(a+b) - 5][(a+b) + 2] = (a + b - 5)(a + b + 2)$

101. $x^6 + 7x^3 - 8 = (x^3 + 8)(x^3 - 1) = (x + 2)(x^2 - 2x + 4)(x - 1)(x^2 + x + 1)$

103. $$\begin{aligned}x^4 + x^2 + 1 &= x^4 + 2x^2 + 1 - x^2 \\ &= \left(x^2+1\right)\left(x^2+1\right) - x^2 \\ &= \left(x^2+1\right)^2 - x^2 \\ &= \left(x^2 + 1 + x\right)\left(x^2 + 1 - x\right)\end{aligned}$$

105. $$\begin{aligned}x^4 + 7x^2 + 16 &= x^4 + 8x^2 + 16 - x^2 \\ &= \left(x^2+4\right)\left(x^2+4\right) - x^2 \\ &= \left(x^2+4\right)^2 - x^2 \\ &= \left(x^2 + 4 + x\right)\left(x^2 + 4 - x\right)\end{aligned}$$

107. $$\begin{aligned}4a^4 + 1 + 3a^2 = 4a^4 + 4a^2 + 1 - a^2 = \left(2a^2+1\right)\left(2a^2+1\right) - a^2 &= \left(2a^2+1\right)^2 - a^2 \\ &= \left(2a^2 + 1 + a\right)\left(2a^2 + 1 - a\right)\end{aligned}$$

109. $V = \dfrac{4}{3}\pi r_1^3 - \dfrac{4}{3}\pi r_2^3 = \dfrac{4}{3}\pi\left(r_1^3 - r_2^3\right) = \dfrac{4}{3}\pi(r_1 - r_2)\left(r_1^2 + r_1r_2 + r_2^2\right)$

111. Answers may vary.

113. Answers may vary.

115. $3x + 2 = 2\left(\frac{3x}{2} + \frac{2}{2}\right) = 2\left(\frac{3}{2}x + 1\right)$

117. $$\begin{aligned}x^2 + 2x + 4 &= 2\left(\tfrac{x^2}{2} + \tfrac{2x}{2} + \tfrac{4}{2}\right) \\ &= 2\left(\tfrac{1}{2}x^2 + x + 2\right)\end{aligned}$$

119. $a + b = a\left(\frac{a}{a} + \frac{b}{a}\right) = a\left(1 + \frac{b}{a}\right)$

121. $$\begin{aligned}x + x^{1/2} &= x^{1/2}\left(x^{1-1/2} + x^{1/2-1/2}\right) \\ &= x^{1/2}\left(x^{1/2} + 1\right)\end{aligned}$$

123. $2x + \sqrt{2}y = \sqrt{2}\left(\frac{2x}{\sqrt{2}} + \frac{\sqrt{2}y}{\sqrt{2}}\right) = \sqrt{2}\left(\sqrt{2}x + y\right)$

125. $ab^{3/2} - a^{3/2}b = ab\left(\frac{ab^{3/2}}{ab} - \frac{a^{3/2}b}{ab}\right) = ab\left(b^{1/2} - a^{1/2}\right)$

127. $x^2 + x - 6 + xy - 2y = (x+3)(x-2) + y(x-2) = (x-2)(x+3+y)$

129. $a^4 + 2a^3 + a^2 + a + 1 = a^2\left(a^2 + 2a + 1\right) + a + 1 = a^2(a+1)(a+1) + 1(a+1) = (a+1)\left[a^2(a+1) + 1\right] = (a+1)\left(a^3 + a^2 + 1\right)$

131. The number 1 is neither prime nor composite.

133. $\left(x^3x^2\right)^4 = \left(x^5\right)^4 = x^{20}$

135. $\left(\frac{3x^4x^3}{6x^{-2}x^4}\right)^0 = 1$

137. $\sqrt{20x} - \sqrt{125x} = \sqrt{4}\sqrt{5x} - \sqrt{25}\sqrt{5x} = 2\sqrt{5x} - 5\sqrt{5x} = -3\sqrt{5x}$

Exercises 0.6 (page 71)

1. numerator

3. $ad = bc$

5. $\frac{ac}{bd}$

7. $\frac{a+c}{b}$

9. $\frac{8x}{3y} \stackrel{?}{=} \frac{16x}{6y}$

$8x \cdot 6y \stackrel{?}{=} 3y \cdot 16x$

$48xy = 48xy$

EQUAL

11. $\frac{25xyz}{12ab^2c} \stackrel{?}{=} \frac{50a^2bc}{24xyz}$

$25xyz \cdot 24xyz \stackrel{?}{=} 12ab^2c \cdot 50a^2bc$

$600x^2y^2z^2 \neq 600a^3b^3c^2$

NOT EQUAL

13. $\frac{7a^2b}{21ab^2} = \frac{a \cdot 7ab}{3b \cdot 7ab} = \frac{a}{3b} \cdot \frac{7ab}{7ab} = \frac{a}{3b}$

15. $\frac{4x}{7} \cdot \frac{2}{5a} = \frac{4x \cdot 2}{7 \cdot 5a} = \frac{8x}{35a}$

17. $\frac{8m}{5n} \div \frac{3m}{10n} = \frac{8m}{5n} \cdot \frac{10n}{3m} = \frac{80mn}{15mn} = \frac{16}{3}$

19. $\frac{3z}{5c} + \frac{2z}{5c} = \frac{3z + 2z}{5c} = \frac{5z}{5c} = \frac{z}{c}$

21. $\frac{15x^2y}{7a^2b^3} - \frac{x^2y}{7a^2b^3} = \frac{14x^2y}{7a^2b^3} = \frac{2x^2y}{a^2b^3}$

23. $\frac{2x-4}{x^2-4} = \frac{2(x-2)}{(x+2)(x-2)} = \frac{2}{x+2}$

25. $\frac{4-x^2}{x^2-5x+6} = \frac{(2+x)(2-x)}{(x-3)(x-2)} = -\frac{x+2}{x-3}$

27. $\frac{6x^3 + x^2 - 12x}{4x^3 + 4x^2 - 3x} = \frac{x(6x^2 + x - 12)}{x(4x^2 + 4x - 3)} = \frac{x(2x+3)(3x-4)}{x(2x+3)(2x-1)} = \frac{3x-4}{2x-1}$

SECTION 0.6

29. $$\frac{x^3-8}{x^2+ax-2x-2a}=\frac{x^3-2^3}{x(x+a)-2(x+a)}=\frac{(x-2)(x^2+2x+4)}{(x+a)(x-2)}=\frac{x^2+2x+4}{x+a}$$

31. $$\frac{x^2-1}{x}\cdot\frac{x^2}{x^2+2x+1}=\frac{(x+1)(x-1)}{x}\cdot\frac{x^2}{(x+1)(x+1)}=\frac{x(x-1)}{x+1}$$

33. $$\frac{3x^2+7x+2}{x^2+2x}\cdot\frac{x^2-x}{3x^2+x}=\frac{(3x+1)(x+2)}{x(x+2)}\cdot\frac{x(x-1)}{x(3x+1)}=\frac{x-1}{x}$$

35. $$\frac{x^2+x}{x-1}\cdot\frac{x^2-1}{x+2}=\frac{x(x+1)}{x-1}\cdot\frac{(x+1)(x-1)}{x+2}=\frac{x(x+1)^2}{x+2}$$

37. $$\frac{2x^2+32}{8}\div\frac{x^2+16}{2}=\frac{2x^2+32}{8}\cdot\frac{2}{x^2+16}=\frac{2(x^2+16)}{8}\cdot\frac{2}{x^2+16}=\frac{1}{2}$$

39. $$\frac{z^2+z-20}{z^2-4}\div\frac{z^2-25}{z-5}=\frac{z^2+z-20}{z^2-4}\cdot\frac{z-5}{z^2-25}=\frac{(z+5)(z-4)}{(z+2)(z-2)}\cdot\frac{z-5}{(z+5)(z-5)}$$
$$=\frac{z-4}{(z+2)(z-2)}$$

41. $$\frac{3x^2+5x-2}{x^3+2x^2}\div\frac{6x^2+13x-5}{2x^3+5x^2}=\frac{3x^2+5x-2}{x^3+2x^2}\cdot\frac{2x^3+5x^2}{6x^2+13x-5}$$
$$=\frac{(3x-1)(x+2)}{x^2(x+2)}\cdot\frac{x^2(2x+5)}{(3x-1)(2x+5)}=1$$

43. $$\frac{x^2+7x+12}{x^3-x^2-6x}\cdot\frac{x^2-3x-10}{x^2+2x-3}\cdot\frac{x^3-4x^2+3x}{x^2-x-20}$$
$$=\frac{(x+3)(x+4)}{x(x-3)(x+2)}\cdot\frac{(x-5)(x+2)}{(x+3)(x-1)}\cdot\frac{x(x-3)(x-1)}{(x-5)(x+4)}=1$$

45. $$\frac{x^2-2x-3}{21x^2-50x-16}\cdot\frac{3x-8}{x-3}\div\frac{x^2+6x+5}{7x^2-33x-10}=\frac{x^2-2x-3}{21x^2-50x-16}\cdot\frac{3x-8}{x-3}\cdot\frac{7x^2-33x-10}{x^2+6x+5}$$
$$=\frac{(x-3)(x+1)}{(7x+2)(3x-8)}\cdot\frac{3x-8}{x-3}\cdot\frac{(7x+2)(x-5)}{(x+5)(x+1)}$$
$$=\frac{x-5}{x+5}$$

47. $$\frac{3}{x+3}+\frac{x+2}{x+3}=\frac{3+x+2}{x+3}=\frac{x+5}{x+3}$$

49. $$\frac{4x}{x-1}-\frac{4}{x-1}=\frac{4x-4}{x-1}=\frac{4(x-1)}{x-1}=4$$

51. $$\frac{2}{5-x}+\frac{1}{x-5}=\frac{-2}{x-5}+\frac{1}{x-5}=\frac{-1}{x-5}$$

53. $$\frac{3}{x+1}+\frac{2}{x-1}=\frac{3(x-1)}{(x+1)(x-1)}+\frac{2(x+1)}{(x-1)(x+1)}=\frac{3x-3}{(x+1)(x-1)}+\frac{2x+2}{(x-1)(x+1)}$$
$$=\frac{5x-1}{(x+1)(x-1)}$$

55. $$\frac{a+3}{a^2+7a+12}+\frac{a}{a^2-16}=\frac{a+3}{(a+3)(a+4)}+\frac{a}{(a+4)(a-4)}$$
$$=\frac{1}{a+4}+\frac{a}{(a+4)(a-4)}$$
$$=\frac{1(a-4)}{(a+4)(a-4)}+\frac{a}{(a+4)(a-4)}$$
$$=\frac{a-4}{(a+4)(a-4)}+\frac{a}{(a+4)(a-4)}$$
$$=\frac{2a-4}{(a+4)(a-4)}=\frac{2(a-2)}{(a+4)(a-4)}$$

57. $$\frac{x}{x^2-4}-\frac{1}{x+2}=\frac{x}{(x+2)(x-2)}-\frac{1}{x+2}=\frac{x}{(x+2)(x-2)}-\frac{1(x-2)}{(x+2)(x-2)}$$
$$=\frac{x}{(x+2)(x-2)}-\frac{x-2}{(x+2)(x-2)}$$
$$=\frac{2}{(x+2)(x-2)}$$

59. $$\frac{3x-2}{x^2+2x+1}-\frac{x}{x^2-1}=\frac{3x-2}{(x+1)(x+1)}-\frac{x}{(x+1)(x-1)}$$
$$=\frac{(3x-2)(x-1)}{(x+1)(x+1)(x-1)}-\frac{x(x+1)}{(x+1)(x-1)(x+1)}$$
$$=\frac{3x^2-5x+2}{(x+1)(x+1)(x-1)}-\frac{x^2+x}{(x+1)(x+1)(x-1)}$$
$$=\frac{2x^2-6x+2}{(x+1)(x+1)(x-1)}=\frac{2(x^2-3x+1)}{(x+1)^2(x-1)}$$

61. $$\frac{2}{y^2-1}+3+\frac{1}{y+1}=\frac{2}{(y+1)(y-1)}+\frac{3}{1}+\frac{1}{y+1}$$
$$=\frac{2}{(y+1)(y-1)}+\frac{3(y+1)(y-1)}{1(y+1)(y-1)}+\frac{1(y-1)}{(y+1)(y-1)}$$
$$=\frac{2}{(y+1)(y-1)}+\frac{3y^2-3}{(y+1)(y-1)}+\frac{y-1}{(y+1)(y-1)}$$
$$=\frac{3y^2+y-2}{(y+1)(y-1)}=\frac{(3y-2)(y+1)}{(y+1)(y-1)}=\frac{3y-2}{y-1}$$

63. $$\frac{1}{x-2}+\frac{3}{x+2}-\frac{3x-2}{x^2-4}=\frac{1}{x-2}+\frac{3}{x+2}-\frac{3x-2}{(x+2)(x-2)}$$
$$=\frac{1(x+2)}{(x-2)(x+2)}+\frac{3(x-2)}{(x+2)(x-2)}-\frac{3x-2}{(x+2)(x-2)}$$
$$=\frac{x+2}{(x+2)(x-2)}+\frac{3x-6}{(x+2)(x-2)}-\frac{3x-2}{(x+2)(x-2)}$$
$$=\frac{x-2}{(x+2)(x-2)}=\frac{1}{x+2}$$

65. $$\left(\frac{1}{x-2}+\frac{1}{x-3}\right)\cdot\frac{x-3}{2x}=\left(\frac{1(x-3)}{(x-2)(x-3)}+\frac{1(x-2)}{(x-3)(x-2)}\right)\cdot\frac{x-3}{2x}$$
$$=\left(\frac{x-3}{(x-2)(x-3)}+\frac{x-2}{(x-2)(x-3)}\right)\cdot\frac{x-3}{2x}$$
$$=\frac{2x-5}{(x-2)(x-3)}\cdot\frac{x-3}{2x}=\frac{2x-5}{2x(x-2)}$$

67. $$\frac{3x}{x-4}-\frac{x}{x+4}-\frac{3x+1}{16-x^2}=\frac{3x}{x-4}-\frac{x}{x+4}-\frac{3x+1}{(4+x)(4-x)}$$
$$=\frac{3x}{x-4}-\frac{x}{x+4}+\frac{3x+1}{(x+4)(x-4)}$$
$$=\frac{3x(x+4)}{(x-4)(x+4)}-\frac{x(x-4)}{(x+4)(x-4)}+\frac{3x+1}{(x+4)(x-4)}$$
$$=\frac{3x^2+12x}{(x+4)(x-4)}-\frac{x^2-4x}{(x+4)(x-4)}+\frac{3x+1}{(x+4)(x-4)}$$
$$=\frac{2x^2+19x+1}{(x+4)(x-4)}$$

69. $$\frac{1}{x^2+3x+2}-\frac{2}{x^2+4x+3}+\frac{1}{x^2+5x+6}$$
$$=\frac{1}{(x+2)(x+1)}-\frac{2}{(x+3)(x+1)}+\frac{1}{(x+2)(x+3)}$$
$$=\frac{1(x+3)}{(x+2)(x+1)(x+3)}-\frac{2(x+2)}{(x+3)(x+1)(x+2)}+\frac{1(x+1)}{(x+2)(x+3)(x+1)}$$
$$=\frac{x+3}{(x+2)(x+1)(x+3)}-\frac{2x+4}{(x+2)(x+1)(x+3)}+\frac{x+1}{(x+2)(x+1)(x+3)}$$
$$=\frac{x+3-2x-4+x+1}{(x+2)(x+1)(x+3)}=\frac{0}{(x+2)(x+1)(x+3)}=0$$

71. $\dfrac{3x-2}{x^2+x-20}-\dfrac{4x^2+2}{x^2-25}+\dfrac{3x^2-25}{x^2-16}$

$$=\frac{3x-2}{(x+5)(x-4)}-\frac{4x^2+2}{(x+5)(x-5)}+\frac{3x^2-25}{(x+4)(x-4)}$$

$$=\frac{(3x-2)(x-5)(x+4)}{(x+5)(x-4)(x-5)(x+4)}-\frac{(4x^2+2)(x-4)(x+4)}{(x+5)(x-5)(x-4)(x+4)}\cdots$$

$$+\frac{(3x^2-25)(x+5)(x-5)}{(x+4)(x-4)(x+5)(x-5)}$$

$$=\frac{3x^3-5x^2-58x+40}{(x+5)(x-4)(x-5)(x+4)}-\frac{4x^4-62x^2-32}{(x+5)(x-4)(x-5)(x+4)}\cdots$$

$$+\frac{3x^4-100x^2+625}{(x+5)(x-4)(x-5)(x+4)}$$

$$=\frac{3x^3-5x^2-58x+40-4x^4+62x^2+32+3x^4-100x^2+625}{(x+5)(x-4)(x-5)(x+4)}$$

$$=\frac{-x^4+3x^3-43x^2-58x+697}{(x+5)(x-4)(x-5)(x+4)}$$

73. $\dfrac{\frac{3a}{b}}{\frac{6ac}{b^2}}=\dfrac{3a}{b}\div\dfrac{6ac}{b^2}=\dfrac{3a}{b}\cdot\dfrac{b^2}{6ac}=\dfrac{b}{2c}$

75. $\dfrac{3a^2b}{\frac{ab}{27}}=\dfrac{3a^2b}{1}\div\dfrac{ab}{27}=\dfrac{3a^2b}{1}\cdot\dfrac{27}{ab}=81a$

77. $\dfrac{\frac{x-y}{ab}}{\frac{y-x}{ab}}=\dfrac{x-y}{ab}\div\dfrac{y-x}{ab}=\dfrac{x-y}{ab}\cdot\dfrac{ab}{y-x}=-1$

79. $\dfrac{\frac{1}{x}+\frac{1}{y}}{xy}=\dfrac{xy\left(\frac{1}{x}+\frac{1}{y}\right)}{xy(xy)}=\dfrac{xy\left(\frac{1}{x}\right)+xy\left(\frac{1}{y}\right)}{x^2y^2}$

$$=\frac{y+x}{x^2y^2}$$

81. $\dfrac{\frac{1}{x}+\frac{1}{y}}{\frac{1}{x}-\frac{1}{y}}=\dfrac{xy\left(\frac{1}{x}+\frac{1}{y}\right)}{xy\left(\frac{1}{x}-\frac{1}{y}\right)}=\dfrac{xy\left(\frac{1}{x}\right)+xy\left(\frac{1}{y}\right)}{xy\left(\frac{1}{x}\right)-xy\left(\frac{1}{y}\right)}$

$$=\frac{y+x}{y-x}$$

83. $\dfrac{\frac{3a}{b}-\frac{4a^2}{x}}{\frac{1}{b}+\frac{1}{ax}}=\dfrac{abx\left(\frac{3a}{b}-\frac{4a^2}{x}\right)}{abx\left(\frac{1}{b}+\frac{1}{ax}\right)}=\dfrac{abx\left(\frac{3a}{b}\right)-abx\left(\frac{4a^2}{x}\right)}{abx\left(\frac{1}{b}\right)+abx\left(\frac{1}{ax}\right)}=\dfrac{3a^2x-4a^3b}{ax+b}=\dfrac{a^2(3x-4ab)}{ax+b}$

85. $\dfrac{x+1-\frac{6}{x}}{x+5+\frac{6}{x}}=\dfrac{x\left(x+1-\frac{6}{x}\right)}{x\left(x+5+\frac{6}{x}\right)}=\dfrac{x(x)+x(1)-x\left(\frac{6}{x}\right)}{x(x)+x(5)+x\left(\frac{6}{x}\right)}=\dfrac{x^2+x-6}{x^2+5x+6}=\dfrac{(x+3)(x-2)}{(x+2)(x+3)}=\dfrac{x-2}{x+2}$

87. $\dfrac{3xy}{1-\frac{1}{xy}}=\dfrac{xy(3xy)}{xy\left(1-\frac{1}{xy}\right)}=\dfrac{3x^2y^2}{xy(1)-xy\left(\frac{1}{xy}\right)}$

$$=\frac{3x^2y^2}{xy-1}$$

89. $\dfrac{3x}{x+\frac{1}{x}}=\dfrac{x(3x)}{x\left(x+\frac{1}{x}\right)}=\dfrac{3x^2}{x^2+1}$

91. $\dfrac{\frac{x}{x+2}-\frac{2}{x-1}}{\frac{3}{x+2}+\frac{x}{x-1}} = \dfrac{(x+2)(x-1)\left(\frac{x}{x+2}-\frac{2}{x-1}\right)}{(x+2)(x-1)\left(\frac{3}{x+2}+\frac{x}{x-1}\right)} = \dfrac{(x+2)(x-1)\left(\frac{x}{x+2}\right)-(x+2)(x-1)\left(\frac{2}{x-1}\right)}{(x+2)(x-1)\left(\frac{3}{x+2}\right)+(x+2)(x-1)\left(\frac{x}{x-1}\right)}$

$= \dfrac{(x-1)(x)-(x+2)(2)}{(x-1)(3)+(x+2)(x)}$

$= \dfrac{x^2-x-2x-4}{3x-3+x^2+2x} = \dfrac{x^2-3x-4}{x^2+5x-3}$

93. $\dfrac{1}{1+x^{-1}} = \dfrac{1}{1+\frac{1}{x}} = \dfrac{x(1)}{x\left(1+\frac{1}{x}\right)} = \dfrac{x}{x+1}$

95. $\dfrac{3(x+2)^{-1}+2(x-1)^{-1}}{(x+2)^{-1}} = \dfrac{\frac{3}{x+2}+\frac{2}{x-1}}{\frac{1}{x+2}} = \dfrac{(x+2)(x-1)\left(\frac{3}{x+2}+\frac{2}{x-1}\right)}{(x+2)(x-1)\left(\frac{1}{x+2}\right)}$

$= \dfrac{(x+2)(x-1)\left(\frac{3}{x+2}\right)+(x+2)(x-1)\left(\frac{2}{x-1}\right)}{x-1}$

$= \dfrac{(x-1)(3)+(x+2)(2)}{x-1}$

$= \dfrac{3x-3+2x+4}{x-1} = \dfrac{5x+1}{x-1}$

97. $\dfrac{1}{\frac{1}{k_1}+\frac{1}{k_2}} = \dfrac{k_1k_2(1)}{k_1k_2\left(\frac{1}{k_1}+\frac{1}{k_2}\right)} = \dfrac{k_1k_2}{k_1k_2\left(\frac{1}{k_1}\right)+k_1k_2\left(\frac{1}{k_2}\right)} = \dfrac{k_1k_2}{k_2+k_1}$

99. $\dfrac{x}{1+\frac{1}{3x^{-1}}} = \dfrac{x}{1+\frac{1}{\frac{3}{x}}} = \dfrac{x}{1+\frac{x(1)}{x\left(\frac{3}{x}\right)}} = \dfrac{x}{1+\frac{x}{3}} = \dfrac{3x}{3\left(1+\frac{x}{3}\right)} = \dfrac{3x}{3+x}$

101. $\dfrac{1}{1+\frac{1}{1+\frac{1}{x}}} = \dfrac{1}{1+\frac{x(1)}{x\left(1+\frac{1}{x}\right)}} = \dfrac{1}{1+\frac{x}{x+1}} = \dfrac{(x+1)1}{(x+1)\left(1+\frac{x}{x+1}\right)} = \dfrac{x+1}{x+1+x} = \dfrac{x+1}{2x+1}$

103-105. Answers may vary.

107. Since $-6 < 0$, $|-6| = -(-6) = 6$.

109. $\left(\dfrac{x^3y^{-2}}{x^{-1}y}\right)^{-3} = \left(\dfrac{x^{-1}y}{x^3y^{-2}}\right)^3 = \left(\dfrac{y^2y}{x^3x}\right)^3 = \left(\dfrac{y^3}{x^4}\right)^3 = \dfrac{y^9}{x^{12}}$

111. $\sqrt{20}-\sqrt{45} = \sqrt{4}\sqrt{5}-\sqrt{9}\sqrt{5} = 2\sqrt{5}-3\sqrt{5} = -\sqrt{5}$

Chapter 0 Review (page 75)

1. natural: 3, 6, 8

2. whole: 0, 3, 6, 8

3. integers: $-6, -3, 0, 3, 6, 8$

4. rational: $-6, -3, 0, \frac{1}{2}, 3, 6, 8$

5. irrational: π, $\sqrt{5}$

6. real: $-6, -3, 0, \frac{1}{2}, 3, \pi, \sqrt{5}, 6, 8$

CHAPTER 0 REVIEW

7. prime: 3

8. composite: 6, 8

9. even integers: $-6, 0, 6, 8$

10. odd integers: $-3, 3$

11. associative property of addition

12. commutative property of addition

13. associative property of multiplication

14. distributive property

15. commutative property of multiplication

16. commutative property of addition

17. double negative rule

18.

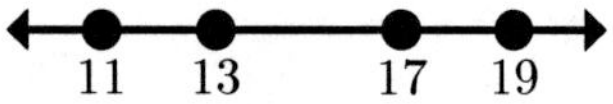

19.

20. $-3 < x \le 5$

21. $x \ge 0$ or $x < -1$

22. $(-2, 4]$

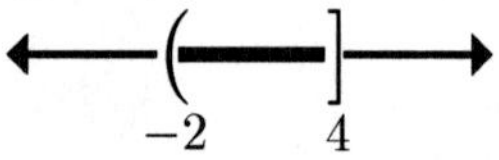

23. $(-\infty, 2) \cap (-5, \infty)$

$(-\infty, 2)$

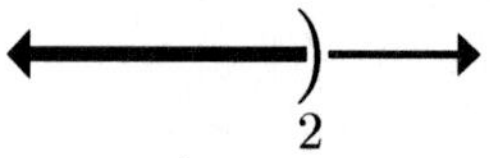

$(-5, \infty)$

$(-\infty, 2) \cap (-5, \infty)$

24. $(-\infty, -4) \cup [6, \infty)$

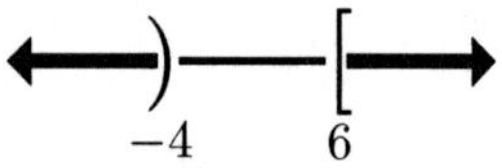

25. Since $6 \ge 0$, $|6| = 6$.

26. Since $-25 < 0$, $|-25| = -(-25) = 25$.

27. Since $1 - \sqrt{2} < 0$,
$\left|1 - \sqrt{2}\right| = -\left(1 - \sqrt{2}\right) = \sqrt{2} - 1$.

28. Since $\sqrt{3} - 1 \ge 0$,
$\left|\sqrt{3} - 1\right| = \sqrt{3} - 1$.

29. distance $= |7 - (-5)| = |12| = 12$

30. $-5a^3 = -5aaa$

31. $(-5a)^2 = (-5a)(-5a)$

CHAPTER 0 REVIEW

32. $3ttt = 3t^3$

33. $(-2b)(3b) = (-2)(3)bb = -6b^2$

34. $n^2n^4 = n^{2+4} = n^6$

35. $(p^3)^2 = p^{3\cdot 2} = p^6$

36. $(x^3y^2)^4 = (x^3)^4(y^2)^4 = x^{12}y^8$

37. $\left(\dfrac{a^4}{b^2}\right)^3 = \dfrac{(a^4)^3}{(b^2)^3} = \dfrac{a^{12}}{b^6}$

38. $(m^{-3}n^0)^2 = (m^{-3}\cdot 1)^2 = m^{-6} = \dfrac{1}{m^6}$

39. $\left(\dfrac{p^{-2}q^2}{2}\right)^3 = \left(\dfrac{q^2}{2p^2}\right)^3 = \dfrac{(q^2)^3}{(2p^2)^3} = \dfrac{q^6}{8p^6}$

40. $\dfrac{a^5}{a^8} = a^{5-8} = a^{-3} = \dfrac{1}{a^3}$

41. $\left(\dfrac{a^2}{b^3}\right)^{-2} = \left(\dfrac{b^3}{a^2}\right)^2 = \dfrac{b^6}{a^4}$

42. $\left(\dfrac{3x^2y^{-2}}{x^2y^2}\right)^{-2} = \left(\dfrac{x^2y^2}{3x^2y^{-2}}\right)^2 = \left(\dfrac{x^2y^2y^2}{3x^2}\right)^2 = \left(\dfrac{y^4}{3}\right)^2 = \dfrac{y^8}{9}$

43. $\left(\dfrac{a^{-3}b^2}{ab^{-3}}\right)^{-2} = \left(\dfrac{ab^{-3}}{a^{-3}b^2}\right)^2 = \left(\dfrac{aa^3}{b^2b^3}\right)^2 = \left(\dfrac{a^4}{b^5}\right)^2 = \dfrac{a^8}{b^{10}}$

44. $\left(\dfrac{-3x^3y}{xy^3}\right)^{-2} = \left(\dfrac{xy^3}{-3x^3y}\right)^2 = \left(\dfrac{y^2}{-3x^2}\right)^2 = \dfrac{y^4}{9x^4}$

45. $\left(-\dfrac{2m^{-2}n^0}{4m^2n^{-1}}\right)^{-3} = \left(-\dfrac{4m^2n^{-1}}{2m^{-2}n^0}\right)^3 = \left(-\dfrac{2m^2m^2}{n^1n^0}\right)^3 = \left(-\dfrac{2m^4}{n}\right)^3 = -\dfrac{8m^{12}}{n^3}$

46. $-x^2 - xy^2 = -(-3)^2 - (-3)(3)^2 = -(+9) - (-3)(9) = -9 - (-27) = -9 + 27 = 18$

47. $6{,}750 = 6.750 \times 10^3$

48. $0.00023 = 2.3 \times 10^{-4}$

49. $4.8 \times 10^2 = 480$

50. $0.25 \times 10^{-3} = 0.00025$

51. $\dfrac{(45{,}000)(350{,}000)}{0.000105} = \dfrac{(4.5\times 10^4)(3.5\times 10^5)}{1.05\times 10^{-4}} = \dfrac{4.5\times 3.5\times 10^4\times 10^5}{1.05\times 10^{-4}} = \dfrac{15.75\times 10^9}{1.05\times 10^{-4}}$
$= 15\times 10^{13}$
$= 1.5\times 10^{14}$

52. $121^{1/2} = (11^2)^{1/2} = 11$

53. $\left(\dfrac{27}{125}\right)^{1/3} = \left[\left(\dfrac{3}{5}\right)^3\right]^{1/3} = \dfrac{3}{5}$

54. $(32x^5)^{1/5} = 32^{1/5}(x^5)^{1/5} = 2x$

55. $(81a^4)^{1/4} = 81^{1/4}(a^4)^{1/4} = 3|a|$

56. $(-1{,}000x^6)^{1/3} = (-1{,}000)^{1/3}(x^6)^{1/3}$
$= -10x^2$

57. $(-25x^2)^{1/2} = (-25)^{1/2}(x^2)^{1/2}$
$\Rightarrow$ not a real number

58. $(x^{12}y^2)^{1/2} = (x^{12})^{1/2}(y^2)^{1/2} = x^6|y|$

59. $\left(\dfrac{x^{12}}{y^4}\right)^{-1/2} = \left(\dfrac{y^4}{x^{12}}\right)^{1/2} = \dfrac{y^2}{x^6}$

60. $\left(\dfrac{-c^{2/3}c^{5/3}}{c^{-2/3}}\right)^{1/3} = \left(\dfrac{-c^{7/3}}{c^{-2/3}}\right)^{1/3} = (-c^{9/3})^{1/3} = (-c^3)^{1/3} = -c$

61. $\left(\dfrac{a^{-1/4}a^{3/4}}{a^{9/2}}\right)^{-1/2} = \left(\dfrac{a^{9/2}}{a^{-1/4}a^{3/4}}\right)^{1/2} = \left(\dfrac{a^{9/2}}{a^{2/4}}\right)^{1/2} = \left(\dfrac{a^{9/2}}{a^{1/2}}\right)^{1/2} = (a^{8/2})^{1/2} = (a^4)^{1/2} = a^2$

62. $64^{2/3} = (64^{1/3})^2 = 4^2 = 16$

63. $32^{-3/5} = \dfrac{1}{32^{3/5}} = \dfrac{1}{(32^{1/5})^3} = \dfrac{1}{2^3} = \dfrac{1}{8}$

64. $\left(\dfrac{16}{81}\right)^{3/4} = \dfrac{16^{3/4}}{81^{3/4}} = \dfrac{(16^{1/4})^3}{(81^{1/4})^3} = \dfrac{2^3}{3^3} = \dfrac{8}{27}$

65. $\left(\dfrac{32}{243}\right)^{2/5} = \dfrac{32^{2/5}}{243^{2/5}} = \dfrac{(32^{1/5})^2}{(243^{1/5})^2} = \dfrac{2^2}{3^2} = \dfrac{4}{9}$

66. $\left(\dfrac{8}{27}\right)^{-2/3} = \left(\dfrac{27}{8}\right)^{2/3} = \dfrac{27^{2/3}}{8^{2/3}} = \dfrac{(27^{1/3})^2}{(8^{1/3})^2} = \dfrac{3^2}{2^2} = \dfrac{9}{4}$

67. $\left(\dfrac{16}{625}\right)^{-3/4} = \left(\dfrac{625}{16}\right)^{3/4} = \dfrac{625^{3/4}}{16^{3/4}} = \dfrac{(625^{1/4})^3}{(16^{1/4})^3} = \dfrac{5^3}{2^3} = \dfrac{125}{8}$

68. $(-216x^3)^{2/3} = (-216)^{2/3}(x^3)^{2/3} = 36x^2$

69. $\dfrac{p^{a/2}p^{a/3}}{p^{a/6}} = \dfrac{p^{3a/6}p^{2a/6}}{p^{a/6}} = \dfrac{p^{5a/6}}{p^{a/6}} = p^{4a/6} = p^{2a/3}$

70. $\sqrt{36} = 6$

71. $-\sqrt{49} = -7$

72. $\sqrt{\dfrac{9}{25}} = \dfrac{\sqrt{9}}{\sqrt{25}} = \dfrac{3}{5}$

73. $\sqrt[3]{\dfrac{27}{125}} = \dfrac{\sqrt[3]{27}}{\sqrt[3]{125}} = \dfrac{3}{5}$

74. $\sqrt{x^2y^4} = \sqrt{x^2}\sqrt{y^4} = |x|y^2$

75. $\sqrt[3]{x^3} = x$

76. $\sqrt[4]{\dfrac{m^8n^4}{p^{16}}} = \dfrac{\sqrt[4]{m^8}\sqrt[4]{n^4}}{\sqrt[4]{p^{16}}} = \dfrac{m^2|n|}{p^4}$

77. $\sqrt[5]{\dfrac{a^{15}b^{10}}{c^5}} = \dfrac{\sqrt[5]{a^{15}}\sqrt[5]{b^{10}}}{\sqrt[5]{c^5}} = \dfrac{a^3b^2}{c}$

78. $\sqrt{50} + \sqrt{8} = \sqrt{25}\sqrt{2} + \sqrt{4}\sqrt{2} = 5\sqrt{2} + 2\sqrt{2} = 7\sqrt{2}$

79. $\sqrt{12} + \sqrt{3} - \sqrt{27} = \sqrt{4}\sqrt{3} + \sqrt{3} - \sqrt{9}\sqrt{3} = 2\sqrt{3} + \sqrt{3} - 3\sqrt{3} = 3\sqrt{3} - 3\sqrt{3} = 0$

80. $\sqrt[3]{24x^4} - \sqrt[3]{3x^4} = \sqrt[3]{8x^3}\sqrt[3]{3x} - \sqrt[3]{x^3}\sqrt[3]{3x} = 2x\sqrt[3]{3x} - x\sqrt[3]{3x} = x\sqrt[3]{3x}$

CHAPTER 0 REVIEW

81. $\frac{\sqrt{7}}{\sqrt{5}} = \frac{\sqrt{7}}{\sqrt{5}} \cdot \frac{\sqrt{5}}{\sqrt{5}} = \frac{\sqrt{35}}{5}$

82. $\frac{8}{\sqrt{8}} = \frac{8}{\sqrt{8}} \cdot \frac{\sqrt{2}}{\sqrt{2}} = \frac{8\sqrt{2}}{\sqrt{16}} = \frac{8\sqrt{2}}{4} = 2\sqrt{2}$

83. $\frac{1}{\sqrt[3]{2}} = \frac{1}{\sqrt[3]{2}} \cdot \frac{\sqrt[3]{4}}{\sqrt[3]{4}} = \frac{\sqrt[3]{4}}{\sqrt[3]{8}} = \frac{\sqrt[3]{4}}{2}$

84. $\frac{2}{\sqrt[3]{25}} = \frac{2}{\sqrt[3]{25}} \cdot \frac{\sqrt[3]{5}}{\sqrt[3]{5}} = \frac{2\sqrt[3]{5}}{\sqrt[3]{125}} = \frac{2\sqrt[3]{5}}{5}$

85. $\frac{\sqrt{2}}{5} = \frac{\sqrt{2}}{5} \cdot \frac{\sqrt{2}}{\sqrt{2}} = \frac{2}{5\sqrt{2}}$

86. $\frac{\sqrt{5}}{5} = \frac{\sqrt{5}}{5} \cdot \frac{\sqrt{5}}{\sqrt{5}} = \frac{5}{5\sqrt{5}} = \frac{1}{\sqrt{5}}$

87. $\frac{\sqrt{2x}}{3} = \frac{\sqrt{2x}}{3} \cdot \frac{\sqrt{2x}}{\sqrt{2x}} = \frac{2x}{3\sqrt{2x}}$

88. $\frac{3\sqrt[3]{7x}}{2} = \frac{3\sqrt[3]{7x}}{2} \cdot \frac{\sqrt[3]{49x^2}}{\sqrt[3]{49x^2}} = \frac{3\sqrt[3]{343x^3}}{2\sqrt[3]{49x^2}} = \frac{21x}{2\sqrt[3]{49x^2}}$

89. 3rd degree, binomial

90. 2nd degree, trinomial

91. 2nd degree, monomial

92. 4th degree, trinomial

93. $2(x+3) + 3(x-4) = 2x + 6 + 3x - 12 = 5x - 6$

94. $3x^2(x-1) - 2x(x+3) - x^2(x+2) = 3x^3 - 3x^2 - 2x^2 - 6x - x^3 - 2x^2 = 2x^3 - 7x^2 - 6x$

95. $(3x+2)(3x+2) = 9x^2 + 6x + 6x + 4 = 9x^2 + 12x + 4$

96. $(3x+y)(2x-3y) = 6x^2 - 9xy + 2xy - 3y^2 = 6x^2 - 7xy - 3y^2$

97. $(4a+2b)(2a-3b) = 8a^2 - 12ab + 4ab - 6b^2 = 8a^2 - 8ab - 6b^2$

98. $(z+3)(3z^2+z-1) = 3z^3 + z^2 - z + 9z^2 + 3z - 3 = 3z^3 + 10z^2 + 2z - 3$

99. $(a^n+2)(a^n-1) = a^{2n} - a^n + 2a^n - 2 = a^{2n} + a^n - 2$

100. $\left(\sqrt{2}+x\right)^2 = \left(\sqrt{2}+x\right)\left(\sqrt{2}+x\right) = \left(\sqrt{2}\right)^2 + x\sqrt{2} + x\sqrt{2} + x^2 = 2 + 2x\sqrt{2} + x^2$

101. $\left(\sqrt{2}+1\right)\left(\sqrt{3}+1\right) = \sqrt{6} + \sqrt{2} + \sqrt{3} + 1$

102. $\left(\sqrt[3]{3}-2\right)\left(\sqrt[3]{9}+2\sqrt[3]{3}+4\right) = \sqrt[3]{27} + 2\sqrt[3]{9} + 4\sqrt[3]{3} - 2\sqrt[3]{9} - 4\sqrt[3]{3} - 8 = 3 - 8 = -5$

103. $\frac{2}{\sqrt{3}-1} = \frac{2}{\sqrt{3}-1} \cdot \frac{\sqrt{3}+1}{\sqrt{3}+1} = \frac{2\left(\sqrt{3}+1\right)}{\left(\sqrt{3}\right)^2 - 1^2} = \frac{2\left(\sqrt{3}+1\right)}{3-1} = \frac{2\left(\sqrt{3}+1\right)}{2} = \sqrt{3}+1$

104. $\dfrac{-2}{\sqrt{3}-\sqrt{2}} = \dfrac{-2}{\sqrt{3}-\sqrt{2}} \cdot \dfrac{\sqrt{3}+\sqrt{2}}{\sqrt{3}+\sqrt{2}} = \dfrac{-2\left(\sqrt{3}+\sqrt{2}\right)}{\left(\sqrt{3}\right)^2-\left(\sqrt{2}\right)^2} = \dfrac{-2\left(\sqrt{3}+\sqrt{2}\right)}{3-2}$

$= \dfrac{-2\left(\sqrt{3}+\sqrt{2}\right)}{1}$

$= -2\left(\sqrt{3}+\sqrt{2}\right)$

105. $\dfrac{2x}{\sqrt{x}-2} = \dfrac{2x}{\sqrt{x}-2} \cdot \dfrac{\sqrt{x}+2}{\sqrt{x}+2} = \dfrac{2x\left(\sqrt{x}+2\right)}{\left(\sqrt{x}\right)^2-2^2} = \dfrac{2x\left(\sqrt{x}+2\right)}{x-4}$

106. $\dfrac{\sqrt{x}-\sqrt{y}}{\sqrt{x}+\sqrt{y}} = \dfrac{\sqrt{x}-\sqrt{y}}{\sqrt{x}+\sqrt{y}} \cdot \dfrac{\sqrt{x}-\sqrt{y}}{\sqrt{x}-\sqrt{y}} = \dfrac{\sqrt{x^2}-\sqrt{xy}-\sqrt{xy}+y}{\left(\sqrt{x}\right)^2-\left(\sqrt{y}\right)^2} = \dfrac{x-2\sqrt{xy}+y}{x-y}$

107. $\dfrac{\sqrt{x}+2}{5} = \dfrac{\sqrt{x}+2}{5} \cdot \dfrac{\sqrt{x}-2}{\sqrt{x}-2} = \dfrac{\left(\sqrt{x}\right)^2-2^2}{5\left(\sqrt{x}-2\right)} = \dfrac{x-4}{5\left(\sqrt{x}-2\right)}$

108. $\dfrac{1-\sqrt{a}}{a} = \dfrac{1-\sqrt{a}}{a} \cdot \dfrac{1+\sqrt{a}}{1+\sqrt{a}} = \dfrac{1^2-\left(\sqrt{a}\right)^2}{a\left(1+\sqrt{a}\right)} = \dfrac{1-a}{a\left(1+\sqrt{a}\right)}$

109. $\dfrac{3x^2y^2}{6x^3y} = \dfrac{y}{2x}$

110. $\dfrac{4a^2b^3+6ab^4}{2b^2} = \dfrac{4a^2b^3}{2b^2} + \dfrac{6ab^4}{2b^2}$

$= 2a^2b + 3ab^2$

111.

$$\begin{array}{r} x^2 + 2x + 1 \\ 2x+3\,\big)\overline{2x^3+7x^2+8x+3} \\ \underline{2x^3+3x^2} \\ 4x^2+8x \\ \underline{4x^2+6x} \\ 2x+3 \\ \underline{2x+3} \\ 0 \end{array}$$

112.

$$\begin{array}{r} x^3 + 2x - 3 + \frac{-6}{x^2-1} \\ x^2-1\,\big)\overline{x^5+0x^4+x^3-3x^2-2x-3} \\ \underline{x^5-x^3} \\ 2x^3-3x^2-2x \\ \underline{2x^3-2x} \\ -3x^2-3 \\ \underline{-3x^2+3} \\ -6 \end{array}$$

113. $3t^3-3t = 3t(t^2-1) = 3t(t+1)(t-1)$

114. $5r^3-5 = 5(r^3-1) = 5(r^3-1^3) = 5(r-1)(r^2+r+1)$

115. $6x^2+7x-24 = (3x+8)(2x-3)$

116. $3a^2+ax-3a-x = a(3a+x)-1(3a+x) = (3a+x)(a-1)$

117. $8x^3-125 = (2x)^3-5^3 = (2x-5)\left[(2x)^2+(2x)(5)+5^2\right] = (2x-5)(4x^2+10x+25)$

118. $6x^2-20x-16 = 2(3x^2-10x-8) = 2(3x+2)(x-4)$

CHAPTER 0 REVIEW

119. $x^2 + 6x + 9 - t^2 = (x+3)(x+3) - t^2 = (x+3)^2 - t^2 = (x+3+t)(x+3-t)$

120. $3x^2 - 1 + 5x = 3x^2 + 5x - 1 \Rightarrow$ prime

121. $8z^3 + 343 = (2z)^3 + 7^3 = (2z+7)\left[(2z)^2 - (2z)(7) + 7^2\right] = (2z+7)(4z^2 - 14z + 49)$

122. $1 + 14b + 49b^2 = 49b^2 + 14b + 1 = (7b+1)(7b+1) = (7b+1)^2$

123. $121z^2 + 4 - 44z = 121z^2 - 44z + 4 = (11z-2)(11z-2) = (11z-2)^2$

124. $64y^3 - 1{,}000 = 8(8y^3 - 125) = 8\left[(2y)^3 - 5^3\right] = 8(2y-5)(4y^2 + 10y + 25)$

125. $2xy - 4zx - wy + 2zw = 2x(y-2z) - w(y-2z) = (y-2z)(2x-w)$

126.
$$\begin{aligned} x^8 + x^4 + 1 = x^8 + 2x^4 + 1 - x^4 &= (x^4+1)(x^4+1) - x^4 \\ &= (x^4+1)^2 - (x^2)^2 \\ &= (x^4+1+x^2)(x^4+1-x^2) \\ &= (x^4+2x^2+1-x^2)(x^4-x^2+1) \\ &= \left[(x^2+1)(x^2+1) - x^2\right](x^4-x^2+1) \\ &= (x^2+1+x)(x^2+1-x)(x^4-x^2+1) \end{aligned}$$

127. $\dfrac{2-x}{x^2-4x+4} = \dfrac{-(x-2)}{(x-2)(x-2)} = \dfrac{-1}{x-2}$

128. $\dfrac{a^2-9}{a^2-6a+9} = \dfrac{(a+3)(a-3)}{(a-3)(a-3)} = \dfrac{a+3}{a-3}$

129. $\dfrac{x^2-4x+4}{x+2} \cdot \dfrac{x^2+5x+6}{x-2} = \dfrac{(x-2)(x-2)}{x+2} \cdot \dfrac{(x+2)(x+3)}{x-2} = (x-2)(x+3)$

130. $\dfrac{2y^2-11y+15}{y^2-6y+8} \cdot \dfrac{y^2-2y-8}{y^2-y-6} = \dfrac{(2y-5)(y-3)}{(y-4)(y-2)} \cdot \dfrac{(y-4)(y+2)}{(y-3)(y+2)} = \dfrac{2y-5}{y-2}$

131.
$$\begin{aligned} \frac{2t^2+t-3}{3t^2-7t+4} \div \frac{10t+15}{3t^2-t-4} &= \frac{2t^2+t-3}{3t^2-7t+4} \cdot \frac{3t^2-t-4}{10t+15} \\ &= \frac{(2t+3)(t-1)}{(3t-4)(t-1)} \cdot \frac{(3t-4)(t+1)}{5(2t+3)} = \frac{t+1}{5} \end{aligned}$$

132.
$$\begin{aligned} \frac{p^2+7p+12}{p^3+8p^2+4p} \div \frac{p^2-9}{p^2} = \frac{p^2+7p+12}{p^3+8p^2+4p} \cdot \frac{p^2}{p^2-9} &= \frac{(p+3)(p+4)}{p(p^2+8p+4)} \cdot \frac{p^2}{(p+3)(p-3)} \\ &= \frac{p(p+4)}{(p^2+8p+4)(p-3)} \end{aligned}$$

133. $\dfrac{x^2+x-6}{x^2-x-6}\cdot\dfrac{x^2-x-6}{x^2+x-2}\div\dfrac{x^2-4}{x^2-5x+6}$

$$=\frac{x^2+x-6}{x^2-x-6}\cdot\frac{x^2-x-6}{x^2+x-2}\cdot\frac{x^2-5x+6}{x^2-4}$$

$$=\frac{(x+3)(x-2)}{(x-3)(x+2)}\cdot\frac{(x-3)(x+2)}{(x+2)(x-1)}\cdot\frac{(x-2)(x-3)}{(x+2)(x-2)}=\frac{(x+3)(x-2)(x-3)}{(x+2)^2(x-1)}$$

134. $\left(\dfrac{2x+6}{x+5}\div\dfrac{2x^2-2x-4}{x^2-25}\right)\dfrac{x^2-x-2}{x^2-2x-15}=\dfrac{2x+6}{x+5}\cdot\dfrac{x^2-25}{2(x^2-x-2)}\cdot\dfrac{x^2-x-2}{x^2-2x-15}$

$$=\frac{2(x+3)}{x+5}\cdot\frac{(x+5)(x-5)}{2(x-2)(x+1)}\cdot\frac{(x-2)(x+1)}{(x-5)(x+3)}=1$$

135. $\dfrac{2}{x-4}+\dfrac{3x}{x+5}=\dfrac{2(x+5)}{(x-4)(x+5)}+\dfrac{3x(x-4)}{(x+5)(x-4)}=\dfrac{2x+10}{(x-4)(x+5)}+\dfrac{3x^2-12x}{(x-4)(x+5)}$

$$=\frac{3x^2-10x+10}{(x-4)(x+5)}$$

136. $\dfrac{5x}{x-2}-\dfrac{3x+7}{x+2}+\dfrac{2x+1}{x+2}=\dfrac{5x}{x-2}+\dfrac{-x-6}{x+2}=\dfrac{5x(x+2)}{(x-2)(x+2)}+\dfrac{(-x-6)(x-2)}{(x+2)(x-2)}$

$$=\frac{5x^2+10x}{(x-2)(x+2)}+\frac{-x^2-4x+12}{(x-2)(x+2)}$$

$$=\frac{4x^2+6x+12}{(x-2)(x+2)}=\frac{2(2x^2+3x+6)}{(x-2)(x+2)}$$

137. $\dfrac{x}{x-1}+\dfrac{x}{x-2}+\dfrac{x}{x-3}$

$$=\frac{x(x-2)(x-3)}{(x-1)(x-2)(x-3)}+\frac{x(x-1)(x-3)}{(x-2)(x-1)(x-3)}+\frac{x(x-1)(x-2)}{(x-3)(x-1)(x-2)}$$

$$=\frac{x^3-5x^2+6x}{(x-1)(x-2)(x-3)}+\frac{x^3-4x^2+3x}{(x-1)(x-2)(x-3)}+\frac{x^3-3x^2+2x}{(x-1)(x-2)(x-3)}$$

$$=\frac{3x^3-12x^2+11x}{(x-1)(x-2)(x-3)}=\frac{x(3x^2-12x+11)}{(x-1)(x-2)(x-3)}$$

138. $\dfrac{x}{x+1}-\dfrac{3x+7}{x+2}+\dfrac{2x+1}{x+2}=\dfrac{x}{x+1}+\dfrac{-3x-7}{x+2}+\dfrac{2x+1}{x+2}$

$$=\frac{x}{x+1}+\frac{-x-6}{x+2}$$

$$=\frac{x(x+2)}{(x+1)(x+2)}+\frac{(-x-6)(x+1)}{(x+2)(x+1)}$$

$$=\frac{x^2+2x}{(x+1)(x+2)}+\frac{-x^2-7x-6}{(x+1)(x+2)}=\frac{-5x-6}{(x+1)(x+2)}$$

139. $$\frac{3(x+1)}{x} - \frac{5(x^2+3)}{x^2} + \frac{x}{x+1} = \frac{3x(x+1)(x+1)}{x^2(x+1)} - \frac{5(x^2+3)(x+1)}{x^2(x+1)} + \frac{x(x^2)}{x^2(x+1)}$$
$$= \frac{3x^3+6x^2+3x}{x^2(x+1)} - \frac{5x^3+5x^2+15x+15}{x^2(x+1)} + \frac{x^3}{x^2(x+1)}$$
$$= \frac{-x^3+x^2-12x-15}{x^2(x+1)}$$

140. $$\frac{3x}{x+1} + \frac{x^2+4x+3}{x^2+3x+2} - \frac{x^2+x-6}{x^2-4} = \frac{3x}{x+1} + \frac{(x+3)(x+1)}{(x+1)(x+2)} - \frac{(x+3)(x-2)}{(x+2)(x-2)}$$
$$= \frac{3x}{x+1} + \frac{x+3}{x+2} - \frac{x+3}{x+2} = \frac{3x}{x+1}$$

141. $$\frac{\frac{5x}{2}}{\frac{3x^2}{8}} = \frac{5x}{2} \div \frac{3x^2}{8} = \frac{5x}{2} \cdot \frac{8}{3x^2} = \frac{20}{3x}$$

142. $$\frac{\frac{3x}{y}}{\frac{6x}{y^2}} = \frac{3x}{y} \div \frac{6x}{y^2} = \frac{3x}{y} \cdot \frac{y^2}{6x} = \frac{y}{2}$$

143. $$\frac{\frac{1}{x}+\frac{1}{y}}{x-y} = \frac{xy\left(\frac{1}{x}+\frac{1}{y}\right)}{xy(x-y)} = \frac{xy\left(\frac{1}{x}\right)+xy\left(\frac{1}{y}\right)}{xy(x-y)} = \frac{y+x}{xy(x-y)}$$

144. $$\frac{x^{-1}+y^{-1}}{y^{-1}-x^{-1}} = \frac{\frac{1}{x}+\frac{1}{y}}{\frac{1}{y}-\frac{1}{x}} = \frac{xy\left(\frac{1}{x}+\frac{1}{y}\right)}{xy\left(\frac{1}{y}-\frac{1}{x}\right)} = \frac{xy\left(\frac{1}{x}\right)+xy\left(\frac{1}{y}\right)}{xy\left(\frac{1}{y}\right)-xy\left(\frac{1}{x}\right)} = \frac{y+x}{x-y}$$

Chapter 0 Test (page 82)

1. odd integers: $-7, 1, 3$

2. prime numbers: 3

3. commutative property of addition

4. distributive property

5. $-4 < x \leq 2 \Rightarrow$ 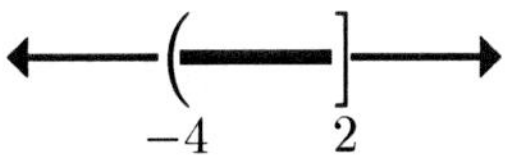

6. $(-\infty, -3) \cup [6, \infty) \Rightarrow$ 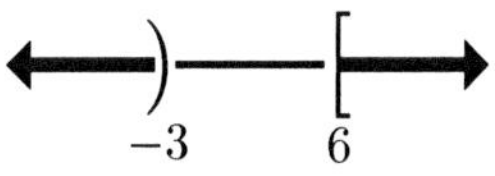

7. Since $-17 < 0$, $|-17| = -(-17) = 17$

8. If $x < 0$, then $x - 7 < 0$. Then $|x-7| = -(x-7)$.

9. distance $= |12 - (-4)| = |16| = 16$

10. distance $= |-12 - (-20)| = |8| = 8$

11. $x^4x^5x^2 = x^{4+5+2} = x^{11}$

12. $\frac{r^2r^3s}{r^4s^2} = \frac{r^5s}{r^4s^2} = \frac{r}{s}$

13. $\frac{(a^{-1}a^2)^{-2}}{a^{-3}} = \frac{(a^1)^{-2}}{a^{-3}} = \frac{a^{-2}}{a^{-3}} = a$

14. $\left(\frac{x^0x^2}{x^{-2}}\right)^6 = \left(\frac{x^2}{x^{-2}}\right)^6 = (x^4)^6 = x^{24}$

15. $450{,}000 = 4.5 \times 10^5$

16. $0.000345 = 3.45 \times 10^{-4}$

CHAPTER 0 TEST

17. $3.7 \times 10^3 = 3{,}700$

18. $1.2 \times 10^{-3} = 0.0012$

19. $\left(25a^4\right)^{1/2} = 25^{1/2}\left(a^4\right)^{1/2} = 5a^2$

20. $\left(\dfrac{36}{81}\right)^{3/2} = \dfrac{36^{3/2}}{81^{3/2}} = \dfrac{\left(36^{1/2}\right)^3}{\left(81^{1/2}\right)^3} = \dfrac{216}{729}$

21. $\left(\dfrac{8t^6}{27s^9}\right)^{-2/3} = \left(\dfrac{27s^9}{8t^6}\right)^{2/3} = \dfrac{27^{2/3}\left(s^9\right)^{2/3}}{8^{2/3}\left(t^6\right)^{2/3}} = \dfrac{\left(27^{1/3}\right)^2 s^6}{\left(8^{1/3}\right)^2 t^4} = \dfrac{9s^6}{4t^4}$

22. $\sqrt[3]{27a^6} = \sqrt[3]{27}\sqrt[3]{a^6} = 3a^2$

23. $\sqrt{12} + \sqrt{27} = \sqrt{4}\sqrt{3} + \sqrt{9}\sqrt{3}$
$= 2\sqrt{3} + 3\sqrt{3} = 5\sqrt{3}$

24. $2\sqrt[3]{3x^4} - 3x\sqrt[3]{24x} = 2\sqrt[3]{x^3}\sqrt[3]{3x} - 3x\sqrt[3]{8}\sqrt[3]{3x} = 2x\sqrt[3]{3x} - 3x(2)\sqrt[3]{3x} = 2x\sqrt[3]{3x} - 6x\sqrt[3]{3x}$
$= -4x\sqrt[3]{3x}$

25. $\dfrac{x}{\sqrt{x}-2} = \dfrac{x}{\sqrt{x}-2}\cdot\dfrac{\sqrt{x}+2}{\sqrt{x}+2} = \dfrac{x\left(\sqrt{x}+2\right)}{\left(\sqrt{x}\right)^2 - 2^2} = \dfrac{x\left(\sqrt{x}+2\right)}{x-4}$

26. $\dfrac{\sqrt{x}-\sqrt{y}}{\sqrt{x}+\sqrt{y}} = \dfrac{\sqrt{x}-\sqrt{y}}{\sqrt{x}+\sqrt{y}}\cdot\dfrac{\sqrt{x}+\sqrt{y}}{\sqrt{x}+\sqrt{y}} = \dfrac{\left(\sqrt{x}\right)^2 - \left(\sqrt{y}\right)^2}{\sqrt{x^2}+\sqrt{xy}+\sqrt{xy}+\sqrt{y^2}} = \dfrac{x-y}{x+2\sqrt{xy}+y}$

27. $\left(a^2+3\right) - \left(2a^2-4\right) = a^2 + 3 - 2a^2 + 4$
$= -a^2 + 7$

28. $(3a^3b^2)(-2a^3b^4) = -6a^6b^6$

29. $(3x-4)(2x+7) = 6x^2 + 21x - 8x - 28$
$= 6x^2 + 13x - 28$

30. $(a^n+2)(a^n-3) = a^{2n} - 3a^n + 2a^n - 6$
$= a^{2n} - a^n - 6$

31. $(x^2+4)(x^2-4) = x^4 - 4x^2 + 4x^2 - 16 = x^4 - 16$

32. $(x^2 - x + 2)(2x - 3) = 2x^3 - 3x^2 - 2x^2 + 3x + 4x - 6 = 2x^3 - 5x^2 + 7x - 6$

33.
$$\begin{array}{r|l} & 6x + 19 + \frac{34}{x-3} \\ \hline x-3 & 6x^2 + x - 23 \\ & \underline{6x^2 - 18x} \\ & 19x - 23 \\ & \underline{19x - 57} \\ & 34 \end{array}$$

34.
$$\begin{array}{r|l} & x^2 + 2x + 1 \\ \hline 2x-1 & 2x^3 + 3x^2 + 0x - 1 \\ & \underline{2x^3 - x^2} \\ & 4x^2 + 0x \\ & \underline{4x^2 - 2x} \\ & 2x - 1 \\ & \underline{2x - 1} \\ & 0 \end{array}$$

35. $3x + 6y = 3(x + 2y)$

36. $x^2 - 100 = x^2 - 10^2 = (x+10)(x-10)$

37. $10t^2 - 19tw + 6w^2 = (5t - 2w)(2t - 3w)$

38. $3a^3 - 648 = 3\left(a^3 - 216\right)$
$= 3(a-6)\left(a^2 + 6a + 36\right)$

39. $x^4 - x^2 - 12 = (x^2 - 4)(x^2 + 3)$
$= (x + 2)(x - 2)(x^2 + 3)$

40. $6x^4 + 11x^2 - 10 = (3x^2 - 2)(2x^2 + 5)$

41. $\frac{x}{x+2} + \frac{2}{x+2} = \frac{x+2}{x+2} = 1$

42. $\frac{x}{x+1} - \frac{x}{x-1} = \frac{x(x-1)}{(x+1)(x-1)} - \frac{x(x+1)}{(x+1)(x-1)} = \frac{x^2 - x - x^2 - x}{(x+1)(x-1)} = \frac{-2x}{(x+1)(x-1)}$

43. $\frac{x^2 + x - 20}{x^2 - 16} \cdot \frac{x^2 - 25}{x - 5} = \frac{(x+5)(x-4)}{(x+4)(x-4)} \cdot \frac{(x+5)(x-5)}{x-5} = \frac{(x+5)^2}{x+4}$

44. $\frac{x+2}{x^2 + 2x + 1} \div \frac{x^2 - 4}{x+1} = \frac{x+2}{(x+1)(x+1)} \cdot \frac{x+1}{(x+2)(x-2)} = \frac{1}{(x+1)(x-2)}$

45. $\frac{\frac{1}{a} + \frac{1}{b}}{\frac{1}{b}} = \frac{ab\left(\frac{1}{a} + \frac{1}{b}\right)}{ab\left(\frac{1}{b}\right)} = \frac{ab\left(\frac{1}{a}\right) + ab\left(\frac{1}{b}\right)}{a} = \frac{b+a}{a}$

46. $\frac{x^{-1}}{x^{-1} + y^{-1}} = \frac{\frac{1}{x}}{\frac{1}{x} + \frac{1}{y}} = \frac{xy\left(\frac{1}{x}\right)}{xy\left(\frac{1}{x} + \frac{1}{y}\right)} = \frac{y}{xy\left(\frac{1}{x}\right) + xy\left(\frac{1}{y}\right)} = \frac{y}{y+x}$

Exercises 1.1 (page 93)

1. root, solution

3. no

5. linear

7. one

9. $x + 3 = 1$; no restrictions

11. $\frac{1}{x} = 12$; $x \neq 0$

13.
$$\frac{8}{x-3} = \frac{5}{x+2}$$
$$x - 3 \neq 0 \quad x + 2 \neq 0$$
$$x \neq 3 \quad x \neq -2$$
$$x \neq 3, x \neq -2$$

15.
$$\frac{1}{x-3} = \frac{5x}{x^2 - 16}$$
$$\frac{1}{x-3} = \frac{5x}{(x+4)(x-4)}$$
$$x - 3 \neq 0 \quad x + 4 \neq 0 \quad x - 4 \neq 0$$
$$x \neq 3 \quad x \neq -4 \quad x \neq 4$$
$$x \neq 3, x \neq -4, x \neq 4$$

17.
$$2x + 5 = 15$$
$$2x + 5 - 5 = 15 - 5$$
$$2x = 10$$
$$\frac{2x}{2} = \frac{10}{2}$$
$$x = 5$$
conditional equation

19.
$$2(n + 2) - 5 = 2n$$
$$2n + 4 - 5 = 2n$$
$$2n - 1 = 2n$$
$$2n - 2n - 1 = 2n - 2n$$
$$-1 \neq 0$$
contradiction

21.
$$\frac{x+7}{2} = 7$$
$$2 \cdot \frac{x+7}{2} = 2(7)$$
$$x + 7 = 14$$
$$x + 7 - 7 = 14 - 7$$
$$x = 7$$
conditional equation

23.
$$\begin{aligned}2(a+1) &= 3(a-2)-a\\ 2a+2 &= 3a-6-a\\ 2a+2 &= 2a-6\\ 2a-2a+2 &= 2a-2a-6\\ 2 &\neq -6\end{aligned}$$
contradiction

25.
$$\begin{aligned}3(x-3) &= \frac{6x-18}{2}\\ 3x-9 &= \frac{6x-18}{2}\\ 2(3x-9) &= 2\cdot\frac{6x-18}{2}\\ 6x-18 &= 6x-18\end{aligned}$$
identity

27.
$$\begin{aligned}\frac{3}{b-3} &= 1\\ (b-3)\cdot\frac{3}{b-3} &= (b-3)(1)\\ 3 &= b-3\\ 3+3 &= b-3+3\\ 6 &= b\end{aligned}$$
conditional equation

29.
$$\begin{aligned}2x^2+5x-3 &= (2x-1)(x+3)\\ 2x^2+5x-3 &= 2x^2+5x-3\end{aligned}$$
identity

31.
$$\begin{aligned}2x+7 &= 10-x\\ 3x+7 &= 10\\ 3x &= 3\\ x &= 1\end{aligned}$$

33.
$$\begin{aligned}5(x-2) &= 2(x+4)\\ 5x-10 &= 2x+8\\ 3x-10 &= 8\\ 3x &= 18\\ x &= 6\end{aligned}$$

35.
$$\begin{aligned}7(2x+5)-6(x+8) &= 7\\ 14x+35-6x-48 &= 7\\ 8x-13 &= 7\\ 8x &= 20\\ x &= \frac{20}{8} = \frac{5}{2}\end{aligned}$$

37.
$$\begin{aligned}\frac{5}{3}z-8 &= 7\\ \frac{5}{3}z &= 15\\ 3\cdot\frac{5}{3}z &= 3(15)\\ 5z &= 45\\ z &= 9\end{aligned}$$

39.
$$\begin{aligned}\frac{z}{5}+2 &= 4\\ \frac{z}{5} &= 2\\ 5\cdot\frac{z}{5} &= 5(2)\\ z &= 10\end{aligned}$$

41.
$$\begin{aligned}\frac{3x-2}{3} &= 2x+\frac{7}{3}\\ 3\cdot\frac{3x-2}{3} &= 3\left(2x+\frac{7}{3}\right)\\ 3x-2 &= 6x+7\\ -3x-2 &= 7\\ -3x &= 9\\ x &= -3\end{aligned}$$

43.
$$\begin{aligned}\frac{3x+1}{20} &= \frac{1}{2}\\ 20\cdot\frac{3x+1}{20} &= 20\cdot\frac{1}{2}\\ 3x+1 &= 10\\ 3x &= 9\\ x &= 3\end{aligned}$$

45.

$$\begin{aligned}
\frac{3+x}{3} + \frac{x+7}{2} &= 4x+1 \\
6\left(\frac{3+x}{3} + \frac{x+7}{2}\right) &= 6(4x+1) \\
2(3+x) + 3(x+7) &= 24x+6 \\
6 + 2x + 3x + 21 &= 24x+6 \\
5x + 27 &= 24x+6 \\
-19x + 27 &= 6 \\
-19x &= -21 \\
x &= \frac{21}{19}
\end{aligned}$$

47.

$$\begin{aligned}
\frac{3}{2}(3x-2) - 10x - 4 &= 0 \\
2\left[\frac{3}{2}(3x-2) - 10x - 4\right] &= 2(0) \\
3(3x-2) - 20x - 8 &= 0 \\
9x - 6 - 20x - 8 &= 0 \\
-11x - 14 &= 0 \\
-11x &= 14 \\
x &= -\frac{14}{11}
\end{aligned}$$

49.

$$\begin{aligned}
\frac{(y+2)^2}{3} &= y + 2 + \frac{y^2}{3} \\
3\left[\frac{(y+2)^2}{3}\right] &= 3\left(y + 2 + \frac{y^2}{3}\right) \\
(y+2)^2 &= 3y + 6 + y^2 \\
y^2 + 4y + 4 &= y^2 + 3y + 6 \\
4y + 4 &= 3y + 6 \\
y + 4 &= 6 \\
y &= 2
\end{aligned}$$

51.

$$\begin{aligned}
x(x+2) &= (x+1)^2 - 1 \\
x^2 + 2x &= (x+1)(x+1) - 1 \\
x^2 + 2x &= x^2 + 2x + 1 - 1 \\
x^2 + 2x &= x^2 + 2x \\
0 &= 0 \Rightarrow \text{identity}
\end{aligned}$$

53.

$$\begin{aligned}
2(s+2) + (s+3)^2 &= s(s+5) + 2\left(\frac{17}{2} + s\right) \\
2s + 4 + s^2 + 6s + 9 &= s^2 + 5s + 17 + 2s \\
s^2 + 8s + 13 &= s^2 + 7s + 17 \\
8s + 13 &= 7s + 17 \\
s &= 4
\end{aligned}$$

55.

$$\begin{aligned}
\frac{2}{x+1} + \frac{1}{3} &= \frac{1}{x+1} \\
3(x+1)\left(\frac{2}{x+1} + \frac{1}{3}\right) &= 3(x+1)\cdot\frac{1}{x+1} \\
6 + 1(x+1) &= 3(1) \\
6 + x + 1 &= 3 \\
x + 7 &= 3 \\
x &= -4
\end{aligned}$$

57.

$$\frac{9t+6}{t(t+3)} = \frac{7}{t+3}$$
$$t(t+3)\left[\frac{9t+6}{t(t+3)}\right] = t(t+3)\cdot\frac{7}{t+3}$$
$$9t+6 = 7t$$
$$2t+6 = 0$$
$$2t = -6$$
$$t = -3; \text{ The answer does not check. } \Rightarrow \text{ no solution}$$

59.

$$\frac{2}{(a-7)(a+2)} = \frac{4}{(a+3)(a+2)}$$
$$(a-7)(a+2)(a+3)\cdot\frac{2}{(a-7)(a+2)} = (a-7)(a+2)(a+3)\cdot\frac{4}{(a+3)(a+2)}$$
$$2(a+3) = 4(a-7)$$
$$2a+6 = 4a-28$$
$$-2a = -34$$
$$a = 17$$

61.

$$\frac{2x+3}{x^2+5x+6} + \frac{3x-2}{x^2+x-6} = \frac{5x-2}{x^2-4}$$
$$\frac{2x+3}{(x+3)(x+2)} + \frac{3x-2}{(x+3)(x-2)} = \frac{5x-2}{(x+2)(x-2)}$$
$$(x-2)(2x+3)+(x+2)(3x-2) = (x+3)(5x-2) \quad \{\text{multiply by common denominator}\}$$
$$2x^2-x-6+3x^2+4x-4 = 5x^2+13x-6$$
$$5x^2+3x-10 = 5x^2+13x-6$$
$$3x-10 = 13x-6$$
$$-10x = 4$$
$$x = -\frac{4}{10} = -\frac{2}{5}$$

63.

$$\frac{3x+5}{x^3+8} + \frac{3}{x^2-4} = \frac{2(3x-2)}{(x-2)(x^2-2x+4)}$$
$$\frac{3x+5}{(x+2)(x^2-2x+4)} + \frac{3}{(x+2)(x-2)} = \frac{2(3x-2)}{(x-2)(x^2-2x+4)}$$
$$(x-2)(3x+5)+\left(x^2-2x+4\right)(3) = 2(x+2)(3x-2) \quad \{\text{multiply by common denominator}\}$$
$$3x^2-x-10+3x^2-6x+12 = 6x^2+8x-8$$
$$6x^2-7x+2 = 6x^2+8x-8$$
$$-15x = -10$$
$$x = \frac{-10}{-15} = \frac{2}{3}$$

65.
$$\begin{aligned}
\frac{1}{11-n}-\frac{2(3n-1)}{-7n^2+74n+33}&=\frac{1}{7n+3}\\
\frac{-1}{n-11}+\frac{2(3n-1)}{7n^2-74n-33}&=\frac{1}{7n+3}\\
\frac{-1}{n-11}+\frac{6n-2}{(7n+3)(n-11)}&=\frac{1}{7n+3}\\
-(7n+3)+6n-2&=(n-11)1 \quad \{\text{multiply by common denominator}\}\\
-7n-3+6n-2&=n-11\\
-n-5&=n-11\\
-2n&=-6\\
n&=3
\end{aligned}$$

67.
$$\begin{aligned}
\frac{5}{y+4}+\frac{2}{y+2}&=\frac{6}{y+2}-\frac{1}{y^2+6y+8}\\
\frac{5}{y+4}&=\frac{4}{y+2}-\frac{1}{(y+2)(y+4)}\\
5(y+2)&=4(y+4)-1 \quad \{\text{multiply by common denominator}\}\\
5y+10&=4y+16-1\\
5y+10&=4y+15\\
y&=5
\end{aligned}$$

69.
$$\begin{aligned}
\frac{3y}{6-3y}+\frac{2y}{2y+4}&=\frac{8}{4-y^2}\\
\frac{3y}{3(2-y)}+\frac{2y}{2(y+2)}&=\frac{8}{(2+y)(2-y)}\\
\frac{y}{2-y}+\frac{y}{2+y}&=\frac{8}{(2+y)(2-y)}\\
y(2+y)+y(2-y)&=8 \quad \{\text{multiply by common denominator}\}\\
2y+y^2+2y-y^2&=8\\
4y&=8\\
y&=2 \Rightarrow \text{The solution does not check, so the equation has no solution.}
\end{aligned}$$

71.
$$\begin{aligned}
\frac{a}{a+2}-1&=-\frac{3a+2}{a^2+4a+4}\\
\frac{a}{a+2}-\frac{1}{1}&=-\frac{3a+2}{(a+2)(a+2)}\\
(a+2)(a+2)\left[\frac{a}{a+2}-1\right]&=(a+2)(a+2)\cdot\left[-\frac{3a+2}{(a+2)(a+2)}\right]\\
a(a+2)-(a+2)(a+2)&=-(3a+2)\\
a^2+2a-\left(a^2+4a+4\right)&=-3a-2\\
a^2+2a-a^2-4a-4&=-3a-2\\
-2a-4&=-3a-2\\
a&=2
\end{aligned}$$

73.
$$\begin{aligned} k &= 2.2p \\ \frac{k}{2.2} &= \frac{2.2p}{2.2} \\ \frac{k}{2.2} &= p \end{aligned}$$

75.
$$\begin{aligned} p &= 2l + 2w \\ p - 2l &= 2w \\ \frac{p-2l}{2} &= \frac{2w}{2} \\ \frac{p-2l}{2} &= w \end{aligned}$$

77.
$$\begin{aligned} V &= \frac{1}{3}\pi r^2 h \\ 3V &= 3 \cdot \frac{1}{3}\pi r^2 h \\ 3V &= \pi r^2 h \\ \frac{3V}{\pi h} &= \frac{\pi r^2 h}{\pi h} \\ \frac{3V}{\pi h} &= r^2 \end{aligned}$$

79.
$$\begin{aligned} P_n &= L + \frac{si}{f} \\ P_n - L &= \frac{si}{f} \\ f(P_n - L) &= f \cdot \frac{si}{f} \\ f(P_n - L) &= si \\ \frac{f(P_n - L)}{i} &= \frac{si}{i} \\ \frac{f(P_n - L)}{i} &= s \end{aligned}$$

81.
$$\begin{aligned} F &= \frac{mMg}{r^2} \\ Fr^2 &= \frac{mMg}{r^2} \cdot r^2 \\ Fr^2 &= mMg \\ \frac{Fr^2}{Mg} &= \frac{mMg}{Mg} \\ \frac{Fr^2}{Mg} &= m \end{aligned}$$

83.
$$\begin{aligned} \frac{x}{a} + \frac{y}{b} &= 1 \\ \frac{y}{b} &= 1 - \frac{x}{a} \\ b \cdot \frac{y}{b} &= b\left(1 - \frac{x}{a}\right) \\ y &= b\left(1 - \frac{x}{a}\right) \end{aligned}$$

85.
$$\begin{aligned} \frac{1}{r} &= \frac{1}{r_1} + \frac{1}{r_2} \\ rr_1r_2 \cdot \frac{1}{r} &= rr_1r_2\left(\frac{1}{r_1} + \frac{1}{r_2}\right) \\ r_1r_2 &= rr_2 + rr_1 \\ r_1r_2 &= r(r_2 + r_1) \\ \frac{r_1r_2}{r_2 + r_1} &= \frac{r(r_2 + r_1)}{r_2 + r_1} \\ \frac{r_1r_2}{r_2 + r_1} &= r \end{aligned}$$

87.
$$\begin{aligned} l &= a + (n-1)d \\ l &= a + nd - d \\ l - a + d &= nd \\ \frac{l - a + d}{d} &= \frac{nd}{d} \\ \frac{l - a + d}{d} &= n \end{aligned}$$

89.
$$\begin{aligned} a &= (n-2)\frac{180}{n} \\ an &= (n-2)\frac{180}{n} \cdot n \\ an &= (n-2)180 \\ an &= 180n - 360 \\ 360 &= 180n - an \\ 360 &= n(180 - a) \\ \frac{360}{180 - a} &= n \end{aligned}$$

91.

$$R = \frac{1}{\frac{1}{r_1} + \frac{1}{r_2} + \frac{1}{r_3}}$$
$$R = \frac{r_1 r_2 r_3(1)}{r_1 r_2 r_3\left(\frac{1}{r_1} + \frac{1}{r_2} + \frac{1}{r_3}\right)}$$
$$R = \frac{r_1 r_2 r_3}{r_2 r_3 + r_1 r_3 + r_1 r_2}$$
$$R(r_2 r_3 + r_1 r_3 + r_1 r_2) = r_1 r_2 r_3$$
$$R r_2 r_3 + R r_1 r_3 + R r_1 r_2 = r_1 r_2 r_3$$
$$R r_1 r_3 + R r_1 r_2 - r_1 r_2 r_3 = -R r_2 r_3$$
$$r_1(R r_3 + R r_2 - r_2 r_3) = -R r_2 r_3$$
$$r_1 = \frac{-R r_2 r_3}{R r_3 + R r_2 - r_2 r_3}$$

93. **Answers may vary.**

95. $(25x^2)^{1/2} = \left[(5x)^2\right]^{1/2} = 5|x|$

97. $\left(\frac{125x^3}{8y^6}\right)^{-2/3} = \left(\frac{8y^6}{125x^3}\right)^{2/3} = \frac{8^{2/3}(y^6)^{2/3}}{125^{2/3}(x^3)^{2/3}} = \frac{4y^4}{25x^2}$

99. $\sqrt{25y^2} = \sqrt{(5y)^2} = 5|y|$

101. $\sqrt[4]{\frac{a^4b^{12}}{z^8}} = \sqrt[4]{\left(\frac{ab^3}{z^2}\right)^4} = \frac{|ab^3|}{z^2}$

Exercises 1.2 (page 101)

1. add

3. amount

5. rate, time

7. Let $x =$ the score on the first exam. Then $x + 5 =$ the score on the midterm, and $x + 13 =$ the score on the final.

$$\frac{\boxed{\text{Sum of scores}}}{3} = 90$$
$$\frac{x + x + 5 + x + 13}{3} = 90$$
$$\frac{3x + 18}{3} = 90$$
$$3x + 18 = 270$$
$$3x = 252$$
$$x = 84$$

His score on the first exam was 84.

9. Let $x =$ the program development score.

$$\frac{\boxed{\text{Sum of scores}}}{4} = 86$$
$$\frac{82 + 90 + x + 78}{4} = 86$$
$$\frac{x + 250}{4} = 86$$
$$x + 250 = 344$$
$$x = 94$$

The program development score was 94.

11. Let $x =$ the number of locks replaced.

$$40 + 28 \cdot \boxed{\text{Number of locks}} = 236$$
$$40 + 28x = 236$$
$$28x = 196$$
$$x = 7$$

7 locks can be changed for \$236.

13. Let $x =$ the width.

Then $x + 26 =$ the height.

$$\boxed{\text{Perimeter}} = 92$$
$$2x + 2(x + 26) = 92$$
$$2x + 2x + 52 = 92$$
$$4x = 40$$
$$x = 10$$

The dimensions are 10 ft by 36 ft.

15.

$$\boxed{\text{Perimeter}} = 14$$
$$x + (x + 2) + x + (x + 2) = 14$$
$$4x + 4 = 14$$
$$4x = 10$$
$$x = \frac{5}{2} = 2\frac{1}{2}$$

The width is $2\frac{1}{2}$ feet.

17.

$$\boxed{\text{Total Area}} = 2 \cdot \boxed{\text{Triangular Area}}$$
$$20x + \frac{1}{2}(16)(20) = 2 \cdot \frac{1}{2}(16)(20)$$
$$20x + 160 = 320$$
$$20x = 160$$
$$x = 8$$

The dimensions are 8 feet by 20 feet.

19.

$$\boxed{\text{New Area}} = \boxed{\text{Old Area}} + 0.50 \cdot \boxed{\text{Old Area}}$$
$$12(x + 10) + 12x = 12(x + 10) + 0.50 \cdot 12(x + 10)$$
$$12x + 120 + 12x = 12x + 120 + 6x + 60$$
$$24x + 120 = 18x + 180$$
$$6x = 60$$

$x = 10 \Rightarrow$ The length of the living room is $x + 10 = 20$ feet.

21. Let $x =$ the amount invested at 7%. Then $22000 - x =$ the amount invested at 6%.

$$\boxed{\text{Interest at 7\%}} + \boxed{\text{Interest at 6\%}} = \boxed{\text{Total interest}}$$
$$0.07x + 0.06(22000 - x) = 1420$$
$$0.07x + 1320 - 0.06x = 1420$$
$$0.01x = 100$$
$$x = 10000$$

\$10,000 was invested at 7% and \$12,000 was invested at 6%.

23. Let $x =$ the amount invested at each rate.

$$\boxed{\text{Interest at 6\%}} + \boxed{\text{Interest at 7\%}} + \boxed{\text{Interest at 8\%}} = \boxed{\text{Total interest}}$$
$$0.06x + 0.07x + 0.08x = 2037$$
$$0.21x = 2037$$
$$x = 9700$$

\$9,700 was invested at each rate, for a total investment of \$29,100.

SECTION 1.2

25. Let $x =$ the number of full-price tickets sold. Then $585 - x =$ the number of student tickets sold.

$$\begin{aligned} 2.50 \cdot \boxed{\substack{\text{\# of}\\ \text{full-price}}} + 1.75 \cdot \boxed{\substack{\text{\# of}\\ \text{student}}} &= 1217.25 \\ 2.50x + 1.75(585 - x) &= 1217.25 \\ 2.50x + 1023.75 - 1.75x &= 1217.25 \\ 0.75x &= 193.50 \\ x &= 258 \Rightarrow \text{There were 327 student tickets sold.} \end{aligned}$$

27. Let $p =$ the original price.

$$\begin{aligned} \boxed{\substack{\text{Original}\\ \text{price}}} - \boxed{\text{Discount}} &= \boxed{\substack{\text{New}\\ \text{price}}} \\ p - 0.20p &= 63.96 \\ 0.80p &= 63.96 \\ p &= 79.95 \end{aligned}$$

The original price was \$79.95.

29. Let $x =$ # of plates for equal costs.

$$\begin{aligned} \boxed{\substack{\text{Cost of 1st}\\ \text{machine}}} &= \boxed{\substack{\text{Cost of 2nd}\\ \text{machine}}} \\ 600 + 3x &= 800 + 2x \\ x &= 200 \end{aligned}$$

The break point is 200 plates.

31. Let $x =$ # of computers to break even.

$$\begin{aligned} \boxed{\text{Income}} &= \boxed{\text{Expenses}} \\ 1275x &= 8925 + 850x \\ 425x &= 8925 \\ x &= 21 \end{aligned}$$

21 computers need to be sold to break even.

33. Let $x =$ days for both working together.

$$\begin{aligned} \boxed{\substack{\text{Man in}\\ \text{1 day}}} + \boxed{\substack{\text{Roofer in}\\ \text{1 day}}} &= \boxed{\substack{\text{Total in}\\ \text{1 day}}} \\ \frac{1}{7} + \frac{1}{4} &= \frac{1}{x} \\ 28x\left(\frac{1}{7} + \frac{1}{4}\right) &= 28x\left(\frac{1}{x}\right) \\ 4x + 7x &= 28 \\ 11x &= 28 \\ x &= \frac{28}{11} = 2\frac{6}{11} \end{aligned}$$

They can roof the house in $2\frac{6}{11}$ days.

35. Let $x =$ hours for both working together.

$$\begin{aligned} \boxed{\substack{\text{Woman}\\ \text{in 1 hour}}} + \boxed{\substack{\text{Man in}\\ \text{1 hour}}} &= \boxed{\substack{\text{Total in}\\ \text{1 hour}}} \\ \frac{1}{2} + \frac{1}{4} &= \frac{1}{x} \\ 4x\left(\frac{1}{2} + \frac{1}{4}\right) &= 4x\left(\frac{1}{x}\right) \\ 2x + x &= 4 \\ 3x &= 4 \\ x &= \frac{4}{3} = 1\frac{1}{3} \end{aligned}$$

They can mow the lawn in $1\frac{1}{3}$ hours.

37. Let $x =$ hours for pool to fill with drain open.

$$\begin{aligned} \boxed{\substack{\text{Pipe in}\\ \text{1 hour}}} - \boxed{\substack{\text{Drain in}\\ \text{1 hour}}} &= \boxed{\substack{\text{Total in}\\ \text{1 hour}}} \\ \frac{1}{10} - \frac{1}{19} &= \frac{1}{x} \\ 190x\left(\frac{1}{10} - \frac{1}{19}\right) &= 190x\left(\frac{1}{x}\right) \\ 19x - 10x &= 190 \\ 9x &= 190 \\ x &= \frac{190}{9} = 21\frac{1}{9} \end{aligned}$$

The pool can be filled in $21\frac{1}{9}$ hours.

39. Let $x =$ the ouncesof water added.

$$\boxed{\text{Oz of alc. at start}} + \boxed{\text{Oz of alc. added}} = \boxed{\text{Oz of alc. at end}}$$

$$\begin{aligned} 0.15(20) + 0(x) &= 0.10(20 + x) \\ 3 &= 2 + 0.1x \\ 1 &= 0.1x \\ \frac{1}{0.1} &= x \\ 10 &= x \end{aligned}$$

10 oz of water should be added.

41. Let $x =$ the liters of liquid replaced with pure antifreeze.

$$\boxed{\text{Liters of a.f. at start}} - \boxed{\text{Liters of a.f. removed}} + \boxed{\text{Liters of a.f. replaced}} = \boxed{\text{Liters of a.f. at end}}$$

$$\begin{aligned} 0.40(6) - 0.40x + x &= 0.50(6) \\ 2.4 + 0.6x &= 3 \\ 0.6x &= 0.6 \\ x &= 1 \Rightarrow \text{1 liter should be replaced with pure antifreeze.} \end{aligned}$$

43. Let $x =$ the liters of pure alcohol added.

$$\boxed{\text{Liters of alcohol at start}} + \boxed{\text{Liters of alcohol added}} = \boxed{\text{Liters of alcohol at end}}$$

$$\begin{aligned} 0.20(1) + x &= 0.25(1 + x) \\ 0.20 + x &= 0.25 + 0.25x \\ 0.75x &= 0.05 \\ x &= \frac{0.05}{0.75} = \frac{1}{15} \Rightarrow \frac{1}{15} \text{ of a liter of pure alcohol should be added.} \end{aligned}$$

45. Let $x =$ the gallons of pure chlorine added.

$$\boxed{\text{Gallons of chlorine at start}} + \boxed{\text{Gallons of chlorine added}} = \boxed{\text{Gallons of chlorine at end}}$$

$$\begin{aligned} 0(15000) + x &= 0.0003(15000 + x) \\ x &= 4.5 + 0.0003x \\ 0.9997x &= 4.5 \\ x &\approx 4.5 \Rightarrow \text{About 4.5 gallons of pure chlorine should be added.} \end{aligned}$$

47. Let $x =$ the liters of water evaporated.

$$\boxed{\text{Liters of salt at start}} - \boxed{\text{Liters of salt evaporated}} = \boxed{\text{Liters of salt at end}}$$

$$\begin{aligned} 0.24(12) - 0(x) &= 0.36(12 - x) \\ 2.88 - 0 &= 4.32 - 0.36x \\ 0.36x &= 1.44 \\ x &= 4 \Rightarrow \text{4 liters of water should be evaporated.} \end{aligned}$$

49. Let $x =$ the pounds of extra-lean hamburger used.

$$\boxed{\text{Pounds of fat in hamburger}} + \boxed{\text{Pounds of fat in lean hamburger}} = \boxed{\text{Pounds of fat in mixture}}$$

$$\begin{aligned} 0.15(30) + 0.07(x) &= 0.10(30 + x) \\ 4.5 + 0.07x &= 3 + 0.1x \\ 1.5 &= 0.03x \\ 50 &= x \end{aligned}$$

50 pounds of the extra-lean hamburger should be used.

51. Let $x =$ the gallons of 5% solution used.

$$\boxed{\text{Gallons of alc. in 5\% solution}} + \boxed{\text{Gallons of alc. in 1\% solution}} = \boxed{\text{Gallons of alc. in 2\% solution}}$$

$$\begin{aligned} 0.05(x) + 0.01(90) &= 0.02(x + 90) \\ 0.05x + 0.9 &= 0.02x + 1.8 \\ 0.03x &= 0.9 \\ x &= 30 \end{aligned}$$

30 gallons of the 5% solution should be used.

53. Let $r =$ his first rate. Then $r + 26 =$ his return rate.

$$\boxed{\text{Distance to city}} = \boxed{\text{Return distance}}$$

$$\begin{aligned} 5r &= 3(r + 26) \\ 5r &= 3r + 78 \\ 2r &= 78 \\ r &= 39 \Rightarrow \text{He drove 39 mph going and 65 mph returning.} \end{aligned}$$

55. Let $t =$ the time the cars travel.

$$\boxed{\text{Distance 1st car travels}} + \boxed{\text{Distance 2nd car travels}} = \boxed{\text{Total distance}}$$

$$\begin{aligned} 60t + 64t &= 310 \\ 124t &= 310 \\ t &= 2.5 \Rightarrow \text{They will be 310 miles apart after 2.5 hours.} \end{aligned}$$

57. Let $t =$ the time the runners run.

$$\boxed{\text{Distance 1st runs}} + \boxed{\text{Distance 2nd runs}} = \boxed{\text{Distance between them (in miles)}}$$

$$\begin{aligned} 8t + 10t &= \frac{440}{1760} \\ 18t &= \frac{1}{4} \\ t &= \frac{1}{72} \text{ hour} = \frac{1}{72}(60 \text{ minutes}) = \frac{5}{6} \text{ minute} = 50 \text{ seconds} \end{aligned}$$

They will meet after 50 seconds.

59. Let $r =$ the speed of the boat in still water.
Then the speed of the boat is $r + 2$ downstream and $r - 2$ upstream.

$\boxed{\text{Time upstream}} = \boxed{\text{Time downstream}}$ {Note: Time = Distance ÷ Rate}

$$\frac{5}{r-2} = \frac{7}{r+2}$$
$$(r+2)(r-2)\frac{5}{r-2} = (r+2)(r-2)\frac{7}{r+2}$$
$$5(r+2) = 7(r-2)$$
$$5r + 10 = 7r - 14$$
$$24 = 2r$$
$12 = r \Rightarrow$ The speed of the boat is 12 mph.

61. Since the mixture is to be 25% barley, there will be $0.25(2400) = 600$ pounds of barley used. Thus, the other 1800 pounds will be either oats or soybean meal.
Let $x =$ the number of pounds of oats used. Then $1800 - x =$ the number of pounds of meal used.

$\boxed{\text{Pounds of protein from barley}} + \boxed{\text{Pounds of protein from oats}} + \boxed{\text{Pounds of protein from soybean meal}} = \boxed{\text{Total pounds of protein}}$

$$0.117(600) + 0.118x + 0.445(1800 - x) = 0.14(2400)$$
$$70.2 + 0.118x + 801 - 0.445x = 336$$
$$871.2 - 0.327x = 336$$
$$-0.327x = -535.2$$
$$x \approx 1637$$

The farmer should use 600 pounds of barley, 1,637 pounds of oats and 163 pounds of soybean meal.

63.
$$V = \pi r^2 h$$
$$712.51 = \pi(4.5)^2 d$$
$$\frac{712.51}{\pi(4.5)^2} = d$$
$11.2 \approx d$; The hole is about 11.2 millimeters deep.

65. **Answers may vary.**

67. $x^2 - 2x - 63 = (x-9)(x+7)$

69. $9x^2 - 12x - 5 = (3x+1)(3x-5)$

71. $x^2 + 6x + 9 = (x+3)(x+3) = (x+3)^2$

73. $x^3 + 8 = (x+2)(x^2 - 2x + 2^2) = (x+2)(x^2 - 2x + 4)$

Exercises 1.3 (page 116)

1. $ax^2 + bx + c = 0$

3. $\sqrt{c}, -\sqrt{c}$

5. rational numbers

7.
$$x^2 - x - 6 = 0$$
$$(x+2)(x-3) = 0$$
$$x + 2 = 0 \quad \textbf{or} \quad x - 3 = 0$$
$$x = -2 \qquad\qquad x = 3$$

9.
$$x^2 - 144 = 0$$
$$(x+12)(x-12) = 0$$
$$x + 12 = 0 \quad \textbf{or} \quad x - 12 = 0$$
$$x = -12 \qquad\qquad x = 12$$

SECTION 1.3

11.
$$\begin{aligned} 2x^2 + x - 10 &= 0 \\ (2x+5)(x-2) &= 0 \end{aligned}$$
$2x + 5 = 0$ **or** $x - 2 = 0$
$2x = -5$ $\quad x = 2$
$x = -\frac{5}{2}$ $\quad x = 2$

13.
$$\begin{aligned} 5x^2 - 13x + 6 &= 0 \\ (5x-3)(x-2) &= 0 \end{aligned}$$
$5x - 3 = 0$ **or** $x - 2 = 0$
$5x = 3$ $\quad x = 2$
$x = \frac{3}{5}$ $\quad x = 2$

15.
$$\begin{aligned} 15x^2 + 16x &= 15 \\ 15x^2 + 16x - 15 &= 0 \\ (3x+5)(5x-3) &= 0 \end{aligned}$$
$3x + 5 = 0$ **or** $5x - 3 = 0$
$3x = -5$ $\quad 5x = 3$
$x = -\frac{5}{3}$ $\quad x = \frac{3}{5}$

17.
$$\begin{aligned} 12x^2 + 9 &= 24x \\ 12x^2 - 24x + 9 &= 0 \\ 3\left(4x^2 - 8x + 3\right) &= 0 \\ (2x-1)(2x-3) &= 0 \end{aligned}$$
$2x - 1 = 0$ **or** $2x - 3 = 0$
$2x = 1$ $\quad 2x = 3$
$x = \frac{1}{2}$ $\quad x = \frac{3}{2}$

19.
$$x^2 = 9$$
$x = \sqrt{9}$ **or** $x = -\sqrt{9}$
$x = 3$ $\quad x = -3$

21.
$$\begin{aligned} y^2 - 50 &= 0 \\ y^2 &= 50 \end{aligned}$$
$y = \sqrt{50}$ **or** $y = -\sqrt{50}$
$y = 5\sqrt{2}$ $\quad y = -5\sqrt{2}$

23.
$$\begin{aligned} 2x^2 &= 40 \\ x^2 &= 20 \end{aligned}$$
$x = \sqrt{20}$ **or** $x = -\sqrt{20}$
$x = 2\sqrt{5}$ $\quad x = -2\sqrt{5}$

25.
$$\begin{aligned} 4x^2 &= 7 \\ x^2 &= \tfrac{7}{4} \end{aligned}$$
$x = \sqrt{\frac{7}{4}}$ **or** $x = -\sqrt{\frac{7}{4}}$
$x = \frac{\sqrt{7}}{2}$ $\quad x = -\frac{\sqrt{7}}{2}$

27.
$$\begin{aligned} 2x^2 - 13 &= 0 \\ 2x^2 &= 13 \\ x^2 &= \tfrac{13}{2} \end{aligned}$$
$x = \sqrt{\frac{13}{2}}$ **or** $x = -\sqrt{\frac{13}{2}}$
$x = \frac{\sqrt{13}}{\sqrt{2}} \cdot \frac{\sqrt{2}}{\sqrt{2}}$ $\quad x = -\frac{\sqrt{13}}{\sqrt{2}} \cdot \frac{\sqrt{2}}{\sqrt{2}}$
$x = \frac{\sqrt{26}}{2}$ $\quad x = -\frac{\sqrt{26}}{2}$

29.
$$(x-1)^2 = 4$$
$x - 1 = \sqrt{4}$ **or** $x - 1 = -\sqrt{4}$
$x - 1 = 2$ $\quad x - 1 = -2$
$x = 3$ $\quad x = -1$

31.
$$\begin{aligned} (x+1)^2 - 8 &= 0 \\ (x+1)^2 &= 8 \end{aligned}$$
$x + 1 = \sqrt{8}$ **or** $x + 1 = -\sqrt{8}$
$x + 1 = 2\sqrt{2}$ $\quad x + 1 = -2\sqrt{2}$
$x = -1 + 2\sqrt{2}$ $\quad x = -1 - 2\sqrt{2}$

33.

$$(2x+1)^2 = 27$$

$2x+1 = \sqrt{27}$ **or** $2x+1 = -\sqrt{27}$

$2x+1 = 3\sqrt{3}$ $\quad$ $2x+1 = -3\sqrt{3}$

$2x = -1+3\sqrt{3}$ $\quad$ $2x = -1-3\sqrt{3}$

$x = \frac{-1+3\sqrt{3}}{2}$ $\quad$ $x = \frac{-1-3\sqrt{3}}{2}$

35. $x^2 + 6x + \left[\frac{1}{2}(6)\right]^2 = x^2 + 6x + 3^2$
$= x^2 + 6x + 9$

37. $x^2 - 4x + \left[\frac{1}{2}(-4)\right]^2 = x^2 - 4x + (-2)^2$
$= x^2 - 4x + 4$

39. $a^2 + 5a + \left[\frac{1}{2}(5)\right]^2 = a^2 + 5a + \left(\frac{5}{2}\right)^2 = a^2 + 5a + \frac{25}{4}$

41. $r^2 - 11r + \left[\frac{1}{2}(-11)\right]^2 = r^2 - 11r + \left(\frac{-11}{2}\right)^2 = r^2 - 11r + \frac{121}{4}$

43. $y^2 + \frac{3}{4}y + \left[\frac{1}{2}\left(\frac{3}{4}\right)\right]^2 = y^2 + \frac{3}{4}y + \left(\frac{3}{8}\right)^2 = y^2 + \frac{3}{4}y + \frac{9}{64}$

45. $q^2 - \frac{1}{5}q + \left[\frac{1}{2}\left(-\frac{1}{5}\right)\right]^2 = q^2 - \frac{1}{5}q + \left(\frac{-1}{10}\right)^2 = q^2 - \frac{1}{5}q + \frac{1}{100}$

47.

$$x^2 - 8x + 15 = 0$$
$$x^2 - 8x = -15$$
$$x^2 - 8x + 16 = -15 + 16$$
$$(x-4)^2 = 1$$

$x - 4 = \sqrt{1}$ **or** $x - 4 = -\sqrt{1}$

$x - 4 = 1$ $\quad$ $x - 4 = -1$

$x = 5$ $\quad$ $x = 3$

49.

$$x^2 + 12x = -8$$
$$x^2 + 12x + 36 = -8 + 36$$
$$(x+6)^2 = 28$$

$x + 6 = \sqrt{28}$ **or** $x + 6 = -\sqrt{28}$

$x + 6 = 2\sqrt{7}$ $\quad$ $x + 6 = -2\sqrt{7}$

$x = -6 + 2\sqrt{7}$ $\quad$ $x = -6 - 2\sqrt{7}$

51.

$$\begin{aligned} x^2+5 &= -5x \\ x^2+5x &= -5 \\ x^2+5x+\frac{25}{4} &= -5+\frac{25}{4} \\ \left(x+\frac{5}{2}\right)^2 &= \frac{5}{4} \end{aligned}$$

$$x+\frac{5}{2}=\sqrt{\frac{5}{4}} \quad \textbf{or} \quad x+\frac{5}{2}=-\sqrt{\frac{5}{4}}$$

$$x+\frac{5}{2}=\frac{\sqrt{5}}{2} \qquad x+\frac{5}{2}=-\frac{\sqrt{5}}{2}$$

$$x=\frac{-5+\sqrt{5}}{2} \qquad x=\frac{-5-\sqrt{5}}{2}$$

53.

$$\begin{aligned} 2x^2-20x &= -49 \\ x^2-10x &= -\frac{49}{2} \\ x^2-10x+25 &= -\frac{49}{2}+\frac{50}{2} \\ (x-5)^2 &= \frac{1}{2} \end{aligned}$$

$$x-5=\sqrt{\frac{1}{2}} \quad \textbf{or} \quad x-5=-\sqrt{\frac{1}{2}}$$

$$x-5=\frac{1}{\sqrt{2}} \qquad x-5=-\frac{1}{\sqrt{2}}$$

$$x-\frac{10}{2}=\frac{\sqrt{2}}{2} \qquad x-\frac{10}{2}=-\frac{\sqrt{2}}{2}$$

$$x=\frac{10+\sqrt{2}}{2} \qquad x=\frac{10-\sqrt{2}}{2}$$

55.

$$\begin{aligned} 3x^2 &= 1-4x \\ 3x^2+4x &= 1 \\ x^2+\frac{4}{3}x &= \frac{1}{3} \\ x^2+\frac{4}{3}x+\frac{4}{9} &= \frac{1}{3}+\frac{4}{9} \\ \left(x+\frac{2}{3}\right)^2 &= \frac{7}{9} \end{aligned}$$

$$x+\frac{2}{3}=\sqrt{\frac{7}{9}} \quad \textbf{or} \quad x+\frac{2}{3}=-\sqrt{\frac{7}{9}}$$

$$x+\frac{2}{3}=\frac{\sqrt{7}}{3} \qquad x+\frac{2}{3}=-\frac{\sqrt{7}}{3}$$

$$x=\frac{-2+\sqrt{7}}{3} \qquad x=\frac{-2-\sqrt{7}}{3}$$

57.

$$\begin{aligned}2x^2 &= 3x+1\\ 2x^2-3x &= 1\\ x^2-\frac{3}{2}x &= \frac{1}{2}\\ x^2-\frac{3}{2}x+\frac{9}{16} &= \frac{1}{2}+\frac{9}{16}\\ \left(x-\frac{3}{4}\right)^2 &= \frac{17}{16}\end{aligned}$$

$$x-\frac{3}{4}=\sqrt{\frac{17}{16}} \quad \textbf{or} \quad x-\frac{3}{4}=-\sqrt{\frac{17}{16}}$$

$$x-\frac{3}{4}=\frac{\sqrt{17}}{4} \qquad x-\frac{3}{4}=-\frac{\sqrt{17}}{4}$$

$$x=\frac{3+\sqrt{17}}{4} \qquad x=\frac{3-\sqrt{17}}{4}$$

59. $x^2-12=0 \Rightarrow a=1, b=0, c=-12$

$$x=\frac{-b\pm\sqrt{b^2-4ac}}{2a}=\frac{-(0)\pm\sqrt{(0)^2-4(1)(-12)}}{2(1)}=\frac{0\pm\sqrt{0+48}}{2}=\frac{0\pm\sqrt{48}}{2}=\frac{\pm\sqrt{48}}{2}$$

$$x=\frac{\sqrt{48}}{2}=\frac{4\sqrt{3}}{2}=2\sqrt{3} \text{ or } x=\frac{-\sqrt{48}}{2}=\frac{-4\sqrt{3}}{2}=-2\sqrt{3}$$

61. $x^2-25x=0 \Rightarrow a=1, b=-25, c=0$

$$x=\frac{-b\pm\sqrt{b^2-4ac}}{2a}=\frac{-(-25)\pm\sqrt{(-25)^2-4(1)(0)}}{2(1)}=\frac{25\pm\sqrt{625-0}}{2}=\frac{25\pm\sqrt{625}}{2}$$

$$x=\frac{25+25}{2}=\frac{50}{2}=25 \text{ or } x=\frac{25-25}{2}=\frac{0}{2}=0$$

63. $2x^2-x-15=0 \Rightarrow a=2, b=-1, c=-15$

$$x=\frac{-b\pm\sqrt{b^2-4ac}}{2a}=\frac{-(-1)\pm\sqrt{(-1)^2-4(2)(-15)}}{2(2)}=\frac{1\pm\sqrt{1+120}}{4}=\frac{1\pm\sqrt{121}}{4}=\frac{1\pm 11}{4}$$

$$x=\frac{1+11}{4}=\frac{12}{4}=3 \text{ or } x=\frac{1-11}{4}=\frac{-10}{4}=-\frac{5}{2}$$

65. $3x^2=-5x-1 \Rightarrow 3x^2+5x+1=0 \Rightarrow a=3, b=5, c=1$

$$x=\frac{-b\pm\sqrt{b^2-4ac}}{2a}=\frac{-(5)\pm\sqrt{(5)^2-4(3)(1)}}{2(3)}=\frac{-5\pm\sqrt{25-12}}{6}=\frac{-5\pm\sqrt{13}}{6}$$

67. $x^2+1=-7x \Rightarrow x^2+7x+1=0 \Rightarrow a=1, b=7, c=1$

$$x=\frac{-b\pm\sqrt{b^2-4ac}}{2a}=\frac{-(7)\pm\sqrt{(7)^2-4(1)(1)}}{2(1)}=\frac{-7\pm\sqrt{49-4}}{2}=\frac{-7\pm\sqrt{45}}{2}=\frac{-7\pm3\sqrt{5}}{2}$$

69. $3x^2+6x=-1 \Rightarrow 3x^2+6x+1=0 \Rightarrow a=3, b=6, c=1$

$$x=\frac{-b\pm\sqrt{b^2-4ac}}{2a}=\frac{-(6)\pm\sqrt{(6)^2-4(3)(1)}}{2(3)}=\frac{-6\pm\sqrt{36-12}}{6}=\frac{-6\pm\sqrt{24}}{6}$$

$$x=\frac{-6\pm\sqrt{24}}{6}=\frac{-6\pm2\sqrt{6}}{6}=\frac{2\left(-3\pm\sqrt{6}\right)}{6}=\frac{-3\pm\sqrt{6}}{3}$$

71. $5x\left(x+\frac{1}{5}\right)=3 \Rightarrow 5x^2+x-3=0 \Rightarrow a=5, b=1, c=-3$

$$x=\frac{-b\pm\sqrt{b^2-4ac}}{2a}=\frac{-(1)\pm\sqrt{(1)^2-4(5)(-3)}}{2(5)}=\frac{-1\pm\sqrt{1+60}}{10}=\frac{-1\pm\sqrt{61}}{10}$$

73.

$$h=\tfrac{1}{2}gt^2$$
$$2h=gt^2$$
$$\tfrac{2h}{g}=t^2$$
$$\pm\sqrt{\frac{2h}{g}}=t$$

75.

$$h=64t-16t^2$$
$$16t^2-64t+h=0;\ a=16, b=-64, c=h$$
$$t=\frac{-b\pm\sqrt{b^2-4ac}}{2a}$$
$$=\frac{-(-64)\pm\sqrt{(-64)^2-4(16)h}}{2(16)}$$
$$=\frac{64\pm\sqrt{4096-64h}}{32}$$
$$=\frac{64\pm\sqrt{64(64-h)}}{32}$$
$$=\frac{64\pm8\sqrt{64-h}}{32}=\frac{8\pm\sqrt{64-h}}{4}$$

77.

$$\frac{x^2}{a^2}+\frac{y^2}{b^2}=1$$
$$\frac{y^2}{b^2}=1-\frac{x^2}{a^2}$$
$$\frac{y^2}{b^2}=\frac{a^2-x^2}{a^2}$$
$$y^2=\frac{b^2(a^2-x^2)}{a^2}$$
$$y=\pm\sqrt{\frac{b^2(a^2-x^2)}{a^2}}=\pm\frac{b\sqrt{a^2-x^2}}{a}$$

79.
$$\frac{x^2}{a^2} - \frac{y^2}{b^2} = 1$$
$$a^2b^2\left(\frac{x^2}{a^2} - \frac{y^2}{b^2}\right) = a^2b^2(1)$$
$$b^2x^2 - a^2y^2 = a^2b^2$$
$$b^2x^2 = a^2b^2 + a^2y^2$$
$$b^2x^2 = a^2(b^2 + y^2)$$
$$\frac{b^2x^2}{b^2 + y^2} = a^2$$
$$\pm\sqrt{\frac{b^2x^2}{b^2 + y^2}} = a$$
$$\pm\frac{bx\sqrt{b^2 + y^2}}{b^2 + y^2} = a$$

81. $x^2 + xy - y^2 = 0 \Rightarrow a = 1, b = y, c = -y^2$
$$x = \frac{-b \pm \sqrt{b^2 - 4ac}}{2a}$$
$$= \frac{-(y) \pm \sqrt{(y)^2 - 4(1)(-y^2)}}{2(1)}$$
$$= \frac{-y \pm \sqrt{y^2 + 4y^2}}{2}$$
$$= \frac{-y \pm \sqrt{5y^2}}{2} = \frac{-y \pm y\sqrt{5}}{2}$$

83. $x^2 + 6x + 9 = 0 \Rightarrow a = 1, b = 6, c = 9$
$b^2 - 4ac = 6^2 - 4(1)(9) = 36 - 36 = 0$
one repeated rational root

85. $3x^2 - 2x + 5 = 0 \Rightarrow a = 3, b = -2, c = 5$
$b^2 - 4ac = (-2)^2 - 4(3)(5)$
$= 4 - 60 = -56$
no real roots

87. $10x^2 + 29x = 21 \Rightarrow 10x^2 + 29x - 21 = 0$
$a = 10, b = 29, c = -21$
$b^2 - 4ac = (29)^2 - 4(10)(-21)$
$= 841 + 840 = 1681$
two different rational roots

89. $x^2 - 5x + 2 = 0 \Rightarrow a = 1, b = -5, c = 2$
$b^2 - 4ac = (-5)^2 - 4(1)(2) = 25 - 8 = 17$
two different irrational roots

91. $1492x^2 + 1984x - 1776 = 0$
$a = 1492, b = 1984, c = -1776$
$b^2 - 4ac = (1984)^2 - 4(1492)(-1776)$
$= 3{,}936{,}256 + 10{,}599{,}168$
$= 14{,}535{,}424$
The solutions are real numbers.

93. $x^2 + kx + 3k - 5 = 0$
$a = 1, b = k, c = 3k - 5$
Set the discriminant equal to 0:
$$b^2 - 4ac = 0$$
$$k^2 - 4(1)(3k - 5) = 0$$
$$k^2 - 4(3k - 5) = 0$$
$$k^2 - 12k + 20 = 0$$
$$(k - 2)(k - 10) = 0$$
$k = 2$ or $k = 10$

SECTION 1.3

95.

$$\begin{aligned} x+1 &= \frac{12}{x} \\ x(x+1) &= x\left(\frac{12}{x}\right) \\ x^2+x &= 12 \\ x^2+x-12 &= 0 \\ (x+4)(x-3) &= 0 \end{aligned}$$

$$x+4=0 \quad \textbf{or} \quad x-3=0$$
$$x=-4 \qquad\qquad x=3$$

97.

$$\begin{aligned} 8x-\frac{3}{x} &= 10 \\ x\left(8x-\frac{3}{x}\right) &= x(10) \\ 8x^2-3 &= 10x \\ 8x^2-10x-3 &= 0 \\ (4x+1)(2x-3) &= 0 \end{aligned}$$

$$4x+1=0 \quad \textbf{or} \quad 2x-3=0$$
$$4x=-1 \qquad\qquad 2x=3$$
$$x=-\tfrac{1}{4} \qquad\qquad x=\tfrac{3}{2}$$

99.

$$\begin{aligned} \frac{5}{x} &= \frac{4}{x^2}-6 \\ x^2\left(\frac{5}{x}\right) &= x^2\left(\frac{4}{x^2}-6\right) \\ 5x &= 4-6x^2 \\ 6x^2+5x-4 &= 0 \\ (3x+4)(2x-1) &= 0 \end{aligned}$$

$$3x+4=0 \quad \textbf{or} \quad 2x-1=0$$
$$3x=-4 \qquad\qquad 2x=1$$
$$x=-\tfrac{4}{3} \qquad\qquad x=\tfrac{1}{2}$$

101.

$$\begin{aligned} x\left(30-\frac{13}{x}\right) &= \frac{10}{x} \\ 30x-13 &= \frac{10}{x} \\ x(30x-13) &= x\left(\frac{10}{x}\right) \\ 30x^2-13x &= 10 \\ 30x^2-13x &= 10 \\ 30x^2-13x-10 &= 0 \\ (5x+2)(6x-5) &= 0 \end{aligned}$$

$$5x+2=0 \quad \textbf{or} \quad 6x-5=0$$
$$5x=-2 \qquad\qquad 6x=5$$
$$x=-\tfrac{2}{5} \qquad\qquad x=\tfrac{5}{6}$$

103.

$$\begin{aligned} \frac{1}{x}+\frac{3}{x+2} &= 2 \\ x(x+2)\left(\frac{1}{x}+\frac{3}{x+2}\right) &= x(x+2)(2) \\ 1(x+2)+3x &= 2x(x+2) \\ x+2+3x &= 2x^2+4x \\ 0 &= 2x^2-2 \\ 0 &= 2(x+1)(x-1) \end{aligned}$$

$$x+1=0 \quad \textbf{or} \quad x-1=0$$
$$x=-1 \qquad\qquad x=1$$

SECTION 1.3

105.

$$\frac{1}{x+1}+\frac{5}{2x-4}=1$$

$$(x+1)(2x-4)\left(\frac{1}{x+1}+\frac{5}{2x-4}\right)=(x+1)(2x-4)1$$

$$1(2x-4)+5(x+1)=(x+1)(2x-4)$$

$$2x-4+5x+5=2x^2-2x-4$$

$$0=2x^2-9x-5$$

$$0=(2x+1)(x-5)$$

$2x+1=0$ **or** $x-5=0$

$2x=-1$ $\quad x=5$

$x=-\frac{1}{2}$ $\quad x=5$

107.

$$x+1+\frac{x+2}{x-1}=\frac{3}{x-1}$$

$$(x-1)\left(\frac{x+1}{1}+\frac{x+2}{x-1}\right)=(x-1)\frac{3}{x-1}$$

$$(x-1)(x+1)+x+2=3$$

$$x^2-1+x+2=3$$

$$x^2+x-2=0$$

$$(x+2)(x-1)=0$$

$x+2=0$ **or** $x-1=0$

$x=-2$ $\quad x=1$

Since $x=1$ does not check, the only solution is $x=-2$.

109.

$$\frac{24}{a}-11=\frac{-12}{a+1}$$

$$a(a+1)\left(\frac{24}{a}-11\right)=a(a+1)\frac{-12}{a+1}$$

$$24(a+1)-11a(a+1)=a(-12)$$

$$24a+24-11a^2-11a=-12a$$

$$0=11a^2-25a-24$$

$$0=(11a+8)(a-3)$$

$11a+8=0$ **or** $a-3=0$

$11a=-8$ $\quad a=3$

$a=-\frac{8}{11}$ $\quad a=3$

111.

$$\frac{4+a}{2a}=\frac{a-2}{3}$$

$$6a\left(\frac{4+a}{2a}\right)=6a\left(\frac{a-2}{3}\right)$$

$$12+3a=2a^2-4a$$

$$0=2a^2-7a-12$$

$a=2,\ b=-7,\ c=-12$

$$a=\frac{-b\pm\sqrt{b^2-4ac}}{2a}=\frac{-(-7)\pm\sqrt{(-7)^2-4(2)(-12)}}{2(2)}=\frac{7\pm\sqrt{145}}{4}$$

113. If r_1 and r_2 are the roots of $ax^2+bx+c=0$, then their values are

$$r_1=\frac{-b+\sqrt{b^2-4ac}}{2a}\text{ and } r_2=\frac{-b-\sqrt{b^2-4ac}}{2a}.$$

$$r_1+r_2=\frac{-b+\sqrt{b^2-4ac}}{2a}+\frac{-b-\sqrt{b^2-4ac}}{2a}=\frac{-2b}{2a}=-\frac{b}{a}$$

115. Rewrite the equation as $16t^2 - v_0 t + h = 0$ and solve for t using the quadratic formula.
$a = 16$, $b = -v_0$, $c = h$

$$t = \frac{-b \pm \sqrt{b^2 - 4ac}}{2a} = \frac{-(-v_0) \pm \sqrt{(-v_0)^2 - 4(16)(h)}}{2(16)} = \frac{v_0 \pm \sqrt{v_0^2 - 64h}}{32}$$

Since t_1 and t_2 are the solutions to the equation, we have

$t_1 = \dfrac{v_0 - \sqrt{v_0^2 - 64h}}{32}$ and $t_2 = \dfrac{v_0 + \sqrt{v_0^2 - 64h}}{32}$. Calculate $16t_1t_2$:

$$16t_1t_2 = 16 \cdot \frac{v_0 - \sqrt{v_0^2 - 64h}}{32} \cdot \frac{v_0 + \sqrt{v_0^2 - 64h}}{32} = 16 \cdot \frac{v_0^2 - (v_0^2 - 64h)}{1024}$$
$$= \frac{16 \cdot 64h}{1024} = \frac{1024h}{1024} = h$$

Thus, $h = 16t_1t_2$.

117. Answers may vary.

119. Answers may vary.

121. $5x(x-2) - x(3x-2) = 5x^2 - 10x - 3x^2 + 2x = 2x^2 - 8x$

123.
$$\begin{aligned}(m+3)^2 - (m-3)^2 &= (m+3)(m+3) - (m-3)(m-3)\\ &= m^2 + 6m + 9 - \left(m^2 - 6m + 9\right)\\ &= m^2 + 6m + 9 - m^2 + 6m - 9 = 12m\end{aligned}$$

125. $\sqrt{50x^3} - x\sqrt{8x} = \sqrt{25x^2}\sqrt{2x} - x\sqrt{4}\sqrt{2x} = 5x\sqrt{2x} - 2x\sqrt{2x} = 3x\sqrt{2x}$

Exercises 1.4 (page 125)

1. $A = lw$

3. Let $w =$ the width of the rectangle. Then $w + 4 =$ the length.

$\boxed{\text{Width}} \cdot \boxed{\text{Length}} = \boxed{\text{Area}}$

$$\begin{aligned} w(w+4) &= 32\\ w^2 + 4w &= 32\\ w^2 + 4w - 32 &= 0\\ (w+8)(w-4) &= 0 \end{aligned}$$

$w + 8 = 0$ **or** $w - 4 = 0$
$w = -8$ $\qquad$ $w = 4$

Since the width cannot be negative, the only reasonable solution is $w = 4$. The dimensions are 4 feet by 8 feet.

5. Let $w =$ the width of the screen. Then $w + 88 =$ the length.

$\boxed{\text{Width}} \cdot \boxed{\text{Length}} = \boxed{\text{Area}}$

$$w(w+88) = 11520$$
$$w^2 + 88w = 11520$$
$$w^2 + 88w - 11520 = 0 \Rightarrow a = 1, b = 88, c = -11520$$
$$w = \frac{-b \pm \sqrt{b^2 - 4ac}}{2a} = \frac{-(88) \pm \sqrt{(88)^2 - 4(1)(-11520)}}{2(1)} = \frac{-88 \pm \sqrt{53824}}{2}$$

$w = 72$ or $w = -160$; Since the screen cannot have a negative width, the solution is $w = 72$ and the dimensions of the screen are 160 ft by 72 ft.

7. Let $s =$ the side of the second square.
Then $s - 4 =$ the side of the first square.

$\boxed{\text{Area of first}} + \boxed{\text{Area of second}} = 106$

$$(s-4)^2 + s^2 = 106$$
$$s^2 - 8s + 16 + s^2 = 106$$
$$2s^2 - 8s - 90 = 0$$
$$2(s^2 - 4s - 45) = 0$$
$$2(s+5)(s-9) = 0$$
$$s + 5 = 0 \quad \textbf{or} \quad s - 9 = 0$$
$$s = -5 \qquad\qquad s = 9$$

Since the side cannot be negative, the only reasonable solution is $s = 9$.
The larger square has a side of length 9 cm.

9. Let the dimensions be x and $1.9x$.

$\boxed{\text{Width}} \cdot \boxed{\text{Length}} = \boxed{\text{Area}}$

$$x(1.9x) = 100$$
$$1.9x^2 = 100$$
$$x^2 = \frac{100}{1.9}$$
$$x = \pm\sqrt{\frac{100}{1.9}}$$
$$x \approx \pm 7.25$$

$1.9x \approx 1.9(7.25) \approx 13.75$
Since the dimensions cannot be negative, the only reasonable solution $7\frac{1}{4}$ ft by $13\frac{3}{4}$ ft.

11. The floor area of the box is a square with a side of length $12 - 2x$.

$\boxed{\text{Floor area}} = 64$

$$(12 - 2x)^2 = 64$$
$$144 - 48x + 4x^2 = 64$$
$$4x^2 - 48x + 80 = 0$$
$$4(x^2 - 12x + 20) = 0$$
$$4(x-2)(x-10) = 0$$
$$x - 2 = 0 \quad \textbf{or} \quad x - 10 = 0$$
$$x = 2 \qquad\qquad x = 10$$

The solution $x = 10$ does not make sense in the problem, so the depth is 2 inches.

13. Let $h =$ the height of the triangle.
Then $\frac{1}{3}h =$ the base of the triangle.

$\frac{1}{2} \cdot \boxed{\text{Base}} \cdot \boxed{\text{Height}} = \boxed{\text{Area}}$

$$\frac{1}{2} \cdot \frac{1}{3}h \cdot h = 24$$
$$\frac{1}{6}h^2 = 24$$
$$h^2 = 144$$
$$h = \sqrt{144} \quad \textbf{or} \quad h = -\sqrt{144}$$
$$h = 12 \qquad\qquad h = -12$$

Since the height cannot be negative, the only reasonable solution is $h = 12$.
The base has a length of 4 meters.

15. Let the legs have lengths x and $x - 14$.

$$x^2 + (x-14)^2 = 26^2$$
$$x^2 + x^2 - 28x + 196 = 676$$
$$2x^2 - 28x - 480 = 0$$
$$2(x^2 - 14x - 240) = 0$$
$$2(x-24)(x+10) = 0$$
$$x - 24 = 0 \quad \textbf{or} \quad x + 10 = 0$$
$$x = 24 \qquad\qquad x = -10$$

Since lengths are positive, the answer is $x = 24$, and the legs have length 10 meters and 24 meters.

17. Let $x =$ the height of the screen. Then $x + 17.5 =$ the width. Use the Pythagorean Theorem:

$$\text{height}^2 + \text{width}^2 = \text{diagonal}^2$$
$$x^2 + (x+17.5)^2 = 46^2$$
$$x^2 + x^2 + 35x + 306.25 = 2116$$
$$2x^2 + 35x - 1809.75 = 0$$
$$a = 2,\ b = 35,\ c = -1809.75$$

$$x = \frac{-b \pm \sqrt{b^2 - 4ac}}{2a}$$
$$= \frac{-35 \pm \sqrt{35^2 - 4(2)(-1809.75)}}{2(2)}$$
$$= \frac{-35 \pm \sqrt{15703}}{4} \approx \frac{-35 \pm 125.312}{4}$$

The only positive solution is 22.6. The dimensions are 22.6 inches by 40.1 inches.

19. Let $r =$ the cyclist's rate from DeKalb to Rockford. Then his return rate is $r - 10$.

$$\boxed{\text{Return time}} = \boxed{\text{First time}} + 2$$

	Rate	Time	Dist.
First trip	r	$\frac{40}{r}$	40
Return trip	$r - 10$	$\frac{40}{r-10}$	40

$$\frac{40}{r-10} = \frac{40}{r} + 2$$
$$r(r-10)\frac{40}{r-10} = r(r-10)\left(\frac{40}{r} + 2\right)$$
$$40r = 40(r-10) + 2r(r-10)$$
$$40r = 40r - 400 + 2r^2 - 20r$$
$$0 = 2r^2 - 20r - 400$$
$$0 = 2(r-20)(r+10)$$
$$r - 20 = 0 \quad \textbf{or} \quad r + 10 = 0$$
$$r = 20 \qquad\qquad r = -10$$

Since $r = -10$ does not make sense, the solution is $r = 20$. The cyclist rides 20 mph going and 10 mph returning.

21. Let $r =$ the slower rate. Then the faster rate is $r + 10$.

$$\boxed{\text{Faster time}} = \boxed{\text{Slower time}} - 1$$

	Rate	Time	Dist.
Slower trip	r	$\frac{420}{r}$	420
Faster trip	$r+10$	$\frac{420}{r+10}$	420

$$\frac{420}{r+10} = \frac{420}{r} - 1$$
$$r(r+10)\frac{420}{r+10} = r(r+10)\left(\frac{420}{r} - 1\right)$$
$$420r = 420(r+10) - r(r+10)$$
$$420r = 420r + 4200 - r^2 - 10r$$
$$r^2 + 10r - 4200 = 0$$
$$(r-60)(r+70) = 0$$
$$r - 60 = 0 \quad \textbf{or} \quad r + 70 = 0$$
$$r = 60 \qquad\qquad r = -70$$

Since $r = -70$ does not make sense, the solution is $r = 60$. The slower speed results in a trip of length 7 hours.

23. Set $h = 0$:

$$h = -16t^2 + 400t$$
$$0 = -16t^2 + 400t$$
$$16t^2 - 400t = 0$$
$$16t(t - 25) = 0$$
$$16t = 0 \quad \textbf{or} \quad t - 25 = 0$$
$$t = 0 \qquad\qquad t = 25$$

$t = 0$ represents when the projectile was fired, so it returns to earth after 25 seconds.

25. Set $s = 1454$:

$$s = 16t^2$$
$$1454 = 16t^2$$
$$\frac{1454}{16} = t^2$$
$$t = \sqrt{\frac{1454}{16}} \quad \textbf{or} \quad t = -\sqrt{\frac{1454}{16}}$$
$$t \approx 9.5 \qquad\qquad t \approx -9.5$$

$t = -9.5$ does not make sense, so it takes it about 9.5 seconds to hit the ground.

27. Set $h = 5$ and $s = 48$.

$$h = s - 16t^2$$
$$5 = 48 - 16t^2$$
$$16t^2 = 43$$
$$t = \sqrt{\frac{43}{16}} \quad \textbf{or} \quad t = -\sqrt{\frac{43}{16}}$$
$$t \approx 1.6 \qquad\qquad t \approx -1.6$$

$t = -1.6$ does not make sense, so she has about 1.6 seconds to get out of the way.

29. Let $x =$ the number of nickel increases. The new fare $= 25 + 5x$ (in cents), while the number of passengers $= 3000 - 80x$.

$$\boxed{\text{Number of passengers}} \cdot \boxed{\text{Fare}} = \boxed{\text{Revenue}}$$

$$(3000 - 80x)(25 + 5x) = 99400$$
$$75000 + 13000x - 400x^2 = 99400$$
$$400x^2 - 13000x + 24400 = 0$$
$$200(2x^2 - 65x + 122) = 0$$
$$200(2x - 61)(x - 2) = 0$$
$$2x - 61 = 0 \quad \textbf{or} \quad x - 2 = 0$$
$$x = 30.5 \qquad\qquad x = 2$$

Since you cannot have half of a nickel increase, $x = 30.5$ does not make sense. Thus, there should be 2 nickel increases, for a fare increase of 10 cents.

31. Let $x =$ the number of \$0.50 decreases. The new price $= 15 - 0.5x$, while the number attending $= 1200 + 40x$.

$$\boxed{\text{Number attending}} \cdot \boxed{\text{Price}} = \boxed{\text{Revenue}}$$

$$(1200 + 40x)(15 - 0.5x) = 17280$$
$$18000 - 20x^2 = 17280$$
$$20x^2 = 720$$
$$x^2 = 36$$
$$x = \sqrt{36} \quad \textbf{or} \quad x = -\sqrt{36}$$
$$x = 6 \qquad\qquad x = -6$$

$x = -6$ does not make sense. Thus, there should be six 50-cent decreases, for a ticket price of \$12 and an attendance of 1440 people.

33. Let $x =$ Chloe's principal.

	I	P	r
Chloe	240	x	$\frac{240}{x}$
Morgan	280	$x + 1000$	$\frac{280}{x+1000}$

$$\boxed{\text{Morgan's rate}} = \boxed{\text{Chloe's rate}} - 0.01$$

$$\frac{280}{x + 1000} = \frac{240}{x} - 0.01$$
$$x(x + 1000)\frac{280}{x + 1000} = x(x + 1000)\left(\frac{240}{x} - 0.01\right)$$
$$280x = 240(x + 1000) - 0.01x(x + 1000)$$
$$0.01x^2 + 50x - 240{,}000 = 0$$
$$x^2 + 5000x - 24{,}000{,}000 = 0$$
$$(x - 3000)(x + 8000) = 0$$
$$x - 3000 = 0 \quad \textbf{or} \quad x + 8000 = 0$$
$$x = 3000 \qquad\qquad x = -8000 \Rightarrow$$

$x = -8000$ does not make sense. The principal amounts were \$3000 and \$4000. The interest rates were 8% for Chloe and 7% for Morgan.

35. Let $x =$ the total number of professors.

$$\boxed{\text{New share with lower number}} = \boxed{\text{Original share}} + 10$$

$$\frac{150}{x - 4} = \frac{150}{x} + 10$$
$$x(x - 4)\frac{150}{x - 4} = x(x - 4)\left(\frac{150}{x} + 10\right)$$
$$150x = 150(x - 4) + 10x(x - 4)$$
$$150x = 150x - 600 + 10x^2 - 40x$$

$$150x = 150x - 600 + 10x^2 - 40x$$
$$0 = 10x^2 - 40x - 600$$
$$0 = x^2 - 4x - 60$$
$$0 = (x - 10)(x + 6)$$
$$x - 10 = 0 \quad \textbf{or} \quad x + 6 = 0$$
$$x = 10 \qquad\qquad x = -6$$

$x = -6$ does not make sense, so there are 10 professors in the department.

37. Let $x =$ time for the second pipe to fill tank.

$$\boxed{\text{First in 1 hour}} + \boxed{\text{Second in 1 hour}} = \boxed{\text{Total in 1 hour}}$$

$$\frac{1}{4} + \frac{1}{x} = \frac{1}{x-2}$$

$$4x(x-2)\left(\frac{1}{4} + \frac{1}{x}\right) = 4x(x-2) \cdot \frac{1}{x-2}$$

$$x(x-2) + 4(x-2) = 4x$$

$$x^2 - 2x + 4x - 8 = 4x$$

$$x^2 - 2x + 4x - 8 = 4x$$

$$x^2 - 2x - 8 = 0$$

$$(x-4)(x+2) = 0$$

$$x - 4 = 0 \quad \textbf{or} \quad x + 2 = 0$$

$$x = 4 \qquad\qquad x = -2$$

Since $x = -2$ does not make sense, the solution is $x = 4$. It takes the second pipe 4 hours to fill the tank alone.

39. Let $x =$ time for the Steven to mow lawn.

$$\boxed{\text{Steven in 1 hour}} + \boxed{\text{Kristy in 1 hour}} = \boxed{\text{Total in 1 hour}}$$

$$\frac{1}{x} + \frac{1}{x-1} = \frac{1}{5}$$

$$5x(x-1)\left(\frac{1}{x} + \frac{1}{x-1}\right) = 5x(x-1) \cdot \frac{1}{5}$$

$$5(x-1) + 5x = x(x-1)$$

$$5x - 5 + 5x = x^2 - x$$

$$0 = x^2 - 11x + 5$$

$a = 1, b = -11, c = 5$

$$x = \frac{-b \pm \sqrt{b^2 - 4ac}}{2a}$$

$$= \frac{-(-11) \pm \sqrt{(-11)^2 - 4(1)(5)}}{2(1)}$$

$$= \frac{11 \pm \sqrt{101}}{2} \approx 10.5 \text{ or } 0.5$$

$x = 0.5$ does not make sense, so Kristy could mow the lawn in about 9.5 hours alone.

41. The number of trees is the length of the row divided by the space between the trees, plus 1. Let $x =$ the original spacing.

$$\boxed{\text{Original number}} - 44 = \boxed{\text{New number}}$$

$$\frac{1320}{x} + 1 - 44 = \frac{1320}{x+1} + 1$$

$$\frac{1320}{x} - 44 = \frac{1320}{x+1}$$

$$x(x+1)\left(\frac{1320}{x} - 44\right) = x(x+1)\frac{1320}{x+1}$$

$$1320(x+1) - 44x(x+1) = 1320x$$

$$44x^2 + 44x - 1320 = 0$$

$$x^2 + x - 30 = 0$$

$$(x-5)(x+6) = 0$$

$$x - 5 = 0 \quad \textbf{or} \quad x + 6 = 0$$

$$x = 5 \qquad\qquad x = -6$$

$x = -6$ does not make sense. The original spacing was 5 feet, resulting in 265 trees, so the new spacing will require 221 trees.

43. Let $h = 15$:

$$\frac{l}{h} = \frac{h}{l-h}$$

$$\frac{l}{15} = \frac{15}{l-15}$$

$$15(l-15) \cdot \frac{l}{15} = 15(l-15) \cdot \frac{15}{l-15}$$

$$l(l-15) = 15^2$$

$$l^2 - 15l - 225 = 0$$

$a = 1, b = -15, c = -225$

$$l = \frac{-b \pm \sqrt{b^2 - 4ac}}{2a}$$

$$= \frac{-(-15) \pm \sqrt{(-15)^2 - 4(1)(-225)}}{2(1)}$$

$$= \frac{15 \pm \sqrt{1125}}{2} \approx \frac{15 \pm 33.541}{2}$$

The only positive solution is $l = 24.3$ ft.

45.
$$\begin{aligned} V &= \pi r^2 h \\ 47.75 &= \pi r^2 (5.25) \\ \frac{47.75}{5.25\pi} &= r^2 \\ \pm\sqrt{\frac{47.75}{5.25\pi}} &= r \\ \pm 1.70 &\approx r \end{aligned}$$

The radius is about 1.70 in.

47. Let $x =$ the length of the diagonal.

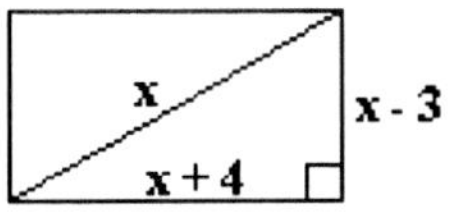

Using the Pythagorean Theorem:

$$\begin{aligned} (x+4)^2 + (x-3)^2 &= x^2 \\ x^2 + 8x + 16 + x^2 - 6x + 9 &= x^2 \\ x^2 + 2x + 25 &= 0 \end{aligned}$$

$a = 1, b = 2, c = 25$

$$\begin{aligned} x &= \frac{-b \pm \sqrt{b^2 - 4ac}}{2a} \\ &= \frac{-2 \pm \sqrt{2^2 - 4(1)(25)}}{2(1)} = \frac{-2 \pm \sqrt{-96}}{2} \end{aligned}$$

This does not equal a real number, so it is not possible.

49. $$\frac{2}{x} - \frac{1}{x-3} = \frac{2(x-3)}{x(x-3)} - \frac{1(x)}{(x-3)(x)} = \frac{2x-6}{x(x-3)} - \frac{x}{x(x-3)} = \frac{x-6}{x(x-3)}$$

51. $$\frac{x+3}{x^2-x-6} \div \frac{x^2+3x}{x^2-9} = \frac{x+3}{x^2-x-6} \cdot \frac{x^2-9}{x^2+3x} = \frac{x+3}{(x-3)(x+2)} \cdot \frac{(x+3)(x-3)}{x(x+3)} = \frac{x+3}{x(x+2)}$$

53. $$\frac{\frac{1}{x}+\frac{1}{y}}{\frac{1}{x}-\frac{1}{y}} = \frac{xy\left(\frac{1}{x}+\frac{1}{y}\right)}{xy\left(\frac{1}{x}-\frac{1}{y}\right)} = \frac{xy\left(\frac{1}{x}\right)+xy\left(\frac{1}{y}\right)}{xy\left(\frac{1}{x}\right)-xy\left(\frac{1}{y}\right)} = \frac{y+x}{y-x}$$

Exercises 1.5 (page 138)

1. imaginary

3. imaginary

5. $2 - 5i$

7. real

9. $\sqrt{-144} = \sqrt{-1}\sqrt{144} = 12i$

11. $$\begin{aligned} -2\sqrt{-24} = -2\sqrt{-1}\sqrt{24} &= -2i \cdot 2\sqrt{6} \\ &= -4i\sqrt{6} \end{aligned}$$

13. $$\sqrt{-\frac{50}{9}} = \sqrt{-1} \cdot \frac{\sqrt{50}}{\sqrt{9}} = i \cdot \frac{5\sqrt{2}}{3} = \frac{5}{3}i\sqrt{2}$$

15. $$-7\sqrt{-\frac{3}{8}} = -7\sqrt{-1} \cdot \frac{\sqrt{3}}{\sqrt{8}} = -7i \cdot \frac{\sqrt{3}}{\sqrt{8}} \cdot \frac{\sqrt{2}}{\sqrt{2}} = -7i \cdot \frac{\sqrt{6}}{\sqrt{16}} = -\frac{7}{4}i\sqrt{6}$$

17. Equate real parts: $\boxed{x = 3}$ Equate imaginary parts:
$$\begin{aligned} x + y &= 8 \\ 3 + y &= 8 \\ \boxed{y = 5} \end{aligned}$$

SECTION 1.5

19. Equate real parts: $3x = 2$ $\boxed{x = \frac{2}{3}}$

Equate imaginary parts: $-2y = x + y$

$-3y = x$

$y = -\frac{1}{3}x$

$y = -\frac{1}{3} \cdot \frac{2}{3}$

$\boxed{y = -\frac{2}{9}}$

21. $(2 - 7i) + (3 + i) = 2 - 7i + 3 + i$
$= 5 - 6i$

23. $(5 - 6i) - (7 + 4i) = 5 - 6i - 7 - 4i$
$= -2 - 10i$

25. $(14i + 2) + \left(2 - \sqrt{-16}\right) = (14i + 2) + (2 - 4i) = 14i + 2 + 2 - 4i = 4 + 10i$

27. $\left(3 + \sqrt{-4}\right) - \left(2 + \sqrt{-9}\right) = (3 + 2i) - (2 + 3i) = 3 + 2i - 2 - 3i = 1 - i$

29. $-5(3 + 5i) = -15 - 25i$

31. $7i(4 - 8i) = 28i - 56i^2 = 28i - 56(-1) = 28i + 56 = 56 + 28i$

33. $(2 + 3i)(3 + 5i) = 6 + 19i + 15(-1) = 6 + 19i - 15 = -9 + 19i$

35. $(2 + 3i)^2 = (2 + 3i)(2 + 3i) = 4 + 12i + 9i^2 = 4 + 12i + 9(-1) = 4 + 12i - 9 = -5 + 12i$

37. $\left(11 + \sqrt{-25}\right)\left(2 - \sqrt{-36}\right) = (11 + 5i)(2 - 6i) = 22 - 56i - 30i^2 = 22 - 56i - 30(-1)$
$= 22 - 56i + 30 = 52 - 56i$

39. $\left(\sqrt{-16} + 3\right)\left(2 + \sqrt{-9}\right) = (4i + 3)(2 + 3i) = 6 + 17i + 12i^2 = 6 + 17i + 12(-1)$
$= 6 + 17i - 12 = -6 + 17i$

41. $\dfrac{1}{-i} = \dfrac{1}{-i} \cdot \dfrac{i}{i} = \dfrac{i}{-i^2} = \dfrac{i}{1} = 0 + i$

43. $\dfrac{-4}{3i} = \dfrac{-4}{3i} \cdot \dfrac{i}{i} = \dfrac{-4i}{3i^2} = \dfrac{-4i}{-3} = 0 + \dfrac{4}{3}i$

45. $\dfrac{1}{2 + i} = \dfrac{1(2 - i)}{(2 + i)(2 - i)} = \dfrac{2 - i}{2^2 - i^2} = \dfrac{2 - i}{4 - (-1)} = \dfrac{2 - i}{5} = \dfrac{2}{5} - \dfrac{1}{5}i$

47. $\dfrac{2i}{7 + i} = \dfrac{2i(7 - i)}{(7 + i)(7 - i)} = \dfrac{14i - 2i^2}{7^2 - i^2} = \dfrac{14i - 2(-1)}{49 - (-1)} = \dfrac{14i + 2}{50} = \dfrac{7i + 1}{25} = \dfrac{1}{25} + \dfrac{7}{25}i$

49. $\dfrac{2 + i}{3 - i} = \dfrac{(2 + i)(3 + i)}{(3 - i)(3 + i)} = \dfrac{6 + 5i + i^2}{9 - i^2} = \dfrac{5 + 5i}{10} = \dfrac{5}{10} + \dfrac{5}{10}i = \dfrac{1}{2} + \dfrac{1}{2}i$

51. $\dfrac{4 - 5i}{2 + 3i} = \dfrac{(4 - 5i)(2 - 3i)}{(2 + 3i)(2 - 3i)} = \dfrac{8 - 22i + 15i^2}{4 - 9i^2} = \dfrac{-7 - 22i}{13} = -\dfrac{7}{13} - \dfrac{22}{13}i$

SECTION 1.5

53. $\dfrac{5-\sqrt{-16}}{-8+\sqrt{-4}} = \dfrac{5-4i}{-8+2i} = \dfrac{(5-4i)(-8-2i)}{(-8+2i)(-8-2i)} = \dfrac{-40+22i+8i^2}{64-4i^2} = \dfrac{-48+22i}{68} = -\dfrac{48}{68}+\dfrac{22}{68}i$
$= -\dfrac{12}{17}+\dfrac{11}{34}i$

55. $\dfrac{2+i\sqrt{3}}{3+i} = \dfrac{\left(2+i\sqrt{3}\right)(3-i)}{(3+i)(3-i)} = \dfrac{6-2i+3i\sqrt{3}-i^2\sqrt{3}}{9-i^2} = \dfrac{6+\sqrt{3}+\left(3\sqrt{3}-2\right)i}{10}$
$= \dfrac{6+\sqrt{3}}{10}+\dfrac{3\sqrt{3}-2}{10}i$

57. $i^9 = i^8 i = \left(i^4\right)^2 i = 1^2 i = i$

59. $i^{38} = i^{36}i^2 = \left(i^4\right)^9 i^2 = 1^9 i^2 = i^2 = -1$

61. $i^{87} = i^{84}i^3 = \left(i^4\right)^{21} i^3 = 1^{21}i^3 = i^3 = -i$

63. $i^{100} = \left(i^4\right)^{25} = 1^{25} = 1$

65. $i^{-6} = \dfrac{1}{i^6} = \dfrac{1\cdot i^2}{i^6\cdot i^2} = \dfrac{i^2}{i^8} = \dfrac{i^2}{1} = i^2 = -1$

67. $i^{-10} = \dfrac{1}{i^{10}} = \dfrac{1\cdot i^2}{i^{10}\cdot i^2} = \dfrac{i^2}{i^{12}} = \dfrac{i^2}{1} = i^2 = -1$

69. $\dfrac{1}{i^3} = \dfrac{1\cdot i}{i^3\cdot i} = \dfrac{i}{i^4} = \dfrac{i}{1} = i = i$

71. $\dfrac{-4}{i^{10}} = \dfrac{-4\cdot i^2}{i^{10}\cdot i^2} = \dfrac{-4i^2}{i^{12}} = \dfrac{-4i^2}{1} = -4(-1)$
$= 4$

73. $|3+4i| = \sqrt{3^2+4^2} = \sqrt{9+16}$
$= \sqrt{25} = 5$

75. $|2+3i| = \sqrt{2^2+3^2} = \sqrt{4+9} = \sqrt{13}$

77. $\left|-7+\sqrt{-49}\right| = |-7+7i| = \sqrt{(-7)^2+7^2} = \sqrt{49+49} = \sqrt{98} = 7\sqrt{2}$

79. $\left|\dfrac{1}{2}+\dfrac{1}{2}i\right| = \sqrt{\left(\dfrac{1}{2}\right)^2+\left(\dfrac{1}{2}\right)^2} = \sqrt{\dfrac{1}{4}+\dfrac{1}{4}} = \sqrt{\dfrac{1}{2}} = \dfrac{\sqrt{2}}{2}$

81. $|-6i| = |0-6i| = \sqrt{0^2+(-6)^2} = \sqrt{0+36} = \sqrt{36} = 6$

83. $\left|\dfrac{2}{1+i}\right| = \left|\dfrac{2(1-i)}{(1+i)(1-i)}\right| = \left|\dfrac{2(1-i)}{1-i^2}\right| = \left|\dfrac{2(1-i)}{2}\right| = |1-i| = \sqrt{1^2+(-1)^2} = \sqrt{2}$

85. $\left|\dfrac{-3i}{2+i}\right| = \left|\dfrac{-3i(2-i)}{(2+i)(2-i)}\right| = \left|\dfrac{-3i(2-i)}{4-i^2}\right| = \left|\dfrac{-3i(2-i)}{5}\right| = \left|\dfrac{-6i+3i^2}{5}\right|$
$= \left|-\tfrac{3}{5}-\tfrac{6}{5}i\right|$
$= \sqrt{\left(-\tfrac{3}{5}\right)^2+\left(-\tfrac{6}{5}\right)^2}$
$= \sqrt{\tfrac{9}{25}+\tfrac{36}{25}}$
$= \sqrt{\tfrac{45}{25}} = \tfrac{\sqrt{45}}{5} = \tfrac{3\sqrt{5}}{5}$

SECTION 1.5

87. $\left|\dfrac{i+2}{i-2}\right| = \left|\dfrac{(i+2)(i+2)}{(i-2)(i+2)}\right| = \left|\dfrac{i^2+4i+4}{i^2-4}\right| = \left|\dfrac{3+4i}{5}\right| = \left|\dfrac{3}{5}+\dfrac{4}{5}i\right| = \sqrt{\left(\dfrac{3}{5}\right)^2+\left(\dfrac{4}{5}\right)^2}$

$$= \sqrt{\frac{9}{25}+\frac{16}{25}}$$

$$= \sqrt{\frac{25}{25}} = \sqrt{1} = 1$$

89.

$$x^2 = -169$$

$x = \sqrt{-169}$ **or** $x = -\sqrt{-169}$

$x = 13i$ $\quad x = -13i$

91.

$$y^2 + 54 = 0$$

$$y^2 = -54$$

$y = \sqrt{-54}$ **or** $y = -\sqrt{-54}$

$y = i\sqrt{9}\sqrt{6}$ $\quad y = -i\sqrt{9}\sqrt{6}$

$y = 3i\sqrt{6}$ $\quad y = -3i\sqrt{6}$

93.

$$2x^2 = -90$$

$$x^2 = -45$$

$x = \sqrt{-45}$ **or** $x = -\sqrt{-45}$

$x = i\sqrt{9}\sqrt{5}$ $\quad x = -i\sqrt{9}\sqrt{5}$

$x = 3i\sqrt{5}$ $\quad x = -3i\sqrt{5}$

95.

$$9x^2 = -7$$

$$x^2 = -\frac{7}{9}$$

$x = \sqrt{-\dfrac{7}{9}}$ **or** $x = -\sqrt{-\dfrac{7}{9}}$

$x = i\dfrac{\sqrt{7}}{\sqrt{9}}$ $\quad x = -i\dfrac{\sqrt{7}}{\sqrt{9}}$

$x = \dfrac{\sqrt{7}}{3}i$ $\quad x = -\dfrac{\sqrt{7}}{3}i$

97.

$$2x^2 + 15 = 0$$

$$2x^2 = -15$$

$$x^2 = -\frac{15}{2}$$

$x = \sqrt{-\dfrac{15}{2}}$ **or** $x = -\sqrt{-\dfrac{15}{2}}$

$x = i\dfrac{\sqrt{15}}{\sqrt{2}} \cdot \dfrac{\sqrt{2}}{\sqrt{2}}$ $\quad x = -i\dfrac{\sqrt{15}}{\sqrt{2}} \cdot \dfrac{\sqrt{2}}{\sqrt{2}}$

$x = \dfrac{\sqrt{30}}{2}i$ $\quad x = -\dfrac{\sqrt{30}}{2}i$

99.

$$(x+1)^2 + 12 = 0$$

$$(x+1)^2 = -12$$

$x + 1 = \sqrt{-12}$ **or** $x + 1 = -\sqrt{-12}$

$x = -1 + i\sqrt{4}\sqrt{3}$ $\quad x = -1 - i\sqrt{4}\sqrt{3}$

$x = -1 + 2i\sqrt{3}$ $\quad x = -1 - 2i\sqrt{3}$

101.

$$(5x+1)^2 = -8$$

$$5x+1 = \sqrt{-8} \qquad \textbf{or} \qquad 5x+1 = -\sqrt{-8}$$

$$5x = -1 + i\sqrt{4}\sqrt{2} \qquad\qquad 5x = -1 - i\sqrt{4}\sqrt{2}$$

$$x = -\frac{1}{5} + \frac{2i\sqrt{2}}{5} \qquad\qquad x = -\frac{1}{5} - \frac{2i\sqrt{2}}{5}$$

103.

$$x^2 - 10x + 37 = 0$$

$$x^2 - 10x = -37$$

$$x^2 - 10x + 25 = -37 + 25$$

$$(x-5)^2 = -12$$

$$x - 5 = \sqrt{-12} \qquad \textbf{or} \qquad x - 5 = -\sqrt{-12}$$

$$x = 5 + i\sqrt{4}\sqrt{3} \qquad\qquad x = 5 - i\sqrt{4}\sqrt{3}$$

$$x = 5 + 2i\sqrt{3} \qquad\qquad x = 5 - 2i\sqrt{3}$$

105.

$$y^2 + 11y = -49$$

$$y^2 + 11y + \frac{121}{4} = -\frac{196}{4} + \frac{121}{4}$$

$$\left(y + \frac{11}{2}\right)^2 = -\frac{75}{4}$$

$$y + \frac{11}{2} = \sqrt{-\frac{75}{4}} \qquad \textbf{or} \qquad y + \frac{11}{2} = -\sqrt{-\frac{75}{4}}$$

$$y = -\frac{11}{2} + i\frac{\sqrt{75}}{\sqrt{4}} \qquad\qquad y = -\frac{11}{2} - i\frac{\sqrt{75}}{\sqrt{4}}$$

$$y = -\frac{11}{2} + \frac{5\sqrt{3}}{2}i \qquad\qquad y = -\frac{11}{2} - \frac{5\sqrt{3}}{2}i$$

107.

$$9x^2 = 18x - 14$$

$$9x^2 - 18x = -14$$

$$x^2 - 2x = -\frac{14}{9}$$

$$x^2 - 2x + 1 = -\frac{14}{9} + \frac{9}{9}$$

$$(x-1)^2 = -\frac{5}{9}$$

$$x - 1 = \sqrt{-\frac{5}{9}} \qquad \textbf{or} \qquad x - 1 = -\sqrt{-\frac{5}{9}}$$

$$x = 1 + i\frac{\sqrt{5}}{\sqrt{9}} \qquad\qquad x = 1 - i\frac{\sqrt{5}}{\sqrt{9}}$$

$$x = 1 + \frac{\sqrt{5}}{3}i \qquad\qquad x = 1 - \frac{\sqrt{5}}{3}i$$

109.
$$\begin{aligned} 2x^2 &= 14x - 30 \\ 2x^2 - 14x &= -30 \\ x^2 - 7x &= -15 \\ x^2 - 7x + \frac{49}{4} &= -\frac{60}{4} + \frac{49}{4} \\ \left(x - \frac{7}{2}\right)^2 &= -\frac{11}{4} \end{aligned}$$

$$x - \frac{7}{2} = \sqrt{-\frac{11}{4}} \quad \textbf{or} \quad x - \frac{7}{2} = -\sqrt{-\frac{11}{4}}$$

$$x = \frac{7}{2} + i\frac{\sqrt{11}}{\sqrt{4}} \qquad x = \frac{7}{2} - i\frac{\sqrt{11}}{\sqrt{4}}$$

$$x = \frac{7}{2} + \frac{\sqrt{11}}{2}i \qquad x = \frac{7}{2} - \frac{\sqrt{11}}{2}i$$

111. $x^2 + 2x + 2 = 0 \Rightarrow a = 1, b = 2, c = 2$

$$x = \frac{-b \pm \sqrt{b^2 - 4ac}}{2a} = \frac{-2 \pm \sqrt{2^2 - 4(1)(2)}}{2(1)} = \frac{-2 \pm \sqrt{4 - 8}}{2} = \frac{-2 \pm \sqrt{-4}}{2} = \frac{-2 \pm 2i}{2}$$
$$= -1 \pm i$$

113. $y^2 + 4y + 5 = 0 \Rightarrow a = 1, b = 4, c = 5$

$$y = \frac{-b \pm \sqrt{b^2 - 4ac}}{2a} = \frac{-4 \pm \sqrt{4^2 - 4(1)(5)}}{2(1)} = \frac{-4 \pm \sqrt{16 - 20}}{2} = \frac{-4 \pm \sqrt{-4}}{2} = \frac{-4 \pm 2i}{2}$$
$$= -2 \pm i$$

115. $x^2 - 2x = -5 \Rightarrow x^2 - 2x + 5 = 0 \Rightarrow a = 1, b = -2, c = 5$

$$x = \frac{-b \pm \sqrt{b^2 - 4ac}}{2a} = \frac{-(-2) \pm \sqrt{(-2)^2 - 4(1)(5)}}{2(1)} = \frac{2 \pm \sqrt{4 - 20}}{2} = \frac{2 \pm \sqrt{-16}}{2} = \frac{2 \pm 4i}{2}$$
$$= 1 \pm 2i$$

117. $x^2 - \frac{2}{3}x = -\frac{2}{9} \Rightarrow x^2 - \frac{2}{3}x + \frac{2}{9} = 0 \Rightarrow 9x^2 - 6x + 2 = 0 \Rightarrow a = 9, b = -6, c = 2$

$$x = \frac{-b \pm \sqrt{b^2 - 4ac}}{2a} = \frac{-(-6) \pm \sqrt{(-6)^2 - 4(9)(2)}}{2(9)} = \frac{6 \pm \sqrt{36 - 72}}{18} = \frac{6 \pm \sqrt{-36}}{18}$$
$$= \frac{6 \pm 6i}{18} = \frac{1}{3} \pm \frac{1}{3}i$$

119. $x^2 + 4 = x^2 - (-4) = x^2 - (2i)^2 = (x + 2i)(x - 2i)$

121. $25p^2 + 36q^2 = (5p)^2 - (-36q^2) = (5p)^2 - (6qi)^2 = (5p + 6qi)(5p - 6qi)$

123. $2y^2 + 8z^2 = 2(y^2 + 4z^2) = 2[y^2 - (-4z^2)] = 2\left[y^2 - (2zi)^2\right] = 2(y + 2zi)(y - 2zi)$

125. $50m^2 + 2n^2 = 2(25m^2 + n^2) = 2\left[(5m)^2 - (-n^2)\right] = 2\left[(5m)^2 - (ni)^2\right] = 2(5m + ni)(5m - ni)$

127. $V = IR = (3-2i)(3+6i) = 9+18i-6i-12i^2 = 9+12i+12 = 21+12i$

129. $V = IZ = (0.5+2.0i)(0.4-3.0i) = 0.2-1.5i+0.8i-6i^2 = 0.2-0.7i+6 = 6.2-0.7i$

131.
$$\begin{aligned}(a+bi)+(c+di) &= a+bi+c+di\\ &= a+c+bi+di\\ &= (a+c)+(b+d)i\end{aligned}$$
$$\begin{aligned}(c+di)+(a+bi) &= c+di+a+bi\\ &= c+a+di+bi\\ &= (c+a)+(d+b)i\\ &= (a+c)+(b+d)i\end{aligned}$$

133.
$$\begin{aligned}[(a+bi)+(c+di)]+(e+fi) &= a+bi+c+di+e+fi\\ &= a+c+e+bi+di+fi\\ &= (a+c+e)+(b+d+f)i\\ (a+bi)+[(c+di)+(e+fi)] &= a+bi+c+di+e+fi\\ &= a+c+e+bi+di+fi\\ &= (a+c+e)+(b+d+f)i\end{aligned}$$

135. Answers may vary.

137. $\sqrt{8x^3}\sqrt{4x} = \sqrt{32x^4} = \sqrt{16x^4}\sqrt{2} = 4x^2\sqrt{2}$

139. $\left(\sqrt{x+1}-2\right)^2 = \left(\sqrt{x+1}-2\right)\left(\sqrt{x+1}-2\right) = x+1-4\sqrt{x+1}+4 = x-4\sqrt{x+1}+5$

141. $$\frac{4}{\sqrt{5}-1} = \frac{4}{\sqrt{5}-1}\cdot\frac{\sqrt{5}+1}{\sqrt{5}+1} = \frac{4\left(\sqrt{5}+1\right)}{\left(\sqrt{5}\right)^2-1^2} = \frac{4\left(\sqrt{5}+1\right)}{5-1} = \frac{4\left(\sqrt{5}+1\right)}{4} = \sqrt{5}+1$$

Exercises 1.6 (page 147)

1. equal

3. extraneous

5.
$$\begin{aligned}x^3+9x^2+20x &= 0\\ x(x^2+9x+20) &= 0\\ x(x+5)(x+4) &= 0\end{aligned}$$
$x=0$ **or** $x+5=0$ **or** $x+4=0$

$x=0$ $\quad x=-5$ $\quad x=-4$

7.
$$\begin{aligned}6a^3-5a^2-4a &= 0\\ a(6a^2-5a-4) &= 0\\ a(2a+1)(3a-4) &= 0\end{aligned}$$
$a=0$ **or** $2a+1=0$ **or** $3a-4=0$

$a=0$ $\quad 2a=-1$ $\quad 3a=4$

$a=0$ $\quad a=-\frac{1}{2}$ $\quad a=\frac{4}{3}$

9.
$$\begin{aligned}y^4-26y^2+25 &= 0\\ (y^2-25)(y^2-1) &= 0\end{aligned}$$
$y^2-25=0$ **or** $y^2-1=0$

$y^2=25$ $\quad y^2=1$

$y=\pm 5$ $\quad y=\pm 1$

11.
$$\begin{aligned}x^4-37x^2+36 &= 0\\ (x^2-36)(x^2-1) &= 0\end{aligned}$$
$x^2-36=0$ **or** $x^2-1=0$

$x^2=36$ $\quad x^2=1$

$x=\pm 6$ $\quad x=\pm 1$

SECTION 1.6

13.
$$2y^4 - 46y^2 = -180$$
$$2(y^4 - 23y^2 + 90) = 0$$
$$2(y^2 - 18)(y^2 - 5) = 0$$
$$y^2 - 18 = 0 \quad \textbf{or} \quad y^2 - 5 = 0$$
$$y^2 = 18 \qquad y^2 = 5$$
$$y = \pm\sqrt{18} \qquad y = \pm\sqrt{5}$$
$$y = \pm 3\sqrt{2} \qquad y = \pm\sqrt{5}$$

15.
$$z^{3/2} - z^{1/2} = 0$$
$$z^{1/2}(z^{2/2} - 1) = 0$$
$$z^{1/2}(z - 1) = 0$$
$$z^{1/2} = 0 \quad \textbf{or} \quad z - 1 = 0$$
$$(z^{1/2})^2 = 0^2 \qquad z = 1$$
$$z = 0 \qquad z = 1$$
Both answers check.

17.
$$2m^{2/3} + 3m^{1/3} - 2 = 0$$
$$(2m^{1/3} - 1)(m^{1/3} + 2) = 0$$
$$2m^{1/3} - 1 = 0 \quad \textbf{or} \quad m^{1/3} + 2 = 0$$
$$m^{1/3} = \tfrac{1}{2} \qquad m^{1/3} = -2$$
$$(m^{1/3})^3 = \left(\tfrac{1}{2}\right)^3 \qquad (m^{1/3})^3 = (-2)^3$$
$$m = \tfrac{1}{8} \qquad m = -8$$
Both answers check.

19.
$$x - 13x^{1/2} + 12 = 0$$
$$(x^{1/2} - 12)(x^{1/2} - 1) = 0$$
$$x^{1/2} - 12 = 0 \quad \textbf{or} \quad x^{1/2} - 1 = 0$$
$$x^{1/2} = 12 \qquad x^{1/2} = 1$$
$$(x^{1/2})^2 = (12)^2 \qquad (x^{1/2})^2 = (1)^2$$
$$x = 144 \qquad x = 1$$
Both answers check.

21.
$$2t^{1/3} + 3t^{1/6} - 2 = 0$$
$$(2t^{1/6} - 1)(t^{1/6} + 2) = 0$$
$$2t^{1/6} - 1 = 0 \quad \textbf{or} \quad t^{1/6} + 2 = 0$$
$$t^{1/6} = \tfrac{1}{2} \qquad t^{1/6} = -2$$
$$(t^{1/6})^6 = \left(\tfrac{1}{2}\right)^6 \qquad (t^{1/6})^6 = (-2)^6$$
$$t = \tfrac{1}{64} \qquad t = 64$$
$t = 64$ does not check and is extraneous.

23.
$$6p + p^{1/2} - 1 = 0$$
$$(2p^{1/2} + 1)(3p^{1/2} - 1) = 0$$
$$2p^{1/2} + 1 = 0 \quad \textbf{or} \quad 3p^{1/2} - 1 = 0$$
$$p^{1/2} = -\tfrac{1}{2} \qquad p^{1/2} = \tfrac{1}{3}$$
$$(p^{1/2})^2 = \left(-\tfrac{1}{2}\right)^2 \qquad (p^{1/2})^2 = \left(\tfrac{1}{3}\right)^2$$
$$p = \tfrac{1}{4} \qquad p = \tfrac{1}{9}$$
$p = \frac{1}{4}$ does not check and is extraneous.

25.
$$\sqrt{x-2} - 3 = 2$$
$$\sqrt{x-2} = 5$$
$$\left(\sqrt{x-2}\right)^2 = 5^2$$
$$x - 2 = 25$$
$$x = 27$$
The solution checks.

27.
$$3\sqrt{x+1} = \sqrt{6}$$
$$\left(3\sqrt{x+1}\right)^2 = \left(\sqrt{6}\right)^2$$
$$9(x+1) = 6$$
$$9x + 9 = 6$$
$$9x = -3$$
$$x = -\tfrac{1}{3}$$
The solution checks.

29.

$$\begin{aligned}\sqrt{5a-2} &= \sqrt{a+6}\\ \left(\sqrt{5a-2}\right)^2 &= \left(\sqrt{a+6}\right)^2\\ 5a-2 &= a+6\\ 4a &= 8\\ a &= 2\end{aligned}$$

The solution checks.

31.

$$\begin{aligned}2\sqrt{x^2+3} &= \sqrt{-16x-3}\\ \left(2\sqrt{x^2+3}\right)^2 &= \left(\sqrt{-16x-3}\right)^2\\ 4\left(x^2+3\right) &= -16x-3\\ 4x^2+12 &= -16x-3\\ 4x^2+16x+15 &= 0\\ (2x+3)(2x+5) &= 0\end{aligned}$$

$2x+3=0$ **or** $2x+5=0$

$x=-\frac{3}{2}$ $\quad$ $x=-\frac{5}{2}$

Both solutions check.

33.

$$\begin{aligned}\sqrt[3]{7x+1} &= 4\\ \left(\sqrt[3]{7x+1}\right)^3 &= 4^3\\ 7x+1 &= 64\\ 7x &= 63\\ x &= 9\end{aligned}$$

The solution checks.

35.

$$\begin{aligned}\sqrt[4]{30t+25} &= 5\\ \left(\sqrt[4]{30t+25}\right)^4 &= 5^4\\ 30t+25 &= 625\\ 30t &= 600\\ t &= 20\end{aligned}$$

The solution checks.

37.

$$\begin{aligned}\sqrt{x^2+21} &= x+3\\ \left(\sqrt{x^2+21}\right)^2 &= (x+3)^2\\ x^2+21 &= x^2+6x+9\\ 21 &= 6x+9\\ 12 &= 6x\\ 2 &= x\end{aligned}$$

The solution checks.

39.

$$\begin{aligned}\sqrt{y+2} &= 4-y\\ \left(\sqrt{y+2}\right)^2 &= (4-y)^2\\ y+2 &= 16-8y+y^2\\ 0 &= y^2-9y+14\\ 0 &= (y-2)(y-7)\end{aligned}$$

$y-2=0$ **or** $y-7=0$

$y=2$ $\quad$ $y=7$

$y=7$ does not check and is extraneous.

41.
$$\begin{aligned} x-\sqrt{7x-12}&=0\\ x&=\sqrt{7x-12}\\ x^2&=\left(\sqrt{7x-12}\right)^2\\ x^2&=7x-12\\ x^2-7x+12&=0\\ (x-4)(x-3)&=0\\ x-4=0 \quad &\textbf{or} \quad x-3=0\\ x=4 \quad & \quad\quad x=3 \end{aligned}$$
Both solutions check.

43.
$$\begin{aligned} x+4&=\sqrt{\frac{6x+6}{5}}+3\\ x+1&=\sqrt{\frac{6x+6}{5}}\\ (x+1)^2&=\left(\sqrt{\frac{6x+6}{5}}\right)^2\\ x^2+2x+1&=\frac{6x+6}{5}\\ 5x^2+10x+5&=6x+6\\ 5x^2+4x-1&=0\\ (5x-1)(x+1)&=0\\ 5x-1=0 \quad &\textbf{or} \quad x+1=0\\ x=\tfrac{1}{5} \quad & \quad\quad x=-1 \end{aligned}$$
Both solutions check.

45.
$$\begin{aligned} \sqrt{\frac{x^2-1}{x-2}}&=2\sqrt{2}\\ \left(\sqrt{\frac{x^2-1}{x-2}}\right)^2&=\left(2\sqrt{2}\right)^2\\ \frac{x^2-1}{x-2}&=8\\ x^2-1&=8x-16\\ x^2-8x+15&=0\\ (x-3)(x-5)&=0\\ x-3=0 \quad &\textbf{or} \quad x-5=0\\ x=3 \quad & \quad\quad x=5 \end{aligned}$$
Both solutions check.

47.
$$\begin{aligned} \sqrt[3]{x^3+7}&=x+1\\ \left(\sqrt[3]{x^3+7}\right)^3&=(x+1)^3\\ x^3+7&=x^3+3x^2+3x+1\\ 0&=3x^2+3x-6\\ 0&=3(x+2)(x-1)\\ x+2=0 \quad &\textbf{or} \quad x-1=0\\ x=-2 \quad & \quad\quad x=1 \end{aligned}$$
Both solutions check.

49.
$$\begin{aligned} \sqrt[3]{8x^3+61}&=2x+1\\ \left(\sqrt[3]{8x^3+61}\right)^3&=(2x+1)^3\\ 8x^3+61&=8x^3+12x^2+6x+1\\ 0&=12x^2+6x-60\\ 0&=6(2x+5)(x-2)\\ 2x+5=0 \quad &\textbf{or} \quad x-2=0\\ x=-\tfrac{5}{2} \quad & \quad\quad x=2 \end{aligned}$$
Both solutions check.

51.
$$\begin{aligned} \sqrt{2p+1}-1&=\sqrt{p}\\ \left(\sqrt{2p+1}-1\right)^2&=\left(\sqrt{p}\right)^2\\ 2p+1-2\sqrt{2p+1}+1&=p\\ p+2&=2\sqrt{2p+1}\\ (p+2)^2&=\left(2\sqrt{2p+1}\right)^2\\ p^2+4p+4&=4(2p+1)\\ p^2+4p+4&=8p+4\\ p^2-4p&=0\\ p(p-4)&=0\\ p=0 \quad &\textbf{or} \quad p-4=0\\ p=0 \quad & \quad\quad p=4 \end{aligned}$$
Both solutions check.

53.
$$\begin{aligned}
\sqrt{x+3} &= \sqrt{2x+8}-1\\
\left(\sqrt{x+3}\right)^2 &= \left(\sqrt{2x+8}-1\right)^2\\
x+3 &= 2x+8-2\sqrt{2x+8}+1\\
2\sqrt{2x+8} &= x+6\\
\left(2\sqrt{2x+8}\right)^2 &= (x+6)^2\\
4(2x+8) &= x^2+12x+36\\
8x+32 &= x^2+12x+36\\
0 &= x^2+4x+4\\
0 &= (x+2)(x+2)\\
x+2=0 \quad &\textbf{or} \quad x+2=0\\
x=-2 \quad &\quad\quad x=-2
\end{aligned}$$
The solution checks.

55.
$$\begin{aligned}
\sqrt{y+8}-\sqrt{y-4} &= -2\\
\sqrt{y+8} &= \sqrt{y-4}-2\\
\left(\sqrt{y+8}\right)^2 &= \left(\sqrt{y-4}-2\right)^2\\
y+8 &= y-4-4\sqrt{y-4}+4\\
4\sqrt{y-4} &= -8\\
\left(4\sqrt{y-4}\right)^2 &= (-8)^2\\
16(y-4) &= 64\\
16y-64 &= 64\\
16y &= 128\\
y &= 8
\end{aligned}$$
The solution does not check. ⇒ No solution.

57.
$$\begin{aligned}
\sqrt{2b+3}-\sqrt{b+1} &= \sqrt{b-2}\\
\left(\sqrt{2b+3}-\sqrt{b+1}\right)^2 &= \left(\sqrt{b-2}\right)^2\\
2b+3-2\sqrt{(2b+3)(b+1)}+b+1 &= b-2\\
3b+4-2\sqrt{2b^2+5b+3} &= b-2\\
2b+6 &= 2\sqrt{2b^2+5b+3}\\
(2b+6)^2 &= \left(2\sqrt{2b^2+5b+3}\right)^2\\
4b^2+24b+36 &= 4\left(2b^2+5b+3\right)\\
4b^2+24b+36 &= 8b^2+20b+12\\
0 &= 4b^2-4b-24\\
0 &= 4(b-3)(b+2)\\
b-3=0 \quad &\textbf{or} \quad b+2=0\\
b=3 \quad &\quad\quad b=-2
\end{aligned}$$
$b=-2$ does not check, so it is an extraneous solution.

59.
$$\begin{aligned}
\sqrt{\sqrt{b}+\sqrt{b+8}} &= 2\\
\left(\sqrt{\sqrt{b}+\sqrt{b+8}}\right)^2 &= 2^2\\
\sqrt{b}+\sqrt{b+8} &= 4\\
\sqrt{b+8} &= 4-\sqrt{b}\\
\left(\sqrt{b+8}\right)^2 &= \left(4-\sqrt{b}\right)^2
\end{aligned}$$

$$\begin{aligned}
\left(\sqrt{b+8}\right)^2 &= \left(4-\sqrt{b}\right)^2\\
b+8 &= 16-8\sqrt{b}+b\\
8\sqrt{b} &= 8\\
\sqrt{b} &= 1\\
\left(\sqrt{b}\right)^2 &= 1^2
\end{aligned}$$
$b=1$; The solution checks.

61.
$$t = \sqrt{\frac{d}{16}}$$
$$5 = \sqrt{\frac{d}{16}}$$
$$5^2 = \left(\sqrt{\frac{d}{16}}\right)^2$$
$$25 = \frac{d}{16}$$
$400 = d \Rightarrow$ The bridge is 400 feet high.

63.
$$l = \sqrt{f^2 + h^2}$$
$$10 = \sqrt{f^2 + 6^2}$$
$$10^2 = \left(\sqrt{f^2 + 36}\right)^2$$
$$100 = f^2 + 36$$
$$64 = f^2$$
$$\pm 8 = f$$
He should nail the brace to the floor 8 feet from the wall.

65.
$$r = \sqrt[n]{\frac{A}{P}} - 1$$
$$0.065 = \sqrt[4]{\frac{4000}{P}} - 1$$
$$1.065 = \sqrt[4]{\frac{4000}{P}}$$
$$(1.065)^4 = \left(\sqrt[4]{\frac{4000}{P}}\right)^4$$
$$1.286466 \approx \frac{4000}{P}$$
$$1.286466P \approx 4000$$
$$P \approx 3109$$
The original price was about \$3109.

67. **Answers may vary.**

69. Natural numbers between -4 and 4:

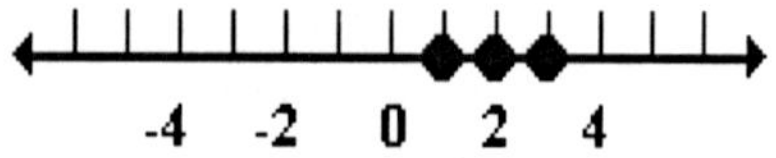

71. $x \geq 3 \Rightarrow [3, \infty)$

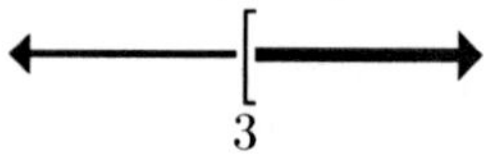

73. $-2 \leq x < 1 \Rightarrow [-2, 1)$

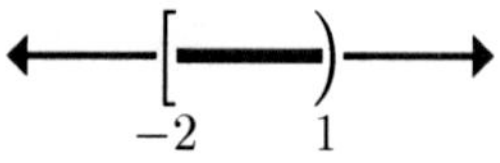

75. $x < 1$ or $x \geq 2$
$(-\infty, 1) \cup [2, \infty)$

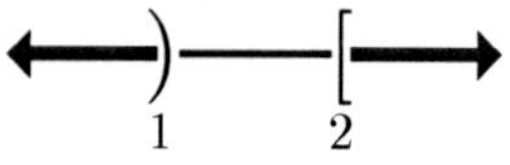

Exercises 1.7 (page 162)

1. right

3. $a < c$

5. $b - c$

7. $>$

9. linear

11. equivalent

13.
$$3x + 2 < 5$$
$$3x < 3$$
$$x < 1 \Rightarrow (-\infty, 1)$$

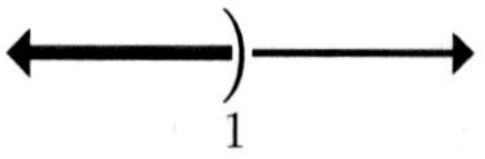

15.
$$3x + 2 \geq 5$$
$$3x \geq 3$$
$$x \geq 1 \Rightarrow [1, \infty)$$

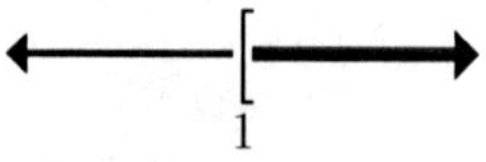

SECTION 1.7

17. $-5x + 3 > -2$

$$-5x > -5$$

$$x < 1 \Rightarrow (-\infty, 1)$$

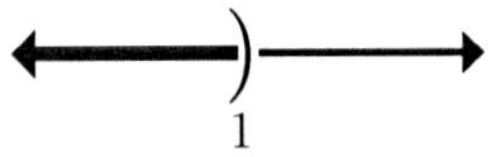

19. $-5x + 3 \le -2$

$$-5x \le -5$$

$$x \ge 1 \Rightarrow [1, \infty)$$

21. $2(x - 3) \le -2(x - 3)$

$$2x - 6 \le -2x + 6$$

$$4x \le 12$$

$$x \le 3 \Rightarrow (-\infty, 3]$$

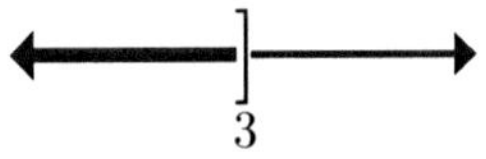

23. $\dfrac{3}{5}x + 4 > 2$

$$5\left(\frac{3}{5}x + 4\right) > 5(2)$$

$$3x + 20 > 10$$

$$3x > -10$$

$$x > -\tfrac{10}{3} \Rightarrow \left(-\tfrac{10}{3}, \infty\right)$$

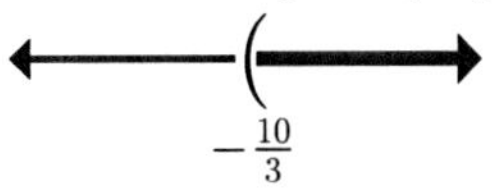

25. $\dfrac{x+3}{4} < \dfrac{2x-4}{3}$

$$12 \cdot \frac{x+3}{4} < 12 \cdot \frac{2x-4}{3}$$

$$3(x + 3) < 4(2x - 4)$$

$$3x + 9 < 8x - 16$$

$$-5x < -25$$

$$x > 5 \Rightarrow (5, \infty)$$

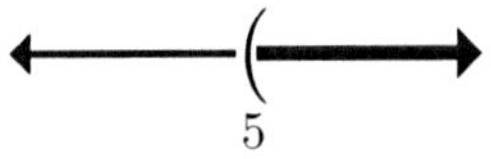

27. $\dfrac{6(x-4)}{5} \ge \dfrac{3(x+2)}{4}$

$$20 \cdot \frac{6x - 24}{5} \ge 20 \cdot \frac{3x + 6}{4}$$

$$4(6x - 24) \ge 5(3x + 6)$$

$$24x - 96 \ge 15x + 30$$

$$9x \ge 126$$

$$x \ge 14 \Rightarrow [14, \infty)$$

29. $\dfrac{5}{9}(a + 3) - a \ge \dfrac{4}{3}(a - 3) - 1$

$$9\left[\frac{5}{9}(a + 3) - a\right] \ge 9\left[\frac{4}{3}(a - 3) - 1\right]$$

$$5(a + 3) - 9a \ge 12(a - 3) - 9$$

$$5a + 15 - 9a \ge 12a - 36 - 9$$

$$-16a \ge -60$$

$$a \le \frac{-60}{-16}$$

$$a \le \tfrac{15}{4} \Rightarrow \left(-\infty, \tfrac{15}{4}\right]$$

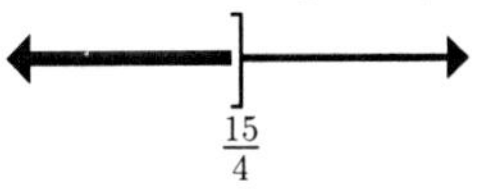

31. $\dfrac{2}{3}a - \dfrac{3}{4}a < \dfrac{3}{5}\left(a + \dfrac{2}{3}\right) + \dfrac{1}{3}$

$$60\left(\frac{2}{3}a - \frac{3}{4}a\right) < 60\left[\frac{3}{5}\left(a + \frac{2}{3}\right) + \frac{1}{3}\right]$$

$$40a - 45a < 36\left(a + \frac{2}{3}\right) + 20$$

$$-5a < 36a + 24 + 20$$

$$-41a < 44$$

$$a > -\tfrac{44}{41} \Rightarrow \left(-\tfrac{44}{41}, \infty\right)$$

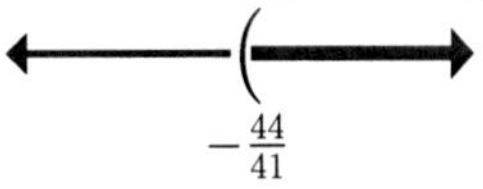

33. $4 < 2x - 8 \leq 10$
$12 < 2x \leq 18$
$6 < x \leq 9 \Rightarrow (6, 9]$

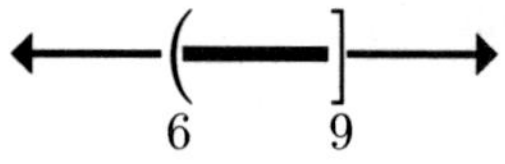

35. $9 \geq \dfrac{x-4}{2} > 2$
$18 \geq x - 4 > 4$
$22 \geq x > 8$
$8 < x \leq 22 \Rightarrow (8, 22]$

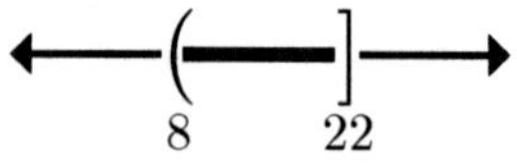

37. $0 \leq \dfrac{4-x}{3} \leq 5$
$0 \leq 4 - x \leq 15$
$-4 \leq -x \leq 11$
$4 \geq x \geq -11$
$-11 \leq x \leq 4 \Rightarrow [-11, 4]$

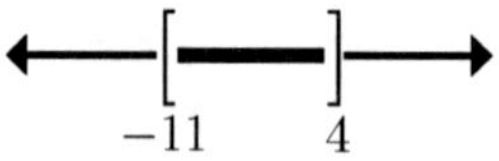

39. $-2 \geq \dfrac{1-x}{2} \geq -10$
$-4 \geq 1 - x \geq -20$
$-5 \geq -x \geq -21$
$5 \leq x \leq 21 \Rightarrow [5, 21]$

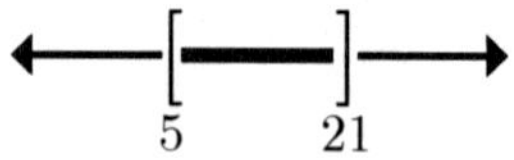

41. $-3x > -2x > -x$
$-3x > -2x$ **and** $-2x > -x$
$-x > 0$ $\quad$ $-x > 0$
$x < 0$ $\quad$ $x < 0$
$x < 0 \Rightarrow (-\infty, 0)$

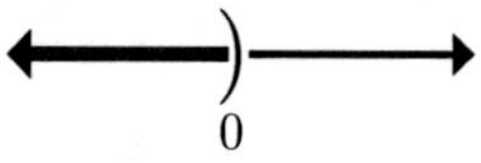

43. $x < 2x < 3x$
$x < 2x$ **and** $2x < 3x$
$-x < 0$ $\quad$ $-x < 0$
$x > 0$ $\quad$ $x > 0$
$x > 0 \Rightarrow (0, \infty)$

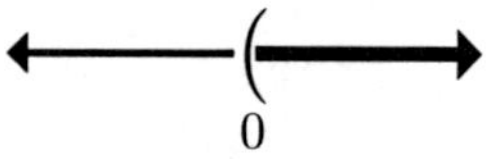

45. $2x + 1 < 3x - 2 < 12$
$2x + 1 < 3x - 2$ **and** $3x - 2 < 12$
$-x < -3$ $\quad$ $3x < 14$
$x > 3$ $\quad$ $x < \frac{14}{3}$

$x > 3$
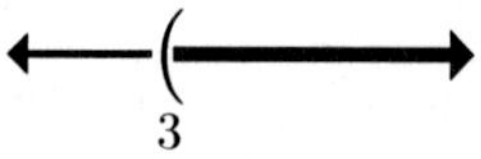

$x < \frac{14}{3}$
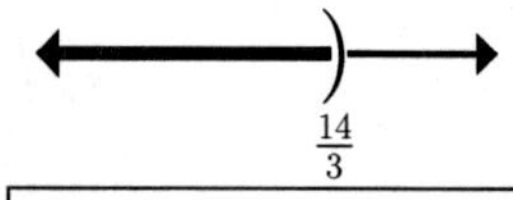

$x > 3$
and
$x < \frac{14}{3}$
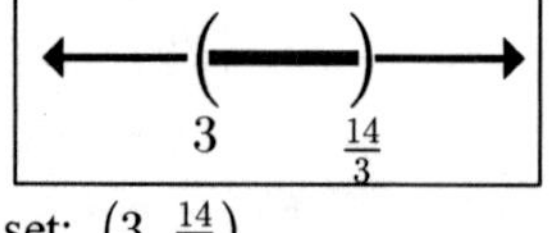

Solution set: $\left(3, \frac{14}{3}\right)$

47. $2 + x < 3x - 2 < 5x + 2$
$2 + x < 3x - 2$ **and** $3x - 2 < 5x + 2$
$-2x < -4$ $\quad$ $-2x < 4$
$x > 2$ $\quad$ $x > -2$

$x > 2$
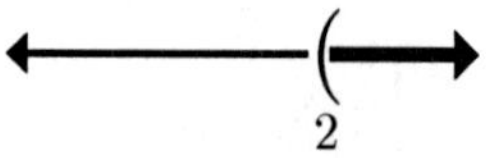

$x > -2$
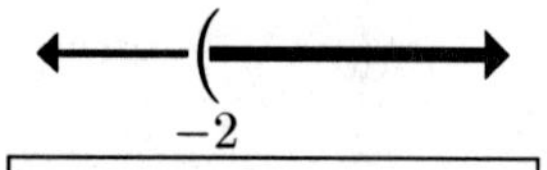

$x > 2$
and
$x > -2$
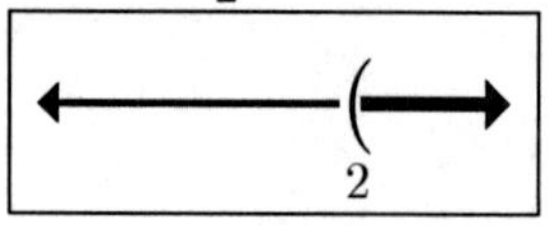

Solution set: $(2, \infty)$

49. $3 + x > 7x - 2 > 5x - 10$

$3 + x > 7x - 2$ **and** $7x - 2 > 5x - 10$

$-6x > -5$ $\qquad 2x > -8$

$x < \frac{5}{6}$ $\qquad x > -4$

$x < \frac{5}{6}$

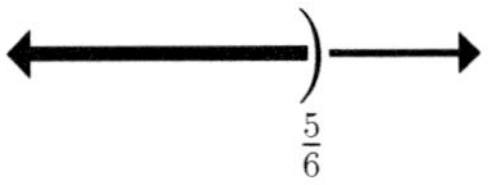

$x > -4$

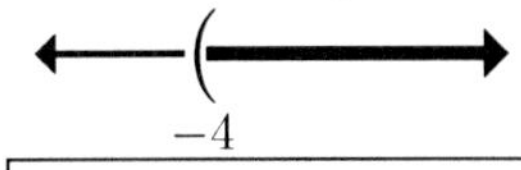

$x < \frac{5}{6}$ **and** $x > -4$

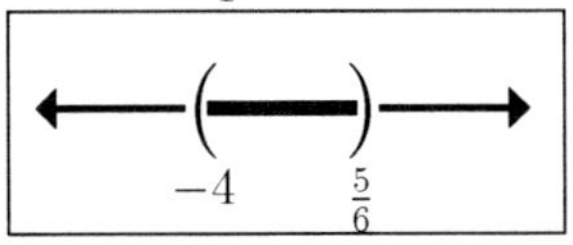

Solution set: $\left(-4, \frac{5}{6}\right)$

51. $x \leq x + 1 \leq 2x + 3$

$x \leq x + 1$ **and** $x + 1 \leq 2x + 3$

$0 \leq 1$ $\qquad -x \leq 2$

true for all real numbers x $\qquad x \geq -2$

$0 \leq 1$

$x \geq -2$

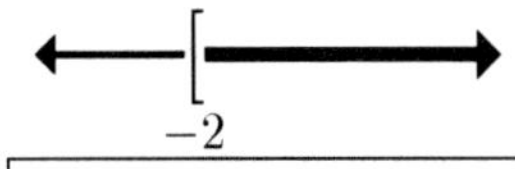

$0 \leq 1$ **and** $x \geq -2$

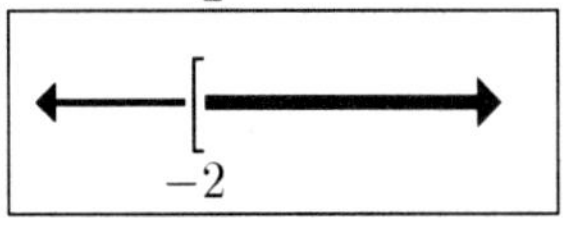

Solution set: $[-2, \infty)$

53. $x^2 + 7x + 12 < 0$

$(x + 3)(x + 4) < 0$

factors $= 0$: $x = -3$, $x = -4$

intervals: $(-\infty, -4)$, $(-4, -3)$, $(-3, \infty)$

interval	test number	value of $x^2+7x+12$
$(-\infty, -4)$	-5	$+2$
$(-4, -3)$	-3.5	-0.25
$(-3, \infty)$	0	$+12$

Solution: $(-4, -3)$

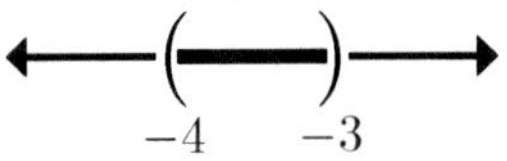

55. $x^2 - 5x + 6 \geq 0$

$(x - 3)(x - 2) \geq 0$

factors $= 0$: $x = 3$, $x = 2$

intervals: $(-\infty, 2)$, $(2, 3)$, $(3, \infty)$

interval	test number	value of x^2-5x+6
$(-\infty, 2)$	0	$+6$
$(2, 3)$	2.5	-0.25
$(3, \infty)$	4	$+2$

Solution: $(-\infty, 2] \cup [3, \infty)$

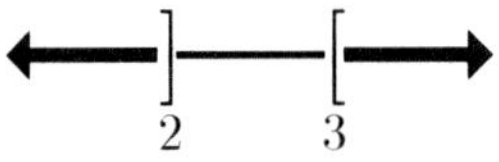

57. $x^2 + 5x + 6 < 0$

$(x + 3)(x + 2) < 0$

factors $= 0$: $x = -3$, $x = -2$

intervals: $(-\infty, -3)$, $(-3, -2)$, $(-2, \infty)$

interval	test number	value of x^2+5x+6
$(-\infty, -3)$	-4	$+2$
$(-3, -2)$	-2.5	-0.25
$(-2, \infty)$	0	$+6$

Solution: $(-3, -2)$

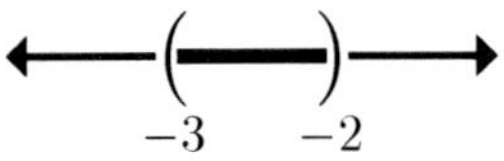

59. $6x^2 + 5x + 1 \geq 0$

$(2x + 1)(3x + 1) \geq 0$

factors $= 0$: $x = -\frac{1}{2}$, $x = -\frac{1}{3}$

intervals: $\left(-\infty, -\frac{1}{2}\right)$, $\left(-\frac{1}{2}, -\frac{1}{3}\right)$, $\left(-\frac{1}{3}, \infty\right)$

interval	test number	value of $6x^2+5x+1$
$\left(-\infty, -\frac{1}{2}\right)$	-1	$+2$
$\left(-\frac{1}{2}, -\frac{1}{3}\right)$	-0.4	-0.04
$\left(-\frac{1}{3}, \infty\right)$	0	$+1$

Solution: $\left(-\infty, -\frac{1}{2}\right] \cup \left[-\frac{1}{3}, \infty\right)$

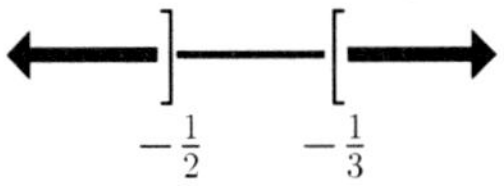

61.

$$6x^2 - 5x < -1$$
$$6x^2 - 5x + 1 < 0$$
$$(2x-1)(3x-1) < 0$$

factors $= 0$: $x = \frac{1}{2}$, $x = \frac{1}{3}$

intervals: $\left(-\infty, \frac{1}{3}\right)$, $\left(\frac{1}{3}, \frac{1}{2}\right)$, $\left(\frac{1}{2}, \infty\right)$

interval	test number	value of $6x^2-5x+1$
$\left(-\infty, \frac{1}{3}\right)$	0	+1
$\left(\frac{1}{3}, \frac{1}{2}\right)$	0.4	−0.04
$\left(\frac{1}{2}, \infty\right)$	1	+2

Solution: $\left(\frac{1}{3}, \frac{1}{2}\right)$

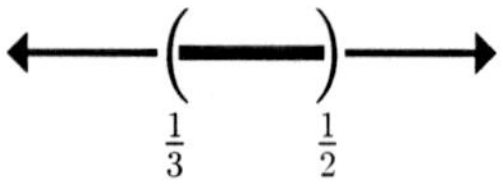

63.

$$2x^2 \geq 3 - x$$
$$2x^2 + x - 3 \geq 0$$
$$(2x+3)(x-1) \geq 0$$

factors $= 0$: $x = -\frac{3}{2}$, $x = 1$

intervals: $\left(-\infty, -\frac{3}{2}\right)$, $\left(-\frac{3}{2}, 1\right)$, $(1, \infty)$

interval	test number	value of $2x^2+x-3$
$\left(-\infty, -\frac{3}{2}\right)$	−2	+3
$\left(-\frac{3}{2}, 1\right)$	0	−3
$(1, \infty)$	2	+7

Solution: $\left(-\infty, -\frac{3}{2}\right] \cup [1, \infty)$

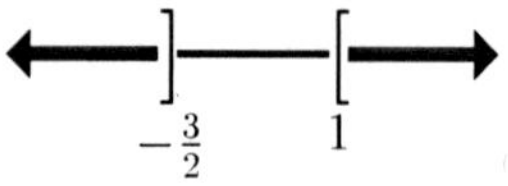

65.

$$x^2 - 3 \geq 0$$
$$x^2 - 3 = 0$$
$$x^2 = 3$$
$$x = \pm\sqrt{3}$$

intervals: $\left(-\infty, -\sqrt{3}\right)$, $\left(-\sqrt{3}, \sqrt{3}\right)$, $\left(\sqrt{3}, \infty\right)$

interval	test number	value of x^2-3
$\left(-\infty, -\sqrt{3}\right)$	−2	+1
$\left(-\sqrt{3}, \sqrt{3}\right)$	0	−3
$\left(\sqrt{3}, \infty\right)$	2	+1

Solution: $\left(-\infty, -\sqrt{3}\right] \cup \left[\sqrt{3}, \infty\right)$

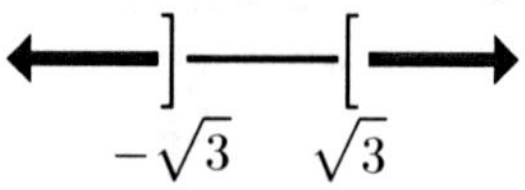

67.

$$x^2 - 11 < 0$$
$$x^2 - 11 = 0$$
$$x^2 = 11$$
$$x = \pm\sqrt{11}$$

intervals: $\left(-\infty, -\sqrt{11}\right)$, $\left(-\sqrt{11}, \sqrt{11}\right)$, $\left(\sqrt{11}, \infty\right)$

interval	test number	value of x^2-11
$\left(-\infty, -\sqrt{11}\right)$	−4	+5
$\left(-\sqrt{11}, \sqrt{11}\right)$	0	−11
$\left(\sqrt{11}, \infty\right)$	4	+5

Solution: $\left(-\sqrt{11}, \sqrt{11}\right)$

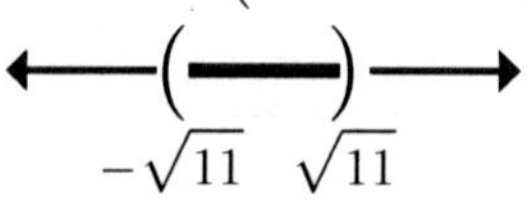

69. $\dfrac{x+3}{x-2} < 0$

factors $= 0$: $x = -3, x = 2$

intervals: $(-\infty, -3), (-3, 2), (2, \infty)$

interval	test number	sign of $\frac{x+3}{x-2}$
$(-\infty, -3)$	-4	$+$
$(-3, 2)$	0	$-$
$(2, \infty)$	3	$+$

Solution: $(-3, 2)$

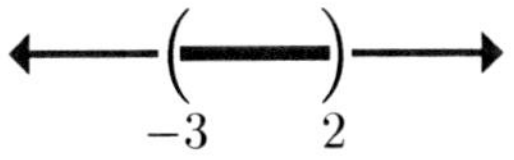

71. $\dfrac{x^2+x}{x^2-1} > 0$

$\dfrac{x(x+1)}{(x+1)(x-1)} > 0$

factors $= 0$: $x = 0, x = -1, x = 1$

intervals: $(-\infty, -1), (-1, 0), (0, 1), (1, \infty)$

interval	test number	sign of $\frac{x^2+x}{x^2-1}$
$(-\infty, -1)$	-2	$+$
$(-1, 0)$	$-\frac{1}{2}$	$+$
$(0, 1)$	$\frac{1}{2}$	$-$
$(1, \infty)$	2	$+$

Solution: $(-\infty, -1) \cup (-1, 0) \cup (1, \infty)$

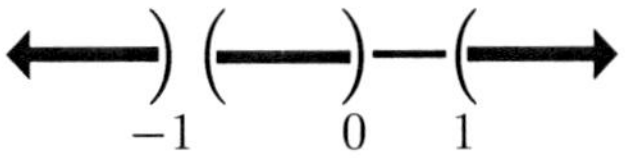

73. $\dfrac{x^2+5x+6}{x^2+x-6} \geq 0 \Rightarrow \dfrac{(x+3)(x+2)}{(x+3)(x-2)} \geq 0$

factors $= 0$: $x = -3, x = \pm 2$

intervals: $(-\infty, -3), (-3, -2), (-2, 2),$
$(2, \infty)$

interval	test number	sign of $\frac{x^2+5x+6}{x^2+x-6}$
$(-\infty, -3)$	-4	$+$
$(-3, -2)$	-2.5	$+$
$(-2, 2)$	0	$-$
$(2, \infty)$	3	$+$

Include endpoints which make the numerator equal to 0. Do not include endpoints which make the denominator equal to 0.

Solution: $(-\infty, -3) \cup (-3, -2] \cup (2, \infty)$

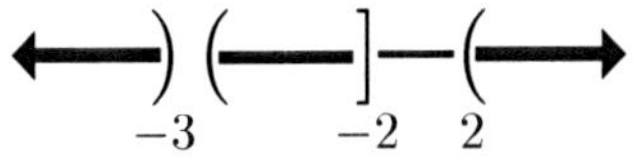

75. $\dfrac{6x^2-x-1}{x^2+4x+4} > 0 \Rightarrow \dfrac{(2x-1)(3x+1)}{(x+2)(x+2)} > 0$

factors $= 0$: $x = \frac{1}{2}, x = -\frac{1}{3}, x = -2$

intervals: $(-\infty, -2), \left(-2, -\frac{1}{3}\right), \left(-\frac{1}{3}, \frac{1}{2}\right),$
$\left(\frac{1}{2}, \infty\right)$

interval	test number	sign of $\frac{6x^2-x-1}{x^2+4x+4}$
$(-\infty, -2)$	-3	$+$
$\left(-2, -\frac{1}{3}\right)$	-1	$+$
$\left(-\frac{1}{3}, \frac{1}{2}\right)$	0	$-$
$\left(\frac{1}{2}, \infty\right)$	1	$+$

Solution: $(-\infty, -2) \cup \left(-2, -\frac{1}{3}\right) \cup \left(\frac{1}{2}, \infty\right)$

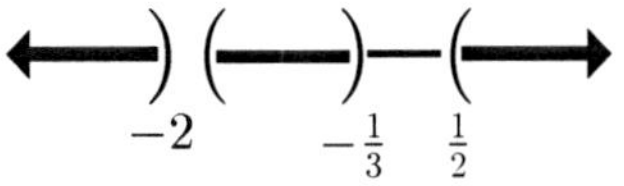

77. $\dfrac{3}{x} > 2$

$\dfrac{3}{x} - 2 > 0$

$\dfrac{3-2x}{x} > 0$

factors $= 0$: $x = \frac{3}{2}, x = 0$

intervals: $(-\infty, 0), \left(0, \frac{3}{2}\right), \left(\frac{3}{2}, \infty\right)$

interval	test number	sign of $\frac{3-2x}{x}$
$(-\infty, 0)$	-1	$-$
$\left(0, \frac{3}{2}\right)$	1	$+$
$\left(\frac{3}{2}, \infty\right)$	2	$-$

Solution: $\left(0, \frac{3}{2}\right)$

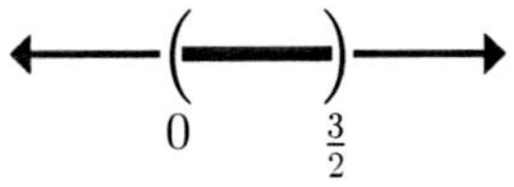

79.

$$\frac{6}{x} < 4$$
$$\frac{6}{x} - 4 < 0$$
$$\frac{6-4x}{x} < 0$$

factors = 0: $x = \frac{3}{2}, x = 0$

intervals: $(-\infty, 0), \left(0, \frac{3}{2}\right), \left(\frac{3}{2}, \infty\right)$

interval	test number	sign of $\frac{6-4x}{x}$
$(-\infty, 0)$	-1	$-$
$\left(0, \frac{3}{2}\right)$	1	$+$
$\left(\frac{3}{2}, \infty\right)$	2	$-$

Solution: $(-\infty, 0) \cup \left(\frac{3}{2}, \infty\right)$

81.

$$\frac{3}{x-2} \le 5$$
$$\frac{3}{x-2} - 5 \le 0$$
$$\frac{3}{x-2} - \frac{5(x-2)}{x-2} \le 0$$
$$\frac{3-5x+10}{x-2} \le 0$$
$$\frac{13-5x}{x-2} \le 0$$

factors = 0: $x = \frac{13}{5}, x = 2$

intervals: $(-\infty, 2), \left(2, \frac{13}{5}\right), \left(\frac{13}{5}, \infty\right)$

interval	test number	sign of $\frac{13-5x}{x-2}$
$(-\infty, 2)$	0	$-$
$\left(2, \frac{13}{5}\right)$	$\frac{11}{5}$	$+$
$\left(\frac{13}{5}, \infty\right)$	3	$-$

Solution: $(-\infty, 2) \cup \left[\frac{13}{5}, \infty\right)$

Include endpoints which make the numerator equal to 0. Do not include endpoints which make the denominator equal to 0.

83.

$$\frac{6}{x^2-1} < 1$$
$$\frac{6}{x^2-1} - 1 < 0$$
$$\frac{6}{x^2-1} - \frac{x^2-1}{x^2-1} < 0$$
$$\frac{7-x^2}{x^2-1} < 0$$
$$\frac{7-x^2}{(x+1)(x-1)} < 0$$

factors = 0: $x = \pm\sqrt{7}, x = \pm 1$

intervals: $\left(-\infty, -\sqrt{7}\right), \left(-\sqrt{7}, -1\right), (-1, 1),$ $\left(1, \sqrt{7}\right), \left(\sqrt{7}, \infty\right)$

interval	test number	sign of $\frac{7-x^2}{x^2-1}$
$\left(-\infty, -\sqrt{7}\right)$	-3	$-$
$\left(-\sqrt{7}, -1\right)$	-2	$+$
$(-1, 1)$	0	$-$
$\left(1, \sqrt{7}\right)$	2	$+$
$\left(\sqrt{7}, \infty\right)$	3	$-$

Sol'n: $\left(-\infty, -\sqrt{7}\right) \cup (-1, 1) \cup \left(\sqrt{7}, \infty\right)$

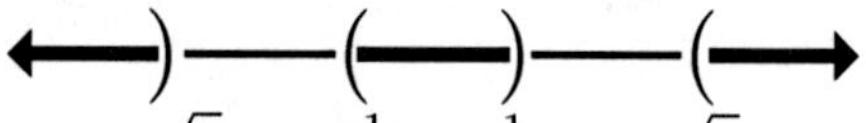

85. Let $x =$ the number of minutes after 3 minutes. The total cost $= 40 + 10x$ cents.

$$\boxed{\text{Total cost}} < 200$$
$$40 + 10x < 200$$
$$10x < 160$$
$$x < 16$$

A person can talk for up to 16 minutes after the initial 3 minutes, for a total of up to 19 minutes for less than \$2.

87. Let $x =$ the number of CDs. Then the total cost $= 150 + 9.75x$.

$$\boxed{\text{Total cost}} < 275$$
$$150 + 9.75x < 275$$
$$9.75x < 125$$
$$x < 12.8$$

He can buy up to 12 disks.

89. Let $p =$ the price of the refrigerator. Then the total cost $= p + 0.065p + 0.0025p$.

$$\boxed{\text{Total cost}} < 1200$$
$$p + 0.065p + 0.0025p < 1200$$
$$1.0675p < 1200$$
$$p < 1124.122$$
$$p \leq \$1124.12$$

91. Let $a =$ the assessed value. Find when Method 1 < Method 2:

$$2200 + 0.04a < 1200 + 0.06a$$
$$1000 < 0.02a$$
$$50000 < a$$

The first method will benefit the taxpayer when $a >$ \$50,000.

93. Let $b =$ the hospital bill. Find when Cost of Plan 1 > Cost of Plan 2:

$$200 + 0.30(b - 200) > 400 + 0.20(b - 400)$$
$$200 + 0.30b - 60 > 400 + 0.20b - 80$$
$$0.1b > 180$$
$$b > 1800$$

Plan 2 is better for bills over \$1,800.

95. Let $P =$ the perimeter. Then the length of one side is equal to $\frac{P}{3}$.

$$50 < P < 60$$
$$\frac{50}{3} < \frac{P}{3} < \frac{60}{3}$$
$$16\tfrac{2}{3} < \text{length} < 20$$

The length of a side is between $16\frac{2}{3}$ and 20 cm.

97.
$$20 < l < 30$$
$$40 < 2l < 60$$
$$40 + 2w < 2l + 2w < 60 + 2w$$
$$40 + 2w < P < 60 + 2w$$

99. **Answers may vary.**

101. even integers: $0, -2, 2$

103. prime numbers: 2

105. real numbers: all in the list

Exercises 1.8 (page 171)

1. x

3. $x = k$ or $x = -k$

5. $-k < x < k$

7. $x \leq -k$ or $x \geq k$

9. $|7| = 7$

11. $|0| = 0$

13. $|5| - |-3| = 5 - 3 = 2$

15. $|\pi - 2| = +(\pi - 2) = \pi - 2$

17. $x \geq 5 \Rightarrow |x - 5| = x - 5$

SECTION 1.8

19. $|x^3| = \begin{cases} x^3 & \text{if } x \geq 0 \\ -x^3 & \text{if } x < 0 \end{cases}$

21.
$$|x+2| = 2$$
$$x+2 = 2 \quad \textbf{or} \quad x+2 = -2$$
$$x = 0 \qquad\qquad x = -4$$

23.
$$|3x-1| - 7 = -2$$
$$|3x-1| = 5$$
$$3x-1 = 5 \quad \textbf{or} \quad 3x-1 = -5$$
$$3x = 6 \qquad\qquad 3x = -4$$
$$x = 2 \qquad\qquad x = -\tfrac{4}{3}$$

25.
$$\left|\frac{3x-4}{2}\right| = 5$$
$$\frac{3x-4}{2} = 5 \quad \textbf{or} \quad \frac{3x-4}{2} = -5$$
$$3x-4 = 10 \qquad\qquad 3x-4 = -10$$
$$3x = 14 \qquad\qquad 3x = -6$$
$$x = \tfrac{14}{3} \qquad\qquad x = -2$$

27.
$$\left|\frac{2x-4}{5}\right| + 6 = 8$$
$$\left|\frac{2x-4}{5}\right| = 2$$
$$\frac{2x-4}{5} = 2 \quad \textbf{or} \quad \frac{2x-4}{5} = -2$$
$$2x-4 = 10 \qquad\qquad 2x-4 = -10$$
$$2x = 14 \qquad\qquad 2x = -6$$
$$x = 7 \qquad\qquad x = -3$$

29.
$$\left|\frac{x-3}{4}\right| = -2$$
An absolute value can never equal a negative number.
no solution

31.
$$\left|\frac{x-5}{3}\right| = 0$$
$$\frac{x-5}{3} = 0 \quad \textbf{or} \quad \frac{x-5}{3} = -0$$
$$x-5 = 0 \qquad\qquad x-5 = 0$$
$$x = 5 \qquad\qquad x = 5$$

33.
$$\left|\frac{4x-2}{x}\right| = 3$$
$$\frac{4x-2}{x} = 3 \quad \textbf{or} \quad \frac{4x-2}{x} = -3$$
$$4x-2 = 3x \qquad\qquad 4x-2 = -3x$$
$$x = 2 \qquad\qquad 7x = 2$$
$$x = 2 \qquad\qquad x = \tfrac{2}{7}$$

35.
$$|x| = x$$
True for all $x \geq 0$.

37.
$$|x+3| = |x|$$
$$x+3 = x \quad \textbf{or} \quad x+3 = -x$$
$$0 = 3 \qquad\qquad 2x = -3$$
$$\text{not true} \qquad\qquad x = -\tfrac{3}{2}$$

39.
$$|x-3| = |2x+3|$$
$$x-3 = 2x+3 \quad \textbf{or} \quad x-3 = -(2x+3)$$
$$-x = 6 \qquad\qquad x-3 = -2x-3$$
$$x = -6 \qquad\qquad 3x = 0$$
$$x = 0$$

41.
$$|x+2| = |x-2|$$
$$x+2 = x-2 \quad \textbf{or} \quad x+2 = -(x-2)$$
$$0 = -4 \qquad\qquad x+2 = -x+2$$
$$\text{not true} \qquad\qquad 2x = 0$$
$$x = 0$$

43.

$$\left|\frac{x+3}{2}\right| = |2x-3|$$

$$\begin{aligned} \frac{x+3}{2} &= 2x-3 \quad \textbf{or} \quad & \frac{x+3}{2} &= -(2x-3) \\ x+3 &= 4x-6 & \frac{x+3}{2} &= -2x+3 \\ -3x &= -9 & x+3 &= -4x+6 \\ x &= 3 & 5x &= 3 \\ & & x &= \tfrac{3}{5} \end{aligned}$$

45.

$$\left|\frac{3x-1}{2}\right| = \left|\frac{2x+3}{3}\right|$$

$$\begin{aligned} \frac{3x-1}{2} &= \frac{2x+3}{3} \quad \textbf{or} \quad & \frac{3x-1}{2} &= -\frac{2x+3}{3} \\ 6\left(\frac{3x-1}{2}\right. &= \left.\frac{2x+3}{3}\right) & 6\left(\frac{3x-1}{2}\right. &= \left.-\frac{2x+3}{3}\right) \\ 3(3x-1) &= 2(2x+3) & 3(3x-1) &= -2(2x+3) \\ 9x-3 &= 4x+6 & 9x-3 &= -4x-6 \\ 5x &= 9 & 13x &= -3 \\ x &= \tfrac{9}{5} & x &= -\tfrac{3}{13} \end{aligned}$$

47.

$$\begin{aligned} |x-3| &< 6 \\ -6 < x-3 &< 6 \\ -3 < \quad x \quad &< 9 \\ (-3, 9)& \end{aligned}$$

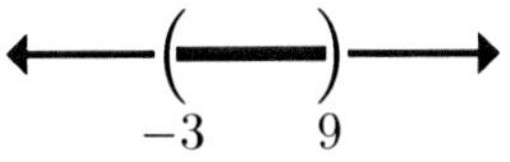

49.

$$\begin{aligned} & |x+3| > 6 \\ x+3 > 6 \quad &\textbf{or} \quad x+3 < -6 \\ x > 3 \quad & \qquad\quad x < -9 \\ (-\infty, -9)& \cup (3, \infty) \end{aligned}$$

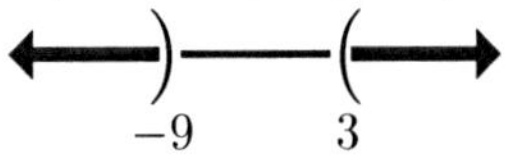

51.

$$\begin{aligned} & |2x+4| \geq 10 \\ 2x+4 \geq 10 \quad &\textbf{or} \quad 2x+4 \leq -10 \\ 2x \geq 6 \quad & \qquad\quad 2x \leq -14 \\ x \geq 3 \quad & \qquad\quad x \leq -7 \\ (-\infty, -7]& \cup [3, \infty) \end{aligned}$$

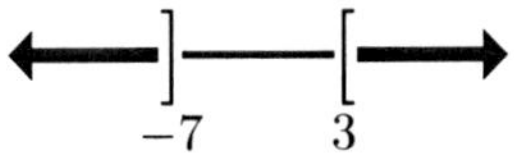

53.

$$\begin{aligned} |3x+5|+1 &\leq 9 \\ |3x+5| &\leq 8 \\ -8 \leq 3x+5 &\leq 8 \\ -13 \leq \quad 3x \quad &\leq 3 \\ -\tfrac{13}{3} \leq \quad x \quad &\leq 1 \\ \left[-\tfrac{13}{3}, 1\right]& \end{aligned}$$

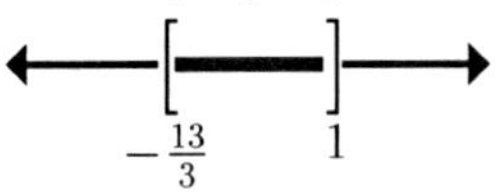

55.

$$|x+3| > 0$$

$$x+3>0 \quad \textbf{or} \quad x+3<-0$$

$$x>-3 \qquad\qquad x<-3$$

$$(-\infty, -3) \cup (-3, \infty)$$

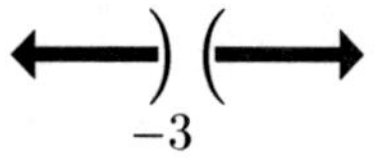

57.

$$\left|\tfrac{5x+2}{3}\right| < 1$$

$$-1 < \tfrac{5x+2}{3} < 1$$

$$-3 < 5x+2 < 3$$

$$-5 < 5x < 1$$

$$-1 < x < \tfrac{1}{5}$$

$$\left(-1, \tfrac{1}{5}\right)$$

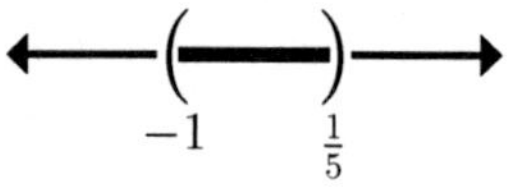

59.

$$3\left|\tfrac{3x-1}{2}\right| > 5$$

$$\left|\tfrac{3x-1}{2}\right| > \tfrac{5}{3}$$

$$\tfrac{3x-1}{2} > \tfrac{5}{3} \quad \textbf{or} \quad \tfrac{3x-1}{2} < -\tfrac{5}{3}$$

$$6\cdot\tfrac{3x-1}{2} > 6\cdot\tfrac{5}{3} \qquad 6\cdot\tfrac{3x-1}{2} < 6\left(-\tfrac{5}{3}\right)$$

$$9x-3>10 \qquad 9x-3<-10$$

$$9x>13 \qquad 9x<-7$$

$$x>\tfrac{13}{9} \qquad x<-\tfrac{7}{9}$$

$$\left(-\infty, -\tfrac{7}{9}\right) \cup \left(\tfrac{13}{9}, \infty\right)$$

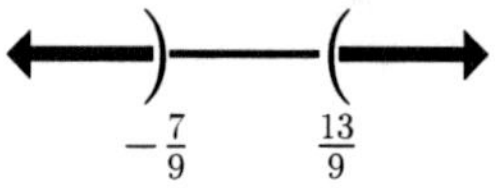

61.

$$\frac{|x-1|}{-2} > -3$$

$$|x-1| < 6$$

$$-6 < x-1 < 6$$

$$-5 < x < 7$$

$$(-5, 7)$$

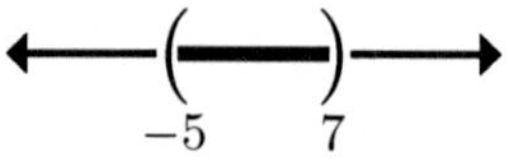

63.

$$0 < |2x+1| < 3$$

$$0 < |2x+1| \quad \textbf{and} \quad |2x+1| < 3$$

(1) $|2x+1| > 0$

$$2x+1>0 \quad \textbf{or} \quad 2x+1<-0$$

$$2x>-1 \qquad 2x<-1$$

$$x>-\tfrac{1}{2} \qquad x<-\tfrac{1}{2}$$

(2) $|2x+1| < 3$

$$-3 < 2x+1 < 3$$

$$-4 < 2x < 2$$

$$-2 < x < 1$$

(1) 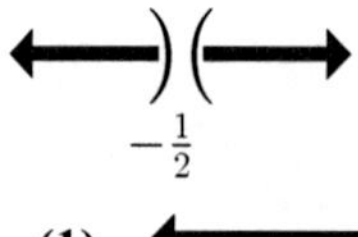

(2) 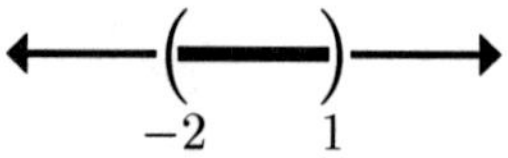

(1) 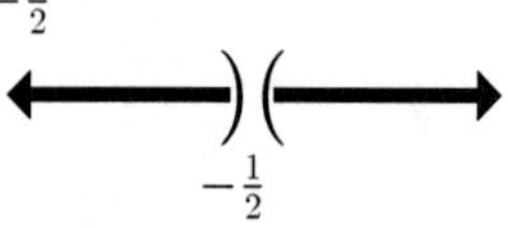

(2) 

(1) and (2) 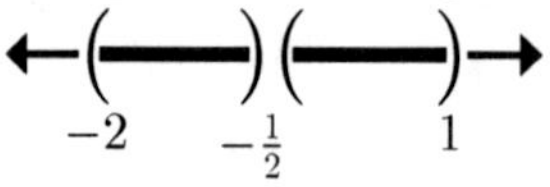

$\Rightarrow \left(-2, -\tfrac{1}{2}\right) \cup \left(-\tfrac{1}{2}, 1\right)$

SECTION 1.8

65. $$8 > |3x - 1| > 3$$

$|3x - 1| > 3$ **and** $8 > |3x - 1|$

(1) $|3x - 1| > 3$ **(2)** $|3x - 1| < 8$

$3x - 1 > 3$ **or** $3x - 1 < -3$ $\qquad -8 < 3x - 1 < 8$

$3x > 4$ $\qquad 3x < -2$ $\qquad -7 < 3x < 9$

$x > \frac{4}{3}$ $\qquad x < -\frac{2}{3}$ $\qquad -\frac{7}{3} < x < 3$

(1) 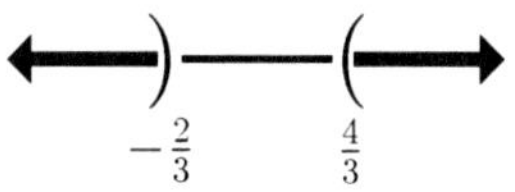

(2) 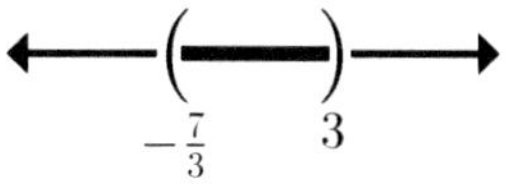

(1) 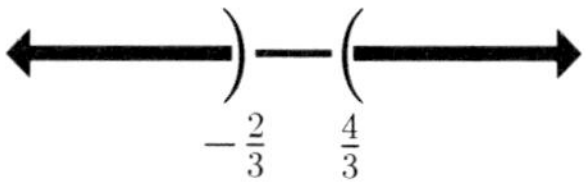

(2) 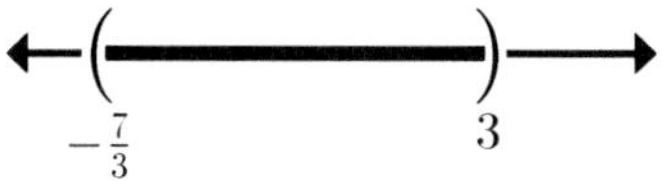

(1) and (2) 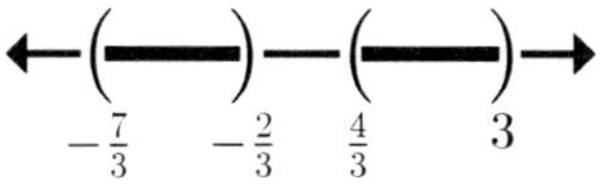 $\Rightarrow \left(-\frac{7}{3}, -\frac{2}{3}\right) \cup \left(\frac{4}{3}, 3\right)$

67. $$2 < \left|\frac{x-5}{3}\right| < 4$$

$2 < \left|\frac{x-5}{3}\right|$ **and** $\left|\frac{x-5}{3}\right| < 4$

(1) $\left|\frac{x-5}{3}\right| > 2$ **(2)** $\left|\frac{x-5}{3}\right| < 4$

$\frac{x-5}{3} > 2$ **or** $\frac{x-5}{3} < -2$ $\qquad -4 < \frac{x-5}{3} < 4$

$x - 5 > 6$ $\qquad x - 5 < -6$ $\qquad -12 < x - 5 < 12$

$x > 11$ $\qquad x < -1$ $\qquad -7 < x < 17$

(1) 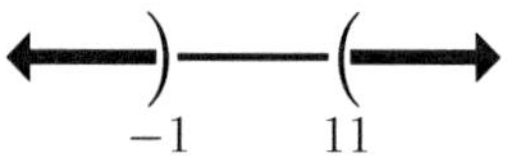

(2) 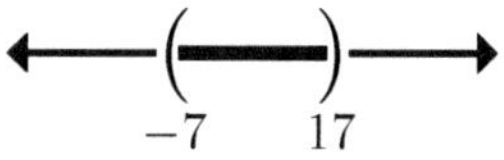

(1) 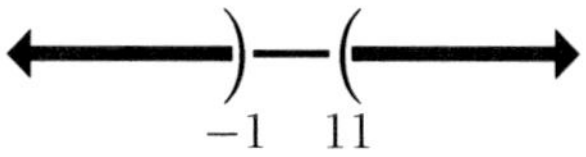

(2) 

(1) and (2) 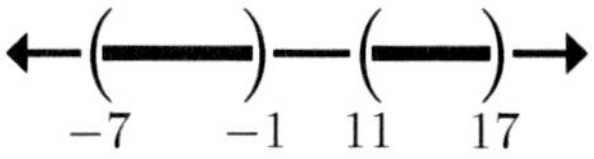 $\Rightarrow (-7, -1) \cup (11, 17)$

69. $$10 > \left|\frac{x-2}{2}\right| > 4$$

$\left|\frac{x-2}{2}\right| > 4$ **and** $10 > \left|\frac{x-2}{2}\right|$

(1) $\left|\frac{x-2}{2}\right| > 4$

$\frac{x-2}{2} > 4$ **or** $\frac{x-2}{2} < -4$

$x-2 > 8 \qquad x-2 < -8$

$x > 10 \qquad x < -6$

(2) $\left|\frac{x-2}{2}\right| < 10$

$-10 < \frac{x-2}{2} < 10$

$-20 < x-2 < 20$

$-18 < x < 22$

(1) 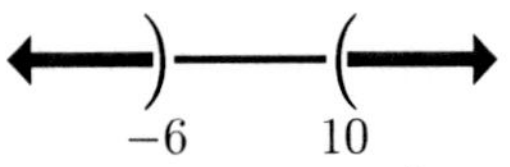

(2) 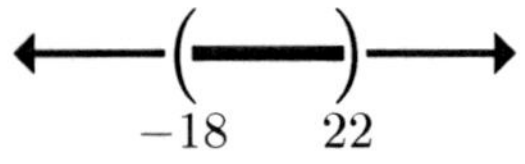

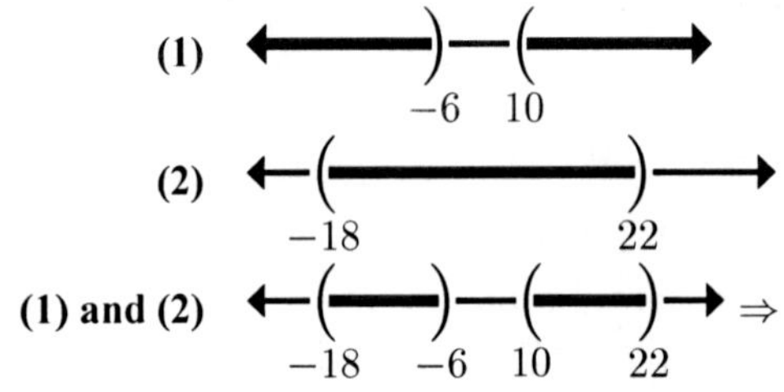

$\Rightarrow (-18, -6) \cup (10, 22)$

71. $$2 \le \left|\frac{x+1}{3}\right| < 3$$

$2 \le \left|\frac{x+1}{3}\right|$ **and** $\left|\frac{x+1}{3}\right| < 3$

(1) $\left|\frac{x+1}{3}\right| \ge 2$

$\frac{x+1}{3} \ge 2$ **or** $\frac{x+1}{3} \le -2$

$x+1 \ge 6 \qquad x+1 \le -6$

$x \ge 5 \qquad x \le -7$

(2) $\left|\frac{x+1}{3}\right| < 3$

$-3 < \frac{x+1}{3} < 3$

$-9 < x+1 < 9$

$-10 < x < 8$

(1) 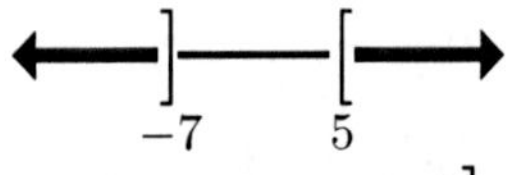

(2) 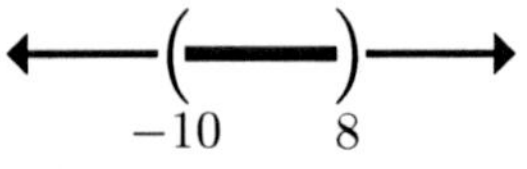

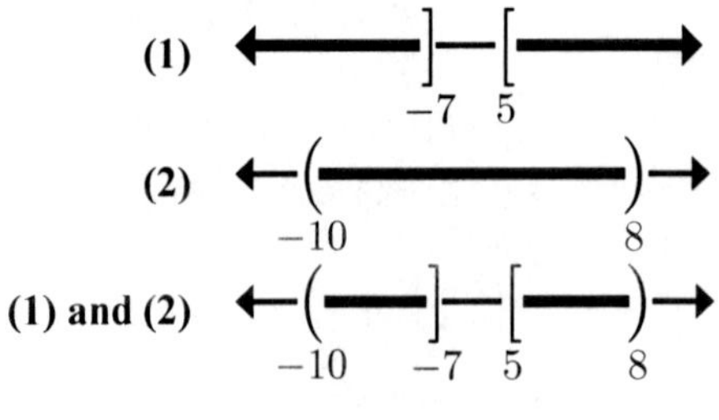

$\Rightarrow (-10, -7] \cup [5, 8)$

73.
$$\begin{aligned} |x+1| &\geq |x| \\ \sqrt{(x+1)^2} &\geq \sqrt{x^2} \\ (x+1)^2 &\geq x^2 \\ x^2+2x+1 &\geq x^2 \\ 2x &\geq -1 \\ x &\geq -\tfrac{1}{2} \end{aligned}$$
Solution: $\left[-\frac{1}{2}, \infty\right)$

75.
$$\begin{aligned} |2x+1| &< |2x-1| \\ \sqrt{(2x+1)^2} &< \sqrt{(2x-1)^2} \\ (2x+1)^2 &< (2x-1)^2 \\ 4x^2+4x+1 &< 4x^2-4x+1 \\ 8x &< 0 \\ x &< 0 \end{aligned}$$
Solution: $(-\infty, 0)$

77.
$$\begin{aligned} |x+1| &< |x| \\ \sqrt{(x+1)^2} &< \sqrt{x^2} \\ (x+1)^2 &< x^2 \\ x^2+2x+1 &< x^2 \\ 2x &< -1 \\ x &< -\tfrac{1}{2} \end{aligned}$$
Solution: $\left(-\infty, -\frac{1}{2}\right)$

79.
$$\begin{aligned} |2x+1| &\geq |2x-1| \\ \sqrt{(2x+1)^2} &\geq \sqrt{(2x-1)^2} \\ (2x+1)^2 &\geq (2x-1)^2 \\ 4x^2+4x+1 &\geq 4x^2-4x+1 \\ 8x &\geq 0 \\ x &\geq 0 \end{aligned}$$
Solution: $[0, \infty)$

81.
$$\begin{aligned} |t-78^\circ| &\leq 8^\circ \\ -8^\circ \leq t-78^\circ &\leq 8^\circ \\ 70^\circ \leq \quad t \quad &\leq 86^\circ \end{aligned}$$

83.
$$\begin{aligned} 0.6^\circ + 0.5^\circ &= 1.1^\circ \\ 0.6^\circ - 0.5^\circ &= 0.1^\circ \\ 0.1^\circ \leq \quad c \quad &\leq 1.1^\circ \\ 0.6^\circ - 0.5^\circ \leq \quad c \quad &\leq 0.6^\circ + 0.5^\circ \\ -0.5^\circ \leq c - 0.6^\circ &\leq 0.5^\circ \\ |c-0.6^\circ| &\leq 0.5^\circ \end{aligned}$$

85.
$$\begin{aligned} \frac{38+72}{2} = \frac{110}{2} &= 55 \\ 38 &= 55-17 \\ 72 &= 55+17 \\ 38 < \quad h \quad &< 72 \\ 55-17 < \quad h \quad &< 55+17 \\ -17 < h-55 &< 17 \\ |h-55| &< 17 \end{aligned}$$

87.
$$\begin{aligned} |p-25.46| &\leq 1.00 \\ -1.00 \leq p-25.46 &\leq 1.00 \\ 24.46 \leq \quad p \quad &\leq 26.46 \end{aligned}$$

a. 24.76% and 26.45% are within the range.

b. The error is less than 1%.

89-93. Answers may vary.

95. $37{,}250 = 3.725 \times 10^4$

97. $5.23 \times 10^5 = 523{,}000$

99.
$$\begin{aligned} (x-y)^2 - (x+y)^2 = [(x-y)(x-y)] - [(x+y)(x+y)] &= \left[x^2 - 2xy + y^2\right] - \left[x^2 + 2xy + y^2\right] \\ &= x^2 - 2xy + y^2 - x^2 - 2xy - y^2 \\ &= -4xy \end{aligned}$$

Chapter 1 Review (page 175)

1. $3x + 7 = 4$
no restrictions on x

2. $x + \dfrac{1}{x} = 2$
restriction: $x \neq 0$

3. $\dfrac{1}{x-1} = 4$
restriction: $x \neq 1$

4. $\dfrac{1}{x-2} = \dfrac{2}{x-3}$
restriction: $x \neq 2$, $x \neq 3$

5.
$$\begin{aligned} 3(9x+4) &= 28 \\ 27x + 12 &= 28 \\ 27x &= 16 \\ x &= \tfrac{16}{27} \end{aligned}$$
conditional equation

6.
$$\begin{aligned} \tfrac{3}{2}a &= 7(a+11) \\ 2 \cdot \tfrac{3}{2}a &= 2 \cdot 7(a+11) \\ 3a &= 14a + 154 \\ -11a &= 154 \\ a &= -\tfrac{154}{11} = -14 \end{aligned}$$
conditional equation

7.
$$\begin{aligned} 8(3x-5) - 4(x+3) &= 12 \\ 24x - 40 - 4x - 12 &= 12 \\ 20x - 52 &= 12 \\ 20x &= 64 \\ x &= \tfrac{64}{20} = \tfrac{16}{5} \Rightarrow \text{conditional equation} \end{aligned}$$

8.
$$\begin{aligned} \frac{x+3}{x+4} + \frac{x+3}{x+2} &= 2 \\ (x+4)(x+2)\left(\frac{x+3}{x+4} + \frac{x+3}{x+2}\right) &= (x+4)(x+2) \cdot 2 \\ (x+2)(x+3) + (x+4)(x+3) &= \left(x^2 + 6x + 8\right) \cdot 2 \\ x^2 + 5x + 6 + x^2 + 7x + 12 &= 2x^2 + 12x + 16 \\ 2x^2 + 12x + 18 &= 2x^2 + 12x + 16 \\ 18 &\neq 16 \Rightarrow \text{contradiction} \end{aligned}$$

9.
$$\begin{aligned} \frac{3}{x-1} &= \frac{1}{2} \\ 2(x-1) \cdot \frac{3}{x-1} &= 2(x-1) \cdot \frac{1}{2} \\ 6 &= x - 1 \\ 7 &= x \end{aligned}$$
conditional equation

10.
$$\begin{aligned} \frac{8x^2 + 72x}{9+x} &= 8x \\ (9+x) \cdot \frac{8x^2 + 72x}{9+x} &= (9+x) \cdot 8x \\ 8x^2 + 72x &= 72x + 8x^2 \end{aligned}$$
identity

11.
$$\begin{aligned}
\frac{3x}{x-1}-\frac{5}{x+3}&=3\\
(x-1)(x+3)\left(\frac{3x}{x-1}-\frac{5}{x+3}\right)&=(x-1)(x+3)\cdot 3\\
3x(x+3)-5(x-1)&=\left(x^2+2x-3\right)\cdot 3\\
3x^2+9x-5x+5&=3x^2+6x-9\\
4x+5&=6x-9\\
-2x&=-14\\
x&=7\Rightarrow \text{conditional equation}
\end{aligned}$$

12.
$$\begin{aligned}
x+\frac{1}{2x-3}&=\frac{2x^2}{2x-3}\\
(2x-3)\left(x+\frac{1}{2x-3}\right)&=(2x-3)\cdot\frac{2x^2}{2x-3}\\
(2x-3)x+1&=2x^2\\
2x^2-3x+1&=2x^2\\
-3x&=-1\\
x&=\tfrac{1}{3}\Rightarrow \text{conditional equation}
\end{aligned}$$

13.
$$\begin{aligned}
\frac{4}{x^2-13x-48}-\frac{1}{x^2+x-6}&=\frac{2}{x^2-18x+32}\\
\frac{4}{(x-16)(x+3)}-\frac{1}{(x+3)(x-2)}&=\frac{2}{(x-16)(x-2)}\\
4(x-2)-(x-16)&=2(x+3)\qquad \{\text{multiply by common denominator}\}\\
4x-8-x+16&=2x+6\\
3x+8&=2x+6\\
x&=-2\Rightarrow \text{conditional equation}
\end{aligned}$$

14.
$$\begin{aligned}
\frac{a-1}{a+3}+\frac{2a-1}{3-a}&=\frac{2-a}{a-3}\\
\frac{a-1}{a+3}+\frac{1-2a}{a-3}&=\frac{2-a}{a-3}\\
(a-1)(a-3)+(1-2a)(a+3)&=(2-a)(a+3)\qquad \{\text{multiply by common denominator}\}\\
a^2-3a-a+3+a+3-2a^2-6a&=2a+6-a^2-3a\\
-a^2-9a+6&=-a^2-a+6\\
-9a+6&=-a+6\\
0&=8a\\
0&=a\Rightarrow \text{conditional equation}
\end{aligned}$$

15.

$$C = \frac{5}{9}(F - 32)$$
$$\frac{9}{5}C = \frac{9}{5} \cdot \frac{5}{9}(F - 32)$$
$$\frac{9}{5}C = F - 32$$
$$\frac{9}{5}C + 32 = F$$

16.

$$P_n = l + \frac{si}{f}$$
$$P_n - l = \frac{si}{f}$$
$$f(P_n - l) = f \cdot \frac{si}{f}$$
$$f(P_n - l) = si$$
$$\frac{f(P_n - l)}{P_n - l} = \frac{si}{P_n - l}$$
$$f = \frac{si}{P_n - l}$$

17.

$$\frac{1}{f} = \frac{1}{f_1} + \frac{1}{f_2}$$
$$ff_1f_2 \cdot \frac{1}{f} = ff_1f_2\left(\frac{1}{f_1} + \frac{1}{f_2}\right)$$
$$f_1f_2 = ff_2 + ff_1$$
$$f_1f_2 - ff_1 = ff_2$$
$$f_1(f_2 - f) = ff_2$$
$$\frac{f_1(f_2 - f)}{f_2 - f} = \frac{ff_2}{f_2 - f}$$
$$f_1 = \frac{ff_2}{f_2 - f}$$

18.

$$S = \frac{a - lr}{1 - r}$$
$$S(1 - r) = \frac{a - lr}{1 - r}(1 - r)$$
$$S(1 - r) = a - lr$$
$$S - Sr = a - lr$$
$$lr = a - S + Sr$$
$$\frac{lr}{r} = \frac{a - S + Sr}{r}$$
$$l = \frac{a - S + Sr}{r}$$

19. Let $x =$ the score on the first exam. Then his scores on the following tests were $x + 4$, $x + 8$ and $x + 12$.

$$\frac{\boxed{\text{Sum of scores}}}{4} = 66$$
$$\frac{x + x + 4 + x + 8 + x + 12}{4} = 66$$
$$\frac{4x + 24}{4} = 66$$
$$4x + 24 = 264$$
$$4x = 240$$
$$x = 60$$

His score on the first test was 60%.

20. Let $w =$ the width. Then $w + 5 =$ the length.

$$\boxed{\text{Perimeter}} = 100$$
$$2w + 2(w + 5) = 100$$
$$2w + 2w + 10 = 100$$
$$4w + 10 = 100$$
$$4w = 90$$
$$w = 22.5$$

The dimensions are 22.5 ft by 27.5 ft.

21. Let $t =$ the time the cars travel.

$$\boxed{\text{Distance 1st car travels}} + \boxed{\text{Distance 2nd car travels}} = \boxed{\text{Total distance}}$$
$$45t + 50t = 285$$
$$95t = 285$$
$$t = 3 \Rightarrow$$ They will be 285 miles apart after 3 hours.

CHAPTER 1 REVIEW

22. Let $t =$ the time the cars travel.

$$\boxed{\text{Distance 1st car travels}} - \boxed{\text{Distance 2nd car travels}} = \boxed{\text{Distance between them}}$$

$$46t - 40t = 3$$
$$6t = 3$$
$$t = 0.5 \Rightarrow \text{They will be 3 miles apart after 0.5 hours.}$$

23. Let $x =$ the liters of water added.

$$\boxed{\text{Liters of alcohol at start}} + \boxed{\text{Liters of alcohol added}} = \boxed{\text{Liters of alcohol at end}}$$

$$0.50(1) + 0 = 0.20(1 + x)$$
$$0.50 = 0.20 + 0.20x$$
$$0.30 = 0.20x$$
$$1.5 = x \Rightarrow \text{1.5 liters of water should be added.}$$

24. Let $x =$ hours for both working together.

$$\boxed{\text{Number Scott washes in 1 hour}} \cdot \boxed{\text{Number of hours}} + \boxed{\text{Number Bill washes in 1 hour}} \cdot \boxed{\text{Number of hours}} = \boxed{\text{100 windows}}$$

$$\frac{37}{3}x + \frac{27}{2}x = 100$$
$$6\left(\frac{37}{3}x + \frac{27}{2}x\right) = 6(100)$$
$$74x + 81x = 600$$
$$155x = 600$$
$$x = \frac{600}{155} \approx 3.9$$

They can wash 100 windows together in about 3.9 hours.

25. Let $x =$ hours for both pipes to fill the tank.

$$\boxed{\text{1st pipe in 1 hour}} + \boxed{\text{2nd pipe in 1 hour}} = \boxed{\text{Total in 1 hour}}$$

$$\frac{1}{9} + \frac{1}{12} = \frac{1}{x}$$
$$36x\left(\frac{1}{9} + \frac{1}{12}\right) = 36x\left(\frac{1}{x}\right)$$
$$4x + 3x = 36$$
$$7x = 36$$
$$x = \frac{36}{7} = 5\frac{1}{7}$$

The tank can be filled in $5\frac{1}{7}$ hours.

26. Let $x =$ the ounces of pure zinc added.

$$\boxed{\text{Ounces of zinc at start}} + \boxed{\text{Ounces of zinc added}} = \boxed{\text{Ounces of zinc at end}}$$

$$0.30(20) + x = 0.40(20 + x)$$
$$6 + x = 8 + 0.40x$$
$$0.60x = 2$$
$$6x = 20$$
$$x = \frac{20}{6} = 3\frac{1}{3}$$

$3\frac{1}{3}$ ounces of zinc should be added.

27. Let x = the amount invested at 11%. Then $10{,}000 - x$ = the amount invested at 14%.

$$\boxed{\text{Interest at 11\%}} + \boxed{\text{Interest at 14\%}} = \boxed{\text{Total interest}}$$

$$\begin{aligned} 0.11x + 0.14(10{,}000 - x) &= 1{,}265 \\ 0.11x + 1{,}400 - 0.14x &= 1{,}265 \\ -0.03x &= -135 \\ x &= 4{,}500 \end{aligned}$$

\$4,500 was invested at 11% and \$5,500 was invested at 14%.

28. Let x = # of rugs for equal costs.

$$\boxed{\text{Cost of 1st loom}} = \boxed{\text{Cost of 2nd loom}}$$

$$\begin{aligned} 750 + 115x &= 950 + 95x \\ 20x &= 200 \\ x &= 10 \end{aligned}$$

The costs are the same on either loom for 10 rugs.

29.

$$\begin{aligned} 2x^2 - x - 6 &= 0 \\ (2x+3)(x-2) &= 0 \end{aligned}$$

$$\begin{array}{lcl} 2x + 3 = 0 & \textbf{or} & x - 2 = 0 \\ 2x = -3 & & x = 2 \\ x = -\frac{3}{2} & & x = 2 \end{array}$$

30.

$$\begin{aligned} 12x^2 + 13x &= 4 \\ 12x^2 + 13x - 4 &= 0 \\ (4x-1)(3x+4) &= 0 \end{aligned}$$

$$\begin{array}{lcl} 4x - 1 = 0 & \textbf{or} & 3x + 4 = 0 \\ 4x = 1 & & 3x = -4 \\ x = \frac{1}{4} & & x = -\frac{4}{3} \end{array}$$

31.

$$\begin{aligned} 5x^2 - 8x &= 0 \\ x(5x-8) &= 0 \end{aligned}$$

$$\begin{array}{lcl} x = 0 & \textbf{or} & 5x - 8 = 0 \\ x = 0 & & 5x = 8 \\ x = 0 & & x = \frac{8}{5} \end{array}$$

32.

$$\begin{aligned} 27x^2 &= 30x - 8 \\ 27x^2 - 30x + 8 &= 0 \\ (9x-4)(3x-2) &= 0 \end{aligned}$$

$$\begin{array}{lcl} 9x - 4 = 0 & \textbf{or} & 3x - 2 = 0 \\ 9x = 4 & & 3x = 2 \\ x = \frac{4}{9} & & x = \frac{2}{3} \end{array}$$

33.

$$\begin{aligned} 2x^2 &= 16 \\ x^2 &= 8 \\ \sqrt{x^2} &= \pm\sqrt{8} \\ x &= \pm 2\sqrt{2} \end{aligned}$$

34.

$$\begin{aligned} 12x^2 &= 60 \\ x^2 &= 5 \\ \sqrt{x^2} &= \pm\sqrt{5} \\ x &= \pm\sqrt{5} \end{aligned}$$

35.

$$\begin{aligned} (4z-5)^2 &= 32 \\ \sqrt{(4z-5)^2} &= \pm\sqrt{32} \\ 4z - 5 &= \pm 4\sqrt{2} \\ 4z &= 5 \pm 4\sqrt{2} \\ z &= \frac{5 \pm 4\sqrt{2}}{4} \end{aligned}$$

36.

$$\begin{aligned} (5x-7)^2 &= 45 \\ \sqrt{(5x-7)^2} &= \pm\sqrt{45} \\ 5x - 7 &= \pm 3\sqrt{5} \\ 5x &= 7 \pm 3\sqrt{5} \\ x &= \frac{7 \pm 3\sqrt{5}}{5} \end{aligned}$$

37.

$$\begin{aligned}
x^2 - 8x + 15 &= 0 \\
x^2 - 8x &= -15 \\
x^2 - 8x + 16 &= -15 + 16 \\
(x-4)^2 &= 1
\end{aligned}$$

$x - 4 = \sqrt{1}$ **or** $x - 4 = -\sqrt{1}$

$x - 4 = 1$ $\qquad x - 4 = -1$

$x = 5$ $\qquad x = 3$

38.

$$\begin{aligned}
3x^2 + 18x &= -24 \\
\frac{3x^2 + 18x}{3} &= \frac{-24}{3} \\
x^2 + 6x &= -8 \\
x^2 + 6x + 9 &= -8 + 9 \\
(x+3)^2 &= 1
\end{aligned}$$

$x + 3 = \sqrt{1}$ **or** $x + 3 = -\sqrt{1}$

$x + 3 = 1$ $\qquad x + 3 = -1$

$x = -2$ $\qquad x = -4$

39.

$$\begin{aligned}
5x^2 - x - 1 &= 0 \\
5x^2 - x &= 1 \\
x^2 - \frac{1}{5}x &= \frac{1}{5} \\
x^2 - \frac{1}{5}x + \frac{1}{100} &= \frac{1}{5} + \frac{1}{100} \\
\left(x - \frac{1}{10}\right)^2 &= \frac{21}{100}
\end{aligned}$$

$x - \frac{1}{10} = \sqrt{\frac{21}{100}}$ **or** $x - \frac{1}{10} = -\sqrt{\frac{21}{100}}$

$x - \frac{1}{10} = \frac{\sqrt{21}}{10}$ $\qquad x - \frac{1}{10} = -\frac{\sqrt{21}}{10}$

$x = \frac{1 + \sqrt{21}}{10}$ $\qquad x = \frac{1 - \sqrt{21}}{10}$

40.

$$\begin{aligned}
5x^2 - x &= 0 \\
x^2 - \frac{1}{5}x &= 0 \\
x^2 - \frac{1}{5}x + \frac{1}{100} &= 0 + \frac{1}{100} \\
\left(x - \frac{1}{10}\right)^2 &= \frac{1}{100}
\end{aligned}$$

$x - \frac{1}{10} = \sqrt{\frac{1}{100}}$ **or** $x - \frac{1}{10} = -\sqrt{\frac{1}{100}}$

$x - \frac{1}{10} = \frac{1}{10}$ $\qquad x - \frac{1}{10} = -\frac{1}{10}$

$x = \frac{2}{10} = \frac{1}{5}$ $\qquad x = \frac{0}{10} = 0$

41. $x^2 + 5x - 14 = 0 \Rightarrow a = 1, b = 5, c = -14$

$$x = \frac{-b \pm \sqrt{b^2 - 4ac}}{2a} = \frac{-(5) \pm \sqrt{(5)^2 - 4(1)(-14)}}{2(1)} = \frac{-5 \pm \sqrt{25 + 56}}{2} = \frac{-5 \pm \sqrt{81}}{2} = \frac{-5 \pm 9}{2}$$

$$x = \frac{-5 + 9}{2} = \frac{4}{2} = 2 \text{ or } x = \frac{-5 - 9}{2} = \frac{-14}{2} = -7$$

42. $3x^2 - 25x = 18 \Rightarrow 3x^2 - 25x - 18 = 0 \Rightarrow a = 3, b = -25, c = -18$

$$x = \frac{-b \pm \sqrt{b^2 - 4ac}}{2a} = \frac{-(-25) \pm \sqrt{(-25)^2 - 4(3)(-18)}}{2(3)} = \frac{25 \pm \sqrt{625 + 216}}{6} = \frac{25 \pm \sqrt{841}}{6} = \frac{25 \pm 29}{6}$$

$$x = \frac{25 + 29}{6} = \frac{54}{6} = 9 \text{ or } x = \frac{25 - 29}{6} = \frac{-4}{6} = -\frac{2}{3}$$

43. $5x^2 = 1 - x \Rightarrow 5x^2 + x - 1 = 0 \Rightarrow a = 5, b = 1, c = -1$

$$x = \frac{-b \pm \sqrt{b^2 - 4ac}}{2a} = \frac{-(1) \pm \sqrt{(1)^2 - 4(5)(-1)}}{2(5)} = \frac{-1 \pm \sqrt{1 + 20}}{10} = \frac{-1 \pm \sqrt{21}}{10}$$

44. $-5 = a^2 + 2a \Rightarrow a^2 + 2a + 5 = 0 \Rightarrow a = 1, b = 2, c = 5$

$$a = \frac{-b \pm \sqrt{b^2 - 4ac}}{2a} = \frac{-(2) \pm \sqrt{(2)^2 - 4(1)(5)}}{2(1)} = \frac{-2 \pm \sqrt{4 - 20}}{2} = \frac{-2 \pm \sqrt{-16}}{2} = \frac{-2 \pm 4i}{2} = -1 \pm 2i$$

45. $6x^2 + 5x + 1 = 0$

$a = 6, b = 5, c = 1$

$b^2 - 4ac = (5)^2 - 4(6)(1) = 25 - 24 = 1$

46. The roots are unequal rational numbers.

47. $kx^2 + 4x + 12 = 0$

$a = k, b = 4, c = 12$

Set the discriminant equal to 0:

$$\begin{aligned} b^2 - 4ac &= 0 \\ 4^2 - 4(k)(12) &= 0 \\ 16 - 48k &= 0 \\ -48k &= -16 \\ k &= \frac{1}{3} \end{aligned}$$

48.

$$\begin{aligned} 4y^2 + (k+2)y &= 1 - k \\ 4y^2 + (k+2)y - 1 + k &= 0 \end{aligned}$$

$a = 4, b = k + 2, c = -1 + k$

Set the discriminant equal to 0:

$$\begin{aligned} b^2 - 4ac &= 0 \\ (k+2)^2 - 4(4)(-1+k) &= 0 \\ k^2 + 4k + 4 + 16 - 16k &= 0 \\ k^2 - 12k + 20 &= 0 \\ (k-10)(k-2) &= 0 \end{aligned}$$

$k - 10 = 0$ **or** $k - 2 = 0$

$k = 10$ $\quad$ $k = 2$

49.
$$\begin{aligned}\frac{1}{a}-\frac{1}{5}&=\frac{3}{2a}\\ 10a\left(\tfrac{1}{a}-\tfrac{1}{5}\right)&=10a\left(\tfrac{3}{2a}\right)\\ 10-2a&=15\\ -2a&=5\\ a&=-\tfrac{5}{2}\end{aligned}$$

50.
$$\begin{aligned}\frac{4}{a-4}+\frac{4}{a-1}&=5\\ (a-4)(a-1)\left(\tfrac{4}{a-4}+\tfrac{4}{a-1}\right)&=(a-4)(a-1)5\\ 4(a-1)+4(a-4)&=5\left(a^2-5a+4\right)\\ 4a-4+4a-16&=5a^2-25a+20\\ 0&=5a^2-33a+40\\ 0&=(5a-8)(a-5)\end{aligned}$$

$5a-8=0$ **or** $a-5=0$

$5a=8 \qquad a=5$

$a=\frac{8}{5} \qquad a=5$

51. Let x = one side of the garden.

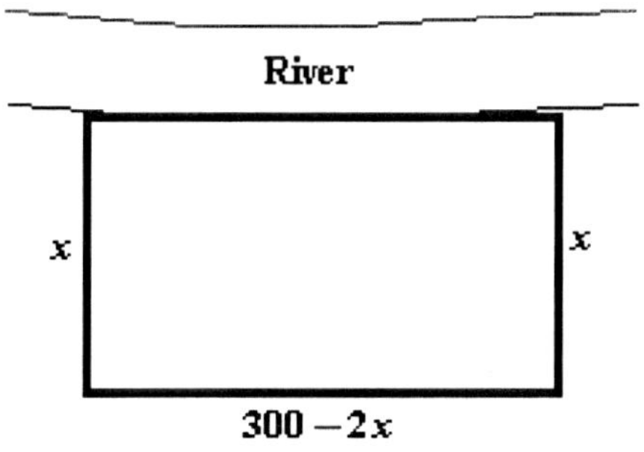

$$\begin{aligned}\text{Area}&=10450\\ x(300-2x)&=10450\\ -2x^2+300x&=10450\\ 0&=2x^2-300x+10450\\ 0&=2\left(x^2-150x+5225\right)\\ 0&=2(x-95)(x-55)\end{aligned}$$

$x-95=0$ **or** $x-55=0$

$x=95 \qquad x=55$

The dimensions are 95 yards by 110 yards or 55 yards by 190 yards.

52. Let r = the rate of the propeller-driven plane. Then the rate of the jet plane is $r+120$.

$$\boxed{\text{Jet time}}=\boxed{\text{Propeller time}}-3$$

	Rate	Time	Dist.
Propeller	r	$\frac{3520}{r}$	3520
Jet	$r+120$	$\frac{3520}{r+120}$	3520

$$\begin{aligned}\frac{3520}{r+120}&=\frac{3520}{r}-3\\ r(r+120)\frac{3520}{r+120}&=r(r+120)\left(\frac{3520}{r}-3\right)\\ 3520r&=3520(r+120)-3r(r+120)\\ 3520r&=3520r+422400-3r^2-360r\end{aligned}$$

$$\begin{aligned}3r^2+360r+422{,}400&=0\\ 3(r-320)(r+440)&=0\end{aligned}$$

$r-320=0$ **or** $r+440=0$

$r=320 \qquad r=-440$

Since $r=-440$ does not make sense, the solution is $r=320$. The prop. plane's rate is 320 mph, while the jet plane's rate is 440 mph.

53. Set $h = 48$:

$$h = -16t^2 + 64t$$
$$48 = -16t^2 + 64t$$
$$16t^2 - 64t + 48 = 0$$
$$16(t-1)(t-3) = 0$$
$$t - 1 = 0 \quad \textbf{or} \quad t - 3 = 0$$
$$t = 1 \qquad\qquad t = 3$$

The shortest time required for the ball to reach a height of 48 feet is 1 second.

54. Let $x =$ the width of the walk. Then the total dimensions are $16 + 2x$ by $20 + 2x$.

$$\boxed{\text{Total area}} - \boxed{\text{Area of pool}} = \boxed{\text{Area of walk}}$$
$$(16 + 2x)(20 + 2x) - (16)(20) = 117$$
$$320 + 72x + 4x^2 - 320 = 117$$
$$4x^2 + 72x - 117 = 0$$
$$(2x + 39)(2x - 3) = 0$$
$$2x + 39 = 0 \quad \textbf{or} \quad 2x - 3 = 0$$
$$x = -\tfrac{39}{2} \qquad\qquad x = \tfrac{3}{2}$$

Since $x = -\frac{39}{2}$ does not make sense, the only solution is $x = \frac{3}{2}$. The walk is $1\frac{1}{2}$ feet wide.

55. $(2 - 3i) + (-4 + 2i) = 2 - 3i - 4 + 2i$
$= -2 - i$

56. $(2 - 3i) - (4 + 2i) = 2 - 3i - 4 - 2i$
$= -2 - 5i$

57. $\left(3 - \sqrt{-36}\right) + \left(\sqrt{-16} + 2\right) = (3 - 6i) + (4i + 2) = 3 - 6i + 4i + 2 = 5 - 2i$

58. $\left(3 + \sqrt{-9}\right)\left(2 - \sqrt{-25}\right) = (3 + 3i)(2 - 5i) = 6 - 9i - 15i^2 = 6 - 9i + 15 = 21 - 9i$

59. $\dfrac{3}{i} = \dfrac{3i}{ii} = \dfrac{3i}{i^2} = \dfrac{3i}{-1} = 0 - 3i$

60. $-\dfrac{2}{i^3} = -\dfrac{2i}{i^3 i} = -\dfrac{2i}{i^4} = -\dfrac{2i}{1} = 0 - 2i$

61. $\dfrac{3}{1+i} = \dfrac{3(1-i)}{(1+i)(1-i)} = \dfrac{3(1-i)}{1^2 - i^2} = \dfrac{3 - 3i}{2} = \dfrac{3}{2} - \dfrac{3}{2}i$

62. $\dfrac{2i}{2-i} = \dfrac{2i(2+i)}{(2-i)(2+i)} = \dfrac{4i + 2i^2}{2^2 - i^2} = \dfrac{-2 + 4i}{5} = -\dfrac{2}{5} + \dfrac{4}{5}i$

63. $\dfrac{3+i}{3-i} = \dfrac{(3+i)(3+i)}{(3-i)(3+i)} = \dfrac{9 + 6i + i^2}{3^2 - i^2} = \dfrac{8 + 6i}{10} = \dfrac{8}{10} + \dfrac{6}{10}i = \dfrac{4}{5} + \dfrac{3}{5}i$

64. $\dfrac{3 - 2i}{1+i} = \dfrac{(3-2i)(1-i)}{(1+i)(1-i)} = \dfrac{3 - 5i + 2i^2}{1^2 - i^2} = \dfrac{1 - 5i}{2} = \dfrac{1}{2} - \dfrac{5}{2}i$

65. $i^{53} = i^{52}i = \left(i^4\right)^{13} i = 1^{13} i = 0 + i$

66. $i^{103} = i^{100} i^3 = \left(i^4\right)^{25} i^3 = 1^{25} i^3 = 0 - i$

67. $|3 - i| = \sqrt{3^2 + (-1)^2} = \sqrt{9 + 1} = \sqrt{10}$

68. $\left|\dfrac{1+i}{1-i}\right| = \left|\dfrac{(1+i)(1+i)}{(1-i)(1+i)}\right| = \left|\dfrac{1 + 2i + i^2}{1^2 - i^2}\right| = \left|\dfrac{2i}{2}\right| = |0 + i| = \sqrt{0^2 + 1^2} = 1$

69. $3x^2 - 2x + 1 = 0 \Rightarrow a = 3, b = -2, c = 1$

$$x = \frac{-b \pm \sqrt{b^2 - 4ac}}{2a} = \frac{-(-2) \pm \sqrt{(-2)^2 - 4(3)(1)}}{2(3)} = \frac{2 \pm \sqrt{4 - 12}}{6} = \frac{2 \pm \sqrt{-8}}{6}$$

$$= \frac{2}{6} \pm \frac{2\sqrt{2}}{6}i$$

$$= \frac{1}{3} \pm \frac{\sqrt{2}}{3}i$$

70. $3x^2 + 4 = 2x \Rightarrow 3x^2 - 2x + 4 = 0 \Rightarrow a = 3, b = -2, c = 4$

$$x = \frac{-b \pm \sqrt{b^2 - 4ac}}{2a} = \frac{-(-2) \pm \sqrt{(-2)^2 - 4(3)(4)}}{2(3)} = \frac{2 \pm \sqrt{4 - 48}}{6} = \frac{2 \pm \sqrt{-44}}{6}$$

$$= \frac{2}{6} \pm \frac{2\sqrt{11}}{6}i$$

$$= \frac{1}{3} \pm \frac{\sqrt{11}}{3}i$$

71.

$$\frac{3x}{2} - \frac{2x}{x-1} = x - 3$$

$$2(x-1)\left(\frac{3x}{2} - \frac{2x}{x-1}\right) = 2(x-1)(x-3)$$

$$3x(x-1) - 2(2x) = 2\left(x^2 - 4x + 3\right)$$

$$3x^2 - 3x - 4x = 2x^2 - 8x + 6$$

$$x^2 + x - 6 = 0$$

$$(x+3)(x-2) = 0$$

$x + 3 = 0$ **or** $x - 2 = 0$

$x = -3$ $\quad$ $x = 2$

72.

$$\frac{12}{x} - \frac{x}{2} = x - 3$$

$$2x\left(\frac{12}{x} - \frac{x}{2}\right) = 2x(x-3)$$

$$2(12) - x(x) = 2x^2 - 6x$$

$$24 - x^2 = 2x^2 - 6x$$

$$0 = 3x^2 - 6x - 24$$

$$0 = 3(x+2)(x-4)$$

$x + 2 = 0$ **or** $x - 4 = 0$

$x = -2$ $\quad$ $x = 4$

73.

$$x^4 - 2x^2 + 1 = 0$$

$$\left(x^2 - 1\right)\left(x^2 - 1\right) = 0$$

$x^2 - 1 = 0$ **or** $x^2 - 1 = 0$

$x^2 = 1$ $\quad$ $x^2 = 1$

$x = \pm 1$ $\quad$ $x = \pm 1$

74.

$$x^4 + 36 = 37x^2$$

$$x^4 - 37x^2 + 36 = 0$$

$$\left(x^2 - 36\right)\left(x^2 - 1\right) = 0$$

$x^2 - 36 = 0$ **or** $x^2 - 1 = 0$

$x^2 = 36$ $\quad$ $x^2 = 1$

$x = \pm 6$ $\quad$ $x = \pm 1$

75.

$$a - a^{1/2} - 6 = 0$$

$$\left(a^{1/2} + 2\right)\left(a^{1/2} - 3\right) = 0$$

$a^{1/2} + 2 = 0$ **or** $a^{1/2} - 3 = 0$

$a^{1/2} = -2$ $\quad$ $a^{1/2} = 3$

$\left(a^{1/2}\right)^2 = (-2)^2$ $\quad$ $\left(a^{1/2}\right)^2 = (3)^2$

$a = 4$ $\quad$ $a = 9$

$a = 4$ does not check and is extraneous.

76.
$$x^{2/3} + x^{1/3} - 6 = 0$$
$$\left(x^{1/3} - 2\right)\left(x^{1/3} + 3\right) = 0$$

$$x^{1/3} - 2 = 0 \quad \textbf{or} \quad x^{1/3} + 3 = 0$$
$$x^{1/3} = 2 \qquad x^{1/3} = -3$$
$$\left(x^{1/3}\right)^3 = (2)^3 \qquad \left(x^{1/3}\right)^3 = (-3)^3$$
$$x = 8 \qquad x = -27$$

Both answers check.

77.
$$\sqrt{x-1} + x = 7$$
$$\sqrt{x-1} = 7 - x$$
$$\left(\sqrt{x-1}\right)^2 = (7-x)^2$$
$$x - 1 = 49 - 14x + x^2$$
$$0 = x^2 - 15x + 50$$
$$0 = (x-5)(x-10)$$
$$x - 5 = 0 \quad \textbf{or} \quad x - 10 = 0$$
$$x = 5 \qquad x = 10$$

$x = 10$ does not check and is extraneous.

78.
$$\sqrt{a+9} - \sqrt{a} = 3$$
$$\sqrt{a+9} = 3 + \sqrt{a}$$
$$\left(\sqrt{a+9}\right)^2 = \left(3 + \sqrt{a}\right)^2$$
$$a + 9 = 9 + 6\sqrt{a} + a$$
$$0 = 6\sqrt{a}$$
$$0^2 = \left(6\sqrt{a}\right)^2$$
$$0 = 36a \Rightarrow a = 0$$

The solution checks.

79.
$$\sqrt{5-x} + \sqrt{5+x} = 4$$
$$\sqrt{5+x} = 4 - \sqrt{5-x}$$
$$\left(\sqrt{5+x}\right)^2 = \left(4 - \sqrt{5-x}\right)^2$$
$$5 + x = 16 - 8\sqrt{5-x} + 5 - x$$
$$8\sqrt{5-x} = 16 - 2x$$
$$\left(8\sqrt{5-x}\right)^2 = (16-2x)^2$$
$$64(5-x) = 256 - 64x + 4x^2$$
$$320 - 64x = 4x^2 - 64x + 256$$
$$0 = 4x^2 - 64$$
$$0 = 4(x+4)(x-4)$$
$$x + 4 = 0 \quad \textbf{or} \quad x - 4 = 0$$
$$x = -4 \qquad x = 4$$

Both solutions check.

80.
$$\sqrt{y+5} + \sqrt{y} = 1$$
$$\sqrt{y+5} = 1 - \sqrt{y}$$
$$\left(\sqrt{y+5}\right)^2 = \left(1 - \sqrt{y}\right)^2$$
$$y + 5 = 1 - 2\sqrt{y} + y$$
$$2\sqrt{y} = -4$$
$$\left(2\sqrt{y}\right)^2 = (-4)^2$$
$$4y = 16$$
$$y = 4$$

The solution does not check. $\Rightarrow$ No solution.

81.
$$2x - 9 < 5$$
$$2x < 14$$
$$x < 7 \Rightarrow (-\infty, 7)$$

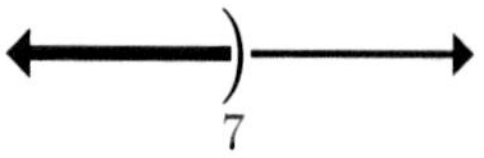

82.
$$5x + 3 \geq 2$$
$$5x \geq -1$$
$$x \geq -\tfrac{1}{5} \Rightarrow \left[-\tfrac{1}{5}, \infty\right)$$

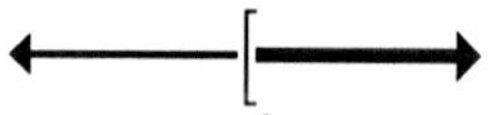

83. $\dfrac{5(x-1)}{2} < x$

$5(x-1) < 2x$

$5x - 5 < 2x$

$3x < 5$

$x < \frac{5}{3} \Rightarrow \left(-\infty, \frac{5}{3}\right)$

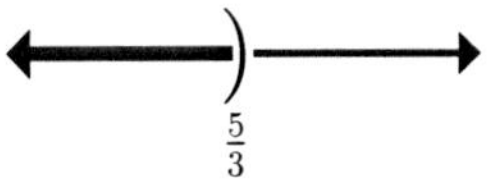

84. $\dfrac{1}{4}x + \dfrac{2}{3}x - x > \dfrac{1}{2} + \dfrac{1}{2}(x+1)$

$12\left(\dfrac{1}{4}x + \dfrac{2}{3}x - x\right) > 12\left(\dfrac{1}{2} + \dfrac{1}{2}(x+1)\right)$

$3x + 8x - 12x > 6 + 6(x+1)$

$-x > 6 + 6x + 6$

$-12 > 7x$

$-\dfrac{12}{7} > x \Rightarrow \left(-\infty, -\dfrac{12}{7}\right)$

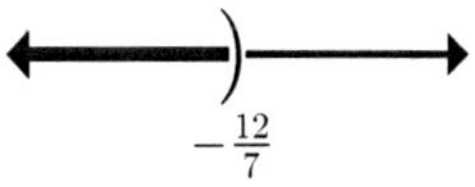

85. $0 \le \dfrac{3+x}{2} < 4$

$0 \le 3 + x < 8$

$-3 \le x < 5 \Rightarrow [-3, 5)$

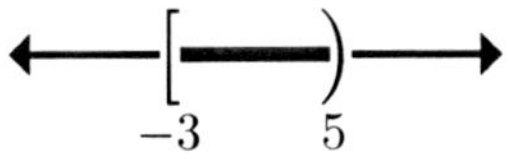

86. $2 + a < 3a - 2 \le 5a + 2$

$2 + a < 3a - 2$ **and** $3a - 2 \le 5a + 2$

$4 < 2a$ $\quad$ $-4 \le 2a$

$a > 2$ $\quad$ $a \ge -2$

$a > 2$

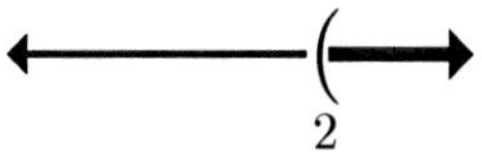

$a > -2$

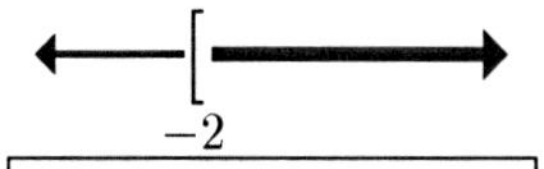

$a > 2$
and
$a \ge -2$

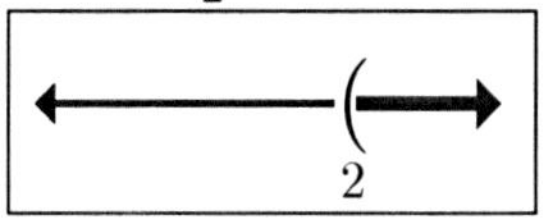

Solution set: $(2, \infty)$

87. $(x+2)(x-4) > 0$

factors $= 0$: $x = -2, x = 4$

intervals: $(-\infty, -2), (-2, 4), (4, \infty)$

interval	test number	value of $(x+2)(x-4)$
$(-\infty, -2)$	-3	$+7$
$(-2, 4)$	0	-8
$(4, \infty)$	5	$+7$

Solution: $(-\infty, -2) \cup (4, \infty)$

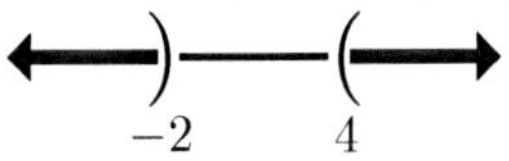

88. $(x-1)(x+4) < 0$

factors $= 0$: $x = 1, x = -4$

intervals: $(-\infty, -4), (-4, 1), (1, \infty)$

interval	test number	value of $(x-1)(x+4)$
$(-\infty, -4)$	-5	$+6$
$(-4, 1)$	0	-4
$(1, \infty)$	2	$+8$

Solution: $(-4, 1)$

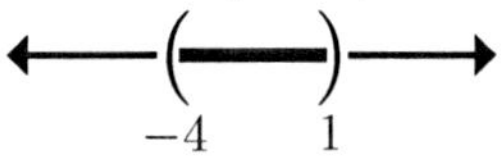

89. $x^2 - 2x - 3 < 0$
$(x - 3)(x + 1) < 0$
factors $= 0$: $x = 3$, $x = -1$
intervals: $(-\infty, -1)$, $(-1, 3)$, $(3, \infty)$

interval	test number	value of x^2-2x-3
$(-\infty, -1)$	-2	$+5$
$(-1, 3)$	0	-3
$(3, \infty)$	4	$+5$

Solution: $(-1, 3)$

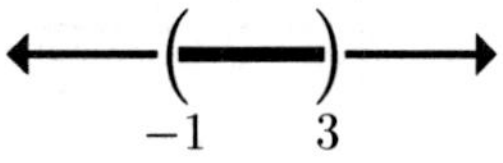

90. $2x^2 + x - 3 > 0$
$(2x + 3)(x - 1) > 0$
factors $= 0$: $x = -\frac{3}{2}$, $x = 1$
intervals: $(-\infty, -\frac{3}{2})$, $(-\frac{3}{2}, 1)$, $(1, \infty)$

interval	test number	value of $2x^2+x-3$
$(-\infty, -\frac{3}{2})$	-2	$+3$
$(-\frac{3}{2}, 1)$	0	-3
$(1, \infty)$	2	$+7$

Solution: $(-\infty, -\frac{3}{2}) \cup (1, \infty)$

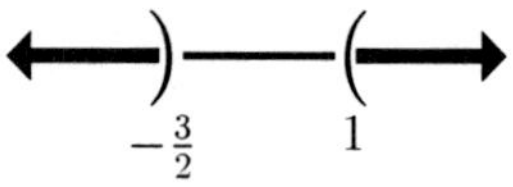

91. $\frac{x+2}{x-3} \geq 0$
factors $= 0$: $x = -2$, $x = 3$
intervals: $(-\infty, -2)$, $(-2, 3)$, $(3, \infty)$

interval	test number	sign of $\frac{x+2}{x-3}$
$(-\infty, -2)$	-3	$+$
$(-2, 3)$	0	$-$
$(3, \infty)$	4	$+$

Include endpoints which make the numerator equal to 0. Do not include endpoints which make the denominator equal to 0.
Solution: $(-\infty, -2] \cup (3, \infty)$

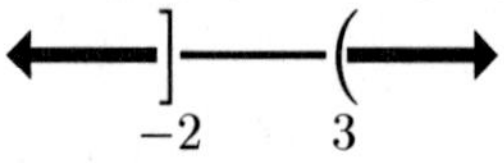

92. $\frac{x-1}{x+4} \leq 0$
factors $= 0$: $x = 1$, $x = -4$
intervals: $(-\infty, -4)$, $(-4, 1)$, $(1, \infty)$

interval	test number	sign of $\frac{x-1}{x+4}$
$(-\infty, -4)$	-5	$+$
$(-4, 1)$	0	$-$
$(1, \infty)$	2	$+$

Include endpoints which make the numerator equal to 0. Do not include endpoints which make the denominator equal to 0.
Solution: $(-4, 1]$

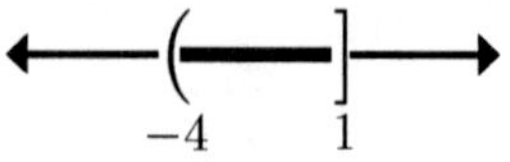

93. $\frac{x^2 + x - 2}{x - 3} \geq 0$
$\frac{(x + 2)(x - 1)}{x - 3} \geq 0$
factors $= 0$: $x = -2$, $x = 1$, $x = 3$
int.: $(-\infty, -2)$, $(-2, 1)$, $(1, 3)$, $(3, \infty)$

interval	test number	sign of $\frac{x^2+x-2}{x-3}$
$(-\infty, -2)$	-3	$-$
$(-2, 1)$	0	$+$
$(1, 3)$	2	$-$
$(3, \infty)$	4	$+$

Include endpoints which make the numerator equal to 0. Do not include endpoints which make the denominator equal to 0.
Solution: $[-2, 1] \cup (3, \infty)$

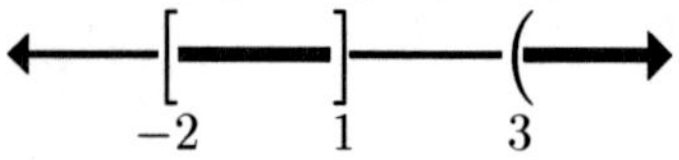

94.
$$\frac{5}{x} < 2$$
$$\frac{5}{x} - 2 < 0$$
$$\frac{5-2x}{x} < 0$$
factors $= 0$: $x = \frac{5}{2}$, $x = 0$

intervals: $(-\infty, 0)$, $(0, \frac{5}{2})$, $(\frac{5}{2}, \infty)$

interval	test number	value of $\frac{5-2x}{x}$
$(-\infty, 0)$	-1	-7
$(0, \frac{5}{2})$	1	$+3$
$(\frac{5}{2}, \infty)$	3	$-\frac{1}{3}$

Solution: $(-\infty, 0) \cup (\frac{5}{2}, \infty)$

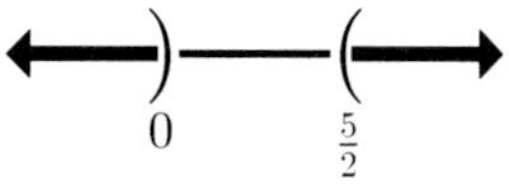

95.
$$|x+1| = 6$$
$x + 1 = 6$ **or** $x + 1 = -6$

$x = 5$ $\quad$ $x = -7$

96.
$$|2x-1| = |2x+1|$$
$2x - 1 = 2x + 1$ **or** $2x - 1 = -(2x+1)$

$0 = 2$ $\quad$ $2x - 1 = -2x - 1$

never true $\quad$ $4x = 0$

$x = 0$

97.
$$\left|\frac{3x+11}{7}\right| - 1 = 0$$
$$\left|\frac{3x+11}{7}\right| = 1$$
$\frac{3x+11}{7} = 1$ **or** $\frac{3x+11}{7} = -1$

$3x + 11 = 7$ $\quad$ $3x + 11 = -7$

$3x = -4$ $\quad$ $3x = -18$

$x = -\frac{4}{3}$ $\quad$ $x = -6$

98.
$$\left|\frac{2a-6}{3a}\right| - 6 = 0$$
$$\left|\frac{2a-6}{3a}\right| = 6$$
$\frac{2a-6}{3a} = 6$ **or** $\frac{2a-6}{3a} = -6$

$2a - 6 = 18a$ $\quad$ $2a - 6 = -18a$

$-16a = 6$ $\quad$ $20a = 6$

$a = -\frac{6}{16}$ $\quad$ $a = \frac{6}{20}$

$a = -\frac{3}{8}$ $\quad$ $a = \frac{3}{10}$

99.
$$|x+3| < 3$$
$$-3 < x + 3 < 3$$
$$-6 < x < 0$$
$$(-6, 0)$$

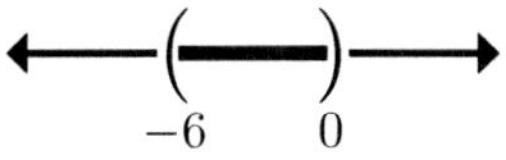

100.
$$|3x-7| \geq 1$$
$3x - 7 \geq 1$ **or** $3x - 7 \leq -1$

$3x \geq 8$ $\quad$ $3x \leq 6$

$x \geq \frac{8}{3}$ $\quad$ $x \leq 2$

$$(-\infty, 2] \cup [\tfrac{8}{3}, \infty)$$

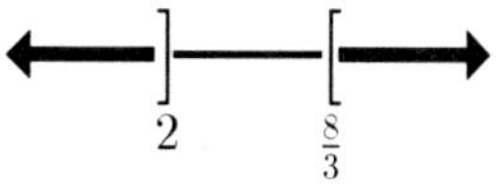

CHAPTER 1 REVIEW

101.

$$\left|\frac{x+2}{3}\right| < 1$$
$$-1 < \frac{x+2}{3} < 1$$
$$-3 < x+2 < 3$$
$$-5 < x < 1$$
$$(-5, 1)$$

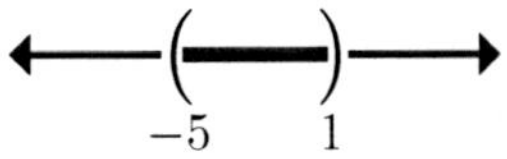

102.

$$\left|\frac{x-3}{4}\right| > 8$$

$\frac{x-3}{4} > 8$ **or** $\frac{x-3}{4} < -8$

$x - 3 > 32$ $\quad$ $x - 3 < -32$

$x > 35$ $\quad$ $x < -29$

$$(-\infty, -29) \cup (35, \infty)$$

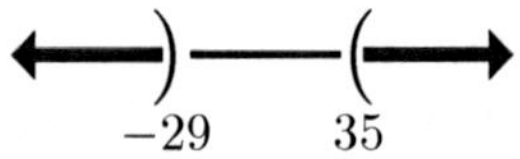

103.

$$1 < |2x+3| < 4$$

$1 < |2x+3|$ **and** $|2x+3| < 4$

(1) $|2x+3| > 1$

$2x + 3 > 1$ **or** $2x + 3 < -1$

$2x > -2$ $\quad$ $2x < -4$

$x > -1$ $\quad$ $x < -2$

(2) $|2x+3| < 4$

$$-4 < 2x+3 < 4$$
$$-7 < 2x < 1$$
$$-\tfrac{7}{2} < x < \tfrac{1}{2}$$

(1) 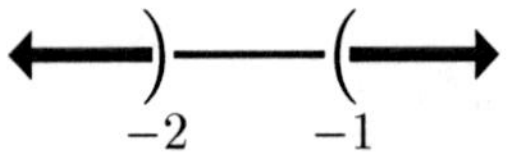

(2) 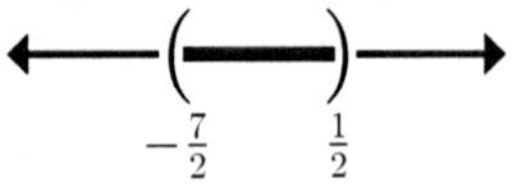

(1) 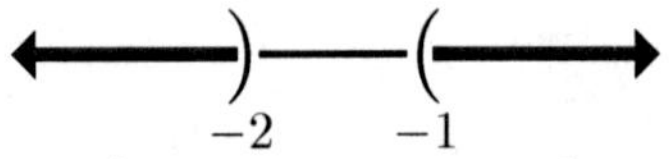

(2) 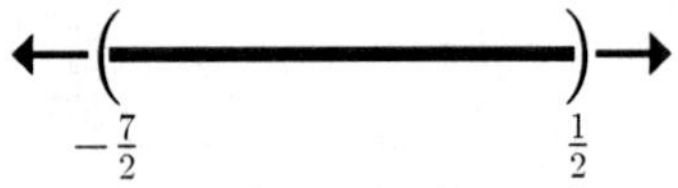

(1) and (2) 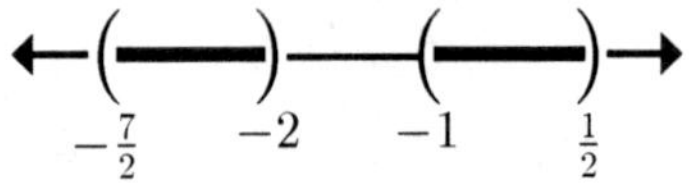

$\Rightarrow \left(-\frac{7}{2}, -2\right) \cup \left(-1, \frac{1}{2}\right)$

104.

$$0 < |3x-4| < 7$$

$0 < |3x-4|$ **and** $|3x-4| < 7$

(1) $|3x-4| > 0$

$3x - 4 > 0$ **or** $3x - 4 < -0$

$3x > 4$ $\quad$ $3x < 4$

$x > \frac{4}{3}$ $\quad$ $x < \frac{4}{3}$

(2) $|3x-4| < 7$

$$-7 < 3x-4 < 7$$
$$-3 < 3x < 11$$
$$-1 < x < \tfrac{11}{3}$$

(1) 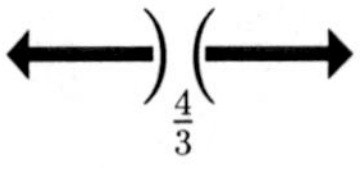

(2) 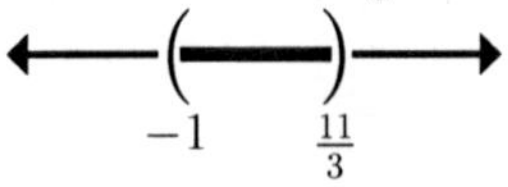

(1) 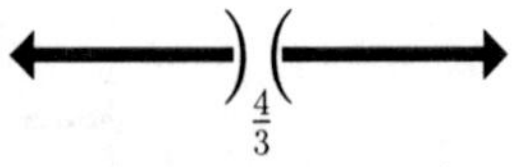

(2) 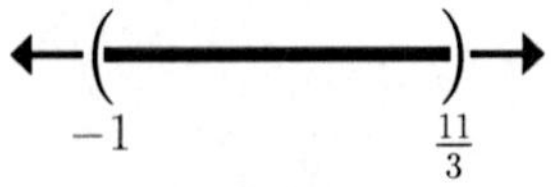

(1) and (2) 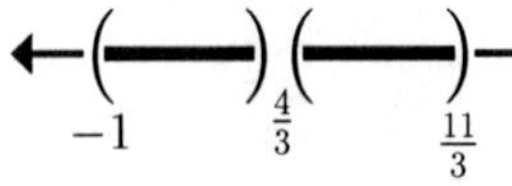

$\Rightarrow \left(-1, \frac{4}{3}\right) \cup \left(\frac{4}{3}, \frac{11}{3}\right)$

Chapter 1 Test (page 185)

1. $\dfrac{x}{x(x-1)} = 2$

restrictions: $x \neq 0$, $x \neq 1$

2. $\dfrac{4}{3x-2} + 3 = 7$

restrictions: $x \neq \frac{2}{3}$

3.
$$\begin{aligned} 7(2a+5) - 7 &= 6(a+8) \\ 14a + 35 - 7 &= 6a + 48 \\ 8a &= 20 \\ a &= \frac{20}{8} = \frac{5}{2} \end{aligned}$$

4.
$$\begin{aligned} \frac{3}{x^2 - 5x - 14} &= \frac{4}{x^2 + 5x + 6} \\ \frac{3}{(x-7)(x+2)} &= \frac{4}{(x+2)(x+3)} \\ (x-7)(x+2)(x+3)\frac{3}{(x-7)(x+2)} &= (x-7)(x+2)(x+3)\frac{4}{(x+2)(x+3)} \\ 3(x+3) &= 4(x-7) \\ 3x + 9 &= 4x - 28 \\ 37 &= x \end{aligned}$$

5.
$$\begin{aligned} z &= \frac{x-\mu}{\sigma} \\ \sigma z &= \sigma \cdot \frac{x-\mu}{\sigma} \\ z\sigma &= x - \mu \\ z\sigma + \mu &= x \end{aligned}$$

6.
$$\begin{aligned} \frac{1}{a} &= \frac{1}{b} + \frac{1}{c} \\ abc \cdot \frac{1}{a} &= abc\left(\frac{1}{b} + \frac{1}{c}\right) \\ bc &= ac + ab \\ bc &= a(c+b) \\ \frac{bc}{c+b} &= \frac{a(c+b)}{c+b} \\ \frac{bc}{c+b} &= a \end{aligned}$$

7. Let $x =$ the score on the final.
Note: This score is counted twice.

$$\begin{aligned} \frac{\boxed{\text{Sum of scores}}}{5} &= 80 \\ \frac{75 + 75 + 75 + x + x}{5} &= 80 \\ \frac{2x + 225}{5} &= 80 \\ 2x + 225 &= 400 \\ 2x &= 175 \\ x &= 87.5 \end{aligned}$$

The student needs to score 87.5.

8. Let $x =$ the amount invested at 6%. Then $20{,}000 - x =$ the amount invested at 7%.

$$\boxed{\text{Interest at 6\%}} + \boxed{\text{Interest at 7\%}} = \boxed{\text{Total interest}}$$

$$\begin{aligned} 0.06x + 0.07(20{,}000 - x) &= 1{,}260 \\ 0.06x + 1{,}400 - 0.07x &= 1{,}260 \\ -0.01x &= -140 \\ x &= 14{,}000 \end{aligned}$$

\$14,000 was invested at 6%.

CHAPTER 1 TEST

9.
$$\begin{aligned} 4x^2 - 8x + 3 &= 0 \\ (2x-3)(2x-1) &= 0 \\ 2x - 3 = 0 \quad &\textbf{or} \quad 2x - 1 = 0 \\ 2x = 3 \quad &\quad\quad 2x = 1 \\ x = \tfrac{3}{2} \quad &\quad\quad x = \tfrac{1}{2} \end{aligned}$$

10.
$$\begin{aligned} 2b^2 - 12 &= -5b \\ 2b^2 + 5b - 12 &= 0 \\ (2b-3)(b+4) &= 0 \\ 2b - 3 = 0 \quad &\textbf{or} \quad b + 4 = 0 \\ 2b = 3 \quad &\quad\quad b = -4 \\ b = \tfrac{3}{2} \quad &\quad\quad b = -4 \end{aligned}$$

11. $x = \dfrac{-b \pm \sqrt{b^2 - 4ac}}{2a}$

12. $3x^2 - 5x - 9 = 0 \Rightarrow a = 3, b = -5, c = -9$

$$x = \frac{-b \pm \sqrt{b^2 - 4ac}}{2a} = \frac{-(-5) \pm \sqrt{(-5)^2 - 4(3)(-9)}}{2(3)} = \frac{5 \pm \sqrt{25 + 108}}{6} = \frac{5 \pm \sqrt{133}}{6}$$

13. $x^2 + (k+1)x + k + 4 = 0$

$a = 1, b = k + 1, c = k + 4$

Set the discriminant equal to 0:

$$\begin{aligned} b^2 - 4ac &= 0 \\ (k+1)^2 - 4(1)(k+4) &= 0 \\ k^2 + 2k + 1 - 4k - 16 &= 0 \\ k^2 - 2k - 15 &= 0 \\ (k-5)(k+3) &= 0 \\ k - 5 = 0 \quad &\textbf{or} \quad k + 3 = 0 \\ k = 5 \quad &\quad\quad k = -3 \end{aligned}$$

14. Set $h = 0$:

$$\begin{aligned} h &= -16t^2 + 128t \\ 0 &= -16t^2 + 128t \\ 0 &= -16t(t - 8) \\ -16t = 0 \quad &\textbf{or} \quad t - 8 = 0 \\ t = 0 \quad &\quad\quad t = 8 \end{aligned}$$

The ball will return after 8 seconds.

15. $(4 - 5i) - (-3 + 7i) = 4 - 5i + 3 - 7i$
$= 7 - 12i$

16. $(4 - 5i)(3 - 7i) = 12 - 43i + 35i^2$
$= 12 - 43i - 35$
$= -23 - 43i$

17. $\dfrac{2}{2-i} = \dfrac{2(2+i)}{(2-i)(2+i)} = \dfrac{4+2i}{2^2 - i^2} = \dfrac{4+2i}{5} = \dfrac{4}{5} + \dfrac{2}{5}i$

18. $\dfrac{1+i}{1-i} = \dfrac{(1+i)(1+i)}{(1-i)(1+i)} = \dfrac{1+2i+i^2}{1^2 - i^2} = \dfrac{2i}{2} = 0 + i$

19. $i^{13} = i^{12}i = (i^4)^3 i = 1^3 i = i$

20. $i^0 = 1$

21. $|5 - 12i| = \sqrt{5^2 + (-12)^2} = \sqrt{25 + 144} = \sqrt{169} = 13$

CHAPTER 1 TEST

22. $\left|\frac{1}{3+i}\right| = \left|\frac{1(3-i)}{(3+i)(3-i)}\right| = \left|\frac{3-i}{3^2-i^2}\right| = \left|\frac{3-i}{10}\right| = \left|\frac{3}{10} - \frac{1}{10}i\right| = \sqrt{\left(\frac{3}{10}\right)^2 + \left(-\frac{1}{10}\right)^2}$

$$= \sqrt{\frac{9}{100} + \frac{1}{100}}$$
$$= \sqrt{\frac{10}{100}} = \frac{\sqrt{10}}{10}$$

23.
$$z^4 - 13z^2 + 36 = 0$$
$$(z^2 - 4)(z^2 - 9) = 0$$

$z^2 - 4 = 0$ **or** $z^2 - 9 = 0$

$z^2 = 4$ $\qquad z^2 = 9$

$z = \pm 2$ $\qquad z = \pm 3$

24.
$$2p^{2/5} - p^{1/5} - 1 = 0$$
$$(2p^{1/5} + 1)(p^{1/5} - 1) = 0$$

$2p^{1/5} + 1 = 0$ **or** $p^{1/5} - 1 = 0$

$p^{1/5} = -\frac{1}{2}$ $\qquad p^{1/5} = 1$

$(p^{1/5})^5 = (-\frac{1}{2})^5$ $\qquad (p^{1/5})^5 = (1)^5$

$p = -\frac{1}{32}$ $\qquad p = 1$

Both answers check.

25.
$$\sqrt{x+5} = 12$$
$$\left(\sqrt{x+5}\right)^2 = 12^2$$
$$x + 5 = 144$$
$$x = 139$$

The answer checks.

26.
$$\sqrt{2z+3} = 1 - \sqrt{z+1}$$
$$\left(\sqrt{2z+3}\right)^2 = \left(1 - \sqrt{z+1}\right)^2$$
$$2z + 3 = 1 - 2\sqrt{z+1} + z + 1$$
$$2\sqrt{z+1} = -z - 1$$
$$\left(2\sqrt{z+1}\right)^2 = (-z-1)^2$$
$$4(z+1) = z^2 + 2z + 1$$
$$4z + 4 = z^2 + 2z + 1$$
$$0 = z^2 - 2z - 3$$
$$0 = (z+1)(z-3)$$

$z + 1 = 0$ **or** $z - 3 = 0$

$z = -1$ $\qquad z = 3$

The answer $z = 3$ is extraneous.

27.
$$5x - 3 \le 7$$
$$5x \le 10$$
$$x \le 2 \Rightarrow (-\infty, 2]$$

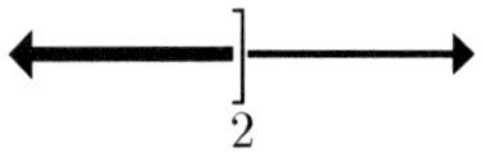

28.
$$\frac{x+3}{4} > \frac{2x-4}{3}$$
$$12 \cdot \tfrac{x+3}{4} > 12 \cdot \tfrac{2x-4}{3}$$
$$3(x+3) > 4(2x-4)$$
$$3x + 9 > 8x - 16$$
$$-5x > -25$$
$$x < 5 \Rightarrow (-\infty, 5)$$

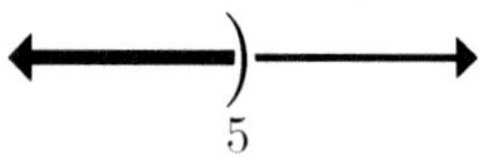

29. $5 \le 2x - 1 < 7$

$6 \le 2x < 8$

$3 \le x < 4 \Rightarrow [3, 4)$

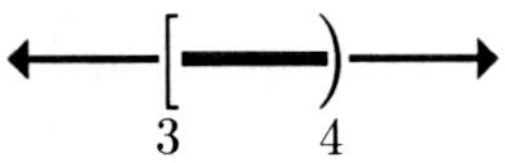

30. $1 + x < 3x - 3 < 4x - 2$

$1 + x < 3x - 3$ **and** $3x - 3 < 4x - 2$

$-2x < -4$ $\quad$ $-x < 1$

$x > 2$ $\quad$ $x > -1$

$x > 2$

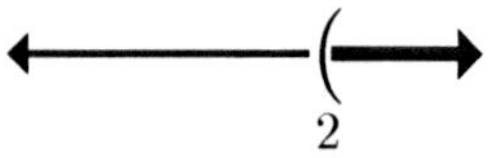

$x > -1$

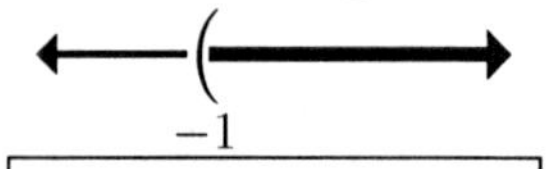

$x > 2$
and
$x > -1$

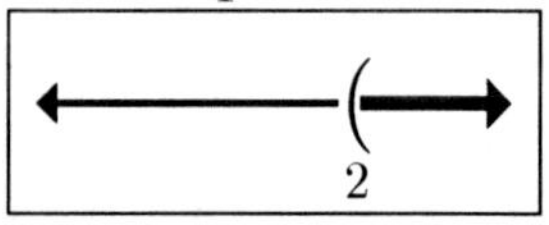

Solution set: $(2, \infty)$

31. $x^2 - 7x - 8 \ge 0$

$(x + 1)(x - 8) \ge 0$

factors $= 0$: $x = -1, x = 8$

intervals: $(-\infty, -1), (-1, 8), (8, \infty)$

interval	test number	value of x^2-7x-8
$(-\infty, -1)$	-2	$+10$
$(-1, 8)$	0	-8
$(8, \infty)$	9	$+10$

Solution: $(-\infty, -1] \cup [8, \infty)$

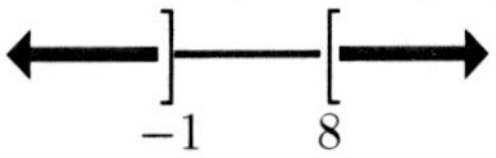

32. $\dfrac{x + 2}{x - 1} \le 0$

factors $= 0$: $x = -2, x = 1$

intervals: $(-\infty, -2), (-2, 1), (1, \infty)$

interval	test number	sign of $\frac{x+2}{x-1}$
$(-\infty, -2)$	-3	$+$
$(-2, 1)$	0	$-$
$(1, \infty)$	2	$+$

Include endpoints which make the numerator equal to 0. Do not include endpoints which make the denominator equal to 0.

Solution: $[-2, 1)$

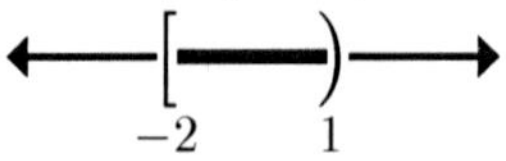

33. $\left|\dfrac{3x + 2}{2}\right| = 4$

$\frac{3x+2}{2} = 4$ **or** $\frac{3x+2}{2} = -4$

$3x + 2 = 8$ $\quad$ $3x + 2 = -8$

$3x = 6$ $\quad$ $3x = -10$

$x = 2$ $\quad$ $x = -\frac{10}{3}$

34. $|x + 3| = |x - 3|$

$x + 3 = x - 3$ **or** $x + 3 = -(x - 3)$

$0 = -6$ $\quad$ $x + 3 = -x + 3$

not true $\quad$ $2x = 0$

$x = 0$

35.
$$|2x-5|>2$$
$$2x-5>2 \quad \textbf{or} \quad 2x-5<-2$$
$$2x>7 \qquad\qquad 2x<3$$
$$x>\tfrac{7}{2} \qquad\qquad x<\tfrac{3}{2}$$
$$\left(-\infty,\tfrac{3}{2}\right)\cup\left(\tfrac{7}{2},\infty\right)$$

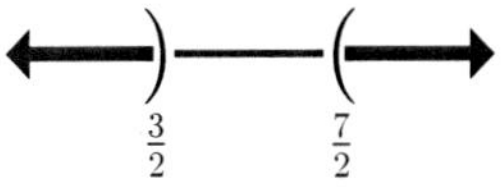

36.
$$\left|\frac{2x+3}{3}\right|\le 5$$
$$-5 \le \tfrac{2x+3}{3} \le 5$$
$$-15 \le 2x+3 \le 15$$
$$-18 \le 2x \le 12$$
$$-9 \le x \le 6$$
$$[-9,6]$$

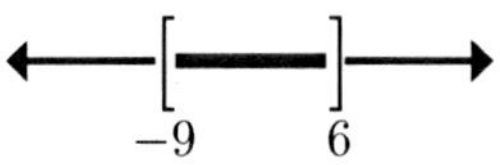

Cumulative Review Exercises (page 186)

1. even integers: $-2, 0, 2, 6$

2. prime numbers: $2, 5, 11$

3. $-4 \le x < 7 \Rightarrow [-4, 7)$

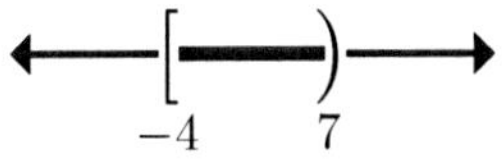

4. $x \ge 2$ or $x < 0 \Rightarrow (-\infty, 0)\cup[2,\infty)$

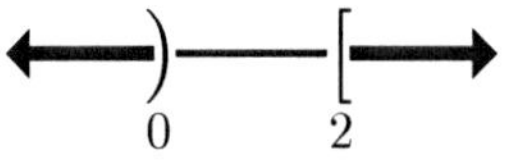

5. commutative property of addition

6. transitive property

7. $(81a^4)^{1/2} = \left[(9a^2)^2\right]^{1/2} = 9a^2$

8. $81(a^4)^{1/2} = 81\left[(a^2)^2\right]^{1/2} = 81a^2$

9. $(a^{-3}b^{-2})^{-2} = (a^{-3})^{-2}(b^{-2})^{-2} = a^6b^4$

10. $\left(\dfrac{4x^4}{12x^2y}\right)^{-2} = \left(\dfrac{12x^2y}{4x^4}\right)^2 = \left(\dfrac{3y}{x^2}\right)^2 = \dfrac{9y^2}{x^4}$

11. $\left(\dfrac{4x^0y^2}{x^2y}\right)^{-2} = \left(\dfrac{x^2y}{4x^0y^2}\right)^2 = \left(\dfrac{x^2}{4y}\right)^2 = \dfrac{x^4}{16y^2}$

12. $\left(\dfrac{4x^{-5}y^2}{6x^{-2}y^{-3}}\right)^2 = \left(\dfrac{2y^5}{3x^3}\right)^2 = \dfrac{4y^{10}}{9x^6}$

13. $(a^{1/2}b)^2(ab^{1/2})^2 = (ab^2)(a^2b) = a^3b^3$

14. $(a^{1/2}b^{1/2}c)^2 = abc^2$

15. $\dfrac{3}{\sqrt{3}} = \dfrac{3\sqrt{3}}{\sqrt{3}\sqrt{3}} = \dfrac{3\sqrt{3}}{3} = \sqrt{3}$

16. $\dfrac{2}{\sqrt[3]{4x}} = \dfrac{2\sqrt[3]{2x^2}}{\sqrt[3]{4x}\sqrt[3]{2x^2}} = \dfrac{2\sqrt[3]{2x^2}}{\sqrt[3]{8x^3}} = \dfrac{2\sqrt[3]{2x^2}}{2x} = \dfrac{\sqrt[3]{2x^2}}{x}$

17. $\dfrac{3}{y-\sqrt{3}} = \dfrac{3\left(y+\sqrt{3}\right)}{\left(y-\sqrt{3}\right)\left(y+\sqrt{3}\right)} = \dfrac{3\left(y+\sqrt{3}\right)}{y^2-\left(\sqrt{3}\right)^2} = \dfrac{3\left(y+\sqrt{3}\right)}{y^2-3}$

CUMULATIVE REVIEW EXERCISES

18. $\dfrac{3x}{\sqrt{x}-1} = \dfrac{3x\left(\sqrt{x}+1\right)}{\left(\sqrt{x}-1\right)\left(\sqrt{x}+1\right)} = \dfrac{3x\left(\sqrt{x}+1\right)}{\left(\sqrt{x}\right)^2-1^2} = \dfrac{3x\left(\sqrt{x}+1\right)}{x-1}$

19. $\sqrt{75}-3\sqrt{5}=\sqrt{25}\sqrt{3}-3\sqrt{5}=5\sqrt{3}-3\sqrt{5}$

20. $\sqrt{18}+\sqrt{8}-2\sqrt{2}=\sqrt{9}\sqrt{2}+\sqrt{4}\sqrt{2}-2\sqrt{2}=3\sqrt{2}+2\sqrt{2}-2\sqrt{2}=3\sqrt{2}$

21. $\left(\sqrt{2}-\sqrt{3}\right)^2=\left(\sqrt{2}-\sqrt{3}\right)\left(\sqrt{2}-\sqrt{3}\right)=\sqrt{4}-2\sqrt{6}+\sqrt{9}=5-2\sqrt{6}$

22. $\left(3-\sqrt{5}\right)\left(3+\sqrt{5}\right)=9-\sqrt{25}=9-5=4$

23. $(3x^2-2x+5)-3(x^2+2x-1)=3x^2-2x+5-3x^2-6x+3=-8x+8$

24. $5x^2(2x^2-x)+x(x^2-x^3)=10x^4-5x^3+x^3-x^4=9x^4-4x^3$

25. $(3x-5)(2x+7)=6x^2+21x-10x-35=6x^2+11x-35$

26. $(z+2)(z^2-z+2)=z^3-z^2+2z+2z^2-2z+4=z^3+z^2+4$

27.
$$\begin{array}{r|l} & 2x^2 - x + 1 \\ \hline 3x+2 & 6x^3 + x^2 + x + 2 \\ & \underline{6x^3 + 4x^2} \\ & \quad -3x^2 + x \\ & \quad \underline{-3x^2 - 2x} \\ & \qquad 3x + 2 \\ & \qquad \underline{3x + 2} \\ & \qquad\quad 0 \end{array}$$

28.
$$\begin{array}{r|l} & 3x^2 + 1 + \frac{-x}{x^2+2} \\ \hline x^2+2 & 3x^4 + 0x^3 + 7x^2 - x + 2 \\ & \underline{3x^4 \qquad\quad + 6x^2} \\ & \qquad\qquad x^2 - x + 2 \\ & \qquad\qquad \underline{x^2 \qquad + 2} \\ & \qquad\qquad\quad -x \end{array}$$

29. $3t^2-6t=3t(t-2)$

30. $3x^2-10x-8=(3x+2)(x-4)$

31.
$$\begin{aligned} x^8-2x^4+1 &= \left(x^4-1\right)\left(x^4-1\right)=\left(x^2+1\right)\left(x^2-1\right)\left(x^2+1\right)\left(x^2-1\right) \\ &= \left(x^2+1\right)^2(x+1)(x-1)(x+1)(x-1) \\ &= \left(x^2+1\right)^2(x+1)^2(x-1)^2 \end{aligned}$$

32. $x^6-1=\left(x^3\right)^2-1^2=(x^3+1)(x^3-1)=(x+1)(x^2-x+1)(x-1)(x^2+x+1)$

33. $\dfrac{x^2-4}{x^2+5x+6}\cdot\dfrac{x^2-2x-15}{x^2+3x-10}=\dfrac{(x+2)(x-2)}{(x+2)(x+3)}\cdot\dfrac{(x-5)(x+3)}{(x+5)(x-2)}=\dfrac{x-5}{x+5}$

34.
$$\begin{aligned} \frac{6x^3+x^2-x}{x+2}\div\frac{3x^2-x}{x^2+4x+4} &= \frac{x(6x^2+x-1)}{x+2}\cdot\frac{x^2+4x+4}{3x^2-x} \\ &= \frac{x(2x+1)(3x-1)}{x+2}\cdot\frac{(x+2)(x+2)}{x(3x-1)}=(2x+1)(x+2) \end{aligned}$$

CUMULATIVE REVIEW EXERCISES

35. $\dfrac{2}{x+3}+\dfrac{5x}{x-3}=\dfrac{2(x-3)}{(x+3)(x-3)}+\dfrac{5x(x+3)}{(x-3)(x+3)}=\dfrac{2x-6}{(x+3)(x-3)}+\dfrac{5x^2+15x}{(x+3)(x-3)}$

$$=\frac{5x^2+17x-6}{(x+3)(x-3)}$$

36. $\dfrac{x-2}{x+3}\left(\dfrac{x+3}{x^2-4}-1\right)=\dfrac{x-2}{x+3}\left(\dfrac{x+3}{x^2-4}-\dfrac{x^2-4}{x^2-4}\right)=\dfrac{x-2}{x+3}\left(\dfrac{-x^2+x+7}{(x+2)(x-2)}\right)=\dfrac{-x^2+x+7}{(x+3)(x+2)}$

37. $\dfrac{\frac{1}{a}+\frac{1}{b}}{\frac{1}{ab}}=\dfrac{ab\left(\frac{1}{a}+\frac{1}{b}\right)}{ab\left(\frac{1}{ab}\right)}=\dfrac{b+a}{1}=b+a$

38. $\dfrac{x^{-1}-y^{-1}}{x-y}=\dfrac{\frac{1}{x}-\frac{1}{y}}{x-y}=\dfrac{xy\left(\frac{1}{x}-\frac{1}{y}\right)}{xy(x-y)}$

$$=\frac{y-x}{xy(x-y)}=-\frac{1}{xy}$$

39.
$$\begin{aligned}\frac{3x}{x+5}&=\frac{x}{x-5}\\ 3x(x-5)&=x(x+5)\\ 3x^2-15x&=x^2+5x\\ 2x^2-20x&=0\\ 2x(x-10)&=0\end{aligned}$$
$2x=0$ **or** $x-10=0$

$x=0$ $\quad\quad x=10$

40.
$$\begin{aligned}8(2x-3)-3(5x+2)&=4\\ 16x-24-15x-6&=4\\ x&=34\end{aligned}$$

41.
$$\begin{aligned}\frac{1}{R}&=\frac{1}{R_1}+\frac{1}{R_2}\\ RR_1R_2\cdot\frac{1}{R}&=RR_1R_2\left(\frac{1}{R_1}+\frac{1}{R_2}\right)\\ R_1R_2&=RR_2+RR_1\\ R_1R_2&=R(R_2+R_1)\\ \frac{R_1R_2}{R_2+R_1}&=\frac{R(R_2+R_1)}{R_2+R_1}\\ \frac{R_1R_2}{R_2+R_1}&=R\end{aligned}$$

42.
$$\begin{aligned}S&=\frac{a-lr}{1-r}\\ S(1-r)&=\frac{a-lr}{1-r}(1-r)\\ S(1-r)&=a-lr\\ S-Sr&=a-lr\\ S-a&=Sr-lr\\ S-a&=r(S-l)\\ \frac{S-a}{S-l}&=r\end{aligned}$$

43.

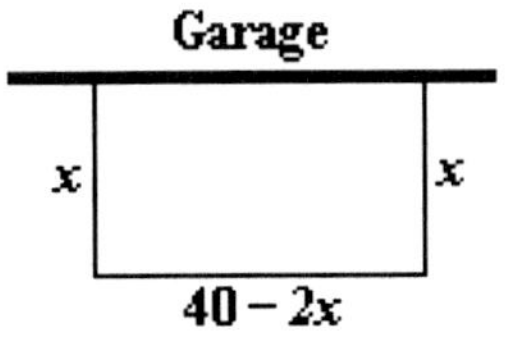

$$\begin{aligned}\text{Area}&=192\\ x(40-2x)&=192\\ 40x-2x^2&=192\\ 0&=2x^2-40x+192\\ 0&=2(x-8)(x-12)\end{aligned}$$
$x-8=0$ **or** $x-12=0$

$x=8$ $\quad\quad x=12$

If $x=8$, then the dimensions are 8 feet by 24 feet.

If $x=12$, then the dimensions are 12 feet by 16 feet.

CUMULATIVE REVIEW EXERCISES

44. Let x = the amount invested at 6%. Then $25{,}000 - x$ = the amount invested at 7%.

$$\boxed{\text{Interest at 6\%}} + \boxed{\text{Interest at 7\%}} = \boxed{\text{Total interest}}$$

$$\begin{aligned} 0.06x + 0.07(25{,}000 - x) &= 1{,}670 \\ 0.06x + 1{,}750 - 0.07x &= 1{,}670 \\ -0.01x &= -80 \\ x &= 8{,}000 \Rightarrow \$8{,}000 \text{ was invested at 6\%.} \end{aligned}$$

45. $$\frac{2+i}{2-i} = \frac{(2+i)(2+i)}{(2-i)(2+i)} = \frac{4+4i+i^2}{4-i^2} = \frac{3+4i}{5} = \frac{3}{5} + \frac{4}{5}i$$

46. $$\begin{aligned} \frac{i(3-i)}{(1+i)(1+i)} &= \frac{i(3-i)(1-i)(1-i)}{(1+i)(1-i)(1+i)(1-i)} = \frac{(3i-i^2)(1-2i+i^2)}{(1-i^2)(1-i^2)} \\ &= \frac{(1+3i)(-2i)}{1-2i^2+i^4} = \frac{-2i-6i^2}{4} = \frac{6-2i}{4} = \frac{3}{2} - \frac{1}{2}i \end{aligned}$$

47. $|3+4i| = \sqrt{3^2+4^2} = \sqrt{9+16} = \sqrt{25} = 5$

48. $$\frac{5}{i^7} + 5i = \frac{5i}{i^7 i} + 5i = \frac{5i}{i^8} + 5i = \frac{5i}{(i^4)^2} + 5i = \frac{5i}{1^2} + 5i = 5i + 5i = 10i = 0 + 10i$$

49. $$\begin{aligned} \frac{x+3}{x-1} - \frac{6}{x} &= 1 \\ x(x-1)\left(\frac{x+3}{x-1} - \frac{6}{x}\right) &= x(x-1)(1) \\ x(x+3) - 6(x-1) &= x^2 - x \\ x^2 + 3x - 6x + 6 &= x^2 - x \\ -2x &= -6 \\ x &= 3 \end{aligned}$$

50. $$\begin{aligned} x^4 + 36 &= 13x^2 \\ x^4 - 13x^2 + 36 &= 0 \\ (x^2-4)(x^2-9) &= 0 \end{aligned}$$

$$\begin{aligned} x^2 - 4 &= 0 \quad \textbf{or} \quad & x^2 - 9 &= 0 \\ x^2 &= 4 & x^2 &= 9 \\ x &= \pm 2 & x &= \pm 3 \end{aligned}$$

51. $$\begin{aligned} \sqrt{y+2} + \sqrt{11-y} &= 5 \\ \sqrt{y+2} - 5 &= -\sqrt{11-y} \\ \left(\sqrt{y+2} - 5\right)^2 &= \left(-\sqrt{11-y}\right)^2 \\ y + 2 - 10\sqrt{y+2} + 25 &= 11 - y \\ -10\sqrt{y+2} &= -2y - 16 \\ \left(-10\sqrt{y+2}\right)^2 &= (-2y-16)^2 \\ 100(y+2) &= 4y^2 + 64y + 256 \\ 100y + 200 &= 4y^2 + 64y + 256 \\ 0 &= 4y^2 - 36y + 56 \\ 0 &= 4(y-2)(y-7) \end{aligned}$$

$$\begin{aligned} y - 2 &= 0 \quad \textbf{or} \quad & y - 7 &= 0 \\ y &= 2 & y &= 7 \end{aligned}$$

Both solutions check.

52. $$\begin{aligned} z^{2/3} - 13z^{1/3} + 36 &= 0 \\ \left(z^{1/3} - 4\right)\left(z^{1/3} - 9\right) &= 0 \end{aligned}$$

$$\begin{aligned} z^{1/3} - 4 &= 0 \quad \textbf{or} \quad & z^{1/3} - 9 &= 0 \\ z^{1/3} &= 4 & z^{1/3} &= 9 \\ \left(z^{1/3}\right)^3 &= 4^3 & \left(z^{1/3}\right)^3 &= 9^3 \\ z &= 64 & z &= 729 \end{aligned}$$

Both solutions check.

CUMULATIVE REVIEW EXERCISES

53. $5x - 7 \le 4$

$5x \le 11$

$x \le \frac{11}{5} \Rightarrow \left(-\infty, \frac{11}{5}\right]$

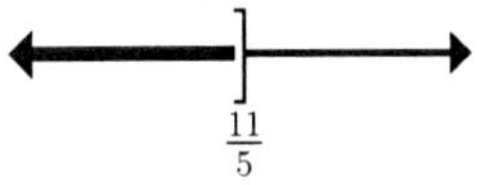

54. $x^2 - 8x + 15 > 0$

$(x-3)(x-5) > 0$

factors $= 0$: $x = 3$, $x = 5$

intervals: $(-\infty, 3)$, $(3, 5)$, $(5, \infty)$

interval	test number	value of $x^2-8x+15$
$(-\infty, 3)$	0	$+15$
$(3, 5)$	4	-1
$(5, \infty)$	6	$+3$

Solution: $(-\infty, 3) \cup (5, \infty)$

55. $\dfrac{x^2 + 4x + 3}{x - 2} \ge 0$

$\dfrac{(x+3)(x+1)}{x-2} \ge 0$

factors $= 0$: $x = -3$, $x = -1$, $x = 2$

int.: $(-\infty, -3)$, $(-3, -1)$, $(-1, 2)$, $(2, \infty)$

interval	test number	sign of $\frac{x^2+4x+3}{x-2}$
$(-\infty, -3)$	-4	$-$
$(-3, -1)$	-2	$+$
$(-1, 2)$	0	$-$
$(2, \infty)$	3	$+$

Include endpoints which make the numerator equal to 0. Do not include endpoints which make the denominator equal to 0.

Solution: $[-3, -1] \cup (2, \infty)$

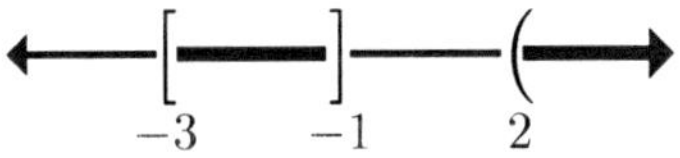

56. $\dfrac{9}{x} > x$

$\dfrac{9}{x} - x > 0$

$\dfrac{9 - x^2}{x} > 0$

$\dfrac{(3+x)(3-x)}{x} > 0$

factors $= 0$: $x = -3$, $x = 3$, $x = 0$

int.: $(-\infty, -3)$, $(-3, 0)$, $(0, 3)$, $(3, \infty)$

interval	test number	sign of $\frac{9-x^2}{x}$
$(-\infty, -3)$	-4	$+$
$(-3, 0)$	-1	$-$
$(0, 3)$	1	$+$
$(3, \infty)$	4	$-$

Solution: $(-\infty, -3) \cup (0, 3)$

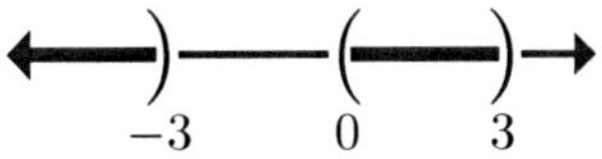

57. $|2x - 3| \ge 5$

$2x - 3 \ge 5$ **or** $2x - 3 \le -5$

$2x \ge 8$ $\quad$ $2x \le -2$

$x \ge 4$ $\quad$ $x \le -1$

$(-\infty, -1] \cup [4, \infty)$

58. $\left|\dfrac{3x - 5}{2}\right| < 2$

$-2 < \dfrac{3x-5}{2} < 2$

$-4 < 3x - 5 < 4$

$1 < 3x < 9$

$\frac{1}{3} < x < 3$

$\left(\frac{1}{3}, 3\right)$

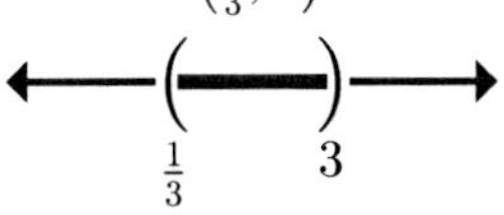

Exercises 2.1 (page 202)

1. quadrants **3.** to the right **5.** first **7.** linear

9. x-intercept **11.** horizontal **13.** midpoint **15.** $A(2, 3)$

17. $C(-2, -3)$ **19.** $E(0, 0)$ **21.** $G(-5, -5)$

23, 25, 27, 29.

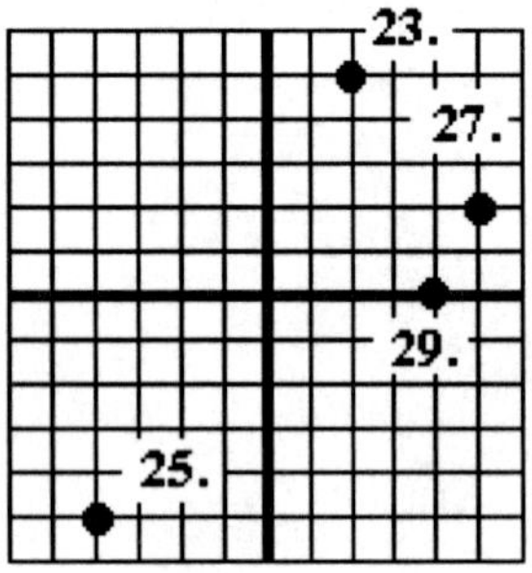

23. QI

25. QIII

27. QI

29. $+\ x$-axis

31. $y - 2x = 7$
$y = 2x + 7$

x	y
0	7
-2	3

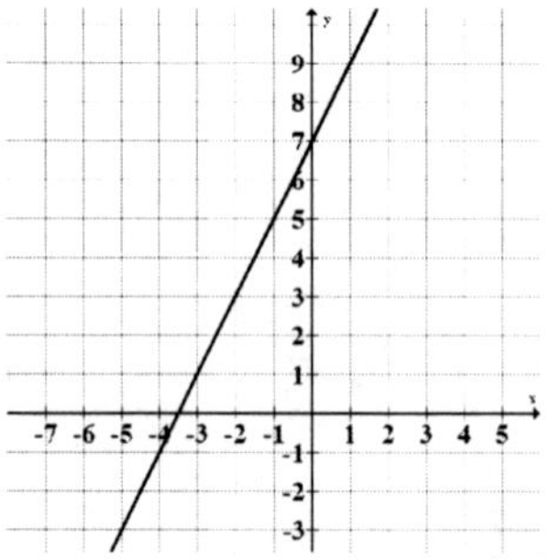

33. $y + 5x = 5$
$y = -5x + 5$

x	y
0	5
1	0

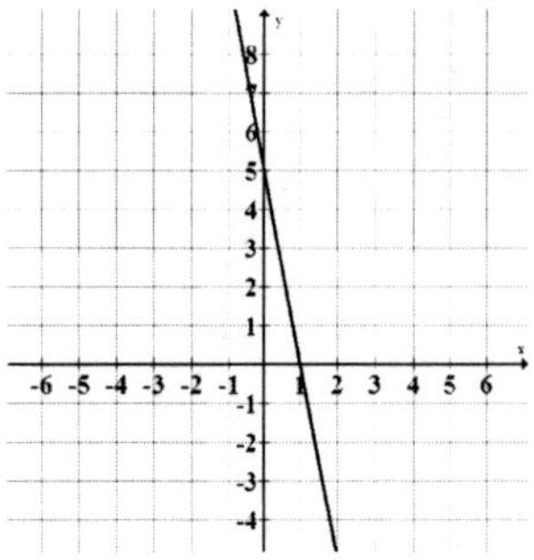

35. $6x - 3y = 10$
$-3y = -6x + 10$
$y = 2x - \frac{10}{3}$

x	y
0	$-\frac{10}{3}$
2	$\frac{2}{3}$

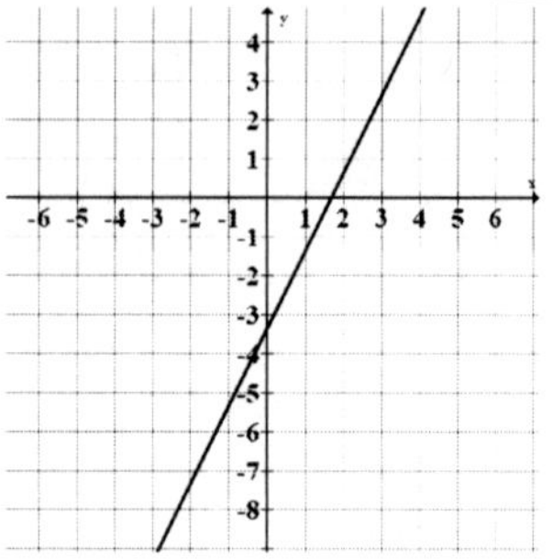

SECTION 2.1

37.

$$3x = 6y - 1$$
$$-6y = -3x - 1$$
$$y = \tfrac{1}{2}x + \tfrac{1}{6}$$

x	y
0	$\frac{1}{6}$
-2	$-\frac{5}{6}$

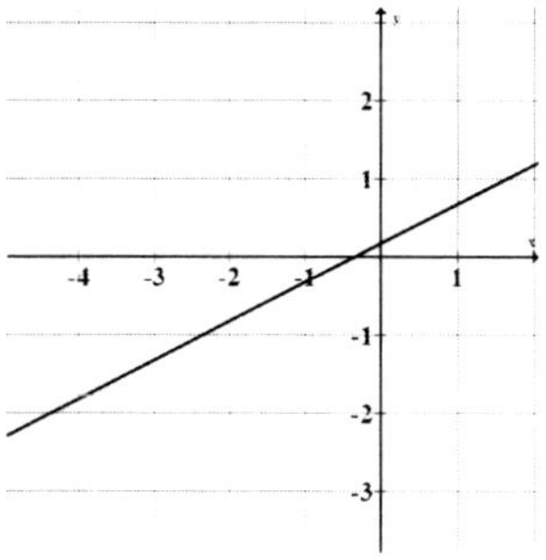

39.

$$2(x + y + 1) = x + 2$$
$$2x + 2y + 2 = x + 2$$
$$2y = -x$$
$$y = -\tfrac{1}{2}x$$

x	y
0	0
-2	1

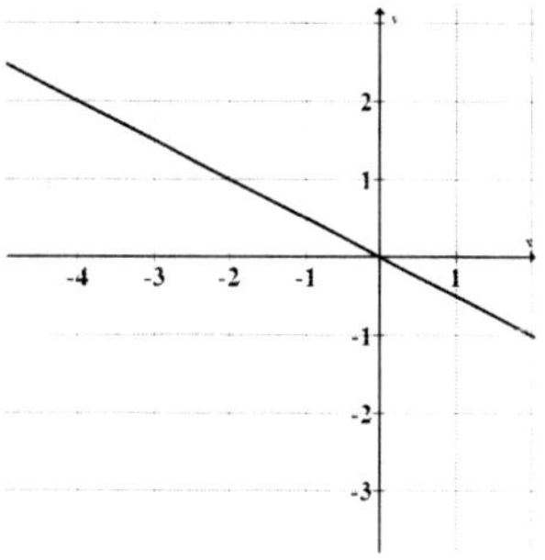

41.

$x + y = 5$ $x + y = 5$

$x + 0 = 5$ $0 + y = 5$

$x = 5$ $y = 5$

$(5, 0)$ $(0, 5)$

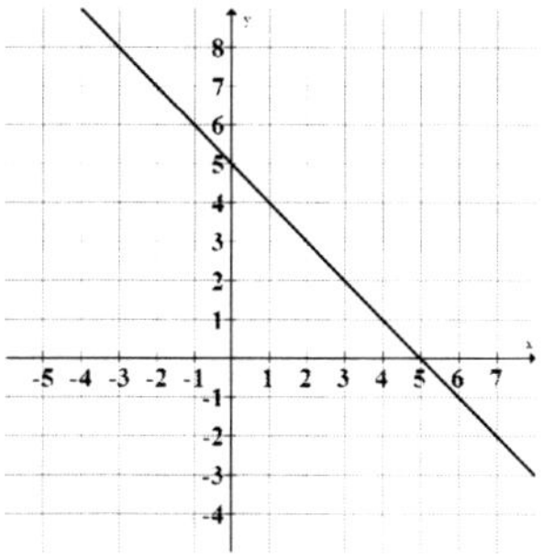

43.

$2x - y = 4$ $2x - y = 4$

$2x - 0 = 4$ $2(0) - y = 4$

$2x = 4$ $-y = 4$

$x = 2$ $y = -4$

$(2, 0)$ $(0, -4)$

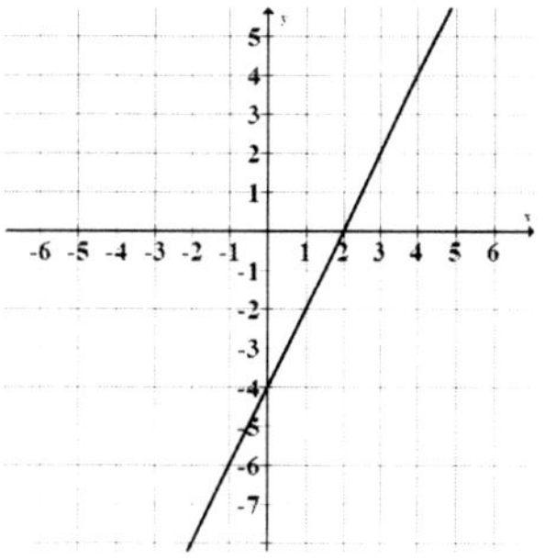

45.

$3x + 2y = 6$ $3x + 2y = 6$

$3x + 2(0) = 6$ $3(0) + 2y = 6$

$3x = 6$ $2y = 6$

$x = 2$ $y = 3$

$(2, 0)$ $(0, 3)$

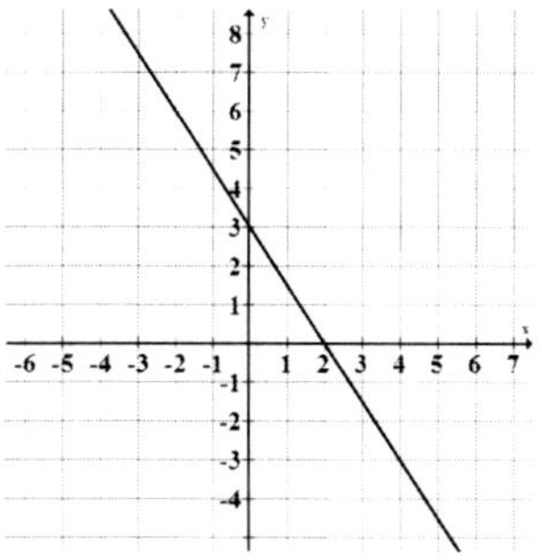

47.

$4x - 5y = 20$ $4x - 5y = 20$

$4x - 5(0) = 20$ $4(0) - 5y = 20$

$4x = 20$ $-5y = 20$

$x = 5$ $y = -4$

$(5, 0)$ $(0, -4)$

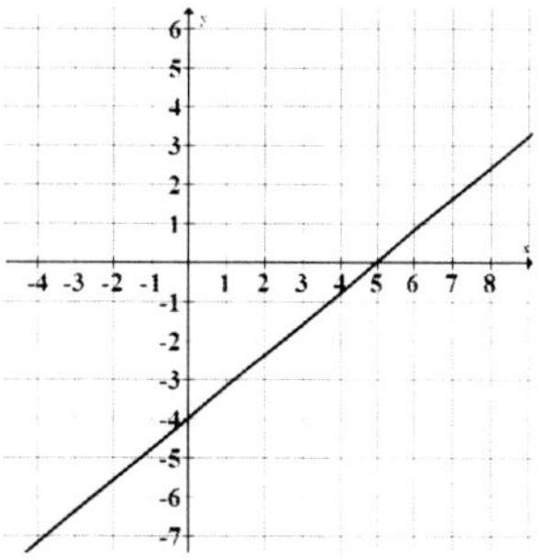

49. $y = 3$

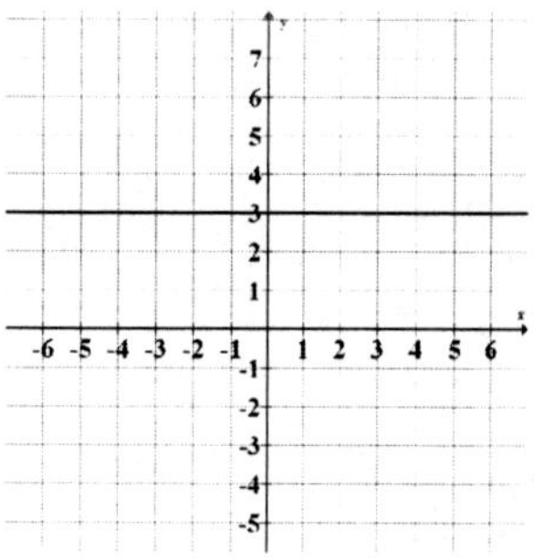

51. $3x + 5 = -1$

$3x = -6 \Rightarrow x = -2$

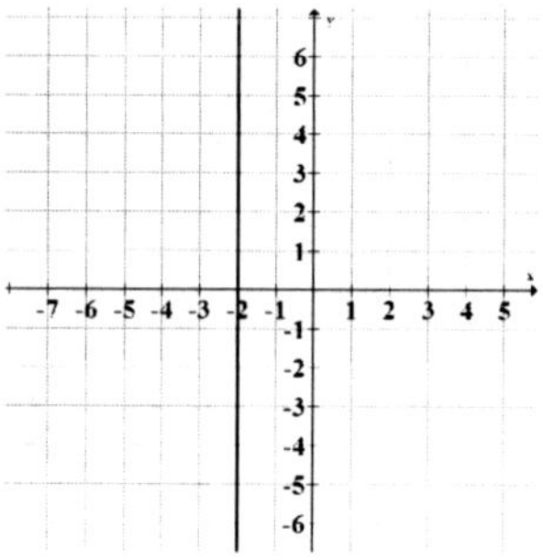

53. $3(y+2) = y$

$3y + 6 = y$

$2y = -6 \Rightarrow y = -3$

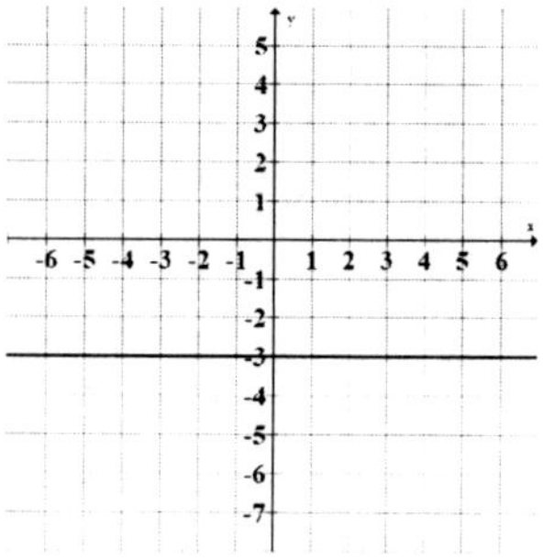

55. $3(y+2x) = 6x + y$

$3y + 6x = 6x + y$

$2y = 0 \Rightarrow y = 0$

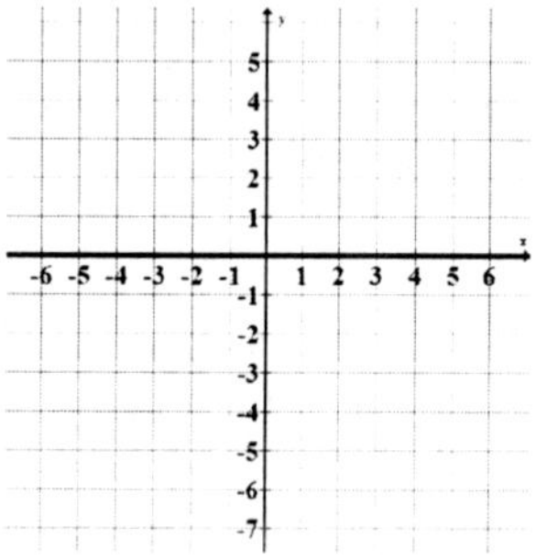

57. $y = 3.7x - 4.5$

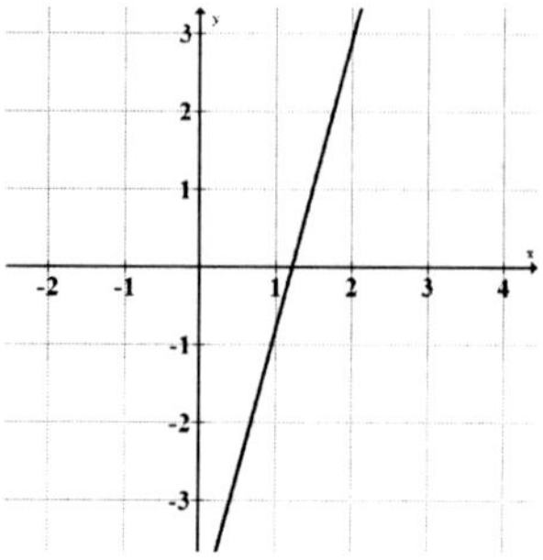

x-int: $x = 1.22$

59. $1.5x - 3y = 7$

$-3y = -1.5x + 7$

$y = 0.5x - \frac{7}{3}$

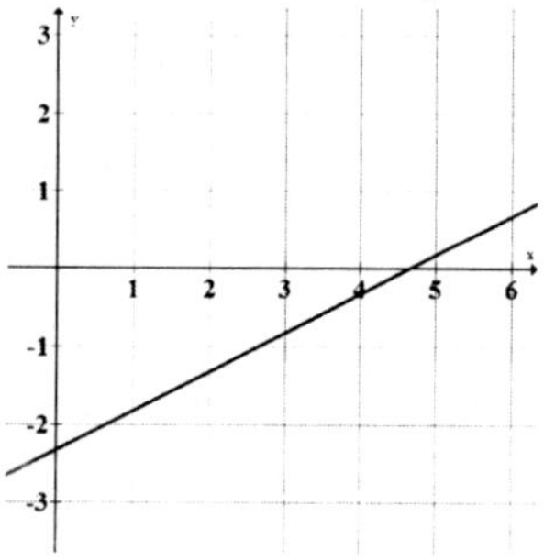

x-int: $x = 4.67$

61.
$$\begin{aligned} d &= \sqrt{(x_2 - x_1)^2 + (y_2 - y_1)^2} \\ &= \sqrt{(4-0)^2 + (-3-0)^2} \\ &= \sqrt{4^2 + (-3)^2} \\ &= \sqrt{16+9} = \sqrt{25} = 5 \end{aligned}$$

63.
$$\begin{aligned} d &= \sqrt{(x_2 - x_1)^2 + (y_2 - y_1)^2} \\ &= \sqrt{(-3-0)^2 + (2-0)^2} \\ &= \sqrt{(-3)^2 + (2)^2} \\ &= \sqrt{9+4} = \sqrt{13} \end{aligned}$$

65.
$$\begin{aligned} d &= \sqrt{(x_2 - x_1)^2 + (y_2 - y_1)^2} \\ &= \sqrt{(1-0)^2 + (1-0)^2} \\ &= \sqrt{(1)^2 + (1)^2} \\ &= \sqrt{1+1} = \sqrt{2} \end{aligned}$$

67.
$$\begin{aligned} d &= \sqrt{(x_2 - x_1)^2 + (y_2 - y_1)^2} \\ &= \sqrt{\left(\sqrt{3}-0\right)^2 + (1-0)^2} \\ &= \sqrt{\left(\sqrt{3}\right)^2 + (1)^2} \\ &= \sqrt{3+1} = \sqrt{4} = 2 \end{aligned}$$

SECTION 2.1

69. $d = \sqrt{(x_2 - x_1)^2 + (y_2 - y_1)^2}$
$= \sqrt{(3-6)^2 + (7-3)^2}$
$= \sqrt{(-3)^2 + (4)^2}$
$= \sqrt{9+16} = \sqrt{25} = 5$

71. $d = \sqrt{(x_2 - x_1)^2 + (y_2 - y_1)^2}$
$= \sqrt{[4-(-1)]^2 + [-6-6]^2}$
$= \sqrt{(5)^2 + (-12)^2}$
$= \sqrt{25+144} = \sqrt{169} = 13$

73. $d = \sqrt{(x_2 - x_1)^2 + (y_2 - y_1)^2}$
$= \sqrt{[-2-(-6)]^2 + [-15-(-21)]^2}$
$= \sqrt{(4)^2 + (6)^2}$
$= \sqrt{16+36} = \sqrt{52} = 2\sqrt{13}$

75. $d = \sqrt{(x_2 - x_1)^2 + (y_2 - y_1)^2}$
$= \sqrt{[3-(-5)]^2 + [-3-5]^2}$
$= \sqrt{(8)^2 + (-8)^2}$
$= \sqrt{64+64} = \sqrt{128} = 8\sqrt{2}$

77. $d = \sqrt{(x_2 - x_1)^2 + (y_2 - y_1)^2} = \sqrt{[\pi - \pi]^2 + [-2-5]^2} = \sqrt{(0)^2 + (-7)^2} = \sqrt{0+49}$
$= \sqrt{49} = 7$

79. $M\left(\frac{x_1+x_2}{2}, \frac{y_1+y_2}{2}\right) = M\left(\frac{2+6}{2}, \frac{4+8}{2}\right) = M\left(\frac{8}{2}, \frac{12}{2}\right) = M(4, 6)$

81. $M\left(\frac{x_1+x_2}{2}, \frac{y_1+y_2}{2}\right) = M\left(\frac{2+(-2)}{2}, \frac{-5+7}{2}\right) = M\left(\frac{0}{2}, \frac{2}{2}\right) = M(0, 1)$

83. $M\left(\frac{x_1+x_2}{2}, \frac{y_1+y_2}{2}\right) = M\left(\frac{-8+6}{2}, \frac{5+(-4)}{2}\right) = M\left(\frac{-2}{2}, \frac{1}{2}\right) = M\left(-1, \frac{1}{2}\right)$

85. $M\left(\frac{x_1+x_2}{2}, \frac{y_1+y_2}{2}\right) = M\left(\frac{0+\sqrt{5}}{2}, \frac{0+\sqrt{5}}{2}\right) = M\left(\frac{\sqrt{5}}{2}, \frac{\sqrt{5}}{2}\right)$

87. Let Q have coordinates (x, y):
$M\left(\frac{x_1+x_2}{2}, \frac{y_1+y_2}{2}\right) = (3, 5)$
$\frac{x_1+x_2}{2} = 3 \qquad \frac{y_1+y_2}{2} = 5$
$\frac{1+x}{2} = 3 \qquad \frac{4+y}{2} = 5$
$1+x = 6 \qquad 4+y = 10$
$x = 5 \qquad y = 6$
$Q(5, 6)$

89. Let Q have coordinates (x, y):
$M\left(\frac{x_1+x_2}{2}, \frac{y_1+y_2}{2}\right) = (5, 5)$
$\frac{x_1+x_2}{2} = 5 \qquad \frac{y_1+y_2}{2} = 5$
$\frac{5+x}{2} = 5 \qquad \frac{-5+y}{2} = 5$
$5+x = 10 \qquad -5+y = 10$
$x = 5 \qquad y = 15$
$Q(5, 15)$

SECTION 2.1

91. Let the points be identified as $A(13, -2)$, $B(9, -8)$ and $C(5, -2)$.

$AB = \sqrt{(x_2 - x_1)^2 + (y_2 - y_1)^2} = \sqrt{(13-9)^2 + (-2-(-8))^2} = \sqrt{16+36} = \sqrt{52} = 2\sqrt{13}$

$BC = \sqrt{(x_2 - x_1)^2 + (y_2 - y_1)^2} = \sqrt{(9-5)^2 + (-8-(-2))^2} = \sqrt{16+36} = \sqrt{52} = 2\sqrt{13}$

Since AB and BC have the same length, the triangle is isosceles.

93. $M = \left(\frac{2+6}{2}, \frac{4+10}{2}\right) = \left(\frac{8}{2}, \frac{14}{2}\right) = (4, 7)$; $N = \left(\frac{4+6}{2}, \frac{6+10}{2}\right) = \left(\frac{10}{2}, \frac{16}{2}\right) = (5, 8)$

$MN = \sqrt{(x_2 - x_1)^2 + (y_2 - y_1)^2} = \sqrt{(4-5)^2 + (7-8)^2} = \sqrt{1+1} = \sqrt{2}$

95. $M = \left(\frac{0+a}{2}, \frac{b+0}{2}\right) = \left(\frac{a}{2}, \frac{b}{2}\right)$; $L = \left(\frac{a}{2}, 0\right)$; $N = \left(0, \frac{b}{2}\right)$

Area of $AOB = \frac{1}{2} \cdot \text{base} \cdot \text{height} = \frac{1}{2}(OA)(OB) = \frac{1}{2}(a)(b) = \frac{1}{2}ab$

Area of $OLMN = \text{length} \cdot \text{width} = (OL)(ON) = \frac{a}{2} \cdot \frac{b}{2} = \frac{1}{4}ab = \frac{1}{2}(\text{Area of } AOB)$

97.
$y = 17500x + 225000$
$y = 17500(5) + 225000$
$y = 87500 + 225000$
$y = 312500$
The value will be \$312,500.

99.
$p = -\frac{1}{10}q + 170$
$150 = -\frac{1}{10}q + 170$
$\frac{1}{10}q = 20$
$q = 200$
200 scanners will be sold.

101.
$V = \frac{nv}{N}$
$60 = \frac{12v}{20}$
$1200 = 12v$
$100 = v$
The smaller gear is spinning at 100 rpm.

103.

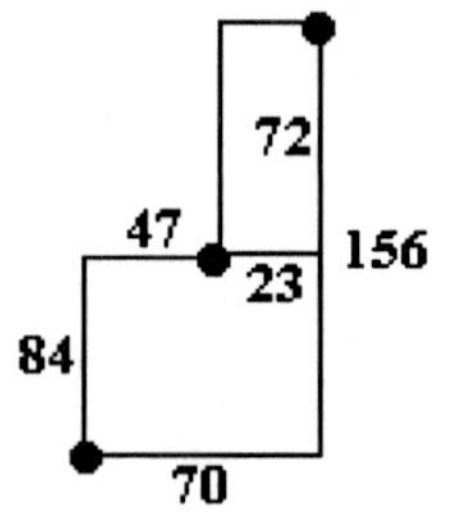

$d^2 = 70^2 + 156^2$
$d^2 = 4900 + 24{,}336$
$d^2 = 29{,}236$
$d = \sqrt{29{,}236}$
$d \approx 171$ miles

105. Answers may vary.

107. Answers may vary.

109. $[-3, 2)$

$(-2, 3]$

$[-3, 2) \cup (-2, 3]$

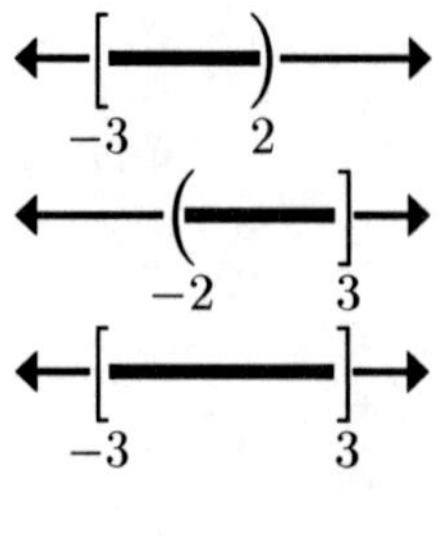

111. $[-3, -2)$

$(2, 3]$

$[-3, 2) \cap (-2, 3]$ no intersection

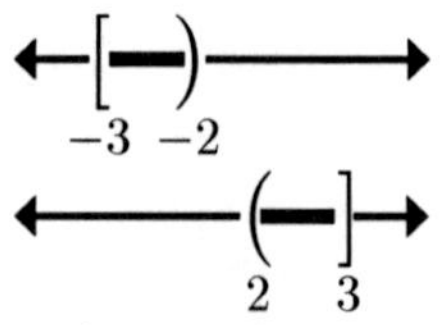

113.
$$\begin{aligned}\frac{3}{y+6} &= \frac{4}{y+4}\\ (y+6)(y+4)\cdot\frac{3}{y+6} &= (y+6)(y+4)\cdot\frac{4}{y+4}\\ 3(y+4) &= 4(y+6)\\ 3y+12 &= 4y+24\\ -12 &= y\end{aligned}$$

Exercises 2.2 (page 214)

1. divided

3. run

5. the change in

7. vertical

9. perpendicular

11. $m = \frac{y_2 - y_1}{x_2 - x_1} = \frac{2-(-1)}{2-(-1)} = \frac{3}{3} = 1$

13. $m = \frac{y_2 - y_1}{x_2 - x_1} = \frac{-2-3}{6-(-6)} = \frac{-5}{12} = -\frac{5}{12}$

15. $m = \frac{y_2 - y_1}{x_2 - x_1} = \frac{5-(-2)}{-1-3} = \frac{7}{-4} = -\frac{7}{4}$

17. $m = \frac{y_2 - y_1}{x_2 - x_1} = \frac{1-(-7)}{4-8} = \frac{8}{-4} = -2$

19. $m = \frac{y_2 - y_1}{x_2 - x_1} = \frac{-3-3}{-4-(-4)} = \frac{-6}{0} \Rightarrow$ und.

21. $m = \frac{y_2 - y_1}{x_2 - x_1} = \frac{\frac{7}{3}-\frac{2}{3}}{\frac{5}{2}-\frac{3}{2}} = \frac{\frac{5}{3}}{\frac{2}{2}} = \frac{\frac{5}{3}}{1} = \frac{5}{3}$

23. $m = \frac{y_2 - y_1}{x_2 - x_1} = \frac{a-c}{(b+c)-(a+b)} = \frac{a-c}{c-a} = -1$

25. $y = 3x + 2$

x	y
0	2
1	5

$m = \frac{y_2 - y_1}{x_2 - x_1} = \frac{5-2}{1-0} = \frac{3}{1} = 3$

27. $5x - 10y = 3$

x	y
0	$-\frac{3}{10}$
1	$\frac{1}{5}$

$m = \frac{y_2 - y_1}{x_2 - x_1} = \frac{\frac{1}{5} - \left(-\frac{3}{10}\right)}{1-0} = \frac{\frac{5}{10}}{1} = \frac{1}{2}$

29. $3(y+2) = 2x - 3$
$3y - 2x = -9$

x	y
0	-3
3	-1

$m = \frac{y_2 - y_1}{x_2 - x_1} = \frac{-1-(-3)}{3-0} = \frac{2}{3}$

31. $3(y+x) = 3(x-1)$
$3y = -3$
$y = -1$

x	y
0	-1
1	-1

$m = \frac{y_2 - y_1}{x_2 - x_1} = \frac{-1-(-1)}{1-0} = \frac{0}{1} = 0$

33. horizontal $\Rightarrow m = 0$

35. vertical $\Rightarrow m$ is undefined.

37. The slope is negative.

39. The slope is positive.

41. The slope is undefined.

43. $m_1 m_2 = 3\left(-\frac{1}{3}\right) = -1$
perpendicular

45. $m_1 = \sqrt{8} = 2\sqrt{2} = m_2$
parallel

47. $m_1 m_2 = -\sqrt{2}\left(\frac{\sqrt{2}}{2}\right) = -1$
perpendicular

49. $m_1 m_2 = -0.125(8) = -1$
perpendicular

51. $m_1 m_2 = ab^{-1}\left(-a^{-1}b\right) = -a^0 b^0 = -1$
perpendicular

For Exercises 53-57 use the slope of line through R and S calculated below:

$$m_{RS} = \frac{y_2 - y_1}{x_2 - x_1} = \frac{7-5}{2-(-3)} = \frac{2}{5}$$

53. $m_{PQ} = \frac{y_2 - y_1}{x_2 - x_1} = \frac{6-4}{7-2} = \frac{2}{5} = m_{RS} \Rightarrow$ parallel

55. $m_{PQ} = \frac{y_2 - y_1}{x_2 - x_1} = \frac{1-6}{-2-(-4)} = \frac{-5}{2} = -\frac{5}{2} \Rightarrow$ perpendicular

57. $m_{PQ} = \frac{y_2 - y_1}{x_2 - x_1} = \frac{6a-a}{3a-a} = \frac{5a}{2a} = \frac{5}{2} \Rightarrow$ neither

59. $m_{PQ} = \frac{y_2 - y_1}{x_2 - x_1} = \frac{9-7}{2-(-3)} = \frac{2}{5}$; $m_{RS} = \frac{y_2 - y_1}{x_2 - x_1} = \frac{-6-(-4)}{x-10} = \frac{-2}{x-10}$
$\frac{2}{5} \cdot \frac{-1}{-1} = \frac{-2}{-5}$; $x - 10 = -5 \Rightarrow \boxed{x = 5}$

61. $m_{PQ} = \frac{y_2 - y_1}{x_2 - x_1} = \frac{0-(-7)}{1-2} = \frac{7}{-1} = -7$; $m_{RS} = \frac{y_2 - y_1}{x_2 - x_1} = \frac{y-5}{-2-(-9)} = \frac{y-5}{7}$
$-7 = \frac{-7}{1}$; Perp. slope $= \frac{1}{7}$; $y - 5 = 1 \Rightarrow \boxed{y = 6}$

63. $m_{PQ} = \frac{y_2 - y_1}{x_2 - x_1} = \frac{9-8}{-6-(-2)} = \frac{1}{-4} = -\frac{1}{4}$
$m_{PR} = \frac{y_2 - y_1}{x_2 - x_1} = \frac{5-8}{2-(-2)} = \frac{-3}{4} = -\frac{3}{4} \Rightarrow$ not on same line

65. $m_{PQ} = \frac{y_2 - y_1}{x_2 - x_1} = \frac{0-a}{0-(-a)} = \frac{-a}{a} = -1$
$m_{PR} = \frac{y_2 - y_1}{x_2 - x_1} = \frac{-a-a}{a-(-a)} = \frac{-2a}{2a} = -1 \Rightarrow$ on same line

67. $m_{PQ} = \frac{y_2 - y_1}{x_2 - x_1} = \frac{-5-4}{2-5} = \frac{-9}{-3} = 3$
$m_{PR} = \frac{y_2 - y_1}{x_2 - x_1} = \frac{-3-4}{8-5} = \frac{-7}{3} = -\frac{7}{3}$
$m_{QR} = \frac{y_2 - y_1}{x_2 - x_1} = \frac{-3-(-5)}{8-2} = \frac{2}{6} = \frac{1}{3} \Rightarrow$ None are perpendicular.

69. $m_{PQ} = \frac{y_2 - y_1}{x_2 - x_1} = \frac{9-3}{1-1} = \frac{6}{0} \Rightarrow$ undefined $\Rightarrow$ vertical

$m_{PR} = \frac{y_2 - y_1}{x_2 - x_1} = \frac{3-3}{7-1} = \frac{0}{6} = 0 \Rightarrow$ horizontal

$m_{QR} = \frac{y_2 - y_1}{x_2 - x_1} = \frac{3-9}{7-1} = \frac{-6}{6} = -1 \Rightarrow PQ$ and PR are perpendicular.

71. $m_{PQ} = \frac{y_2 - y_1}{x_2 - x_1} = \frac{b-0}{a-0} = \frac{b}{a}$

$m_{PR} = \frac{y_2 - y_1}{x_2 - x_1} = \frac{a-0}{-b-0} = \frac{a}{-b} = -\frac{a}{b}$

$m_{QR} = \frac{y_2 - y_1}{x_2 - x_1} = \frac{a-b}{-b-a} = \frac{a-b}{-b-a} \Rightarrow PQ$ and PR are perpendicular.

73. $m_{AB} = \frac{y_2 - y_1}{x_2 - x_1} = \frac{4-(-1)}{-3-(-1)} = \frac{5}{-2} = -\frac{5}{2}$

$m_{AC} = \frac{y_2 - y_1}{x_2 - x_1} = \frac{1-(-1)}{4-(-1)} = \frac{2}{5} \Rightarrow AB$ and AC are perpendicular. $\Rightarrow$ right triangle

75. $m_{AB} = \frac{y_2 - y_1}{x_2 - x_1} = \frac{0-(-1)}{3-1} = \frac{1}{2}$; $d(A, B) = \sqrt{(1-3)^2 + (-1-0)^2} = \sqrt{5}$

$m_{BC} = \frac{y_2 - y_1}{x_2 - x_1} = \frac{2-0}{2-3} = \frac{2}{-1} = -2$; $d(B, C) = \sqrt{(3-2)^2 + (0-2)^2} = \sqrt{5}$

$m_{CD} = \frac{y_2 - y_1}{x_2 - x_1} = \frac{1-2}{0-2} = \frac{-1}{-2} = \frac{1}{2}$; $d(C, D) = \sqrt{(2-0)^2 + (2-1)^2} = \sqrt{5}$

$m_{DA} = \frac{y_2 - y_1}{x_2 - x_1} = \frac{1-(-1)}{0-1} = \frac{2}{-1} = -2$; $d(D, A) = \sqrt{(1-0)^2 + (-1-1)^2} = \sqrt{5}$

Adjacent sides are perpendicular and congruent, so the figure is a square.

77. $m_{AB} = \frac{y_2 - y_1}{x_2 - x_1} = \frac{3-(-2)}{3-(-2)} = \frac{5}{5} = 1$; $m_{BC} = \frac{y_2 - y_1}{x_2 - x_1} = \frac{6-3}{2-3} = \frac{3}{-1} = -3$

$m_{CD} = \frac{y_2 - y_1}{x_2 - x_1} = \frac{1-6}{-3-2} = \frac{-5}{-5} = 1$; $m_{DA} = \frac{y_2 - y_1}{x_2 - x_1} = \frac{1-(-2)}{-3-(-2)} = \frac{3}{-1} = -3$

Opposite sides are parallel, so the figure is a parallelogram.

79. $M\left(\frac{5+7}{2}, \frac{9+5}{2}\right) = M\left(\frac{12}{2}, \frac{14}{2}\right) = M(6, 7)$; $N\left(\frac{1+7}{2}, \frac{3+5}{2}\right) = N\left(\frac{8}{2}, \frac{8}{2}\right) = N(4, 4)$

$m_{MN} = \frac{y_2 - y_1}{x_2 - x_1} = \frac{4-7}{4-6} = \frac{-3}{-2} = \frac{3}{2}$; $m_{AC} = \frac{y_2 - y_1}{x_2 - x_1} = \frac{9-3}{5-1} = \frac{6}{4} = \frac{3}{2} \Rightarrow MN \parallel AC$

81. $m = \frac{y_2 - y_1}{x_2 - x_1} = \frac{26-12}{5-1} = \frac{14}{4} = 3.5$

The rate of growth was 3.5 students per year.

83. $m = \frac{y_2 - y_1}{x_2 - x_1} = \frac{6700-2200}{10-3}$

$= \frac{4500}{7} \approx 642.86$

The cost decreased about \$642.86 per year.

85. $\dfrac{\Delta T}{\Delta t}$ = the hourly rate of change of temperature.
(Let $t = x$ and $T = y$.)

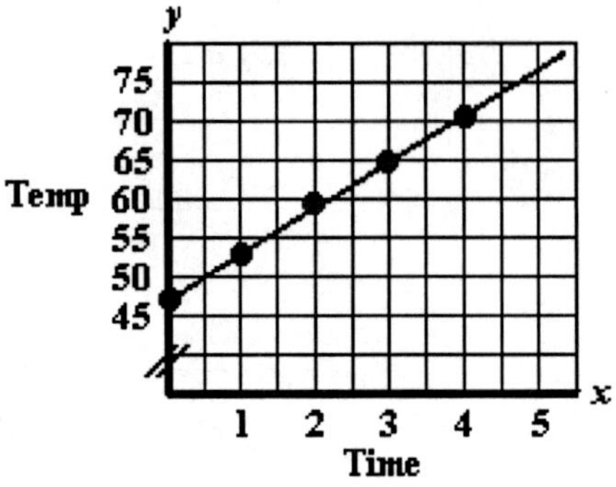

87. $D = 590t$; The slope is the speed of the plane.

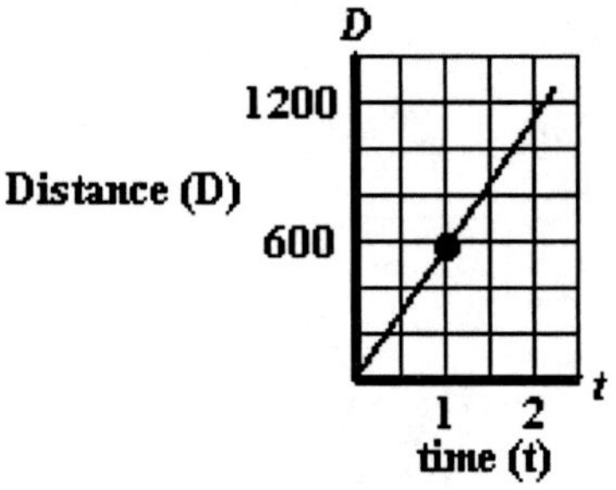

89. **Answers may vary.**

91.
$$\begin{aligned} 3x + 7y &= 21 \\ 7y &= -3x + 21 \\ \frac{7y}{7} &= \frac{-3x}{7} + \frac{21}{7} \\ y &= -\frac{3}{7}x + 3 \end{aligned}$$

93.
$$\begin{aligned} \frac{x}{5} + \frac{y}{2} &= 1 \\ 2\left(\frac{x}{5} + \frac{y}{2}\right) &= 2(1) \\ \frac{2}{5}x + y &= 2 \\ y &= -\frac{2}{5}x + 2 \end{aligned}$$

95. $6p^2 + p - 12 = (2p + 3)(3p - 4)$

97.
$$\begin{aligned} mp + mq + np + nq &= m(p + q) + n(p + q) \\ &= (p + q)(m + n) \end{aligned}$$

Exercises 2.3 (page 229)

1. $y - y_1 = m(x - x_1)$

3. slope-intercept

5. $-\dfrac{A}{B}$

7.
$$\begin{aligned} y - y_1 &= m(x - x_1) \\ y - 4 &= 2(x - 2) \\ y - 4 &= 2x - 4 \\ -2x + y &= 0 \\ 2x - y &= 0 \end{aligned}$$

9.
$$\begin{aligned} y - y_1 &= m(x - x_1) \\ y - \frac{1}{2} &= 2\left(x + \frac{3}{2}\right) \\ y - \frac{1}{2} &= 2x + 3 \\ 2y - 1 &= 4x + 6 \\ -4x + 2y &= 7 \\ 4x - 2y &= -7 \end{aligned}$$

11.
$$\begin{aligned} y - y_1 &= m(x - x_1) \\ y - 1 &= \frac{2}{5}(x + 1) \\ 5(y - 1) &= 2(x + 1) \\ 5y - 5 &= 2x + 2 \\ -2x + 5y &= 7 \\ 2x - 5y &= -7 \end{aligned}$$

SECTION 2.3

13.
$$\begin{aligned} y - y_1 &= m(x - x_1) \\ y + 3 &= 0(x + 6) \\ y + 3 &= 0 \\ y &= -3 \end{aligned}$$

15. m is und $\Rightarrow$ vertical

$x =$ constant

$x = -6$

17.
$$\begin{aligned} y - y_1 &= m(x - x_1) \\ y - 0 &= \pi(x - \pi) \\ y &= \pi x - \pi^2 \\ -\pi x + y &= -\pi^2 \\ \pi x - y &= \pi^2 \end{aligned}$$

19. From the graph, $m = \frac{2}{3}$ and the line passes through $(2, 5)$.
$$\begin{aligned} y - y_1 &= m(x - x_1) \\ y - 5 &= \frac{2}{3}(x - 2) \\ 3(y - 5) &= 3 \cdot \frac{2}{3}(x - 2) \\ 3y - 15 &= 2(x - 2) \\ 3y - 15 &= 2x - 4 \\ -2x + 3y &= 11 \\ 2x - 3y &= -11 \end{aligned}$$

21. $m = \dfrac{y_2 - y_1}{x_2 - x_1} = \dfrac{4 - 0}{4 - 0} = \dfrac{4}{4} = 1$
$$\begin{aligned} y - y_1 &= m(x - x_1) \\ y - 0 &= 1(x - 0) \\ y &= x \end{aligned}$$

23. $m = \dfrac{y_2 - y_1}{x_2 - x_1} = \dfrac{-3 - 4}{0 - 3} = \dfrac{-7}{-3} = \dfrac{7}{3}$
$$\begin{aligned} y - y_1 &= m(x - x_1) \\ y + 3 &= \frac{7}{3}(x - 0) \\ y &= \frac{7}{3}x - 3 \end{aligned}$$

25. From the graph, $m = -\frac{9}{5}$ and the line passes through $(-2, 4)$
$$\begin{aligned} y - y_1 &= m(x - x_1) \\ y - 4 &= -\frac{9}{5}(x + 2) \\ y - 4 &= -\frac{9}{5}x - \frac{18}{5} \\ y &= -\frac{9}{5}x - \frac{18}{5} + 4 \\ y &= -\frac{9}{5}x + \frac{2}{5} \end{aligned}$$

27.
$$\begin{aligned} y &= mx + b \\ y &= 3x - 2 \end{aligned}$$

29.
$$\begin{aligned} y &= mx + b \\ y &= 5x - \frac{1}{5} \end{aligned}$$

31.
$$\begin{aligned} y &= mx + b \\ y &= ax + \frac{1}{a} \end{aligned}$$

33.
$$\begin{aligned} y &= mx + b \\ y &= ax + a \end{aligned}$$

35.
$$\begin{aligned} y &= mx + b \\ 0 &= \frac{3}{2}(0) + b \\ 0 &= b \end{aligned}$$
$$\begin{aligned} y &= mx + b \\ y &= \frac{3}{2}x + 0 \\ 2y &= 3x \\ -3x + 2y &= 0 \\ 3x - 2y &= 0 \end{aligned}$$

37.
$$\begin{aligned} y &= mx + b \\ 5 &= -3(-3) + b \\ 5 &= 9 + b \\ -4 &= b \end{aligned}$$
$$\begin{aligned} y &= mx + b \\ y &= -3x - 4 \\ 3x + y &= -4 \end{aligned}$$

39.
$$y = mx + b$$
$$\sqrt{2} = \sqrt{2}(0) + b$$
$$\sqrt{2} = b$$
$$y = mx + b$$
$$y = \sqrt{2}x + \sqrt{2}$$
$$-\sqrt{2}x + y = \sqrt{2}$$
$$\sqrt{2}x - y = -\sqrt{2}$$

41.
$$x - y = 1$$
$$y = x - 1$$
$$m = 1,\ (0, -1)$$

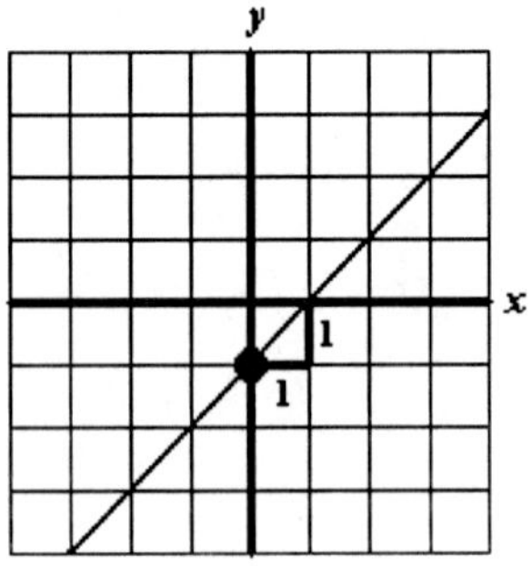

43.
$$x = \frac{3}{2}y - 3$$
$$2x = 3y - 6$$
$$-3y = -2x - 6$$
$$y = \frac{2}{3}x + 2$$
$$m = \frac{2}{3},\ (0, 2)$$

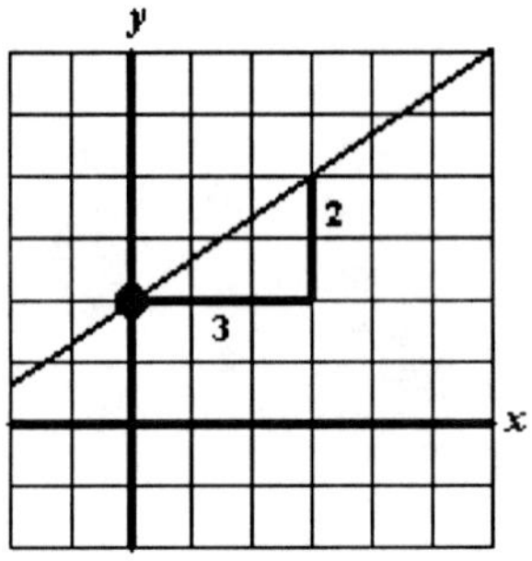

45.
$$3(y - 4) = -2(x - 3)$$
$$3y - 12 = -2x + 6$$
$$3y = -2x + 18$$
$$y = -\frac{2}{3}x + 6$$
$$m = -\frac{2}{3},\ (0, 6)$$

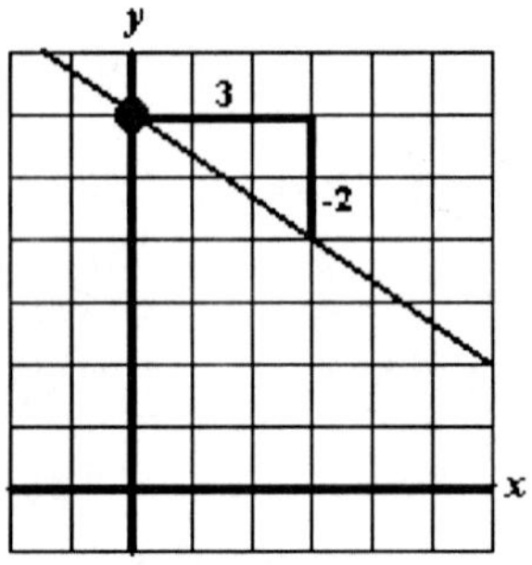

47.
$$3x - 2y = 8$$
$$-2y = -3x + 8$$
$$y = \frac{3}{2}x - 4$$
$$m = \frac{3}{2},\ (0, -4)$$

49.
$$-2(x + 3y) = 5$$
$$-2x - 6y = 5$$
$$-6y = 2x + 5$$
$$y = -\frac{1}{3}x - \frac{5}{6}$$
$$m = -\frac{1}{3},\ \left(0, -\frac{5}{6}\right)$$

51.
$$x = \frac{2y - 4}{7}$$
$$7x = 2y - 4$$
$$-2y = -7x - 4$$
$$y = \frac{7}{2}x + 2$$
$$m = \frac{7}{2},\ (0, 2)$$

53.
$$y = 3x + 4 \qquad y = 3x - 7$$
$$m = 3 \qquad m = 3$$
The lines are parallel.

55.
$$x + y = 2 \qquad y = x + 5$$
$$y = -x + 2 \qquad m = 1$$
$$m = -1$$
The lines are perpendicular.

57.
$$y = 3x + 7 \qquad 2y = 6x - 9$$
$$m = 3 \qquad y = 3x - \frac{9}{2}$$
$$m = 3$$
The lines are parallel.

59.
$$x = 3y + 4 \qquad y = -3x + 7$$
$$-3y = -x + 4 \qquad m = -3$$
$$y = \frac{1}{3}x - \frac{4}{3}$$
$$m = \frac{1}{3}$$
The lines are perpendicular.

SECTION 2.3

61. $y = 3$ $\quad x = 4$

horizontal $\quad$ vertical

The lines are perpendicular.

63. $x = \dfrac{y-2}{3}$

$3x = y - 2$

$-y = -3x - 2$

$y = 3x + 2$

$m = 3$

$3(y-3) + x = 0$

$3y - 9 + x = 0$

$3y = -x + 9$

$y = -\frac{1}{3}x + 3$

$m = -\frac{1}{3}$

The lines are perpendicular.

65. $y = 4x - 7$

$m = 4$

Use $m = 4$.

$y - y_1 = m(x - x_1)$

$y - 0 = 4(x - 0)$

$\boxed{y = 4x}$

67. $4x - y = 7$

$-y = -4x + 7$

$y = 4x - 7$

$m = 4$

Use $m = 4$.

$y - y_1 = m(x - x_1)$

$y - 5 = 4(x - 2)$

$y - 5 = 4x - 8$

$\boxed{y = 4x - 3}$

69. $x = \frac{5}{4}y - 2$

$4x = 5y - 8$

$-5y = -4x - 8$

$y = \frac{4}{5}x + \frac{8}{5}$

$m = \frac{4}{5}$

Use $m = \frac{4}{5}$.

$y - y_1 = m(x - x_1)$

$y + 2 = \frac{4}{5}(x - 4)$

$y + 2 = \frac{4}{5}x - \frac{16}{5}$

$\boxed{y = \frac{4}{5}x - \frac{26}{5}}$

71. $y = 4x - 7$

$m = 4$

Use $m = -\frac{1}{4}$.

$y - y_1 = m(x - x_1)$

$y - 0 = -\frac{1}{4}(x - 0)$

$\boxed{y = -\frac{1}{4}x}$

73. $4x - y = 7$

$-y = -4x + 7$

$y = 4x - 7$

$m = 4$

Use $m = -\frac{1}{4}$.

$y - y_1 = m(x - x_1)$

$y - 5 = -\frac{1}{4}(x - 2)$

$y - 5 = -\frac{1}{4}x + \frac{1}{2}$

$\boxed{y = -\frac{1}{4}x + \frac{11}{2}}$

75. $x = \frac{5}{4}y - 2$

$4x = 5y - 8$

$-5y = -4x - 8$

$y = \frac{4}{5}x + \frac{8}{5}$

$m = \frac{4}{5}$

Use $m = -\frac{5}{4}$.

$y - y_1 = m(x - x_1)$

$y + 2 = -\frac{5}{4}(x - 4)$

$y + 2 = -\frac{5}{4}x + 5$

$\boxed{y = -\frac{5}{4}x + 3}$

77. $4x + 5y = 20 \Rightarrow A = 4, B = 5, C = 20$

$m = -\dfrac{A}{B} = -\dfrac{4}{5}$

$b = \dfrac{C}{B} = \dfrac{20}{5} = 4 \Rightarrow (0, 4)$

79. $2x + 3y = 12 \Rightarrow A = 2, B = 3, C = 12$

$m = -\dfrac{A}{B} = -\dfrac{2}{3}$

$b = \dfrac{C}{B} = \dfrac{12}{3} = 4 \Rightarrow (0, 4)$

81. Since $y = 3$ is the equation of a horizontal line, any perpendicular line will be vertical.
Find the midpoint:
$x = \frac{2 + (-6)}{2} = -2; y = \frac{4 + 10}{2} = 7$
The vertical line through $(-2, 7)$ is $x = -2$.

83. Since $x = 3$ is the equation of a vertical line, any parallel line will be vertical.
Find the midpoint:
$x = \frac{2 + 8}{2} = 5; y = \frac{-4 + 12}{2} = 4$
The vertical line through $(5, 4)$ is $x = 5$.

85. Let $x =$ the number of years the truck has been owned and let $y =$ the value of the truck. Then two points on the line are given: $(0, 24300)$ and $(7, 1900)$.

$$m = \frac{24300 - 1900}{0 - 7} = \frac{22400}{-7} = -3200$$

$$\begin{aligned} y - y_1 &= m(x - x_1) \\ y - 24300 &= -3200(x - 0) \\ y - 24300 &= -3200x \\ y &= -3200x + 24300 \end{aligned}$$

87. Let $x =$ the number of years the building has been owned and let $y =$ the value of the building. Then two points on the line are given: $(0, 475000)$ and $(10, 950000)$.

$$\begin{aligned} m &= \frac{950000 - 475000}{10 - 0} = \frac{475000}{10} \\ &= 47500 \end{aligned}$$

$$\begin{aligned} y - y_1 &= m(x - x_1) \\ y - 475000 &= 47500(x - 0) \\ y - 475000 &= 47500x \\ y &= 47500x + 475000 \end{aligned}$$

89. Let $x =$ the number of years the TV has been owned and let $y =$ the value of the TV. Then two points on the line are given: $(0, 1900)$ and $(3, 1190)$.

$$m = \frac{1900 - 1190}{0 - 3} = \frac{710}{-3} = -\frac{710}{3}$$

$$\begin{aligned} y - y_1 &= m(x - x_1) \\ y - 1900 &= -\frac{710}{3}(x - 0) \\ y - 1900 &= -\frac{710}{3}x \\ y &= -\frac{710}{3}x + 1900 \end{aligned}$$

91. Let $x =$ the number of years the copier has been owned and let $y =$ the value of the copier. Then one point on the line is given: $(0, 1050)$. Since the copier depreciates by \$120 per year, $m = -120$.

$$\begin{aligned} y - y_1 &= m(x - x_1) \\ y - 1050 &= -120(x - 0) \\ y - 1050 &= -120x \\ y &= -120x + 1050 \end{aligned}$$

Let $x = 8$ and find the value of y:

$$\begin{aligned} y &= -120x + 1050 \\ &= -120(8) + 1050 = 90 \end{aligned}$$

The salvage value will be \$90.

93. Let $x =$ the number of years the table has been owned and let $y =$ the value of the table. Then one point on the line is given: $(2, 450)$. Since the table appreciates by \$40 per year, $m = 40$.

$$\begin{aligned} y - y_1 &= m(x - x_1) \\ y - 450 &= 40(x - 2) \\ y - 450 &= 40x - 80 \\ y &= 40x + 370 \end{aligned}$$

Let $x = 13$ and find the value of y:

$$\begin{aligned} y &= 40x + 370 \\ &= 40(13) + 370 = 890 \end{aligned}$$

The value will be \$890.

95. Let $x =$ the number of years the cottage has been owned and let $y =$ the value of the cottage. Then one point on the line is given: $(3, 47700)$. Since the cottage appreciates by \$3500 per year, $m = 3500$.

$$y - y_1 = m(x - x_1)$$
$$y - 47700 = 3500(x - 3)$$
$$y - 47700 = 3500x - 10500$$
$$y = 3500x + 37200$$

Let $x = 0$ and find the value of y:

$$y = 3500x + 37200$$
$$= 3500(0) + 37200 = 37200$$

The purchase price was \$37,200.

97. Let $x =$ the hours of labor and let $y =$ the labor charge. Then $m =$ the hourly charge.

$$y = mx \qquad y = 46x$$
$$69 = m(1.5) \qquad y = 46(5) = 230$$
$$46 = m$$

The charge will be \$230.

99. Let $x =$ the number of fires and let $y =$ the population. Then two points on the line are given: $(300, 57000)$ and $(325, 59000)$.

$$m = \frac{59000 - 57000}{325 - 300} = \frac{2000}{25} = 80$$
$$y - y_1 = m(x - x_1)$$
$$y - 57000 = 80(x - 300)$$
$$y - 57000 = 80x - 24000$$
$$y = 80x + 33000$$

Let $y = 100000$ and find the value of x:

$$y = 80x + 33000$$
$$100000 = 80x + 33000$$
$$67000 = 80x$$

$837.5 = x \Rightarrow$ There will be about 838 fires when the population is 100,000.

101. Let F replace x and C replace y. Then two points on the line are given: $(32, 0)$ and $(212, 100)$.

$$m = \frac{100 - 0}{212 - 32} = \frac{100}{180} = \frac{5}{9}$$
$$C - C_1 = m(F - F_1)$$
$$C - 0 = \frac{5}{9}(F - 32)$$
$$C = \frac{5}{9}(F - 32)$$

103. Let $y =$ the percent who smoke and let $x =$ the # of years since 1974. Two points are given: $(0, 47)$ and $(20, 29)$.

$$m = \frac{29 - 47}{20 - 0} = \frac{-18}{20} = -\frac{9}{10}$$
$$y - y_1 = m(x - x_1)$$
$$y - 47 = -\frac{9}{10}(x - 0)$$
$$\boxed{y = -\frac{9}{10}x + 47}$$

Let $x = 40$:

$$y = -\frac{9}{10}(40) + 47 = -36 + 47 = 11$$

11% will smoke in 2014.

105. Two points on the line are given: $(0, 37.5)$ and $(2, 45)$.

$$m = \frac{45 - 37.5}{2 - 0} = \frac{7.5}{2} = 3.75$$
$$y - y_1 = m(x - x_1)$$
$$y - 37.5 = 3.75(x - 0)$$
$$y = 3.75x + 37.5$$

Let $x = 4$ and find the value of y:

$$y = 3.75x + 37.5$$
$$= 3.75(4) + 37.5$$
$$= 15 + 37.5 = 52.5$$

The price will be \$52.50 in the year 2014.

107. The equation describing the production is $y = -70x + 1900$, where x represents the number of years and y is the level of production. Let $x = 3\frac{1}{2} = \frac{7}{2}$.

$y = -70x + 1900 = -70\left(\frac{7}{2}\right) + 1900 = 1655$; The production will be 1655 barrels per day.

109. a.

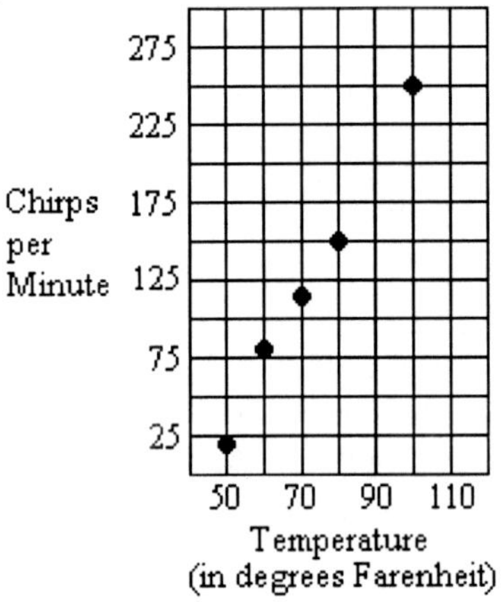

b. Use (50, 20) and (100, 250) for the regression line.

$$m = \frac{250 - 20}{100 - 50} = \frac{230}{50} = \frac{23}{5}$$
$$y - y_1 = m(x - x_1)$$
$$y - 20 = \frac{23}{5}(x - 50)$$
$$y - 20 = \frac{23}{5}x - 230$$
$$y = \frac{23}{5}x - 210$$

c. $y = \frac{23}{5}(90) - 210 = 204$

The rate will be about 204 chirps per minute.

111. $y = 4.44x - 196.62$

113. Answers may vary.

115.
$$m = \frac{b - 0}{0 - a} = -\frac{b}{a}$$
$$y - y_1 = m(x - x_1)$$
$$y - b = -\frac{b}{a}(x - 0)$$
$$y - b = -\frac{b}{a}x$$
$$ay - ab = -bx$$
$$bx + ay = ab$$
$$\frac{bx + ay}{ab} = \frac{ab}{ab}$$
$$\frac{x}{a} + \frac{y}{b} = 1$$

117-121. Answers may vary.

123. $x^7x^3x^{-5} = x^{7+3-5} = x^5$

125. $\left(\frac{81}{25}\right)^{-3/2} = \left(\frac{25}{81}\right)^{3/2} = \left[\frac{5}{9}\right]^3 = \frac{125}{729}$

127. $\sqrt{27} - 2\sqrt{12} = \sqrt{9}\sqrt{3} - 2\sqrt{4}\sqrt{3} = 3\sqrt{3} - 2(2)\sqrt{3} = 3\sqrt{3} - 4\sqrt{3} = -\sqrt{3}$

129. $\frac{5}{\sqrt{x} + 2} = \frac{5}{\sqrt{x} + 2} \cdot \frac{\sqrt{x} - 2}{\sqrt{x} - 2} = \frac{5(\sqrt{x} - 2)}{(\sqrt{x})^2 - 2^2} = \frac{5(\sqrt{x} - 2)}{x - 4}$

Exercises 2.4 (page 250)

1. x-intercept **3.** axis of symmetry **5.** x-axis **7.** circle, center

SECTION 2.4

9. $x^2 + y^2 = r^2$

11.
$y = x^2 - 4$
$0 = (x+2)(x-2)$
$x = -2,\ x = 2$
x-int: $(-2, 0), (2, 0)$

$y = x^2 - 4$
$y = 0^2 - 4$
$y = -4$
y-int: $(0, -4)$

13.
$y = 4x^2 - 2x$
$0 = 2x(2x - 1)$
$x = 0,\ x = \frac{1}{2}$
x-int: $(0, 0), \left(\frac{1}{2}, 0\right)$

$y = 4x^2 - 2x$
$y = 4(0)^2 - 2(0)$
$y = 0$
y-int: $(0, 0)$

15.
$y = x^2 - 4x - 5$
$0 = (x+1)(x-5)$
$x = -1,\ x = 5$
x-int: $(-1, 0), (5, 0)$

$y = x^2 - 4x - 5$
$y = 0^2 - 4(0) - 5$
$y = -5$
y-int: $(0, -5)$

17.
$y = x^2 + x - 2$
$0 = (x+2)(x-1)$
$x = -2,\ x = 1$
x-int: $(-2, 0), (1, 0)$

$y = x^2 + x - 2$
$y = 0^2 + 0 - 2$
$y = -2$
y-int: $(0, -2)$

19.
$y = x^3 - 9x$
$0 = x\left(x^2 - 9\right)$
$0 = x(x+3)(x-3)$
$x = 0,\ x = -3,\ x = 3$
x-int: $(0, 0), (-3, 0), (3, 0)$

$y = x^3 - 9x$
$y = 0^3 - 9(0)$
$y = 0$
y-int: $(0, 0)$

21.
$y = x^4 - 1$
$0 = \left(x^2 + 1\right)\left(x^2 - 1\right)$
$0 = \left(x^2 + 1\right)(x+1)(x-1)$
$x = -1,\ x = 1$
x-int: $(-1, 0), (1, 0)$

$y = x^4 - 1$
$y = 0^4 - 1$
$y = -1$
y-int: $(0, -1)$

23. $y = x^2$
x-int: $(0, 0)$
y-int: $(0, 0)$

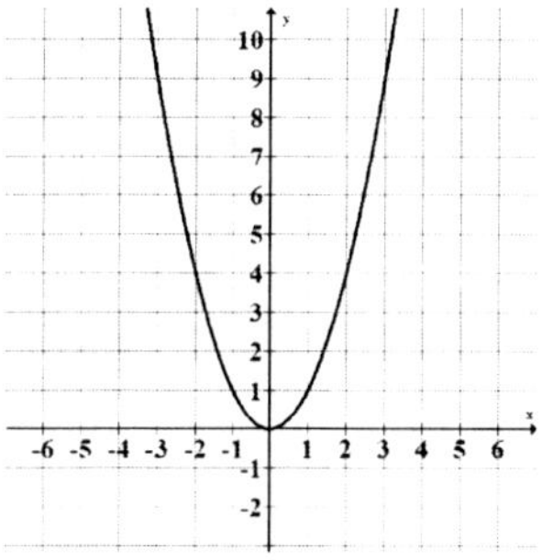

25. $y = -x^2 + 2$
x-int: $\left(\sqrt{2}, 0\right), \left(-\sqrt{2}, 0\right)$
y-int: $(0, 2)$

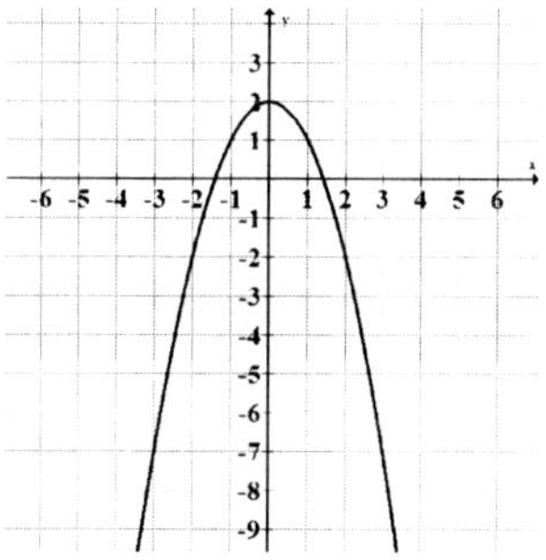

27. $y = x^2 - 4x$
x-int: $(0, 0), (4, 0)$
y-int: $(0, 0)$

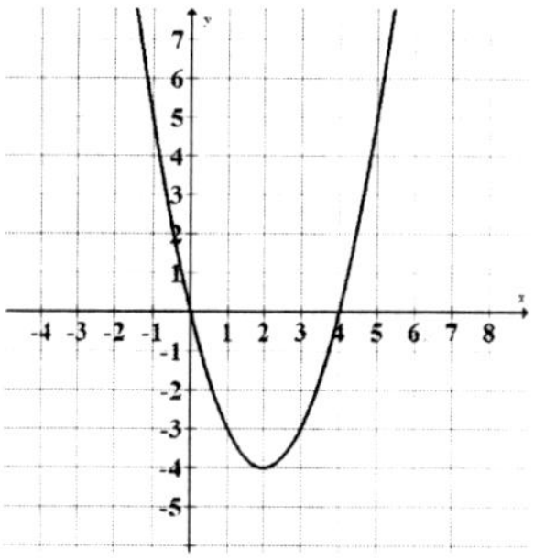

29. $y = \dfrac{1}{2}x^2 - 2x$
x-int: $(0, 0), (4, 0)$
y-int: $(0, 0)$

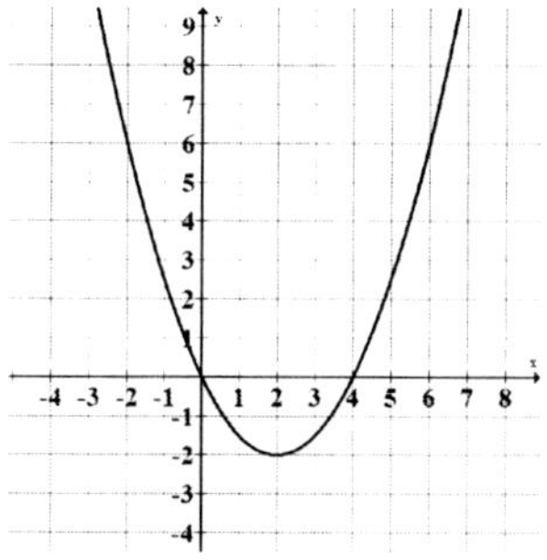

SECTION 2.4

31. $y = x^2 + 2$

x-axis	y-axis	origin
$-y = x^2 + 2$	$y = (-x)^2 + 2$	$-y = (-x)^2 + 2$
not equivalent: no symmetry	$y = x^2 + 2$	$-y = x^2 + 2$
	equivalent: $\boxed{\text{symmetry}}$	not equivalent: no symmetry

33. $y^2 + 1 = x$

x-axis	y-axis	origin
$(-y)^2 + 1 = x$	$y^2 + 1 = -x$	$(-y)^2 + 1 = -x$
$y^2 + 1 = x$	not equivalent: no symmetry	$y^2 + 1 = -x$
equivalent: $\boxed{\text{symmetry}}$		not equivalent: no symmetry

35. $y^2 = x^2$

x-axis	y-axis	origin
$(-y)^2 = x^2$	$y^2 = (-x)^2$	$(-y)^2 = (-x)^2$
$y^2 = x^2$	$y^2 = x^2$	$y^2 = x^2$
equivalent: $\boxed{\text{symmetry}}$	equivalent: $\boxed{\text{symmetry}}$	equivalent: $\boxed{\text{symmetry}}$

37. $y = 3x^2 + 7$

x-axis	y-axis	origin
$-y = 3x^2 + 7$	$y = 3(-x)^2 + 7$	$-y = 3(-x)^2 + 7$
not equivalent: no symmetry	$y = 3x^2 + 7$	$-y = 3x^2 + 7$
	equivalent: $\boxed{\text{symmetry}}$	not equivalent: no symmetry

39. $y = 3x^3 + 7$

x-axis	y-axis	origin
$-y = 3x^3 + 7$	$y = 3(-x)^3 + 7$	$-y = 3(-x)^3 + 7$
not equivalent: no symmetry	$y = -3x^3 + 7$	$-y = -3x^3 + 7$
	not equivalent: no symmetry	$y = 3x^3 - 7$
		not equivalent: no symmetry

41. $y^2 = 3x$

x-axis	y-axis	origin
$(-y)^2 = 3x$	$y^2 = 3(-x)$	$(-y)^2 = 3(-x)$
$y^2 = 3x$	$y^2 = -3x$	$y^2 = -3x$
equivalent: $\boxed{\text{symmetry}}$	not equivalent: no symmetry	not equivalent: no symmetry

SECTION 2.4

43.

$$y = |x|$$

x-axis	y-axis	origin
$-y = \|x\|$	$y = \|-x\|$	$-y = \|-x\|$
not equivalent: no symmetry	$y = \|-1\|\|x\|$	$-y = \|-1\|\|x\|$
	$y = \|x\|$	$-y = \|x\|$
	equivalent: **symmetry**	not equivalent: no symmetry

45.

$$|y| = x$$

x-axis	y-axis	origin
$\|-y\| = x$	$\|y\| = -x$	$\|-y\| = -x$
$\|-1\|\|y\| = x$	not equivalent: no symmetry	$\|-1\|\|y\| = -x$
$\|y\| = x$		$\|y\| = -x$
equivalent: **symmetry**		not equivalent: no symmetry

47. $y = x^2 + 4x$
x-int: $(0, 0), (-4, 0)$
y-int: $(0, 0)$
symmetry: none

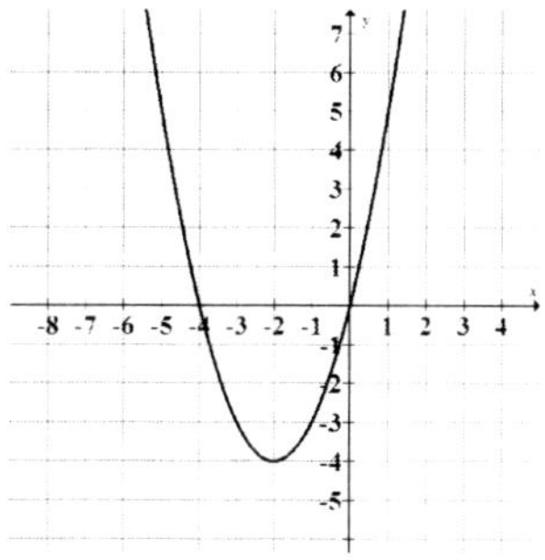

49. $y = x^3$
x-int: $(0, 0)$
y-int: $(0, 0)$
symmetry: origin

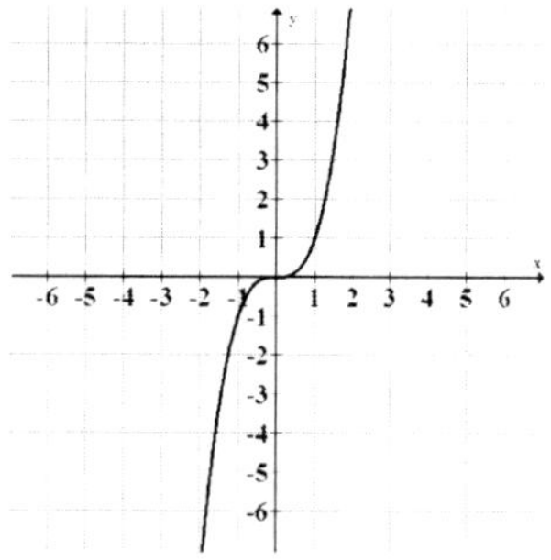

51. $y = |x - 2|$
x-int: $(2, 0)$
y-int: $(0, 2)$
symmetry: none

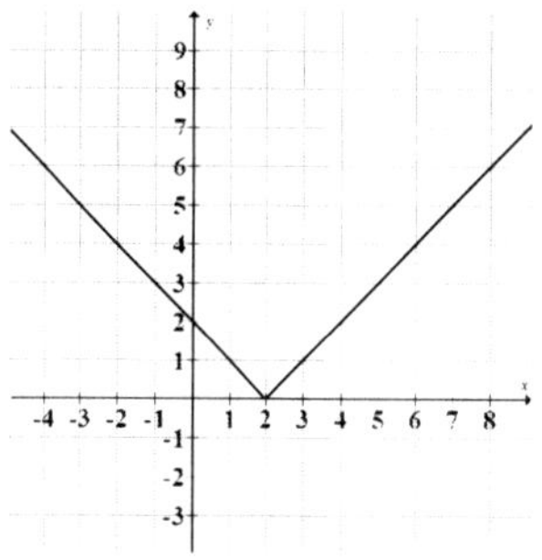

53. $y = 3 - |x|$
x-int: $(-3, 0), (3, 0)$
y-int: $(0, 3)$
symmetry: y-axis

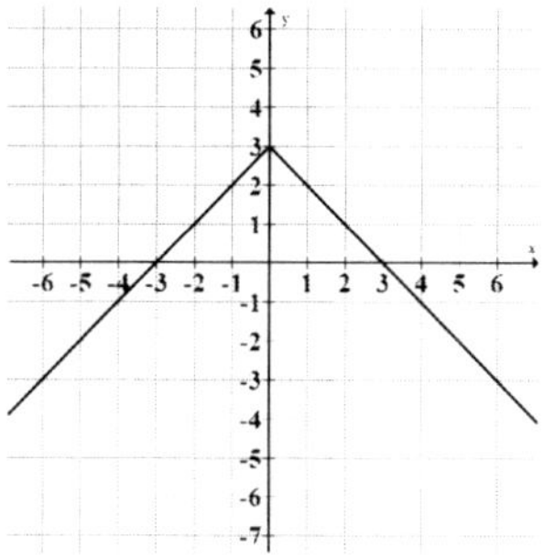

55. $y^2 = -x$
x-int: $(0, 0)$
y-int: $(0, 0)$
symmetry: x-axis

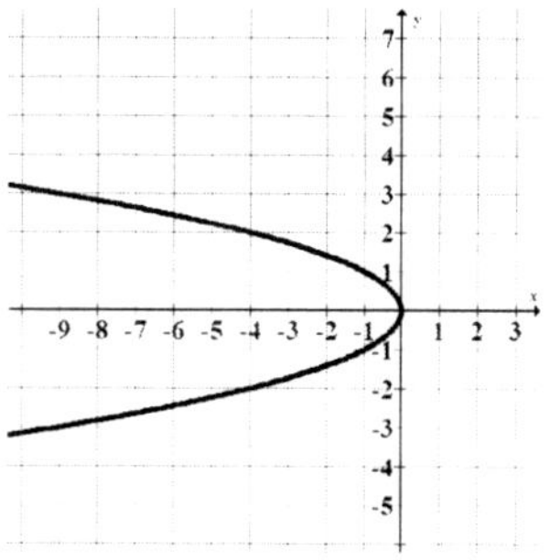

57. $y^2 = 9x$
x-int: $(0, 0)$
y-int: $(0, 0)$
symmetry: x-axis

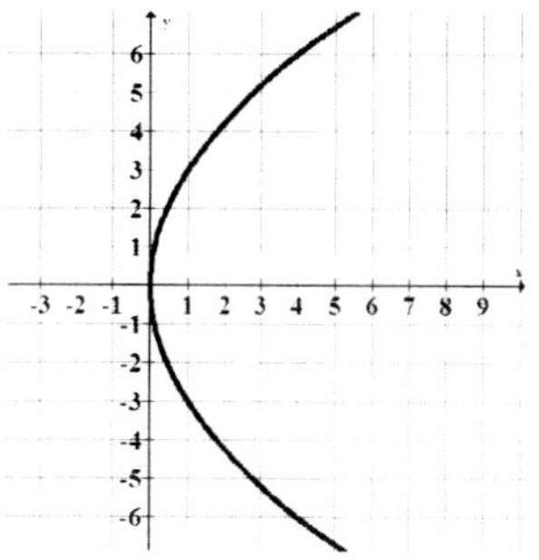

59. $y = \sqrt{x} - 1$
x-int: $(1, 0)$
y-int: $(0, -1)$
symmetry: none

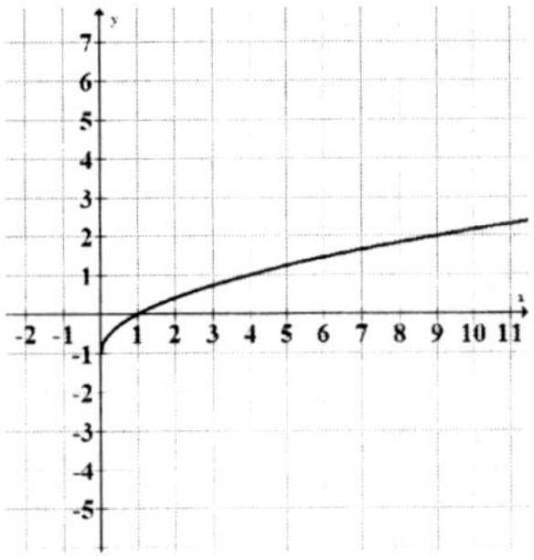

61. $xy = 4$
x-int: none
y-int: none
symmetry: origin

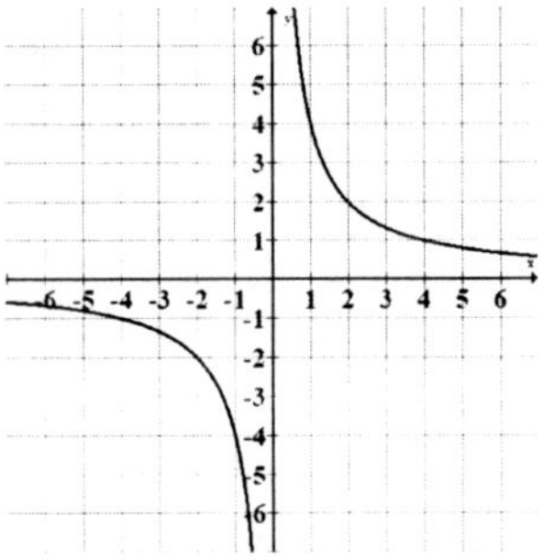

63.
$$x^2 + y^2 = 100$$
$$(x-0)^2 + (y-0)^2 = 10^2$$
C: $(0, 0)$; $r = 10$

65.
$$x^2 + (y-5)^2 = 49$$
$$(x-0)^2 + (y-5)^2 = 7^2$$
C: $(0, 5)$; $r = 7$

67.
$$(x+6)^2 + y^2 = \tfrac{1}{4}$$
$$(x-(-6))^2 + (y-0)^2 = \left(\tfrac{1}{2}\right)^2$$
C: $(-6, 0)$; $r = \frac{1}{2}$

69.
$$(x-4)^2 + (y-1)^2 = 9$$
$$(x-4)^2 + (y-1)^2 = 3^2$$
C: $(4, 1)$; $r = 3$

71.
$$\left(x - \tfrac{1}{4}\right)^2 + (y+2)^2 = 45$$
$$\left(x - \tfrac{1}{4}\right)^2 + (y-(-2))^2 = \left(\sqrt{45}\right)^2$$
$$\left(x - \tfrac{1}{4}\right)^2 + (y-(-2))^2 = \left(3\sqrt{5}\right)^2$$
C: $\left(\frac{1}{4}, -2\right)$; $r = 3\sqrt{5}$

73.
$$(x-0)^2 + (y-0)^2 = 5^2$$
$$x^2 + y^2 = 25$$

75.
$$(x-0)^2 + (y-(-6))^2 = 6^2$$
$$x^2 + (y+6)^2 = 36$$

77.
$$(x-8)^2 + (y-0)^2 = \left(\tfrac{1}{5}\right)^2$$
$$(x-8)^2 + y^2 = \tfrac{1}{25}$$

79.
$$(x-(-2))^2 + (y-12)^2 = 13^2$$
$$(x+2)^2 + (y-12)^2 = 169$$

81. $x^2 + y^2 = 1^2 \Rightarrow x^2 + y^2 - 1 = 0$

83.
$$(x-6)^2 + (y-8)^2 = 4^2$$
$$x^2 - 12x + 36 + y^2 - 16y + 64 = 16$$
$$x^2 + y^2 - 12x - 16y + 84 = 0$$

85.
$$(x-3)^2 + (y+4)^2 = \left(\sqrt{2}\right)^2$$
$$x^2 - 6x + 9 + y^2 + 8y + 16 = 2$$
$$x^2 + y^2 - 6x + 8y + 23 = 0$$

87. Center: $x = \dfrac{3+3}{2} = 3,\ y = \dfrac{-2+8}{2} = 3$

$r =$ distance from center to endpoint

$$= \sqrt{(3-3)^2 + (3-8)^2} = 5$$
$$(x-3)^2 + (y-3)^2 = 5^2$$
$$x^2 - 6x + 9 + y^2 - 6y + 9 = 25$$
$$x^2 + y^2 - 6x - 6y - 7 = 0$$

89. $r =$ distance from center to origin

$$= \sqrt{(0-(-3))^2 + (0-4)^2} = 5$$
$$(x+3)^2 + (y-4)^2 = 5^2$$
$$x^2 + 6x + 9 + y^2 - 8y + 16 = 25$$
$$x^2 + y^2 + 6x - 8y = 0$$

91.
$$x^2 + y^2 - 6x + 4y + 4 = 0$$
$$x^2 - 6x + y^2 + 4y = -4$$
$$x^2 - 6x + 9 + y^2 + 4y + 4 = -4 + 9 + 4$$
$$(x-3)^2 + (y+2)^2 = 9$$

93.
$$x^2 + y^2 - 10x - 12y + 57 = 0$$
$$x^2 - 10x + y^2 - 12y = -57$$
$$x^2 - 10x + 25 + y^2 - 12y + 36 = -57 + 25 + 36$$
$$(x-5)^2 + (y-6)^2 = 4$$

95.
$$2x^2 + 2y^2 - 8x - 16y + 22 = 0$$
$$x^2 + y^2 - 4x - 8y + 11 = 0$$
$$x^2 - 4x + y^2 - 8y = -11$$
$$x^2 - 4x + 4 + y^2 - 8y + 16 = -11 + 4 + 16$$
$$(x-2)^2 + (y-4)^2 = 9$$

97. $x^2 + y^2 - 25 = 0$

$$x^2 + y^2 = 25$$

$C(0, 0),\ r = 5$

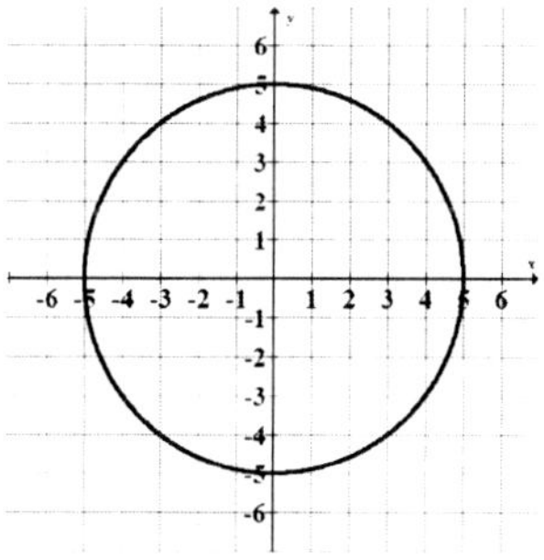

99. $(x-1)^2 + (y+2)^2 = 4$

$C(1, -2),\ r = 2$

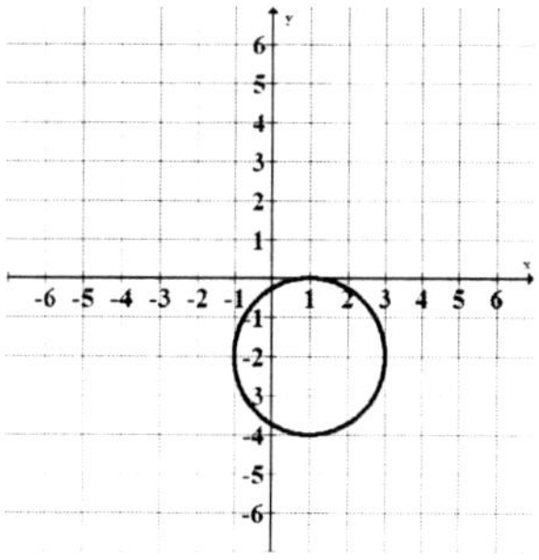

SECTION 2.4

101.
$$x^2 + y^2 + 2x - 24 = 0$$
$$x^2 + 2x + y^2 = 24$$
$$x^2 + 2x + 1 + y^2 = 24 + 1$$
$$(x+1)^2 + y^2 = 25$$
$$C(-1, 0),\ r = 5$$

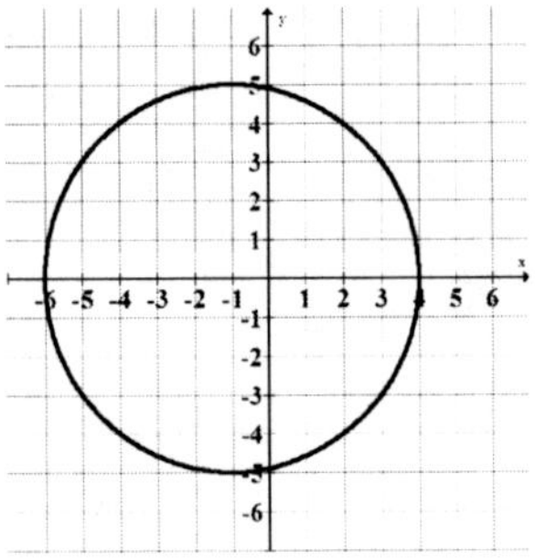

103.
$$x^2 + y^2 + 4x + 2y - 11 = 0$$
$$x^2 + 4x + y^2 + 2y = 11$$
$$x^2 + 4x + 4 + y^2 + 2y + 1 = 11 + 4 + 1$$
$$(x+2)^2 + (y+1)^2 = 16$$
$$C(-2, -1),\ r = 4$$

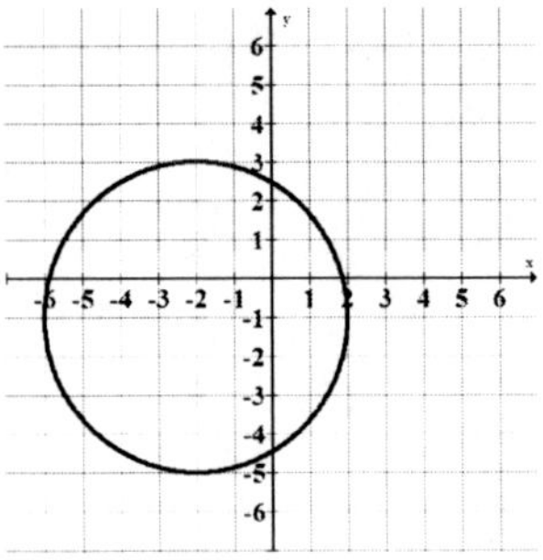

105.
$$9x^2 + 9y^2 - 12y = 5$$
$$x^2 + y^2 - \frac{4}{3}y = \frac{5}{9}$$
$$x^2 + y^2 - \frac{4}{3}y + \frac{4}{9} = \frac{5}{9} + \frac{4}{9}$$
$$x^2 + \left(y - \frac{2}{3}\right)^2 = 1$$
$$C\left(0, \frac{2}{3}\right),\ r = 1$$

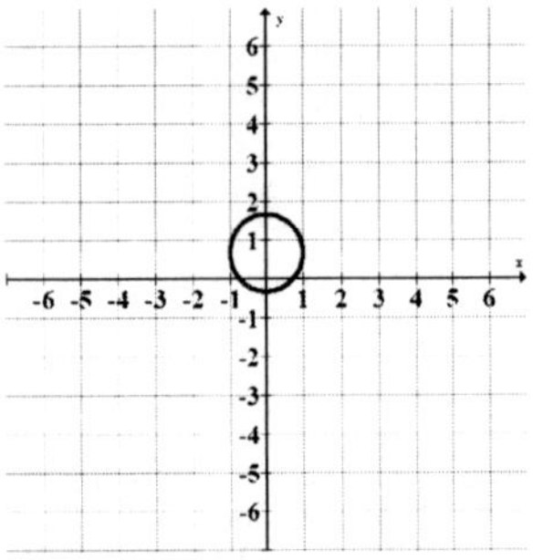

107.
$$4x^2 + 4y^2 - 4x + 8y + 1 = 0$$
$$x^2 + y^2 - x + 2y = -\frac{1}{4}$$
$$x^2 - x + \frac{1}{4} + y^2 + 2y + 1 = -\frac{1}{4} + \frac{1}{4} + 1$$
$$\left(x - \frac{1}{2}\right)^2 + (y+1)^2 = 1$$
$$C\left(\frac{1}{2}, -1\right),\ r = 1$$

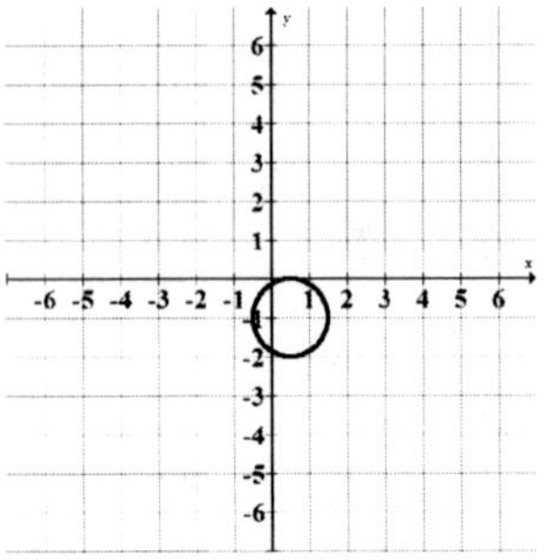

SECTION 2.4

109. $y = 2x^2 - x + 1$
Vertex: $(0.25, 0.88)$

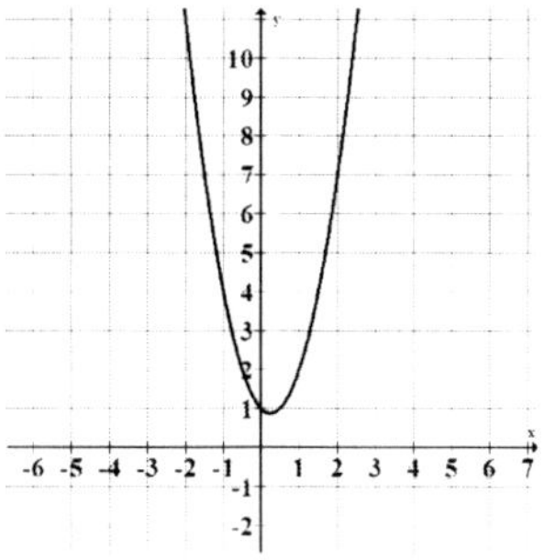

111. $y = 7 + x - x^2$
Vertex: $(0.50, 7.25)$

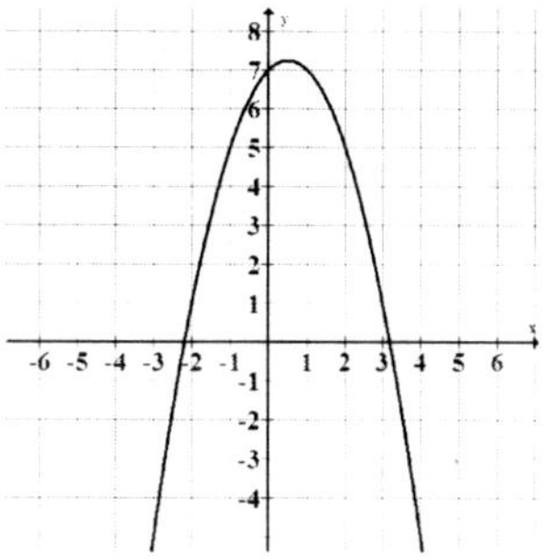

113. Graph $y = x^2 - 7$.
Find the x-intercepts.
$x = -2.65,\ x = 2.65$

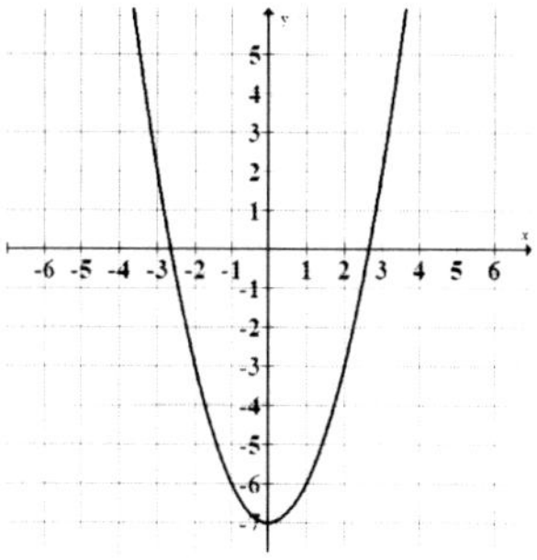

115. Graph $y = x^3 - 3$.
Find the x-intercepts.
$x = 1.44$

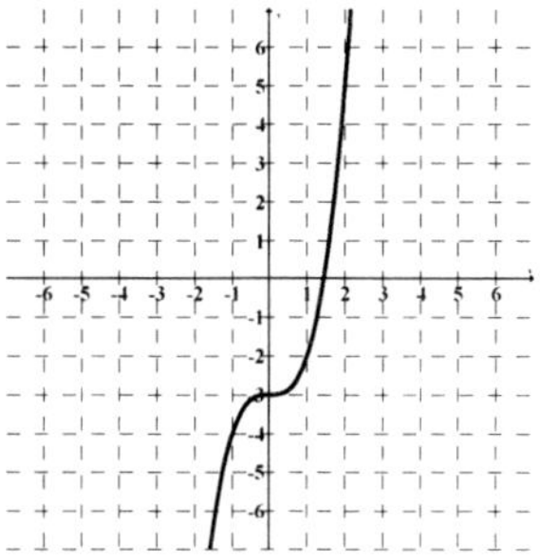

117. Let $y = 0$:
$y = 64t - 16t^2$
$0 = 16t(4 - t)$
$t = 0$ or $t = 4$
It strikes the ground after 4 seconds.

119. $D = 0.08V^2 + 0.9V$;

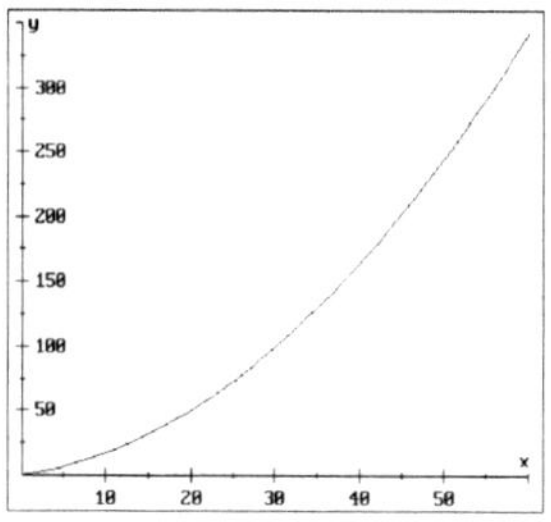

121. $r = \frac{12}{2} = 6$

$$(x - 0)^2 + (y - 0)^2 = 6^2$$
$$x^2 + y^2 = 36$$

123. $r = \frac{60}{2} = 30$

$$(x - 0)^2 + (y - 35)^2 = 30^2$$
$$x^2 + (y - 35)^2 = 900$$

125. $r = \sqrt{(10 - 7)^2 + (0 - 4)^2} = 5$

$$(x - 7)^2 + (y - 4)^2 = 5^2$$
$$x^2 - 14x + 49 + y^2 - 8y + 16 = 25$$
$$x^2 + y^2 - 14x - 8y + 40 = 0$$

127. Graph $y = x^2 + x - 6$. Find the x-coordinates where the graph is below the x-axis.
Solution: $(-3, 2)$

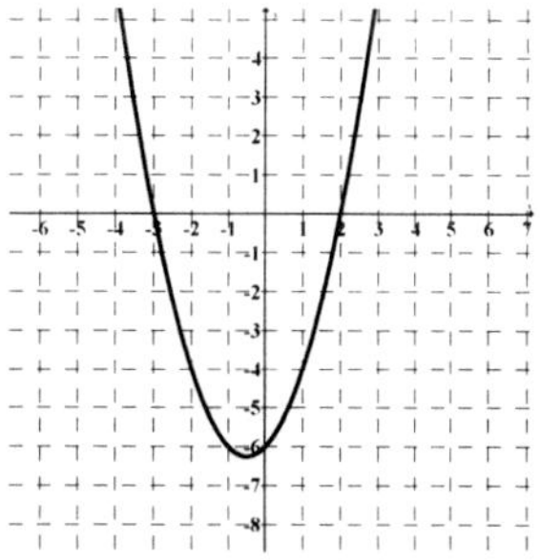

129.
$$x^2 - 4x + y^2 - 6y + 13 = 0$$
$$x^2 - 4x + 4 + y^2 - 6y + 9 = -13 + 4 + 9$$
$$(x+2)^2 + (y-3)^2 = 0 \Rightarrow \text{a single point}$$

131.
$$3(x+2) + x = 5x$$
$$3x + 6 + x = 5x$$
$$4x + 6 = 5x$$
$$6 = x$$

133.
$$\frac{5(2-x)}{3} - 1 = x + 5$$
$$3\left[\frac{10-5x}{3} - 1\right] = 3(x+5)$$
$$10 - 5x - 3 = 3x + 15$$
$$-5x + 7 = 3x + 15$$
$$-8x = 8$$
$$x = -1$$

135. Let $x =$ the ounces of copper added. Since 1 ounce out of 4 ounces are to be gold, the final result should be 25% gold. [Ounces of gold at start] + [Ounces of gold added] = [Ounces of gold at end]
$$20 + 0 = 0.25(60 + x)$$
$$20 = 15 + 0.25x$$
$$5 = 0.25x$$
$20 = x$; 20 ounces of copper should be added.

Exercises 2.5 (page 261)

1. quotient

3. means

5. extremes, means

7. inverse

9. joint

11.
$$\frac{4}{x} = \frac{2}{7}$$
$$4 \cdot 7 = 2 \cdot x$$
$$28 = 2x$$
$$14 = x$$

13.
$$\frac{x}{2} = \frac{3}{x+1}$$
$$x(x+1) = 3 \cdot 2$$
$$x^2 + x = 6$$
$$x^2 + x - 6 = 0$$
$$(x+3)(x-2) = 0$$
$x = -3$ or $x = 2$

15. Let $x =$ the number of women.
$$\tfrac{3}{5} = \tfrac{x}{30}$$
$$3 \cdot 30 = 5 \cdot x$$
$$90 = 5x$$
$18 = x \Rightarrow$ There are 18 women.

17.
$$y = kx$$
$$15 = k(30)$$
$$\tfrac{1}{2} = k$$

19.
$$I = \frac{k}{R}$$
$$50 = \frac{k}{20}$$
$$1000 = k$$

21.
$$E = kIR$$
$$125 = k(5)(25)$$
$$125 = 125k$$
$$1 = k$$

SECTION 2.5

23.
$$y = kx \qquad y = \tfrac{15}{4}x$$
$$15 = k(4) \qquad y = \tfrac{15}{4} \cdot \tfrac{7}{5}$$
$$\tfrac{15}{4} = k \qquad y = \tfrac{21}{4}$$

25.
$$w = \frac{k}{z} \qquad w = \frac{30}{z}$$
$$10 = \frac{k}{3} \qquad w = \frac{30}{5}$$
$$30 = k \qquad w = 6$$

27.
$$P = krs \qquad P = -\tfrac{2}{5}rs$$
$$16 = k(5)(-8) \qquad P = -\tfrac{2}{5}(2)(10)$$
$$16 = -40k \qquad P = -8$$
$$-\tfrac{16}{40} = k$$
$$-\tfrac{2}{5} = k$$

29. direct

31. neither

33. Let $x =$ the amount of caffeine.
$$\tfrac{55}{12} = \tfrac{x}{44} \qquad \tfrac{47}{12} = \tfrac{x}{44} \qquad \tfrac{37}{12} = \tfrac{x}{44}$$
$$55 \cdot 44 = 12 \cdot x \qquad 47 \cdot 44 = 12 \cdot x \qquad 37 \cdot 44 = 12 \cdot x$$
$$2420 = 12x \qquad 2068 = 12x \qquad 1628 = 12x$$
$$202 \text{ mg} \approx x \qquad 172 \text{ mg} \approx x \qquad 136 \text{ mg} \approx x$$

35. Let $x =$ the amount of adhesive needed.
$$\frac{\frac{1}{2}}{140} = \frac{x}{500}$$
$$\frac{1}{2} \cdot 500 = 140 \cdot x$$
$$250 = 140x$$
$$1.79 \approx x$$
About 2 gallons of adhesive will be needed.

37.
$$V = \frac{kT}{P} \qquad V = \frac{\frac{80}{33}T}{P}$$
$$20 = \frac{k(330)}{40} \qquad V = \frac{\frac{80}{33}(300)}{50}$$
$$800 = 330k \qquad V = \frac{\frac{8000}{11}}{50}$$
$$\frac{800}{330} = k \qquad V = \frac{160}{11} = 14\frac{6}{11} \text{ ft}^3$$
$$\frac{80}{33} = k$$

39.
$$d = kt^2 \qquad d = 16t^2$$
$$16 = k(1)^2 \qquad 144 = 16t^2$$
$$16 = k \qquad 9 = t^2$$
$$3 = t \Rightarrow 3 \text{ seconds}$$

41.
$$t = kl^2 \qquad t = l^2$$
$$1 = k(1)^2 \qquad 2 = l^2$$
$$1 = k \qquad \sqrt{2} = l \Rightarrow \sqrt{2} \text{ meters}$$

43.
$$I = \frac{k}{d^2} \qquad I = \frac{6000}{d^2}$$
$$60 = \frac{k}{10^2} \qquad I = \frac{6000}{20^2}$$
$$60 = \frac{k}{100} \qquad I = \frac{6000}{400}$$
$$6000 = k \qquad I = 15 \Rightarrow 15 \text{ lumens}$$

45.
$$E = kmv^2 = k(2m)(3v)^2$$
$$= k(2m)(9v^2)$$
$$= 18 \cdot kmv^2$$
The energy is multiplied by 18.

47. $G = \dfrac{km_1m_2}{d^2} = \dfrac{k(3m_1)(3m_2)}{(2d)^2}$
$= \dfrac{k \cdot 9m_1m_2}{4d^2}$
$= \dfrac{9}{4} \cdot \dfrac{km_1m_2}{d^2}$; The force is multiplied by $\dfrac{9}{4}$.

49. Consider this figure:

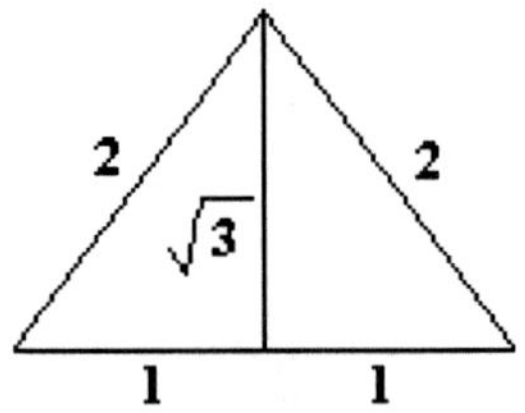

$h = \sqrt{3}$ can be computed using the Pythagorean Theorem.
$A = \dfrac{1}{2}bh = \dfrac{1}{2}(2)\sqrt{3} = \sqrt{3}$
$A = ks^2$
$\sqrt{3} = k(2)^2$
$\sqrt{3} = 4k$
$\dfrac{\sqrt{3}}{4} = k$

51-55. Answers may vary.

57. $\dfrac{1}{x+2} + \dfrac{2}{x+1} = \dfrac{1(x+1)}{(x+2)(x+1)} + \dfrac{2(x+2)}{(x+1)(x+2)} = \dfrac{x+1+2x+4}{(x+1)(x+2)} = \dfrac{3x+5}{(x+1)(x+2)}$

59. $\dfrac{x^2+3x-4}{x^2-5x+4} \div \dfrac{x-1}{x^2-3x-4} = \dfrac{x^2+3x-4}{x^2-5x+4} \cdot \dfrac{x^2-3x-4}{x-1} = \dfrac{(x+4)(x-1)}{(x-4)(x-1)} \cdot \dfrac{(x-4)(x+1)}{x-1}$
$= \dfrac{(x+4)(x+1)}{x-1}$

61. $\dfrac{x^2+4-(x+2)^2}{4x^2} = \dfrac{x^2+4-(x^2+4x+4)}{4x^2} = \dfrac{x^2+4-x^2-4x-4}{4x^2} = \dfrac{-4x}{4x^2} = -\dfrac{1}{x}$

Chapter 2 Review (page 265)

1. $A(2, 0)$ **2.** $B(-2, 1)$ **3.** $C(0, -1)$ **4.** $D(3, -1)$

5. $A(-3, 5)$: QII

6. $B(5, -3)$: QIV

7. $C(0, -7)$: negative y-axis

8. $D\left(-\dfrac{1}{2}, 0\right)$: negative x-axis

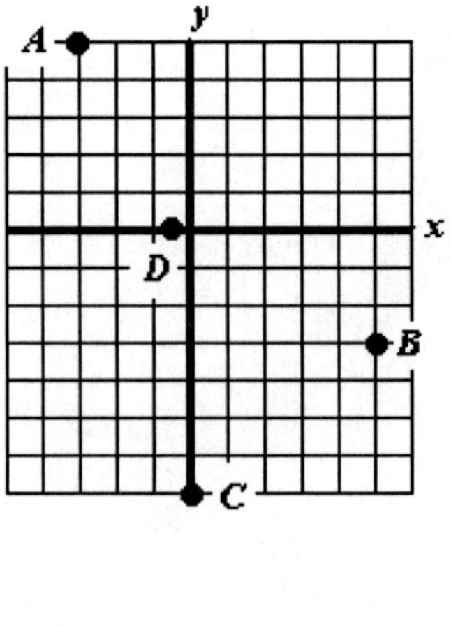

9.

$$\begin{aligned} 2x - y &= 6 \\ -y &= -2x + 6 \\ y &= 2x - 6 \end{aligned}$$

x	y
0	-6
2	-2

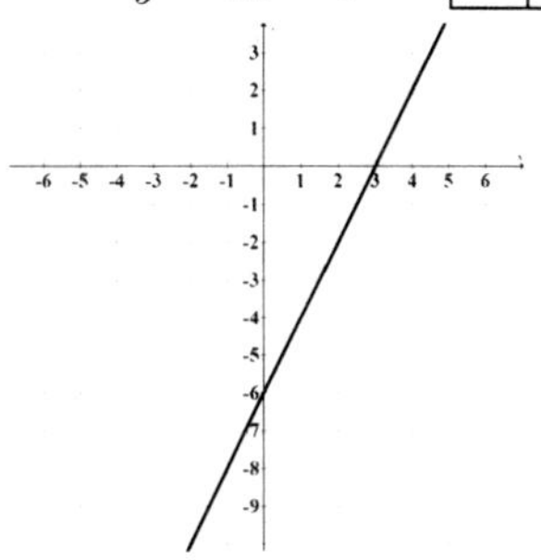

10.

$$\begin{aligned} 2x + 5y &= -10 \\ 5y &= -2x - 10 \\ y &= -\frac{2}{5}x - 2 \end{aligned}$$

x	y
0	-2
-5	0

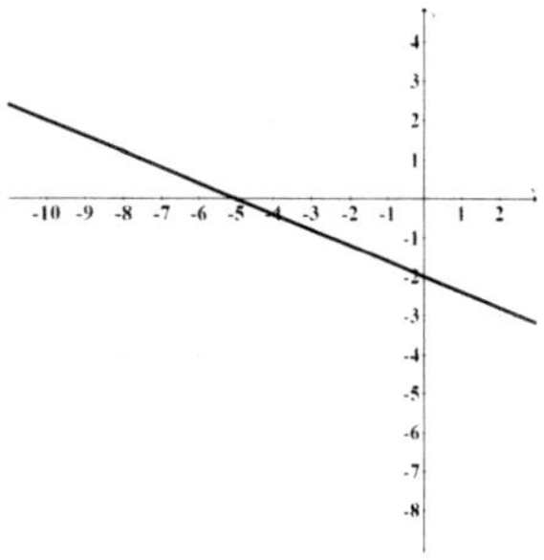

11.

$$\begin{aligned} 3x - 5y &= 15 \\ 3x - 5(0) &= 15 \\ 3x &= 15 \\ x &= 5 \end{aligned} \qquad \begin{aligned} 3x - 5y &= 15 \\ 3(0) - 5y &= 15 \\ -5y &= 15 \\ y &= -3 \end{aligned}$$

$(5, 0)$ $\quad$ $(0, -3)$

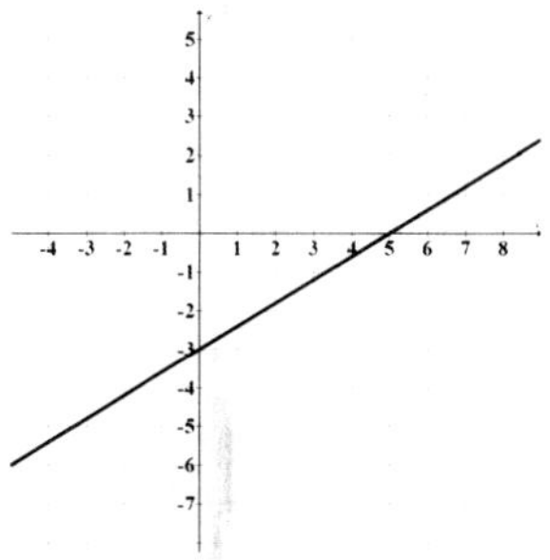

12.

$$\begin{aligned} x + y &= 7 \\ x + 0 &= 7 \\ x &= 7 \end{aligned} \qquad \begin{aligned} x + y &= 7 \\ 0 + y &= 7 \\ y &= 7 \end{aligned}$$

$(7, 0)$ $\quad$ $(0, 7)$

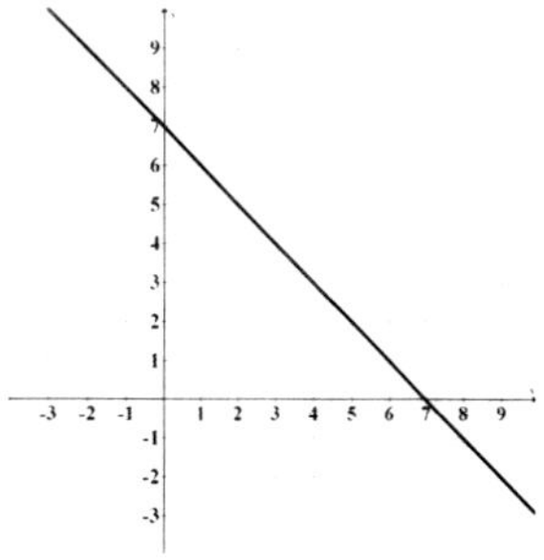

13.

$$\begin{aligned} x + y &= -7 \\ x + 0 &= -7 \\ x &= -7 \end{aligned} \qquad \begin{aligned} x + y &= -7 \\ 0 + y &= -7 \\ y &= -7 \end{aligned}$$

$(-7, 0)$ $\quad$ $(0, -7)$

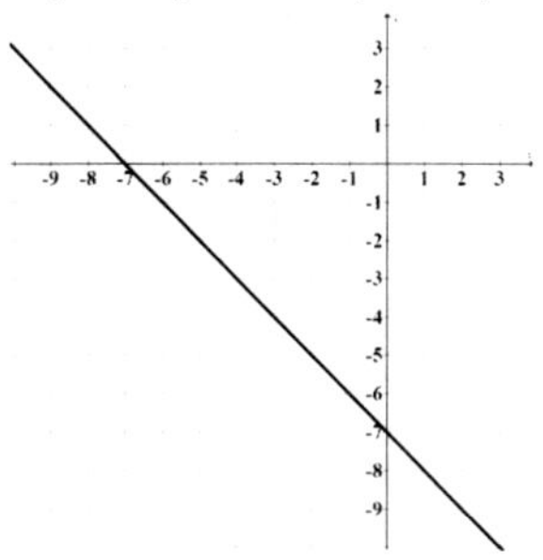

14.

$$\begin{aligned} x - 5y &= 5 \\ x - 5(0) &= 5 \\ x &= 5 \end{aligned} \qquad \begin{aligned} x - 5y &= 5 \\ 0 - 5y &= 5 \\ -5y &= 5 \\ y &= -1 \end{aligned}$$

$(5, 0)$ $\quad$ $(0, -1)$

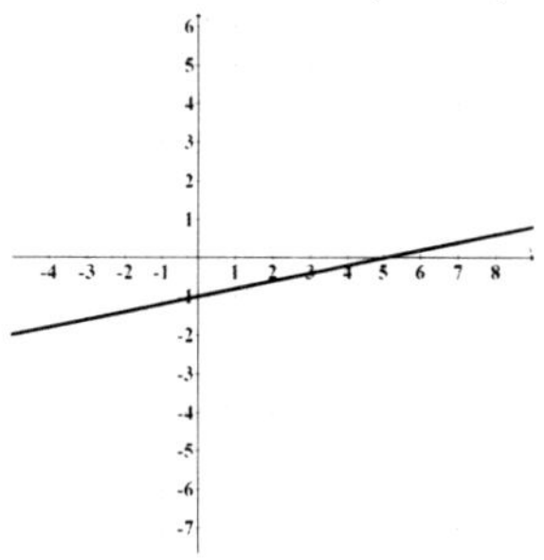

15. $y = 4 \Rightarrow$ horizontal

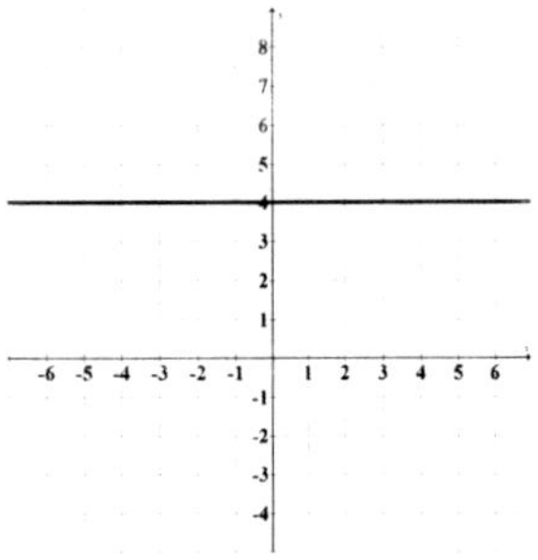

16. $x = -2 \Rightarrow$ vertical

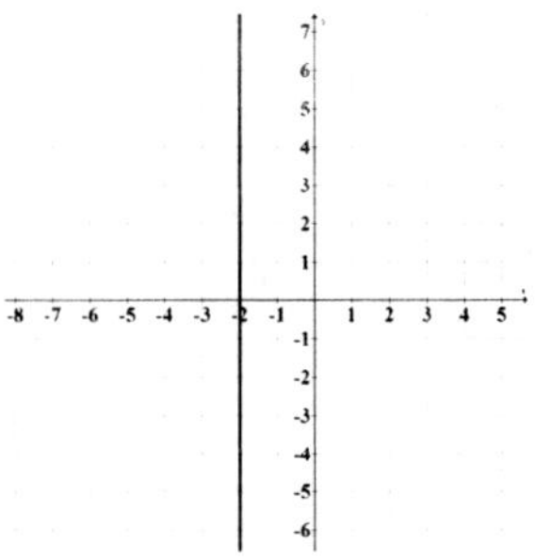

17. Let $x = 3$: $y = -2200x + 18{,}750 = -2200(3) + 18{,}750 = -6600 + 18{,}750 = \$12{,}150$

18. Let $x = 5$: $y = 16{,}500x + 250{,}000 = 16{,}500(5) + 250{,}000 = 82{,}500 + 250{,}000 = \$332{,}500$

19.
$$\begin{aligned} d &= \sqrt{(x_2 - x_1)^2 + (y_2 - y_1)^2} \\ &= \sqrt{(-3-3)^2 + (7-(-1))^2} \\ &= \sqrt{(-6)^2 + (8)^2} \\ &= \sqrt{36+64} = \sqrt{100} = 10 \end{aligned}$$

20.
$$\begin{aligned} d &= \sqrt{(x_2 - x_1)^2 + (y_2 - y_1)^2} \\ &= \sqrt{(-12-(-8))^2 + (10-6)^2} \\ &= \sqrt{(-4)^2 + 4^2} \\ &= \sqrt{16+16} = \sqrt{32} = 4\sqrt{2} \end{aligned}$$

21.
$$\begin{aligned} d &= \sqrt{(x_2 - x_1)^2 + (y_2 - y_1)^2} \\ &= \sqrt{\left(\sqrt{3}-\sqrt{3}\right)^2 + (9-7)^2} \\ &= \sqrt{0^2 + (2)^2} \\ &= \sqrt{0+4} = \sqrt{4} = 2 \end{aligned}$$

22.
$$\begin{aligned} d &= \sqrt{(x_2 - x_1)^2 + (y_2 - y_1)^2} \\ &= \sqrt{(a-(-a))^2 + (-a-a)^2} \\ &= \sqrt{(2a)^2 + (-2a)^2} \\ &= \sqrt{4a^2+4a^2} = \sqrt{8a^2} = 2\sqrt{2}|a| \end{aligned}$$

23. $$M\left(\frac{x_1+x_2}{2}, \frac{y_1+y_2}{2}\right) = M\left(\frac{-3+3}{2}, \frac{7+(-1)}{2}\right) = M\left(\frac{0}{2}, \frac{6}{2}\right) = M(0,3)$$

24. $$M\left(\frac{x_1+x_2}{2}, \frac{y_1+y_2}{2}\right) = M\left(\frac{0+(-12)}{2}, \frac{5+10}{2}\right) = M\left(\frac{-12}{2}, \frac{15}{2}\right) = M\left(-6, \frac{15}{2}\right)$$

25. $$M\left(\frac{x_1+x_2}{2}, \frac{y_1+y_2}{2}\right) = M\left(\frac{\sqrt{3}+\sqrt{3}}{2}, \frac{9+7}{2}\right) = M\left(\frac{2\sqrt{3}}{2}, \frac{16}{2}\right) = M\left(\sqrt{3}, 8\right)$$

26. $$M\left(\frac{x_1+x_2}{2}, \frac{y_1+y_2}{2}\right) = M\left(\frac{a+(-a)}{2}, \frac{-a+a}{2}\right) = M\left(\frac{0}{2}, \frac{0}{2}\right) = M(0,0)$$

27. $$m = \frac{y_2-y_1}{x_2-x_1} = \frac{7-(-5)}{1-3} = \frac{12}{-2} = -6$$

28. $$m = \frac{y_2-y_1}{x_2-x_1} = \frac{-7-7}{-5-2} = \frac{-14}{-7} = 2$$

CHAPTER 2 REVIEW

29. $m = \frac{y_2 - y_1}{x_2 - x_1} = \frac{b - a}{a - b} = -1$

30. $m = \frac{y_2 - y_1}{x_2 - x_1} = \frac{(b - a) - b}{b - (a + b)} = \frac{-a}{-a} = 1$

31. $y = 3x + 6$

x	y
0	6
1	9

$$m = \frac{y_2 - y_1}{x_2 - x_1} = \frac{9 - 6}{1 - 0} = \frac{3}{1} = 3$$

32. $y = 5x - 6$

x	y
0	-6
1	-1

$$m = \frac{y_2 - y_1}{x_2 - x_1} = \frac{-1 - (-6)}{1 - 0} = \frac{5}{1} = 5$$

33. The slope is zero.

34. The slope is undefined.

35. The slope is negative.

36. The slope is positive.

37. $m_1 m_2 = 5\left(-\frac{1}{5}\right) = -1$
perpendicular

38. $m_1 \neq m_2;\ m_1 m_2 = \frac{2}{7} \cdot \frac{7}{2} = 1 \neq -1$
neither

39.
$$\begin{aligned} m &= \frac{y_2 - y_1}{x_2 - x_1} = \frac{10 - 5}{6 - (-2)} = \frac{5}{8} \\ m &= \frac{y_2 - y_1}{x_2 - x_1} = \frac{y - 2}{10 - 2} = \frac{5}{8} \\ 8(y - 2) &= 5(8) \\ 8y - 16 &= 40 \\ 8y &= 56 \\ y &= 7 \end{aligned}$$

40.
$$\begin{aligned} m &= \frac{y_2 - y_1}{x_2 - x_1} = \frac{10 - 5}{6 - (-2)} = \frac{5}{8} \\ m &= \frac{y_2 - y_1}{x_2 - x_1} = \frac{-3 - 5}{x - (-2)} = \frac{-8}{5} \\ 5(-8) &= -8(x + 2) \\ -40 &= -8x - 16 \\ 8x &= 24 \\ x &= 3 \end{aligned}$$

41. $m = \frac{\Delta y}{\Delta x} = \frac{3000}{15} = 200$ ft. per minute

42. $m = \frac{\Delta y}{\Delta x} = \frac{147{,}500 - 50{,}000}{3 - 1} = \frac{97{,}500}{2} = \$48{,}750$ per year

43.
$$\begin{aligned} m &= \frac{y_2 - y_1}{x_2 - x_1} = \frac{7 - 0}{-5 - 0} = -\frac{7}{5} \\ y - y_1 &= m(x - x_1) \\ y - 0 &= -\frac{7}{5}(x - 0) \\ y &= -\frac{7}{5}x \\ 5y &= 5\left(-\frac{7}{5}x\right) \\ 5y &= -7x \\ 7x + 5y &= 0 \end{aligned}$$

44.
$$\begin{aligned} y - y_1 &= m(x - x_1) \\ y - 1 &= -4(x + 2) \\ y - 1 &= -4x - 8 \\ 4x + y &= -7 \end{aligned}$$

45.
$$y - y_1 = m(x - x_1)$$
$$y + 1 = -\frac{1}{5}(x - 2)$$
$$5(y + 1) = 5 \cdot \left[-\frac{1}{5}(x - 2)\right]$$
$$5y + 5 = -(x - 2)$$
$$5y + 5 = -x + 2$$
$$x + 5y = -3$$

46.
$$m = \frac{y_2 - y_1}{x_2 - x_1} = \frac{1 - (-5)}{4 - 7} = \frac{6}{-3} = -2$$
$$y - y_1 = m(x - x_1)$$
$$y + 5 = -2(x - 7)$$
$$y + 5 = -2x + 14$$
$$2x + y = 9$$

47.
$$y = mx + b$$
$$y = \frac{2}{3}x + 3$$

48.
$$y = mx + b$$
$$y = -\frac{3}{2}x - 5$$

49.
$$y = \frac{3}{5}x - 2$$
$$m = \frac{3}{5},\ b = -2$$

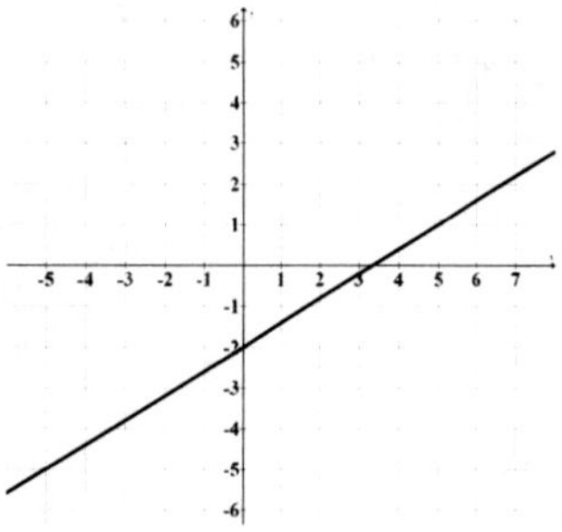

50.
$$y = -\frac{4}{3}x + 3$$
$$m = -\frac{4}{3},\ b = 3$$

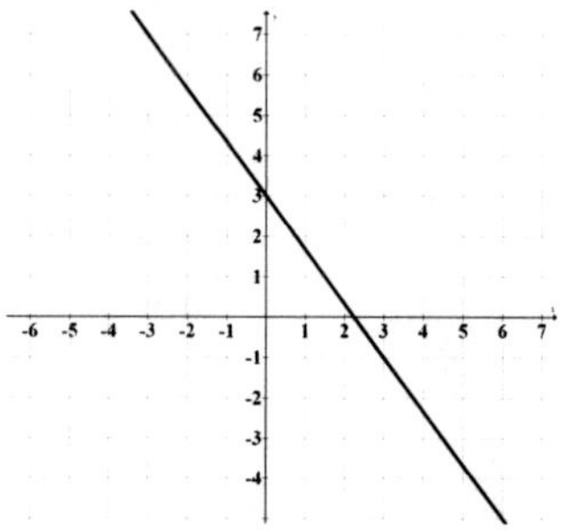

51.
$$3x - 2y = 10$$
$$-2y = -3x + 10$$
$$y = \tfrac{3}{2}x - 5$$
$$m = \tfrac{3}{2},\ (0, -5)$$

52.
$$2x + 4y = -8$$
$$4y = -2x - 8$$
$$y = -\tfrac{1}{2}x - 2$$
$$m = -\tfrac{1}{2},\ (0, -2)$$

53.
$$-2y = -3x + 10$$
$$y = \tfrac{3}{2}x - 5$$
$$m = \tfrac{3}{2},\ (0, -5)$$

54.
$$2x = -4y - 8$$
$$4y = -2x - 8$$
$$y = -\tfrac{1}{2}x - 2$$
$$m = -\tfrac{1}{2},\ (0, -2)$$

55.
$$5x + 2y = 7$$
$$2y = -5x + 7$$
$$y = -\tfrac{5}{2}x + \tfrac{7}{2}$$
$$m = -\tfrac{5}{2},\ \left(0, \tfrac{7}{2}\right)$$

56.
$$3x - 4y = 14$$
$$-4y = -3x + 14$$
$$y = \tfrac{3}{4}x - \tfrac{7}{2}$$
$$m = \tfrac{3}{4},\ \left(0, -\tfrac{7}{2}\right)$$

57. $m = 0 \Rightarrow$ horizontal
$y = 17$

58. m is undefined $\Rightarrow$ vertical
$x = -5$

59. $3x - 4y = 7$
$-4y = -3x + 7$
$y = \frac{3}{4}x - \frac{7}{4}$
$m = \frac{3}{4}$
Use $m = \frac{3}{4}$.
$y - y_1 = m(x - x_1)$
$y - 0 = \frac{3}{4}(x - 2)$
$y = \frac{3}{4}x - \frac{3}{2}$

60. $m = \frac{y_2 - y_1}{x_2 - x_1} = \frac{-10 - 4}{4 - 2} = -7$
$y - y_1 = m(x - x_1)$
$y + 2 = -7(x - 7)$
$y + 2 = -7x + 49$
$y = -7x + 47$

61. $x + 3y = 4$
$3y = -x + 4$
$y = -\frac{1}{3}x + \frac{4}{3}$
$m = -\frac{1}{3}$
Use $m = 3$.
$y - y_1 = m(x - x_1)$
$y - 5 = 3(x - 0)$
$y - 5 = 3x$
$y = 3x + 5$

62. $m = \frac{y_2 - y_1}{x_2 - x_1} = \frac{-10 - 4}{4 - 2} = -7$
$y - y_1 = m(x - x_1)$
$y + 2 = \frac{1}{7}(x - 7)$
$y + 2 = \frac{1}{7}x - 1$
$y = \frac{1}{7}x - 3$

63. $y = 3x + 8$ $\quad$ $2y = 6x - 19$
$m = 3$ $\quad$ $y = 3x - \frac{19}{2}$
$m = 3$
The lines are parallel.

64. $2x + 3y = 6$ $\quad$ $3x - 2y = 15$
$3y = -2x + 6$ $\quad$ $-2y = -3x + 15$
$y = -\frac{2}{3}x + 2$ $\quad$ $y = \frac{3}{2}x - \frac{15}{2}$
$m = -\frac{2}{3}$ $\quad$ $m = \frac{3}{2}$
The lines are perpendicular.

65. $y = 4x - 8x^2$ $\quad$ $y = 4x - 8x^2$
$0 = 4x(1 - 2x)$ $\quad$ $y = 4(0) - 8(0)^2$
$x = 0, x = \frac{1}{2}$ $\quad$ $y = 0$
x-int: $(0, 0), \left(\frac{1}{2}, 0\right)$ $\quad$ y-int: $(0, 0)$

66. $y = x^2 - 10x - 24$ $\quad$ $y = x^2 - 10x - 24$
$0 = (x - 12)(x + 2)$ $\quad$ $y = 0^2 - 10(0) - 24$
$x = 12, x = -2$ $\quad$ $y = -24$
x-int: $(12, 0), (-2, 0)$ $\quad$ y-int: $(0, -24)$

67. $y^2 = 8x$

x-axis	y-axis	origin
$(-y)^2 = 8x$	$y^2 = 8(-x)$	$(-y)^2 = 8(-x)$
$y^2 = 8x$	$y^2 = -8x$	$y^2 = -8x$
equivalent: symmetry	not equivalent: no symmetry	not equivalent: no symmetry

68. $y = 3x^4 + 6$

x-axis	y-axis	origin
$-y = 3x^4 + 6$	$y = 3(-x)^4 + 6$	$-y = 3(-x)^4 + 6$
not equivalent: no symmetry	$y = 3x^4 + 6$	$-y = 3x^4 + 6$
	equivalent: **symmetry**	not equivalent: no symmetry

69. $y = -2|x|$

x-axis	y-axis	origin
$-y = -2\|x\|$	$y = -2\|-x\|$	$-y = -2\|-x\|$
$y = 2\|x\|$	$y = -2\|-1\|\|x\|$	$y = 2\|-x\|$
not equivalent: no symmetry	$y = -2\|x\|$	$y = 2\|-1\|\|x\|$
	equivalent: **symmetry**	$y = 2\|x\|$
		not equivalent: no symmetry

70. $y = |x + 2|$

x-axis	y-axis	origin
$-y = \|x + 2\|$	$y = \|-x + 2\|$	$-y = \|-x + 2\|$
not equivalent: no symmetry	not equivalent: no symmetry	not equivalent: no symmetry

71. $y = x^2 + 2$
x-int: none, y-int: $(0, 2)$
symmetry: y-axis

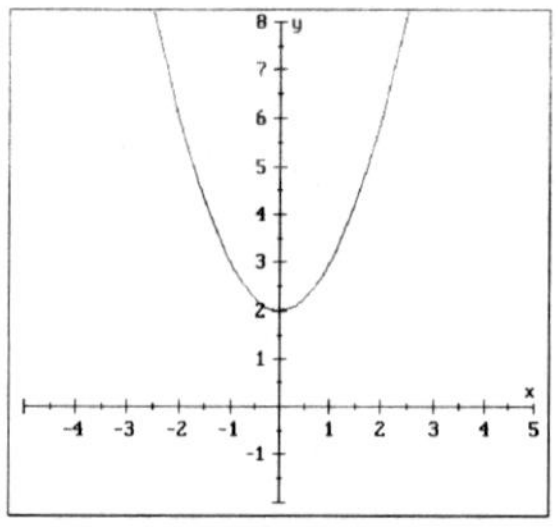

72. $y = x^3 - 2$
x-int: $\left(\sqrt[3]{2}, 0\right)$,
y-int: $(0, -2)$
symmetry: none

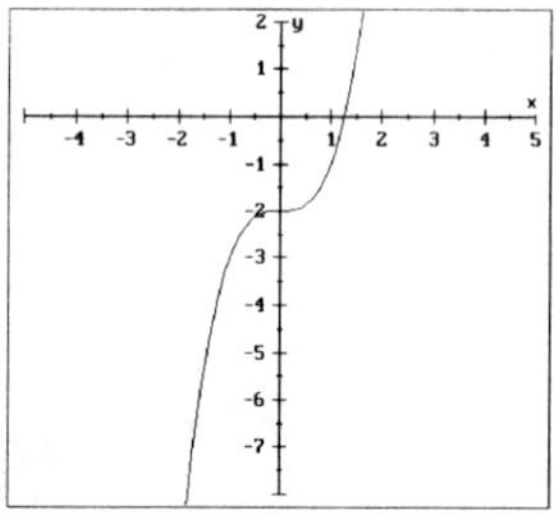

73. $y = \frac{1}{2}|x|$
x-int: $(0, 0)$, y-int: $(0, 0)$
symmetry: y-axis

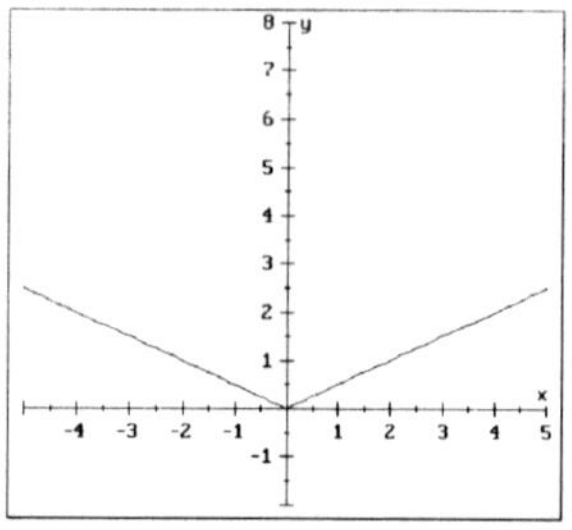

CHAPTER 2 REVIEW

74. $y = -\sqrt{x-4}$
x-int: $(4, 0)$, y-int: none
symmetry: none

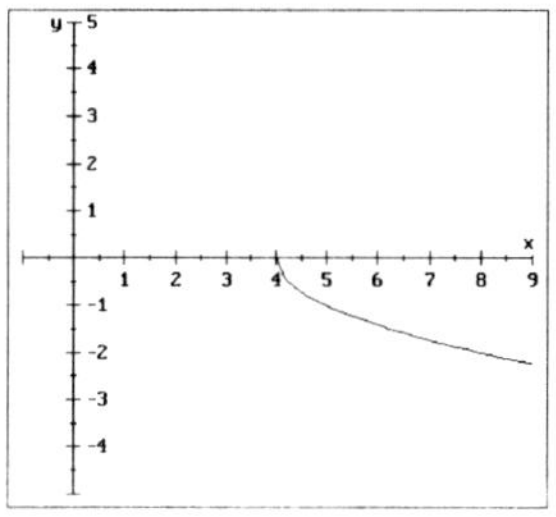

75. $y = \sqrt{x} + 2$
x-int: none, y-int: $(0, 2)$
symmetry: none

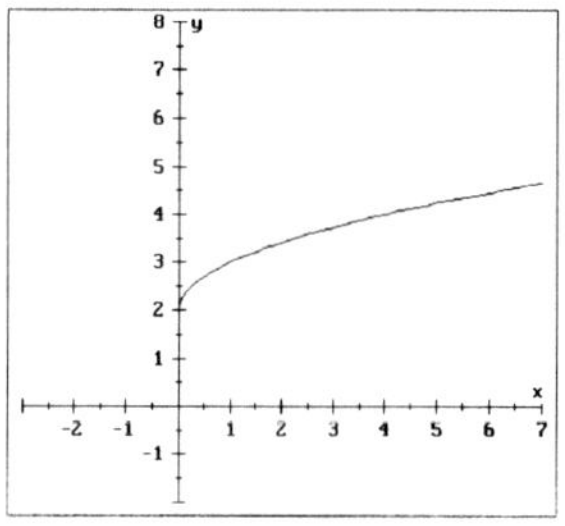

76. $y = |x+1| + 2$
x-int: none, y-int: $(0, 3)$
symmetry: none

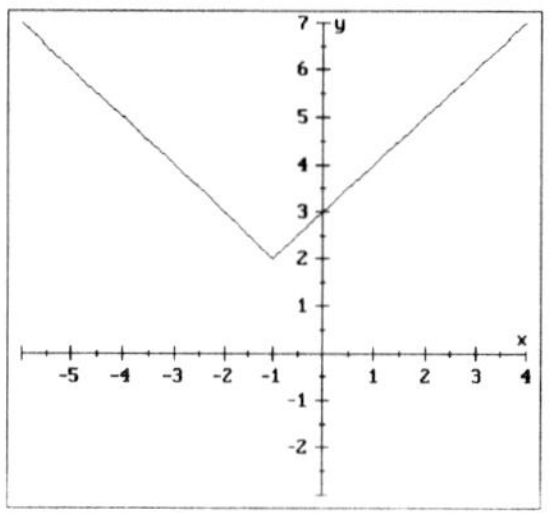

77. $y = |x-4| + 2$

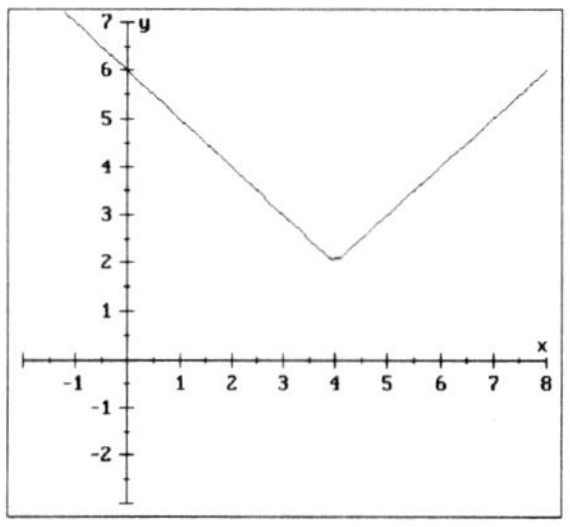

78. $y = -\sqrt{x+2} + 3$

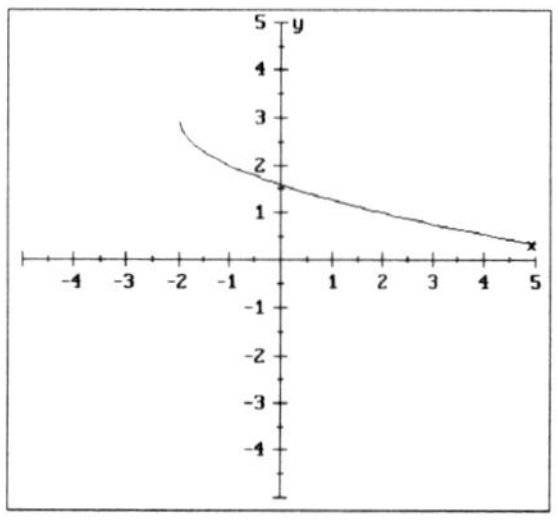

79. $y = x + 2|x|$

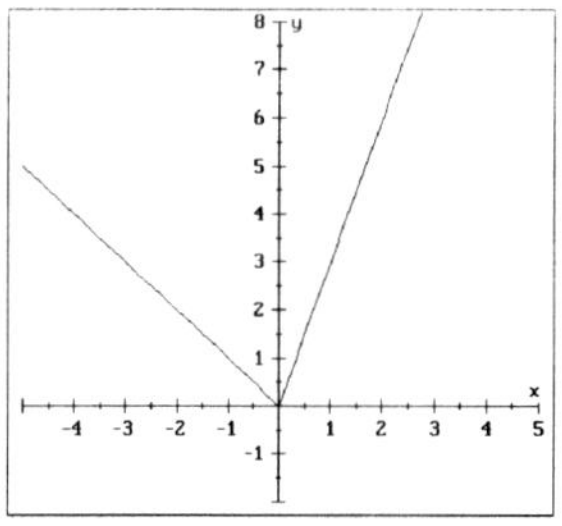

80. $y^2 = x - 3$
Graph $y = \pm\sqrt{x-3}$.

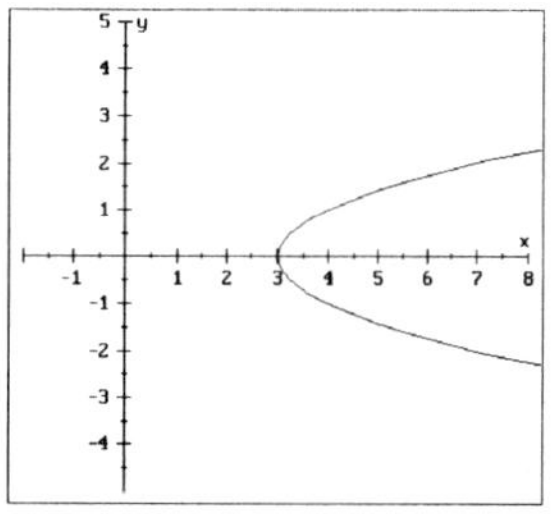

81.
$$x^2 + y^2 = 64$$
$$(x-0)^2 + (y-0)^2 = 8^2$$
$$\text{C: } (0, 0);\ r = 8$$

82.
$$x^2 + (y-6)^2 = 100$$
$$(x-0)^2 + (y-6)^2 = 10^2$$
$$\text{C: } (0, 6);\ r = 10$$

83.
$$(x+7)^2 + y^2 = \tfrac{1}{4}$$
$$(x-(-7))^2 + (y-0)^2 = \left(\tfrac{1}{2}\right)^2$$
$$\text{C: } (-7, 0);\ r = \tfrac{1}{2}$$

84.
$$(x-5)^2 + (y+1)^2 = 9$$
$$(x-5)^2 + (y-(-1))^2 = 3^2$$
$$\text{C: } (5, -1);\ r = 3$$

85. $(x-0)^2+(y-0)^2=7^2$
$x^2+y^2=49$

86. $(x-3)^2+(y-0)^2=\left(\frac{1}{5}\right)^2$
$(x-3)^2+y^2=\frac{1}{25}$

87. $(x-(-2))^2+(y-12)^2=5^2$
$(x+2)^2+(y-12)^2=25$

88. $\left(x-\frac{2}{7}\right)^2+(y-5)^2=9^2$
$\left(x-\frac{2}{7}\right)^2+(y-5)^2=81$

89. $C(-3,4);\ r=12$
$(x-h)^2+(y-k)^2=r^2$
$(x+3)^2+(y-4)^2=144$
or $x^2+y^2+6x-8y-119=0$

90. Center: $x=\dfrac{-6+5}{2}=-\dfrac{1}{2}$
$y=\dfrac{-3+8}{2}=\dfrac{5}{2}$
$r=$ distance from center to endpoint
$=\sqrt{\left(-\frac{1}{2}-5\right)^2+\left(\frac{5}{2}-8\right)^2}=\sqrt{\frac{121}{2}}$
$\left(x+\frac{1}{2}\right)^2+\left(y-\frac{5}{2}\right)^2=\frac{121}{2}$, or
$x^2+y^2+x-5y-54=0$

91. $x^2+y^2+6x-4y+4=0$
$x^2+6x+y^2-4y=-4$
$x^2+6x+9+y^2-4y+4=-4+9+4$
$(x+3)^2+(y-2)^2=9$

92. $2x^2+2y^2-8x-16y-10=0$
$x^2+y^2-4x-8y-5=0$
$x^2-4x+y^2-8y=5$
$x^2-4x+4+y^2-8y+16=5+4+16$
$(x-2)^2+(y-4)^2=25$

93. $x^2+y^2-16=0$
$(x-0)^2+(y-0)^2=16$
$C(0,0),\ r=4$

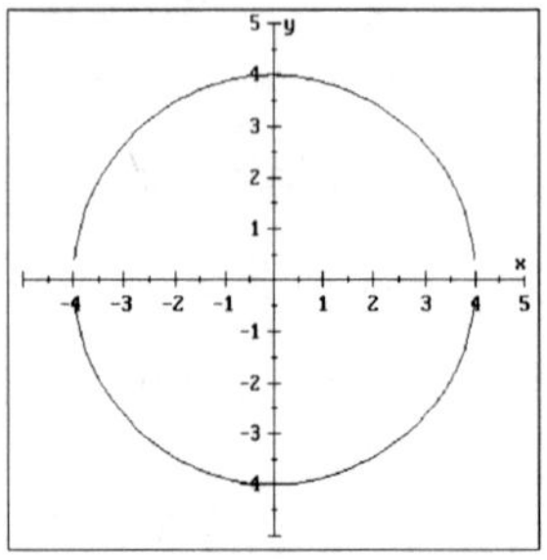

94. $x^2+y^2-4x=5$
$x^2-4x+y^2=5$
$x^2-4x+4+y^2=5+4$
$(x-2)^2+y^2=9$
$C(2,0),\ r=3$

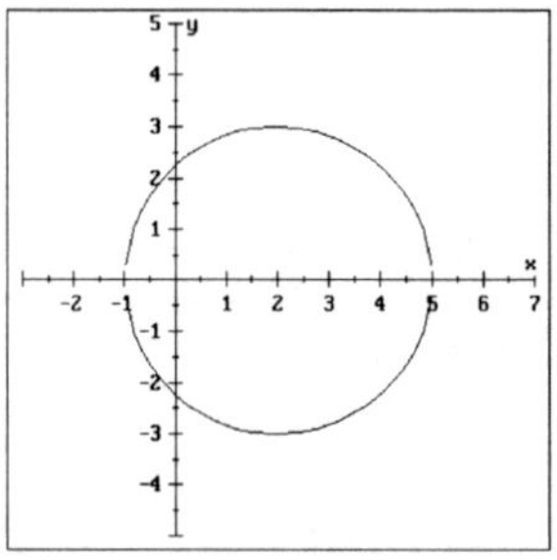

95.
$$x^2 + y^2 - 2y = 15$$
$$x^2 + y^2 - 2y + 1 = 15 + 1$$
$$x^2 + (y - 1)^2 = 16$$
$C(0, 1)$, $r = 4$

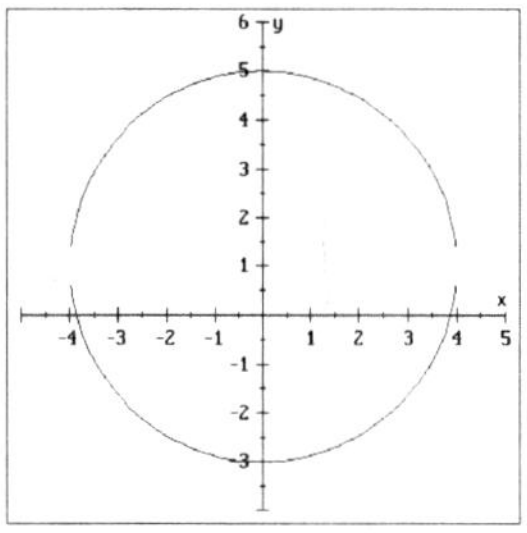

96.
$$x^2 + y^2 - 4x + 2y = 4$$
$$x^2 - 4x + 4 + y^2 + 2y + 1 = 4 + 4 + 1$$
$$(x - 2)^2 + (y + 1)^2 = 9$$
$C(2, -1)$, $r = 3$

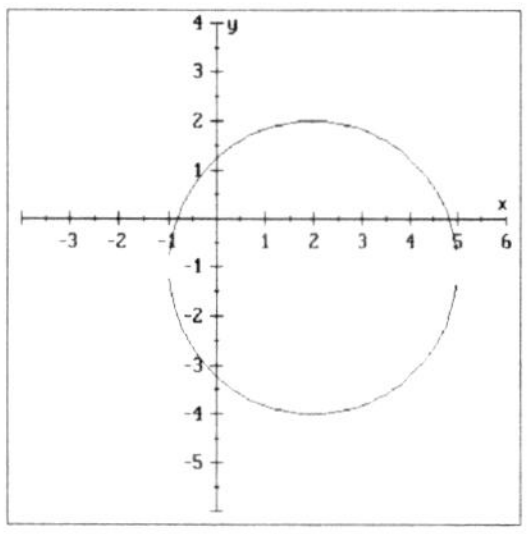

97. Graph $y = x^2 - 11$.
Find the x-intercepts.
$x = -3.32$, $x = 3.32$

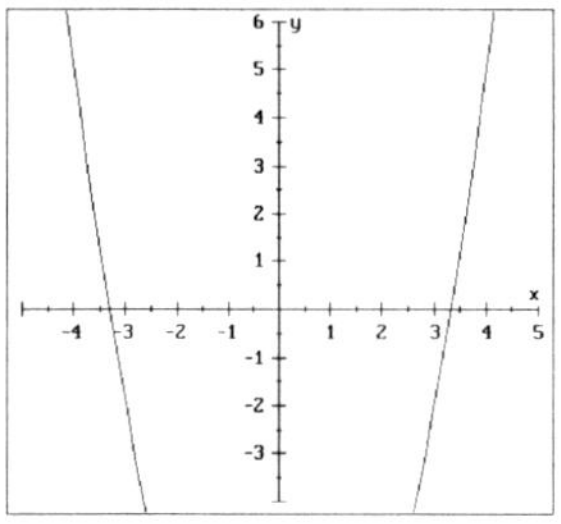

98. Graph $y = x^3 - x$.
Find the x-intercepts.
$x = -1$, $x = 0$, $x = 1$

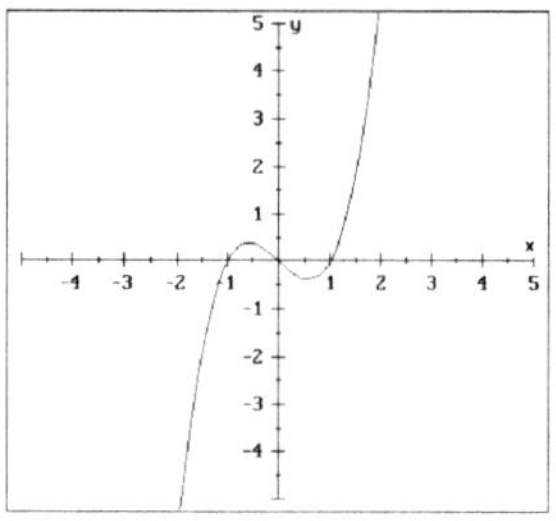

99. Graph $y = |x^2 - 2| - 1$.
Find the x-intercepts.
$x = -1.73$, $x = -1$, $x = 1$, $x = 1.73$

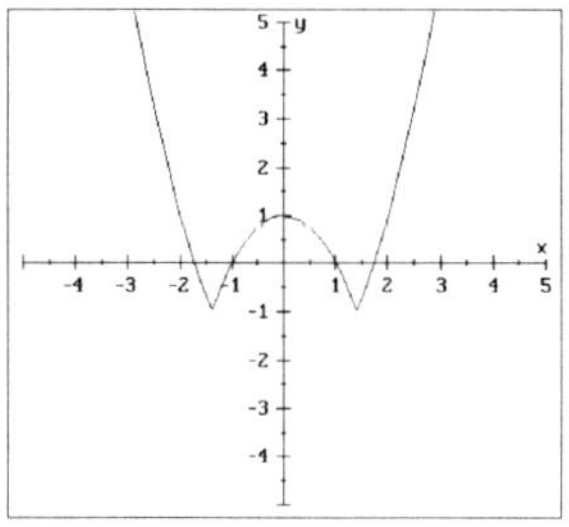

100. Graph $y = x^2 - 3x - 5$.
Find the x-intercepts.
$x = -1.19$, $x = 4.19$

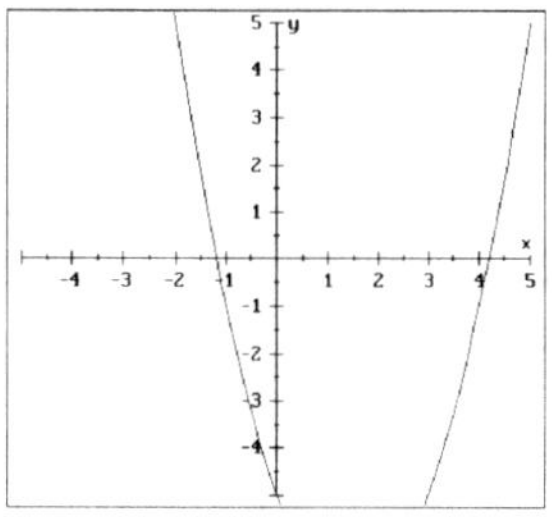

CHAPTER 2 REVIEW

101.
$$\begin{aligned}\frac{x+3}{10} &= \frac{x-1}{x}\\ x(x+3) &= 10(x-1)\\ x^2+3x &= 10x-10\\ x^2-7x+10 &= 0\\ (x-5)(x-2) &= 0\\ x = 5 &\text{ or } x = 2\end{aligned}$$

102.
$$\begin{aligned}\frac{x-1}{2} &= \frac{12}{x+1}\\ (x+1)(x-1) &= 2(12)\\ x^2-1 &= 24\\ x^2 &= 25\\ x &= \pm 5\end{aligned}$$

103.
$$\begin{aligned}f &= ks & f &= \frac{3}{5}s\\ 3 &= k(5) & f &= \frac{3}{5}(3)\\ \frac{3}{5} &= k & f &= \frac{9}{5}\text{ pounds}\end{aligned}$$

104. $E = kv^2$

30 mph	50 mph
$E = k(30)^2$	$E = k(50)^2$
$E = 900k$	$E = 2500k$

$$\text{Factor of increase} = \frac{2500k}{900k} = \frac{25}{9}$$

105.
$$\begin{aligned}V &= \frac{kT}{P} & V &= \frac{\frac{100}{3}T}{P}\\ 400 &= \frac{k(300)}{25} & V &= \frac{\frac{100}{3}(200)}{20}\\ 10000 &= 300k & V &= \frac{1000}{3}\\ \frac{100}{3} &= k & V &= 333\tfrac{1}{3}\text{ cm}^3\end{aligned}$$

106. $A = klw$
$A = 1lw \Rightarrow k = 1$

107.
$$\begin{aligned}R &= \frac{kL}{D^2} & R &= \frac{0.0005L}{D^2}\\ 200 &= \frac{k(1000)}{(.05)^2} & V &= \frac{0.0005(1500)}{(0.08)^2}\\ 200 &= \frac{1000k}{.0025} & V &\approx 117\text{ ohms}\\ 0.0005 &= k\end{aligned}$$

108. Let x = # rolls and y = total charge. Then the equation is $y = mx + b$, where m is the charge per roll and b is the fixed amount.
Given points: (11, 177) and (20, 294)

$$m = \frac{y_2 - y_1}{x_2 - x_1} = \frac{294-177}{20-11} = \frac{117}{9} = 13$$

$$\begin{aligned}y - y_1 &= m(x - x_1) & &\text{Let } x = 27:\\ y - 177 &= 13(x-11) & y &= 13x + 34\\ y &= 13x + 34 & y &= 13(27) + 34\\ & & y &= \$385\end{aligned}$$

109. $14x + 18y = 5040$
Let $x = 180$:
$$\begin{aligned}14(180) + 18y &= 5040\\ 2520 + 18y &= 5040\\ 18y &= 2520\\ y &= 140 \text{ hours of tutoring Spanish.}\end{aligned}$$

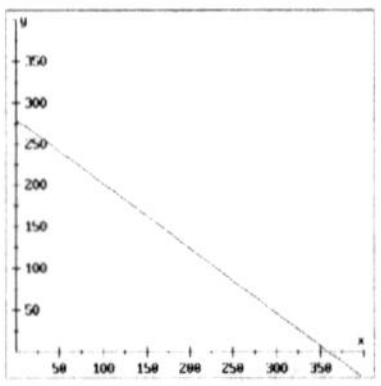

Chapter 2 Test (page 275)

1. $(-3, \pi) \Rightarrow$ QII

2. $(0, -8) \Rightarrow$ negative y-axis

3.
$$\begin{aligned} x + 3y &= 6 \\ x + 3(0) &= 6 \\ x &= 6 \\ &(6, 0) \end{aligned} \qquad \begin{aligned} x + 3y &= 6 \\ 0 + 3y &= 6 \\ y &= 2 \\ &(0, 2) \end{aligned}$$

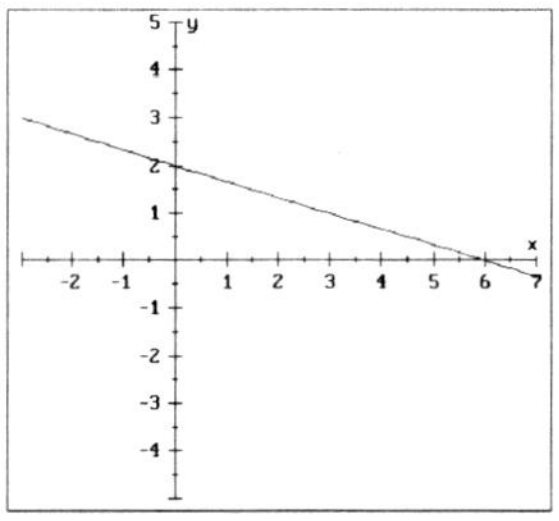

4.
$$\begin{aligned} 2x - 5y &= 10 \\ 2x - 5(0) &= 10 \\ x &= 5 \\ &(5, 0) \end{aligned} \qquad \begin{aligned} 2x - 5y &= 10 \\ 2(0) - 5y &= 10 \\ y &= -2 \\ &(0, -2) \end{aligned}$$

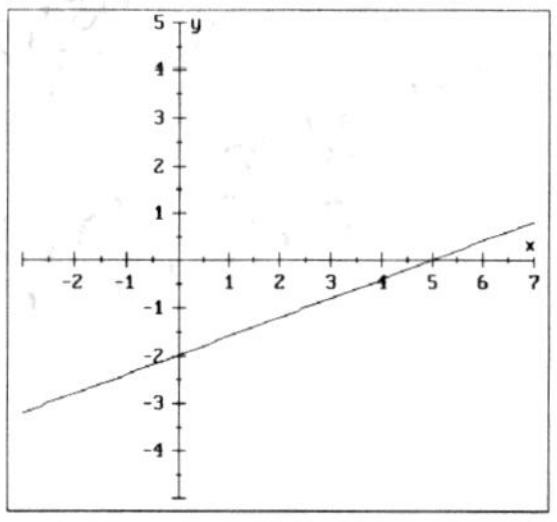

5.
$$\begin{aligned} 2(x + y) &= 3x + 5 \\ 2x + 2y &= 3x + 5 \\ 2y &= x + 5 \\ y &= \tfrac{1}{2}x + \tfrac{5}{2} \end{aligned}$$

x	y
0	$\frac{5}{2}$
1	3

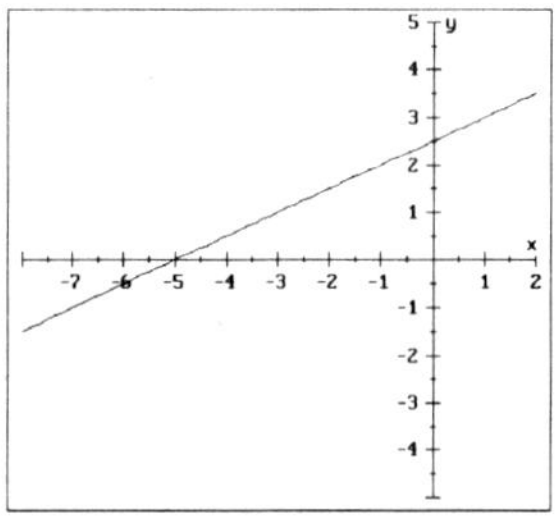

6.
$$\begin{aligned} 3x - 5y &= 3(x - 5) \\ 3x - 5y &= 3x - 15 \\ -5y &= -15 \\ y &= 3 \end{aligned}$$

x	y
0	3
-2	3

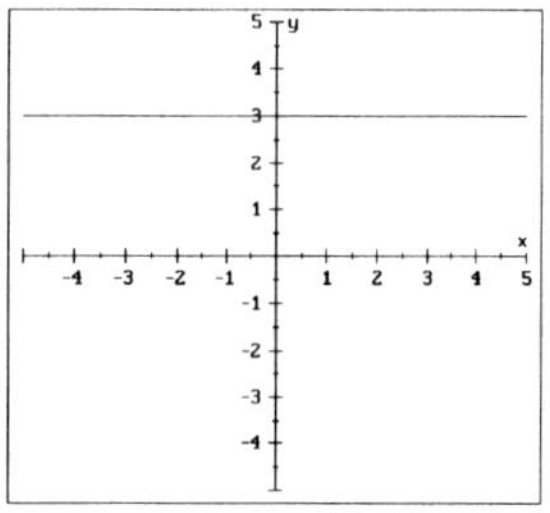

7.

$$\begin{aligned} \tfrac{1}{2}(x-2y) &= y-1 \\ \tfrac{1}{2}x - y &= y-1 \\ x-2y &= 2y-2 \\ -4y &= -x-2 \\ y &= \tfrac{1}{4}x+\tfrac{1}{2} \end{aligned}$$

x	y
0	$\frac{1}{2}$
2	1

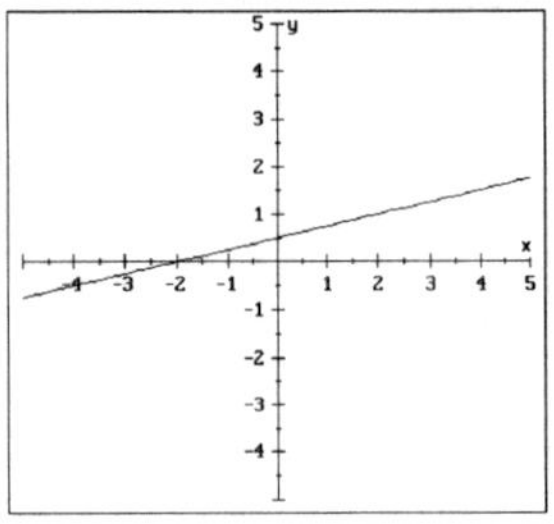

8.

$$\begin{aligned} \frac{x+y-5}{7} &= 3x \\ x+y-5 &= 21x \\ y &= 20x+5 \end{aligned}$$

x	y
0	5
$-\frac{1}{4}$	0

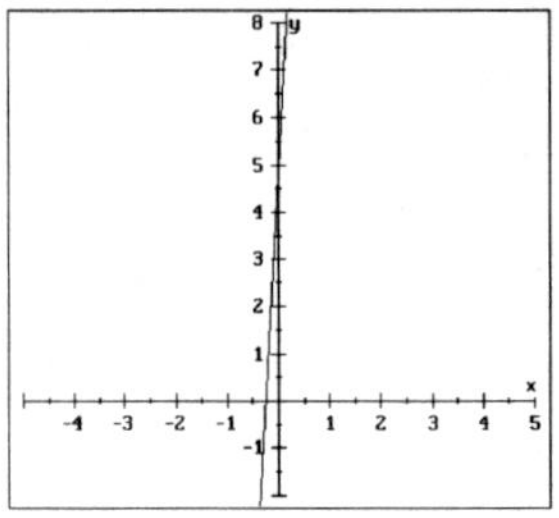

9.

$$\begin{aligned} d &= \sqrt{(x_2-x_1)^2+(y_2-y_1)^2} \\ &= \sqrt{(1-(-3))^2+(-1-4)^2} \\ &= \sqrt{(4)^2+(-5)^2} \\ &= \sqrt{16+25} = \sqrt{41} \end{aligned}$$

10.

$$\begin{aligned} d &= \sqrt{(x_2-x_1)^2+(y_2-y_1)^2} \\ &= \sqrt{(0-(-\pi))^2+(\pi-0)^2} \\ &= \sqrt{\pi^2+\pi^2} \\ &= \sqrt{2\pi^2} = \pi\sqrt{2} \approx 4.44 \end{aligned}$$

11. $M\left(\frac{x_1+x_2}{2}, \frac{y_1+y_2}{2}\right) = M\left(\frac{3+(-3)}{2}, \frac{-7+7}{2}\right) = M\left(\frac{0}{2}, \frac{0}{2}\right) = M(0,0)$

12. $M\left(\frac{x_1+x_2}{2}, \frac{y_1+y_2}{2}\right) = M\left(\frac{0+\sqrt{8}}{2}, \frac{\sqrt{2}+\sqrt{18}}{2}\right) = M\left(\frac{2\sqrt{2}}{2}, \frac{4\sqrt{2}}{2}\right) = M\left(\sqrt{2}, 2\sqrt{2}\right)$

13. $m = \frac{y_2-y_1}{x_2-x_1} = \frac{1-(-9)}{-5-3} = \frac{10}{-8} = -\frac{5}{4}$

14. $m = \frac{y_2-y_1}{x_2-x_1} = \frac{0-3}{-\sqrt{12}-\sqrt{3}} = \frac{-3}{-3\sqrt{3}} = \frac{1}{\sqrt{3}} = \frac{\sqrt{3}}{3}$

15. $y = 3x-2$ $\quad y = 2x-3$

$m = 3$ $\quad m = 2$

neither

16.

$$\begin{aligned} 2x-3y &= 5 \\ -3y &= -2x+5 \\ y &= \tfrac{2}{3}x-\tfrac{5}{3} \\ m &= \tfrac{2}{3} \end{aligned} \qquad \begin{aligned} 3x+2y &= 7 \\ 2y &= -3x+7 \\ y &= -\tfrac{3}{2}x+\tfrac{7}{2} \\ m &= -\tfrac{3}{2} \end{aligned}$$

perpendicular

CHAPTER 2 TEST

17.
$$\begin{aligned} y - y_1 &= m(x - x_1) \\ y + 5 &= 2(x - 3) \\ y + 5 &= 2x - 6 \\ y &= 2x - 11 \end{aligned}$$

18.
$$\begin{aligned} y &= mx + b \\ y &= 3x + \frac{1}{2} \end{aligned}$$

19.
$$\begin{aligned} 2x - y &= 3 \\ -y &= -2x + 3 \\ y &= 2x - 3 \\ m &= 2 \end{aligned} \qquad y = 2x + 5$$

20.
$$\begin{aligned} 2x - y &= 3 \\ -y &= -2x + 3 \\ y &= 2x - 3 \\ m &= 2 \end{aligned} \qquad y = -\tfrac{1}{2}x + 5$$

21.
$$m = \frac{y_2 - y_1}{x_2 - x_1} = \frac{\frac{1}{2} - \left(-\frac{3}{2}\right)}{3 - 2} = \frac{\frac{4}{2}}{1} = 2$$
$$\begin{aligned} y - y_1 &= m(x - x_1) \\ y - \frac{1}{2} &= 2(x - 3) \\ y - \frac{1}{2} &= 2x - 6 \\ y &= 2x - \frac{11}{2} \end{aligned}$$

22. If the line is parallel to the y-axis, then it is a vertical line: $x = 3$

23.
$$\begin{aligned} y &= x^3 - 16x \\ 0 &= x\left(x^2 - 16\right) \\ 0 &= x(x + 4)(x - 4) \end{aligned}$$
$x = 0,\ x = -4,\ x = 4$

x-int: $(0, 0),\ (-4, 0),\ (4, 0)$

$$\begin{aligned} y &= x^3 - 16x \\ y &= 0^3 - 16(0) \\ y &= 0 \end{aligned}$$
y-int: $(0, 0)$

24.
$$\begin{aligned} y &= |x - 4| \\ 0 &= |x - 4| \\ 0 &= x - 4 \\ 4 &= x \end{aligned}$$
x-int: $(4, 0)$

$$\begin{aligned} y &= |x - 4| \\ y &= |0 - 4| \\ y &= |-4| \\ y &= 4 \end{aligned}$$
y-int: $(0, 4)$

25. $y^2 = x - 1$

x-axis	y-axis	origin
$(-y)^2 = x - 1$	$y^2 = -x - 1$	$(-y)^2 = -x - 1$
$y^2 = x - 1$	not equivalent: no symmetry	$y^2 = -x - 1$
equivalent: **symmetry**		not equivalent: no symmetry

26. $y = x^4 + 1$

x-axis	y-axis	origin
$-y = x^4 + 1$	$y = (-x)^4 + 1$	$-y = (-x)^4 + 1$
not equivalent: no symmetry	$y = x^4 + 1$	$-y = x^4 + 1$
	equivalent: **symmetry**	not equivalent: no symmetry

CHAPTER 2 TEST

27. $y = x^2 - 9$
x-int: $(3, 0)$, $(-3, 0)$
y-int: $(0, -9)$
symmetry: y-axis

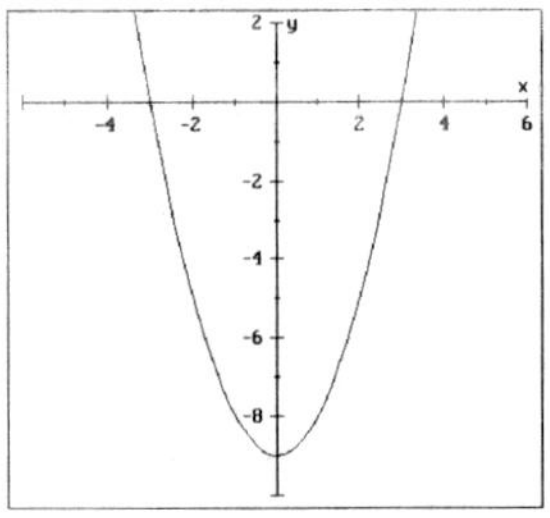

28. $x = |y|$
x-int: $(0, 0)$
y-int: $(0, 0)$
symmetry: x-axis

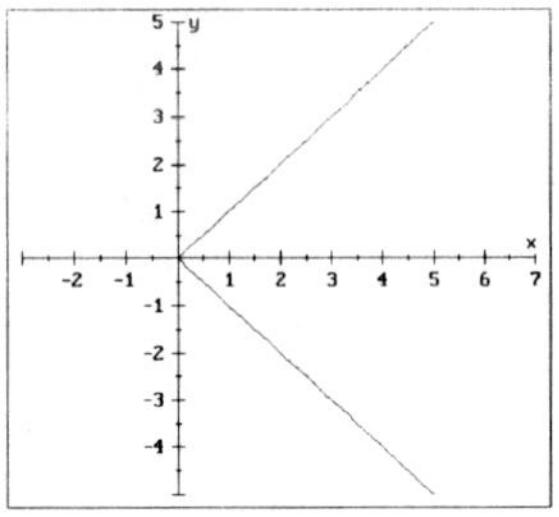

29. $y = 2\sqrt{x}$
x-int: $(0, 0)$
y-int: $(0, 0)$
symmetry: none

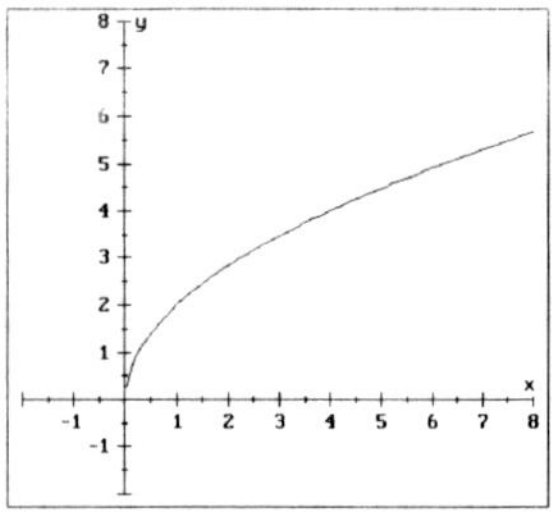

30. $x = y^3$
x-int: $(0, 0)$
y-int: $(0, 0)$
symmetry: origin

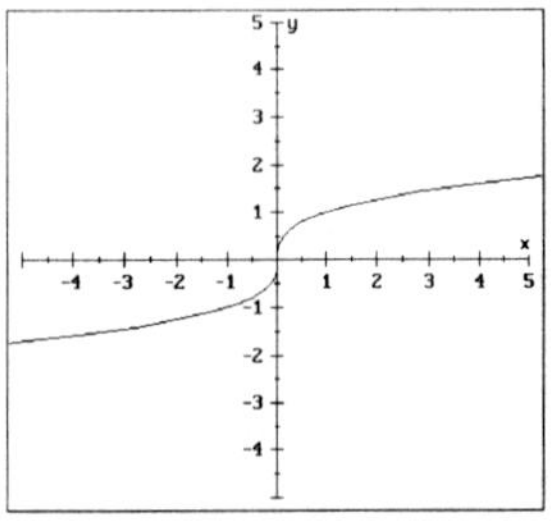

31. $C(5, 7); r = 8$
$(x - h)^2 + (y - k)^2 = r^2$
$(x - 5)^2 + (y - 7)^2 = 64$

32. $r = \sqrt{(2-6)^2 + (4-8)^2}$
$= \sqrt{32}$
$(x - h)^2 + (y - k)^2 = r^2$
$(x - 2)^2 + (y - 4)^2 = 32$

33. $x^2 + y^2 = 9$
$C(0, 0), r = 3$

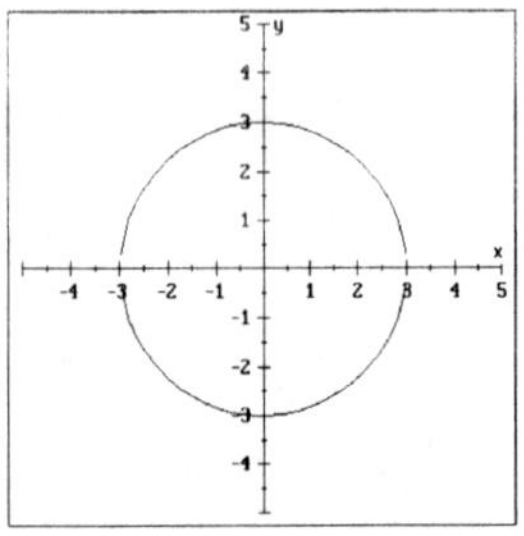

34.
$$\begin{aligned} x^2 - 4x + y^2 + 3 &= 0 \\ x^2 - 4x + y^2 &= -3 \\ x^2 - 4x + 4 + y^2 &= -3 + 4 \\ (x-2)^2 + y^2 &= 1 \\ C(2, 0), r &= 1 \end{aligned}$$

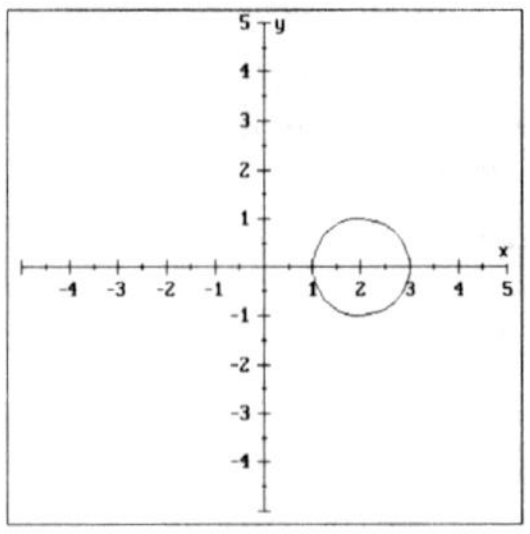

35. $y = kz^2$

36. $w = krs^2$

37.
$$P = kQ \qquad 7 = k(2) \qquad \frac{7}{2} = k$$
$$P = \frac{7}{2}Q \qquad P = \frac{7}{2}(5) \qquad P = \frac{35}{2}$$

38.
$$y = \frac{kx}{z^2} \qquad 16 = \frac{k(3)}{2^2} \qquad 16 = \frac{3k}{4} \qquad 64 = 3k \qquad \frac{64}{3} = k$$
$$y = \frac{\frac{64}{3}x}{z^2} \qquad 2 = \frac{\frac{64}{3}x}{3^2} \qquad 2 = \frac{\frac{64}{3}x}{9} \qquad 18 = \frac{64}{3}x \qquad 54 = 64x \qquad \frac{54}{64} = x \qquad \frac{27}{32} = x$$

39. Graph $y = x^2 - 7$.
Find any positive x-intercepts.
$x = 2.65$

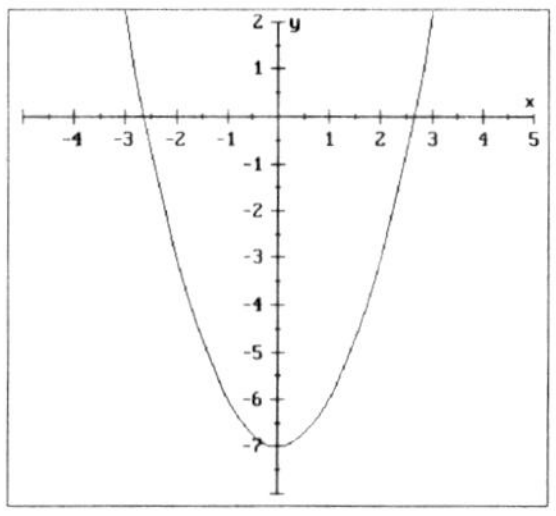

40. Graph $y = x^2 - 5x - 5$.
Find any positive x-intercepts.
$x = 5.85$

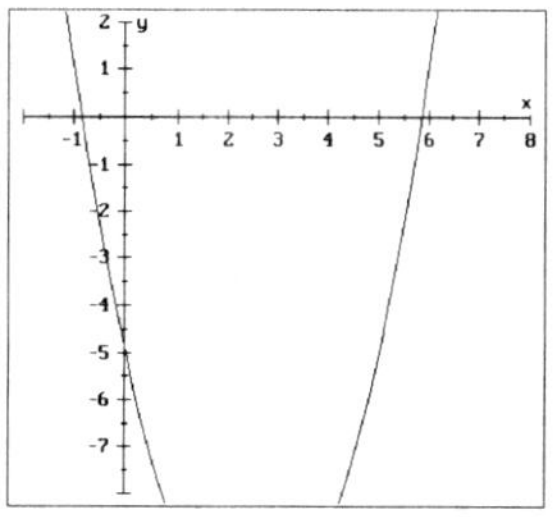

Exercises 3.1 (page 290)

1. function

3. domain

5. $y = f(x)$

7. x

9. vertical, once

11. $y = x$
Each value of x is paired with only one value of y. **function**

13. $y^2 = x$
$y = \pm\sqrt{x}$
At least one value of x is paired with more than one value of y. **not a function**

15. $y = x^2$
Each value of x is paired with only one value of y. **function**

17. $y^2 - 4x = 1$
$y^2 = 4x + 1$
$y = \pm\sqrt{4x+1}$
At least one value of x is paired with more than one value of y. **not a function**

19. $|x| = |y|$
$|y| = |x|$
$y = \pm|x|$
At least one value of x is paired with more than one value of y. **not a function**

21. $y = 7$
Each value of x is paired with only one value of y. **function**

23. $f(x) = 3x + 5 \Rightarrow \text{domain} = (-\infty, \infty)$

25. $f(x) = x^2 - x + 1 \Rightarrow \text{domain} = (-\infty, \infty)$

27. $f(x) = \sqrt{x-2} \Rightarrow x - 2 \geq 0$
domain $= [2, \infty)$

29. $f(x) = \sqrt{4-x} \Rightarrow 4 - x \geq 0$
domain $= (-\infty, 4]$

31. $f(x) = \sqrt{x^2-1} \Rightarrow x^2 - 1 \geq 0$
domain $= (-\infty, -1] \cup [1, \infty)$

33. $f(x) = \sqrt[3]{x+1} \Rightarrow \text{domain} = (-\infty, \infty)$

35. $f(x) = \dfrac{3}{x+1} \Rightarrow x \neq -1$
domain $= (-\infty, -1) \cup (-1, \infty)$

37. $f(x) = \dfrac{x}{x-3} \Rightarrow x \neq 3$
domain $= (-\infty, 3) \cup (3, \infty)$

39. $f(x) = \dfrac{x}{x^2-4} = \dfrac{x}{(x+2)(x-2)}$
$x \neq -2, x \neq 2$
domain $= (-\infty, -2) \cup (-2, 2) \cup (2, \infty)$

41. $f(x) = \dfrac{1}{x^2-4x-5} = \dfrac{1}{(x+1)(x-5)}$
$x \neq -1, x \neq 5$
domain $= (-\infty, -1) \cup (-1, 5) \cup (5, \infty)$

43. $f(x) = 3x - 2$

$f(2) = 3(2) - 2$ $= 6 - 2$ $= 4$	$f(-3) = 3(-3) - 2$ $= -9 - 2$ $= -11$	$f(k) = 3k - 2$	$f(k^2-1) = 3(k^2-1) - 2$ $= 3k^2 - 3 - 2$ $= 3k^2 - 5$

45. $f(x) = \frac{1}{2}x + 3$

$f(2) = \frac{1}{2}(2) + 3$ $= 1 + 3$ $= 4$	$f(-3) = \frac{1}{2}(-3) + 3$ $= -\frac{3}{2} + 3$ $= \frac{3}{2}$	$f(k) = \frac{1}{2}k + 3$	$f(k^2-1) = \frac{1}{2}(k^2-1) + 3$ $= \frac{1}{2}k^2 - \frac{1}{2} + 3$ $= \frac{1}{2}k^2 + \frac{5}{2}$

47. $f(x) = x^2$

$f(2) = 2^2$ $= 4$	$f(-3) = (-3)^2$ $= 9$	$f(k) = k^2$	$f(k^2-1) = (k^2-1)^2$ $= (k^2-1)(k^2-1)$ $= k^4 - 2k^2 + 1$

49. $f(x) = x^2 + 3x - 1$

$f(2) = 2^2 + 3(2) - 1$ $= 4 + 6 - 1$ $= 9$	$f(-3) = (-3)^2 + 3(-3) - 1$ $= 9 - 9 - 1$ $= -1$	$f(k) = k^2 + 3k - 1$

$f(k^2-1) = (k^2-1)^2 + 3(k^2-1) - 1$ $= k^4 - 2k^2 + 1 + 3k^2 - 3 - 1$ $= k^4 + k^2 - 3$

SECTION 3.1

51. $$f(x) = |x^2+1|$$

$f(2) = \|2^2+1\|$ $= \|5\|$ $= 5$	$f(-3) = \|(-3)^2+1\|$ $= \|10\|$ $= 10$	$f(k) = \|k^2+1\|$ $= k^2+1$ $[k^2+1 \geq 0]$	$f(k^2-1) = \|(k^2-1)^2+1\|$ $= (k^2-1)^2+1$ $= k^4-2k^2+1+1$ $= k^4-2k^2+2$ $[(k^2-1)^2+1 \geq 0]$

53. $$f(x) = \frac{2}{x+4}$$

$f(2) = \frac{2}{2+4}$ $= \frac{2}{6} = \frac{1}{3}$	$f(-3) = \frac{2}{-3+4}$ $= \frac{2}{1} = 2$	$f(k) = \frac{2}{k+4}$	$f(k^2-1) = \frac{2}{k^2-1+4}$ $= \frac{2}{k^2+3}$

55. $$f(x) = \frac{1}{x^2-1}$$

$f(2) = \frac{1}{2^2-1}$ $= \frac{1}{4-1}$ $= \frac{1}{3}$	$f(-3) = \frac{1}{(-3)^2-1}$ $= \frac{1}{9-1}$ $= \frac{1}{8}$	$f(k) = \frac{1}{k^2-1}$	$f(k^2-1) = \frac{1}{(k^2-1)^2-1}$ $= \frac{1}{k^4-2k^2+1-1}$ $= \frac{1}{k^4-2k^2}$

57. $$f(x) = \sqrt{x^2+1}$$

$f(2) = \sqrt{2^2+1}$ $= \sqrt{4+1}$ $= \sqrt{5}$	$f(-3) = \sqrt{(-3)^2+1}$ $= \sqrt{9+1}$ $= \sqrt{10}$	$f(k) = \sqrt{k^2+1}$	$f(k^2-1) = \sqrt{(k^2-1)^2+1}$ $= \sqrt{k^4-2k^2+1+1}$ $= \sqrt{k^4-2k^2+2}$

59. $$\frac{f(x+h)-f(x)}{h} = \frac{[3(x+h)+1]-[3x+1]}{h} = \frac{[3x+3h+1]-[3x+1]}{h} = \frac{3x+3h+1-3x-1}{h} = \frac{3h}{h} = 3$$

61. $$\frac{f(x+h)-f(x)}{h} = \frac{[(x+h)^2+1]-[x^2+1]}{h} = \frac{[x^2+2xh+h^2+1]-[x^2+1]}{h} = \frac{x^2+2xh+h^2+1-x^2-1}{h} = \frac{2xh+h^2}{h} = \frac{h(2x+h)}{h} = 2x+h$$

63. $$\frac{f(x+h)-f(x)}{h}=\frac{\left[4(x+h)^2-6\right]-\left[4x^2-6\right]}{h}=\frac{\left[4x^2+8xh+4h^2-6\right]-\left[4x^2-6\right]}{h}$$
$$=\frac{4x^2+8xh+4h^2-6-4x^2+6}{h}$$
$$=\frac{8xh+4h^2}{h}=\frac{h(8x+4h)}{h}=8x+4h$$

65. $$\frac{f(x+h)-f(x)}{h}=\frac{\left[(x+h)^2+3(x+h)-7\right]-\left[x^2+3x-7\right]}{h}$$
$$=\frac{\left[x^2+2xh+h^2+3x+3h-7\right]-\left[x^2+3x-7\right]}{h}$$
$$=\frac{x^2+2xh+h^2+3x+3h-7-x^2-3x+7}{h}$$
$$=\frac{2xh+h^2+3h}{h}=\frac{h(2x+h+3)}{h}=2x+h+3$$

67. $$\frac{f(x+h)-f(x)}{h}=\frac{\left[2(x+h)^2-4(x+h)+2\right]-\left[2x^2-4x+2\right]}{h}$$
$$=\frac{\left[2x^2+4xh+2h^2-4x-4h+2\right]-\left[2x^2-4x+2\right]}{h}$$
$$=\frac{2x^2+4xh+2h^2-4x-4h+2-2x^2+4x-2}{h}$$
$$=\frac{4xh+2h^2-4h}{h}=\frac{h(4x+2h-4)}{h}=4x+2h-4$$

69. $$\frac{f(x+h)-f(x)}{h}=\frac{(x+h)^3-x^3}{h}=\frac{\left[x^3+3x^2h+3xh^2+h^3\right]-\left[x^3\right]}{h}$$
$$=\frac{3x^2h+3xh^2+h^3}{h}$$
$$=\frac{h(3x^2+3xh+h^2)}{h}=3x^2+3xh+h^2$$

71. $f(x)=2x+3$

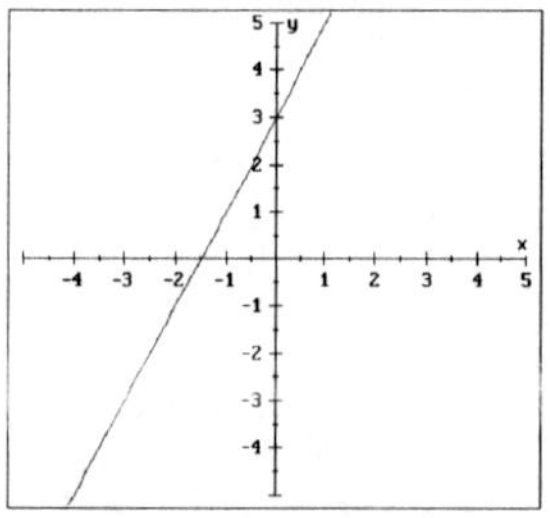

domain $=(-\infty,\infty)$
range $=(-\infty,\infty)$

73. $f(x)=-\frac{3}{4}x+4$

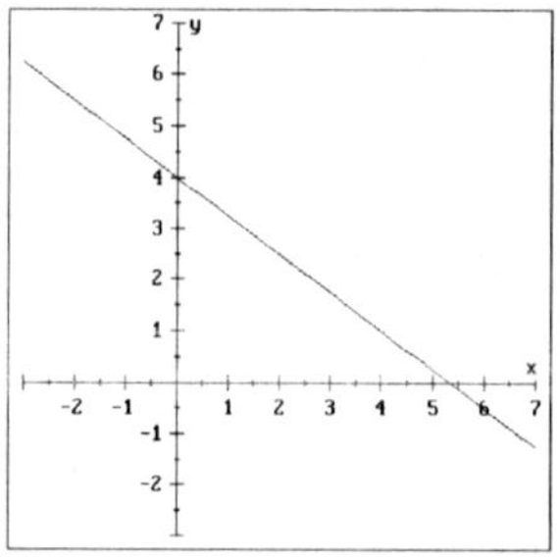

domain $=(-\infty,\infty)$
range $=(-\infty,\infty)$

75. $2x = 3y - 3$
$-3y = -2x - 3$
$y = \dfrac{2}{3}x + 1$

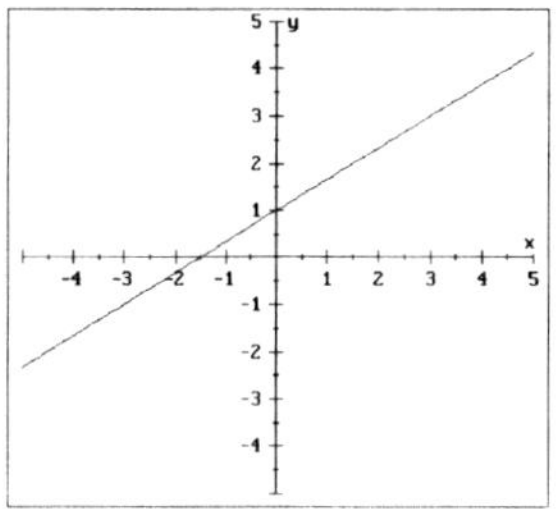

domain $= (-\infty, \infty)$
range $= (-\infty, \infty)$

77. $f(x) = x^2 - 4$

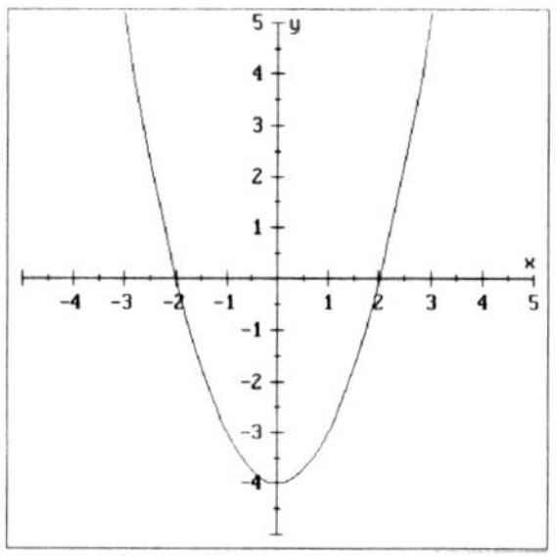

domain $= (-\infty, \infty)$
range $= [-4, \infty)$

79. $f(x) = -x^3 + 2$

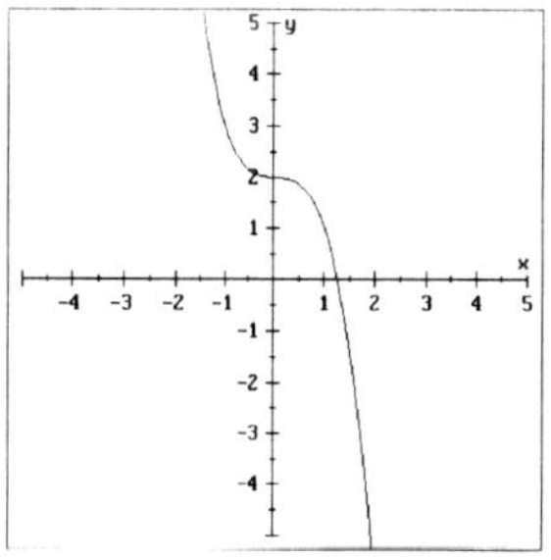

domain $= (-\infty, \infty)$
range $= (-\infty, \infty)$

81. $f(x) = -|x|$

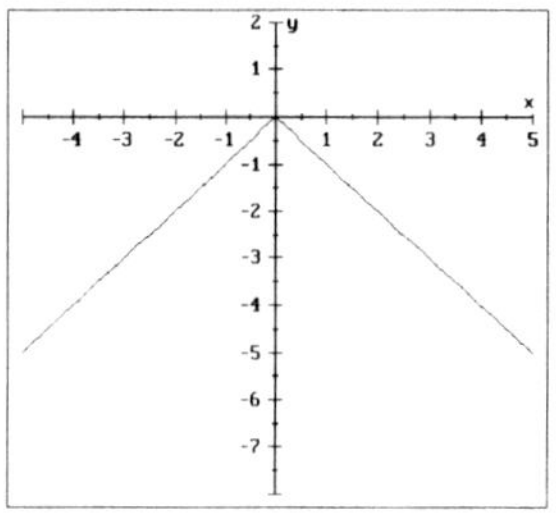

domain $= (-\infty, \infty)$
range $= (-\infty, 0]$

83. $f(x) = |x - 2|$

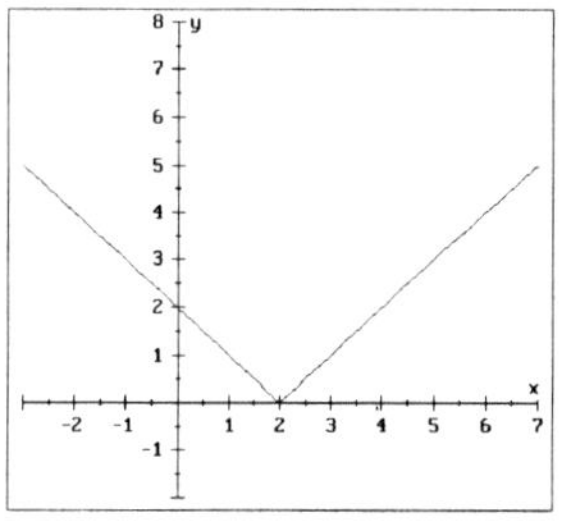

domain $= (-\infty, \infty)$
range $= [0, \infty)$

85. $f(x) = \left|\dfrac{1}{2}x + 3\right|$

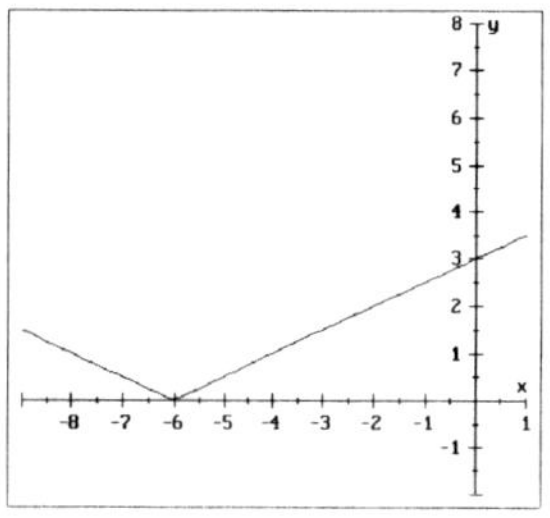

domain $= (-\infty, \infty)$
range $= [0, \infty)$

87. $f(x) = -\sqrt{x + 1}$

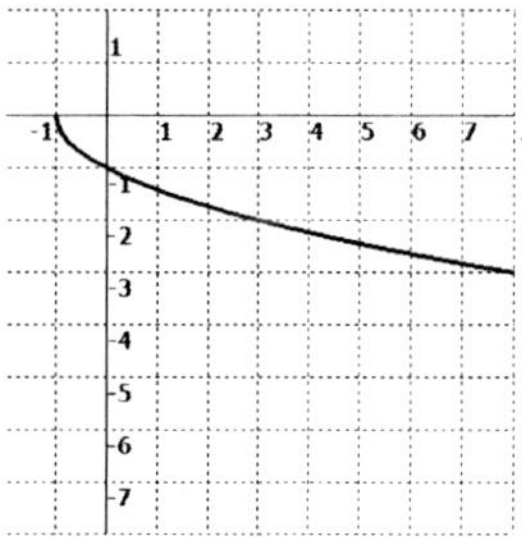

domain $= [-1, \infty)$
range $= [0, \infty)$

89. $f(x) = \sqrt{2x - 4}$

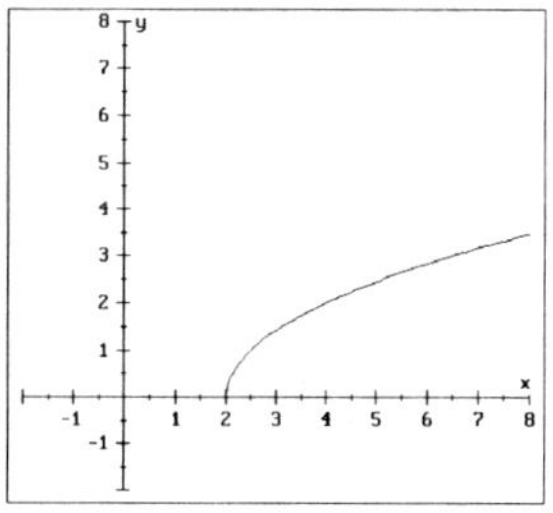

domain $= [2, \infty)$
range $= [0, \infty)$

91. $f(x) = \sqrt[3]{x} + 2$

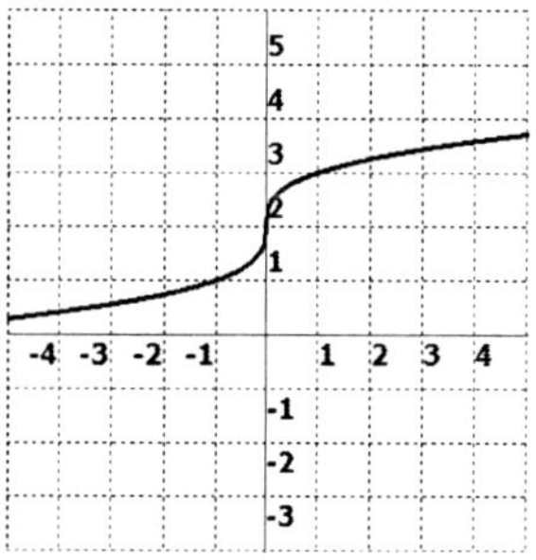

domain $= (-\infty, \infty)$
range $= (-\infty, \infty)$

93.

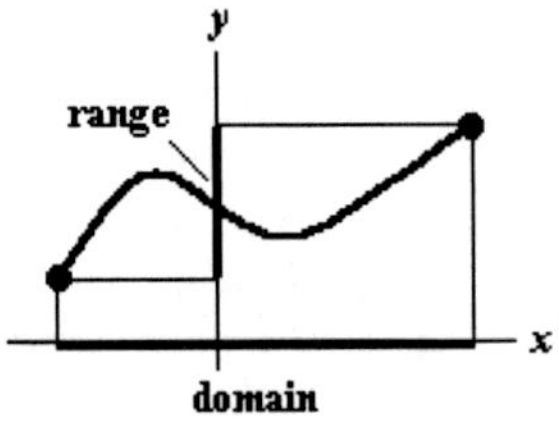

95. function

97. function

99. function

101. $f(x) = |3x + 2|$

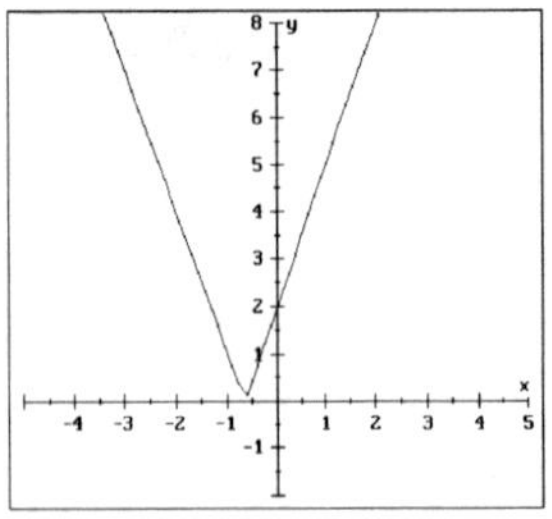

domain: $(-\infty, \infty)$; range: $[0, \infty)$

103. $f(x) = \sqrt[3]{5x - 1}$

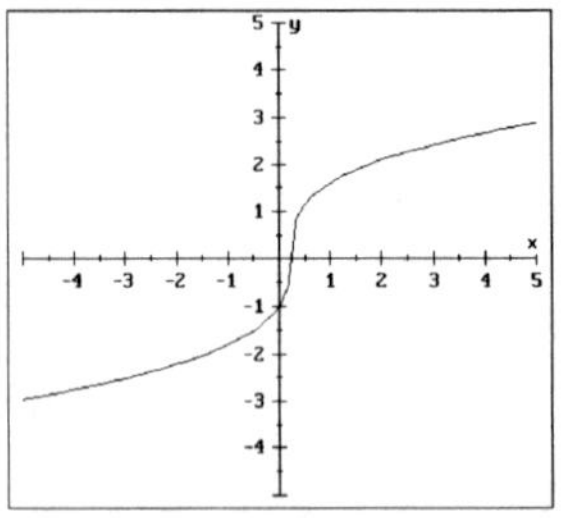

domain: $(-\infty, \infty)$; range: $(-\infty, \infty)$

105. **a.** $C(x) = 8x + 75$
b. $C(85) = 8(85) + 75 = \$755$

107. **a.** $C(t) = 0.07t + 9.99$
b. $C(20) = 0.07(20) + 9.99 = \$11.39$

109. **a.** $C(f) = 95f + 14000$
b. $C(2600) = 95(2600) + 14{,}000 = \$261{,}000$

111. Let (x, c) be a point on the line. Then the points $(100, 17)$ and $(500, 57)$ are on the line.

$$m = \frac{\Delta c}{\Delta x} = \frac{57 - 17}{500 - 100} = \frac{40}{400} = 0.10$$
$$C = mx + b$$
$$17 = 0.10(100) + b$$
$$17 = 10 + b$$
$$7 = b \Rightarrow \boxed{C(x) = 0.10x + 7}$$

113. Let (t, n) be a point on the line. Then the points $(0, 6400)$ and $(4, 13168)$ are on the line.

$$m = \frac{\Delta n}{\Delta t} = \frac{13168 - 6400}{4 - 0} = \frac{6768}{4} = 1692$$
$$n = mt + b$$
$$6400 = 1692(0) + b$$
$$6400 = b \Rightarrow \boxed{n(t) = 1692t + 6400}$$

SECTION 3.1

115. Let (D, E) be a point on the line. Then the points $(50, 69.555)$ and $(125, 173.8875)$ are on the line.

$$m = \frac{\Delta E}{\Delta D} = \frac{173.8875 - 69.555}{125 - 50} = \frac{104.3325}{75} = 1.3911$$

$$E = mD + b$$

$$69.555 = 1.3911(50) + b$$

$$69.555 = 69.555 + b \Rightarrow b = 0 \Rightarrow \boxed{E(D) = 1.3911D}$$

117. $f(x) = 3x + 2$

$$0 = 3x + 2$$

$$-2 = 3x$$

$$-\frac{2}{3} = x$$

119-121. Answers may vary.

123. natural: $1, 7, 8$

125. prime: 7

127. $(-4, 7]$

129. $(-3, 5) \cup [6, \infty)$:

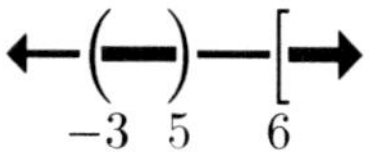

Exercises 3.2 (page 305)

1. $f(x) = ax^2 + bx + c$

3. $(3, 5)$

5. upward

7. $-\frac{b}{2a}$

9. $f(x) = \frac{1}{2}x^2 + 3$
$a = \frac{1}{2} \Rightarrow a > 0$
up, minimum

11. $f(x) = -3(x + 1)^2 + 2$
$a = -3 \Rightarrow a < 0$
down, maximum

13. $f(x) = -2x^2 + 5x - 1$
$a = -2 \Rightarrow a < 0$
down, maximum

15. $y = x^2 - 1 = (x - 0)^2 - 1$
Vertex: $(0, -1)$

17. $f(x) = (x - 3)^2 + 5$
Vertex: $(3, 5)$

19. $f(x) = -2(x + 6)^2 - 4$
Vertex: $(-6, -4)$

21. $f(x) = \frac{2}{3}(x - 3)^2$
Vertex: $(3, 0)$

23. $f(x) = x^2 - 4x + 4;\ a = 1,\ b = -4,\ c = 4$

$$x = -\frac{b}{2a} = -\frac{-4}{2(1)} = 2$$

$$y = x^2 - 4x + 4 = 2^2 - 4(2) + 4 = 0$$

Vertex: $(2, 0)$

25. $y = x^2 + 6x - 3;\ a = 1,\ b = 6,\ c = -3$

$$x = -\frac{b}{2a} = -\frac{6}{2(1)} = -3$$

$$y = x^2 + 6x - 3 = (-3)^2 + 6(-3) - 3 = -12$$

Vertex: $(-3, -12)$

SECTION 3.2

27. $y = -2x^2 + 12x - 17;$

$a = -2, b = 12, c = -17$

$x = -\dfrac{b}{2a} = -\dfrac{12}{2(-2)} = 3$

$y = -2x^2 + 12x - 17$

$= -2(3)^2 + 12(3) - 17 = 1$

Vertex: $(3, 1)$

29. $y = 3x^2 - 4x + 5;$

$a = 3, b = -4, c = 5$

$x = -\dfrac{b}{2a} = -\dfrac{-4}{2(3)} = \dfrac{4}{6} = \dfrac{2}{3}$

$y = 3x^2 - 4x + 5 = 3\left(\dfrac{2}{3}\right)^2 - 4\left(\dfrac{2}{3}\right) + 5$

$= \dfrac{11}{3}$

Vertex: $\left(\dfrac{2}{3}, \dfrac{11}{3}\right)$

31. $y = \dfrac{1}{2}x^2 + 4x - 3;$

$a = \dfrac{1}{2}, b = 4, c = -3$

$x = -\dfrac{b}{2a} = -\dfrac{4}{2\left(\frac{1}{2}\right)} = -4$

$y = \dfrac{1}{2}x^2 + 4x - 3$

$= \dfrac{1}{2}(-4)^2 + 4(-4) - 3 = -11$

Vertex: $(-4, -11)$

33. $f(x) = x^2 - 4 = (x - 0)^2 - 4$

$a = 1 \Rightarrow$ up, vertex: $(0, -4)$

$0 = x^2 - 4$

$0 = (x + 2)(x - 2)$

$x = -2, x = 2 \Rightarrow (-2, 0), (2, 0)$

$f(0) = -4 \Rightarrow (0, -4)$

$f(1) = -3 \Rightarrow (1, -3)$

$(-1, -3)$ on graph by symmetry

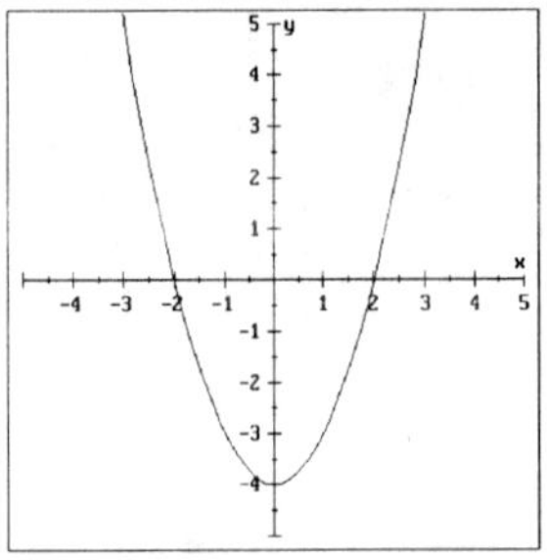

35. $f(x) = -3x^2 + 6 = -3(x - 0)^2 + 6$

$a = -3 \Rightarrow$ down, vertex: $(0, 6)$

$0 = -3x^2 + 6$

$x^2 = 2$

$x = \pm\sqrt{2} \Rightarrow \left(-\sqrt{2}, 0\right), \left(\sqrt{2}, 0\right)$

$f(0) = 6 \Rightarrow (0, 6)$

$f(1) = 3 \Rightarrow (1, 3)$

$(-1, 3)$ on graph by symmetry

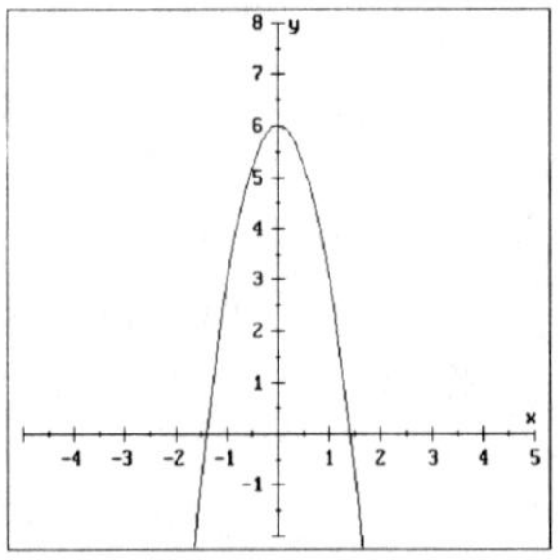

37. $f(x) = -\frac{1}{2}x^2 + 8 = -\frac{1}{2}(x-0)^2 + 8$

$a = -\frac{1}{2} \Rightarrow$ down, vertex: $(0, 8)$

$0 = -\frac{1}{2}x^2 + 8$

$x^2 = 16$

$x = \pm 4 \Rightarrow (-4, 0), (4, 0)$

$f(0) = 8 \Rightarrow (0, 8)$

$f(2) = 6 \Rightarrow (2, 6)$

$(-2, 6)$ on graph by symmetry

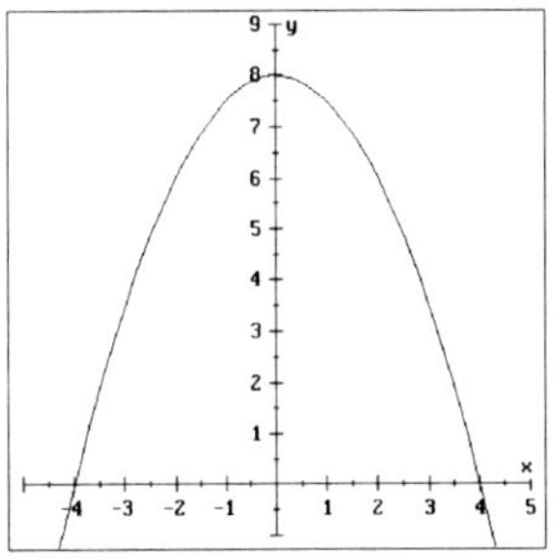

39. $f(x) = (x-3)^2 - 1$

$a = 1 \Rightarrow$ up, vertex: $(3, -1)$

$0 = (x-3)^2 - 1$

$1 = (x-3)^2$

$\pm 1 = x - 3$

$3 \pm 1 = x$

$x = 2, x = 4 \Rightarrow (2, 0), (4, 0)$

$f(0) = 8 \Rightarrow (0, 8)$

$(6, 8)$ on graph by symmetry

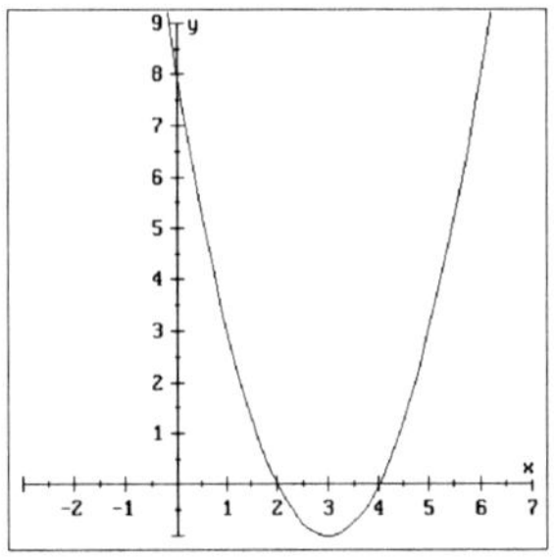

41. $f(x) = 2(x+1)^2 - 2$

$a = 2 \Rightarrow$ up, vertex: $(-1, -2)$

$0 = 2(x+1)^2 - 2$

$2 = 2(x+1)^2$

$1 = (x+1)^2$

$\pm 1 = x + 1$

$-1 \pm 1 = x$

$x = -2, x = 0 \Rightarrow (-2, 0), (0, 0)$

$f(0) = 0 \Rightarrow (0, 0)$

$f(1) = 6 \Rightarrow (1, 6)$

$(-3, 6)$ on graph by symmetry

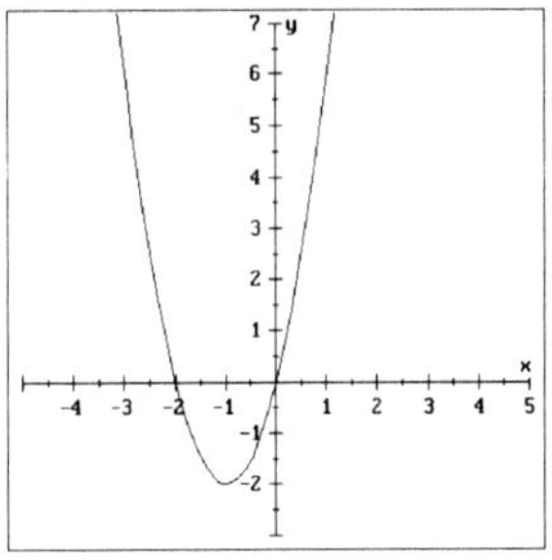

43. $f(x) = -(x+4)^2 + 1$

$a = -1 \Rightarrow$ down, vertex: $(-4, 1)$

$0 = -(x+4)^2 + 1$

$(x+4)^2 = 1$

$x + 4 = \pm 1$

$x = -4 \pm 1$

$x = -5, x = -3 \Rightarrow (-5, 0), (-3, 0)$

$f(0) = -15 \Rightarrow (0, -15)$

$(-8, -15)$ on graph by symmetry

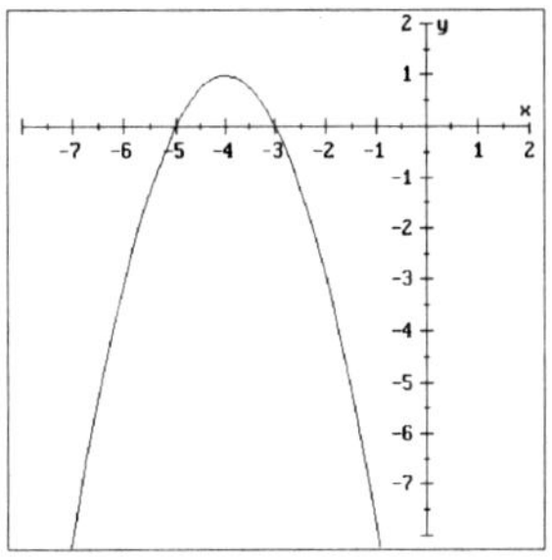

45. $f(x) = -3(x-2)^2 + 6$

$a = -3 \Rightarrow$ down, vertex: $(2, 6)$

$0 = -3(x-2)^2 + 6$

$3(x-2)^2 = 6$

$(x-2)^2 = 2$

$x - 2 = \pm\sqrt{2}$

$x = 2 \pm \sqrt{2}$

$\left(2 - \sqrt{2}, 0\right), \left(2 + \sqrt{2}, 0\right)$

$f(0) = -6 \Rightarrow (0, -6)$

$(4, -6)$ on graph by symmetry

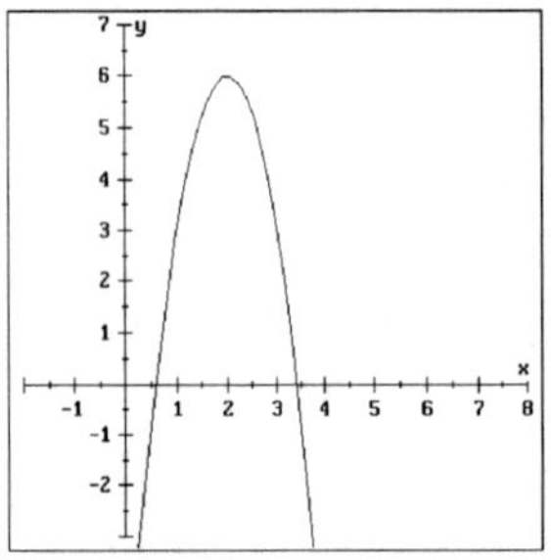

47. $f(x) = x^2 + 2x;\ a = 1, b = 2, c = 0$

$x = -\dfrac{b}{2a} = -\dfrac{2}{2(1)} = -1$

$y = x^2 + 2x = (-1)^2 + 2(-1) = -1$

vertex: $(-1, -1)$, $a = 1 \Rightarrow$ up

$0 = x^2 + 2x$

$0 = x(x+2)$

$x = 0$ or $x = -2 \Rightarrow (0, 0), (-2, 0)$

$f(0) = 0 \Rightarrow (0, 0)$

$f(1) = 3 \Rightarrow (1, 3)$ on graph

$(-3, 3)$ on graph by symmetry

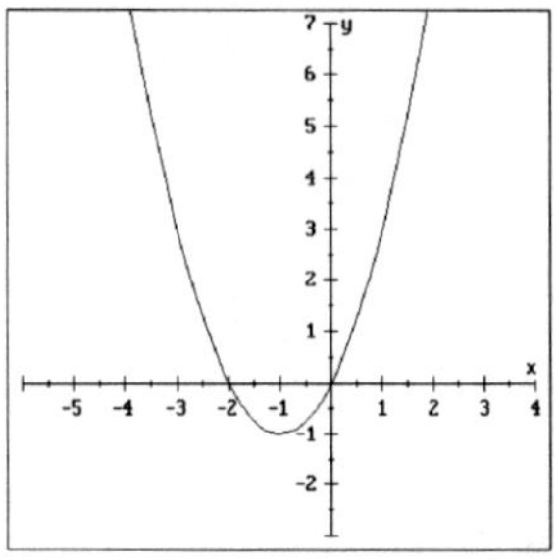

49. $f(x) = x^2 - 4x + 1;\ a = 1, b = -4, c = 1$

$x = -\dfrac{b}{2a} = -\dfrac{-4}{2(1)} = 2$

$y = x^2 - 4x + 1 = 2^2 - 4(2) + 1 = -3$

vertex: $(2, -3)$, $a = 1 \Rightarrow$ up

$0 = x^2 - 4x + 1$

$x = 2 \pm \sqrt{3}$ by quadratic formula

$\left(2 - \sqrt{3}, 0\right), \left(2 + \sqrt{3}, 0\right)$

$f(0) = 0 \Rightarrow (0, 1)$

$(4, 1)$ on graph by symmetry

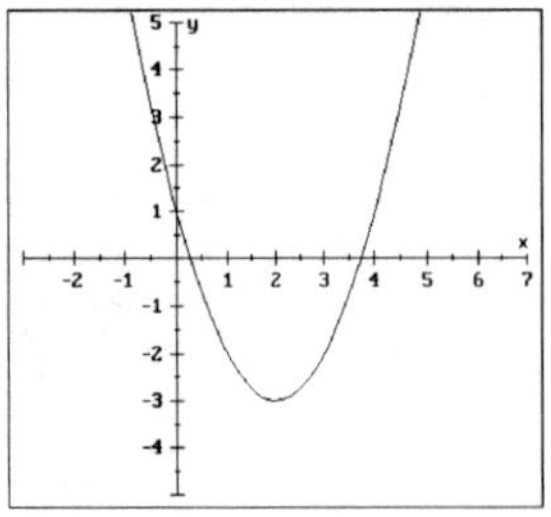

51. $f(x) = 2x^2 - 12x + 10$

$a = 2, b = -12, c = 10$

$x = -\frac{b}{2a} = -\frac{-12}{2(2)} = 3$

$y = 2(3)^2 - 12(3) + 10 = -8$

vertex: $(3, -8)$, $a = 2 \Rightarrow$ up

$0 = 2x^2 - 12x + 10$

$0 = 2\left(x^2 - 6x + 5\right)$

$0 = 2(x-1)(x-5)$

$x = 1$ or $x = 5 \Rightarrow (1, 0), (5, 0)$

$f(0) = 10 \Rightarrow (0, 10)$

$(6, 10)$ on graph by symmetry

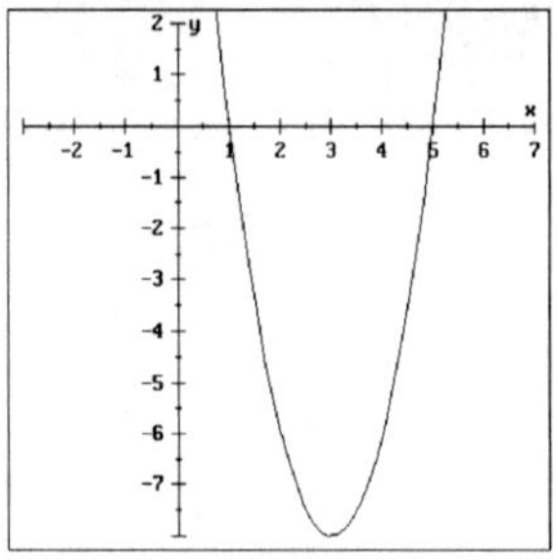

53. $f(x) = -3x^2 - 6x - 9$

$a = -3,\ b = -6,\ c = -9$

$x = -\frac{b}{2a} = -\frac{-6}{2(-3)} = -1$

$y = -3(-1)^2 - 6(-1) - 9 = -6$

vertex: $(-1, -6)$, $a = -3 \Rightarrow$ down

$0 = -3x^2 - 6x - 9$

$0 = -3(x^2 + 2x + 3)$

impossible $\Rightarrow$ no x-intercepts

$f(0) = -9 \Rightarrow (0, -9)$

$(-2, -9)$ on graph by symmetry

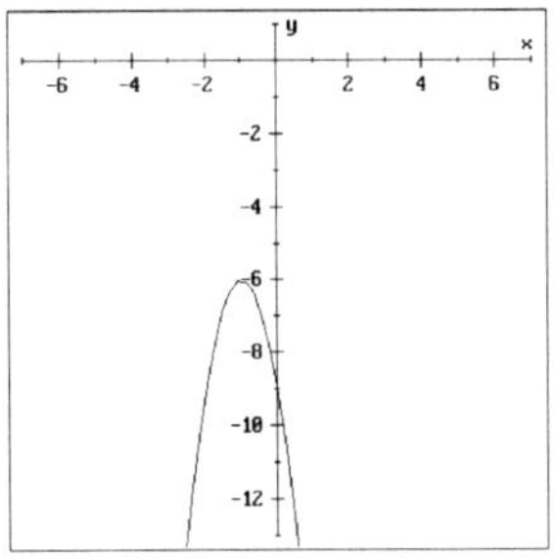

55. Let $x =$ the width of the region.

Then $\dfrac{300 - 2x}{2} = 150 - x =$ the length.

Area = width · length

$y = x(150 - x)$

$y = -x^2 + 150x$

$a = -1,\ b = 150,\ c = 0$

$x = -\dfrac{b}{2a} = -\dfrac{150}{2(-1)} = 75$

$150 - x = 150 - 75 = 75$

$y = 75(150 - 75) = 5625$

The dimensions are 75 ft by 75 ft, with an area of 5625 ft^2.

57. Let $x =$ the width.
Then $50 - x =$ the length.

Area $= lw$

$y = (50 - x)x$

$y = -x^2 + 50x$

$a = -1,\ b = 50,\ c = 0$

$x = -\dfrac{b}{2a} = -\dfrac{50}{2(-1)} = 25$

$y = (50 - 25)25 = 625$

The maximum area occurs when the dimensions are 25 ft by 25 ft.

59. Set up the variables :

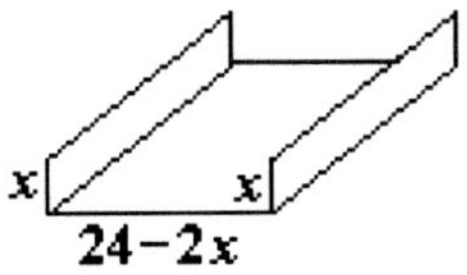

Area $= lw$

$y = x(24 - 2x)$

$y = 24x - 2x^2$

$y = -2x^2 + 24x$

$a = -2,\ b = 24,\ c = 0$

$x = -\dfrac{b}{2a} = -\dfrac{24}{2(-2)} = 6$

$y = -2x^2 + 24x = -2(6^2) + 24(6) = 72$

The maximum area occurs when the depth is 6 inches and the width is 12 inches.

61. $x^2 + 20y - 400 = 0$

$20y = -x^2 + 400$

$y = -\frac{1}{20}x^2 + 20$

$a = -\frac{1}{20},\ b = 0,\ c = 20$

$x = -\dfrac{b}{2a} = -\dfrac{0}{2\left(-\frac{1}{20}\right)} = 0$

$y = -\frac{1}{20}x^2 + 20 = -\frac{1}{20}(0)^2 + 20 = 20$

The maximum height is 20 feet.

63. $f(x) = -0.06x^2 + 1.5x + 6$

$a = -0.06,\ b = 1.5,\ c = 6$

$x = -\dfrac{b}{2a} = -\dfrac{1.5}{2(-0.06)} = 12.5$

$f(12.5) = -0.06(12.5)^2 + 1.5(12.5) + 6$
$= 15.375$

The maximum height is about 15.4 ft.

65. $s(t) = 144 + 64t - 16t^2 = -16t^2 + 64t + 144 \Rightarrow a = -16, b = 64, c = 144$

$$t = -\frac{b}{2a} = -\frac{64}{2(-16)} = 2$$

$s(2) = -16(2)^2 + 64(2) + 144 = 208$

The maximum height is 208 ft.

67. $C(x) = 1.5x^2 - 144x + 5856 \Rightarrow a = 1.5, b = -144, c = 5856$

$$x = -\frac{b}{2a} = -\frac{-144}{2(1.5)} = 48$$

$C(48) = 1.5(48)^2 - 144(48) + 5856 = 2400$

48 cameras should be made, for a minimum cost of \$2400.

69. Let $x =$ # of \$5 increases.

Then Rate $= 90 + 5x$

Rooms $= 200 - 10x$

Revenue $=$ Rate $\cdot$ # Rooms

$y = (90 + 5x)(200 - 10x)$

$y = 18{,}000 + 100x - 50x^2$

$a = -50, b = 100, c = 18{,}000$

Find the vertex:

$$x = -\frac{b}{2a} = -\frac{100}{2(-50)} = 1$$

The maximum revenue occurs when the room rate increases by 1 five-dollar increment, or when the rate is \$95.

71. $s = 80t - 16t^2$

$a = -16, b = 80, c = 0$

Find the x-coord. of the vertex:

$$x = t = -\frac{b}{2a} = -\frac{80}{2(-16)} = \frac{5}{2} = 2.5$$

The max. height occurs after 2.5 seconds.

73. $s = 80t - 16t^2$

$a = -16, b = 80, c = 0$

Find the y-coord. of the vertex.

Note: The x-coord. was found in **#71**.

$y = s = 80t - 16t^2 = 80(2.5) - 16(2.5)^2$
$= 100$

The max. height is 100 ft.

75. $y = 2x^2 + 9x - 56$

Vertex: $(-2.25, -66.13)$

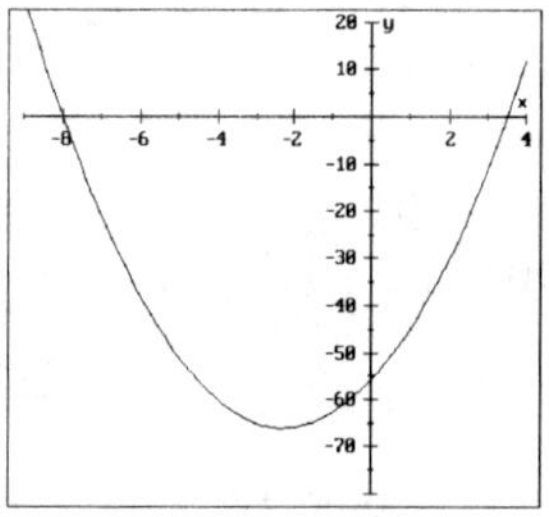

77. $y = (x - 7)(5x + 2)$

Vertex: $(3.3, -68.5)$

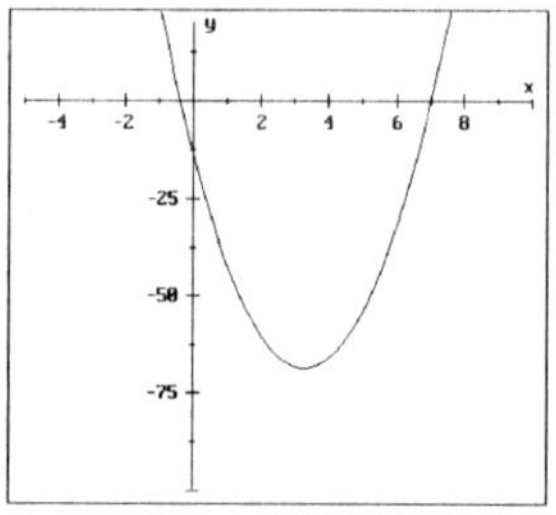

79. $f(x) = x^2 - 5x + 6$
$0 = x^2 - 5x + 6$
$0 = (x-2)(x-3)$
$x = 2$ or $x = 3$

81. The equation of the line is $y = -\frac{3}{4}x + 9$.
Thus the point $(x, y) = \left(x, -\frac{3}{4}x + 9\right)$.
Area $= x\left(-\frac{3}{4}x + 9\right)$
$y = -\frac{3}{4}x^2 + 9x$
$a = -\frac{3}{4}, b = 9, c = 0$
Find the x-coord. of the vertex:
$$x = -\frac{b}{2a} = -\frac{9}{2\left(-\frac{3}{4}\right)} = 6$$
Thus, the dimensions are 6 by $4\frac{1}{2}$ units.

83. Let $x =$ one number. Then $6 - x =$ the other number.
Sum of squares $= x^2 + (6-x)^2 = x^2 + 36 - 12x + x^2 = 2x^2 - 12x + 36$
$a = 2, b = -12, c = 36$; Find the x-coord. of the vertex: $x = -\frac{b}{2a} = -\frac{-12}{2(2)} = 3$
Thus, the numbers are both 3.

85. Answers will vary.

87. $f(a) = a^2 - 3a$
$f(-a) = (-a)^2 - 3(-a)$
$= a^2 + 3a$

89. $f(a) = (5-a)^2$
$f(-a) = [5 - (-a)]^2$
$= (5+a)^2$

91. $f(a) = 7$
$f(-a) = 7$

Exercises 3.3 (page 321)

1. 4

3. $n - 1$

5. odd

7. Piecewise-defined

9. 3

11. $f(x) = x^3 - 9x$
$f(-x) = (-x)^3 - 9(-x)$
$= -x^3 + 9x$
$= -f(x) \Rightarrow$ odd

x-int.	y-int.
$x^3 - 9x = 0$	$f(0) = 0^3 - 9(0)$
$x(x^2 - 9) = 0$	$y = 0$
$x(x+3)(x-3) = 0$	$(0, 0)$
$x = 0, x = -3, x = 3$	
$(0, 0), (-3, 0), (3, 0)$	

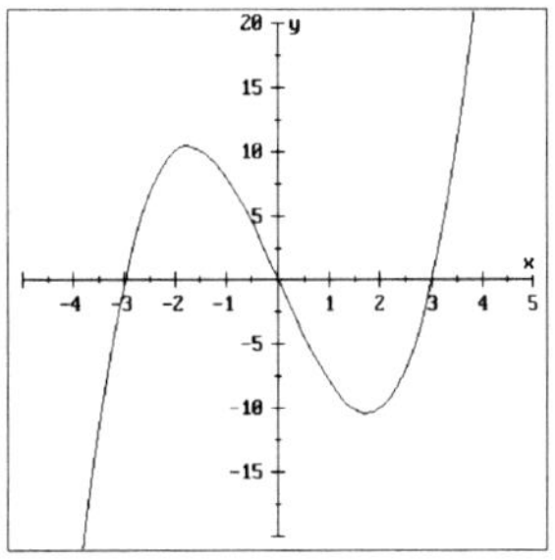

13. $f(x) = -x^3 - 4x^2$

$f(-x) = -(-x)^3 - 4(-x)^2$

$= x^3 - 4x^2$

neither even nor odd

x-int.	y-int.
$-x^3 - 4x^2 = 0$	$f(0) = -(0^3) - 4(0)^2$
$-x^2(x+4) = 0$	$y = 0$
$x = 0,\ x = -4$	$(0, 0)$
$(0, 0), (-4, 0)$	

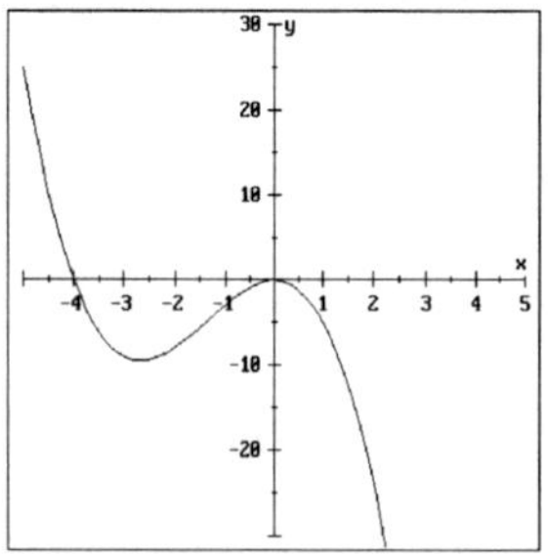

15. $f(x) = x^3 + x^2$

$f(-x) = (-x)^3 + (-x)^2$

$= -x^3 + x^2 \Rightarrow$ neither even nor odd

x-int.	y-int.
$x^3 + x^2 = 0$	$f(0) = 0^3 + 0^2$
$x^2(x+1) = 0$	$y = 0$
$x = 0,\ x = -1$	$(0, 0)$
$(0, 0), (-1, 0)$	

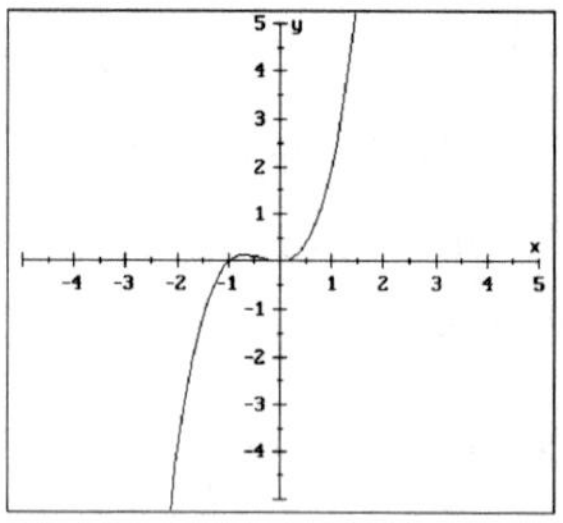

17. $f(x) = x^3 - x^2 - 4x + 4$

$f(-x) = (-x)^3 - (-x)^2 - 4(-x) + 4$

$= -x^3 - x^2 + 4x + 4$

$\Rightarrow$ neither even nor odd

x-int.	y-int.
$x^3 - x^2 - 4x + 4 = 0$	$f(0) = 4$
$x^2(x-1) - 4(x-1) = 0$	$y = 4$
$(x-1)(x^2-4) = 0$	$(0, 4)$
$x = 1$ or $x^2 = 4$	
$x = 1,\ x = 2,\ x = -2$	
$(1, 0), (2, 0), (-2, 0)$	

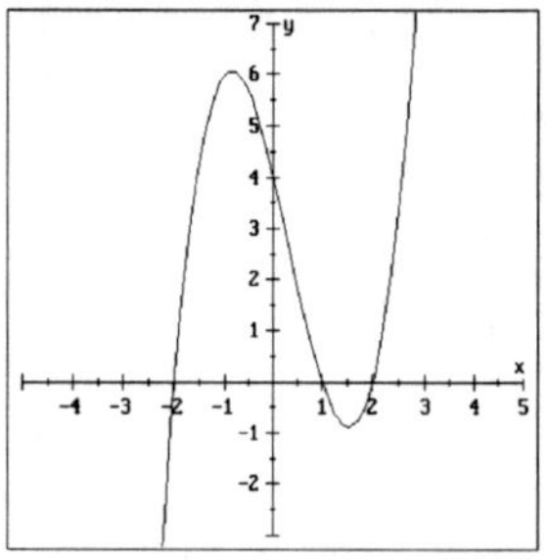

19. $f(x) = x^4 - 2x^2 + 1$

$f(-x) = (-x)^4 - 2(-x)^2 + 1$

$= x^4 - 2x^2 + 1 \Rightarrow$ even

x-int.	y-int.
$x^4 - 2x^2 + 1 = 0$	$f(0) = 1$
$(x^2-1)(x^2-1) = 0$	$y = 1$
$x^2 = 1$	$(0, 1)$
$x = 1,\ x = -1$	
$(1, 0), (-1, 0)$	

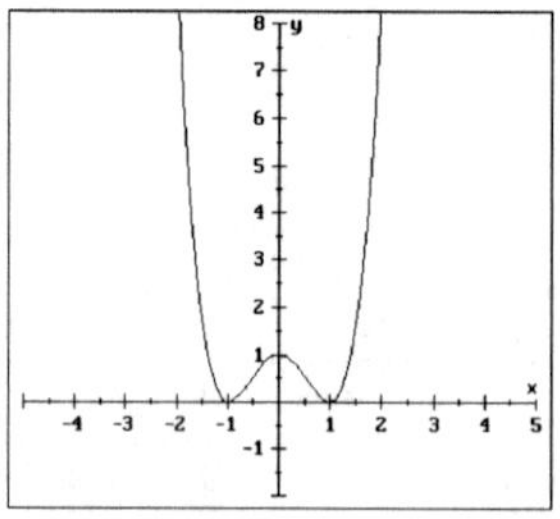

SECTION 3.3

21. $f(x) = -x^4 + 5x^2 - 4$

$$f(-x) = -(-x)^4 + 5(-x)^2 - 4$$
$$= -x^4 + 5x^2 - 4 \Rightarrow \text{even}$$

x-int.	y-int.
$-x^4 + 5x^2 - 4 = 0$	$f(0) = -4$
$-\left(x^4 - 5x^2 + 4\right) = 0$	$y = -4$
$\left(x^2 - 1\right)\left(x^2 - 4\right) = 0$	$(0, -4)$
$x^2 = 1$ or $x^2 = 4$	
$x = \pm 1, x = \pm 2$	
$(\pm 1, 0), (\pm 2, 0)$	

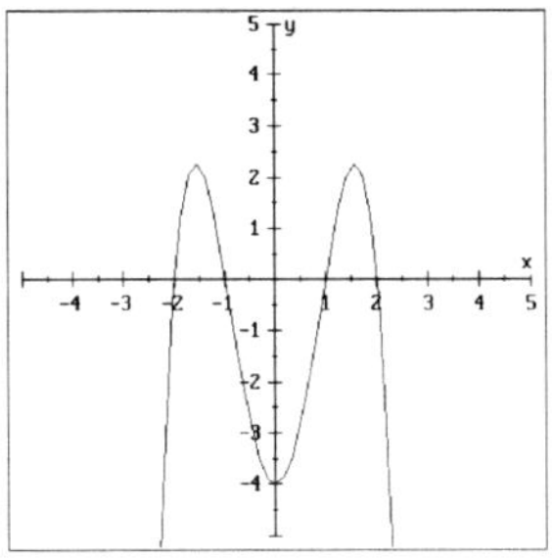

23. $f(x) = x^3 + x^2$

$$f(-x) = (-x)^3 + (-x)^2$$
$$= -x^3 + x^2 \Rightarrow \text{neither}$$

25. $f(x) = x^3 + x^2$

$$f(-x) = (-x)^3 + (-x)^2$$
$$= -x^3 + x^2 \Rightarrow \text{neither}$$

27. $f(x) = x^5 + x^3$

$$f(-x) = (-x)^5 + (-x)^3$$
$$= -x^5 - x^3 = -f(x) \Rightarrow \text{odd}$$

29. $f(x) = 2x^3 - 3x$

$$f(-x) = 2(-x)^3 - 3(-x)$$
$$= -2x^3 + 3x = -f(x) \Rightarrow \text{odd}$$

31. symmetric about x-axis $\Rightarrow$ even

33. symmetric about origin $\Rightarrow$ odd

35. no symmetry $\Rightarrow$ neither

37. decreasing: $(-\infty, 0)$; increasing: $(0, \infty)$

39. increasing: $(-\infty, 0)$; decreasing: $(0, \infty)$

41. decreasing: $(-\infty, -2)$; constant: $(-2, 2)$; increasing: $(2, \infty)$

43. $f(x) = x^2 - 4x + 4$

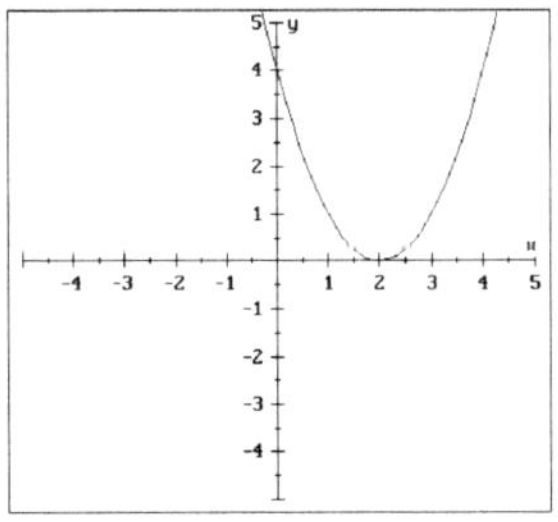

decreasing: $(-\infty, 2)$; increasing: $(2, \infty)$

45. **a.** $f(-2) = 2(-2) + 2 = -2$

b. $f(0) = 3$

SECTION 3.3

47. **a.** $f(-1) = 2$
b. $f(1) = 2 - 1 = 1$
c. $f(2) = 2 + 1 = 3$

49. $f(x) = \begin{cases} x+2 & \text{if } x < 0 \\ 2 & \text{if } x \geq 0 \end{cases}$

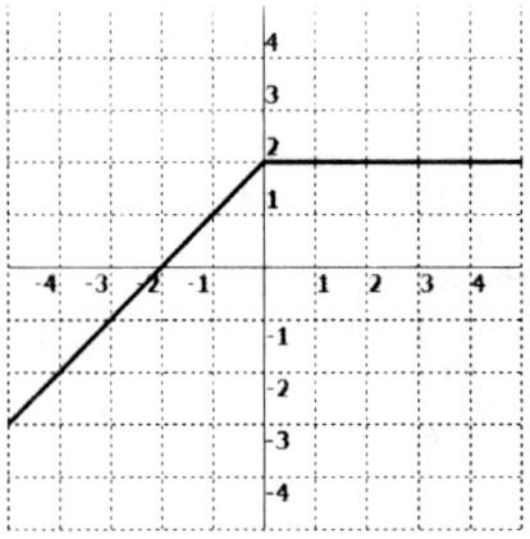

51. $f(x) = \begin{cases} x & \text{if } x \leq 0 \\ 2 & \text{if } x > 0 \end{cases}$

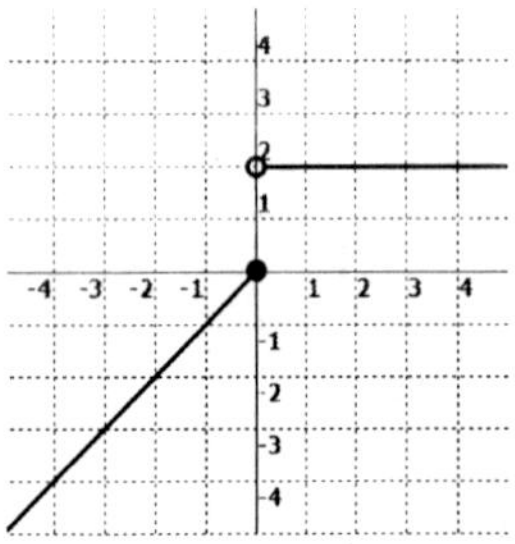

53. $f(x) = \begin{cases} -4-x & \text{if } x < 1 \\ 3 & \text{if } x \geq 1 \end{cases}$

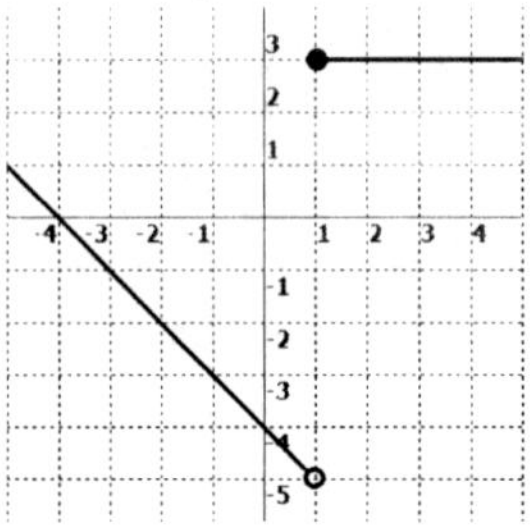

55. $f(x) = \begin{cases} -x & \text{if } x < 0 \\ x^2 & \text{if } x \geq 0 \end{cases}$

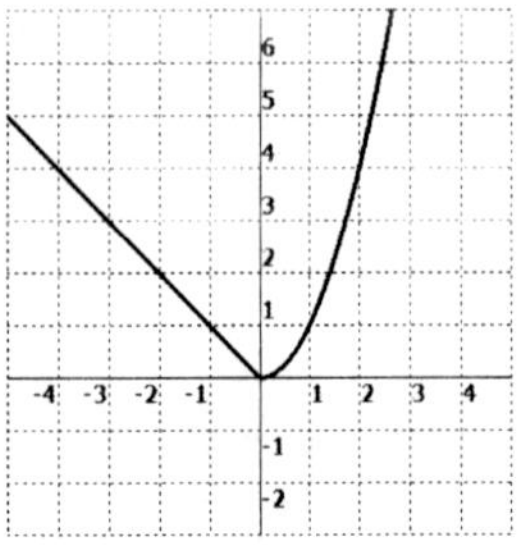

57. $f(x) = \begin{cases} 0 & \text{if } x < 0 \\ x^2 & \text{if } 0 \leq x \leq 2 \\ 4-2x & \text{if } x > 2 \end{cases}$

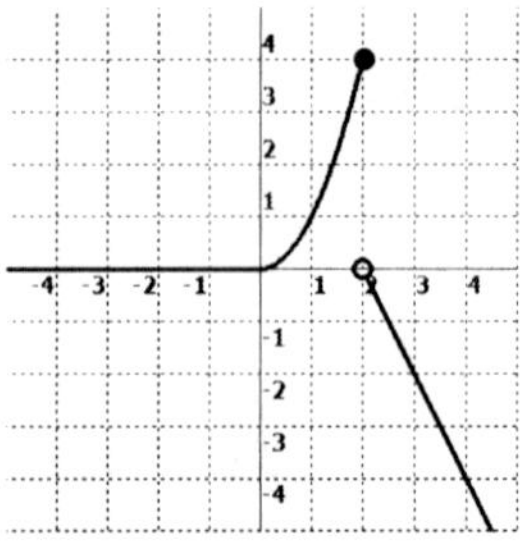

59. **a.** $f(3) = [[3]] = 3$
b. $f(-4) = [[-4]] = -4$
c. $f(-2.3) = [[-2.3]] = -3$

61. **a.** $f(-1) = [[-1+3]] = [[2]] = 2$
b. $f(\frac{2}{3}) = [[\frac{2}{3}+3]] = [[3\frac{2}{3}]] = 3$
c. $f(1.3) = [[1.3+3]] = [[4.3]] = 4$

SECTION 3.3

63. $y = [[2x]]$

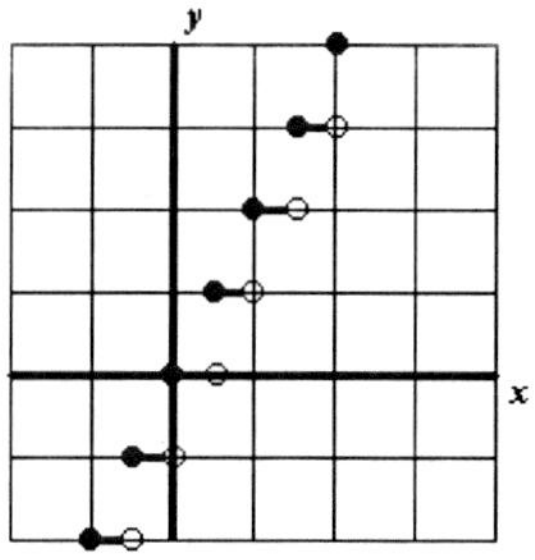

65. $y = [[x]] - 1$

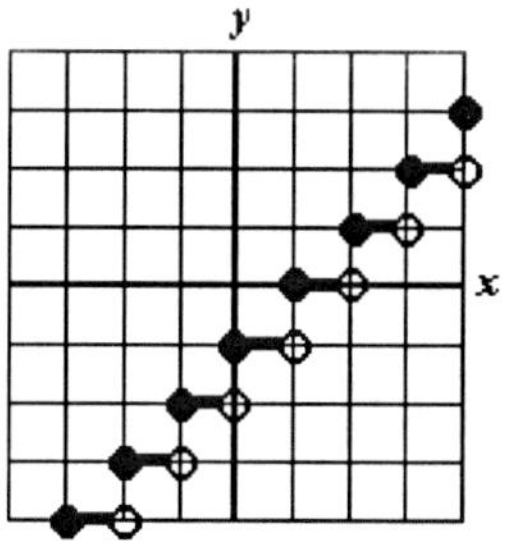

67.

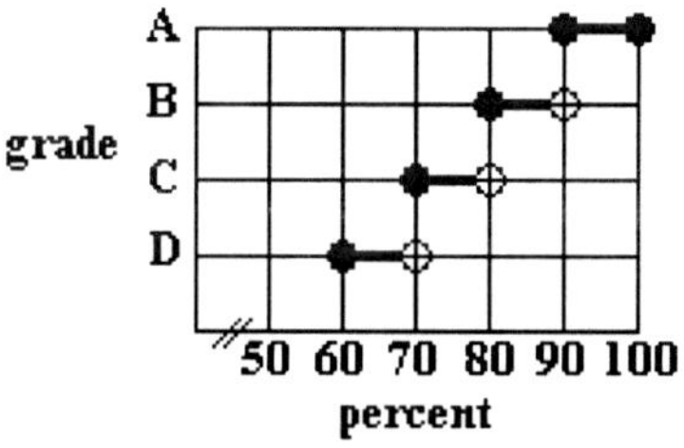

$$\frac{67 + 73 + 84 + 87 + 93}{5} = \frac{404}{5} = 80.8$$

The student's grade is B.

69. \$32 for 275 miles

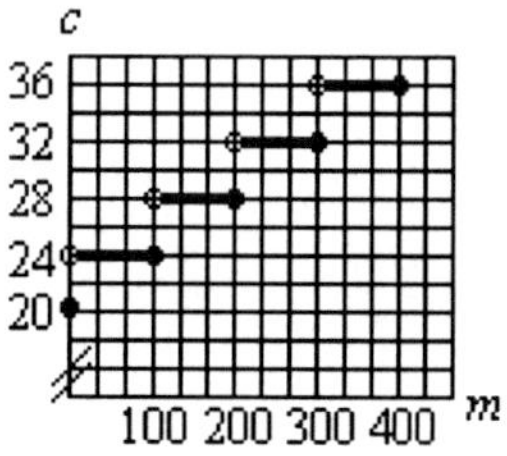

71. \$12 per hour $\Rightarrow$
\$0.20 per minute
\$1.60 for $7\frac{1}{2}$ minutes

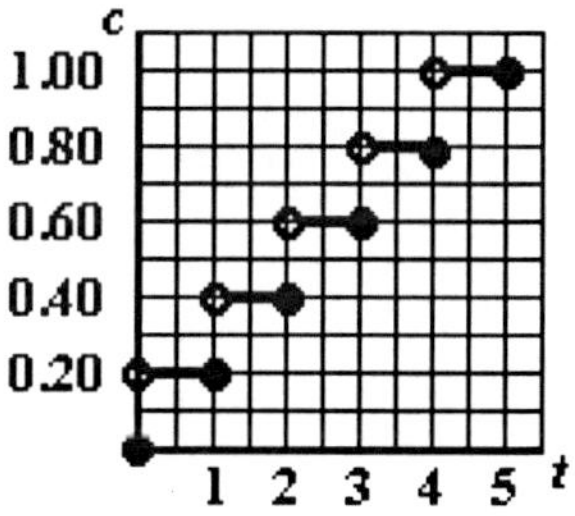

73.

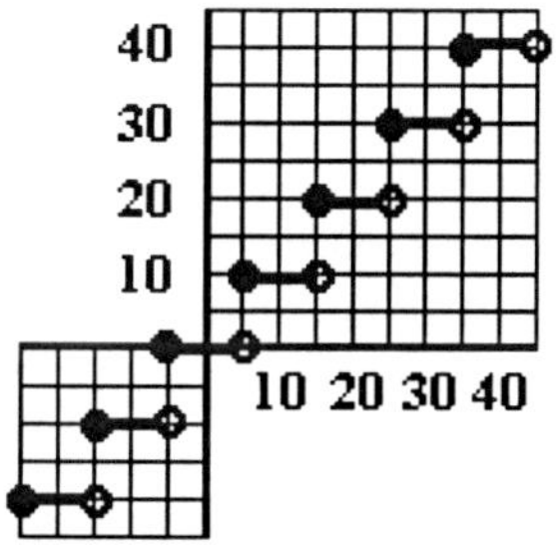

75. $y = \frac{|x|}{x}$, Not defined at $x = 0$, so not the same

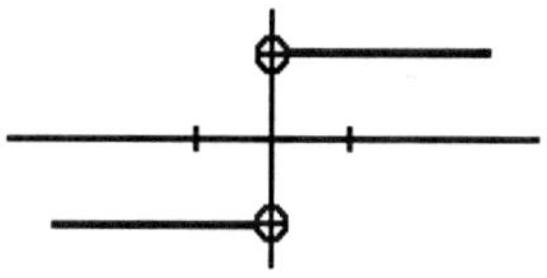

77. Answers may vary.

79. Answers may vary.

81. $f(x + 1) = 3(x + 1) + 2 = 3x + 5$
$f(x) + 1 = (3x + 2) + 1 = 3x + 3$

83. $f(x - 3) = \dfrac{3(x - 3) + 1}{5} = \dfrac{3x - 8}{5}$

$f(x) - 3 = \dfrac{3x + 1}{5} - 3 = \dfrac{3x - 14}{5}$

85.
$$\begin{aligned} 2x^2 - 3 &= x \\ 2x^2 - x - 3 &= 0 \\ (2x-3)(x+1) &= 0 \\ 2x = 3 \text{ or } x &= -1 \\ x = \tfrac{3}{2}, x &= -1 \end{aligned}$$

Exercises 3.4 (page 339)

1. upward

3. to the right

5. 2, downward

7. y-axis

9. horizontally

11. $g(x) = x^2 - 2$
Shift $f(x) = x^2$ D 2

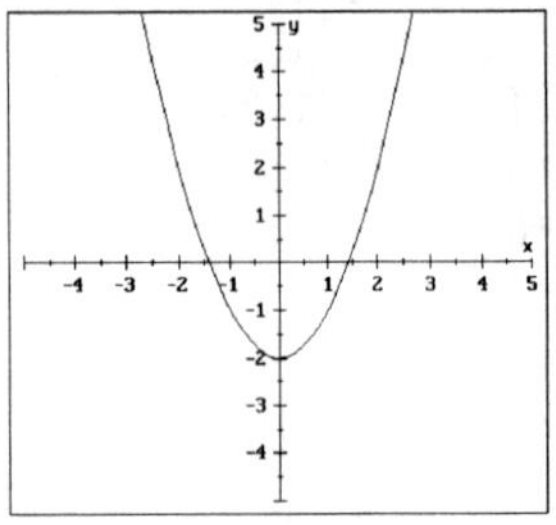

13. $g(x) = (x+3)^2$
Shift $f(x) = x^2$ L 3

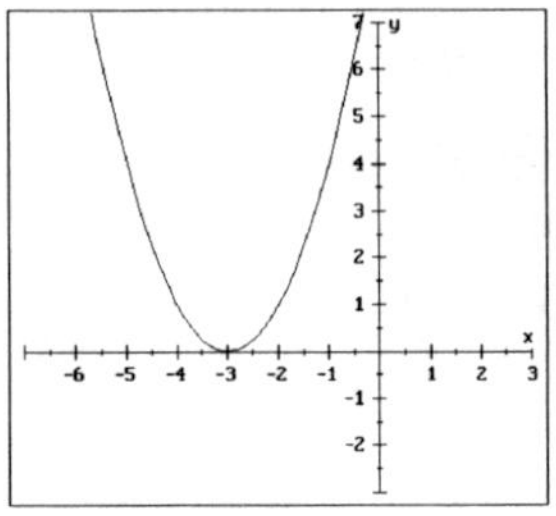

15. $h(x) = (x+1)^2 + 2$
Shift $f(x) = x^2$ U 2, L 1

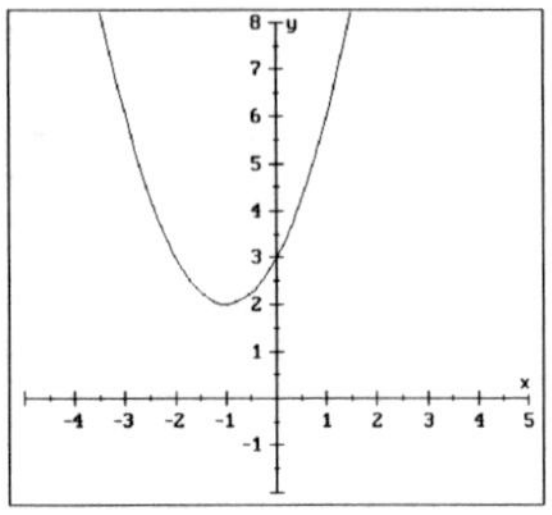

17. $h(x) = \left(x + \dfrac{1}{2}\right)^2 - \dfrac{1}{2}$
Shift $f(x) = x^2$ D $\frac{1}{2}$, L $\frac{1}{2}$

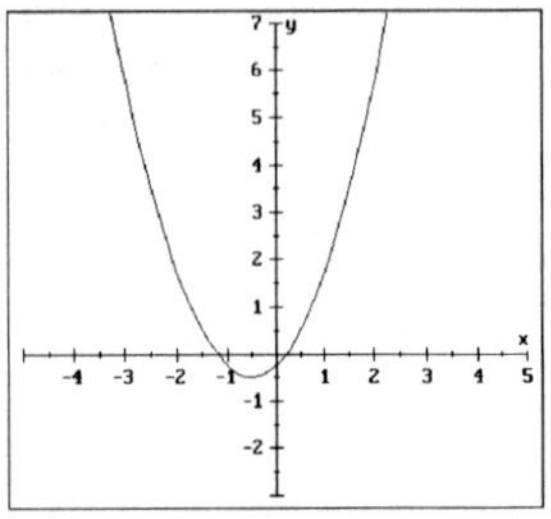

19. $g(x) = x^3 + 1$
Shift $f(x) = x^3$ U 1

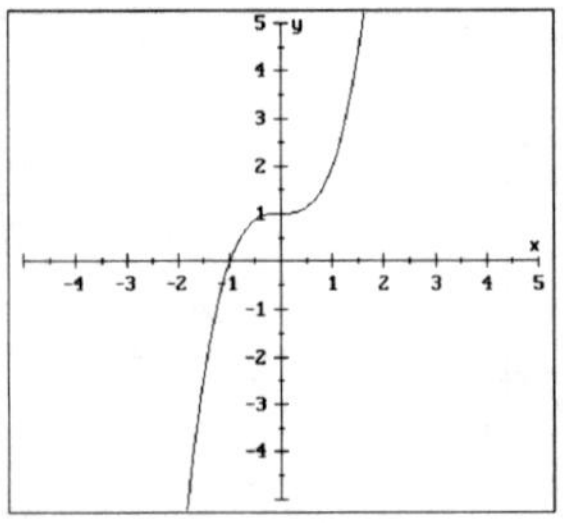

21. $g(x) = (x-2)^3$
Shift $f(x) = x^3$ R 2

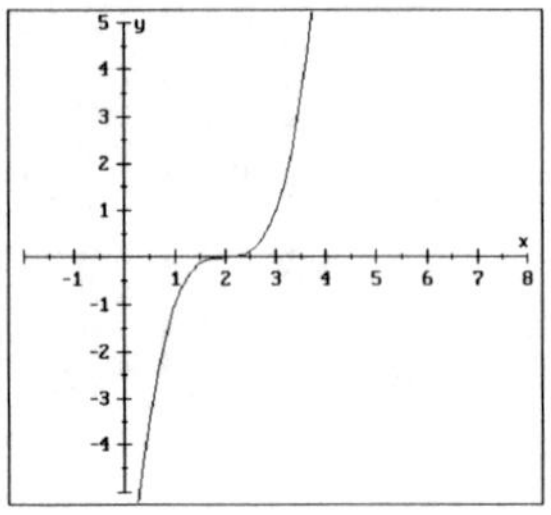

SECTION 3.4

23. $h(x) = (x - 2)^3 - 3$

Shift $f(x) = x^3$ D 3, R 2

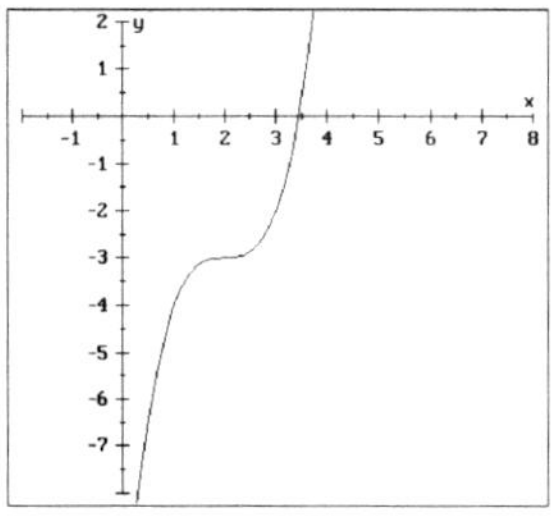

25. $y + 2 = x^3$

$y = x^3 - 2$

Shift $y = x^3$ D 2

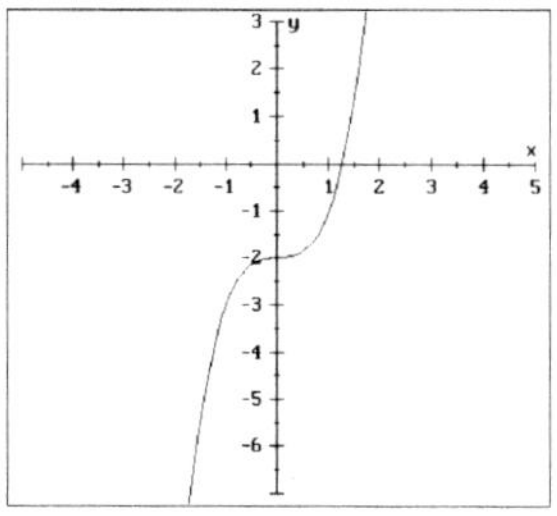

27. $g(x) = |x| + 2$

Shift $f(x) = |x|$ U 2

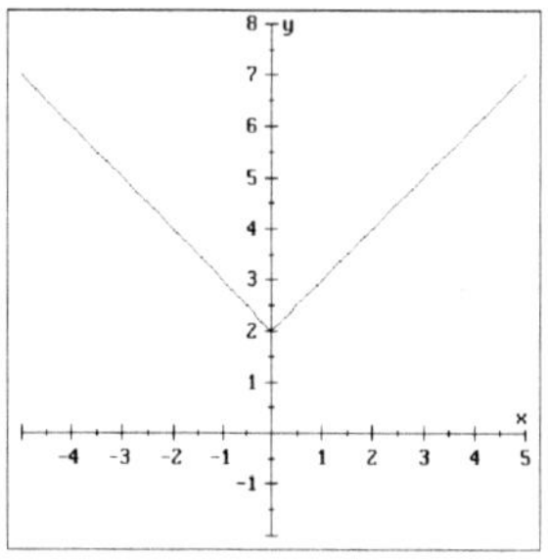

29. $g(x) = |x - 5|$

Shift $f(x) = |x|$ R 5

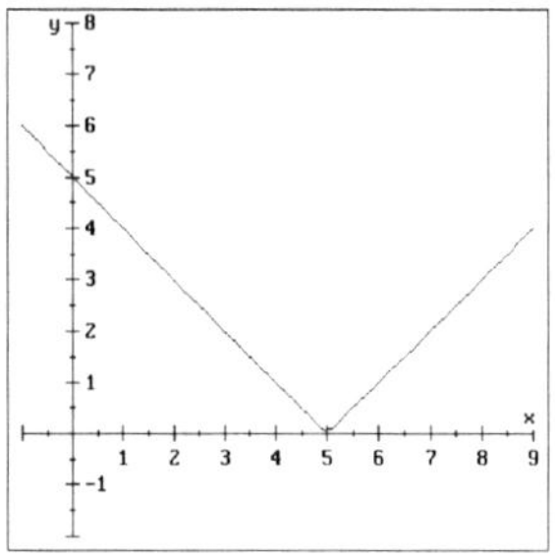

31. $f(x) = |x + 2| - 1$

Shift $f(x) = |x|$ D 1, L 2

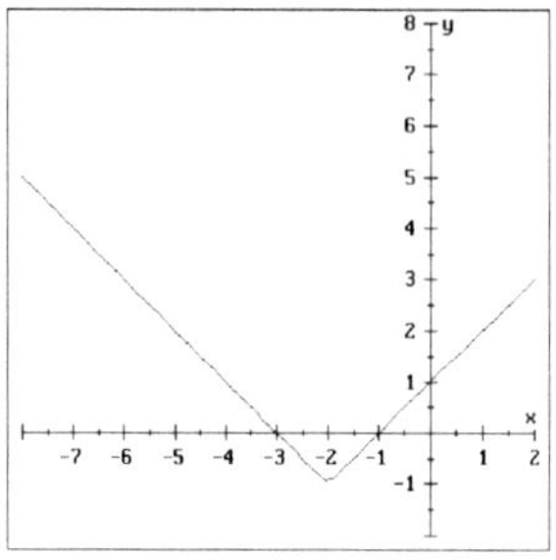

33. $g(x) = \sqrt{x} + 1$

Shift $f(x) = \sqrt{x}$ U 1

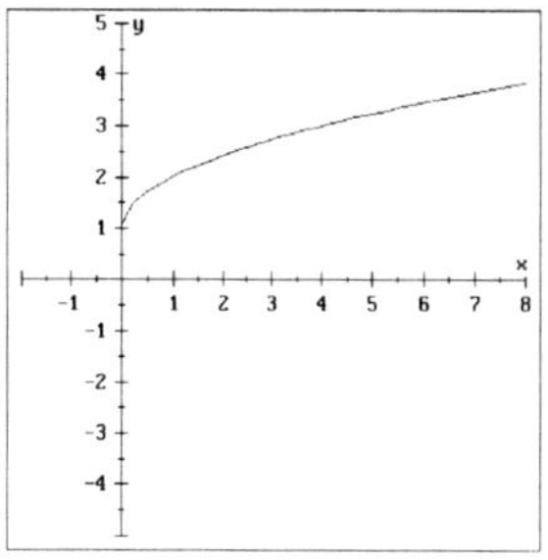

35. $g(x) = \sqrt{x + 2}$

Shift $f(x) = \sqrt{x}$ L 2

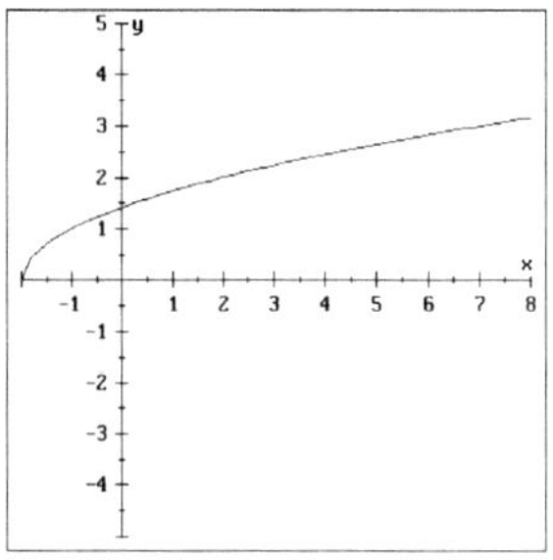

37. $h(x) = \sqrt{x - 2} - 1$

Shift $f(x) = \sqrt{x}$ D 1, R 2

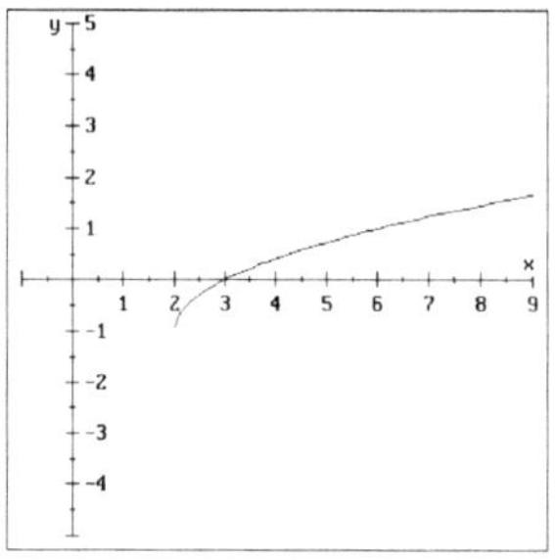

39. $g(x) = \sqrt[3]{x} - 4$

Shift $f(x) = \sqrt[3]{x}$ D 4

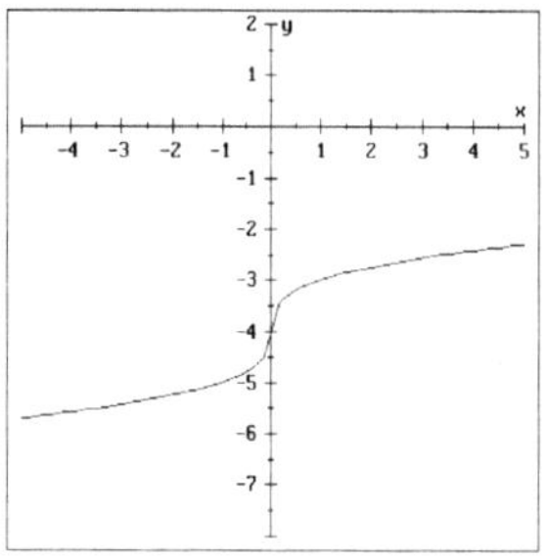

41. $g(x) = \sqrt[3]{x-2}$
Shift $f(x) = \sqrt[3]{x}$ R 2

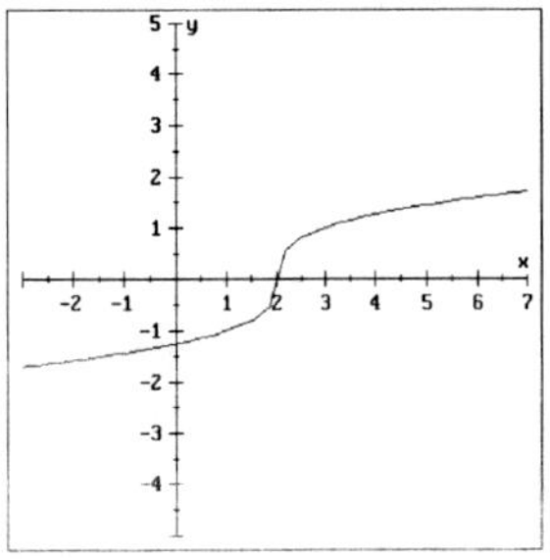

43. $h(x) = \sqrt[3]{x+1} - 1$
Shift $f(x) = \sqrt[3]{x}$ D 1, L 1

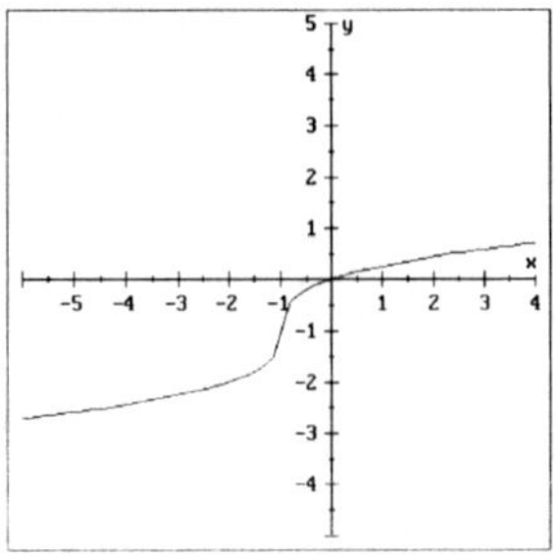

45. $f(x) = -x^2$
Reflect $y = x^2$ about x

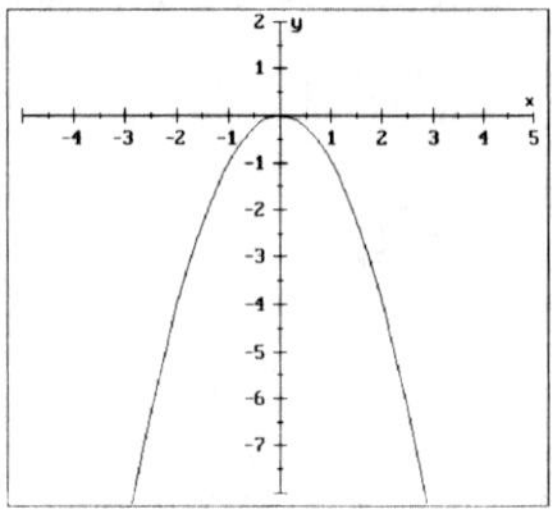

47. $h(x) = -x^3$
Reflect $y = x^3$ about x

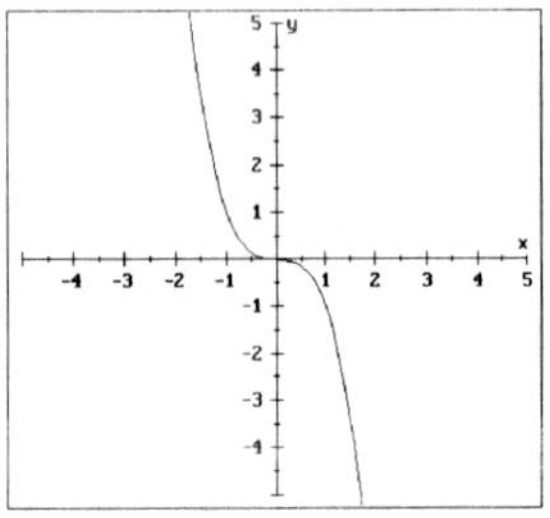

49. $f(x) = -\sqrt{x}$
Reflect $y = \sqrt{x}$ about x

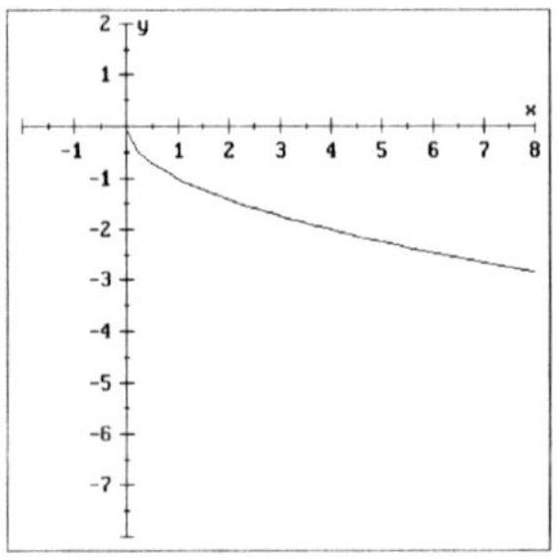

51. $f(x) = |-x|$
Reflect $y = |x|$ about y

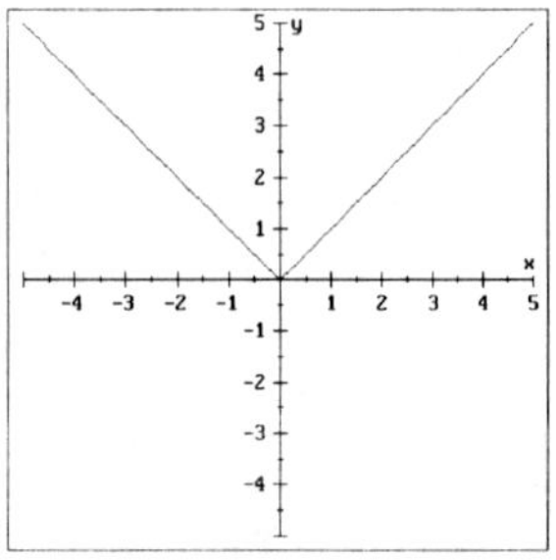

53. $f(x) = 2x^2$: Stretch
$y = x^2$ vert. by a factor of 2

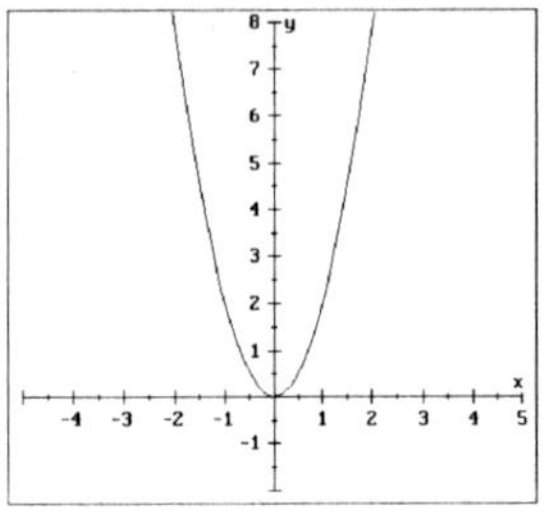

55. $h(x) = -3x^2$: Stretch
$y = x^2$ vert. by a factor of 3
Reflect about x

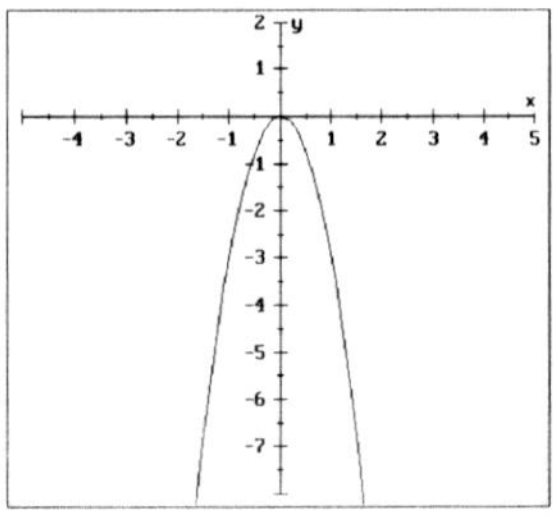

57. $f(x) = \frac{1}{2}x^3$: Shrink
$y = x^3$ vert. by a factor of $\frac{1}{2}$

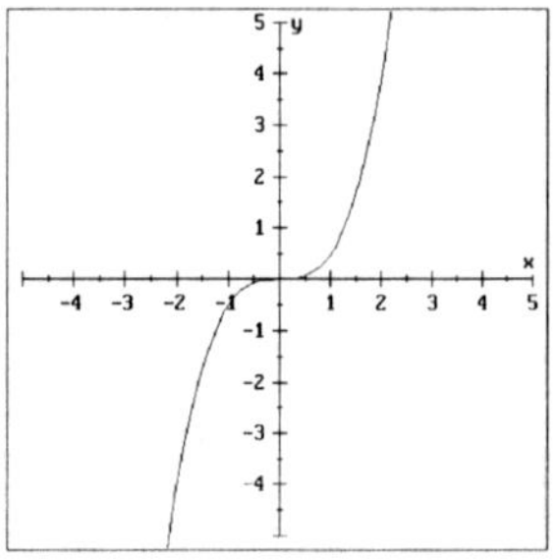

SECTION 3.4

59. $h(x) = -3|x|$: Stretch $y = |x|$ vert. by a factor of 3 Reflect about x

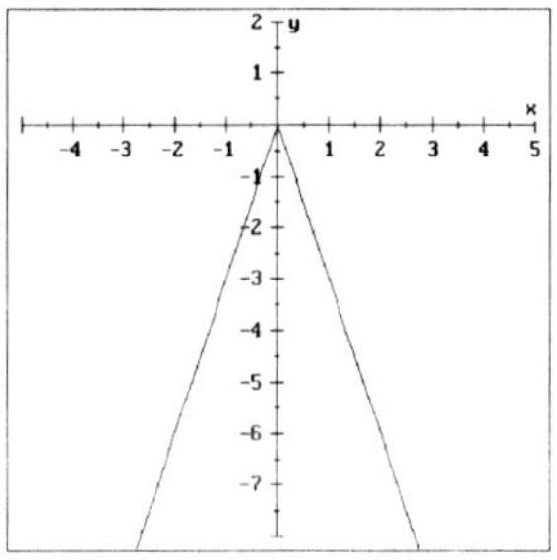

61. $f(x) = \left(\frac{1}{2}x\right)^3$: Stretch $y = x^3$ hor. by a factor of 2

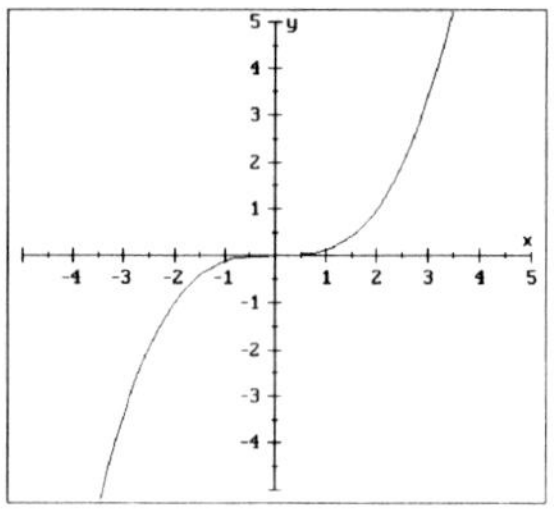

63. $f(x) = (2x)^2$: Shrink $y = x^2$ hor. by a factor of 2

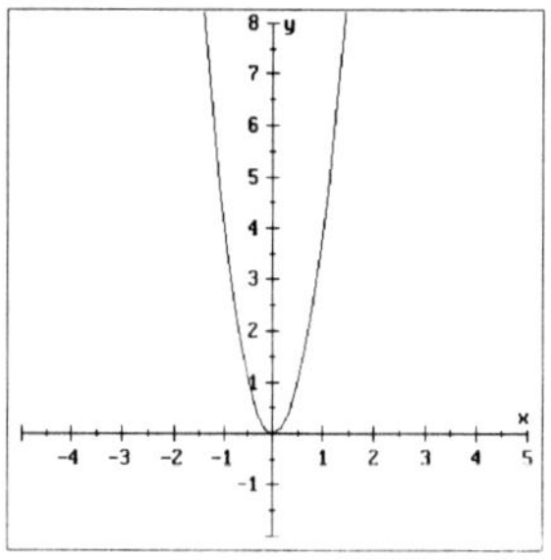

65. $g(x) = 3(x + 2)^2 - 1$
Start with $y = x^2$
Shift L 2, Stretch vert. by a factor of 3, Shift D 1

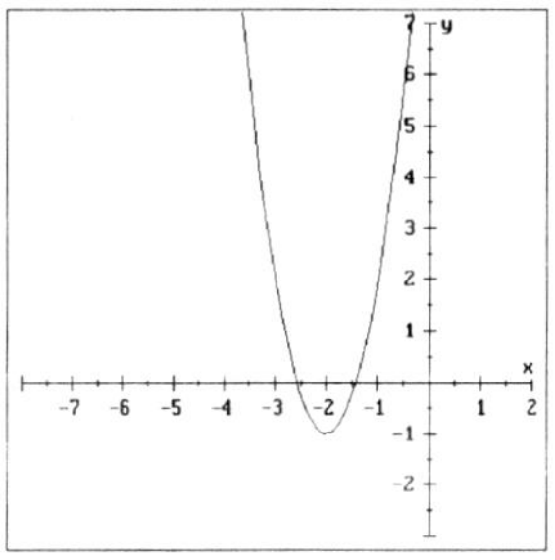

67. $h(x) = -2|x| + 3$
Start with $y = |x|$
Stretch vert. by a factor of 2; Reflect x; Shift U 3

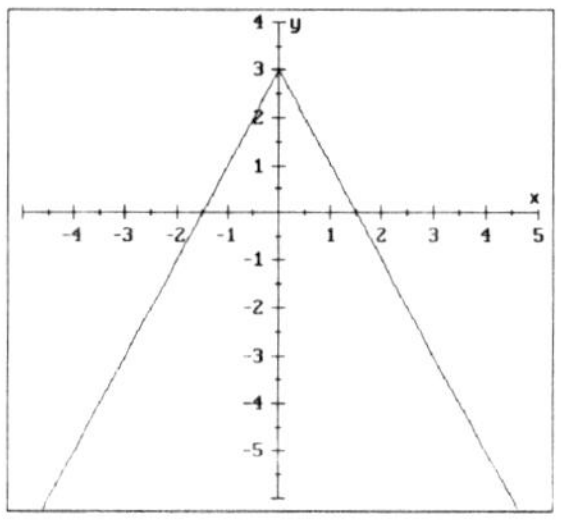

69. $f(x) = 2|x - 2| + 1$
Start with $y = |x|$
Shift R 2, Stretch vert. by a factor of 2, Shift U 1

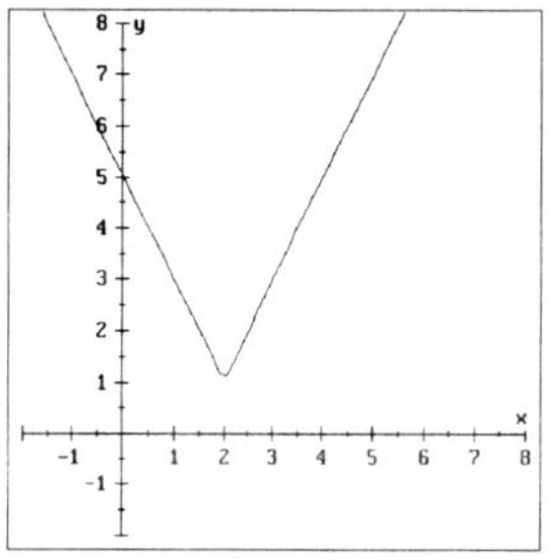

71. $f(x) = 2\sqrt{x} + 3$
Start with $y = \sqrt{x}$
Stretch vert. by a factor of 2, Shift U 3

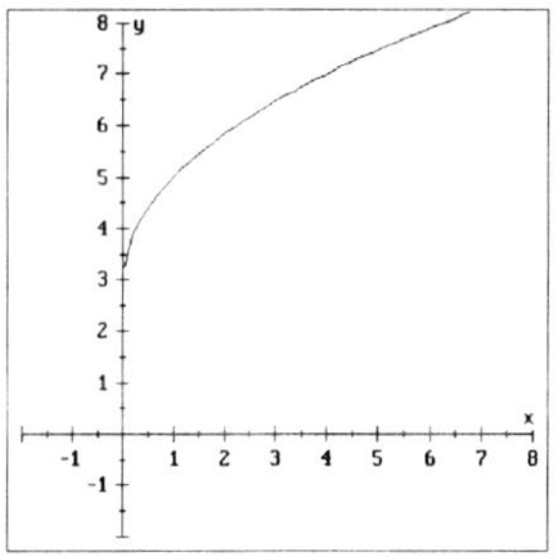

73. $h(x) = 2\sqrt{x - 2} + 1$
Start with $y = \sqrt{x}$
Shift R 2, Stretch vert. by a factor of 2, Shift U 1

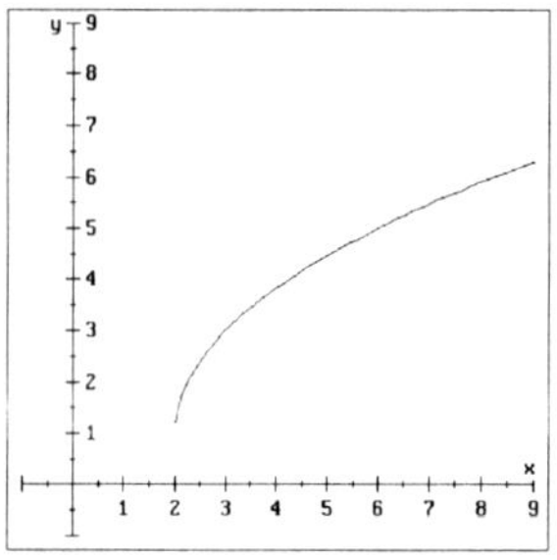

75. $g(x) = -2(x + 2)^3 - 1$
Start with $y = x^3$
Shift L 2, Stretch vert. by a factor of 2, Reflect x, Shift D 1

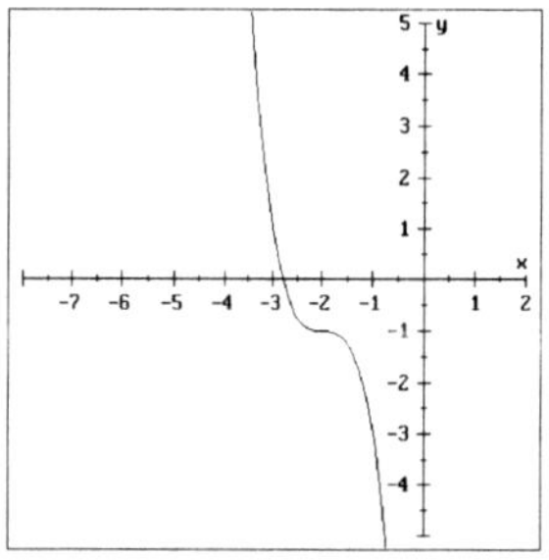

77. $f(x) = 2\sqrt[3]{x} + 4$
Start with $y = \sqrt[3]{x}$
Stretch vert. by a factor of 2, Shift U 4

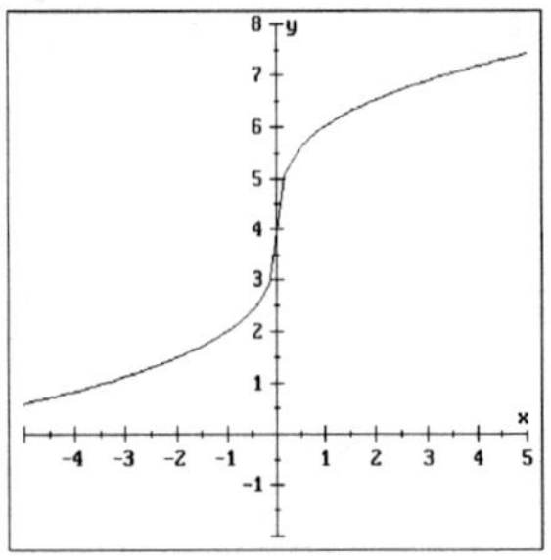

79. Shift $y = f(x)$ U 1

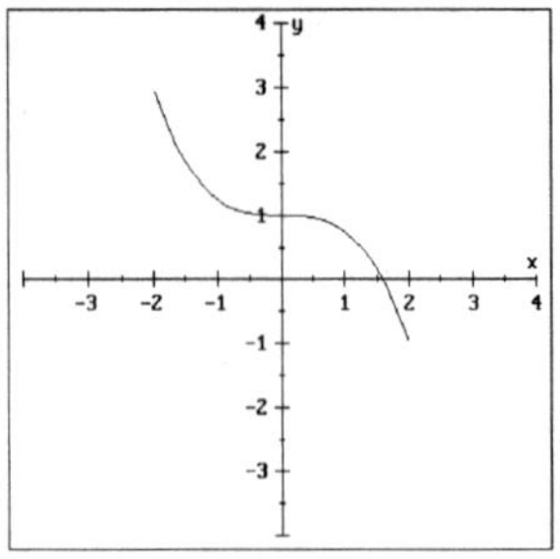

81. Stretch $y = f(x)$ vert. by a factor of 2

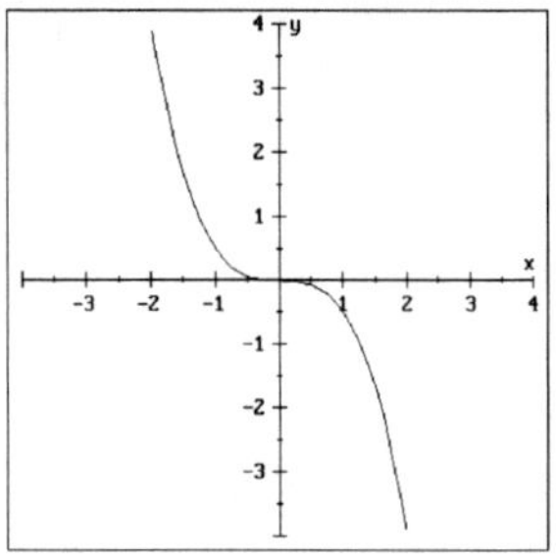

83. Shift $y = f(x)$ U 1, R 2

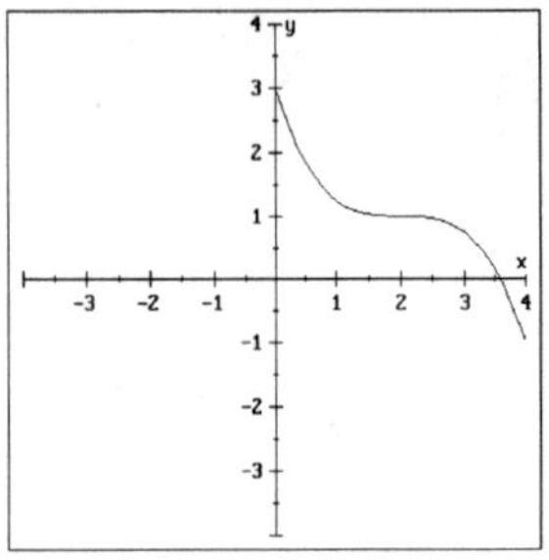

85. Stretch $y = f(x)$ vert. by a factor of 2, reflect about y

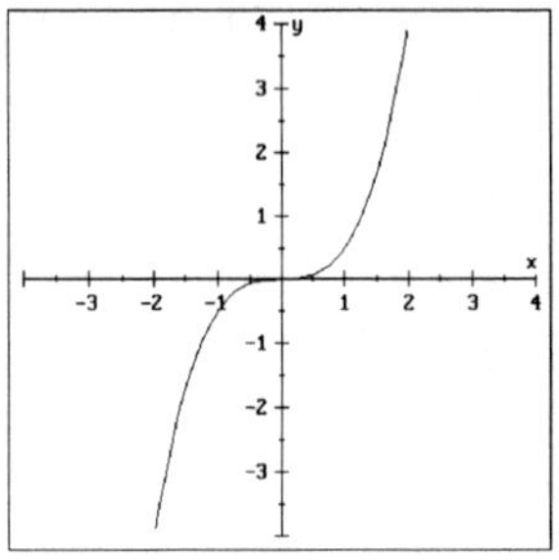

87. **Answers may vary.**

89. **Answers may vary.**

91. **Answers may vary.**

93. $\dfrac{x^2+x-6}{x^2+5x+6} = \dfrac{(x+3)(x-2)}{(x+2)(x+3)} = \dfrac{x-2}{x+2}$

95. $f(x) = \dfrac{x+7}{x-3}$; domain $= (-\infty, 3) \cup (3, \infty)$

97.
$$x + 2 + \tfrac{-2}{x+1}$$
$$x+1 \overline{)\, x^2 + 3x + 0}$$
$$\underline{x^2 + x}$$
$$2x + 0$$
$$\underline{2x + 2}$$
$$-2$$

Exercises 3.5 (page 361)

1. asymptote

3. vertical

5. x-intercept

7. same

9. horizontal or slant; vertical

11. vertical: $x = 2$, horizontal: $y = 1$
domain: $(-\infty, 2) \cup (2, \infty)$
range: $(-\infty, 1) \cup (1, \infty)$

13. $t = f(30) = \frac{600}{30} = 20$ hr

15. $t = f(50) = \frac{600}{50} = 12$ hr

SECTION 3.5

17. $c = f(10) = \dfrac{50{,}000(10)}{100 - 10} \approx \$5555.56$

19. $c = f(50) = \dfrac{50{,}000(50)}{100 - 50} = \$50{,}000.00$

21. $f(x) = \dfrac{x^2}{x - 2}$; den $= 0 \Rightarrow x = 2$
domain $= (-\infty, 2) \cup (2, \infty)$

23. $f(x) = \dfrac{2x^2 + 7x - 2}{x^2 - 25} = \dfrac{2x^2 + 7x - 2}{(x + 5)(x - 5)}$
den $= 0 \Rightarrow x = -5, x = 5$
domain $= (-\infty, -5) \cup (-5, 5) \cup (5, \infty)$

25. $f(x) = \dfrac{x - 1}{x^3 - x} = \dfrac{x - 1}{x(x + 1)(x - 1)}$; den $= 0 \Rightarrow x = 0, x = -1, x = 1$
domain $= (-\infty, -1) \cup (-1, 0) \cup (0, 1) \cup (1, \infty)$

27. $f(x) = \dfrac{3x^2 + 5}{x^2 + 1}$; den $= 0 \Rightarrow$ never true
domain $= (-\infty, \infty)$

29. $f(x) = \dfrac{x}{x - 3}$; den $= 0 \Rightarrow x = 3$
vertical: $x = 3$

31. $f(x) = \dfrac{x + 2}{x^2 - 1} = \dfrac{x + 2}{(x + 1)(x - 1)}$
den $= 0 \Rightarrow x = -1, x = 1$
vertical: $x = -1, x = 1$

33. $f(x) = \dfrac{1}{x^2 - x - 6} = \dfrac{1}{(x + 2)(x - 3)}$
den $= 0 \Rightarrow x = -2, x = 3$
vertical: $x = -2, x = 3$

35. $f(x) = \dfrac{x^2}{x^2 + 5}$; den $= 0 \Rightarrow$ never true
vertical: none

37. $f(x) = \dfrac{2x - 1}{x}$; deg(num) = deg(den)
horizontal: $y = \dfrac{2}{1}$, or $y = 2$

39. $f(x) = \dfrac{x^2 + x - 2}{2x^2 - 4}$; deg(num) = deg(den)
horizontal: $y = \dfrac{1}{2}$

41. $f(x) = \dfrac{x + 1}{x^3 - 4x}$; deg(num) < deg(den)
horizontal: $y = 0$

43. $f(x) = \dfrac{x^2}{x - 2}$; deg(num) > deg(den)
horizontal: none

45. $f(x) = \dfrac{x^2 - 5x - 6}{x - 2} = x - 3 + \dfrac{-12}{x - 2}$
slant: $y = x - 3$

47. $f(x) = \dfrac{2x^2 - 5x + 1}{x - 4} = 2x + 3 + \dfrac{13}{x - 4}$
slant: $y = 2x + 3$

49. $f(x) = \dfrac{x^3 + 2x^2 - x - 1}{x^2 - 1}$
$= x + 2 + \dfrac{1}{x^2 - 1}$
slant: $y = x + 2$

51. $y = \dfrac{1}{x-2}$

Vert: $x = 2$; Horiz: $y = 0$

Slant: none; x-intercepts: none

y-intercepts: $\left(0, -\frac{1}{2}\right)$; Symmetry: none

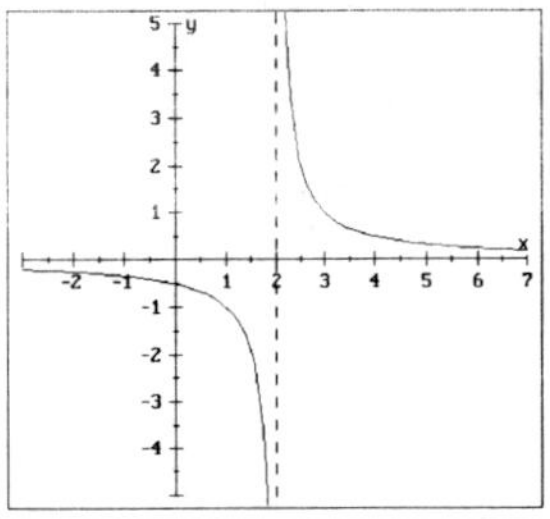

53. $y = \dfrac{x}{x-1}$

Vert: $x = 1$; Horiz: $y = \frac{1}{1} = 1$

Slant: none; x-intercepts: $(0, 0)$

y-intercepts: $(0, 0)$; Symmetry: none

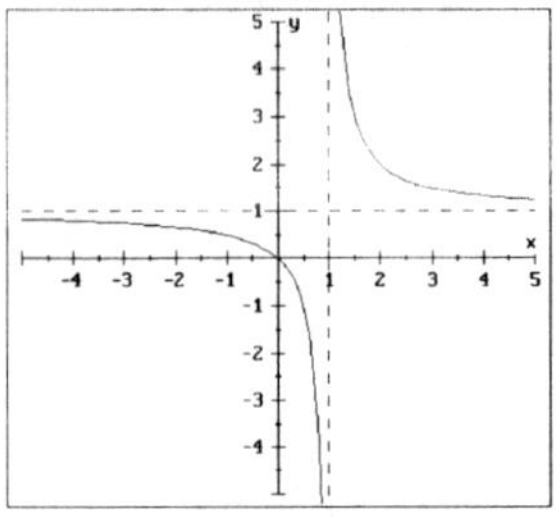

55. $f(x) = \dfrac{x+1}{x+2}$

Vert: $x = -2$; Horiz: $y = \frac{1}{1} = 1$

Slant: none; x-intercepts: $(-1, 0)$

y-intercepts: $\left(0, \frac{1}{2}\right)$; Symmetry: none

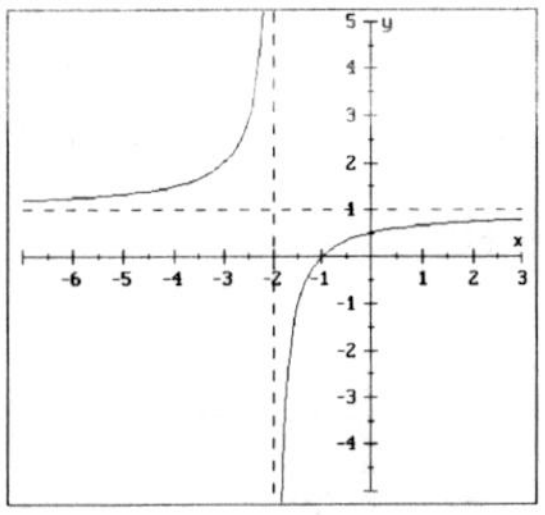

57. $f(x) = \dfrac{2x-1}{x-1}$

Vert: $x = 1$; Horiz: $y = \frac{2}{1} = 2$

Slant: none; x-intercepts: $\left(\frac{1}{2}, 0\right)$

y-intercepts: $(0, 1)$; Symmetry: none

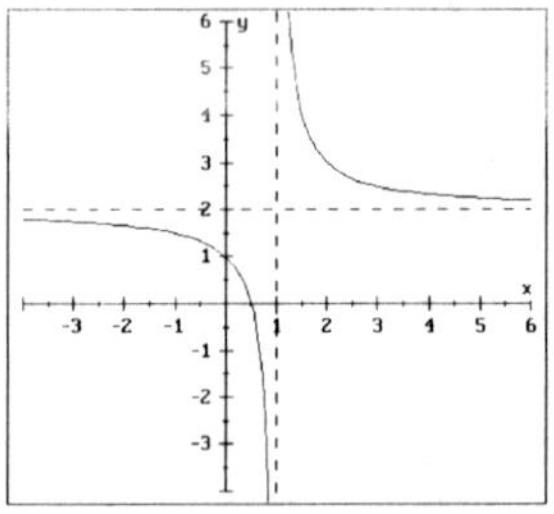

59. $g(x) = \dfrac{x^2-9}{x^2-4} = \dfrac{(x+3)(x-3)}{(x+2)(x-2)}$

Vert: $x = -2, x = 2$; Horiz: $y = \frac{1}{1} = 1$

Slant: none; x-intercepts: $(-3, 0), (3, 0)$

y-intercepts: $\left(0, \frac{9}{4}\right)$; Symmetry: y-axis

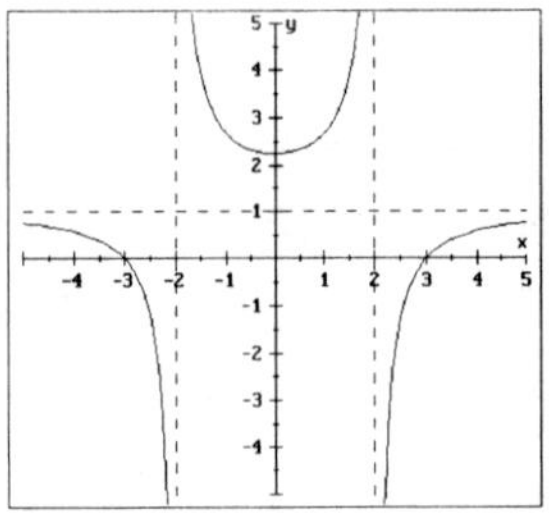

61. $g(x) = \dfrac{x^2-x-2}{x^2-4x+3} = \dfrac{(x+1)(x-2)}{(x-3)(x-1)}$

Vert: $x = 3, x = 1$; Horiz: $y = \frac{1}{1} = 1$

Slant: none; x-intercepts: $(-1, 0), (2, 0)$

y-intercepts: $\left(0, -\frac{2}{3}\right)$; Symmetry: none

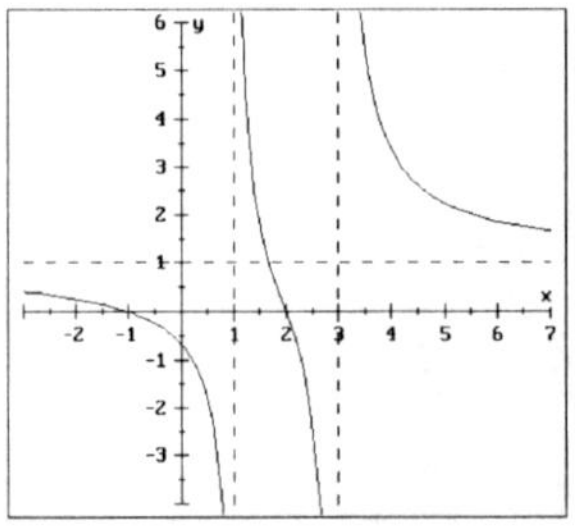

63. $y = \dfrac{x^2+2x-3}{x^3-4x} = \dfrac{(x-1)(x+3)}{x(x+2)(x-2)}$

Vert: $x=0$, $x=-2$, $x=2$; Horiz: $y=0$

Slant: none; x-intercepts: $(1,0)$, $(-3,0)$

y-intercepts: none; Symmetry: none

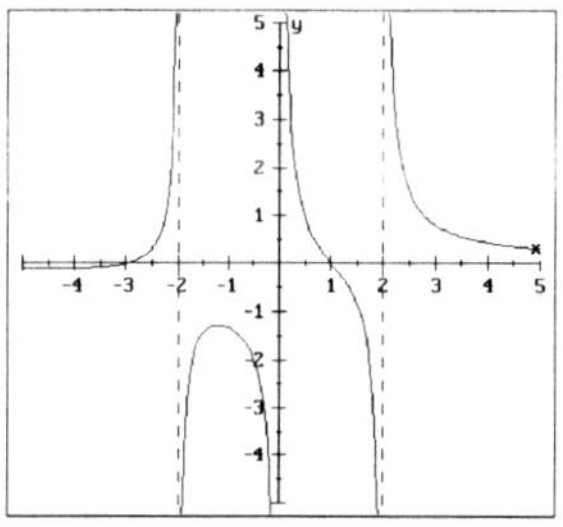

65. $y = \dfrac{x^2-9}{x^2} = \dfrac{(x+3)(x-3)}{x^2}$

Vert: $x=0$; Horiz: $y=\frac{1}{1}=1$

Slant: none; x-intercepts: $(3,0)$, $(-3,0)$

y-intercepts: none; Symmetry: y-axis

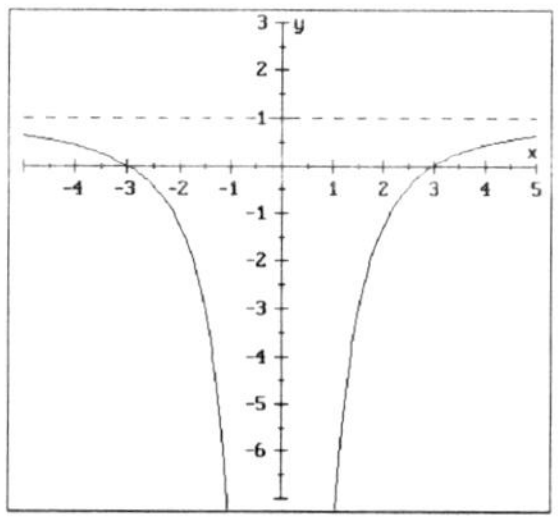

67. $f(x) = \dfrac{x}{(x+3)^2}$

Vert: $x=-3$; Horiz: $y=0$

Slant: none; x-intercepts: $(0,0)$

y-intercepts: $(0,0)$; Symmetry: none

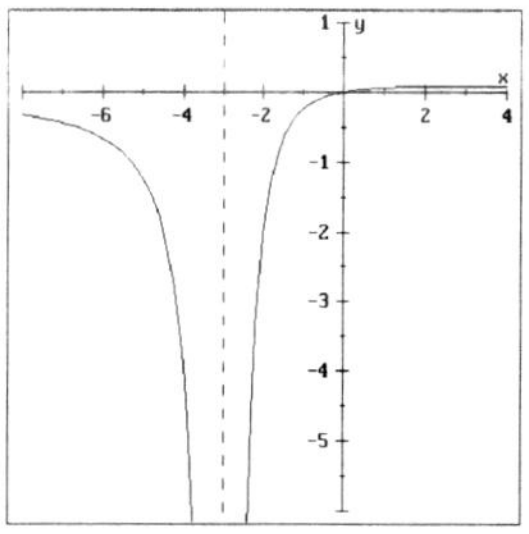

69. $f(x) = \dfrac{x+1}{x^2(x-2)}$

Vert: $x=0$, $x=2$; Horiz: $y=0$

Slant: none; x-intercepts: $(-1,0)$

y-intercepts: none; Symmetry: none

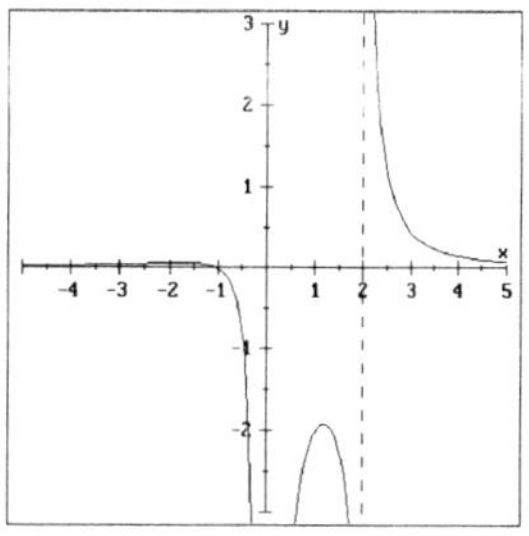

71. $y = \dfrac{x}{x^2+1}$

Vert: none; Horiz: $y=0$

Slant: none; x-intercepts: $(0,0)$

y-intercepts: $(0,0)$; Symmetry: origin

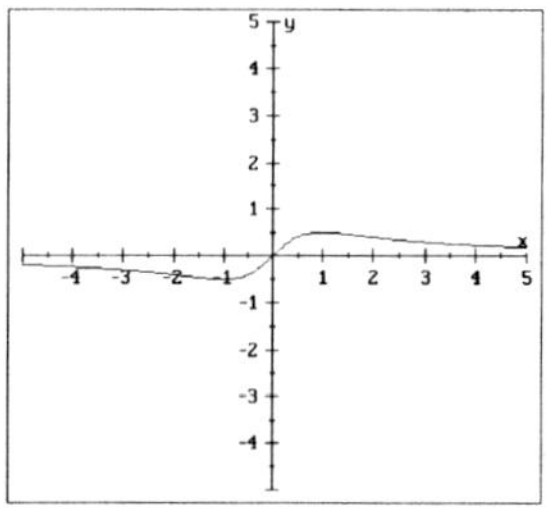

73. $y = \dfrac{3x^2}{x^2+1}$

Vert: none; Horiz: $y=\frac{3}{1}=3$

Slant: none; x-intercepts: $(0,0)$

y-intercepts: $(0,0)$; Symmetry: y-axis

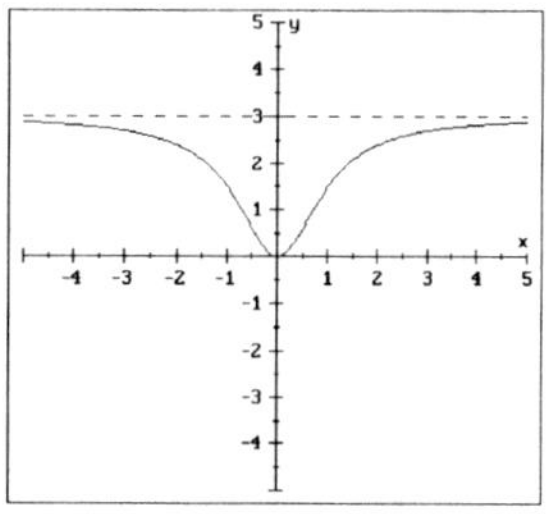

75. $h(x) = \dfrac{x^2 - 2x - 8}{x - 1} = \dfrac{(x+2)(x-4)}{x-1}$
$= x - 1 + \dfrac{-9}{x-1}$

Vert: $x = 1$; Horiz: none; Slant: $y = x - 1$
x-intercepts: $(4, 0)$, $(-2, 0)$
y-intercepts: $(0, 8)$; Symmetry: none

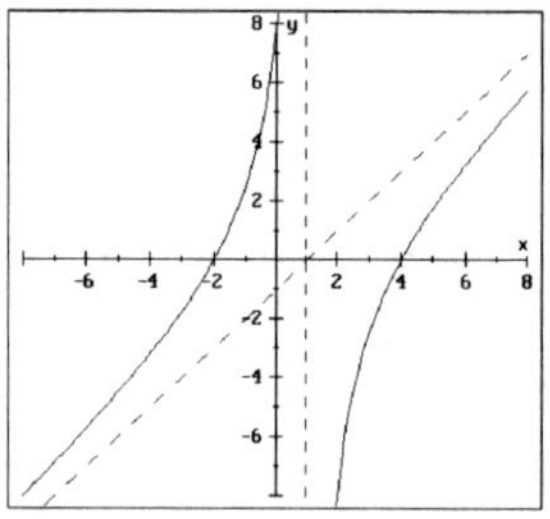

77. $f(x) = \dfrac{x^3 + x^2 + 6x}{x^2 - 1} = \dfrac{x(x^2 + x + 6)}{(x+1)(x-1)}$
$= x + 1 + \dfrac{7x + 1}{x^2 - 1}$

Vert: $x = -1$, $x = 1$; Horiz: none
Slant: $y = x + 1$; x-intercepts: $(0, 0)$
y-intercepts: $(0, 0)$; Symmetry: none

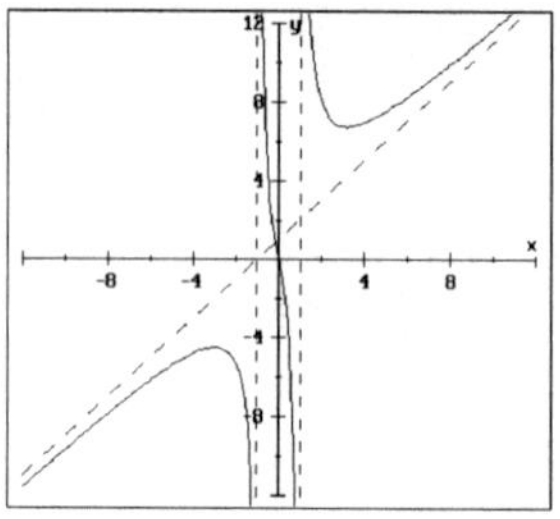

79. $f(x) = \dfrac{x^2}{x} = x$ (if $x \neq 0$)

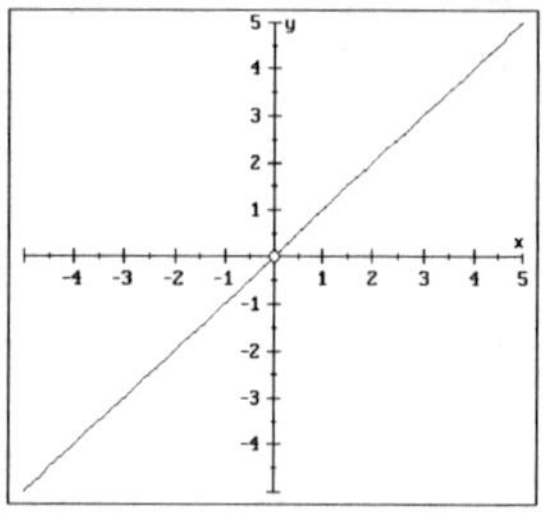

81. $f(x) = \dfrac{x^3 + x}{x} = \dfrac{x(x^2+1)}{x} = x^2 + 1$
(if $x \neq 0$)

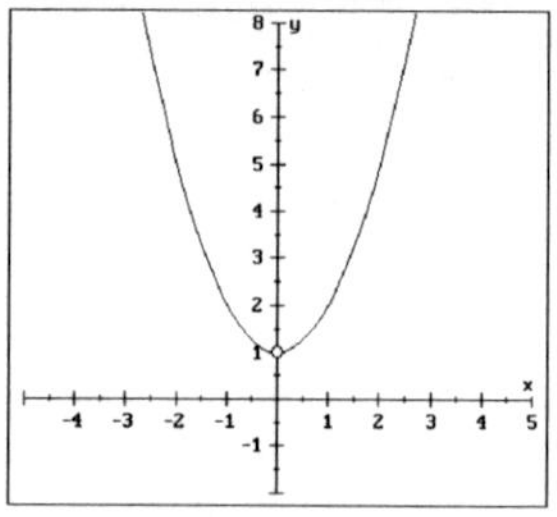

83. $f(x) = \dfrac{x^2 - 2x + 1}{x - 1} = \dfrac{(x-1)(x-1)}{x-1}$
$= x - 1$ (if $x \neq 1$)

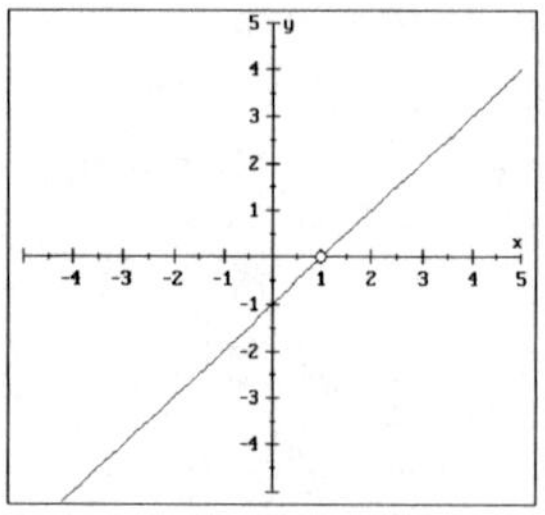

85. $f(x) = \dfrac{x^3 - 1}{x - 1} = \dfrac{(x-1)(x^2+x+1)}{x-1}$
$= x^2 + x + 1$ (if $x \neq 1$)

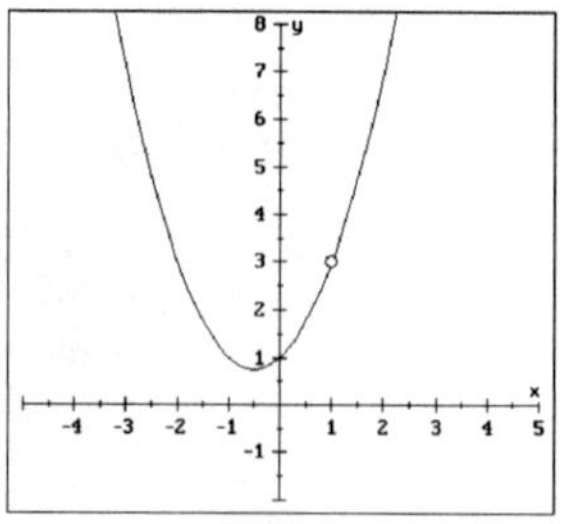

87. **a.** $c(x) = 3.25x + 700$

b. $c(500) = 3.25(500) + 700 = \$2325$

c. $\overline{c}(x) = \frac{3.25x+700}{x}$

d. $\overline{c}(500) = \frac{3.25(500)+700}{500} = \$4.65$

e. $\overline{c}(1000) = \frac{3.25(1000)+700}{1000} = \$3.95$

f. $\overline{c}(2000) = \frac{3.25(2000)+700}{2000} = \$3.60$

89. **a.** $c(n) = 0.095n + 8.50$

b. $\overline{c}(n) = \frac{0.095n+8.50}{n}$

c. $\overline{c}(850) = \frac{0.095(850)+8.50}{850}$
$= \$0.105 = 10.5¢$

91. **Answers may vary.**

93. $$y = \frac{ax+b}{cx^2+d} = \frac{\frac{ax+b}{x^2}}{\frac{cx^2+d}{x^2}} = \frac{\frac{ax}{x^2}+\frac{b}{x^2}}{\frac{cx^2}{x^2}+\frac{d}{x^2}} = \frac{\frac{a}{x}+\frac{b}{x^2}}{c+\frac{d}{x^2}}$$

As x approaches $\pm\infty$, $y \approx \dfrac{0+0}{c+0} = 0$. Thus the horizontal asymptote is $y = 0$.

95. $$y = \frac{ax^2+b}{cx^2+d} = \frac{\frac{ax^2+b}{x^2}}{\frac{cx^2+d}{x^2}} = \frac{\frac{ax^2}{x^2}+\frac{b}{x^2}}{\frac{cx^2}{x^2}+\frac{d}{x^2}} = \frac{a+\frac{b}{x^2}}{c+\frac{d}{x^2}}$$

As x approaches $\pm\infty$, $y \approx \dfrac{a+0}{c+0} = \dfrac{a}{c}$. Thus the horizontal asymptote is $y = \dfrac{a}{c}$.

97. **Answers may vary.**

99. **Answers may vary.**

101. $(2x^2 + 3x) + (x^2 - 2x) = 2x^2 + 3x + x^2 - 2x = 3x^2 + x$

103. $(5x + 2)(2x + 5) = 10x^2 + 25x + 4x + 10 = 10x^2 + 29x + 10$

105. $f(x + 1) = 3(x + 1) + 2 = 3x + 3 + 2 = 3x + 5$

Exercises 3.6 (page 378)

1. $f(x) + g(x)$

3. $f(x)g(x)$

5. intersection

7. $g(f(x))$

9. commutative

11. $(f + g)(x) = f(x) + g(x) = (2x + 1) + (3x - 2) = 5x - 1$; domain $= (-\infty, \infty)$

13. $(f \cdot g)(x) = f(x)g(x) = (2x + 1)(3x - 2) = 6x^2 - x - 2$; domain $= (-\infty, \infty)$

15. $(f - g)(x) = f(x) - g(x) = (x^2 + x) - (x^2 - 1) = x + 1$; domain $= (-\infty, \infty)$

17. $(f/g)(x) = \dfrac{f(x)}{g(x)} = \dfrac{x^2+x}{x^2-1} = \dfrac{x(x+1)}{(x+1)(x-1)} = \dfrac{x}{x-1}$; domain $= (-\infty, -1) \cup (-1, 1) \cup (1, \infty)$

19. $(f+g)(x) = f(x) + g(x) = (x^2 - 7) + \left(\sqrt{x}\right) = x^2 + \sqrt{x} - 7$; domain $= [0, \infty)$

21. $(f/g)(x) = \dfrac{f(x)}{g(x)} = \dfrac{x^2-7}{\sqrt{x}}$; domain $= (0, \infty)$

23. $(f+g)(2) = f(2) + g(2) = \left[(2)^2 - 1\right] + [3(2) - 2] = 3 + 4 = 7$

25. $(f-g)(0) = f(0) - g(0) = \left[(0)^2 - 1\right] - [3(0) - 2] = -1 - (-2) = 1$

27. $(f \cdot g)(2) = f(2) \cdot g(2) = \left[(2)^2 - 1\right] \cdot [3(2) - 2] = (3)(4) = 12$

29. $(f/g)\left(\frac{2}{3}\right) = \dfrac{f\left(\frac{2}{3}\right)}{g\left(\frac{2}{3}\right)} = \dfrac{\left[\left(\frac{2}{3}\right)^2 - 1\right]}{\left[3\left(\frac{2}{3}\right) - 2\right]} = \dfrac{-\frac{5}{9}}{0} \Rightarrow$ undefined

31. Let $f(x) = 3x^2$ and $g(x) = 2x$.
Then $(f+g)(x) = 3x^2 + 2x = h(x)$.

33. Let $f(x) = 3x^2$ and $g(x) = x^2 - 1$.
Then $(f/g)(x) = \dfrac{3x^2}{x^2-1} = h(x)$.

35. Let $f(x) = 3x^3$ and $g(x) = -x$.
Then $(f-g)(x) = 3x^3 + x$
$= x\left(3x^2 + 1\right) = h(x)$.

37. Let $f(x) = x + 9$ and $g(x) = x - 2$.
Then $(f \cdot g)(x) = (x+9)(x-2)$
$= x^2 + 7x - 18 = h(x)$.

39. $(f \circ g)(2) = f(g(2)) = f(5(2) - 2) = f(8) = 2(8) - 5 = 11$

41. $(f \circ f)\left(-\frac{1}{2}\right) = f\left(f\left(-\frac{1}{2}\right)\right) = f\left(2\left(-\frac{1}{2}\right) - 5\right) = f(-6) = 2(-6) - 5 = -17$

43. $(f \circ g)(-3) = f(g(-3)) = f(4(-3) + 4) = f(-8) = 3(-8)^2 - 2 = 190$

45. $(f \circ f)\left(\sqrt{3}\right) = f\left(f\left(\sqrt{3}\right)\right) = f\left(3\left(\sqrt{3}\right)^2 - 2\right) = f(7) = 3(7)^2 - 2 = 145$

47. The domain of $f \circ g$ is the set of all real numbers in the domain of $g(x)$ such that $g(x)$ is in the domain of $f(x)$. Domain of $g(x)$: $(-\infty, \infty)$. Domain of $f(x) = (-\infty, \infty)$. Thus, all values of $g(x)$ are in the domain of $f(x)$. Domain of $f \circ g$: $(-\infty, \infty)$
$(f \circ g)(x) = f(g(x)) = f(x+1) = 3(x+1) = 3x + 3$

49. The domain of $f \circ f$ is the set of all real numbers in the domain of $f(x)$ such that $f(x)$ is in the domain of $f(x)$. Domain of $f(x)$: $(-\infty, \infty)$. Thus, all values of $f(x)$ are in the domain of $f(x)$. Domain of $f \circ f$: $(-\infty, \infty)$ $(f \circ f)(x) = f(f(x)) = f(3x) = 3(3x) = 9x$

SECTION 3.6

51. The domain of $g \circ f$ is the set of all real numbers in the domain of $f(x)$ such that $f(x)$ is in the domain of $g(x)$. Domain of $f(x)$: $(-\infty, \infty)$. Domain of $g(x) = (-\infty, \infty)$. Thus, all values of $f(x)$ are in the domain of $g(x)$. $\boxed{\text{Domain of } g \circ f\text{: } (-\infty, \infty)}$

$(g \circ f)(x) = g(f(x)) = g(x^2) = 2x^2$

53. The domain of $g \circ g$ is the set of all real numbers in the domain of $g(x)$ such that $g(x)$ is in the domain of $g(x)$. Domain of $g(x)$: $(-\infty, \infty)$. Thus, all values of $g(x)$ are in the domain of $g(x)$.

$\boxed{\text{Domain of } g \circ g\text{: } (-\infty, \infty)}$ $(g \circ g)(x) = g(g(x)) = g(2x) = 2(2x) = 4x$

55. The domain of $f \circ g$ is the set of all real numbers in the domain of $g(x)$ such that $g(x)$ is in the domain of $f(x)$. Domain of $g(x)$: $(-\infty, \infty)$. Domain of $f(x) = [0, \infty)$. Thus, we must have $g(x) \geq 0 \Rightarrow x + 1 \geq 0 \Rightarrow x \geq -1$. $\boxed{\text{Domain of } f \circ g\text{: } [-1, \infty)}$

$(f \circ g)(x) = f(g(x)) = f(x+1) = \sqrt{x+1}$

57. The domain of $f \circ f$ is the set of all real numbers in the domain of $f(x)$ such that $f(x)$ is in the domain of $f(x)$. Domain of $f(x)$: $[0, \infty)$. Thus, we must have $f(x) \geq 0 \Rightarrow \sqrt{x} \geq 0$. This is true for all real values of x. $\boxed{\text{Domain of } f \circ f\text{: } [0, \infty)}$

$(f \circ f)(x) = f(f(x)) = f(\sqrt{x}) = \sqrt{\sqrt{x}} = \left((x)^{1/2}\right)^{1/2} = x^{1/4} = \sqrt[4]{x}$

59. The domain of $g \circ f$ is the set of all real numbers in the domain of $f(x)$ such that $f(x)$ is in the domain of $g(x)$. Domain of $f(x)$: $[-1, \infty)$. Domain of $g(x) = (-\infty, \infty)$. Thus, all values of $f(x)$ are in the domain of $g(x)$. $\boxed{\text{Domain of } g \circ f\text{: } [-1, \infty)}$

$(g \circ f)(x) = g(f(x)) = g\left(\sqrt{x+1}\right) = \left(\sqrt{x+1}\right)^2 - 1 = x$

61. The domain of $g \circ g$ is the set of all real numbers in the domain of $g(x)$ such that $g(x)$ is in the domain of $g(x)$. Domain of $g(x)$: $(-\infty, \infty)$. Thus, all values of $g(x)$ are in the domain of $g(x)$.

$\boxed{\text{Domain of } g \circ g\text{: } (-\infty, \infty)}$ $(g \circ g)(x) = g(g(x)) = g(x^2 - 1) = (x^2 - 1)^2 - 1 = x^4 - 2x^2$

63. The domain of $f \circ g$ is the set of all real numbers in the domain of $g(x)$ such that $g(x)$ is in the domain of $f(x)$. Domain of $g(x)$: $(-\infty, 2) \cup (2, \infty)$. Domain of $f(x) = (-\infty, 1) \cup (1, \infty)$. Thus, we must have $g(x) \neq 1 \Rightarrow \dfrac{1}{x-2} \neq 1 \Rightarrow 1 \neq x - 2 \Rightarrow x \neq 3$

$\boxed{\text{Domain of } f \circ g\text{: } (-\infty, 2) \cup (2, 3) \cup (3, \infty)}$

$(f \circ g)(x) = f(g(x)) = f\left(\dfrac{1}{x-2}\right) = \dfrac{1}{\frac{1}{x-2} - 1} = \dfrac{1}{\frac{1}{x-2} - 1} \cdot \dfrac{x-2}{x-2} = \dfrac{x-2}{1-(x-2)} = \dfrac{x-2}{3-x}$

SECTION 3.6

65. The domain of $f \circ f$ is the set of all real numbers in the domain of $f(x)$ such that $f(x)$ is in the domain of $f(x)$. Domain of $f(x)$: $(-\infty, 1) \cup (1, \infty)$. Thus, we must have $f(x) \neq 1 \Rightarrow$

$\dfrac{1}{x-1} \neq 1 \Rightarrow 1 \neq x - 1 \Rightarrow x \neq 2$ $\boxed{\text{Domain of } f \circ f\text{: } (-\infty, 1) \cup (1, 2) \cup (2, \infty)}$

$$(f \circ f)(x) = f(f(x)) = f\left(\frac{1}{x-1}\right) = \frac{1}{\frac{1}{x-1} - 1} = \frac{1}{\frac{1}{x-1} - 1} \cdot \frac{x-1}{x-1} = \frac{x-1}{1-(x-1)} = \frac{x-1}{2-x}$$

67. Let $f(x) = x - 2$ and $g(x) = 3x$.
Then $(f \circ g)(x) = f(g(x))$
$= f(3x) = 3x - 2.$

69. Let $f(x) = x - 2$ and $g(x) = x^2$.
Then $(f \circ g)(x) = f(g(x))$
$= f(x^2) = x^2 - 2.$

71. Let $f(x) = x^2$ and $g(x) = x - 2$.
Then $(f \circ g)(x) = f(g(x))$
$= f(x-2) = (x-2)^2.$

73. Let $f(x) = \sqrt{x}$ and $g(x) = x + 2$.
Then $(f \circ g)(x) = f(g(x))$
$= f(x+2) = \sqrt{x+2}.$

75. Let $f(x) = x + 2$ and $g(x) = \sqrt{x}$.
Then $(f \circ g)(x) = f(g(x))$
$= f(\sqrt{x}) = \sqrt{x} + 2.$

77. Let $f(x) = x$ and $g(x) = x$.
Then $(f \circ g)(x) = f(g(x))$
$= f(x) = x.$

79. $(f+g)(-4) = f(-4) + g(-4)$
$= -2 + 2 = 0$

81. $(f \cdot g)(5) = f(5) \cdot g(5) = -2(0) = 0$

83. $(f \circ g)(3) = f(g(3)) = f(2) = 1$

85. $(f \circ f)(-2) = f(f(-2)) = f(0) = 1$

87. $(f+g)(2) = f(2) + g(2) = 4 + 4 = 8$

89. $(f \circ g)(2) = f(g(2)) = f(4) = 9$

91. **a.** $(R-C)(x) = 300x - (60{,}000 + 40x)$
$= 260x - 60{,}000$

b. $(R-C)(500) = 260(500) - 60{,}000$
$= 70{,}000$

93. $r(t) = \dfrac{d(t)}{2} = \dfrac{3t}{2}$; $A(t) = \pi(r(t))^2 = \pi\left(\dfrac{3t}{2}\right)^2 = \dfrac{9}{4}\pi t^2$

$A(120) = \dfrac{9}{4}\pi(120)^2 \approx 101{,}787.6$ square inches

95. If the area is A and the length of a side is s, then $s^2 = A \Rightarrow s = \sqrt{A}$. Then $P = 4s = 4\sqrt{A}$.

97. $(f+f)(x) = f(x) + f(x) = 3x + 3x = 6x$; $f(x+x) = f(2x) = 3(2x) = 6x$

99. $$(f \circ f)(x) = f(f(x)) = f\left(\frac{x-1}{x+1}\right) = \frac{\frac{x-1}{x+1} - 1}{\frac{x-1}{x+1} + 1} = \frac{(x+1)\left(\frac{x-1}{x+1} - 1\right)}{(x+1)\left(\frac{x-1}{x+1} + 1\right)} = \frac{x-1-(x+1)}{x-1+x+1}$$
$$= \frac{-2}{2x} = -\frac{1}{x}$$

101. **Answers may vary.**

103. **Answers may vary.**

105.
$$\begin{aligned} x &= 3y - 7 \\ x + 7 &= 3y \\ \frac{x+7}{3} &= y \end{aligned}$$

107.
$$\begin{aligned} x &= \frac{y}{y+3} \\ x(y+3) &= y \\ xy + 3x &= y \\ xy - y &= -3x \\ y(x-1) &= -3x \\ y &= \frac{-3x}{x-1} = \frac{3x}{1-x} \end{aligned}$$

Exercises 3.7 (page 390)

1. one-to-one

3. identity

5. $y = 3x$
one-to-one

7. $y = x^2 + 3$
$x = 1$ and $x = -1$ both correspond to $y = 4$.
not one-to-one

9. $y = x^3 - x$
$x = 1$ and $x = -1$ both correspond to $y = 0$.
not one-to-one

11. $y = |x|$
$x = 1$ and $x = -1$ both correspond to $y = 1$.
not one-to-one

13. $y = 5$
$x = 1$ and $x = 2$ both correspond to $y = 5$.
not one-to-one

15. $y = (x-2)^2,\ x \geq 2$
one-to-one

17. one-to-one

19. not one-to-one
(not a function)

21.
$$(f \circ g)(x) = f(g(x)) = f\left(\frac{1}{5}x\right) = 5\left(\frac{1}{5}x\right) = x$$
$$(g \circ f)(x) = g(f(x)) = g(5x) = \frac{1}{5}(5x) = x$$

23.
$$(f \circ g)(x) = f(g(x)) = f\left(\frac{1}{x-1}\right) = \frac{\frac{1}{x-1} + 1}{\frac{1}{x-1}} = \frac{(x-1)\left(\frac{1}{x-1} + 1\right)}{(x-1)\frac{1}{x-1}} = \frac{1 + x - 1}{1} = x$$
$$(g \circ f)(x) = g(f(x)) = g\left(\frac{x+1}{x}\right) = \frac{1}{\frac{x+1}{x} - 1} = \frac{x(1)}{x\left(\frac{x+1}{x} - 1\right)} = \frac{x}{x + 1 - x} = \frac{x}{1} = x$$

25.
$$\begin{aligned} y = f(x) &= 3x \\ x &= 3y \\ \frac{x}{3} &= y \\ f^{-1}(x) &= \frac{x}{3} = \frac{1}{3}x \end{aligned}$$

$$\begin{aligned} (f \circ f^{-1})(x) &= f(f^{-1}(x)) \\ &= f\left(\frac{x}{3}\right) \\ &= 3\left(\frac{x}{3}\right) \\ &= x \end{aligned}$$

$$\begin{aligned} (f^{-1} \circ f)(x) &= f^{-1}(f(x)) \\ &= f^{-1}(3x) \\ &= \frac{3x}{3} \\ &= x \end{aligned}$$

SECTION 3.7

27.
$$\begin{aligned} y = f(x) &= 3x+2 \\ x &= 3y+2 \\ x-2 &= 3y \\ \frac{x-2}{3} &= y \\ f^{-1}(x) &= \frac{x-2}{3} \end{aligned}$$
$$\begin{aligned} (f \circ f^{-1})(x) &= f(f^{-1}(x)) \\ &= f\left(\frac{x-2}{3}\right) \\ &= 3\left(\frac{x-2}{3}\right)+2 \\ &= x-2+2 = x \end{aligned}$$
$$\begin{aligned} (f^{-1} \circ f)(x) &= f^{-1}(f(x)) \\ &= f^{-1}(3x+2) \\ &= \frac{(3x+2)-2}{3} \\ &= \frac{3x}{3} = x \end{aligned}$$

29.
$$\begin{aligned} y = f(x) &= x^3+2 \\ x &= y^3+2 \\ x-2 &= y^3 \\ \sqrt[3]{x-2} &= y \\ f^{-1}(x) &= \sqrt[3]{x-2} \end{aligned}$$
$$\begin{aligned} (f \circ f^{-1})(x) &= f(f^{-1}(x)) \\ &= f\left(\sqrt[3]{x-2}\right) \\ &= \left(\sqrt[3]{x-2}\right)^3+2 \\ &= x-2+2 = x \end{aligned}$$
$$\begin{aligned} (f^{-1} \circ f)(x) &= f^{-1}(f(x)) \\ &= f^{-1}(x^3+2) \\ &= \sqrt[3]{(x^3+2)-2} \\ &= \sqrt[3]{x^3} = x \end{aligned}$$

31.
$$\begin{aligned} y = f(x) &= \sqrt[5]{x} \\ x &= \sqrt[5]{y} \\ x^5 &= y \\ f^{-1}(x) &= x^5 \end{aligned}$$
$$\begin{aligned} (f \circ f^{-1})(x) &= f(f^{-1}(x)) \\ &= f(x^5) \\ &= \sqrt[5]{x^5} = x \end{aligned}$$
$$\begin{aligned} (f^{-1} \circ f)(x) &= f^{-1}(f(x)) \\ &= f^{-1}(\sqrt[5]{x}) \\ &= (\sqrt[5]{x})^5 = x \end{aligned}$$

33.
$$\begin{aligned} y = f(x) &= \frac{1}{x+3} \\ x &= \frac{1}{y+3} \\ x(y+3) &= 1 \\ y+3 &= \frac{1}{x} \\ f^{-1}(x) &= \frac{1}{x}-3 \end{aligned}$$
$$\begin{aligned} (f \circ f^{-1})(x) &= f(f^{-1}(x)) \\ &= f\left(\frac{1}{x}-3\right) \\ &= \frac{1}{\frac{1}{x}-3+3} \\ &= \frac{1}{\frac{1}{x}} \\ &= x \end{aligned}$$
$$\begin{aligned} (f^{-1} \circ f)(x) &= f^{-1}(f(x)) \\ &= f^{-1}\left(\frac{1}{x+3}\right) \\ &= \frac{1}{\frac{1}{x+3}}-3 \\ &= x+3-3 \\ &= x \end{aligned}$$

35.
$$\begin{aligned} y = f(x) &= \frac{1}{2x} \\ x &= \frac{1}{2y} \\ x(2y) &= 1 \\ 2xy &= 1 \\ y &= \frac{1}{2x} \\ f^{-1}(x) &= \frac{1}{2x} \end{aligned}$$
$$\begin{aligned} (f \circ f^{-1})(x) &= f(f^{-1}(x)) \\ &= f\left(\frac{1}{2x}\right) \\ &= \frac{1}{2\left(\frac{1}{2x}\right)} \\ &= \frac{1}{\frac{1}{x}} \\ &= x \end{aligned}$$
$$\begin{aligned} (f^{-1} \circ f)(x) &= f^{-1}(f(x)) \\ &= f^{-1}\left(\frac{1}{2x}\right) \\ &= \frac{1}{2\left(\frac{1}{2x}\right)} \\ &= \frac{1}{\frac{1}{x}} \\ &= x \end{aligned}$$

37.
$$\begin{aligned} y = f(x) &= 5x \\ x &= 5y \\ \frac{x}{5} &= y \\ f^{-1}(x) &= \frac{1}{5}x \end{aligned}$$

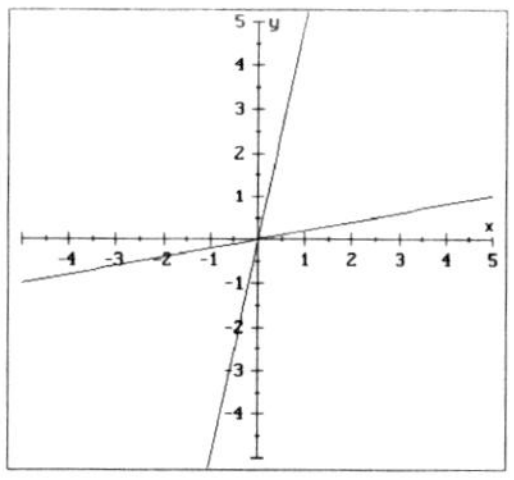

39.
$$\begin{aligned} y = f(x) &= 2x - 4 \\ x &= 2y - 4 \\ x + 4 &= 2y \\ \frac{x+4}{2} &= y \\ f^{-1}(x) &= \frac{x+4}{2} \end{aligned}$$

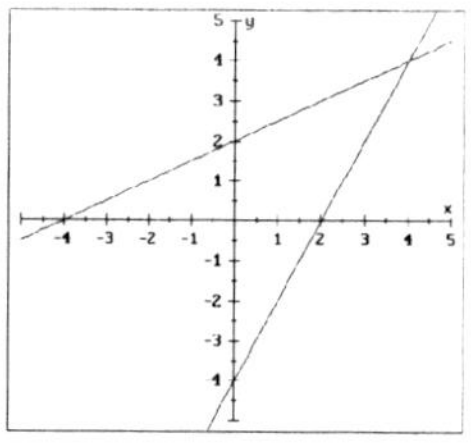

41.
$$\begin{aligned} x - y &= 2 \\ y - x &= 2 \\ y &= x + 2 \\ f^{-1}(x) &= x + 2 \end{aligned}$$

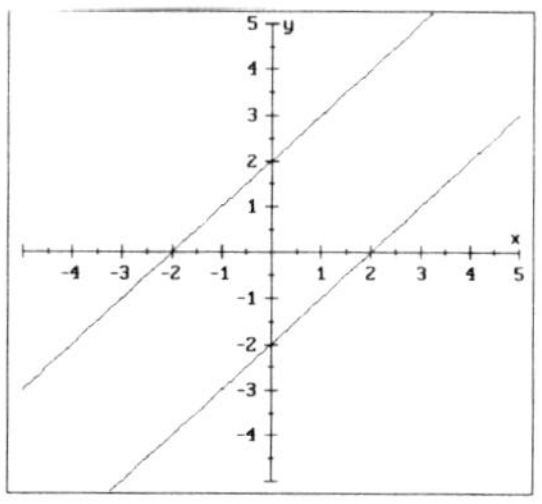

43.
$$\begin{aligned} 2x + y &= 4 \\ 2y + x &= 4 \\ 2y &= 4 - x \\ y &= \frac{4-x}{2} \\ f^{-1}(x) &= \frac{4-x}{2} \end{aligned}$$

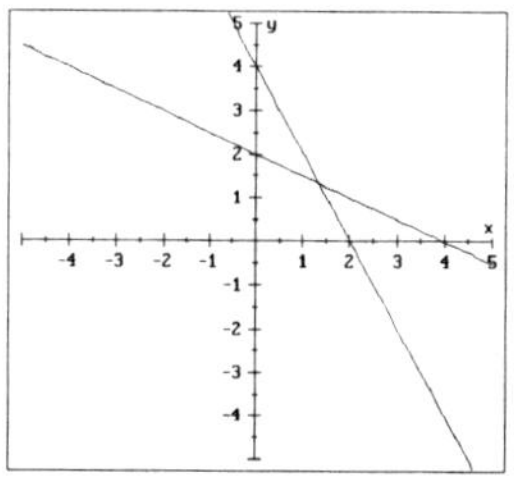

45.
$$\begin{aligned} y = f(x) &= \sqrt[3]{x-4} \\ x &= \sqrt[3]{y-4} \\ x^3 &= y - 4 \\ x^3 + 4 &= y \\ f^{-1}(x) &= x^3 + 4 \end{aligned}$$

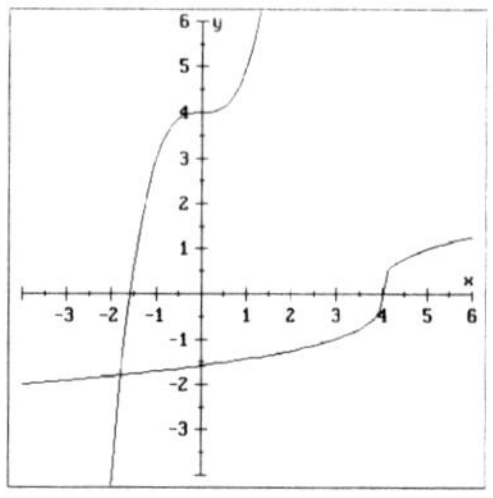

47.
$$\begin{aligned} y = f(x) &= (x-6)^3 \\ x &= (y-6)^3 \\ \sqrt[3]{x} &= y - 6 \\ \sqrt[3]{x} + 6 &= y \\ f^{-1}(x) &= \sqrt[3]{x} + 6 \end{aligned}$$

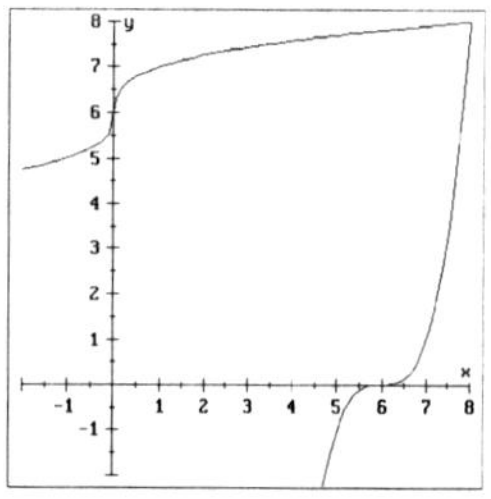

49.
$$\begin{aligned} y &= \frac{1}{2x} \\ x &= \frac{1}{2y} \\ 2xy &= 1 \\ y &= \frac{1}{2x} \\ f^{-1}(x) &= \frac{1}{2x} \end{aligned}$$

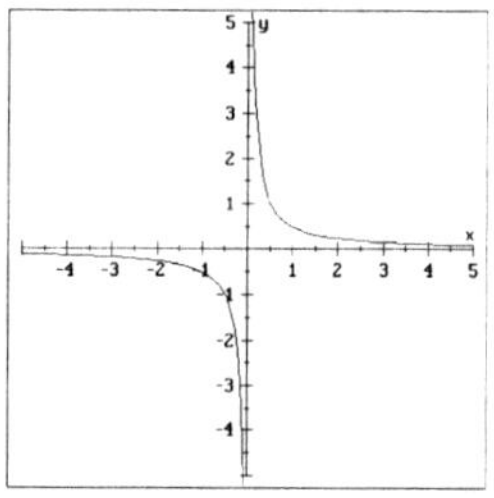

51.
$$\begin{aligned} y &= \frac{x+1}{x-1} \\ x &= \frac{y+1}{y-1} \\ x(y-1) &= y + 1 \\ xy - x &= y + 1 \\ xy - y &= x + 1 \\ y(x-1) &= x + 1 \\ f^{-1}(x) &= \frac{x+1}{x-1} \end{aligned}$$

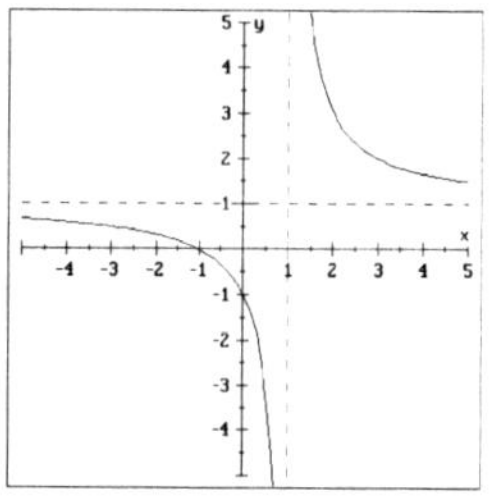

53.
$$\begin{aligned} f(x) &= x^2 - 3 \quad x \le 0 \\ y &= x^2 - 3 \quad x \le 0 \\ x &= y^2 - 3 \quad y \le 0 \\ x + 3 &= y^2 \quad y \le 0 \\ \pm\sqrt{x+3} &= y \quad y \le 0 \end{aligned}$$
Thus, $f^{-1}(x) = -\sqrt{x+3}\ (x \ge -3)$.

55.
$$\begin{aligned} f(x) &= x^4 - 8 \quad x \ge 0 \\ y &= x^4 - 8 \quad x \ge 0 \\ x &= y^4 - 8 \quad y \ge 0 \\ x + 8 &= y^4 \quad y \ge 0 \\ \pm\sqrt[4]{x+8} &= y \quad y \ge 0 \end{aligned}$$
Thus, $f^{-1}(x) = \sqrt[4]{x+8}\ (x \ge -8)$.

57.
$$\begin{aligned} f(x) &= \sqrt{4-x^2} \quad 0 \le x \le 2 \\ y &= \sqrt{4-x^2} \quad 0 \le x \le 2 \\ x &= \sqrt{4-y^2} \quad 0 \le y \le 2 \\ x^2 &= 4 - y^2 \quad 0 \le y \le 2 \\ y^2 &= 4 - x^2 \quad 0 \le y \le 2 \\ y &= \pm\sqrt{4-x^2} \quad 0 \le y \le 2 \end{aligned}$$
Thus, $f^{-1}(x) = \sqrt{4-x^2}\ (0 \le x \le 2)$.

59. $f(x) = \dfrac{x}{x-2}$

Domain of $f = \boxed{(-\infty, 2) \cup (2, \infty)}$

$f^{-1}(x) = \dfrac{2x}{x-1}$

Range of f = Domain of f^{-1}

$= \boxed{(-\infty, 1) \cup (1, \infty)}$

61. $f(x) = \dfrac{1}{x} - 2$

Domain of $f = \boxed{(-\infty, 0) \cup (0, \infty)}$

$f^{-1}(x) = \dfrac{1}{x+2}$

Range of f = Domain of f^{-1}

$= \boxed{(-\infty, -2) \cup (-2, \infty)}$

63. **a.** $y = 0.75x + 8.50$

b. $y = 0.75(4) + 8.50$
$= \$11.50$

c.
$$\begin{aligned} y &= 0.75x + 8.50 \\ x &= 0.75y + 8.50 \\ x - 8.50 &= 0.75y \\ \frac{x - 8.50}{0.75} &= y \end{aligned}$$

d.
$$\begin{aligned} y &= \frac{x - 8.50}{0.75} \\ &= \frac{10 - 8.50}{0.75} \\ &= \frac{1.50}{0.75} = 2 \end{aligned}$$

65. **Answers may vary.**

67. $f(0) = 3$, so $f^{-1}(3) = 0$.

69. $a \ge 0$

71. $16^{3/4} = \left(16^{1/4}\right)^3 = 8$

73. $(-8)^{2/3} = \left((-8)^{1/3}\right)^2 = 4$

75. $\left(\dfrac{64}{125}\right)^{-1/3} = \left(\dfrac{125}{64}\right)^{1/3} = \dfrac{5}{4}$

77. $49^{-1/2} = \dfrac{1}{49^{1/2}} = \dfrac{1}{7}$

Chapter 3 Review (page 394)

1. $y = 3$
Each value of x is paired with only one value of y.
function

2. $y + 5x^2 = 2$
$y = -5x^2 + 2$
Each value of x is paired with only one value of y.
function

3. $y^2 - x = 5$
$y^2 = x + 5$
$y = \pm\sqrt{x+5}$
Each value of x is paired with more than one value of y. **not a function**

4. $y = |x| + x$
Each value of x is paired with only one value of y.
function

5. $f(x) = y = 3x^2 - 5$
domain $= (-\infty, \infty)$

6. $f(x) = y = \dfrac{3x}{x-5}$
domain $= (-\infty, 5) \cup (5, \infty)$

7. $f(x) = y = \sqrt{x-1}$
domain $= [1, \infty)$

8. $f(x) = y = \sqrt{x^2+1}$
$x^2 + 1 \geq 0 \Rightarrow$ domain $= (-\infty, \infty)$

9. $f(x) = 5x - 2$
$f(2) = 5(2) - 2 = 8$
$f(-3) = 5(-3) - 2 = -17$
$f(k) = 5k - 2$

10. $f(x) = \dfrac{6}{x-5}$
$f(2) = \frac{6}{2-5} = \frac{6}{-3} = -2$
$f(-3) = \frac{6}{-3-5} = \frac{6}{-8} = -\frac{3}{4}$
$f(k) = \frac{6}{k-5}$

11. $f(x) = |x - 2|$
$f(2) = |2 - 2| = |0| = 0$
$f(-3) = |-3 - 2| = |-5| = 5$
$f(k) = |k - 2|$

12. $f(x) = \dfrac{x^2-3}{x^2+3}$
$f(2) = \frac{2^2-3}{2^2+3} = \frac{1}{7}$
$f(-3) = \frac{(-3)^2-3}{(-3)^2+3} = \frac{6}{12} = \frac{1}{2}$
$f(k) = \frac{k^2-3}{k^2+3}$

13. $$\frac{f(x+h)-f(x)}{h} = \frac{[5(x+h)-6]-[5x-6]}{h} = \frac{[5x+5h-6]-[5x-6]}{h}$$
$$= \frac{5x+5h-6-5x+6}{h} = \frac{5h}{h} = 5$$

14. $$\frac{f(x+h)-f(x)}{h} = \frac{\left[2(x+h)^2-7(x+h)+3\right]-\left[2x^2-7x+3\right]}{h}$$
$$= \frac{\left[2x^2+4xh+2h^2-7x-7h+3\right]-\left[2x^2-7x+3\right]}{h}$$
$$= \frac{2x^2+4xh+2h^2-7x-7h+3-2x^2+7x-3}{h}$$
$$= \frac{4xh+2h^2-7h}{h} = \frac{h(4x+2h-7)}{h} = 4x+2h-7$$

15. $f(x) = -x^2 + 4$

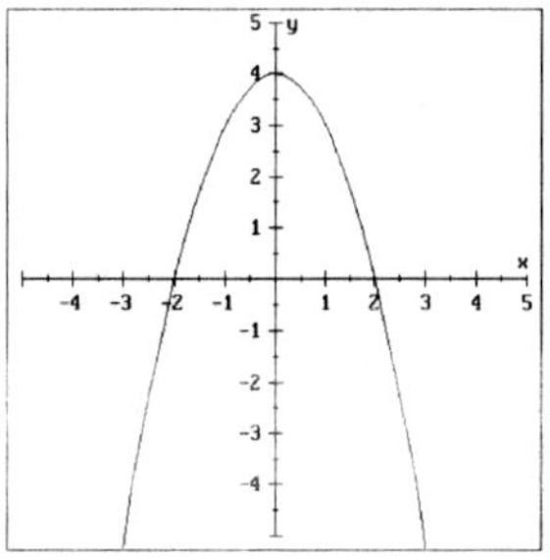

domain $= (-\infty, \infty)$
range $= (-\infty, 4]$

16. $f(x) = 3|x - 2|$

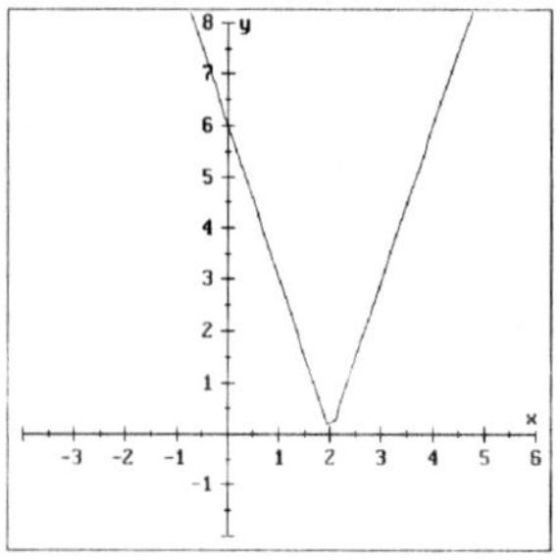

domain $= (-\infty, \infty)$
range $= [0, \infty)$

17. function

18. not a function

19. **a.** $I(h) = 3.5h - 50$
b. $I(200) = 3.5(200) - 50 = \$650$

20. Let (x, C) be a point on the line. Then the points $(100, 20)$ and $(500, 80)$ are on the line.

$$m = \frac{\Delta C}{\Delta x} = \frac{80 - 20}{500 - 100} = \frac{60}{400} = 0.15$$

$$C = mx + b$$
$$20 = 0.15(100) + b$$
$$20 = 15 + b$$
$$5 = b \Rightarrow \boxed{C(x) = 0.15x + 5}$$

21. $f(x) = \frac{1}{2}x^2 + 4$
$a = \frac{1}{2} \Rightarrow a > 0$
upward, minimum

22. $f(x) = -4(x + 1)^2 + 5$
$a = -4 \Rightarrow a < 0$
downward, maximum

23. $f(x) = 2(x - 1)^2 + 6$
Vertex: $(1, 6)$

24. $f(x) = -2(x + 4)^2 - 5$
Vertex: $(-4, -5)$

25. $y = x^2 + 6x - 4;\ a = 1, b = 6, c = -4$

$$x = -\frac{b}{2a} = -\frac{6}{2(1)} = -3$$

$$y = x^2 + 6x - 4 = (-3)^2 + 6(-3) - 4 = -13$$

Vertex: $(-3, -13)$

26. $y = -4x^2 + 4x - 9;\ a = -4, b = 4, c = -9$

$$x = -\frac{b}{2a} = -\frac{4}{2(-4)} = \frac{1}{2}$$

$$y = -4x^2 + 4x - 9 = -4\left(\tfrac{1}{2}\right)^2 + 4\left(\tfrac{1}{2}\right) - 9 = -8$$

Vertex: $\left(\frac{1}{2}, -8\right)$

27. $f(x) = (x-2)^2 - 3$

$a = 1 \Rightarrow$ up, vertex: $(2, -3)$

$0 = (x-2)^2 - 3$

$3 = (x-2)^2$

$\pm\sqrt{3} = x - 2$

$2 \pm \sqrt{3} = x$

$\left(2-\sqrt{3}, 0\right), \left(2+\sqrt{3}, 0\right)$

$f(0) = 1 \Rightarrow (0, 1)$

$f(1) = -2 \Rightarrow (1, -2)$

$(3, -2)$ on graph by symmetry

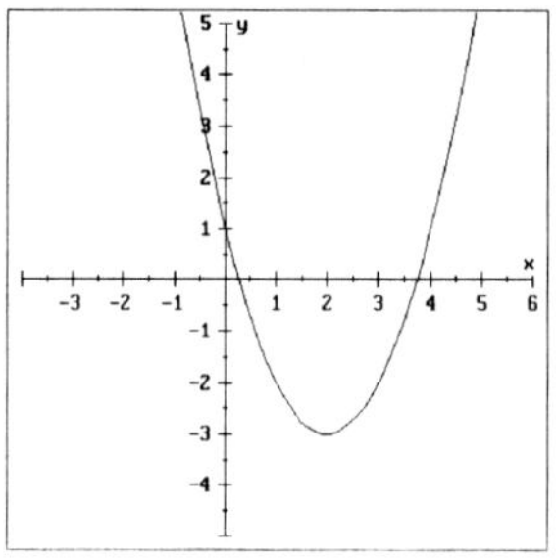

28. $f(x) = -(x-4)^2 + 4$

$a = -1 \Rightarrow$ down, vertex: $(4, 4)$

$0 = -(x-4)^2 + 4$

$(x-4)^2 = 4$

$x - 4 = \pm 2$

$x = 4 \pm 2$

$x = 2$ or $x = 6 \Rightarrow (2, 0), (6, 0)$

$f(0) = -12 \Rightarrow (0, -12)$

$f(3) = 3 \Rightarrow (3, 3)$

$(5, 3)$ on graph by symmetry

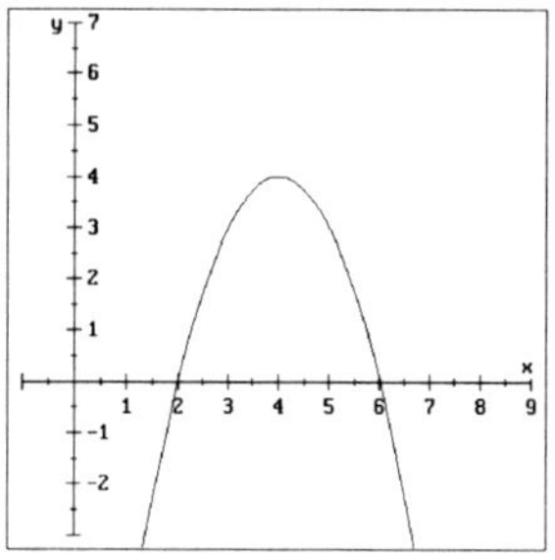

29. $f(x) = x^2 - x;\ a = 1, b = -1, c = 0$

$x = -\frac{b}{2a} = -\frac{-1}{2(1)} = \frac{1}{2}$

$y = x^2 - x = \left(\frac{1}{2}\right)^2 - \frac{1}{2} = -\frac{1}{4}$

vertex: $\left(\frac{1}{2}, -\frac{1}{4}\right)$, $a = 1 \Rightarrow$ up

$0 = x^2 - x$

$0 = x(x-1)$

$x = 0$ or $x = 1 \Rightarrow (0, 0), (1, 0)$

$f(0) = 0 \Rightarrow (0, 0)$

$f(2) = 2 \Rightarrow (2, 2)$ on graph

$(-1, 2)$ on graph by symmetry

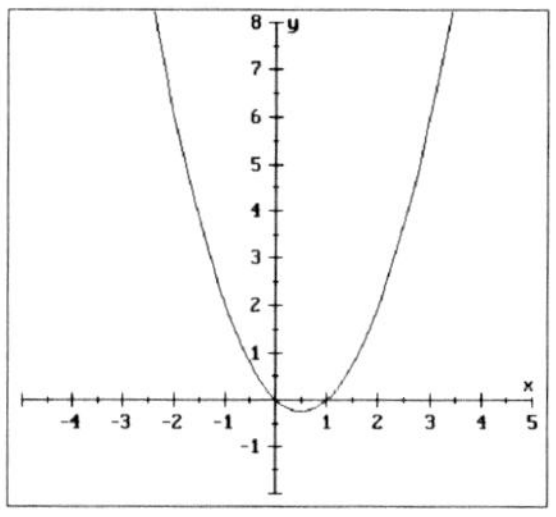

30. $f(x) = x - x^2;\ a = -1, b = 1, c = 0$

$x = -\frac{b}{2a} = -\frac{1}{2(-1)} = \frac{1}{2}$

$y = x - x^2 = \frac{1}{2} - \left(\frac{1}{2}\right)^2 = \frac{1}{4}$

vertex: $\left(\frac{1}{2}, \frac{1}{4}\right)$, $a = -1 \Rightarrow$ down

$0 = x - x^2$

$0 = x(1-x)$

$x = 0$ or $x = 1 \Rightarrow (0, 0), (1, 0)$

$f(0) = 0 \Rightarrow (0, 0)$

$f(2) = -2 \Rightarrow (2, -2)$ on graph

$(-1, -2)$ on graph by symmetry

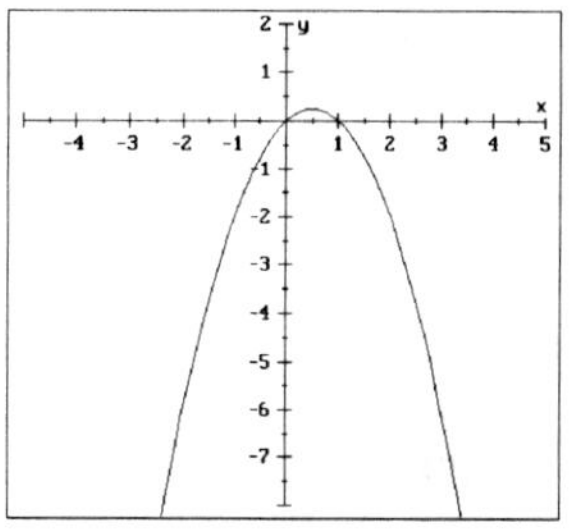

31. $y = x^2 - 3x - 4;\ a = 1, b = -3, c = -4$

$x = -\frac{b}{2a} = -\frac{-3}{2(1)} = \frac{3}{2}$

$y = x^2 - 3x - 4 = \left(\frac{3}{2}\right)^2 - 3\left(\frac{3}{2}\right) - 4 = -\frac{25}{4}$

vertex: $\left(\frac{3}{2}, -\frac{25}{4}\right)$, $a = 1 \Rightarrow$ up

$0 = x^2 - 3x - 4$

$0 = (x+1)(x-4)$

$x = -1$ or $x = 4 \Rightarrow (-1, 0), (4, 0)$

$f(0) = -4 \Rightarrow (0, -4)$

$f(2) = -6 \Rightarrow (2, -6)$ on graph

$(1, -6)$ on graph by symmetry

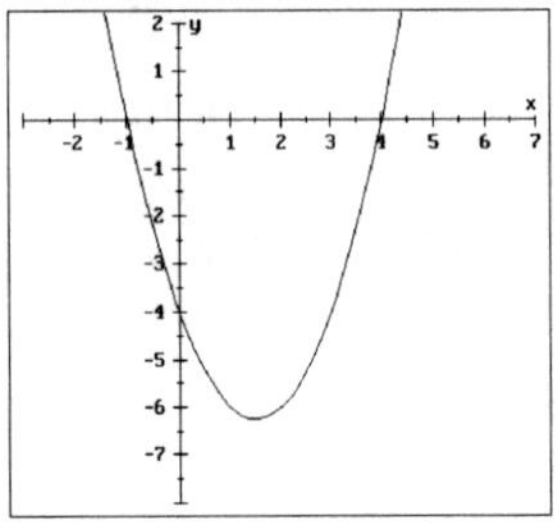

32. $y = 3x^2 - 8x - 3;\ a = 3, b = -8, c = -3$

$x = -\frac{b}{2a} = -\frac{-8}{2(3)} = \frac{4}{3}$

$y = 3x^2 - 8x - 3 = 3\left(\frac{4}{3}\right)^2 - 8\left(\frac{4}{3}\right) - 3 = -\frac{25}{3}$

vertex: $\left(\frac{4}{3}, -\frac{25}{3}\right)$, $a = 3 \Rightarrow$ up

$0 = 3x^2 - 8x - 3$

$0 = (3x+1)(x-3)$

$x = -\frac{1}{3}$ or $x = 3 \Rightarrow \left(-\frac{1}{3}, 0\right), (3, 0)$

$f(0) = -3 \Rightarrow (0, -3)$

$f(1) = -8 \Rightarrow (1, -8)$ on graph

$\left(\frac{5}{3}, -8\right)$ on graph by symmetry

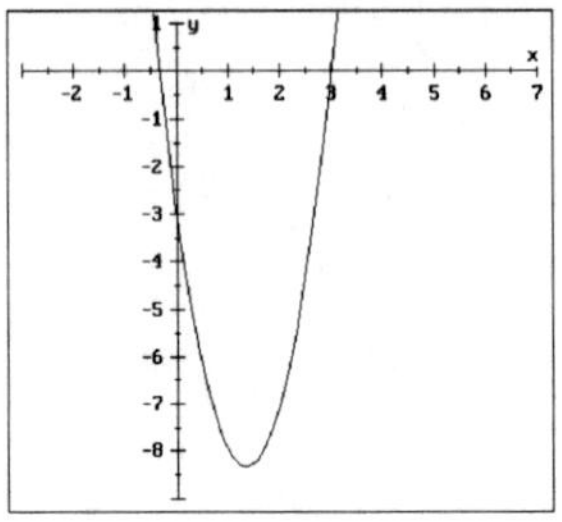

33. $3x^2 + y - 300 = 0$

$y = -3x^2 + 300$

$a = -3, b = 0, c = 300$

$x = -\frac{b}{2a} = -\frac{0}{2(-3)} = 0$

$y = -3(0)^2 + 300 = 300$

The maximum height is 300 units.

34. Let the numbers be x and $1 - x$.

Product $= x(1-x)$

$y = x - x^2$: $a = -1, b = 1, c = 0$

vertex: $x = -\frac{b}{2a} = -\frac{1}{2(-1)} = \frac{1}{2}$

Both numbers are $\frac{1}{2}$.

35. Let $x =$ the width of the region.

Then $\frac{1400 - 2x}{2} = 700 - x =$ the length.

Area = width · length

$y = x(700 - x)$

$y = -x^2 + 700x$

$a = -1, b = 700, c = 0$

$x = -\frac{b}{2a} = -\frac{700}{2(-1)} = 350$

$700 - x = 700 - 350 = 350$

$y = 350(700 - 350) = 122{,}500$

The dimensions are 350 ft by 350 ft, with an area of 122,500 ft^2.

36. $C(x) = 1.5x^2 - 150x + 4850$

$a = 1.5, b = -150, c = 4850$

$x = -\frac{b}{2a} = -\frac{-150}{2(1.5)} = 50$

$C(50) = 1.5(50)^2 - 150(50) + 4850 = 1100$

50 cameras should be made, for a minimum cost of \$1100.

37. $y = f(x) = x^3 - x$

$$f(-x) = (-x)^3 - (-x) = -x^3 + x = -f(x) \Rightarrow \text{odd}$$

x-int.	y-int.
$x^3 - x = 0$	$y = 0^3 - 0$
$x(x^2 - 1) = 0$	$y = 0$
$x(x+1)(x-1) = 0$	$(0, 0)$
$x = 0, x = -1, x = 1$	
$(0, 0), (-1, 0), (1, 0)$	

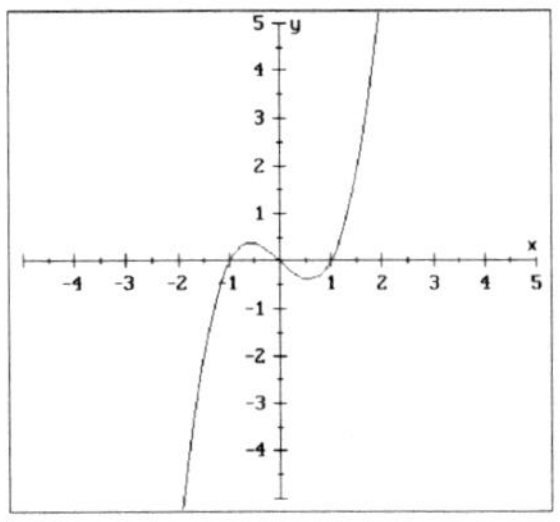

38. $f(x) = x^2 - 4x$

$$f(-x) = (-x)^2 - 4(-x) = x^2 + 4x$$

$\Rightarrow$ neither even nor odd

x-int.	y-int.
$x^2 - 4x = 0$	$y = 0^2 - 4(0)$
$x(x - 4) = 0$	$y = 0$
$x = 0, x = 4$	$(0, 0)$
$(0, 0), (4, 0)$	

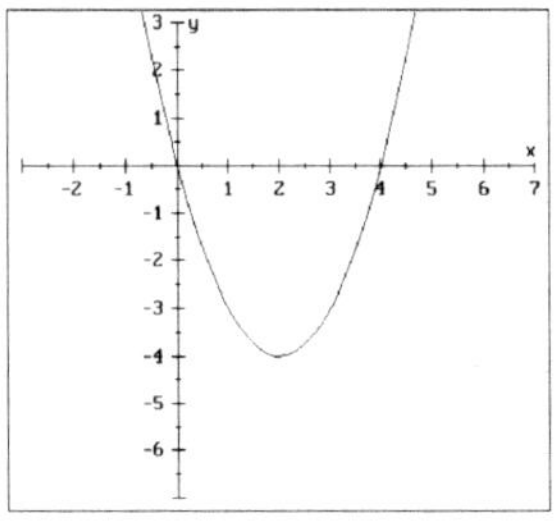

39. $f(x) = x^3 - x^2$

$$f(-x) = (-x)^3 - (-x)^2 = -x^3 - x^2$$

$\Rightarrow$ neither even nor odd

x-int.	y-int.
$x^3 - x^2 = 0$	$y = 0^3 - 0^2$
$x^2(x - 1) = 0$	$y = 0$
$x = 0, x = 1$	$(0, 0)$
$(0, 0), (1, 0)$	

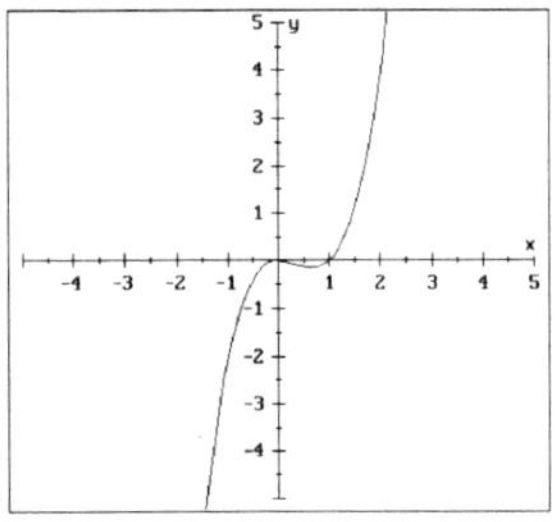

40. $y = f(x) = 1 - x^4$

$$f(-x) = 1 - (-x)^4 = 1 - x^4 = f(x) \Rightarrow \text{even}$$

x-int.	y-int.
$1 - x^4 = 0$	$y = 1 - 0^4$
$(1 + x^2)(1 - x^2) = 0$	$y = 1$
$x^2 = -1, x^2 = 1$	$(0, 1)$
$x = -1, x = 1$	
$(-1, 0), (1, 0)$	

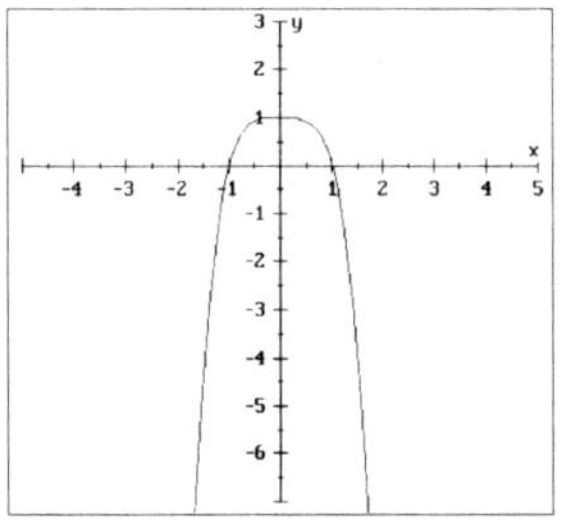

41. **a.** $f(-2) = 2$
b. $f(3) = 3^2 = 9$

42. **a.** $f\left(\frac{3}{2}\right) = 2 - \frac{3}{2} = \frac{1}{2}$
b. $f(2) = 2 + 1 = 3$

43. inc: $(-\infty, 0)$; dec: $(0, \infty)$

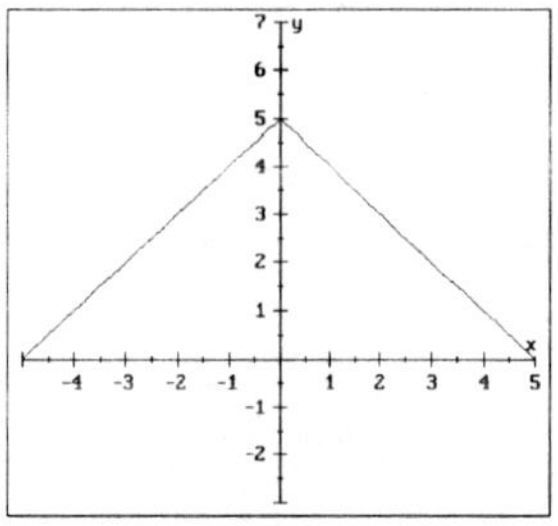

44. inc: $(-\infty, 0)$; const: $(0, \infty)$

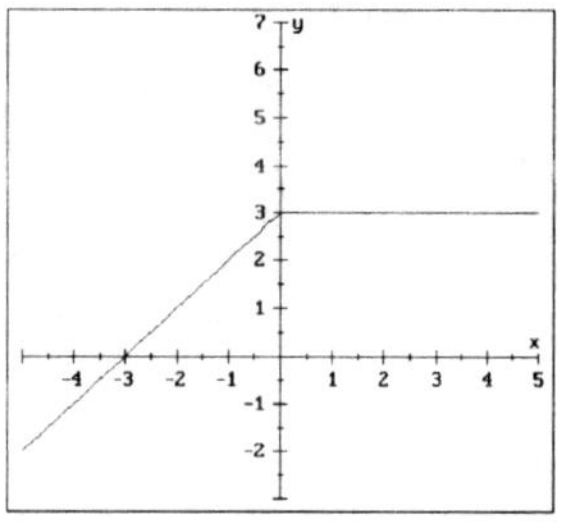

45. $f(1.7) = [[2(1.7)]] = [[3.4]] = 3$

46. $f(4.99) = [[4.99 - 5]] = [[-0.01]] = -1$

47. $f(x) = [[x]] + 2$

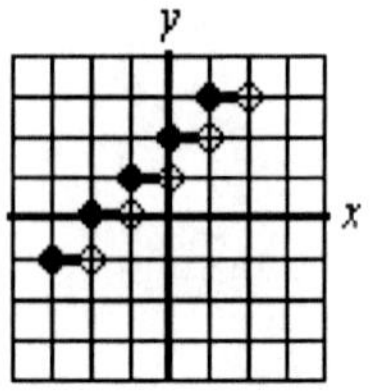

48. $f(x) = [[x - 1]]$

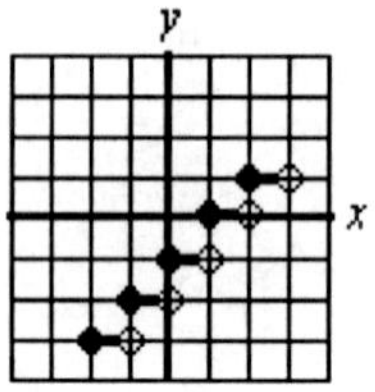

49. $20 + 3(8) = \$44$

50. $4 + 11(2) = \$26$

51. $g(x) = x^2 + 5$
Shift $y = x^2$ U 5

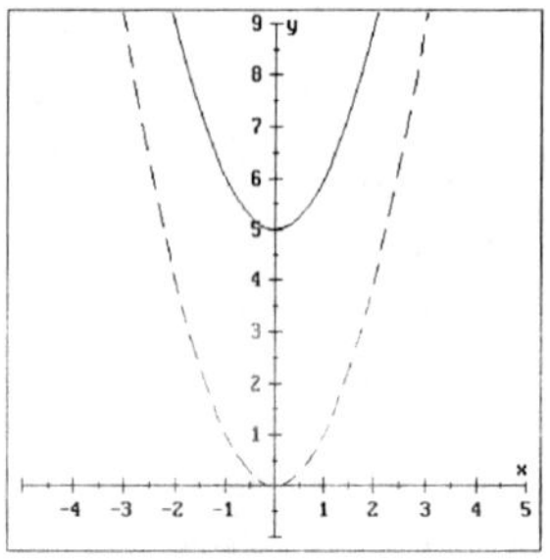

52. $g(x) = (x - 7)^3$
Shift $y = x^3$ R 7

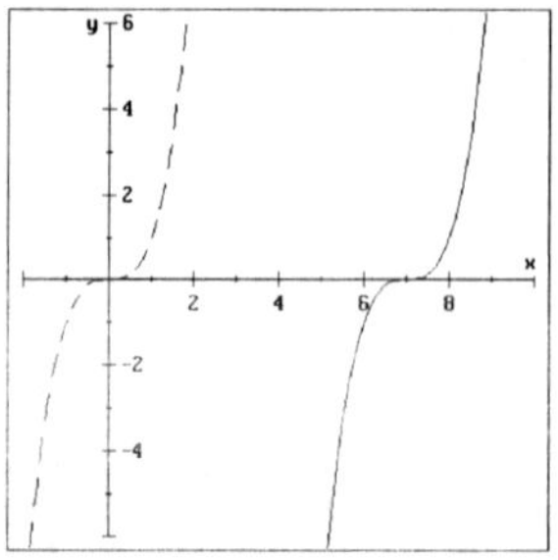

53. $g(x) = \sqrt{x + 2} + 3$
Shift $y = \sqrt{x}$ U 3, L 2

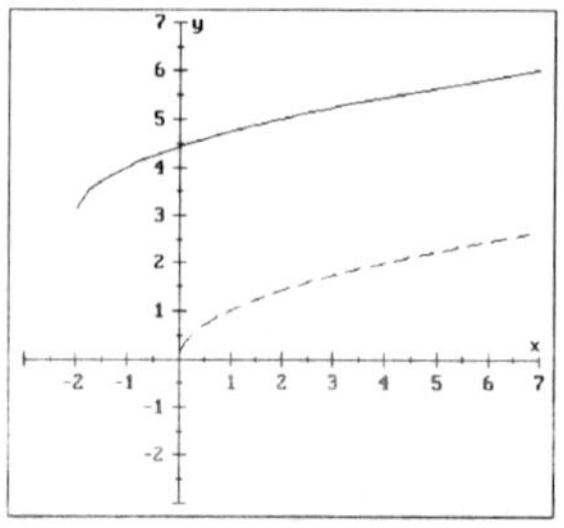

54. $g(x) = |x - 4| + 2$
Shift $y = |x|$ U 2, R 4

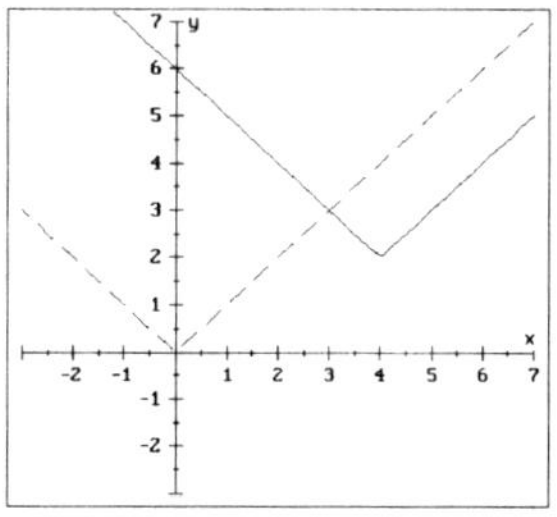

55. $g(x) = \frac{1}{3}x^3$: Shrink $y = x^3$ vert. by a factor of $\frac{1}{3}$

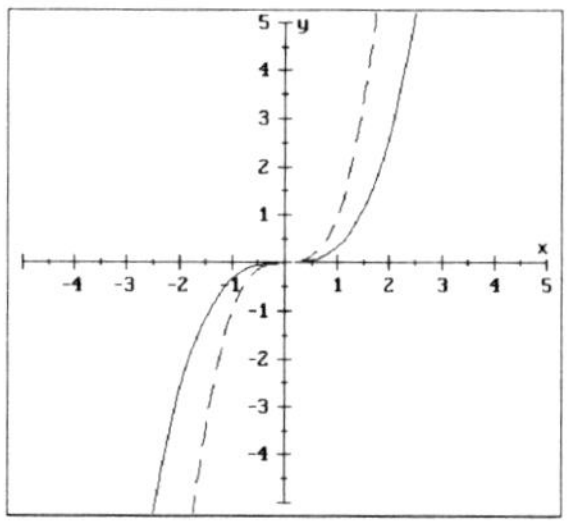

56. $g(x) = (-5x)^3$
Shrink $y = x^3$ horiz. by a factor of $\frac{1}{5}$. Reflect about y.

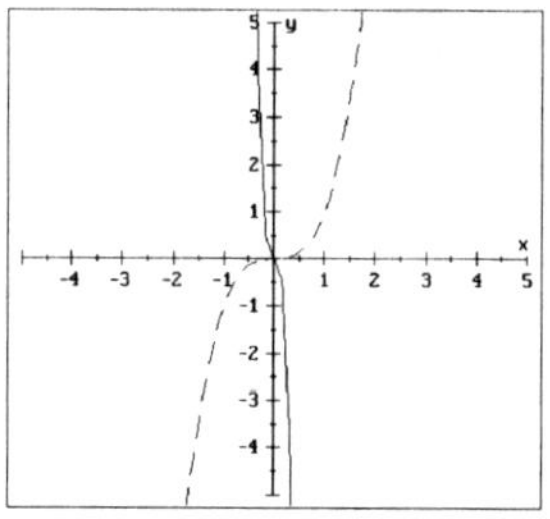

57. $g(x) = -|x - 4| + 3$
Start with $y = |x|$
Shift R 4, Reflect x,
Shift U 3

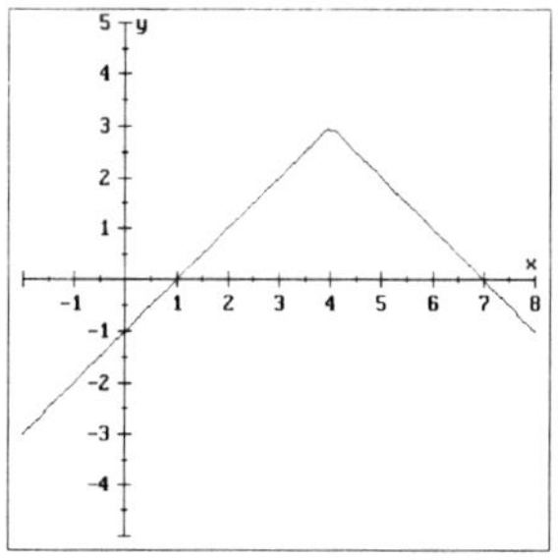

58. $g(x) = \frac{1}{4}|x - 4| + 1$
Start with $y = |x|$
Shift R 4, Shrink vert. by a factor of $\frac{1}{4}$, Shift U 1

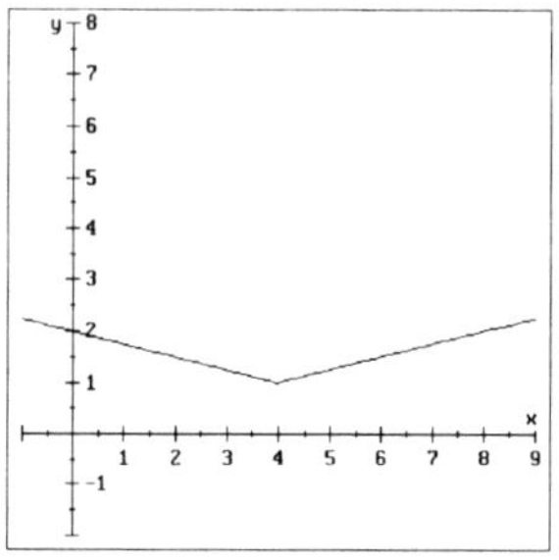

59. $g(x) = 3\sqrt{x + 3} + 2$
Start with $y = \sqrt{x}$
Shift L 3, Stretch vert. by a factor of 3, Shift U 2

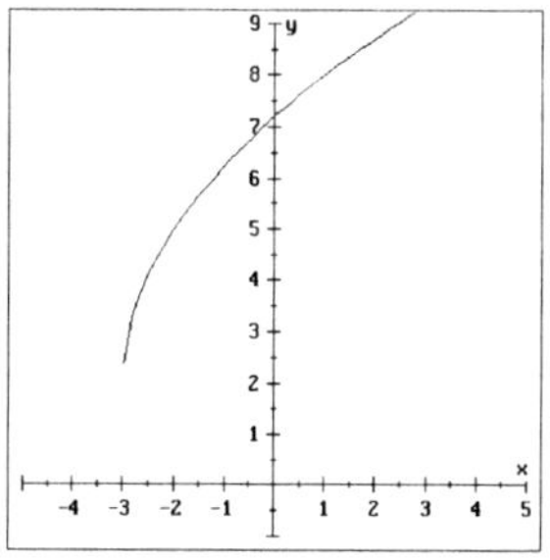

60. $g(x) = \frac{1}{3}(x + 3)^3 + 2$
Start with $y = x^3$
Shift L 3, Shrink vert. by a factor of $\frac{1}{3}$, Shift U 2

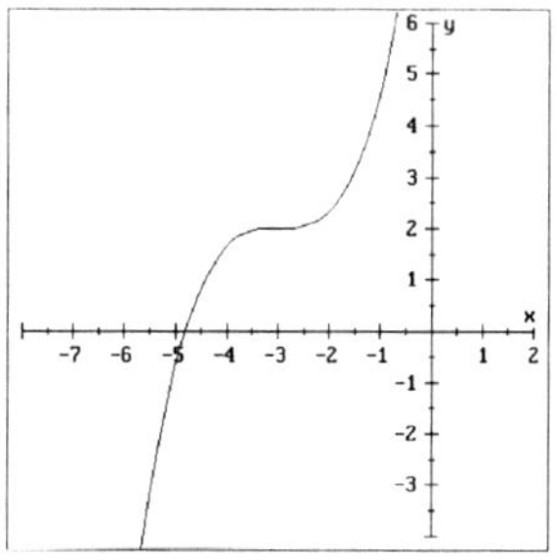

61. $f(x) = \sqrt{-x} + 3$
Start with $y = \sqrt{x}$
Reflect y, Shift U 3

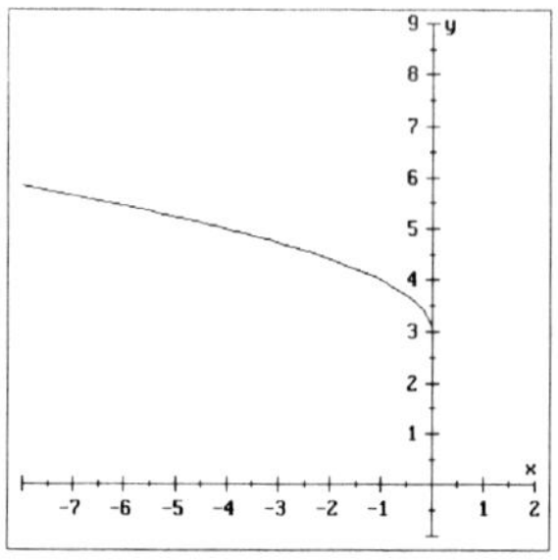

62. $g(x) = 2\sqrt[3]{x} - 5$
Start with $y = \sqrt[3]{x}$
Stretch vert. by a factor of 2, Shift D 5

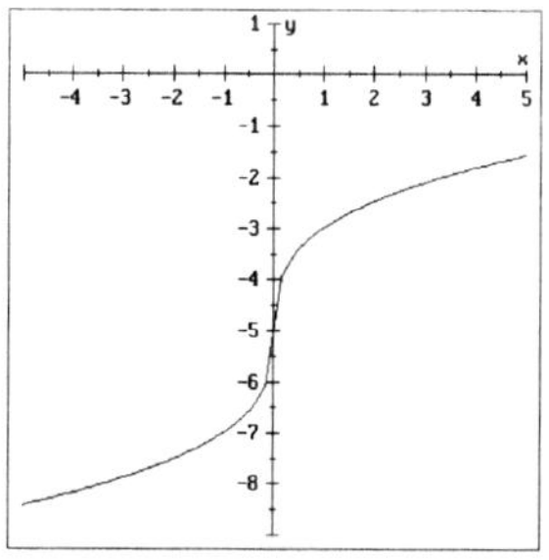

63. $f(x) = \dfrac{3x^2 + x - 2}{x^2 - 25} = \dfrac{(3x-2)(x+1)}{(x+5)(x-5)}$
den $= 0 \Rightarrow x = -5$ or $x = 5$
domain $= (-\infty, -5) \cup (-5, 5) \cup (5, \infty)$

64. $f(x) = \dfrac{2x^2 + 1}{x^2 + 7}$
den $= 0 \Rightarrow$ never true
domain $= (-\infty, \infty)$

65. $f(x) = \dfrac{x+5}{x^2 - 1} = \dfrac{x+5}{(x+1)(x-1)}$
den $= 0 \Rightarrow x = -1$ or $x = 1$
vertical: $x = -1$ or $x = 1$

66. $f(x) = \dfrac{x-7}{x^2 - 49} = \dfrac{x-7}{(x+7)(x-7)} = \dfrac{1}{x+7}$
den $= 0 \Rightarrow x = -7$
vertical: $x = -7$

67. $f(x) = \dfrac{x}{x^2 + x - 6} = \dfrac{x}{(x+3)(x-2)}$
den $= 0 \Rightarrow x = -3$ or $x = 2$
vertical: $x = -3$ or $x = 2$

68. $f(x) = \dfrac{5x+2}{2x^2 - 6x - 8} = \dfrac{5x+2}{(2x+2)(x-4)}$
den $= 0 \Rightarrow x = -1$ or $x = 4$
vertical: $x = -1$ or $x = 4$

69. $f(x) = \dfrac{2x^2 + x - 2}{4x^2 - 4}$; deg(num) = deg(den)
horizontal: $y = \dfrac{2}{4}$, or $y = \dfrac{1}{2}$

70. $f(x) = \dfrac{5x^2 + 4}{4 - x^2}$; deg(num) = deg(den)
horizontal: $y = \dfrac{5}{-1}$, or $y = -5$

71. $f(x) = \dfrac{x+1}{x^3 - 4x}$; deg(num) < deg(den)
horizontal: $y = 0$

72. $f(x) = \dfrac{x^3}{2x^2 - 6x - 8}$; deg(num) > deg(den)
horizontal: none

73. $f(x) = \dfrac{2x^2 - 5x + 1}{x - 4} = 2x + 3 + \dfrac{13}{x-4}$
slant: $y = 2x + 3$

74. $f(x) = \dfrac{5x^3 + 1}{x + 5}$; deg(num) = 3,
deg(den) = 1 $\Rightarrow$ slant: none

75. $f(x) = \dfrac{2x}{x - 4}$
Vert: $x = 4$; Horiz: $y = \frac{2}{1} = 2$
Slant: none; x-intercepts: $(0, 0)$
y-intercepts: $(0, 0)$; Symmetry: none

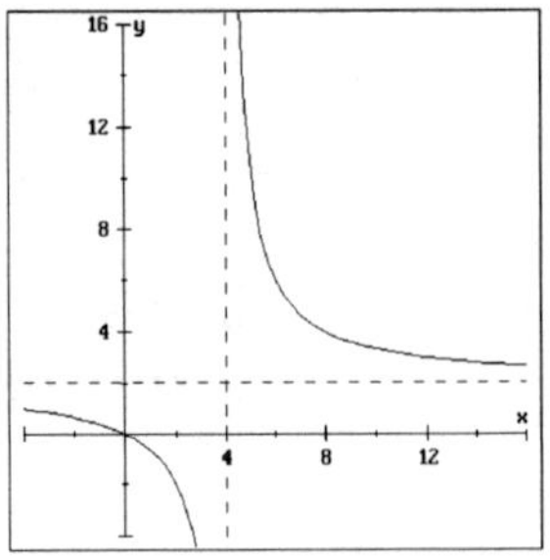

76. $f(x) = \dfrac{-4x}{x + 4}$
Vert: $x = -4$; Horiz: $y = \frac{-4}{1} = -4$
Slant: none; x-intercepts: $(0, 0)$
y-intercepts: $(0, 0)$; Symmetry: none

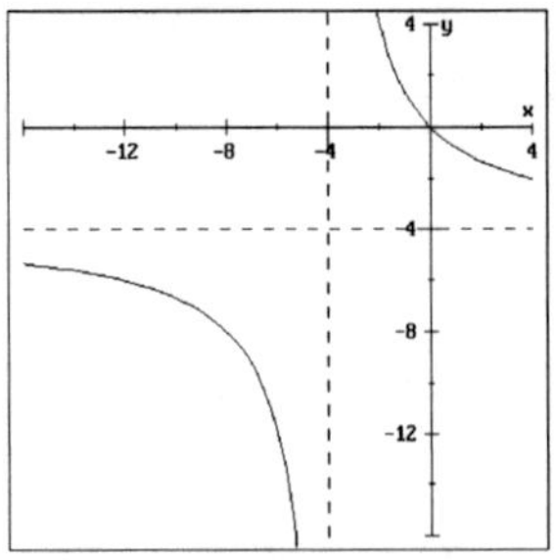

CHAPTER 3 REVIEW

77. $f(x) = \dfrac{x}{(x-1)^2}$
Vert: $x = 1$; Horiz: $y = 0$
Slant: none; x-intercepts: $(0, 0)$
y-intercepts: $(0, 0)$; Symmetry: none

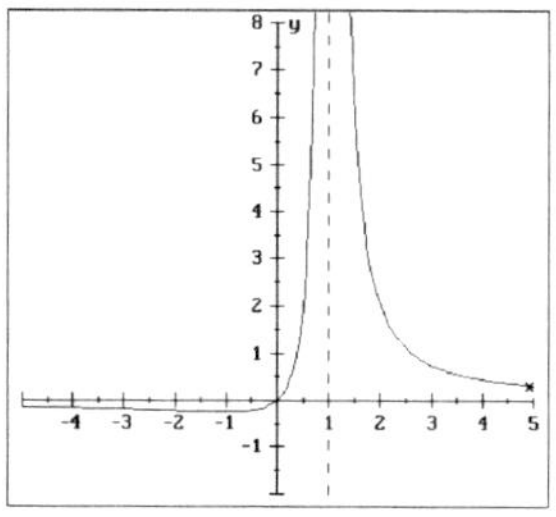

78. $f(x) = \dfrac{(x-1)^2}{x}$: Vert: $x = 0$
Horiz: none; Slant: $y = x - 2$
x-intercepts: $(1, 0)$
y-intercepts: none; Symmetry: none

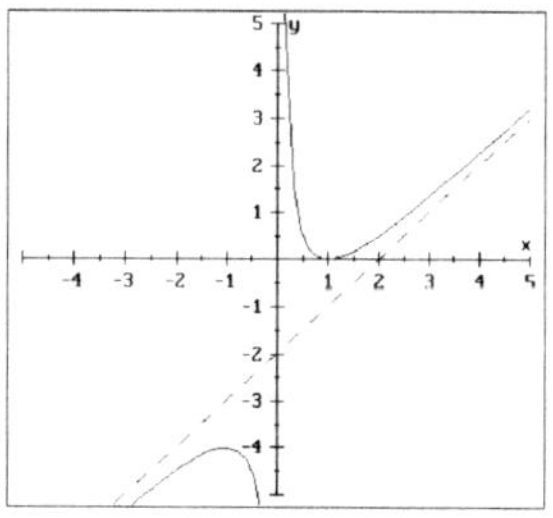

79. $y = \dfrac{x^2 - x - 2}{x^2 + x - 2} = \dfrac{(x+1)(x-2)}{(x+2)(x-1)}$
Vert: $x = -2$, $x = 1$; Horiz: $y = 1$
Slant: none; x-intercepts: $(-1, 0)$, $(2, 0)$
y-intercepts: $(0, 1)$; Symmetry: none

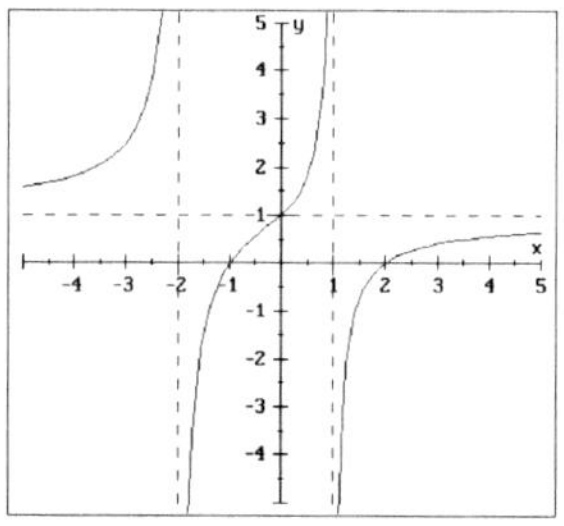

80. $y = \dfrac{x^3 + x}{x^2 - 4} = \dfrac{x(x^2+1)}{(x+2)(x-2)}$
Vert: $x = -2$, $x = 2$;
Slant: $y = x$;
x-int: $(0, 0)$$y$-int: $(0, 0)$;
Symmetry: none

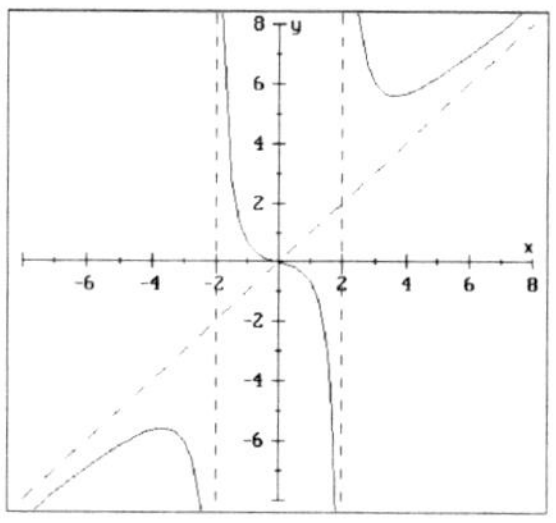

81. $(f+g)(x) = f(x) + g(x) = (x^2 - 1) + (2x + 1) = x^2 + 2x$; domain $= (-\infty, \infty)$

82. $(f \cdot g)(x) = f(x)g(x) = (x^2 - 1)(2x + 1) = 2x^3 + x^2 - 2x - 1$; domain $= (-\infty, \infty)$

83. $(f - g)(x) = f(x) - g(x) = (x^2 - 1) - (2x + 1) = x^2 - 2x - 2$; domain $= (-\infty, \infty)$

84. $(f/g)(x) = \dfrac{f(x)}{g(x)} = \dfrac{x^2 - 1}{2x + 1}$; domain $= \left(-\infty, -\frac{1}{2}\right) \cup \left(-\frac{1}{2}, \infty\right)$

85. $(f+g)(-3) = f(-3) + g(-3) = \left[2(-3)^2 - 1\right] + [2(-3) - 1] = 17 + (-7) = 10$

86. $(f-g)(-5) = f(-5) - g(-5) = \left[2(-5)^2 - 1\right] - [2(-5) - 1] = 49 - (-11) = 60$

87. $(f \cdot g)(2) = f(2) \cdot g(2) = \left[2(2)^2 - 1\right] \cdot [2(2) - 1] = 7 \cdot 3 = 21$

88. $(f/g)\left(\frac{1}{2}\right) = \dfrac{f\left(\frac{1}{2}\right)}{g\left(\frac{1}{2}\right)} = \dfrac{2\left(\frac{1}{2}\right)^2 - 1}{2\left(\frac{1}{2}\right) - 1} = \dfrac{-\frac{1}{2}}{0} \Rightarrow$ undefined

89. The domain of $f \circ g$ is the set of all real numbers in the domain of $g(x)$ such that $g(x)$ is in the domain of $f(x)$. Domain of $g(x)$: $(-\infty, \infty)$. Domain of $f(x) = (-\infty, \infty)$. Thus, all values of $g(x)$ are in the domain of $f(x)$. $\boxed{\text{Domain of } f \circ g\text{: } (-\infty, \infty)}$
$(f \circ g)(x) = f(g(x)) = f(2x+1) = (2x+1)^2 - 1 = 4x^2 + 4x + 1 - 1 = 4x^2 + 4x$

90. The domain of $g \circ f$ is the set of all real numbers in the domain of $f(x)$ such that $f(x)$ is in the domain of $g(x)$. Domain of $f(x)$: $(-\infty, \infty)$. Domain of $g(x) = (-\infty, \infty)$. Thus, all values of $f(x)$ are in the domain of $g(x)$. $\boxed{\text{Domain of } g \circ f\text{: } (-\infty, \infty)}$
$(g \circ f)(x) = g(f(x)) = g\left(x^2 - 1\right) = 2\left(x^2 - 1\right) + 1 = 2x^2 - 2 + 1 = 2x^2 - 1$

91. $(f \circ g)(-2) = f(g(-2)) = f(3(-2)+1) = f(-5) = (-5)^2 - 5 = 20$

92. $(g \circ f)(-2) = g(f(-2)) = g\left((-2)^2 - 5\right) = g(-1) = 3(-1) + 1 = -2$

93. Let $f(x) = x^2$ and $g(x) = x - 5$.
Then $(f \circ g)(x) = f(g(x))$
$= f(x-5) = (x-5)^2$.

94. Let $f(x) = x^3$ and $g(x) = x + 6$.
Then $(f \circ g)(x) = f(g(x))$
$= f(x+6) = (x+6)^3$.

95. $f(x) = x^2 + 7$; $x = 1$ and $x = -1$ both correspond to $f(x) = 8$. Not one-to-one

96. one-to-one

97. one-to-one

98. not one-to-one

99. $(f \circ g)(x) = f(g(x)) = f\left(\dfrac{x+3}{8}\right) = 8\left(\dfrac{x+3}{8}\right) - 3 = x + 3 - 3 = x$

$(g \circ f)(x) = g(f(x)) = g(8x-3) = \dfrac{(8x-3)+3}{8} = \dfrac{8x}{8} = x$

100. $(f \circ g)(x) = f(g(x)) = f\left(2 - \dfrac{1}{x}\right) = \dfrac{1}{2 - \left(2 - \frac{1}{x}\right)} = \dfrac{1}{\frac{1}{x}} = x$

$(g \circ f)(x) = g(f(x)) = g\left(\dfrac{1}{2-x}\right) = 2 - \dfrac{1}{\frac{1}{2-x}} = 2 - (2-x) = x$

CHAPTER 3 REVIEW

101.
$$\begin{aligned} y = f(x) &= 7x - 1 \\ x &= 7y - 1 \\ x + 1 &= 7y \\ \frac{x+1}{7} &= y \\ f^{-1}(x) &= \frac{x+1}{7} \end{aligned}$$

102.
$$\begin{aligned} y = f(x) &= \frac{1}{2-x} \\ x &= \frac{1}{2-y} \\ x(2-y) &= 1 \\ 2 - y &= \frac{1}{x} \\ 2 - \frac{1}{x} &= y \\ f^{-1}(x) &= 2 - \frac{1}{x} \end{aligned}$$

103.
$$\begin{aligned} y = f(x) &= \frac{x}{1-x} \\ x &= \frac{y}{1-y} \\ x(1-y) &= y \\ x - xy &= y \\ x &= xy + y \\ x &= y(x+1) \\ \frac{x}{x+1} &= y \\ f^{-1}(x) &= \frac{x}{x+1} \end{aligned}$$

104.
$$\begin{aligned} y = f(x) &= \frac{3}{x^3} \\ x &= \frac{3}{y^3} \\ xy^3 &= 3 \\ y^3 &= \frac{3}{x} \\ y &= \sqrt[3]{\frac{3}{x}} \\ f^{-1}(x) &= \sqrt[3]{\frac{3}{x}} \end{aligned}$$

105.
$$\begin{aligned} y = f(x) &= 2x - 5 \\ x &= 2y - 5 \\ x + 5 &= 2y \\ \frac{x+5}{2} &= y \\ f^{-1}(x) &= \frac{x+5}{2} \end{aligned}$$

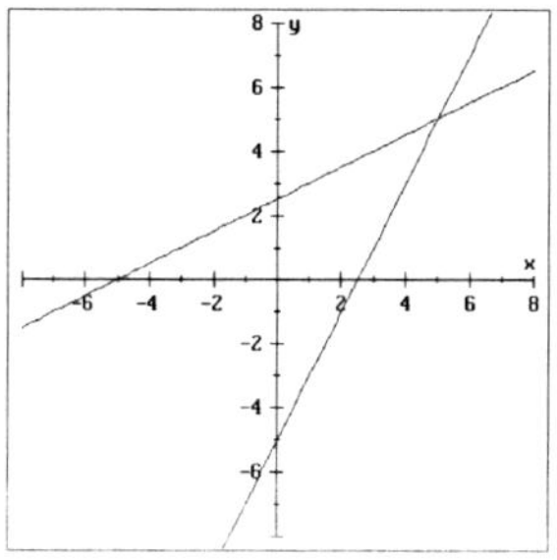

106.
$$\begin{aligned} y &= \frac{2x+3}{5x-10} \\ x &= \frac{2y+3}{5y-10} \\ x(5y-10) &= 2y + 3 \\ 5xy - 10x &= 2y + 3 \\ 5xy - 2y &= 10x + 3 \\ y(5x-2) &= 10x + 3 \\ y &= \frac{10x+3}{5x-2} \end{aligned}$$

Range of f = Domain of f^{-1}

$= \boxed{\left(-\infty, \tfrac{2}{5}\right) \cup \left(\tfrac{2}{5}, \infty\right)}$

Chapter 3 Test (page 409)

1. $f(x) = \dfrac{3}{x-5}$: domain $= (-\infty, 5) \cup (5, \infty)$

2. $f(x) = \sqrt{x+3}$: domain $= [-3, \infty)$

3. $f(-1) = \dfrac{-1}{-1-1} = \dfrac{-1}{-2} = \dfrac{1}{2}$
$f(2) = \dfrac{2}{2-1} = \dfrac{2}{1} = 2$

4. $f(-1) = \sqrt{-1+7} = \sqrt{6}$
$f(2) = \sqrt{2+7} = \sqrt{9} = 3$

5. $y = 3(x-7)^2 - 3$
Vertex: $(7, -3)$

6. $y = x^2 - 2x - 3;\ a = 1, b = -2, c = -3$
vertex: $x = -\frac{b}{2a} = -\frac{-2}{2(1)} = 1$
$y = x^2 - 2x - 3 = 1^2 - 2(1) - 3 = -4$

7. $y = 3x^2 - 24x + 38$
$a = 3, b = -24, c = 38$
vertex: $x = -\frac{b}{2a} = -\frac{-24}{2(3)} = 4$
$y = 3x^2 - 24x + 38 = 3(4)^2 - 24(4) + 38$
$= -10$

8. $y = 5 - 4x - x^2;\ a = -1, b = -4, c = 5$
vertex: $x = -\frac{b}{2a} = -\frac{-4}{2(-1)} = -2$
$y = 5 - 4x - x^2 = 5 - 4(-2) - (-2)^2$
$= 9$

9. $y = f(x) = x^4 - x^2$
$f(-x) = (-x)^4 - (-x)^2$
$= x^4 - x^2 = f(x) \Rightarrow$ even

x-int.	y-int.
$x^4 - x^2 = 0$	$y = 0^4 - 0^2$
$x^2(x^2-1) = 0$	$y = 0$
$x^2 = 0, x^2 = 1$	$(0, 0)$
$x = 0, x = -1, x = 1$	
$(0,0), (-1,0), (1,0)$	

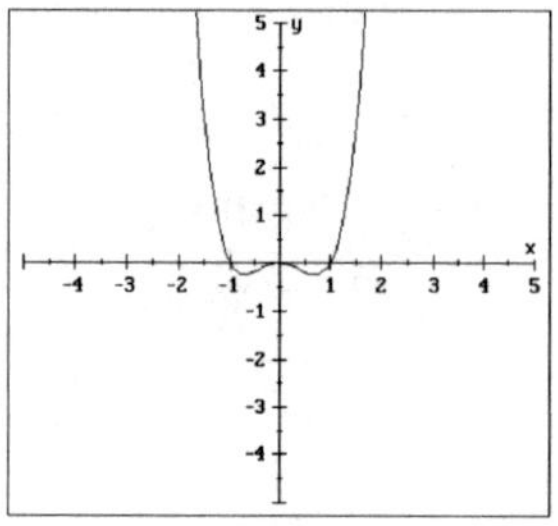

10. $y = f(x) = x^5 - x^3$
$f(-x) = (-x)^5 - (-x)^3$
$= -x^5 - (-x)^3$
$= -x^5 + x^3 = -f(x) \Rightarrow$ odd

x-int.	y-int.
$x^5 - x^3 = 0$	$y = 0^5 - 0^3$
$x^3(x^2-1) = 0$	$y = 0$
$x^3 = 0, x^2 = 1$	$(0, 0)$
$x = 0, x = -1, x = 1$	
$(0,0), (-1,0), (1,0)$	

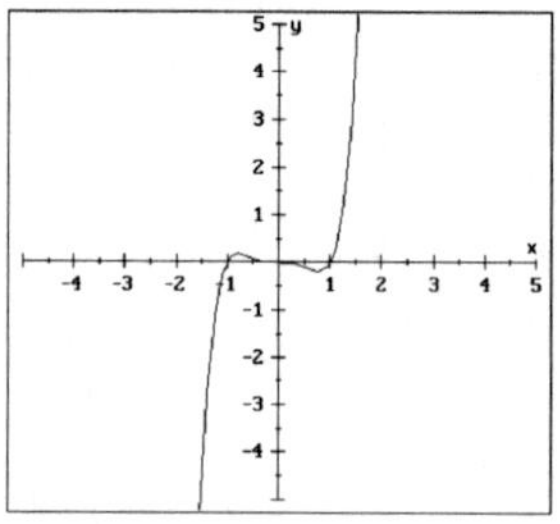

11. $h = 100t - 16t^2$: $a = -16, b = 100, c = 0$
$x = -\dfrac{b}{2a} = -\dfrac{100}{2(-16)} = \dfrac{25}{8}$ seconds

12. $h = 100t - 16t^2$: $a = -16, b = 100, c = 0$
From **#11**, $x = \frac{25}{8}$.
$y = 100\left(\frac{25}{8}\right) - 16\left(\frac{25}{8}\right)^2 = \frac{625}{4}$ feet

CHAPTER 3 TEST

13. The roadway is at $y = 0$, so the distance to the lowest point will be the y-coord. of the vertex:

$$x^2 - 2500y + 25000 = 0$$
$$x^2 + 25000 = 2500y$$
$$\frac{1}{2500}x^2 + 10 = y$$
$$y = c - \frac{b^2}{4a} = 10 - \frac{0^2}{4\left(\frac{1}{2500}\right)} = 10$$

The lowest point is 10 ft above.

14. Let $x = \pm 500$:

$$x^2 - 2500y + 25000 = 0$$
$$(\pm 500)^2 - 2500y + 25000 = 0$$
$$250{,}000 - 2500y + 25{,}000 = 0$$
$$275{,}000 = 2500y$$
$$110 = y$$

The cable attaches to the vertical pillars 110 feet above the roadway.

15. $y = (x - 3)^2 + 1$

Shift $y = x^2$ U 1, R 3

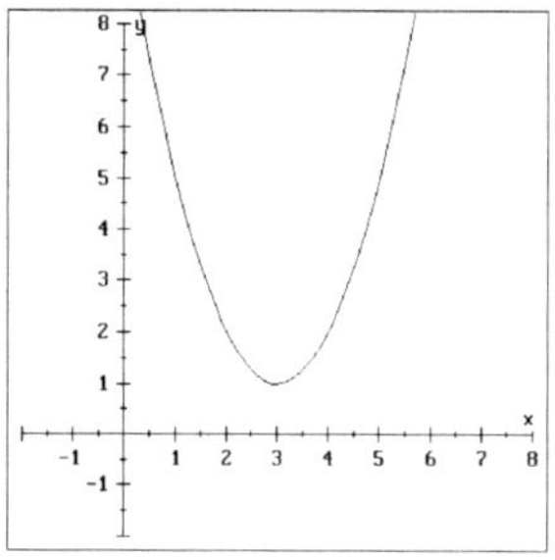

16. $y = \sqrt{x - 1} + 5$

Shift $y = \sqrt{x}$ U 5, R 1

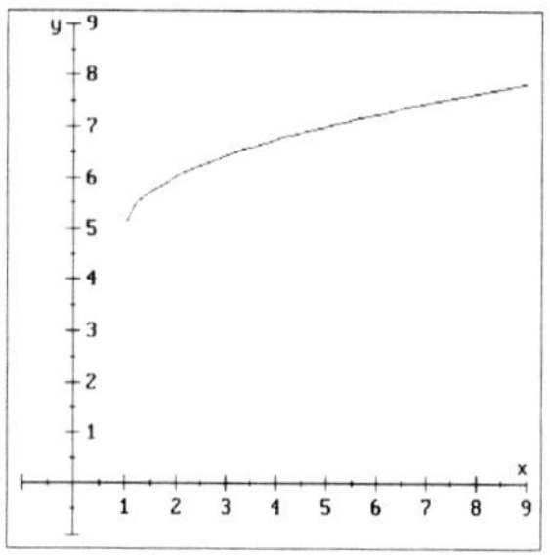

17. $y = \dfrac{x - 1}{x^2 - 9} = \dfrac{x - 1}{(x + 3)(x - 3)}$

Vert: $x = -3$, $x = 3$; Horiz: $y = 0$

18. $y = \dfrac{x^2 - 5x - 14}{x - 3} = \dfrac{(x - 7)(x + 2)}{(x - 3)}$

$= x - 2 + \frac{-20}{x-3}$

Vert: $x = 3$; Slant: $y = x - 2$

19. $y = \dfrac{x^2}{x^2 - 9} = \dfrac{x^2}{(x + 3)(x - 3)}$

Vert: $x = -3$, $x = 3$; Horiz: $y = 1$

Slant: none; x-intercepts: $(0, 0)$

y-intercepts: $(0, 0)$; Symmetry: y-axis

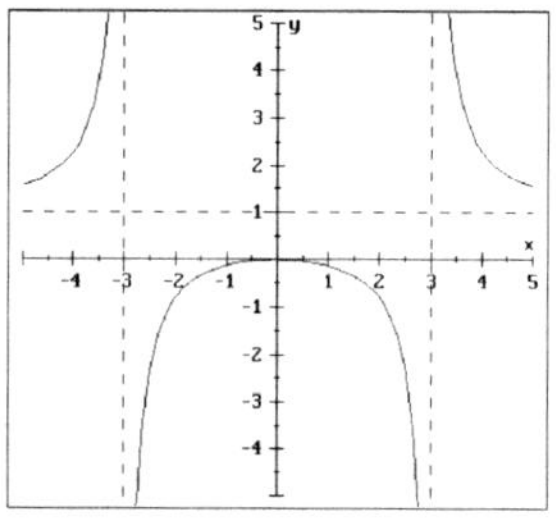

20. $y = \dfrac{x}{x^2 + 1}$

Vert: none; Horiz: $y = 0$

Slant: none; x-intercepts: $(0, 0)$

y-intercepts: $(0, 0)$; Symmetry: origin

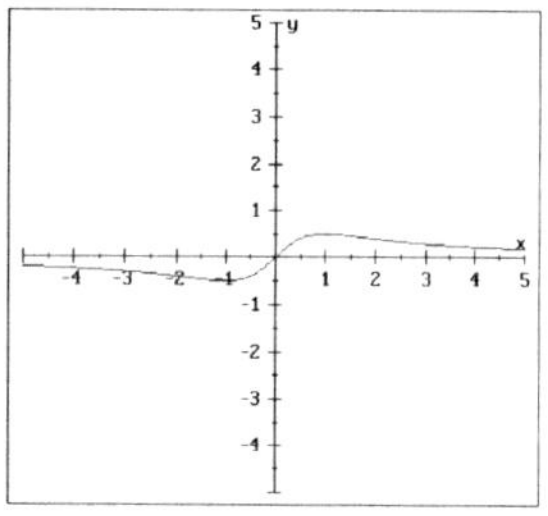

CHAPTER 3 TEST

21. $y = \dfrac{2x^2 - 3x - 2}{x - 2} = \dfrac{(2x+1)(x-2)}{x-2}$
$= 2x + 1 \text{ (if } x \neq 2)$

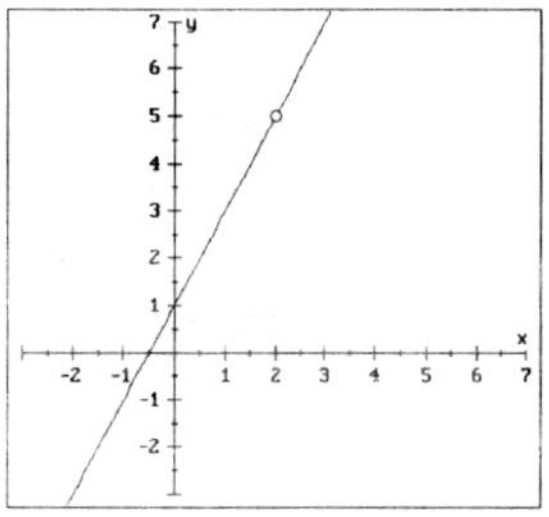

22. $y = \dfrac{x}{x^2 - x} = \dfrac{x}{x(x-1)} = \dfrac{1}{x-1} \text{ (if } x \neq 0)$

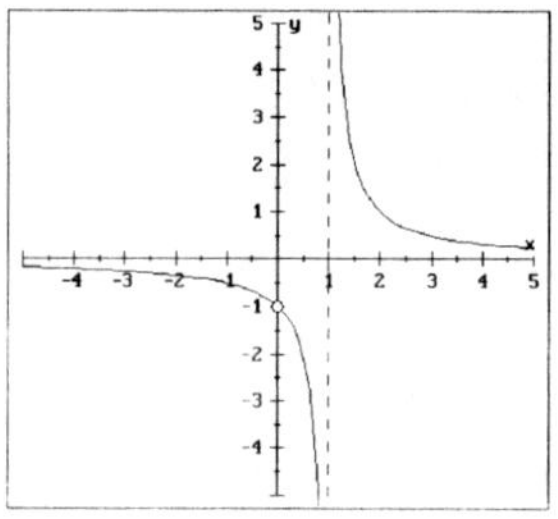

23. $(f+g)(x) = f(x) + g(x) = (3x) + (x^2 + 2) = x^2 + 3x + 2$

24. $(g \circ f)(x) = g(f(x)) = g(3x) = (3x)^2 + 2 = 9x^2 + 2$

25. $(f/g)(x) = \dfrac{f(x)}{g(x)} = \dfrac{3x}{x^2 + 2}$

26. $(f \circ g)(x) = f(g(x)) = f(x^2 + 2) = 3(x^2 + 2) = 3x^2 + 6$

27.
$$\begin{aligned} y &= \frac{x+1}{x-1} \\ x &= \frac{y+1}{y-1} \\ x(y-1) &= y+1 \\ xy - x &= y + 1 \\ xy - y &= x + 1 \\ y(x-1) &= x+1 \\ f^{-1}(x) &= \frac{x+1}{x-1} \end{aligned}$$

28.
$$\begin{aligned} y &= x^3 - 3 \\ x &= y^3 - 3 \\ x + 3 &= y^3 \\ \sqrt[3]{x+3} &= y \\ f^{-1}(x) &= \sqrt[3]{x+3} \end{aligned}$$

29. $f(x) = \dfrac{3}{x} - 2;\ f^{-1}(x) = \dfrac{3}{x+2}$
Range of f = Domain of f^{-1}
$= \boxed{(-\infty, -2) \cup (-2, \infty)}$

30. $f(x) = \dfrac{3x-1}{x-3};\ f^{-1}(x) = \dfrac{3x-1}{x-3}$
Range of f = Domain of f^{-1}
$= \boxed{(-\infty, 3) \cup (3, \infty)}$

Cumulative Review Exercises (page 410)

1.
$$\begin{aligned} 5x - 3y &= 15 \\ 5x - 3(0) &= 15 \\ 5x &= 15 \\ x &= 3 \end{aligned} \qquad \begin{aligned} 5x - 3y &= 15 \\ 5(0) - 3y &= 15 \\ -3y &= 15 \\ y &= -5 \end{aligned}$$
$(3, 0)$ $(0, -5)$

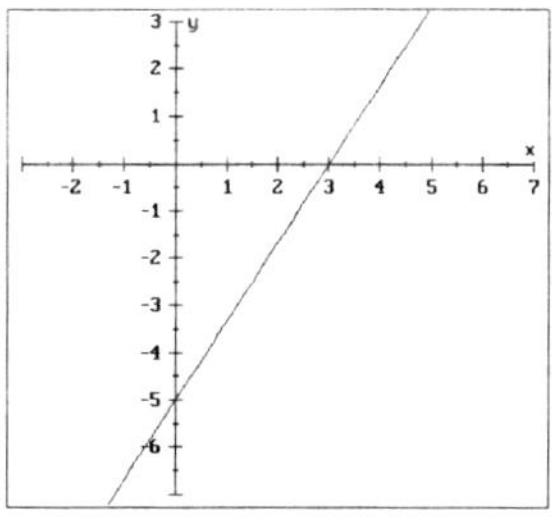

2.
$$\begin{aligned} 3x + 2y &= 12 \\ 3x + 2(0) &= 12 \\ 3x &= 12 \\ x &= 4 \end{aligned} \qquad \begin{aligned} 3x + 2y &= 12 \\ 3(0) + 2y &= 12 \\ 2y &= 12 \\ y &= 6 \end{aligned}$$
$(4, 0)$ $(0, 6)$

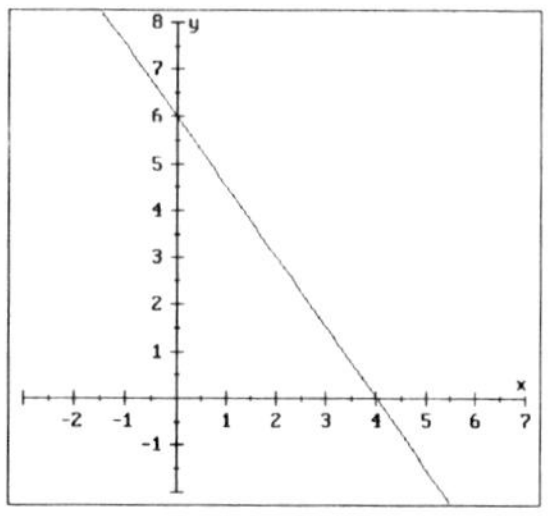

3. **a.** $d = \sqrt{(x_2 - x_1)^2 + (y_2 - y_1)^2} = \sqrt{(-2-3)^2 + \left(\frac{7}{2} - \left(-\frac{1}{2}\right)\right)^2} = \sqrt{25 + 16} = \sqrt{41}$

b. $x = \dfrac{x_1 + x_2}{2} = \dfrac{-2+3}{2} = \dfrac{1}{2}$; $y = \dfrac{\frac{7}{2} + \left(-\frac{1}{2}\right)}{2} = \dfrac{\frac{6}{2}}{2} = \dfrac{3}{2}$ $\left(\dfrac{1}{2}, \dfrac{3}{2}\right)$

c. $m = \dfrac{-\frac{1}{2} - \frac{7}{2}}{3 - (-2)} = \dfrac{-\frac{8}{2}}{5} = -\dfrac{4}{5}$

4. **a.** $d = \sqrt{(x_2 - x_1)^2 + (y_2 - y_1)^2} = \sqrt{(3 - (-7))^2 + (7-3)^2} = \sqrt{100 + 16} = \sqrt{116} = 2\sqrt{29}$

b. $x = \dfrac{x_1 + x_2}{2} = \dfrac{3 + (-7)}{2} = \dfrac{-4}{2} = -2$; $y = \dfrac{7+3}{2} = \dfrac{10}{2} = 5$ $(-2, 5)$

c. $m = \dfrac{3-7}{-7-3} = \dfrac{-4}{-10} = \dfrac{2}{5}$

5. $m = \dfrac{y_2 - y_1}{x_2 - x_1} = \dfrac{-7-5}{3-(-3)} = \dfrac{-12}{6} = -2$
$$\begin{aligned} y - y_1 &= m(x - x_1) \\ y - 5 &= -2(x + 3) \\ y &= -2x - 1 \end{aligned}$$

6.
$$\begin{aligned} y - y_1 &= m(x - x_1) \\ y - \tfrac{5}{2} &= \tfrac{7}{2}\left(x - \tfrac{3}{2}\right) \\ y &= \tfrac{7}{2}x - \tfrac{21}{4} + \tfrac{5}{2} \\ y &= \tfrac{7}{2}x - \tfrac{11}{4} \end{aligned}$$

CUMULATIVE REVIEW EXERCISES

7. $3x - 5y = 7$
$-5y = -3x + 7$
$y = \frac{3}{5}x - \frac{7}{5}$
$m = \frac{3}{5}$
Use $m = \frac{3}{5}$.

$y - y_1 = m(x - x_1)$
$y - 3 = \frac{3}{5}(x + 5)$
$y - 3 = \frac{3}{5}x + 3$
$\boxed{y = \frac{3}{5}x + 6}$

8. $x - 4y = 12$
$-4y = -x + 12$
$y = \frac{1}{4}x - 3$
$m = \frac{1}{4}$
Use $m = -4$.

$y - y_1 = m(x - x_1)$
$y - 0 = -4(x - 0)$
$\boxed{y = -4x}$

9. $x^2 = y - 2$
symmetry: y-axis
x-int: none, y-int: $(0, 2)$

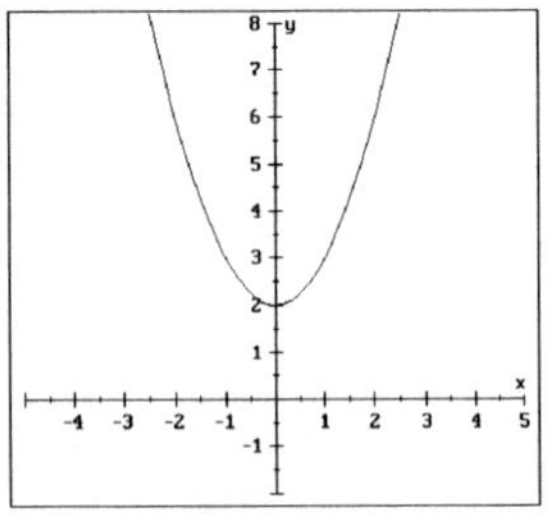

10. $y^2 = x - 2$
symmetry: x-axis
x-int: $(2, 0)$, y-int: none

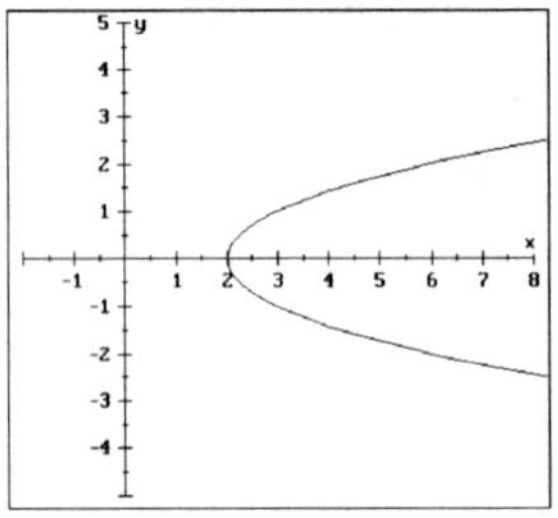

11. $x^2 + y^2 = 100$
Circle: $C(0, 0)$; $r = 10$

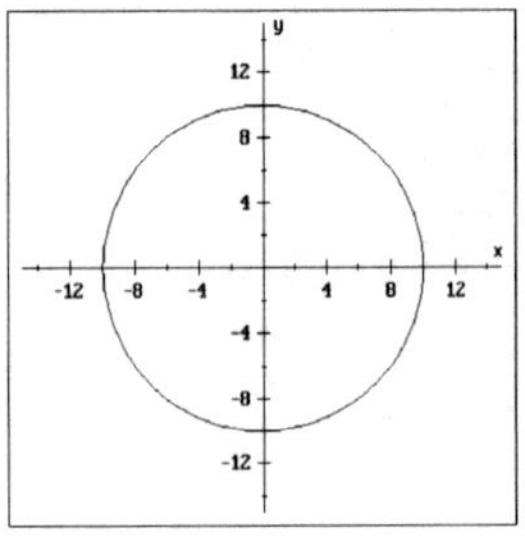

12. $x^2 - 2x + y^2 = 8$
$(x - 1)^2 + y^2 = 9$
Circle: $C(1, 0)$; $r = 3$

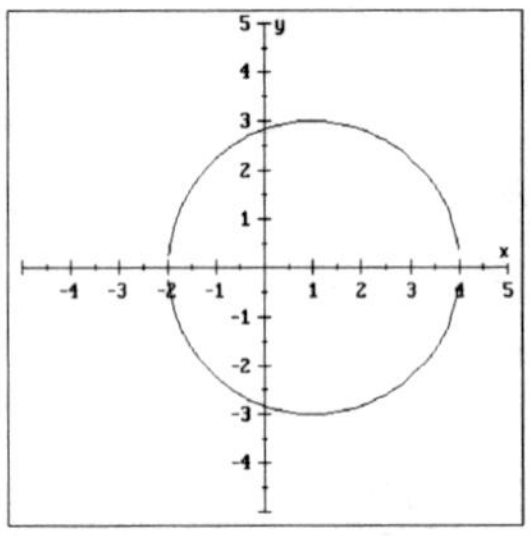

13. $\frac{x - 2}{x} = \frac{x - 6}{5}$
$5(x - 2) = x(x - 6)$
$5x - 10 = x^2 - 6x$
$0 = x^2 - 11x + 10$
$0 = (x - 10)(x - 1)$
$x = 10$ or $x = 1$

14. $\frac{x + 2}{x - 6} = \frac{3x + 1}{2x - 11}$
$(x + 2)(2x - 11) = (3x + 1)(x - 6)$
$2x^2 - 7x - 22 = 3x^2 - 17x - 6$
$0 = x^2 - 10x + 16$
$0 = (x - 8)(x - 2)$
$x = 8$ or $x = 2$

CUMULATIVE REVIEW EXERCISES

15. $m = \dfrac{54-37}{4-2} = \dfrac{17}{2} = 8.5$

$y = mx + b$
$y = 8.5x + b$
$37 = 8.5(2) + b$
$37 = 17 + b$
$20 = b$

$y = 8.5x + 20$
$y = 8.5(5) + 20$
$y = 62.50$
It will cost $62.50.

16. $E = ks^2$

$E = k(50)^2 \quad E = k(20)^2$
$E = 2500k \quad E = 400k$
$\dfrac{50 \text{ mph } E}{20 \text{ mph } E} = \dfrac{2500k}{400k} = \dfrac{25}{4}$

17. $y = 3x - 1$: Each value of x is paired with only one value of y. $\Rightarrow$ **function**

18. $y = x^2 + 3$: Each value of x is paired with only one value of y. $\Rightarrow$ **function**

19. $y = \dfrac{1}{x-2}$: Each value of x is paired with only one value of y. $\Rightarrow$ **function**

20. $y^2 = 4x$
$y = \pm\sqrt{4x}$: Each value of x is paired with more than one value of y. $\Rightarrow$ **not a function**

21. $y = f(x) = x^2 + 5$: domain $= (-\infty, \infty)$

22. $y = f(x) = \dfrac{7}{x+2}$
domain $= (-\infty, -2) \cup (-2, \infty)$

23. $y = f(x) = -\sqrt{x-2} \Rightarrow$ domain $= [2, \infty)$

24. $y = f(x) = \sqrt{x+4} \Rightarrow$ domain $= [-4, \infty)$

25. $y = x^2 + 5x - 6$; $a = 1, b = 5, c = -6$
vertex: $x = -\frac{b}{2a} = -\frac{5}{2(1)} = -\frac{5}{2}$
$y = x^2 + 5x - 6 = \left(-\frac{5}{2}\right)^2 + 5\left(-\frac{5}{2}\right) - 6$
$= -\frac{49}{4}$

26. $y = -x^2 + 5x + 6$; $a = -1, b = 5, c = 6$
vertex: $x = -\frac{b}{2a} = -\frac{5}{2(-1)} = \frac{5}{2}$
$y = -x^2 + 5x + 6 = -\left(\frac{5}{2}\right)^2 + 5\left(\frac{5}{2}\right) + 6$
$= \frac{49}{4}$

27. $y = x^2 - 4$: Shift $y = x^2$ D 4.

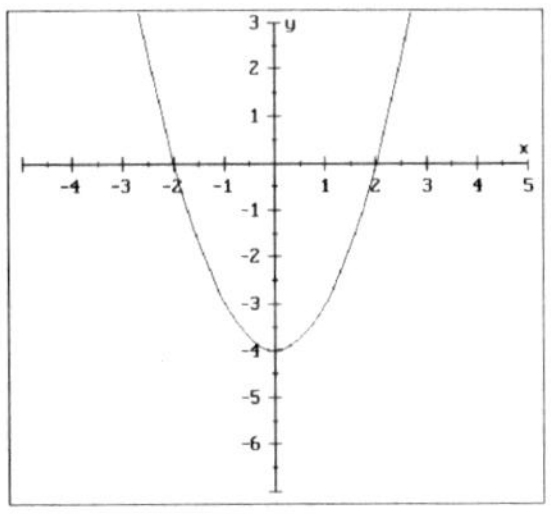

28. $y = -x^2 + 4$: Reflect $y = x^2$ about x and shift U 4.

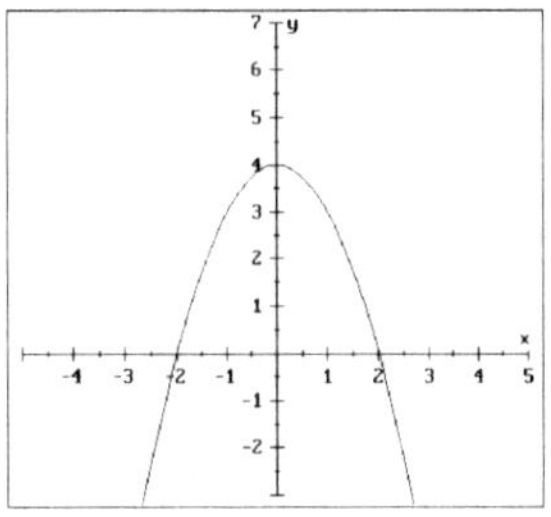

CUMULATIVE REVIEW EXERCISES

29. $y = f(x) = x^3 + x$

$f(-x) = (-x)^3 + (-x)$
$= -x^3 - x$
$= -f(x) \Rightarrow$ odd

x-int.	y-int.
$x^3 + x = 0$	$y = 0^3 + 0$
$x(x^2+1) = 0$	$y = 0$
$x = 0$	$(0, 0)$
$(0, 0)$	

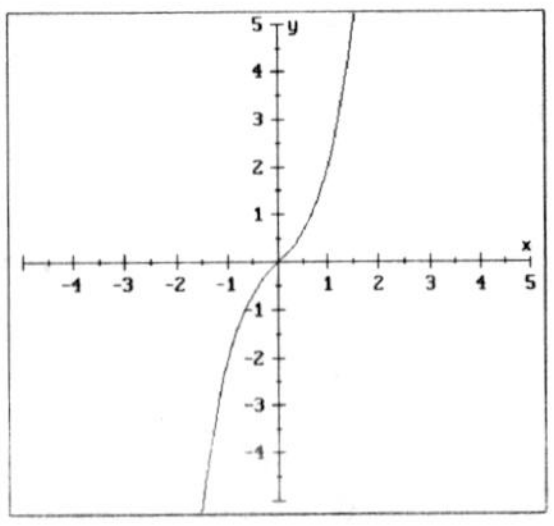

30. $f(x) = -x^4 + 2x^2 + 1$

$f(-x) = -(-x)^4 + 2(-x)^2 + 1$
$= -x^4 + 2x^2 + 1 \Rightarrow$ even

x-int.	y-int.
$-x^4 + 2x^2 + 1 = 0$	$f(0) = 1$
$x^4 - 2x^2 - 1 = 0$	$y = 1$
not rational numbers	$(0, 1)$

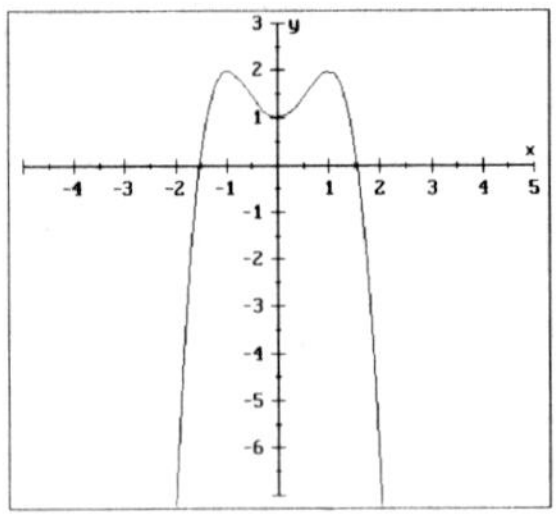

31. $y = \dfrac{x}{x-3}$

Vertical: $x = 3$; Horizontal: $y = 1$
Slant: none; x-intercepts: $(0, 0)$
y-intercepts: $(0, 0)$; Symmetry: none

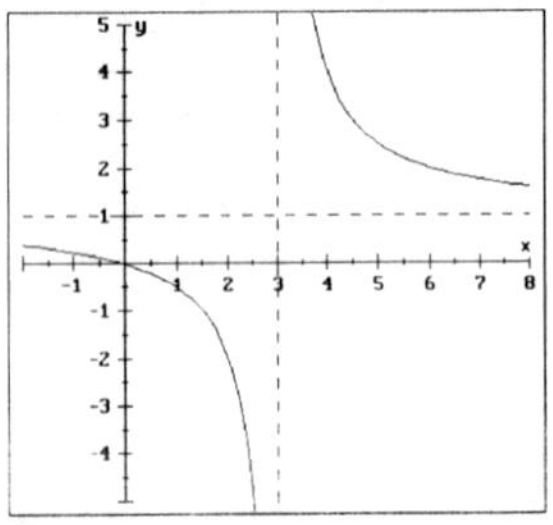

32. $y = \dfrac{x^2-1}{x^2-9} = \dfrac{(x+1)(x-1)}{(x+3)(x-3)}$

Vertical: $x = -3$, $x = 3$; Horizontal: $y = 1$
Slant: none; x-intercepts: $(-1, 0)$, $(1, 0)$
y-intercepts: $\left(0, \frac{1}{9}\right)$; Symmetry: y-axis

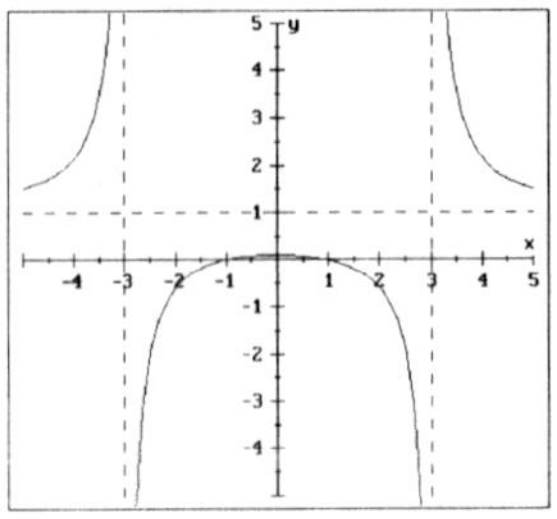

33. $(f+g)(x) = f(x) + g(x) = (3x-4) + (x^2+1) = x^2 + 3x - 3$; domain $= (-\infty, \infty)$

34. $(f-g)(x) = f(x) - g(x) = (3x-4) - (x^2+1) = -x^2 + 3x - 5$; domain $= (-\infty, \infty)$

35. $(f \cdot g)(x) = f(x)g(x) = (3x-4)(x^2+1) = 3x^3 - 4x^2 + 3x - 4$; domain $= (-\infty, \infty)$

36. $(f/g)(x) = \frac{f(x)}{g(x)} = \frac{3x-4}{x^2+1}$; domain $= (-\infty, \infty)$

37. $(f \circ g)(2) = f(g(2)) = f((2^2+1)) = f(5) = 3(5) - 4 = 11$

38. $(g \circ f)(2) = g(f(2)) = g(3(2) - 4) = g(2) = 2^2 + 1 = 5$

SECTION 4.1

39. $(f \circ g)(x) = f(g(x)) = f(x^2+1) = 3(x^2+1) - 4 = 3x^2 - 1$

40. $(g \circ f)(x) = g(f(x)) = g(3x-4) = (3x-4)^2 + 1 = 9x^2 - 24x + 16 + 1 = 9x^2 - 24x + 17$

41.
$$\begin{aligned} y &= 3x + 2 \\ x &= 3y + 2 \\ x - 2 &= 3y \\ \tfrac{x-2}{3} &= y \\ f^{-1}(x) &= \tfrac{x-2}{3} \end{aligned}$$

42.
$$\begin{aligned} y &= \tfrac{1}{x-3} \\ x &= \tfrac{1}{y-3} \\ x(y-3) &= 1 \\ y - 3 &= \tfrac{1}{x} \\ y &= \tfrac{1}{x} + 3 \\ f^{-1}(x) &= \tfrac{1}{x} + 3 \end{aligned}$$

43.
$$\begin{aligned} y &= x^2 + 5 \\ x &= y^2 + 5 \\ x - 5 &= y^2 \\ \pm\sqrt{x-5} &= y \\ \sqrt{x-5} &= y \ \ (y \ge 0) \\ f^{-1}(x) &= \sqrt{x-5} \end{aligned}$$

44.
$$\begin{aligned} 3x - y &= 1 \\ 3y - x &= 1 \\ 3y &= x + 1 \\ y &= \tfrac{x+1}{3} \\ f^{-1}(x) &= \tfrac{x+1}{3} \end{aligned}$$

45. $y = kwz$

46. $y = \dfrac{kx}{t^2}$

Exercise 4.1 (page 425)

1. exponential

3. $(-\infty, \infty)$

5. $(0, \infty)$

7. asymptote

9. increasing

11. 2.72

13. increasing

15. $4^{\sqrt{3}} \approx 11.0357$

17. $7^{\pi} \approx 451.8079$

19. $5^{\sqrt{2}}5^{\sqrt{2}} = 5^{\sqrt{2}+\sqrt{2}} = 5^{2\sqrt{2}} = (5^2)^{\sqrt{2}} = 25^{\sqrt{2}}$

21. $\left(a^{\sqrt{8}}\right)^{\sqrt{2}} = a^{\sqrt{8}\cdot\sqrt{2}} = a^{\sqrt{16}} = a^4$

23. $f(0) = 5^0 = 1,\ f(2) = 5^2 = 25$

25. $f(0) = \left(\frac{1}{3}\right)^{-0} = 1,\ f(2) = \left(\frac{1}{3}\right)^{-2} = 9$

27. $f(x) = 3^x$
points: $(0, 1)$, $(1, 3)$

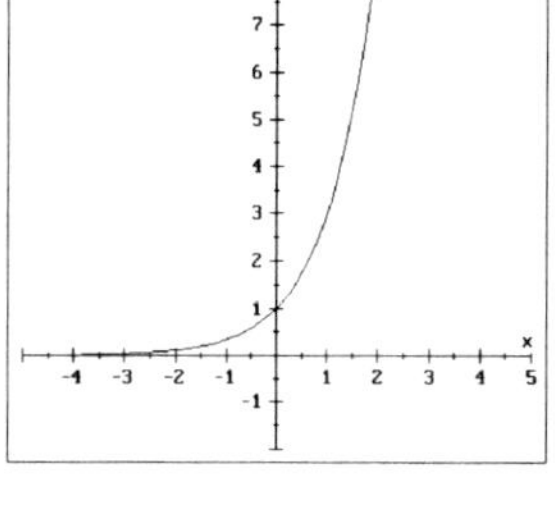

29. $f(x) = \left(\dfrac{1}{5}\right)^x$
points: $(0, 1)$, $\left(1, \frac{1}{5}\right)$

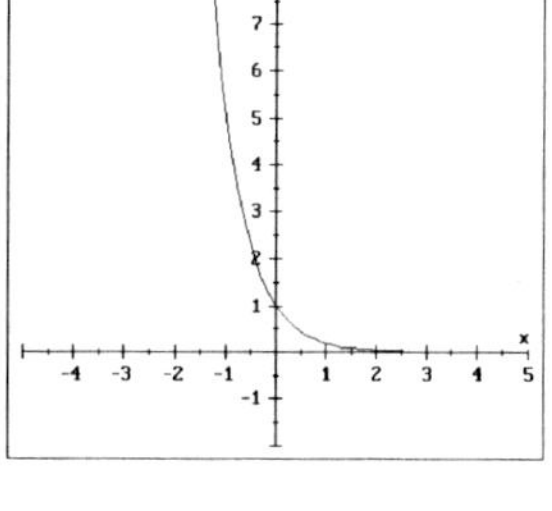

31. $f(x) = \left(\dfrac{3}{4}\right)^x$
points: $(0, 1)$, $\left(1, \frac{3}{4}\right)$

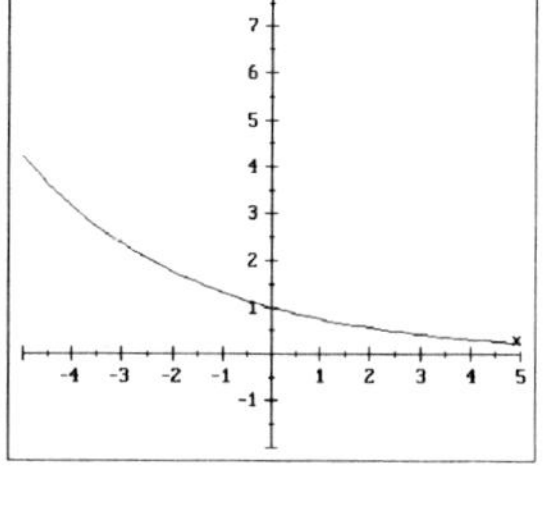

33. $f(x) = (1.5)^x$
points: $(0, 1)$, $(1, 1.5)$

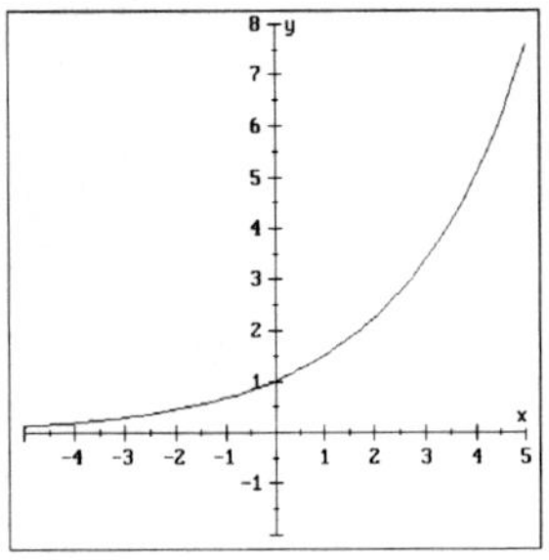

35. $f(x) = 3^{-x}$
points: $(0, 1)$, $\left(1, \frac{1}{3}\right)$

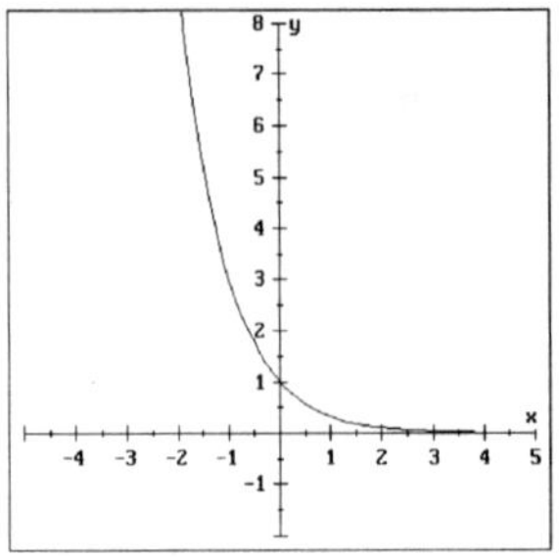

37. $f(x) = -\left(\frac{1}{5}\right)^x$
points: $(0, -1)$, $\left(1, -\frac{1}{5}\right)$

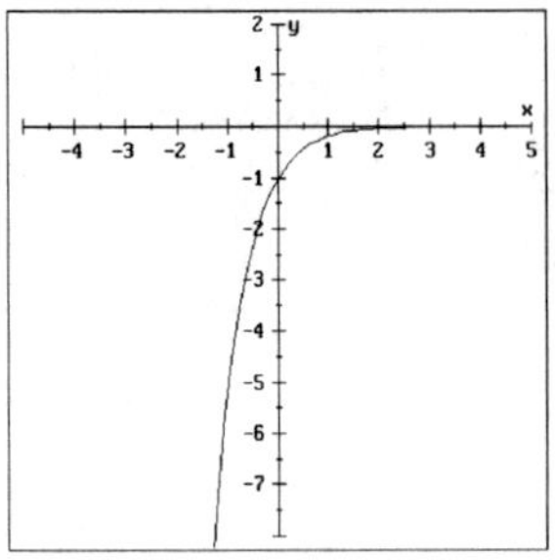

39. The graph passes through $(0, 1)$ and has the x-axis as an asymptote. YES

41. The graph does not pass through $(0, 1)$. NO

43. The graph passes through $(0, 1)$ and has the x-axis as an asymptote, so it could be an exponential function. It passes through the point $\left(1, \frac{1}{2}\right) = (1, b)$. $b = \frac{1}{2}$

45. The graph does not pass through $(0, 1)$. It is not an exponential function.

47. The graph passes through $(0, 1)$ and has the x-axis as an asymptote, so it could be an exponential function. It passes through the point $(1, 2) = (1, b)$. $b = 2$

49. The graph passes through $(0, 1)$ and has the x-axis as an asymptote, so it could be an exponential function.

$$y = b^x$$
$$e^2 = b^2$$
$$e = b$$

51. $f(x) = 3^x - 1$
Shift $y = 3^x$ D1.

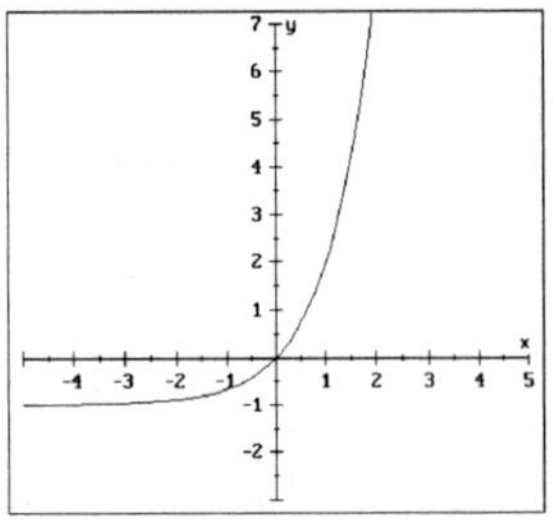

53. $f(x) = 2^x + 1$
Shift $y = 2^x$ U1.

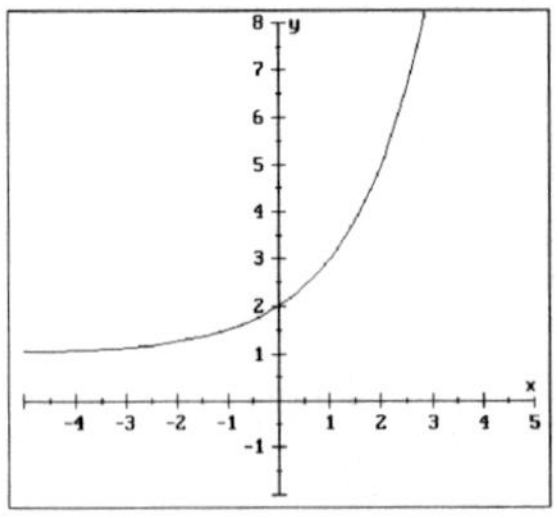

55. $f(x) = 3^{x-1}$
Shift $y = 3^x$ R1.

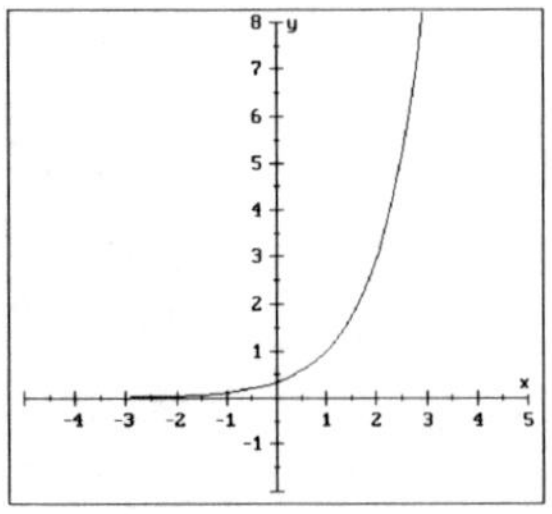

57. $f(x) = 3^{x+1}$
Shift $y = 3^x$ L1.

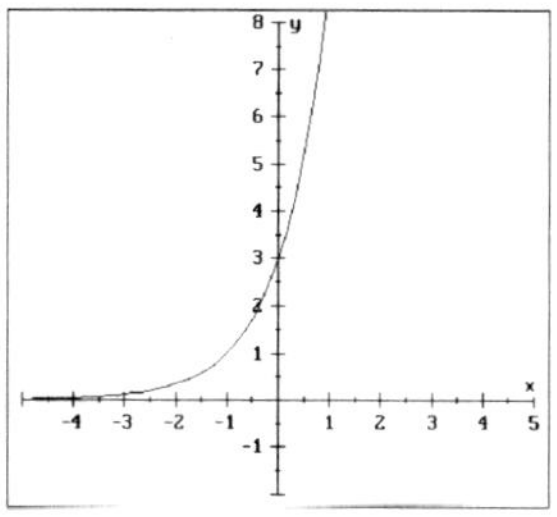

59. $f(x) = e^x - 4$
Shift $y = e^x$ D4.

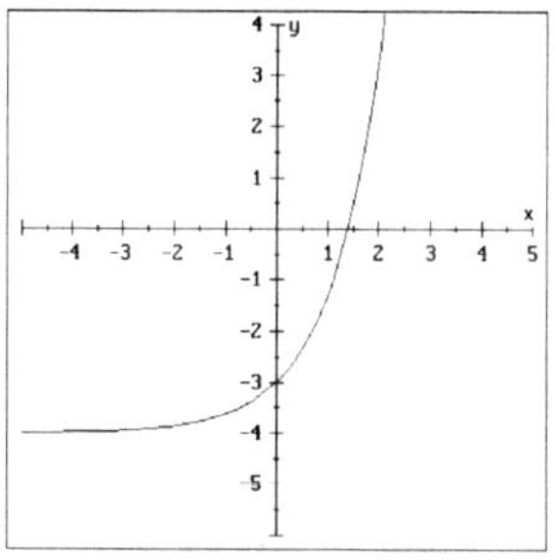

61. $f(x) = e^{x-2}$
Shift $y = e^x$ R2.

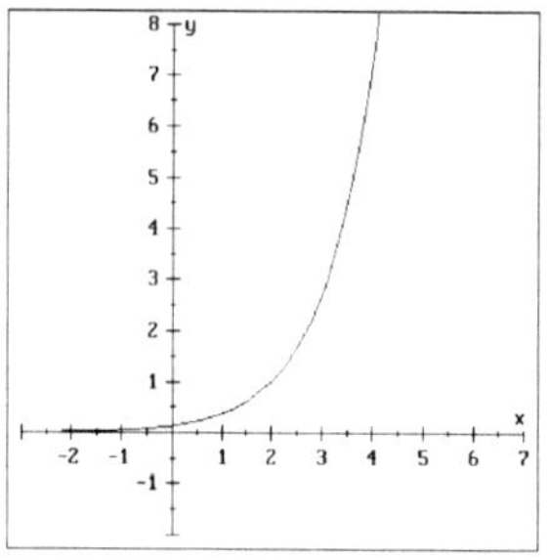

63. $f(x) = 2^{x+1} - 2$
Shift $y = 2^x$ L1, D2.

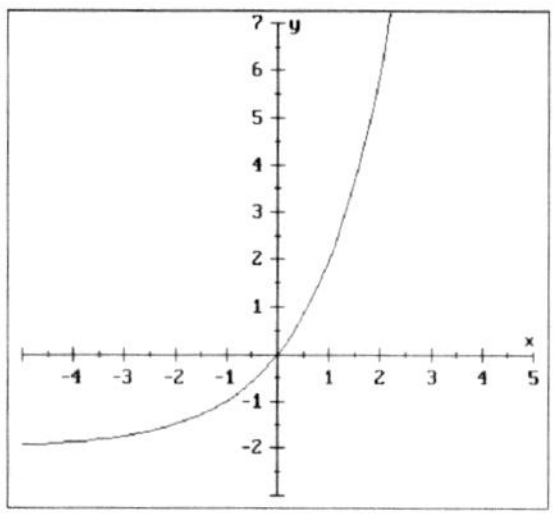

65. $f(x) = 3^{x-2} + 1$
Shift $y = 3^x$ R2, U1.

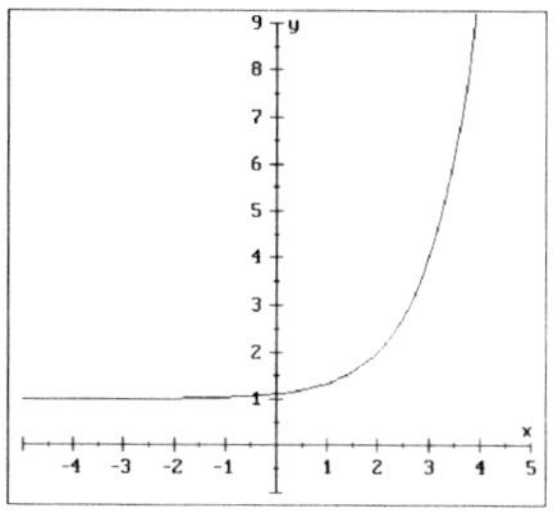

67. $f(x) = -3^x + 1$
Reflect $y = 3^x$ about x,
Shift U1

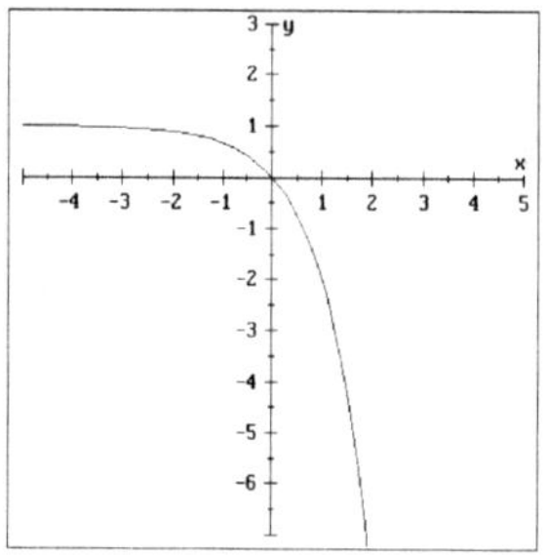

69. $f(x) = 2^{-x} - 3$
Reflect $y = 2^x$ about y,
Shift D3

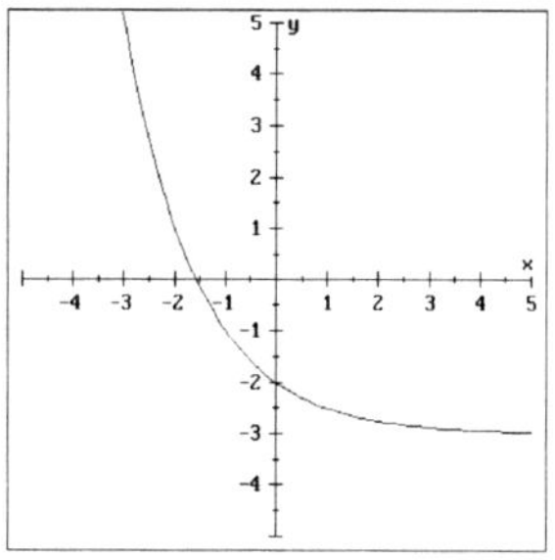

71. $f(x) = -e^x + 2$
Reflect $y = e^x$ about x,
Shift U2

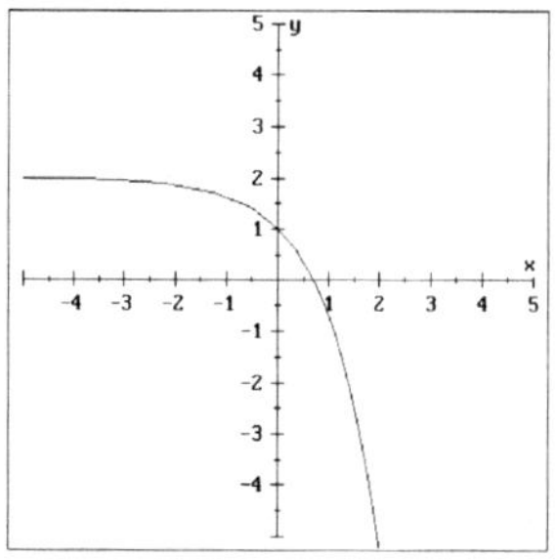

73. $y = 5(2^x)$

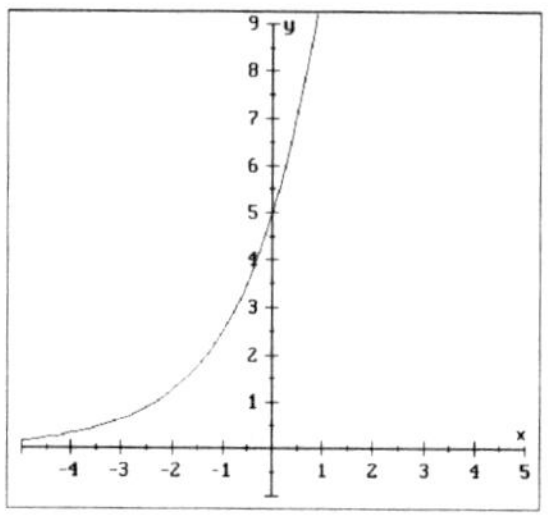

SECTION 4.1

75. $y = 3^{-x}$

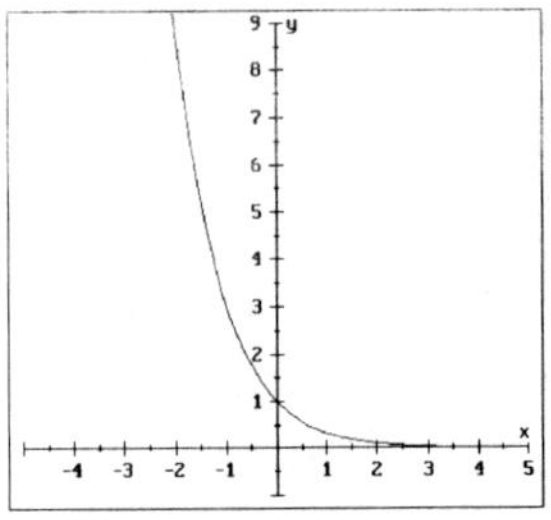

77. $y = 2e^{x}$

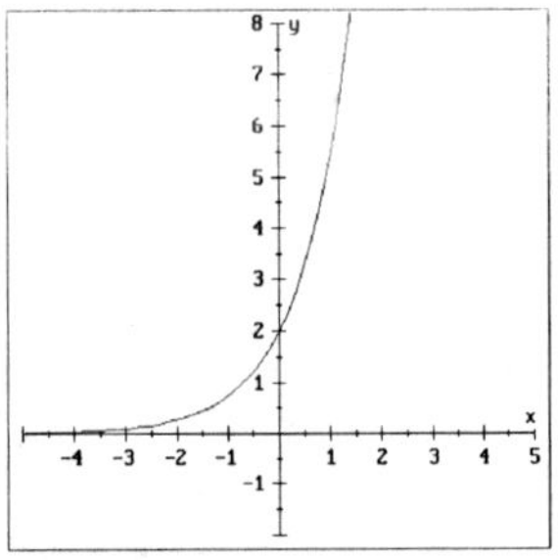

79. $y = 5e^{-0.5x}$

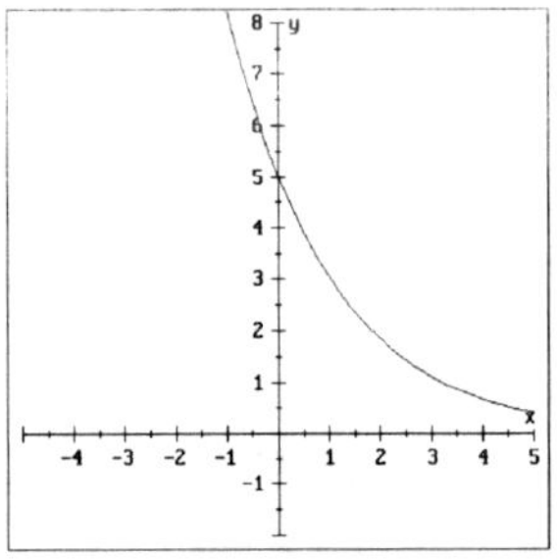

81. $A = P\left(1 + \frac{r}{k}\right)^{kt} = 10000\left(1 + \frac{0.08}{4}\right)^{4(10)} \approx \$22{,}080.40$

83. 5% interest:

$$A = P\left(1 + \frac{r}{k}\right)^{kt} = 500\left(1 + \frac{0.05}{2}\right)^{2(5)} \approx \$640.04$$

$5\frac{1}{2}$% interest:

$$A = P\left(1 + \frac{r}{k}\right)^{kt} = 500\left(1 + \frac{0.055}{2}\right)^{2(5)} \approx \$655.83$$

Difference $= 655.83 - 640.04$
$= \$15.79$ more

85. $$A = P\left(1 + \frac{r}{k}\right)^{kt} = 1\left(1 + \frac{0.05}{1}\right)^{1(300)} \approx \$2{,}273{,}996.13$$

87. $$A = P\left(1 + \frac{r}{k}\right)^{kt} = 1100\left(1 + \frac{0.0175}{1}\right)^{1(8)} \approx \$1263.77$$

89. $$A = Pe^{rt} = 5000e^{0.082(12)} \approx \$13{,}375.68$$

91. Continuous:

$$A = Pe^{rt} = 5000e^{0.085(5)} \approx \$7647.95$$

Annually:

$$A = P\left(1 + \frac{r}{k}\right)^{kt} = 5000\left(1 + \frac{0.085}{1}\right)^{1(5)} \approx \$7518.28$$

93. Quarterly:

$$A = P\left(1 + \frac{r}{k}\right)^{kt} = 10000\left(1 + \frac{0.06}{4}\right)^{4(20)} \approx \$32{,}906.63$$

Daily:

$$A = P\left(1 + \frac{r}{k}\right)^{kt} = 10000\left(1 + \frac{0.06}{365}\right)^{365(20)} \approx \$33{,}197.90$$

Difference
$33{,}197.90 - 32{,}906.63$
$= \boxed{\$291.27}$

95.
$$A = P\left(1+\frac{r}{k}\right)^{kt}$$
$$40{,}000 = P\left(1+\frac{0.06}{4}\right)^{4(20)}$$
$$\frac{40{,}000}{\left(1+\frac{0.06}{4}\right)^{4(20)}} = P$$
$$\$12{,}155.61 \approx P$$

97.
$$P(t) = 1200e^{0.2t}$$
$$P(12) = 1200e^{0.2(12)}$$
$$\approx 13{,}228$$

99.
$$A = P\left(1+\frac{r}{k}\right)^{kt}$$
$$P = A(1+r)^n$$
$$\frac{P}{(1+r)^n} = A$$
$$P(1+r)^{-n} = A$$

101.
$$5^{3t} = k^t$$
$$\left(5^3\right)^t = k^t$$
$$125^t = k^t$$
$$125 = k$$

103. $x^2 + 9x^4 = x^2(1+9x^2)$

105. $x^2 + x - 12 = (x+4)(x-3)$

Exercise 4.2 (page 435)

1. birth, death

3.
$$A = A_0 2^{-t/h}$$
$$= 50 \cdot 2^{-100/(12.4)}$$
$$\approx 0.1868 \text{ grams}$$

5.
$$A = A_0 2^{-t/h}$$
$$= 1000 \cdot 2^{-200/(30.17)}$$
$$\approx 10 \text{ kg}$$

7.
$$A = A_0 2^{-t/h}$$
$$= A_0 \cdot 2^{-60/40}$$
$$\approx A_0(0.354)$$
About 35.4% will remain.

9.
$$A = A_0 2^{-t/h}$$
$$= 1 \cdot 2^{-12/(4.5)}$$
$$\approx 0.1575 \text{ unit}$$

11.
$$I = I_0 k^x$$
$$= 8(0.5)^2$$
$$= 2 \text{ lumens}$$

13.
$$I = I_0 k^x$$
$$1 = I_0(0.5)^3$$
$$\frac{1}{(0.5)^3} = I_0$$
$$8 \text{ lumens} = I_0$$

15.
$$P = P_0 2^{t/2}$$
$$= 10{,}000 \cdot 2^{5/2}$$
$$\approx 56{,}570 \text{ fish}$$

17.
$$T = 40 + 60(0.75)^t$$
$$= 40 + 60(0.75)^{3.5}$$
$$\approx 61.9° \text{ C}$$

19.
$$P = 173e^{0.03t}$$
$$= 173e^{0.03(20)}$$
$$\approx 315$$

21.
$$P = P_0 e^{0.27t}$$
$$= 2e^{0.27(7)}$$
$$\approx 13 \text{ cases}$$

23.
$$P = P_0 e^{kt}$$
$$= 6e^{0.019(30)}$$
$$\approx 10.6 \text{ billion}$$

SECTION 4.2

25. $P = P_0e^{kt}$
$= 6e^{0.019(50)}$
≈ 15.5 billion
$\frac{15.5}{6} \approx$ a factor of 2.6

27. $P = e^{-0.3t}$
$= e^{-0.3(24)}$
≈ 0.0007
$= 0.07\%$

29. Let $t = 0$:
$x = 0.08\left(1 - e^{-0.1t}\right)$
$= 0.08\left(1 - e^{-0.1(0)}\right)$
$= 0.08\left(1 - e^{0}\right)$
$= 0.08(1 - 1) = 0$

31. $N = P\left(1 - e^{-0.1t}\right)$
$= 50{,}000\left(1 - e^{-0.1(10)}\right)$
$\approx 31{,}606$
$50{,}000 - 31{,}606 = 18{,}394$

33. $P = \frac{1{,}200{,}000}{1 + (1200 - 1)e^{-0.4t}}$
$= \frac{1{,}200{,}000}{1 + (1200 - 1)e^{-0.4(8)}}$
$\approx 24{,}060$ people

35. $P = \frac{55{,}000}{1 + (550 - 1)e^{-0.8t}}$
$= \frac{55{,}000}{1 + (550 - 1)e^{-0.8(2)}}$
≈ 492 people

37. $w = 1.54e^{0.503n}$
$= 1.54e^{0.503(5)}$
≈ 19.0 mm

39. $v = 50\left(1 - e^{-0.2t}\right)$
$= 50\left(1 - e^{-0.2(20)}\right)$
≈ 49 meters/second

41.

Males	Females
$P = P_0e^{kt}$	$P = P_0e^{kt}$
$= 133e^{0.01t}$	$= 139e^{0.01t}$
$= 133e^{0.01(20)}$	$= 139e^{0.01(20)}$
≈ 162.4 million	≈ 169.8 million

There will be about 7 million more females.

43. Find where these graphs meet:
$y = 1000e^{0.02t}$, $y = 31x + 2000$

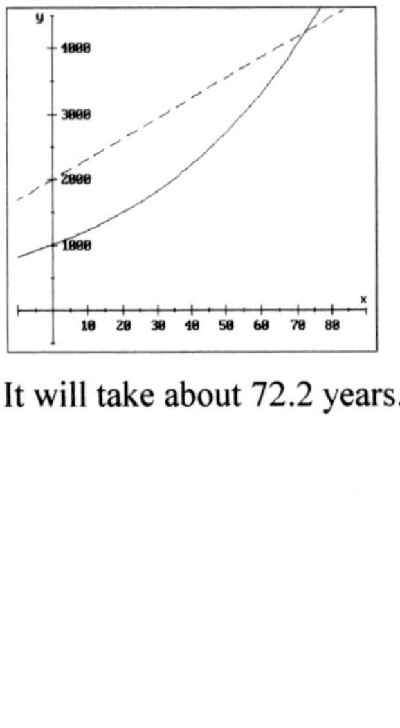

It will take about 72.2 years.

45. $1 + 1 + \frac{1}{2} + \frac{1}{2\cdot3} + \frac{1}{2\cdot3\cdot4} + \frac{1}{2\cdot3\cdot4\cdot5} \approx 2.71\overline{6}$
$e \approx 2.718$: accurate to 2 places

47. $P = \dfrac{1{,}200{,}000}{1 + 1199e^{-0.4t}}$

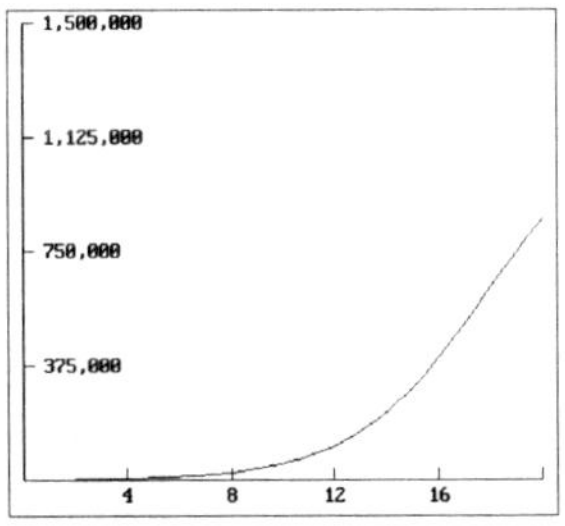

49. $2^3 = x$
$8 = x$

51. $x^3 = 27$
$x = 3$

53. $x^{-3} = \dfrac{1}{8}$
$\dfrac{1}{x^3} = \dfrac{1}{8}$
$x = 2$

55. $9^{1/2} = x$
$3 = x$

Exercise 4.3 (page 448)

1. $x = b^y$

3. range

5. inverse

7. exponent

9. $(b, 1), (1, 0)$

11. $\log_e x$

13. $(-\infty, \infty)$

15. 10

17. $8^2 = 64$
$\log_8 64 = 2$

19. $4^{-2} = \dfrac{1}{16}$
$\log_4 \dfrac{1}{16} = -2$

21. $\left(\dfrac{1}{2}\right)^{-5} = 32$
$\log_{1/2} 32 = -5$

23. $x^y = z$
$\log_x z = y$

25. $\log_3 81 = 4$
$3^4 = 81$

27. $\log_{1/2} \dfrac{1}{8} = 3$
$\left(\dfrac{1}{2}\right)^3 = \dfrac{1}{8}$

29. $\log_4 \dfrac{1}{64} = -3$
$4^{-3} = \dfrac{1}{64}$

31. $\log_\pi \pi = 1$
$\pi^1 = \pi$

33. $\log_2 8 = x$
$2^x = 8$
$x = 3$

35. $\log_4 64 = x$
$4^x = 64$
$x = 3$

37. $\log_{1/2} \dfrac{1}{8} = x$
$\left(\dfrac{1}{2}\right)^x = \dfrac{1}{8}$
$x = 3$

39. $\log_9 3 = x$
$9^x = 3$
$x = \dfrac{1}{2}$

41. $\log_{1/2} 8 = x$
$\left(\dfrac{1}{2}\right)^x = 8$
$x = -3$

43. $\log_8 x = 2$
$8^2 = x$
$64 = x$

45. $\log_7 x = 1$
$7^1 = x$
$7 = x$

47. $\log_{25} x = \dfrac{1}{2}$
$25^{1/2} = x$
$5 = x$

SECTION 4.3

49. $\log_5 x = -2$
$5^{-2} = x$
$\frac{1}{5^2} = x$
$\frac{1}{25} = x$

51. $\log_{36} x = -\frac{1}{2}$
$36^{-1/2} = x$
$\frac{1}{36^{1/2}} = x$
$\frac{1}{6} = x$

53. $\log_x 5^3 = 3$
$x^3 = 5^3$
$x = 5$

55. $\log_x \frac{9}{4} = 2$
$x^2 = \frac{9}{4}$
$x = \frac{3}{2}$

57. $\log_x \frac{1}{64} = -3$
$x^{-3} = \frac{1}{64}$
$\frac{1}{x^3} = \frac{1}{4^3}$
$x = 4$

59. $\log_x \frac{9}{4} = -2$
$x^{-2} = \frac{9}{4}$
$\left(x^{-2}\right)^{-1} = \left(\frac{9}{4}\right)^{-1}$
$x^2 = \frac{4}{9}$
$x = \frac{2}{3}$

61. From the definition:
$2^{\log_2 5} = 5$

63. From the definition:
$x^{\log_4 6} = 6 \Rightarrow x = 4$

65. $\log 3.25 \approx 0.5119$

67. $\log 0.00467 \approx -2.3307$

69. $\ln 45.7 \approx 3.8221$

71. $\ln \frac{2}{3} \approx -0.4055$

73. $\ln 35.15 \approx 3.5596$

75. $\ln 7.896 \approx 2.0664$

77. $\log(\ln 1.7) \approx -0.2752$

79. $\ln(\log 0.1)$: undefined

81. $\log y = 1.4023$
$y \approx 25.2522$

83. $\log y = -3.71$
$y \approx 1.9498 \times 10^{-4}$

85. $\ln y = 1.4023$
$y \approx 4.0645$

87. $\ln y = 4.24$
$y \approx 69.4079$

89. $\ln y = -3.71$
$y \approx 0.0245$

91. $\log y = \ln 8$
$y \approx 120.0719$

93. $\log 10{,}000 = x$
$10^x = 10{,}000$
$x = 4$
$\log 10{,}000 = 4$

95. $\log 0.001 = x$
$10^x = 0.001$
$x = -3$
$\log 0.001 = -3$

97. From the definition:
$e^{\ln 7} = 7$

99. From the definition:
$\ln\left(e^4\right) = 4$

101. The graph passes through the point $(b, 1) = (2, 1)$,
$b = 2$

103. $y = \log_b x$
$b^y = x$
$b^{-1} = \frac{1}{2}$
$\left(b^{-1}\right)^{-1} = \left(\frac{1}{2}\right)^{-1}$
$b = 2$

SECTION 4.3

105. $f(x) = \log_3 x$
points: $(1, 0)$, $(3, 1)$

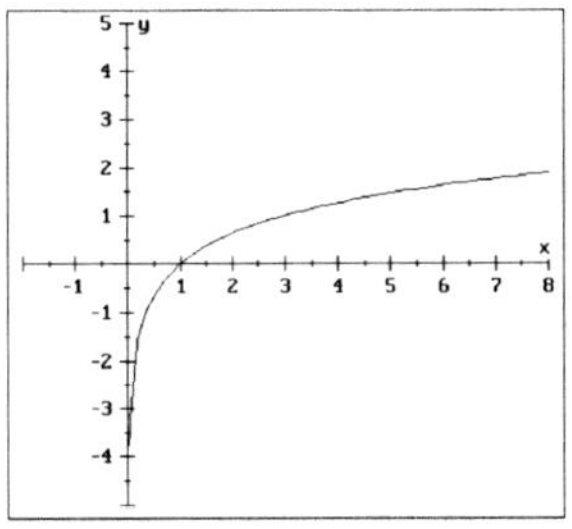

107. $f(x) = \log_{1/3} x$
points: $(1, 0)$, $(\frac{1}{3}, 1)$

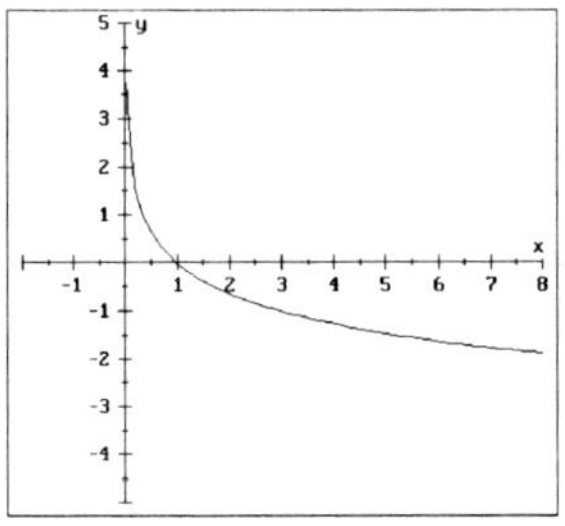

109. $f(x) = -\log_5 x$
Reflect $y = \log_5 x$ about x

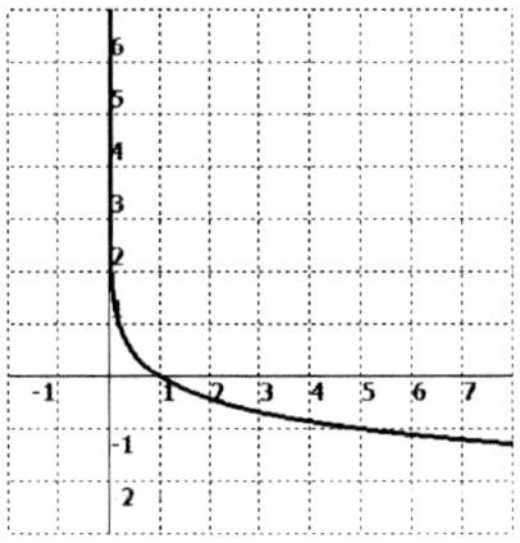

111. $f(x) = 2 + \log_2 x$
Shift $y = \log_2 x$ U2.

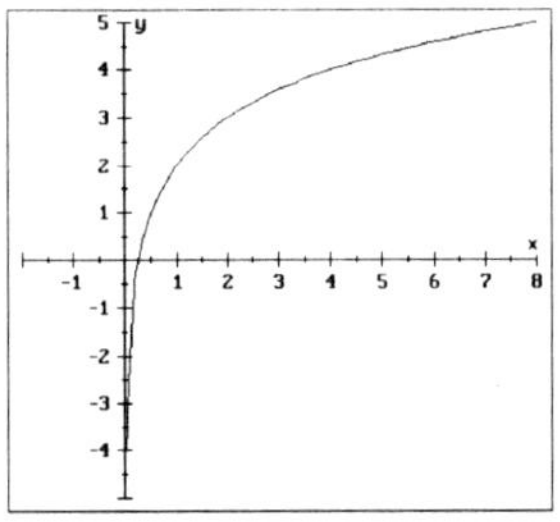

113. $f(x) = \log_3 (x + 2)$
Shift $y = \log_3 x$ L2.

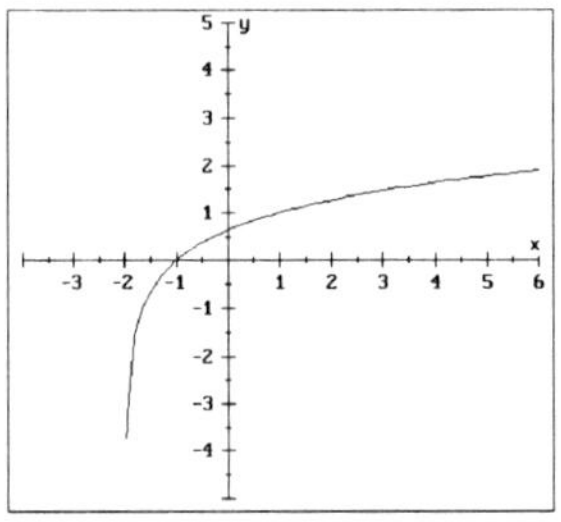

115. $f(x) = 3 + \log_3 (x + 1)$
Shift $y = \log_3 x$ U3, L1.

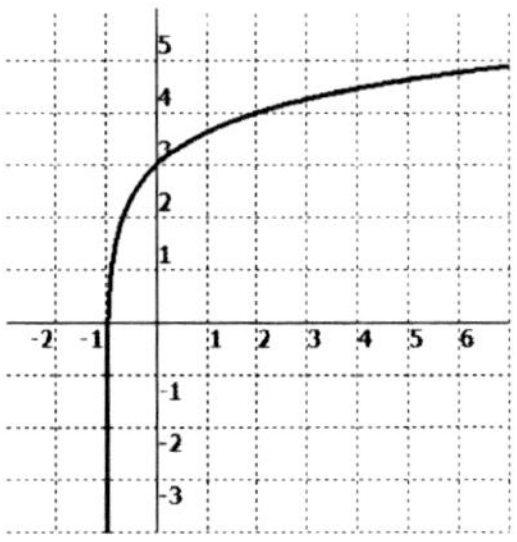

117. $f(x) = -3 + \ln x$
Shift $y = \ln x$ D3.

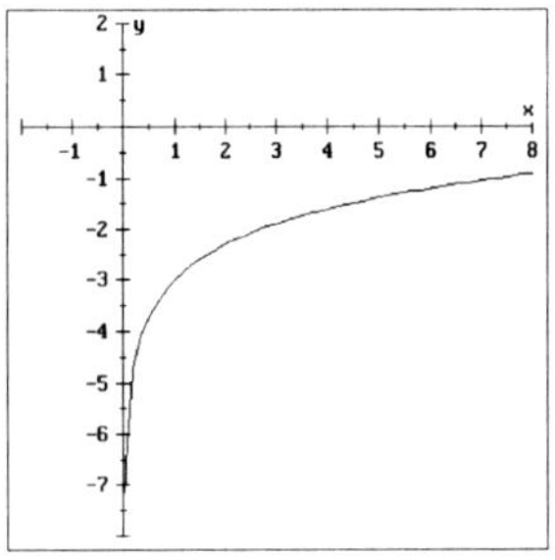

119. $f(x) = \ln (x - 4)$
Shift $y = \ln x$ R4.

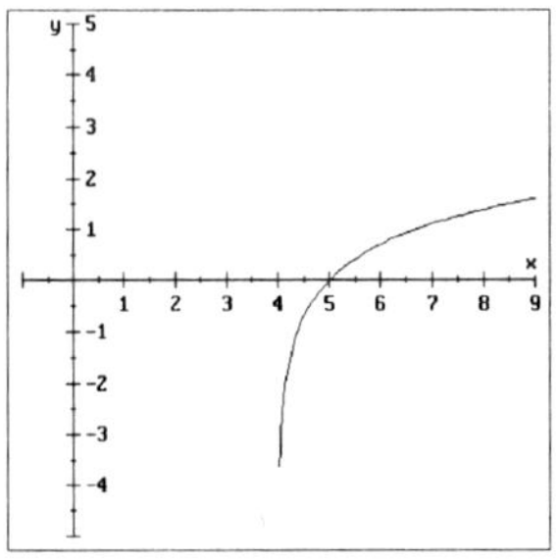

121. $f(x) = 1 - \ln x$
Reflect $y = \ln x$
about x, shift U1.

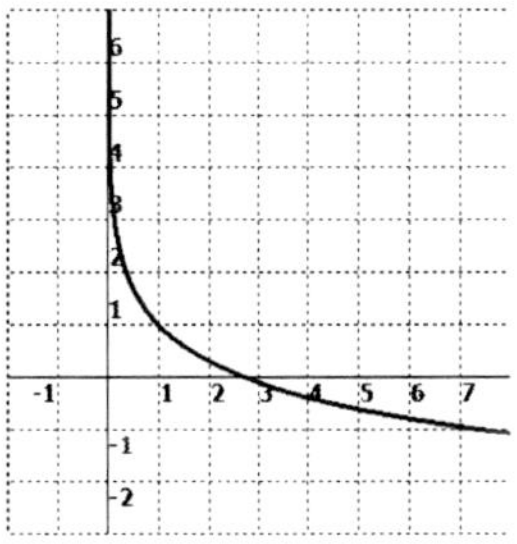

SECTION 4.3

123. $f(x) = \log(3x)$

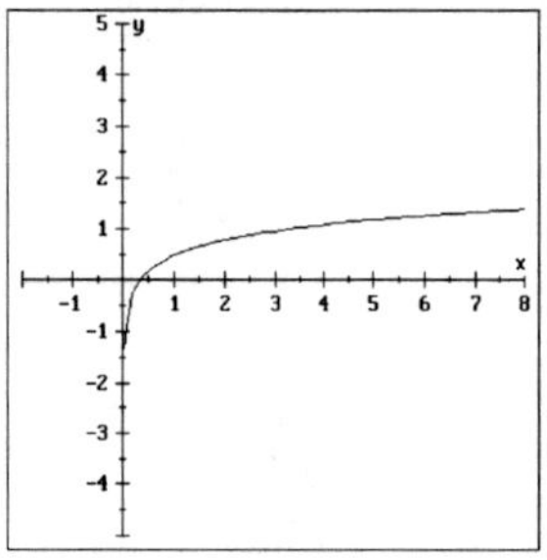

125. $f(x) = \log(-x)$

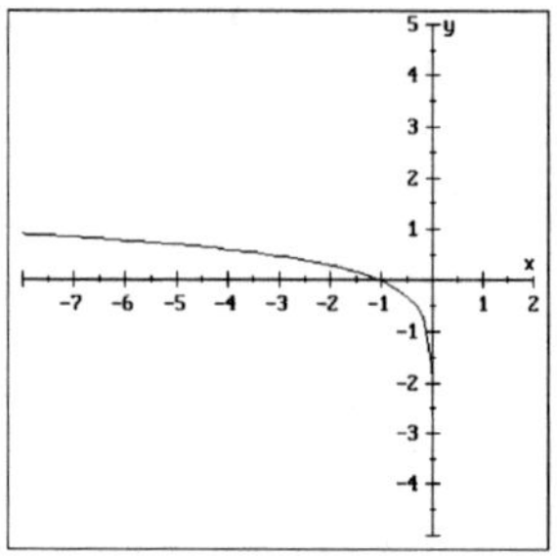

127. $f(x) = \ln\left(\frac{1}{2}x\right)$

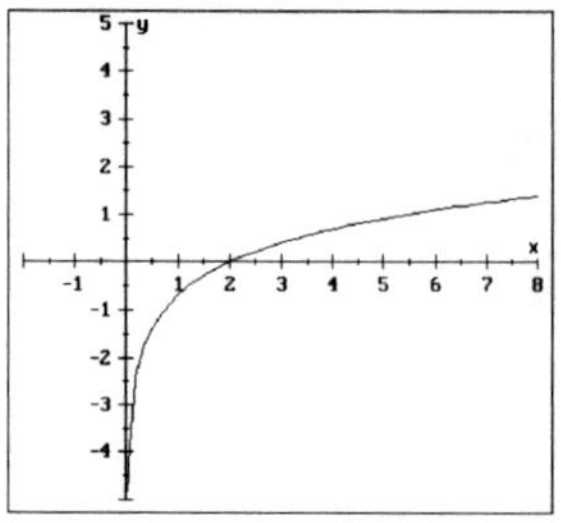

129. $f(x) = \ln(-x)$

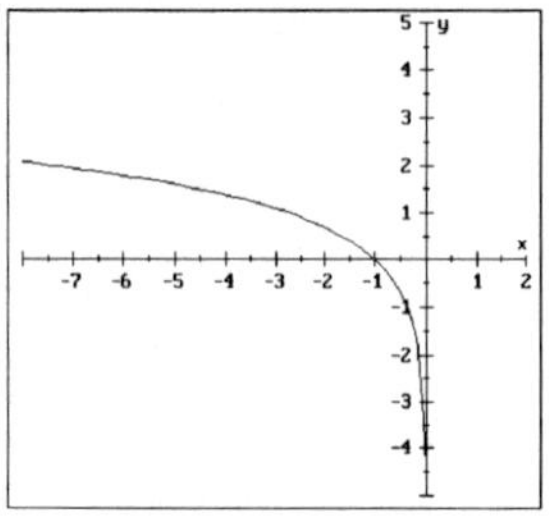

131-133. Answers may vary.

135. $y = x^2 + 7x + 3;\ a = 1, b = 7, c = 3$

$x = -\frac{b}{2a} = -\frac{7}{2(1)} = -\frac{7}{2}$

$y = x^2 + 7x + 3$

$= \left(-\frac{7}{2}\right)^2 + 7\left(-\frac{7}{2}\right) + 3 = -\frac{37}{4}$

Vertex: $\left(-\frac{7}{2}, -\frac{37}{4}\right)$

137. Let $x =$ width. Then $\dfrac{3400 - 4x}{2} =$ length.

$A = x\left(\dfrac{3400 - 4x}{2}\right) = x(1700 - 2x)$

$= 1700x - 2x^2$

$= -2x^2 + 1700x$

$x = -\dfrac{b}{2a} = -\dfrac{1700}{2(-2)} = 425$

The dimensions are 425 feet by 850 feet.

139. $y = 5x - 8:\ m = 5$

Use $m = 5$:

$y - y_1 = m(x - x_1)$

$y - 0 = 5(x - 0)$

$y = 5x$

Exercise 4.4 (page 455)

1. $20 \log \dfrac{E_O}{E_I}$

3. $t = -\dfrac{1}{k} \ln\left(1 - \dfrac{C}{M}\right)$

5. $E = RT \ln\left(\dfrac{V_f}{V_i}\right)$

SECTION 4.4

7. $\text{db gain} = 20 \log \frac{E_0}{E_I}$
$= 20 \log \frac{17}{0.03}$
$\approx 55 \text{ db}$

9. $\text{db gain} = 20 \log \frac{E_0}{E_I}$
$= 20 \log \frac{20}{0.71}$
$\approx 29 \text{ db}$

11. $\text{db gain} = 20 \log \frac{E_0}{E_I}$
$= 20 \log \frac{30}{0.1}$
$\approx 49.5 \text{ db}$

13. $R = \log \frac{A}{P}$
$= \log \frac{5000}{0.2}$
≈ 4.4

15. $R = \log \frac{A}{P}$
$= \log \frac{2500}{\frac{1}{4}}$
$= 4$

17. $R = \log \frac{A}{P}$
$= \log \frac{6000}{0.3}$
$\approx 4.3 \Rightarrow$ no damage

19. $t = -\frac{1}{k} \ln \left(1 - \frac{C}{M}\right)$
$t = -\frac{1}{0.116} \ln \left(1 - \frac{0.9M}{M}\right)$
$t = -\frac{1}{0.116} \ln(1 - 0.9) \approx 19.8$ minutes

21. $t = \frac{\ln 2}{r} = \frac{\ln 2}{0.12} \approx 5.8$ years

23. $t = \frac{\ln 3}{r} = \frac{\ln 3}{0.12} \approx 9.2$ years

25. $E = RT \ln \left(\frac{V_f}{V_i}\right)$
$E = (8.314)(400) \ln \left(\frac{3V_i}{V_i}\right)$
$E = (8.314)(400) \ln (3)$
$E \approx 3654$ joules

27. $r = \frac{1}{t} \ln \frac{P}{P_0}$
$= \frac{1}{7} \ln \frac{10{,}400{,}000}{10{,}400}$
$\approx 0.99 \Rightarrow$ about 99% per year

29. $n = \frac{\log V - \log C}{\log \left(1 - \frac{2}{N}\right)}$
$= \frac{\log 8000 - \log 37{,}000}{\log \left(1 - \frac{2}{5}\right)}$
≈ 3 years old

31. $n = \frac{\log \left[\frac{Ar}{P} + 1\right]}{\log (1 + r)}$
$= \frac{\log \left[\frac{20000(0.12)}{1000} + 1\right]}{\log (1 + 0.12)}$
≈ 10.8 years

33. $V = ER_1 \ln \frac{R_2}{R_1} = (400{,}000)(0.25) \ln \left(\frac{2}{0.25}\right) \approx 208{,}000 \text{ V}$

35. $f(x) = 12 \Rightarrow \ln x = 12 \Rightarrow x \approx 162755 \Rightarrow$ You would go out 162,755 cm.
$162{,}755 \text{ cm} = \frac{162{,}755 \text{ cm}}{1} \cdot \frac{1 \text{ in}}{2.54 \text{ cm}} \cdot \frac{1 \text{ ft}}{12 \text{ in}} \cdot \frac{1 \text{ mi}}{5280 \text{ ft}} \approx 1 \text{ mile}$

37. Set $x = 0$:
$y = \frac{1}{1 + e^{-2(0)}} = \frac{1}{1 + e^0} = \frac{1}{1+1} = \frac{1}{2}$
y-intercept: $\left(0, \frac{1}{2}\right)$

39. $y = mx + b$
$y = 7x + 3$

41. Vertical through $(2, 3)$: $x = 2$

43. $\dfrac{2(x+2)-1}{4x^2-9} = \dfrac{2x+4-1}{(2x+3)(2x-3)} = \dfrac{2x+3}{(2x+3)(2x-3)} = \dfrac{1}{2x-3}$

45. $\dfrac{x^2+3x+2}{3x+9} \cdot \dfrac{x+3}{x^2-4} = \dfrac{(x+2)(x+1)}{3(x+3)} \cdot \dfrac{x+3}{(x+2)(x-2)} = \dfrac{x+1}{3(x-2)}$

Exercise 4.5 (page 468)

1. 0 **3.** M, N **5.** x, y **7.** x

9. $\neq$ **11.** $\log_4 1 = 0$ **13.** $\log_4 4^7 = 7\log_4 4 = 7$

15. $5^{\log_5 10} = 10$ **17.** $\log_5 5 = 1$ **19-23. Answers may vary.**

25. $\log_b 2xy = \log_b 2 + \log_b x + \log_b y$

27. $\log_b \dfrac{2x}{y} = \log_b 2x - \log_b y$
$= \log_b 2 + \log_b x - \log_b y$

29. $\log_b x^2y^3 = \log_b x^2 + \log_b y^3$
$= 2\log_b x + 3\log_b y$

31. $\log_b (xy)^{1/3} = \dfrac{1}{3}\log_b xy$
$= \dfrac{1}{3}(\log_b x + \log_b y)$

33. $\log_b x\sqrt{z} = \log_b xz^{1/2} = \log_b x + \log_b z^{1/2} = \log_b x + \dfrac{1}{2}\log_b z$

35. $\log_b \dfrac{\sqrt[3]{x}}{\sqrt[3]{yz}} = \log_b \sqrt[3]{x} - \log_b \sqrt[3]{yz} = \log_b x^{1/3} - \log_b (yz)^{1/3} = \dfrac{1}{3}\log_b x - \dfrac{1}{3}\log_b yz$
$= \dfrac{1}{3}\log_b x - \dfrac{1}{3}(\log_b y + \log_b z)$
$= \dfrac{1}{3}\log_b x - \dfrac{1}{3}\log_b y - \dfrac{1}{3}\log_b z$

37. $\ln x^7y^8 = \ln x^7 + \ln y^8$
$= 7\ln x + 8\ln y$

39. $\ln \dfrac{x}{y^4z} = \ln x - \ln y^4z$
$= \ln x - \left(\ln y^4 + \ln z\right)$
$= \ln x - (4\ln y + \ln z)$
$= \ln x - 4\ln y - \ln z$

41. $\log_b (x+1) - \log_b x = \log_b \dfrac{x+1}{x}$

43. $2\log_b x + \dfrac{1}{3}\log_b y = \log_b x^2 + \log_b y^{1/3} = \log_b x^2y^{1/3} = \log_b x^2\sqrt[3]{y}$

SECTION 4.5

45. $-3\log_b x - 2\log_b y + \frac{1}{2}\log_b z = \log_b x^{-3} + \log_b y^{-2} + \log_b z^{1/2} = \log_b x^{-3}y^{-2}\sqrt{z} = \log_b \frac{\sqrt{z}}{x^3y^2}$

47. $\log_b\left(\frac{x}{z} + x\right) - \log_b\left(\frac{y}{z} + y\right) = \log_b \frac{\frac{x}{z} + x}{\frac{y}{z} + y} = \log_b \frac{z\left(\frac{x}{z} + x\right)}{z\left(\frac{y}{z} + y\right)} = \log_b \frac{x + xz}{y + yz} = \log_b \frac{x}{y}$

49. $\ln x + \ln(x+5) - \ln 9 = \ln x(x+5) - \ln 9 = \ln \frac{x(x+5)}{9}$

51. $-6\ln x - 2\ln y + \ln z = \ln x^{-6} + \ln y^{-2} + \ln z = \ln x^{-6}y^{-2}z = \ln \frac{z}{x^6y^2}$

53. $\log_b ab = \log_b a + \log_b b = \log_b a + 1$
TRUE

55. $\log_b 0$ is undefined.
FALSE

57. $\log_b(x \cdot y) = \log_b x + \log_b y$, so
$\log_b(x + y) \neq \log_b x + \log_b y$
TRUE (unless $x \cdot y = x + y$)

59. If $\log_a b = c$, then $\log_b a = c$
FALSE

61. $\log_7 7^7 = 7 \Rightarrow 7^7 = 7^7$
TRUE

63. $-\log_b x = \log_b x^{-1} = \log_b \frac{1}{x}$
FALSE

65. $\log_b\left(\frac{A}{B}\right) = \log_b A - \log_b B$, so
$\frac{\log_b(A)}{\log_b(B)} \neq \log_b A - \log_b B$
FALSE

67. $\log_b \frac{1}{5} = \log_b 5^{-1} = -\log_b 5$
TRUE

69. $\frac{1}{3}\log_b a^3 = \frac{1}{3} \cdot 3\log_b a = \log_b a$
TRUE

71. Let $\log_{1/b} y = c$.
Then $\left(\frac{1}{b}\right)^c = y$.
$\left(\left(\frac{b}{1}\right)^{-1}\right)^c = y$
$(b)^{-c} = y \Rightarrow \log_b y = -c.$
$\log_{1/b} y + \log_b y = c + (-c) = 0.$
TRUE

73. $\ln(xy) = \ln x + \ln y$, so
$\ln(xy) \neq (\ln x)(\ln y)$
FALSE

75. $\frac{1}{5}\ln a^5 = \frac{1}{5} \cdot 5\ln a = \ln a$
TRUE

SECTION 4.5

77. $\log_{10} 28 = \log_{10}(4 \cdot 7)$
$= \log_{10} 4 + \log_{10} 7$
$= 0.6021 + 0.8451 = 1.4472$

79. $\log_{10} 2.25 = \log_{10}\left(\frac{9}{4}\right)$
$= \log_{10} 9 - \log_{10} 4$
$= 0.9542 - 0.6021 = 0.3521$

81. $\log_{10}\left(\frac{63}{4}\right) = \log_{10} 63 - \log_{10} 4$
$= \log_{10}(7 \cdot 9) - \log_{10} 4$
$= \log_{10} 7 + \log_{10} 9 - \log_{10} 4$
$= 0.8451 + 0.9542 - 0.6021$
$= 1.1972$

83. $\log_{10} 252 = \log_{10}(4 \cdot 63)$
$= \log_{10}(4 \cdot 7 \cdot 9)$
$= \log_{10} 4 + \log_{10} 7 + \log_{10} 9$
$= 0.6021 + 0.8451 + 0.9542$
$= 2.4014$

85. $\log_{10} 112 = \log_{10}(4 \cdot 28) = \log_{10}(4 \cdot 7 \cdot 4) = \log_{10} 4 + \log_{10} 7 + \log_{10} 4$
$= 0.6021 + 0.8451 + 0.6021$
$= 2.0493$

87. $\log_{10}\left(\frac{144}{49}\right) = \log_{10} 144 - \log_{10} 49 = \log_{10}(4 \cdot 4 \cdot 9) - \log_{10}(7 \cdot 7)$
$= \log_{10} 4 + \log_{10} 4 + \log_{10} 9 - \log_{10} 7 - \log_{10} 7$
$= 0.6021 + 0.6021 + 0.9542 - 0.8451 - 0.8451$
$= 0.4682$

89. $\log_3 7 = \frac{\log_{10} 7}{\log_{10} 3} \approx 1.7712$

91. $\log_\pi 3 = \frac{\log_{10} 3}{\log_{10} \pi} \approx 0.9597$

93. $\log_3 8 = \frac{\log_e 8}{\log_e 3} \approx 1.8928$

95. $\log_{\sqrt{2}} \sqrt{5} = \frac{\log_e \sqrt{5}}{\log_e \sqrt{2}} \approx 2.3219$

97. $\text{pH} = -\log[\text{H}^+]$
$= -\log(6.3 \times 10^{-8})$
≈ 7.20

99. $\text{pH} = -\log[\text{H}^+]$
$= -\log(1.7 \times 10^{-5})$
≈ 4.77

101.

$\text{pH} = -\log[\text{H}^+]$	$\text{pH} = -\log[\text{H}^+]$
$2.9 = -\log[\text{H}^+]$	$3.3 = -\log[\text{H}^+]$
$-2.9 = \log[\text{H}^+]$	$-3.3 = \log[\text{H}^+]$
$1.26 \times 10^{-3} \approx [\text{H}^+]$	$5.01 \times 10^{-4} \approx [\text{H}^+]$

The hydrogen ion concentration can range from 5.01×10^{-4} to 1.26×10^{-3}.

103. $\text{db gain} = 10 \log \frac{P_O}{P_I}$
$= 10 \log \frac{40}{\frac{1}{2}}$
≈ 19 db

105. $L = k \ln I$
$4L = 4k \ln I = k \cdot 4 \ln I = k \ln I^4$
The original intensity must be raised to the fourth power.

107. $E = 8300 \ln V$

$2E = 2 \cdot 8300 \ln V = 8300 \ln V^2$

The original volume is squared.

109. $3^{4\log_3 2} + 5^{\frac{1}{2}\log_5 25} = 3^{\log_3 2^4} + 5^{\log_5 25^{1/2}}$

$= 2^4 + 25^{1/2} = 16 + 5 = 21$

111. Let $\log_b M = x$ and $\log_b N = y$.

Then $b^x = M$ and $b^y = N$.

$\frac{M}{N} = \frac{b^x}{b^y} = b^{x-y}$.

So $\log_b \frac{M}{N} = x - y$, or

$\log_b \frac{M}{N} = \log_b M - \log_b N$.

113. $e^{x \ln a} = e^{\ln a^x} = a^x$

115. $\ln(e^x) = x \ln e = x(1) = x$

117. $\log(0.9) < 0$, so $\ln(\log(0.9))$ is undefined.

119. **Answers may vary.**

121. $y = 3x - 1 \Rightarrow$ function

123. $y^2 = 4x$

$y = \pm\sqrt{4x} \Rightarrow$ not a function

125. $f(x) = x^2 - 4$; domain $= (-\infty, \infty)$

127. $f(x) = \sqrt{x^2 + 4}$; domain $= (-\infty, \infty)$

Exercise 4.6 (page 482)

1. exponential

3. $A_0 2^{-t/h}$

5.
$$\begin{aligned} 2^{3x+2} &= 16^x \\ 2^{3x+2} &= \left(2^4\right)^x \\ 2^{3x+2} &= 2^{4x} \\ 3x + 2 &= 4x \\ 2 &= x \end{aligned}$$

7.
$$\begin{aligned} 27^{x+1} &= 3^{2x+1} \\ \left(3^3\right)^{x+1} &= 3^{2x+1} \\ 3^{3x+3} &= 3^{2x+1} \\ 3x + 3 &= 2x + 1 \\ x &= -2 \end{aligned}$$

9.
$$\begin{aligned} 5^{4x+1} &= 25^{-x-2} \\ 5^{4x+1} &= \left(5^2\right)^{-x-2} \\ 5^{4x+1} &= 5^{-2x-4} \\ 4x + 1 &= -2x - 4 \\ 6x &= -5 \\ x &= -\frac{5}{6} \end{aligned}$$

11.
$$\begin{aligned} 4^{x-2} &= 8^x \\ \left(2^2\right)^{x-2} &= \left(2^3\right)^x \\ 2^{2x-4} &= 2^{3x} \\ 2x - 4 &= 3x \\ -4 &= x \end{aligned}$$

SECTION 4.6

13.
$$\begin{aligned} 81^{2x} &= 27^{2x-5} \\ \left(3^4\right)^{2x} &= \left(3^3\right)^{2x-5} \\ 3^{8x} &= 3^{6x-15} \\ 8x &= 6x - 15 \\ 2x &= -15 \\ x &= -\frac{15}{2} \end{aligned}$$

15.
$$\begin{aligned} 2^{x^2-2x} &= 8 \\ 2^{x^2-2x} &= 2^3 \\ x^2 - 2x &= 3 \\ x^2 - 2x - 3 &= 0 \\ (x-3)(x+1) &= 0 \\ x - 3 = 0 \quad &\textbf{or} \quad x + 1 = 0 \\ x = 3 \quad &\quad\quad x = -1 \end{aligned}$$

17.
$$\begin{aligned} 36^{x^2} &= 216^{x^2-3} \\ \left(6^2\right)^{x^2} &= \left(6^3\right)^{x^2-3} \\ 6^{2x^2} &= 6^{3x^2-9} \\ 2x^2 &= 3x^2 - 9 \\ 9 &= x^2 \\ \pm 3 &= x \end{aligned}$$

19.
$$\begin{aligned} 7^{x^2+3x} &= \frac{1}{49} \\ 7^{x^2+3x} &= 7^{-2} \\ x^2 + 3x &= -2 \\ x^2 + 3x + 2 &= 0 \\ (x+2)(x+1) &= 0 \\ x + 2 = 0 \quad &\textbf{or} \quad x + 1 = 0 \\ x = -2 \quad &\quad\quad x = -1 \end{aligned}$$

21.
$$\begin{aligned} e^{-x+6} &= e^x \\ -x + 6 &= x \\ 6 &= 2x \\ 3 &= x \end{aligned}$$

23.
$$\begin{aligned} e^{x^2-1} &= e^{24} \\ x^2 - 1 &= 24 \\ x^2 - 25 &= 0 \\ (x-5)(x+5) &= 0 \\ x - 5 = 0 \quad &\textbf{or} \quad x + 5 = 0 \\ x = -5 \quad &\quad\quad x = 5 \end{aligned}$$

25.
$$\begin{aligned} 4^x &= 5 \\ \log 4^x &= \log 5 \\ x \log 4 &= \log 5 \\ x &= \frac{\log 5}{\log 4} \\ x &\approx 1.1610 \end{aligned}$$

27.
$$\begin{aligned} 13^{x-1} &= 2 \\ \log 13^{x-1} &= \log 2 \\ (x-1)\log 13 &= \log 2 \\ x \log 13 - \log 13 &= \log 2 \\ x \log 13 &= \log 2 + \log 13 \\ x &= \frac{\log 2 + \log 13}{\log 13} \\ x &\approx 1.2702 \end{aligned}$$

29.
$$\begin{aligned} 2^{x+1} &= 3^x \\ \log 2^{x+1} &= \log 3^x \\ (x+1)\log 2 &= x \log 3 \\ x \log 2 + \log 2 &= x \log 3 \\ x \log 2 - x \log 3 &= -\log 2 \\ x(\log 2 - \log 3) &= -\log 2 \\ x &= \frac{-\log 2}{\log 2 - \log 3} \\ x &\approx 1.7095 \end{aligned}$$

31.
$$\begin{aligned} 2^x &= 3^x \\ \log 2^x &= \log 3^x \\ x \log 2 &= x \log 3 \\ x \log 2 - x \log 3 &= 0 \\ x(\log 2 - \log 3) &= 0 \\ x &= \frac{0}{\log 2 - \log 3} = 0 \end{aligned}$$

33.
$$\begin{aligned} 7^{x^2} &= 10 \\ \log 7^{x^2} &= \log 10 \\ x^2 \log 7 &= \log 10 \\ x^2 &= \frac{\log 10}{\log 7} \\ x &= \pm\sqrt{\frac{\log 10}{\log 7}} \\ x &\approx \pm 1.0878 \end{aligned}$$

35.
$$\begin{aligned} 8^{x^2} &= 9^x \\ \log 8^{x^2} &= \log 9^x \\ x^2 \log 8 &= x \log 9 \\ x^2 \log 8 - x\log 9 &= 0 \\ x(x\log 8 - \log 9) &= 0 \end{aligned}$$
$$\begin{array}{lcl} x = 0 & \textbf{or} & x\log 8 - \log 9 = 0 \\ x = 0 & & x \log 8 = \log 9 \\ x = 0 & & x = \dfrac{\log 9}{\log 8} \\ x = 0 & & x \approx 1.0566 \end{array}$$

37.
$$\begin{aligned} e^x &= 10 \\ \ln e^x &= \ln 10 \\ x \ln e &= \ln 10 \\ x &= \ln 10 \end{aligned}$$

39.
$$\begin{aligned} 4e^{2x} &= 24 \\ e^{2x} &= 6 \\ \ln e^{2x} &= \ln 6 \\ 2x \ln e &= \ln 6 \\ 2x &= \ln 6 \\ x &= \frac{\ln 6}{2} \end{aligned}$$

41.
$$\begin{aligned} 4^{x+2} - 4^x &= 15 \\ 4^x 4^2 - 4^x &= 15 \\ 16 \cdot 4^x - 4^x &= 15 \\ 15 \cdot 4^x &= 15 \\ 4^x &= 1 \\ x &= 0 \end{aligned}$$

43.
$$\begin{aligned} 2(3^x) &= 6^{2x} \\ \log[2(3^x)] &= \log 6^{2x} \\ \log 2 + \log 3^x &= 2x \log 6 \\ \log 2 + x\log 3 &= 2x\log 6 \\ x\log 3 - 2x\log 6 &= -\log 2 \\ x(\log 3 - 2\log 6) &= -\log 2 \\ x &= \frac{-\log 2}{\log 3 - 2\log 6} \\ x &\approx 0.2789 \end{aligned}$$

45.
$$\begin{aligned} 2^{2x} - 10(2^x) + 16 &= 0 \\ y^2 - 10y + 16 &= 0 \\ (y-2)(y-8) &= 0 \end{aligned}$$
$$\begin{array}{lcl} y - 2 = 0 & \textbf{or} & y - 8 = 0 \\ y = 2 & & y = 8 \\ 2^x = 2 & & 2^x = 8 \\ x = 1 & & x = 3 \end{array}$$

47.
$$\begin{aligned} 2^{2x+1} - 2^x &= 1 \\ 2^{2x}2^1 - 2^x - 1 &= 0 \\ 2\left(2^{2x}\right) - 2^x - 1 &= 0 \\ (2(2^x)+1)(2^x - 1) &= 0 \end{aligned}$$
$$\begin{array}{lcl} 2(2^x) + 1 = 0 & \textbf{or} & 2^x - 1 = 0 \\ 2(2^x) = -1 & & 2^x = 1 \\ 2^x = -\frac{1}{2} & & x = 0 \\ \text{impossible} & & \end{array}$$

49.
$$\begin{aligned} \log x^2 &= 2 \\ x^2 &= 10^2 \\ x^2 &= 100 \\ x &= \pm\sqrt{100} = \pm 10 \end{aligned}$$

51. $\log \dfrac{4x+1}{2x+9} = 0$

$$10^0 = \frac{4x+1}{2x+9}$$
$$1 = \frac{4x+1}{2x+9}$$
$$2x+9 = 4x+1$$
$$8 = 2x$$
$$4 = x$$

53. $\ln x = 6$

$$e^6 = x$$

55. $\ln(2x-7) = 4$

$$e^4 = 2x-7$$
$$e^4 + 7 = 2x$$
$$\frac{e^4+7}{2} = x, \text{ or } x = \frac{1}{2}(e^4+7)$$

57. $\log(2x-3) = \log(x+4)$

$$2x-3 = x+4$$
$$x = 7$$

59.

$$\log x + \log(x-48) = 2$$
$$\log x(x-48) = 2$$
$$x(x-48) = 10^2$$
$$x^2 - 48x - 100 = 0$$
$$(x-50)(x+2) = 0$$
$$x - 50 = 0 \quad \textbf{or} \quad x+2 = 0$$
$$x = 50 \qquad x = -2\text{: extraneous}$$

61.

$$\log x + \log(x-15) = 2$$
$$\log x(x-15) = 2$$
$$x(x-15) = 10^2$$
$$x^2 - 15x - 100 = 0$$
$$(x-20)(x+5) = 0$$
$$x - 20 = 0 \quad \textbf{or} \quad x+5 = 0$$
$$x = 20 \qquad x = -5\text{: extraneous}$$

63.

$$\log(x+90) = 3 - \log x$$
$$\log x + \log(x+90) = 3$$
$$\log x(x+90) = 3$$
$$x(x+90) = 10^3$$
$$x^2 + 90x - 1000 = 0$$
$$(x-10)(x+100) = 0$$
$$x - 10 = 0 \quad \textbf{or} \quad x+100 = 0$$
$$x = 10 \qquad x = -100$$

extraneous

65. $\log 5000 - \log(x-2) = 3$

$$\log \frac{5000}{x-2} = 3$$
$$\frac{5000}{x-2} = 10^3$$
$$5000 = 1000(x-2)$$
$$5000 = 1000x - 2000$$
$$7000 = 1000x$$
$$7 = x$$

SECTION 4.6

67.
$$\begin{aligned} \log_7 x + \log_7 (x-5) &= \log_7 6 \\ \log_7 x(x-5) &= \log_7 6 \\ x(x-5) &= 6 \\ x^2 - 5x - 6 &= 0 \\ (x-6)(x+1) &= 0 \end{aligned}$$
$x - 6 = 0$ **or** $x + 1 = 0$
$x = 6$ $\quad x = -1$
extraneous

69.
$$\begin{aligned} \ln 15 - \ln (x-2) &= \ln x \\ \ln \frac{15}{x-2} &= \ln x \\ \frac{15}{x-2} &= x \\ 15 &= x(x-2) \\ 0 &= x^2 - 2x - 15 \\ 0 &= (x-5)(x+3) \end{aligned}$$
$x - 5 = 0$ **or** $x + 3 = 0$
$x = 5$ $\quad x = -3$
extraneous

71.
$$\begin{aligned} \log_6 8 - \log_6 x &= \log_6 (x-2) \\ \log_6 \frac{8}{x} &= \log_6 (x-2) \\ \frac{8}{x} &= x - 2 \\ 8 &= x(x-2) \\ 0 &= x^2 - 2x - 8 \\ 0 &= (x-4)(x+2) \end{aligned}$$
$x - 4 = 0$ **or** $x + 2 = 0$
$x = 4$ $\quad x = -2$
extraneous

73.
$$\begin{aligned} \log (x-1) - \log 6 &= \log (x-2) - \log x \\ \log \tfrac{x-1}{6} &= \log \tfrac{x-2}{x} \\ \tfrac{x-1}{6} &= \tfrac{x-2}{x} \\ x(x-1) &= 6(x-2) \\ x^2 - x &= 6x - 12 \\ x^2 - 7x + 12 &= 0 \\ (x-3)(x-4) &= 0 \end{aligned}$$
$x - 3 = 0$ **or** $x - 4 = 0$
$x = 3$ $\quad x = 4$

75.
$$\begin{aligned} \log (\log x) &= 1 \\ \log x &= 10^1 \\ \log x &= 10 \\ x &= 10^{10} \end{aligned}$$

77.
$$\begin{aligned} \frac{\log (3x-4)}{\log x} &= 2 \\ \log (3x-4) &= 2 \log x \\ \log (3x-4) &= \log x^2 \\ 3x - 4 &= x^2 \\ 0 &= x^2 - 3x + 4 \end{aligned}$$
No real solutions

79.
$$\begin{aligned} \tfrac{\ln (5x+6)}{2} &= \ln x \\ \ln (5x+6) &= 2 \ln x \\ \ln (5x+6) &= \ln x^2 \\ 5x + 6 &= x^2 \\ 0 &= x^2 - 5x - 6 \\ 0 &= (x-6)(x+1) \end{aligned}$$
$x - 6 = 0$ **or** $x + 1 = 0$
$x = 6$ $\quad x = -1$
extraneous

81.
$$\begin{aligned} \log_3 x &= \log_3 \left(\frac{1}{x}\right) + 4 \\ \log_3 x &= \log_3 x^{-1} + 4 \\ \log_3 x &= -\log_3 x + 4 \\ 2 \log_3 x &= 4 \\ \log_3 x &= 2 \\ x &= 9 \end{aligned}$$

83.

$$
\begin{aligned}
2\log_2 x &= 3 + \log_2 (x-2)\\
\log_2 x^2 - \log_2 (x-2) &= 3\\
\log_2 \tfrac{x^2}{x-2} &= 3\\
\tfrac{x^2}{x-2} &= 2^3\\
x^2 &= 8(x-2)\\
x^2 - 8x + 16 &= 0\\
(x-4)(x-4) &= 0\\
x - 4 = 0 \quad \textbf{or} &\quad x - 4 = 0\\
x = 4 \qquad &\qquad x = 4
\end{aligned}
$$

85.

$$
\begin{aligned}
\ln (7y+1) &= 2\ln (y+3) - \ln 2\\
\ln (7y+1) &= \ln \tfrac{(y+3)^2}{2}\\
7y + 1 &= \tfrac{y^2+6y+9}{2}\\
14y + 2 &= y^2 + 6y + 9\\
0 &= y^2 - 8y + 7\\
0 &= (y-7)(y-1)\\
y - 7 = 0 \quad \textbf{or} &\quad y - 1 = 0\\
y = 7 \qquad &\qquad y = 1
\end{aligned}
$$

87. Graph $y = \log x + \log (x - 15)$ and $y = 2$ and find the x-coordinate of the point(s) of intersection: $x = 20$

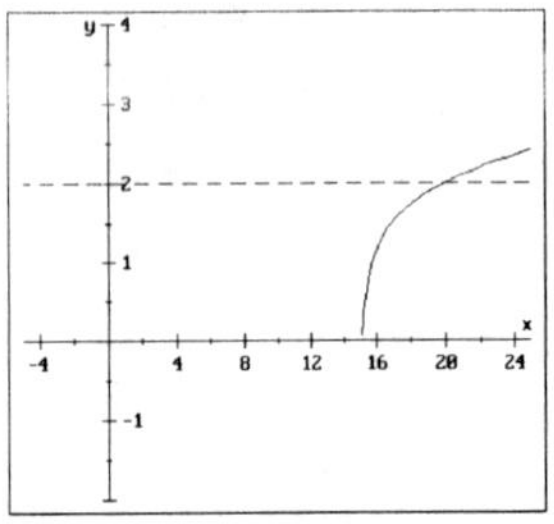

89. Graph $y = 2^{x+1}$ and $y = 7$ and find the x-coordinate of the point(s) of intersection: $x \approx 1.81$

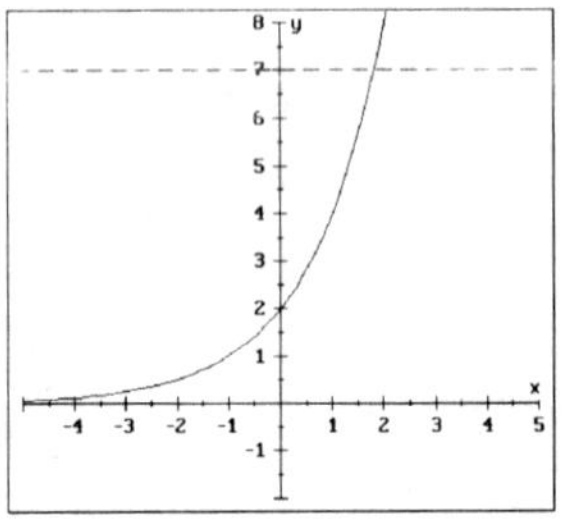

91.

$$
\begin{aligned}
A &= A_0 2^{-t/h}\\
0.75A_0 &= A_0 \cdot 2^{-t/(12.4)}\\
0.75 &= 2^{-t/12.4}\\
\log (0.75) &= \log \left(2^{-t/12.4}\right)\\
\log (0.75) &= -\frac{t}{12.4}\log 2\\
-\frac{12.4\log (0.75)}{\log 2} &= t\\
5.1 \text{ years} &\approx t
\end{aligned}
$$

93.

$$
\begin{aligned}
A &= A_0 2^{-t/h}\\
0.20A_0 &= A_0 \cdot 2^{-t/(18.4)}\\
0.20 &= 2^{-t/18.4}\\
\log (0.20) &= \log \left(2^{-t/18.4}\right)\\
\log (0.20) &= -\frac{t}{18.4}\log 2\\
-\frac{18.4\log (0.20)}{\log 2} &= t\\
42.7 \text{ days} &\approx t
\end{aligned}
$$

95.
$$\begin{aligned} A &= A_0 2^{-t/h} \\ 0.70A_0 &= A_0 \cdot 2^{-t/5700} \\ 0.70 &= 2^{-t/5700} \\ \log(0.70) &= \log\left(2^{-t/5700}\right) \\ \log(0.70) &= -\frac{t}{5700}\log 2 \\ -\frac{5700\log(0.70)}{\log 2} &= t \\ 2900 \text{ years} &\approx t \end{aligned}$$

97.
$$\begin{aligned} A &= P\left(1+\frac{r}{k}\right)^{kt} \\ 800 &= 500\left(1+\frac{0.085}{2}\right)^{2t} \\ \frac{8}{5} &= (1.0425)^{2t} \\ \log\left(\frac{8}{5}\right) &= \log(1.0425)^{2t} \\ \log 8 - \log 5 &= 2t\log(1.0425) \\ \frac{\log 8 - \log 5}{2\log(1.0425)} &= t \\ 5.6 \text{ years} &\approx t \end{aligned}$$

99.
$$\begin{aligned} A &= P\left(1+\frac{r}{k}\right)^{kt} \\ 2100 &= 1300\left(1+\frac{0.09}{4}\right)^{4t} \\ \frac{21}{13} &= (1.0225)^{4t} \\ \log\left(\frac{21}{13}\right) &= \log(1.0225)^{4t} \\ \log 21 - \log 13 &= 4t\log(1.0225) \\ \frac{\log 21 - \log 13}{4\log(1.0225)} &= t \\ 5.4 \text{ years} &\approx t \end{aligned}$$

101.
$$\begin{aligned} A &= Pe^{rt} \\ 2P &= Pe^{rt} \\ 2 &= e^{rt} \\ \ln 2 &= \ln e^{rt} \\ \ln 2 &= rt \\ \frac{\ln 2}{r} &= t \\ \frac{0.70}{r} &\approx t \\ \frac{70}{(100 \cdot r)\%} &\approx t \end{aligned}$$

103.
$$\begin{aligned} P &= P_0 a^t \\ 3P_0 &= P_0 a^5 \\ 3 &= a^5 \\ 3^{1/5} &= \left(a^5\right)^{1/5} \\ 3^{1/5} &= a \end{aligned}$$

$$\begin{aligned} P &= P_0 a^t \\ 2P_0 &= P_0\left(3^{1/5}\right)^t \\ 2 &= 3^{t/5} \\ \log 2 &= \log 3^{t/5} \\ \log 2 &= \frac{t}{5}\log 3 \\ \frac{5\log 2}{\log 3} &= t \\ 3.2 \text{ days} &\approx t \end{aligned}$$

105.
$$\begin{aligned} T &= 70 + 110e^{-0.2t} \\ 80 &= 70 + 110e^{-0.2t} \\ 10 &= 110e^{-0.2t} \\ \tfrac{1}{11} &= e^{-0.2t} \\ \ln\left(\tfrac{1}{11}\right) &= \ln e^{-0.2t} \\ \ln\left(\tfrac{1}{11}\right) &= -0.2t \\ \tfrac{\ln\left(\frac{1}{11}\right)}{-0.2} &= t \\ 12 &\approx t \Rightarrow t \approx 12 \text{ minutes} \end{aligned}$$

107.
$$\begin{aligned} T &= 60 + 40e^{kt} \\ 90 &= 60 + 40e^{k(3)} \\ 30 &= 40e^{3k} \\ 0.75 &= e^{3k} \\ \ln(0.75) &= \ln e^{3k} \\ \ln(0.75) &= 3k \\ \tfrac{\ln(0.75)}{3} &= k \end{aligned}$$

109.
$$\begin{aligned} T &= 300 - 300e^{kt} \\ 100 &= 300 - 300e^{k(5)} \\ -200 &= -300e^{5k} \\ \frac{2}{3} &= e^{5k} \\ \ln\left(\frac{2}{3}\right) &= \ln e^{5k} \\ \ln\left(\frac{2}{3}\right) &= 5k \\ \frac{\ln\left(\frac{2}{3}\right)}{5} &= k \end{aligned}$$

111. **Answers may vary.**

113.
$$\begin{aligned} P &= P_0e^{rt} \\ 3P_0 &= P_0e^{rt} \\ 3 &= e^{rt} \\ \ln 3 &= \ln e^{rt} \\ \ln 3 &= rt \\ \frac{\ln 3}{r} &= t \end{aligned}$$

115.
$$\begin{aligned} \log_2(\log_5(\log_7 x)) &= 2 \\ \log_5(\log_7 x) &= 2^2 = 4 \\ \log_7 x &= 5^4 = 625 \\ x &= 7^{625} \end{aligned}$$

117.
$$\begin{aligned} y &= 3x + 2 \\ x &= 3y + 2 \\ x - 2 &= 3y \\ \frac{x-2}{3} &= f^{-1}(x) = y \end{aligned}$$

119.
$$\begin{aligned} (f \circ g)(2) &= f(g(2)) \\ &= f(2^2) \\ &= f(4) \\ &= 5(4) - 1 \\ &= 19 \end{aligned}$$

121.
$$\begin{aligned} (f \circ g)(x) &= f(g(x)) \\ &= f(x^2) \\ &= 5x^2 - 1 \end{aligned}$$

Chapter 4 Review (page 486)

1. $5^{\sqrt{2}}5^{\sqrt{2}} = 5^{\sqrt{2}+\sqrt{2}} = 5^{2\sqrt{2}}$

2. $\left(2^{\sqrt{5}}\right)^{\sqrt{2}} = 2^{\sqrt{5}\cdot\sqrt{2}} = 2^{\sqrt{10}}$

3. $f(x) = 3^x$: $(0, 1)$, $(1, 3)$

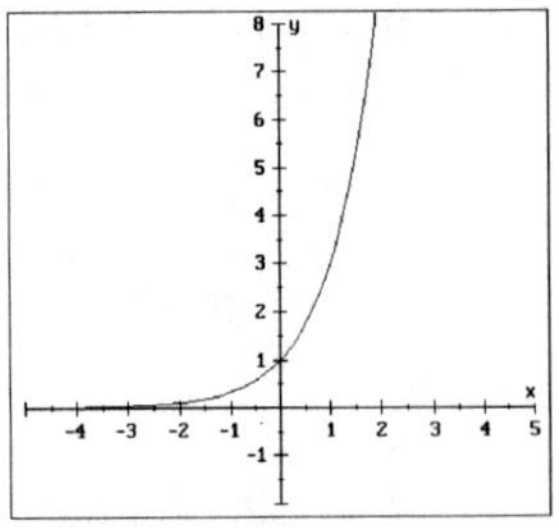

4. $f(x) = \left(\frac{1}{3}\right)^x$: $(0, 1)$, $\left(1, \frac{1}{3}\right)$

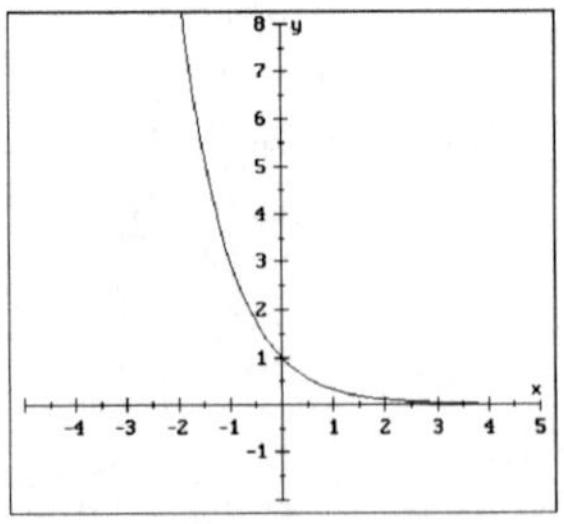

CHAPTER 4 REVIEW

5. $f(x) = 7^x$: goes through $(0, 1)$ and $(1, 7)$
$p = 1, q = 7$

6. $y = b^x$: domain $= (-\infty, \infty)$; range $= (0, \infty)$

7. $f(x) = \left(\frac{1}{2}\right)^x - 2$

Shift $y = \left(\frac{1}{2}\right)^x$ down 2:

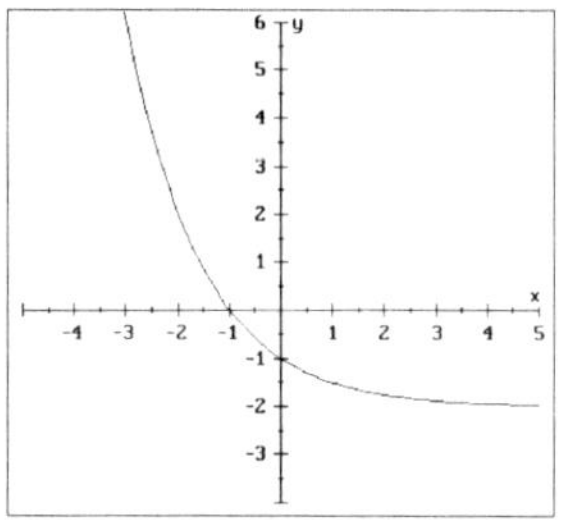

8. $f(x) = \left(\frac{1}{2}\right)^{x+2}$

Shift $y = \left(\frac{1}{2}\right)^x$ left 2:

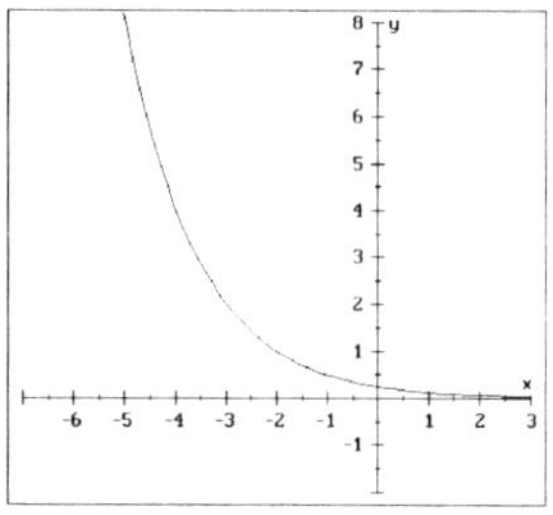

9. $f(x) = -5^x$

Reflect $y = 5^x$ about x-axis.

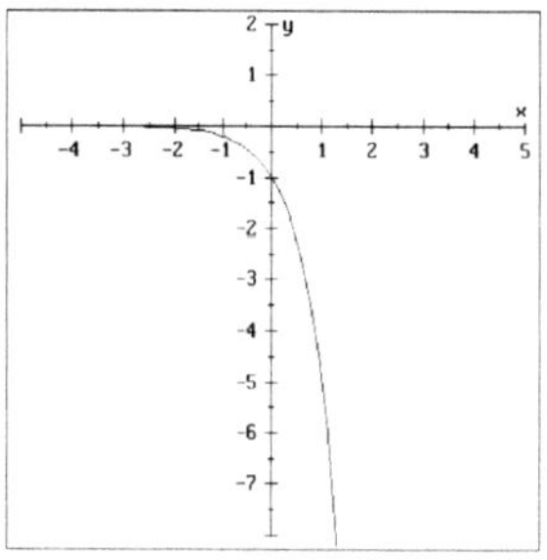

10. $f(x) = -5^x + 4$; Reflect $y = 5^x$ about x-axis.
Shift U4.

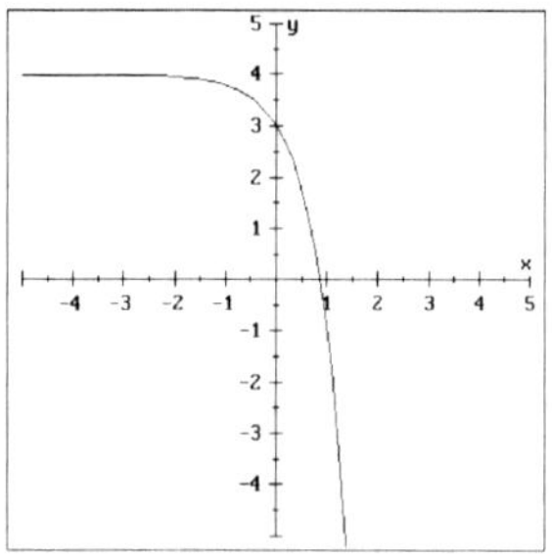

11. $f(x) = e^x + 1$

Shift $y = e^x$ up 1:

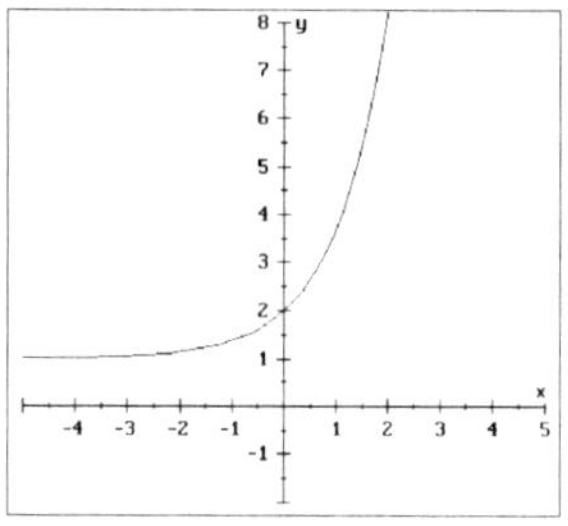

12. $f(x) = e^{x-3}$

Shift $y = e^x$ right 3:

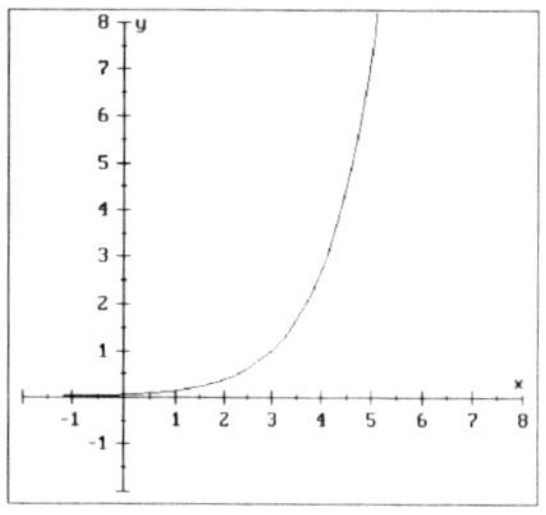

13.
$$\begin{aligned} A &= P\left(1 + \frac{r}{k}\right)^{kt} \\ &= 10{,}500\left(1 + \frac{0.09}{4}\right)^{4(60)} \\ &\approx \$2{,}189{,}703.45 \end{aligned}$$

14.
$$\begin{aligned} A &= Pe^{rt} \\ &= 10{,}500e^{0.09(60)} \\ &\approx 2{,}324{,}767.37 \end{aligned}$$

15.
$$\begin{aligned} A &= A_0 2^{-t/h} \\ &= A_0 \cdot 2^{-20/34.2} \\ &\approx 0.6667A_0 \end{aligned}$$
about $\frac{2}{3}$ of the original

16.
$$\begin{aligned} I &= I_0 k^x \\ &= 14(0.7)^{12} \\ &\approx 0.19 \text{ lumens} \end{aligned}$$

17. $P = P_0 e^{kt}$
$P = 300{,}000{,}000 e^{0.015(50)}$
$P \approx 635{,}000{,}000$ people

18. $P = \dfrac{450{,}000}{1 + (450 - 1)e^{-0.2t}}$
$= \dfrac{450{,}000}{1 + (450 - 1)e^{-0.2(5)}}$
≈ 2708 people

19. domain $= (0, \infty)$; range $= (-\infty, \infty)$

20. domain $= (0, \infty)$; range $= (-\infty, \infty)$

21. $\log_3 9 = ?$
$3^? = 9$
$\log_3 9 = 2$

22. $\log_9 \dfrac{1}{3} = ?$
$9^? = \dfrac{1}{3}$
$\log_9 \dfrac{1}{3} = -\dfrac{1}{2}$

23. $\log_x 1 = ?$
$x^? = 1$
$\log_x 1 = 0$

24. $\log_5 0.04 = ?$
$5^? = 0.04 = \dfrac{1}{25}$
$\log_5 0.04 = -2$

25. $\log_a \sqrt{a} = ?$
$a^? = \sqrt{a}$
$\log_a \sqrt{a} = \dfrac{1}{2}$

26. $\log_a \sqrt[3]{a} = ?$
$a^? = \sqrt[3]{a}$
$\log_a \sqrt[3]{a} = \dfrac{1}{3}$

27. $\log_2 x = 5$
$2^5 = x$
$32 = x$

28. $\log_{\sqrt{3}} x = 4$
$\left(\sqrt{3}\right)^4 = x$
$9 = x$

29. $\log_{\sqrt{2}} x = 6$
$\left(\sqrt{2}\right)^6 = x$
$8 = x$

30. $\log_{0.1} 10 = x$
$(0.1)^x = 10$
$\left(\dfrac{1}{10}\right)^x = 10$
$x = -1$

31. $\log_x 2 = -\frac{1}{3}$
$x^{-1/3} = 2$
$\left(x^{-1/3}\right)^{-3} = 2^{-3}$
$x = \frac{1}{8}$

32. $\log_x 32 = 5$
$x^5 = 32$
$x = 2$

33. $\log_{0.25} x = -1$
$(0.25)^{-1} = x$
$\left(\frac{1}{4}\right)^{-1} = x$
$4 = x$

34. $\log_{0.125} x = -\frac{1}{3}$
$(0.125)^{-1/3} = x$
$\left(\frac{1}{8}\right)^{-1/3} = x$
$2 = x$

35. $\log_{\sqrt{2}} 32 = x$
$\left(\sqrt{2}\right)^x = 32$
$\left(2^{1/2}\right)^x = 2^5$
$\dfrac{1}{2}x = 5$
$x = 10$

36. $\log_{\sqrt{5}} x = -4$

$\left(\sqrt{5}\right)^{-4} = x$

$\dfrac{1}{25} = x$

37. $\log_{\sqrt{3}} 9\sqrt{3} = x$

$\left(\sqrt{3}\right)^{x} = 9\sqrt{3}$

$\left(3^{1/2}\right)^{x} = 3^{5/2}$

$\dfrac{1}{2}x = \dfrac{5}{2}$

$x = 5$

38. $\log_{\sqrt{5}} 5\sqrt{5} = x$

$\left(\sqrt{5}\right)^{x} = 5\sqrt{5}$

$\left(5^{1/2}\right)^{x} = 5^{3/2}$

$\dfrac{1}{2}x = \dfrac{3}{2}$

$x = 3$

39. $f(x) = \log(x-2)$; Shift $y = \log x$ right 2:

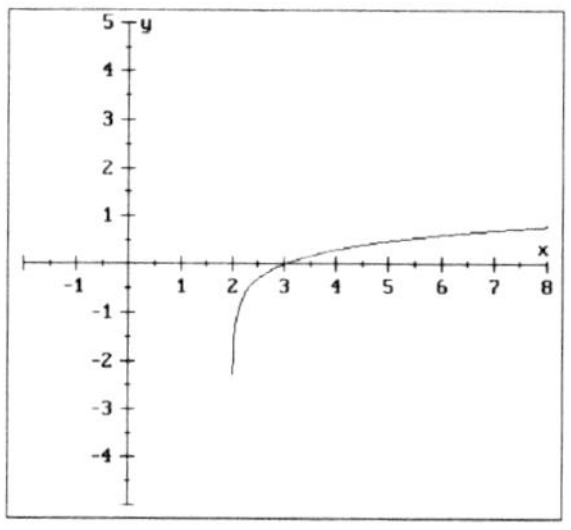

40. $f(x) = 3 + \log x$; Shift $y = \log x$ up 3:

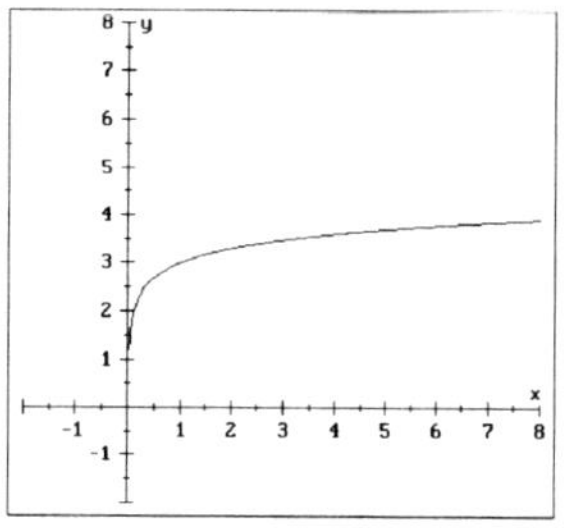

41. $y = 4^x$; $y = \log_4 x$

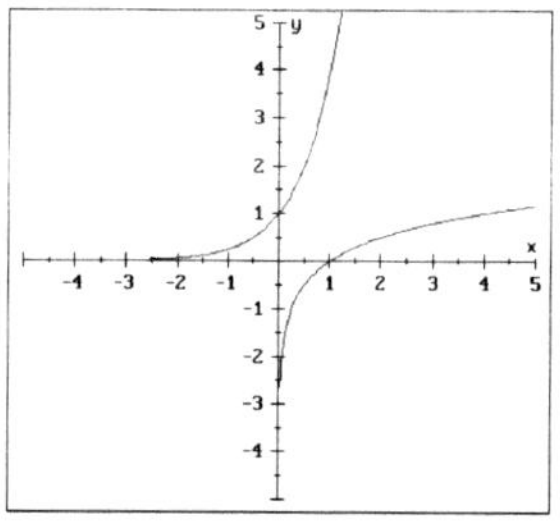

42. $y = \left(\frac{1}{3}\right)^x$; $y = \log_{1/3} x$

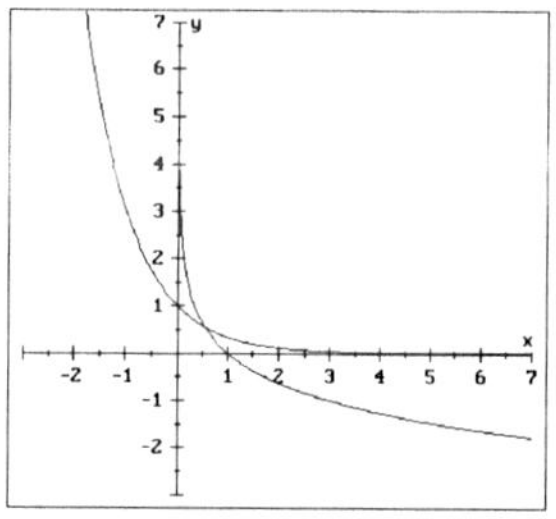

43. $\ln 452 \approx 6.1137$

44. $\ln(\log 7.85) \approx -0.1111$

45. $\ln x = 2.336$

$x \approx 10.3398$

46. $\ln x = \log 8.8$

$x \approx 2.5715$

47. $y = f(x) = 1 + \ln x$

Shift $y = \ln x$ up 1:

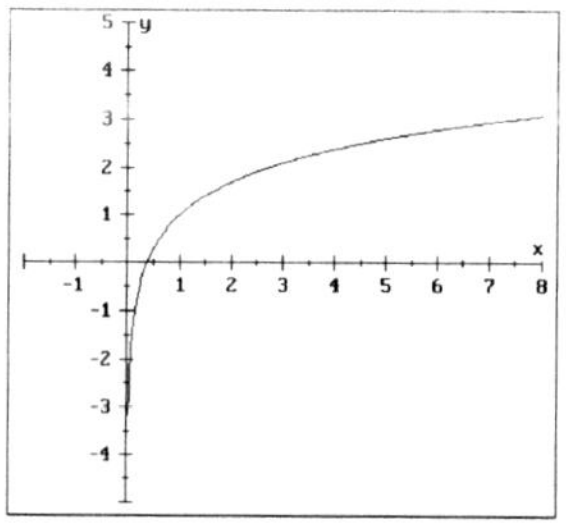

48. $y = f(x) = \ln(x+1)$

Shift $y = \ln x$ left 1:

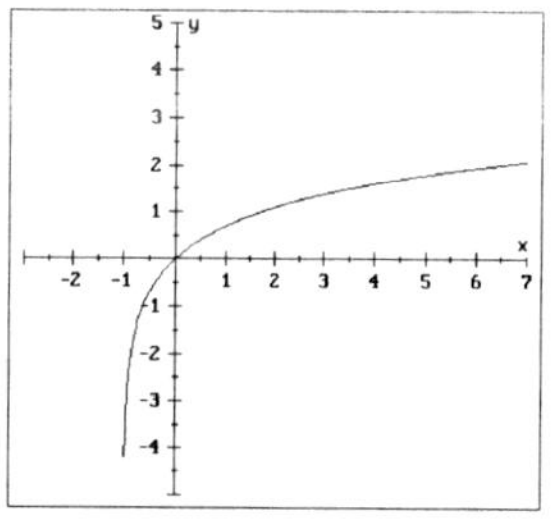

49. $\ln(e^{12}) = 12 \ln e = 12$

50. $e^{\ln 14x} = 14x$

51. $\text{db gain} = 20 \log \frac{E_0}{E_I}$
$= 20 \log \frac{18}{0.04}$
$\approx 53 \text{ db gain}$

52. $R = \log \frac{A}{P}$
$= \log \frac{7500}{0.3}$
≈ 4.4

53. $t = -\frac{1}{k} \ln \left(1 - \frac{C}{M}\right)$
$t = -\frac{1}{0.17} \ln \left(1 - \frac{0.8M}{M}\right)$
$t = -\frac{1}{0.17} \ln(1 - 0.8) \approx 9.5 \text{ minutes}$

54. $t = \frac{\ln 2}{r} = \frac{\ln 2}{0.03} \approx 23 \text{ years}$

55. $E = RT \ln \left(\frac{V_f}{V_i}\right) = (8.314)(350) \ln \left(\frac{2V_i}{V_i}\right) = (8.314)(350) \ln (2) \approx 2017 \text{ joules}$

56. $\log_7 1 = 0$

57. $\log_7 7 = 1$

58. $\log_7 7^3 = 3$

59. $7^{\log_7 4} = 4$

60. $\ln e^4 = 4 \ln e = 4$

61. $\ln 1 = 0$

62. $10^{\log_{10} 7} = 7$

63. $e^{\ln 3} = e^{\log_e 3} = 3$

64. $\log_b b^4 = 4 \log_b b = 4$

65. $\ln e^9 = 9 \ln e = 9$

66. $\log_b \frac{x^2 y^3}{z^4} = \log_b \left(x^2 y^3\right) - \log_b \left(z^4\right)$
$= \log_b x^2 + \log_b y^3 - \log_b z^4$
$= 2 \log_b x + 3 \log_b y - 4 \log_b z$

67. $\log_b \sqrt{\frac{x}{yz^2}} = \log_b \left(\frac{x}{yz^2}\right)^{1/2}$
$= \frac{1}{2} \log_b \left(\frac{x}{yz^2}\right)$
$= \frac{1}{2} (\log_b x - \log_b yz^2)$
$= \frac{1}{2} (\log_b x - \log_b y - 2 \log_b z)$

68. $\ln \frac{x^4}{y^5 z^6} = \ln x^4 - \ln y^5 z^6$
$= \ln x^4 - \left(\ln y^5 + \ln z^6\right)$
$= 4 \ln x - (5 \ln y + 6 \ln z)$
$= 4 \ln x - 5 \ln y - 6 \ln z$

69. $\ln \sqrt[3]{xyz} = \ln (xyz)^{1/3}$
$= \frac{1}{3} \ln xyz$
$= \frac{1}{3} (\ln x + \ln y + \ln z)$

70. $3 \log_b x - 5 \log_b y + 7 \log_b z = \log_b x^3 + \log_b y^{-5} + \log_b z^7 = \log_b \frac{x^3 z^7}{y^5}$

71. $\frac{1}{2} (\log_b x + 3 \log_b y) - 7 \log_b z = \frac{1}{2} \left(\log_b x + \log_b y^3\right) + \log_b z^{-7} = \frac{1}{2} \log_b xy^3 + \log_b z^{-7}$
$= \log_b \sqrt{xy^3} + \log_b z^{-7}$
$= \log_b \frac{\sqrt{xy^3}}{z^7}$

CHAPTER 4 REVIEW

72. $4\ln x - 5\ln y - 6\ln z = \ln x^4 + \ln y^{-5} + \ln z^{-6} = \ln \dfrac{x^4}{y^5 z^6}$

73. $\dfrac{1}{2}\ln x + 3\ln y - \dfrac{1}{3}\ln z = \ln x^{1/2} + \ln y^3 + \ln z^{-1/3} = \ln \dfrac{y^3\sqrt{x}}{\sqrt[3]{z}}$

74. $\log abc = \log a + \log b + \log c = 0.6 + 0.36 + 2.4 = 3.36$

75. $\log a^2 b = \log a^2 + \log b = 2\log a + \log b = 2(0.6) + 0.36 = 1.56$

76. $\log \dfrac{ac}{b} = \log a + \log c - \log b = 0.6 + 2.4 - 0.36 = 2.64$

77. $\log \dfrac{a^2}{c^3 b^2} = \log a^2 - \log c^3 b^2 = \log a^2 - \log c^3 - \log b^2 = 2\log a - 3\log c - 2\log b$

$= 2(0.6) - 3(2.4) - 2(0.36) = -6.72$

78. $\log_5 17 = \dfrac{\log 17}{\log 5} \approx 1.7604$

79.
$$\begin{aligned} \text{pH} &= -\log[\text{H}^+] \\ 3.1 &= -\log[\text{H}^+] \\ -3.1 &= \log[\text{H}^+] \\ 7.94 \times 10^{-4} &\approx [\text{H}^+] \end{aligned}$$

80. $L = k\ln I$

$k\ln \dfrac{I}{2} = k(\ln I - \ln 2) = k\ln I - k\ln 2 \Rightarrow$ The loudness decreases by $k\ln 2$.

81.
$$\begin{aligned} 81^{x+2} &= 27 \\ (3^4)^{x+2} &= 3^3 \\ 3^{4x+8} &= 3^3 \\ 4x + 8 &= 3 \\ 4x &= -5 \\ x &= -\frac{5}{4} \end{aligned}$$

82.
$$\begin{aligned} 2^{x^2+4x} &= \frac{1}{8} \\ 2^{x^2+4x} &= 2^{-3} \\ x^2 + 4x &= -3 \\ x^2 + 4x + 3 &= 0 \\ (x+1)(x+3) &= 0 \end{aligned}$$
$x + 1 = 0$ **or** $x + 3 = 0$

$x = -1$ $\quad x = -3$

83.
$$\begin{aligned} e^x &= e^{-6x+14} \\ x &= -6x + 14 \\ 7x &= 14 \\ x &= 2 \end{aligned}$$

84.
$$\begin{aligned} e^{2x^2} &= e^{18} \\ 2x^2 &= 18 \\ x^2 &= 9 \\ x &= \pm 3 \end{aligned}$$

85.
$$\begin{aligned} 3^x &= 7 \\ \log 3^x &= \log 7 \\ x\log 3 &= \log 7 \\ x &= \frac{\log 7}{\log 3} \\ x &\approx 1.7712 \end{aligned}$$

86.
$$\begin{aligned} 2^x &= 3^{x-1} \\ \log 2^x &= \log 3^{x-1} \\ x\log 2 &= (x-1)\log 3 \\ x\log 2 &= x\log 3 - \log 3 \\ x\log 2 - x\log 3 &= -\log 3 \\ x(\log 2 - \log 3) &= -\log 3 \\ x &= \frac{-\log 3}{\log 2 - \log 3} \\ x &\approx 2.7095 \end{aligned}$$

87.
$$\begin{aligned} 2e^x &= 16 \\ e^x &= 8 \\ \ln e^x &= \ln 8 \\ x\ln e &= \ln 8 \\ x &= \ln 8 \approx 2.0794 \end{aligned}$$

88.
$$\begin{aligned} -5e^x &= -35 \\ e^x &= 7 \\ \ln e^x &= \ln 7 \\ x\ln e &= \ln 7 \\ x &= \ln 7 \approx 1.9459 \end{aligned}$$

89.
$$\begin{aligned} \log_7(-7x+2) &= \log_7(3x+32) \\ -7x+2 &= 3x+32 \\ -30 &= 10x \\ -3 &= x \end{aligned}$$

90.
$$\begin{aligned} \ln(x+3) &= \ln(-5x+51) \\ x+3 &= -5x+51 \\ 6x &= 48 \\ x &= 8 \end{aligned}$$

91.
$$\begin{aligned} \log x + \log(29-x) &= 2 \\ \log x(29-x) &= 2 \\ x(29-x) &= 10^2 \\ -x^2+29x-100 &= 0 \\ x^2-29x+100 &= 0 \\ (x-25)(x-4) &= 0 \end{aligned}$$
$x-25=0$ **or** $x-4=0$

$x=25$ $\quad$ $x=4$

92.
$$\begin{aligned} \log_2 x + \log_2(x-2) &= 3 \\ \log_2 x(x-2) &= 3 \\ x(x-2) &= 2^3 \\ x^2-2x-8 &= 0 \\ (x-4)(x+2) &= 0 \end{aligned}$$
$x-4=0$ **or** $x+2=0$

$x=4$ $\quad$ $x=-2$

extraneous

93.
$$\begin{aligned} \log_2(x+2)+\log_2(x-1) &= 2 \\ \log_2(x+2)(x-1) &= 2 \\ (x+2)(x-1) &= 2^2 \\ x^2+x-2 &= 4 \\ x^2+x-6 &= 0 \\ (x-2)(x+3) &= 0 \end{aligned}$$
$x-2=0$ **or** $x+3=0$

$x=2$ $\quad$ $x=-3$

extraneous

94.
$$\begin{aligned} \frac{\log(7x-12)}{\log x} &= 2 \\ \log(7x-12) &= 2\log x \\ \log(7x-12) &= \log x^2 \\ 7x-12 &= x^2 \\ 0 &= x^2-7x+12 \\ 0 &= (x-3)(x-4) \\ x &= 3,\ x=4 \end{aligned}$$

95.

$$\begin{aligned} \ln x + \ln(x-5) &= \ln 6 \\ \ln x(x-5) &= \ln 6 \\ x(x-5) &= 6 \\ x^2 - 5x &= 6 \\ x^2 - 5x - 6 &= 0 \\ (x-6)(x+1) &= 0 \end{aligned}$$

$x - 6 = 0$ **or** $x + 1 = 0$

$x = 6$ $\quad x = -1$ extraneous

96.

$$\begin{aligned} \log 3 - \log(x-1) &= -1 \\ \log \frac{3}{x-1} &= -1 \\ \frac{3}{x-1} &= 10^{-1} \\ \frac{3}{x-1} &= \frac{1}{10} \\ 30 &= x - 1 \\ 31 &= x \end{aligned}$$

97.

$$\begin{aligned} e^{x \ln 2} &= 9 \\ \ln e^{x \ln 2} &= \ln 9 \\ x \ln 2 \ln e &= \ln 9 \\ x \ln 2 &= \ln 9 \\ x &= \frac{\ln 9}{\ln 2} \approx 3.1699 \end{aligned}$$

98.

$$\begin{aligned} \ln x &= \ln(x-1) \\ x &= x - 1 \end{aligned}$$

no solution

99.

$$\begin{aligned} \ln x - 3 &= 4 \\ \ln x &= 7 \\ x &= e^7 \approx 1096.6332 \end{aligned}$$

100.

$$\begin{aligned} \ln x &= \ln(x-1) + 1 \\ \ln x - \ln(x-1) &= 1 \\ \ln \frac{x}{x-1} &= 1 \\ \frac{x}{x-1} &= e^1 \\ \frac{x}{x-1} &= \frac{e}{1} \\ x &= e(x-1) \\ \frac{e}{e-1} &= x, \text{ or } x \approx 1.5820 \end{aligned}$$

101. Note: $\log_{10} x = \dfrac{\ln x}{\ln 10}$

$$\begin{aligned} \ln x &= \log_{10} x \\ \ln x &= \frac{\ln x}{\ln 10} \\ \ln x \ln 10 &= \ln x \\ \ln x \ln 10 - \ln x &= 0 \\ \ln x (\ln 10 - 1) &= 0 \\ \ln x &= 0 \Rightarrow x = 1 \end{aligned}$$

102.

$$\begin{aligned} A &= A_0 2^{-t/h} \\ \frac{2}{3} A_0 &= A_0 \cdot 2^{-t/5700} \\ \frac{2}{3} &= 2^{-t/5700} \\ \log(2/3) &= \log\left(2^{-t/5700}\right) \\ \log(2/3) &= -\frac{t}{5700} \log 2 \\ -\frac{5700 \log(2/3)}{\log 2} &= t \Rightarrow t \approx 3300 \text{ years} \end{aligned}$$

Chapter 4 Test (page 497)

1. $f(x) = 2^x + 1 \Rightarrow$ Shift $y = 2^x$ up 1.

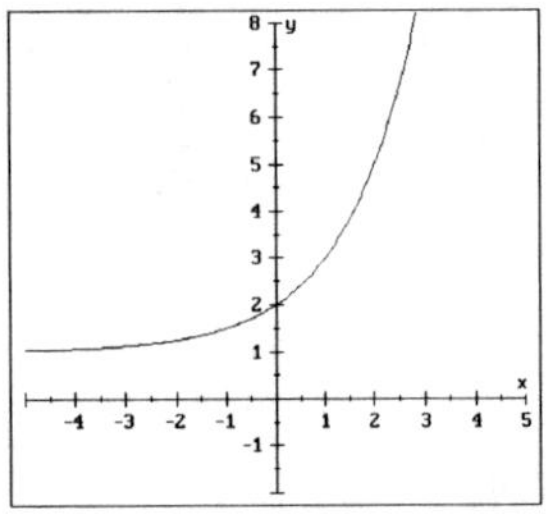

2. $f(x) = e^{x-2} \Rightarrow$ Shift $y = e^x$ right 2.

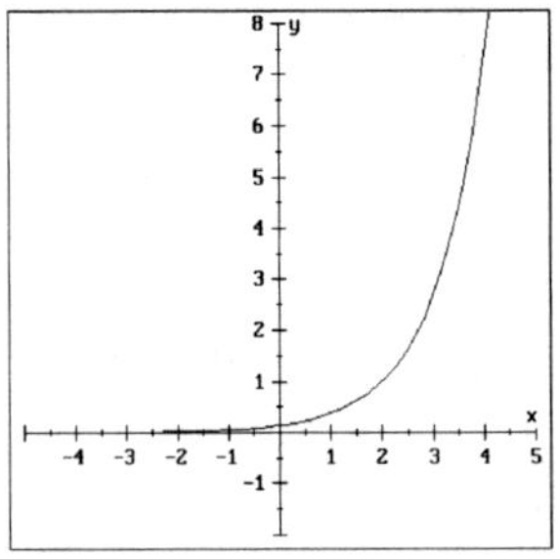

3. $A = 3(2)^{-6} = 3 \cdot \frac{1}{64} = \frac{3}{64}$ gram

4. $A = 1000\left(1 + \frac{0.06}{2}\right)^{2(1)} \approx \$1060.90$

5. $A = 2000e^{0.08(10)} \approx \$4451.08$

6. $\log_7 343 = \log_7 7^3 = 3$

7. $\log_3 \dfrac{1}{27} = \log_3 3^{-3} = -3$

8. $\log_{10} 10^{12} + 10^{\log_{10} 5} = 12 + 5 = 17$

9. $\log_{3/2} \frac{9}{4} = \log_{3/2} \left(\frac{3}{2}\right)^2 = 2$

10. $\log_{2/3} \frac{27}{8} = \log_{2/3} \left(\frac{2}{3}\right)^{-3} = -3$

11. $f(x) = \log(x-1)$; Shift $y = \log x$ right 1.

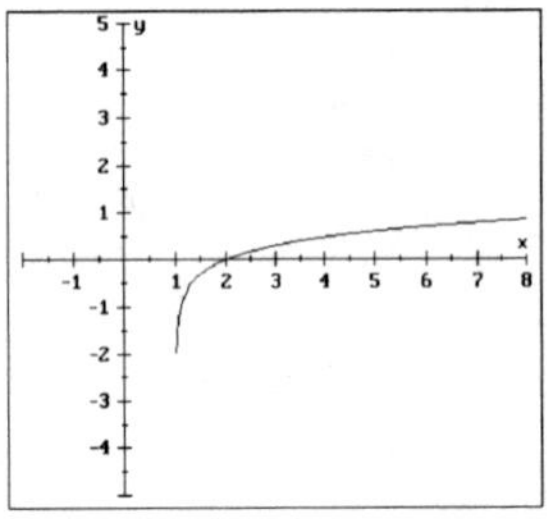

12. $f(x) = 2 + \ln x$; Shift $y = \ln x$ up 2.

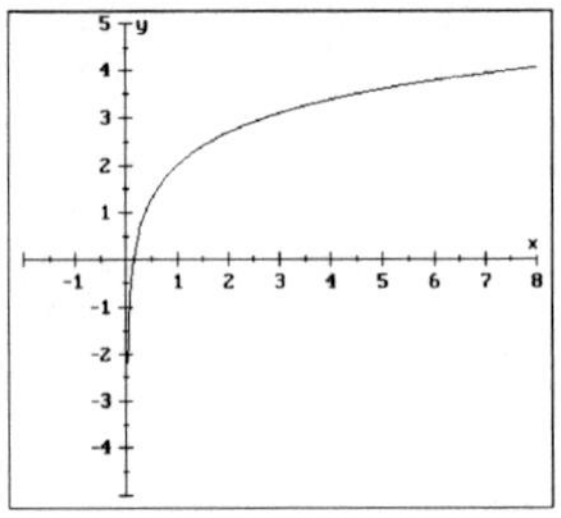

13. $\log a^2bc^3 = \log a^2 + \log b + \log c^3 = 2\log a + \log b + 3\log c$

14. $\ln\sqrt{\dfrac{a}{b^2c}} = \ln\left(\dfrac{a}{b^2c}\right)^{1/2} = \dfrac{1}{2}\ln\dfrac{a}{b^2c} = \dfrac{1}{2}\left(\ln a - \ln b^2 - \ln c\right) = \dfrac{1}{2}(\ln a - 2\ln b - \ln c)$

15. $\dfrac{1}{2}\log(a+2) + \log b - 2\log c = \log\sqrt{a+2} + \log b - \log c^2 = \log\dfrac{b\sqrt{a+2}}{c^2}$

16. $\dfrac{1}{3}(\ln a - 2\ln b) - \ln c = \dfrac{1}{3}\ln\dfrac{a}{b^2} - \ln c = \ln\sqrt[3]{\dfrac{a}{b^2}} - \ln c = \ln\dfrac{\sqrt[3]{\frac{a}{b^2}}}{c}$

17. $\log 24 = \log 8 \cdot 3 = \log 2^3 \cdot 3 = 3\log 2 + \log 3 = 3(0.3010) + 0.4771 = 1.3801$

CHAPTER 4 TEST

18. $\log \frac{8}{3} = \log \frac{2^3}{3} = 3\log 2 - \log 3 = 3(0.3010) - 0.4771 = 0.4259$

19. $\log_7 3 = \frac{\log 3}{\log 7}$ or $\frac{\ln 3}{\ln 7}$

20. $\log_\pi e = \frac{\log e}{\log \pi}$ or $\frac{\ln e}{\ln \pi} = \frac{1}{\ln \pi}$

21. $\log_a ab = \log_a a + \log_a b = 1 + \log_a b$

TRUE

22. $\log \frac{a}{b} = \log a - \log b$

FALSE

23.
$$\begin{aligned} \text{pH} &= -\log [\text{H}^+] \\ &= -\log \left(3.7 \times 10^{-7}\right) \\ &\approx 6.4 \end{aligned}$$

24.
$$\begin{aligned} \text{db gain} &= 20 \log \frac{E_O}{E_I} \\ &= 20 \log \frac{60}{0.3} \\ &\approx 46 \text{ db gain} \end{aligned}$$

25.
$$\begin{aligned} 3^{x^2-2x} &= 27 \\ 3^{x^2-2x} &= 3^3 \\ x^2 - 2x &= 3 \\ x^2 - 2x - 3 &= 0 \\ (x-3)(x+1) &= 0 \end{aligned}$$
$$x - 3 = 0 \quad \textbf{or} \quad x + 1 = 0$$
$$x = 3 \qquad\qquad x = -1$$

26.
$$\begin{aligned} 3^{x-1} &= 100^x \\ \log 3^{x-1} &= \log 100^x \\ (x-1)\log 3 &= x \log 100 \\ x\log 3 - \log 3 &= 2x \\ x \log 3 - 2x &= \log 3 \\ x(\log 3 - 2) &= \log 3 \\ x &= \frac{\log 3}{\log 3 - 2} \\ x &\approx -0.3133 \end{aligned}$$

27.
$$\begin{aligned} 5e^x &= 45 \\ e^x &= 9 \\ \ln e^x &= \ln 9 \\ x \ln e &= \ln 9 \\ x &= \ln 9 \approx 2.1972 \end{aligned}$$

28.
$$\begin{aligned} \ln(5x+2) &= \ln(2x+5) \\ 5x + 2 &= 2x + 5 \\ 3x &= 3 \\ x &= 1 \end{aligned}$$

29.
$$\begin{aligned} \log x + \log(x-9) &= 1 \\ \log x(x-9) &= 1 \\ x(x-9) &= 10^1 \\ x^2 - 9x - 10 &= 0 \\ (x-10)(x+1) &= 0 \end{aligned}$$
$$x - 10 = 0 \quad \textbf{or} \quad x + 1 = 0$$
$$x = 10 \qquad\qquad x = -1$$

extraneous

30.
$$\begin{aligned} \log_6 18 - \log_6 (x-3) &= \log_6 3 \\ \log_6 \frac{18}{x-3} &= \log_6 3 \\ \frac{18}{x-3} &= 3 \\ 18 &= 3x - 9 \\ 27 &= 3x \\ 9 &= x \end{aligned}$$

Exercises 5.1 (page 508)

1. whole

3. any

5. factor

7.

$$\begin{array}{r|l} & 4x^2 + 2x + 1 + \frac{2}{x-1} \\ \hline x-1 & 4x^3 - 2x^2 - x + 1 \\ & \underline{4x^3 - 4x^2} \\ & \quad 2x^2 - x \\ & \quad \underline{2x^2 - 2x} \\ & \qquad x + 1 \\ & \qquad \underline{x - 1} \\ & \qquad\quad 2 \end{array}$$

9.

$$\begin{array}{r|l} & 2x^3 - 3x^2 + 8x - 1 + \frac{-3}{x+2} \\ \hline x+2 & 2x^4 + x^3 + 2x^2 + 15x - 5 \\ & \underline{2x^4 + 4x^3} \\ & \quad -3x^3 + 2x^2 \\ & \quad \underline{-3x^3 - 6x^2} \\ & \qquad 8x^2 + 15x \\ & \qquad \underline{8x^2 + 16x} \\ & \qquad\quad -x - 5 \\ & \qquad\quad \underline{-x - 2} \\ & \qquad\qquad -3 \end{array}$$

11. $P(2) = 3(2)^3 - 2(2)^2 - 5(2) - 7$
$= 3(8) - 2(4) - 5(2) - 7$
$= 24 - 8 - 10 - 7 = \boxed{-1}$

$$\begin{array}{r|l} & 3x^2 + 4x + 3 \\ \hline x-2 & 3x^3 - 2x^2 - 5x - 7 \\ & \underline{3x^3 - 6x^2} \\ & \quad 4x^2 - 5x \\ & \quad \underline{4x^2 - 8x} \\ & \qquad 3x - 7 \\ & \qquad \underline{3x - 6} \\ & \qquad\quad \boxed{-1} \end{array}$$

13. $P(-1) = 7(-1)^4 + 2(-1)^3 + 5(-1)^2 - 1$
$= 7(1) + 2(-1) + 5(1) - 1$
$= 7 - 2 + 5 - 1 = \boxed{9}$

$$\begin{array}{r|l} & 7x^3 - 5x^2 + 10x - 10 \\ \hline x+1 & 7x^4 + 2x^3 + 5x^2 + 0x - 1 \\ & \underline{7x^4 + 7x^3} \\ & \quad -5x^3 + 5x^2 \\ & \quad \underline{-5x^3 - 5x^2} \\ & \qquad 10x^2 + 0x \\ & \qquad \underline{10x^2 + 10x} \\ & \qquad\quad -10x - 1 \\ & \qquad\quad \underline{-10x - 10} \\ & \qquad\qquad \boxed{9} \end{array}$$

15. $P(1) = 2(1)^5 + (1)^4 - (1)^3 - 2(1) + 3$
$= 2(1) + (1) - (1) - 2 + 3$
$= 2 + 1 - 1 - 2 + 3 = \boxed{3}$

$$\begin{array}{r|l} & 2x^4 + 3x^3 + 2x^2 + 2x \\ \hline x-1 & 2x^5 + x^4 - x^3 + 0x^2 - 2x + 3 \\ & \underline{2x^5 - 2x^4} \\ & \quad 3x^4 - x^3 \\ & \quad \underline{3x^4 - 3x^3} \\ & \qquad 2x^3 + 0x^2 \\ & \qquad \underline{2x^3 - 2x^2} \\ & \qquad\quad 2x^2 - 2x \\ & \qquad\quad \underline{2x^2 - 2x} \\ & \qquad\qquad 0x + \boxed{3} \end{array}$$

17. remainder $= P(-2) = 3(-2)^4 + 5(-2)^3 - 4(-2)^2 - 2(-2) + 1$
$= 3(16) + 5(-8) - 4(4) - 2(-2) + 1 = 48 - 40 - 16 + 4 + 1 = -3$

19. remainder $= P(2) = 3(2)^4 + 5(2)^3 - 4(2)^2 - 2(2) + 1$
$= 3(16) + 5(8) - 4(4) - 2(2) + 1 = 48 + 40 - 16 - 4 + 1 = 69$

21. remainder $= P(12) = 3(12)^4 + 5(12)^3 - 4(12)^2 - 2(12) + 1$
$= 3(20{,}736) + 5(1728) - 4(144) - 2(12) + 1$
$= 62{,}208 + 8640 - 576 - 24 + 1 = 70{,}249$

23. remainder $= P(-3.25) = 3(-3.25)^4 + 5(-3.25)^3 - 4(-3.25)^2 - 2(-3.25) + 1$
$= 3(111.56640625) + 5(-34.328125) - 4(10.5625) - 2(-3.25) + 1$
$= 334.69921875 - 171.640625 - 42.25 + 6.5 + 1 = 128.3085938$

25. $(x - 1)$ is a factor if $P(1) = 0$.
$P(1) = 1^7 - 1 = 0$: true

27. $(x - 1)$ is a factor if $P(1) = 0$.
$P(1) = 3(1)^5 + 4(1)^2 - 7 = 0$: true

29. $(x + 3)$ is a factor if $P(-3) = 0$. $P(-3) = 2(-3)^3 - 2(-3)^2 + 1 = -71$: false

31. $(x - 1)$ is a factor if $P(1) = 0$. $P(1) = (1)^{1984} - (1)^{1776} + (1)^{1492} - (1)^{1066} = 0$: true

33.
$$\begin{array}{r|rrrr} 1 & 3 & -2 & -6 & -4 \\ & & 3 & 1 & -5 \\ \hline & 3 & 1 & -5 & -9 \end{array}$$
$(x - 1)(3x^2 + x - 5) - 9$

35.
$$\begin{array}{r|rrrr} 3 & 3 & -2 & -6 & -4 \\ & & 9 & 21 & 45 \\ \hline & 3 & 7 & 15 & 41 \end{array}$$
$(x - 3)(3x^2 + 7x + 15) + 41$

37.
$$\begin{array}{r|rrrr} -1 & 3 & -2 & -6 & -4 \\ & & -3 & 5 & 1 \\ \hline & 3 & -5 & -1 & -3 \end{array}$$
$(x + 1)(3x^2 - 5x - 1) - 3$

39.
$$\begin{array}{r|rrrr} -3 & 3 & -2 & -6 & -4 \\ & & -9 & 33 & -81 \\ \hline & 3 & -11 & 27 & -85 \end{array}$$
$(x + 3)(3x^2 - 11x + 27) - 85$

41.
$$\begin{array}{r|rrrr} 1 & 1 & 1 & 1 & -3 \\ & & 1 & 2 & 3 \\ \hline & 1 & 2 & 3 & 0 \end{array}$$
$x^2 + 2x + 3$

43.
$$\begin{array}{r|rrrr} -1 & 7 & -3 & -5 & 1 \\ & & -7 & 10 & -5 \\ \hline & 7 & -10 & 5 & -4 \end{array}$$
$7x^2 - 10x + 5 + \dfrac{-4}{x + 1}$

45.
$$\begin{array}{r|rrrrr} 3 & 4 & -3 & 0 & -1 & 5 \\ & & 12 & 27 & 81 & 240 \\ \hline & 4 & 9 & 27 & 80 & 245 \end{array}$$
$4x^3 + 9x^2 + 27x + 80 + \dfrac{245}{x - 3}$

47.
$$\begin{array}{r|rrrrrr} 4 & 3 & 0 & 0 & 0 & -768 & 0 \\ & & 12 & 48 & 192 & 768 & 0 \\ \hline & 3 & 12 & 48 & 192 & 0 & 0 \end{array}$$
$3x^4 + 12x^3 + 48x^2 + 192x$

49.
$$\begin{array}{r|rrrr} 2 & 5 & 2 & -1 & 1 \\ & & 10 & 24 & 46 \\ \hline & 5 & 12 & 23 & 47 \end{array}$$
$P(2) = 47$

51.
$$\begin{array}{r|rrrr} -5 & 5 & 2 & -1 & 1 \\ & & -25 & 115 & -570 \\ \hline & 5 & -23 & 114 & -569 \end{array}$$
$P(-5) = -569$

53.

$$\begin{array}{r|rrrr} i & 5 & 2 & -1 & 1 \\ & & 5i & -5+2i & -2-6i \\ \hline & 5 & 2+5i & -6+2i & -1-6i \end{array}$$

$P(i) = -1 - 6i$

55.

$$\begin{array}{r|rrrrr} \frac{1}{2} & 2 & 0 & -1 & 0 & 2 \\ & & 1 & \frac{1}{2} & -\frac{1}{4} & -\frac{1}{8} \\ \hline & 2 & 1 & -\frac{1}{2} & -\frac{1}{4} & \frac{15}{8} \end{array} \quad P\left(\tfrac{1}{2}\right) = \tfrac{15}{8}$$

57.

$$\begin{array}{r|rrrrr} i & 2 & 0 & -1 & 0 & 2 \\ & & 2i & -2 & -3i & 3 \\ \hline & 2 & 2i & -3 & -3i & 5 \end{array} \quad P(i) = 5$$

59.

$$\begin{array}{r|rrrrr} 1 & 1 & -8 & 14 & 8 & -15 \\ & & 1 & -7 & 7 & 15 \\ \hline & 1 & -7 & 7 & 15 & 0 \end{array} \quad P(1) = 0$$

61.

$$\begin{array}{r|rrrrr} -3 & 1 & -8 & 14 & 8 & -15 \\ & & -3 & 33 & -141 & 399 \\ \hline & 1 & -11 & 47 & -133 & 384 \end{array} \quad P(-3) = 384$$

63.

$$\begin{array}{r|rrrrr} -i & 1 & -8 & 14 & 8 & -15 \\ & & -i & -1+8i & 8-13i & -13-16i \\ \hline & 1 & -8-i & 13+8i & 16-13i & -28-16i \end{array} \quad P(-i) = -28 - 16i$$

65.

$$\begin{array}{r|rrrrrr} i & 1 & 0 & -1 & -8 & 0 & 8 \\ & & i & -1 & -2i & 2-8i & 8+2i \\ \hline & 1 & i & -2 & -8-2i & 2-8i & 16+2i \end{array} \quad P(i) = 16 + 2i$$

67.

$$\begin{array}{r|rrrrrr} -2i & 1 & 0 & -1 & -8 & 0 & 8 \\ & & -2i & -4 & 10i & 20+16i & 32-40i \\ \hline & 1 & -2i & -5 & -8+10i & 20+16i & 40-40i \end{array} \quad P(-2i) = 40 - 40i$$

69. $x = -1$ is a solution, so $(x+1)$ is a factor. Use synthetic division to divide by $(x+1)$.

$$\begin{array}{r|rrrr} -1 & 1 & 3 & -13 & -15 \\ & & -1 & -2 & 15 \\ \hline & 1 & 2 & -15 & 0 \end{array}$$

$$x^3 + 3x^2 - 13x - 15 = 0$$
$$(x+1)\left(x^2 + 2x - 15\right) = 0$$
$$(x+1)(x+5)(x-3) = 0$$

Solution set: $\{-1, -5, 3\}$

71. $x = -\frac{1}{2}$ is a solution, so $\left(x + \frac{1}{2}\right)$ is a factor. Use synthetic division to divide by $\left(x + \frac{1}{2}\right)$.

$$\begin{array}{r|rrrr} -\frac{1}{2} & 2 & 1 & -18 & -9 \\ & & -1 & 0 & 9 \\ \hline & 2 & 0 & -18 & 0 \end{array}$$

$$2x^3 + x^2 - 18x - 9 = 0$$
$$\left(x + \tfrac{1}{2}\right)(2x^2 - 18) = 0$$
$$2\left(x + \tfrac{1}{2}\right)(x^2 - 9) = 0$$
$$2\left(x + \tfrac{1}{2}\right)(x+3)(x-3) = 0$$

Solution set: $\left\{-\frac{1}{2}, -3, 3\right\}$

73. $x = 2$ is a solution, so $(x-2)$ is a factor. Use synthetic division to divide by $(x-2)$.

$$\begin{array}{r|rrrr} 2 & 1 & -6 & 7 & 2 \\ & & 2 & -8 & -2 \\ \hline & 1 & -4 & -1 & 0 \end{array}$$

$$x^3 - 6x^2 + 7x + 2 = 0$$
$$(x-2)\left(x^2 - 4x - 1\right) = 0$$

Use the quadratic formula to finish.

Solution set: $\left\{2, 2 + \sqrt{5}, 2 - \sqrt{5}\right\}$

75. $x = -3$ is a solution, so $(x+3)$ is a factor.
Use synthetic division to divide by $(x+3)$.

-3	1	-3	1	57
		-3	18	-57
	1	-6	19	0

$$x^3 - 3x^2 + x + 57 = 0$$
$$(x+3)(x^2 - 6x + 19) = 0$$

Use the quadratic formula to finish.

Solution set: $\left\{-3, 3 \pm \sqrt{10}\, i\right\}$

77. $x = 1$ is a solution, so $(x-1)$ is a factor.
Use synthetic division to divide by $(x-1)$.

1	1	-2	-2	6	-3
		1	-1	-3	3
	1	-1	-3	3	0

$$x^4 - 2x^3 - 2x^2 + 6x - 3 = 0$$
$$(x-1)(x^3 - x^2 - 3x + 3) = 0$$

Use the fact that $x = 1$ is a double root and divide by $(x-1)$:

1	1	-1	-3	3
		1	0	-3
	1	0	-3	0

$$(x-1)(x^3 - x^2 - 3x + 3) = 0$$
$$(x-1)(x-1)(x^2 - 3) = 0$$

Solution set: $\left\{1, 1, \sqrt{3}, -\sqrt{3}\right\}$

79. $x = 2$ is a solution, so $(x-2)$ is a factor.
Use synthetic division to divide by $(x-2)$.

2	1	-5	7	-5	6
		2	-6	2	-6
	1	-3	1	-3	0

$$x^4 - 5x^3 + 7x^2 - 5x + 6 = 0$$
$$(x-2)(x^3 - 3x^2 + x - 3) = 0$$

$x = 3$ is a root, so $(x-3)$ is a factor.
Use synthetic division to divide by $(x-3)$.

3	1	-3	1	-3
		3	0	3
	1	0	1	0

$$(x-2)(x^3 - 3x^2 + x - 3) = 0$$
$$(x-2)(x-3)(x^2 + 1) = 0$$

$x^2 + 1 = 0 \Rightarrow x^2 = -1 \Rightarrow x = \pm i$

Solution set: $\{2, 3, \pm i\}$

81. $(x-4)(x-5) = x^2 - 9x + 20$

83. $(x-1)(x-1)(x-1) = (x^2 - 2x + 1)(x-1) = x^3 - 3x^2 + 3x - 1$

85. $(x-2)(x-4)(x-5) = (x^2 - 6x + 8)(x-5) = x^3 - 11x^2 + 38x - 40$

87. $(x-1)(x+1)\left(x - \sqrt{2}\right)\left(x + \sqrt{2}\right) = (x^2 - 1)(x^2 - 2) = x^4 - 3x^2 + 2$

89. $\left(x - \sqrt{2}\right)(x - i)(x + i) = \left(x - \sqrt{2}\right)(x^2 - i^2) = \left(x - \sqrt{2}\right)(x^2 + 1) = x^3 - \sqrt{2}x^2 + x - \sqrt{2}$

91.
$$\begin{aligned}(x-0)[x-(1+i)][x-(1-i)] &= x\left[x^2 - (1-i)x - (1+i)x + (1+i)(1-i)\right] \\ &= x\left[x^2 - x + ix - x - ix + 1 - i^2\right] \\ &= x\left[x^2 - 2x + 2\right] = x^3 - 2x^2 + 2x\end{aligned}$$

93.
$$\begin{aligned}P(0) &= 0 \\ a_n(0)^n + a_{n-1}(0)^{n-1} + \cdots + a_1(0) + a_0 &= 0 \\ 0 + 0 + \cdots + 0 + a_0 &= 0 \\ a_0 &= 0\end{aligned}$$

95. $P(2) = 0 \Rightarrow (x - 2)$ is a factor. $P(-2) = 0 \Rightarrow (x + 2)$ is a factor. The product of two factors will also be a factor, so $(x - 2)(x + 2) = x^2 - 4$ is a factor of the polynomial $P(x)$.

97. $P(3, -2)$: QIV

99. $R(8, \pi)$: QI

101. $d = \sqrt{(3 - (-5))^2 + (-3 - 3)^2} = \sqrt{(8)^2 + (-6)^2} = \sqrt{64 + 36} = \sqrt{100} = 10$

103. $m = \dfrac{-3 - 5}{-5 - 3} = \dfrac{-8}{-8} = 1$

Exercises 5.2 (page 518)

1. zero

3. conjugate

5. $(-x)^3 - (-x)^2 - 4 = -x^3 - x^2 - 4 \Rightarrow$ 0 variations

7. $7x^4 + 5x^3 - 2x + 1 \Rightarrow 7(-x)^4 + 5(-x)^3 - 2(-x) + 1$
$\Rightarrow 7x^4 - 5x^3 + 2x + 1 \Rightarrow$ 2 variations $\Rightarrow$ at most 2 negative roots

9. lower bound

11.
$x^{10} = 1$
$x^{10} - 1 = 0 \Rightarrow$ 10 roots

13.
$3x^4 - 4x^2 - 2x = -7$
$3x^4 - 4x^2 - 2x + 7 = 0 \Rightarrow$ 4 roots

15.
$x(3x^4 - 2) = 12x$
$3x^5 - 14x = 0$
5 total roots $\Rightarrow$ $\boxed{\text{4 other roots}}$

17. $P(x) = x^4 - 81$
4 linear factors, 4 zeros

19. $P(x) = 4x^5 + 8x^3$
5 linear factors, 5 zeros

21. If $2i$ is a root, then $-2i$ is a root also:
$(x - 2i)(x + 2i) = 0$
$x^2 - 4i^2 = 0$
$x^2 + 4 = 0$

23. If $3 - i$ is a root, then $3 + i$ is a root also:
$[x - (3 - i)][x - (3 + i)] = 0$
$x^2 - (3 + i)x - (3 - i)x + (3 - i)(3 + i) = 0$
$x^2 - 3x - ix - 3x + ix + 9 - i^2 = 0$
$x^2 - 6x + 10 = 0$

25. If $-i$ is a root, then i is a root also:
$(x - 3)(x + i)(x - i) = 0$
$(x - 3)(x^2 - i^2) = 0$
$(x - 3)(x^2 + 1) = 0$
$x^3 - 3x^2 + x - 3 = 0$

27. If $2+i$ is a root, then $2-i$ is a root also:

$$(x-2)[x-(2+i)][x-(2-i)]=0$$
$$(x-2)\left[x^2-(2-i)x-(2+i)x+(2+i)(2-i)\right]=0$$
$$(x-2)\left[x^2-2x+ix-2x-ix+4-i^2\right]=0$$
$$(x-2)\left[x^2-4x+5\right]=0$$
$$x^3-6x^2+13x-10=0$$

29. If i is a root, then $-i$ is a root also:

$$(x-3)(x-2)(x-i)(x+i)=0$$
$$\left(x^2-5x+6\right)\left(x^2-i^2\right)=0$$
$$\left(x^2-5x+6\right)\left(x^2+1\right)=0$$
$$x^4-5x^3+7x^2-5x+6=0$$

31. If i and $1-i$ are roots, then $-i$ and $1+i$ are roots also:

$$(x-i)(x+i)[x-(1-i)][x-(1+i)]=0$$
$$\left(x^2-i^2\right)\left[x^2-(1+i)x-(1-i)x+(1-i)(1+i)\right]=0$$
$$\left(x^2+1\right)\left[x^2-x-ix-x+ix+1-i^2\right]=0$$
$$\left(x^2+1\right)\left[x^2-2x+2\right]=0$$
$$x^4-2x^3+3x^2-2x+2=0$$

33. If $2i$ is a double root, then there are two factors of $(x-2i)$. [The problem does not specify real coefficients, so we do not include $-2i$ as a root.]

$$(x-2i)(x-2i)=0$$
$$x^2-4ix+4i^2=0$$
$$x^2-4ix-4=0$$

35. $P(x)=3x^3+5x^2-4x+3$

2 sign variations $\Rightarrow$ 2 or 0 positive roots

$$P(-x)=3(-x)^3+5(-x)^2-4(-x)+3$$
$$=-3x^3+5x^2+4x+3$$

1 sign variation $\Rightarrow$ 1 negative root

# pos	# neg	# nonreal
2	1	0
0	1	2

37. $P(x)=2x^3+7x^2+5x+5$

0 sign variations $\Rightarrow$ 0 positive roots

$$P(-x)=2(-x)^3+7(-x)^2+5(-x)+3$$
$$=-2x^3+7x^2-5x+3$$

3 sign variations $\Rightarrow$ 3 or 1 negative roots

# pos	# neg	# nonreal
0	3	0
0	1	2

39. $P(x)=8x^4+5$

0 sign variations $\Rightarrow$ 0 positive roots

$$P(-x)=8(-x)^4+5$$
$$=8x^4+5$$

0 sign variations $\Rightarrow$ 0 negative roots

# pos	# neg	# nonreal
0	0	4

SECTION 5.2

41. $P(x) = x^4 + 8x^2 - 5x - 10$: 1 sign variation $\Rightarrow$ 1 positive root

$P(-x) = (-x)^4 + 8(-x)^2 - 5(-x) - 10$

$= x^4 + 8x^2 + 5x - 10$: 1 sign variation $\Rightarrow$ 1 negative root

# pos	# neg	# nonreal
1	1	2

43. $P(x) = -x^{10} - x^8 - x^6 - x^4 - x^2 - 1$: 0 sign variations $\Rightarrow$ 0 positive roots

$P(-x) = -(-x)^{10} - (-x)^8 - (-x)^6 - (-x)^4 - (-x)^2 - 1$

$= -x^{10} - x^8 - x^6 - x^4 - x^2 - 1$: 0 sign variations $\Rightarrow$ 0 negative roots

# pos	# neg	# nonreal
0	0	10

45. $P(x) = x^9 + x^7 + x^5 + x^3 + x = x(x^8 + x^6 + x^4 + x^2 + 1)$: 0 sign variations $\Rightarrow$ 0 positive roots

$P(-x) = (-x)\left[(-x)^8 + (-x)^6 + (-x)^4 + (-x)^2 + 1\right]$

$= -x\left[x^8 + x^6 + x^4 + x^2 + 1\right]$: 0 sign variations $\Rightarrow$ 0 negative roots

# pos	# neg	# zero	# nonreal
0	0	1	8

47. $P(x) = -2x^4 - 3x^2 + 2x + 3$: 1 sign variation $\Rightarrow$ 1 positive root

$P(-x) = -2(-x)^4 - 3(-x)^2 + 2(-x) + 3$

$= -2x^4 - 3x^2 - 2x + 3$: 1 sign variation $\Rightarrow$ 1 negative root

# pos	# neg	# nonreal
1	1	2

49. $P(x) = x^2 - 2x - 4$

4	1	−2	−4
		4	8
	1	2	4

Upper bound: 4

−2	1	−2	−4
		−2	8
	1	−4	4

Lower bound: −2

51. $P(x) = 18x^2 - 6x - 1$

1	18	−6	−1
		18	12
	18	12	11

Upper bound: 1

−1	18	−6	−1
		−18	24
	18	−24	23

Lower bound: −1

53. $P(x) = 6x^3 - 13x^2 - 110x$

6	6	−13	−110	0
		36	138	168
	6	23	28	168

Upper bound: 6

−4	6	−13	−110	0
		−24	148	−152
	6	−37	38	−152

Lower bound: −4

55. $P(x) = x^5 + x^4 - 8x^3 - 8x^2 + 15x + 15$

3	1	1	−8	−8	15	15
		3	12	12	12	81
	1	4	4	4	27	96

Upper bound: 3

−4	1	1	−8	−8	15	15
		−4	12	−16	96	−444
	1	−3	4	−24	111	−429

Lower bound: −4

57. $$P(x) = 3x^5 - 11x^4 - 2x^3 + 38x^2 - 21x - 15$$

$$\begin{array}{r|rrrrrr} 4 & 3 & -11 & -2 & 38 & -21 & -15 \\ & & 12 & 4 & 8 & 184 & 652 \\ \hline & 3 & 1 & 2 & 46 & 163 & 637 \end{array}$$

Upper bound: 4

$$\begin{array}{r|rrrrrr} -2 & 3 & -11 & -2 & 38 & -21 & -15 \\ & & -6 & 34 & -64 & 52 & -62 \\ \hline & 3 & -17 & 32 & -26 & 31 & -77 \end{array}$$

Lower bound: -2

59. **Answers may vary.**

61. The number of nonreal roots must occur in conjugate pairs, so the number of nonreal roots will always be even. Since a polynomial of odd degree has an odd number of roots, at least one root must not be nonreal. Thus, at least one root of such a polynomial will be real.

63. k units to the right

65. reflected about the y-axis

67. stretched vertically by a factor of k

Exercises 5.3 (page 528)

1. -7

3. root

5. num: $\pm 1, \pm 2, \pm 3, \pm 4, \pm 6, \pm 12$; den: ± 1; possible roots: $\pm 1, \pm 2, \pm 3, \pm 4, \pm 6, \pm 12$

7. num: $\pm 1, \pm 2, \pm 3, \pm 6$; den: $\pm 1, \pm 2$; possible roots: $\pm 1, \pm 2, \pm 3, \pm 6, \pm \frac{1}{2}, \pm \frac{3}{2}$

9. num: $\pm 1, \pm 2, \pm 5, \pm 10$; den: $\pm 1, \pm 2, \pm 4$
possible roots: $\pm 1, \pm 2, \pm 5, \pm 10, \pm \frac{1}{2}, \pm \frac{5}{2}, \pm \frac{1}{4}, \pm \frac{5}{4}$

11. $$x^3 - 5x^2 - x + 5 = 0$$

Possible rational roots: $\pm 1, \pm 5$

Descartes' Rule of Signs:

# pos	# neg	# nonreal
2	1	0
0	1	2

Test $x = -1$:

$$\begin{array}{r|rrrr} -1 & 1 & -5 & -1 & 5 \\ & & -1 & 6 & -5 \\ \hline & 1 & -6 & 5 & 0 \end{array}$$

$$x^3 - 5x^2 - x + 5 = 0$$
$$(x+1)(x^2 - 6x + 5) = 0$$
$$(x+1)(x-5)(x-1) = 0$$

Solution set: $\{-1, 5, 1\}$

13. $$x^3 - 2x^2 - x + 2 = 0$$

Possible rational roots

$\pm 1, \pm 2$

Descartes' Rule of Signs

# pos	# neg	# nonreal
2	1	0
0	1	2

Test $x = -1$:

$$\begin{array}{r|rrrr} -1 & 1 & -2 & -1 & 2 \\ & & -1 & 3 & -2 \\ \hline & 1 & -3 & 2 & 0 \end{array}$$

$$x^3 - 2x^2 - x + 2 = 0$$
$$(x+1)(x^2 - 3x + 2) = 0$$
$$(x+1)(x-1)(x-2) = 0$$

Solution set: $\{-1, 1, 2\}$

SECTION 5.3

15. $x^3 - x^2 - 4x + 4 = 0$

Possible rational roots

$\pm 1, \ \pm 2, \ \pm 4$

Descartes' Rule of Signs

# pos	# neg	# nonreal
2	1	0
0	1	2

Test $x = 1$:

$$\begin{array}{r|rrrr} 1 & 1 & -1 & -4 & 4 \\ & & 1 & 0 & -4 \\ \hline & 1 & 0 & -4 & 0 \end{array}$$

$$x^3 - x^2 - 4x + 4 = 0$$
$$(x-1)\left(x^2-4\right) = 0$$
$$(x-1)(x+2)(x-2) = 0$$

Solution set: $\{1, -2, 2\}$

17. $x^3 - 2x^2 - 9x + 18 = 0$

Possible rational roots:

$\pm 1, \ \pm 2, \ \pm 3, \ \pm 6, \ \pm 9, \ \pm 18$

Descartes' Rule of Signs:

# pos	# neg	# nonreal
2	1	0
0	1	2

Test $x = 2$:

$$\begin{array}{r|rrrr} 2 & 1 & -2 & -9 & 18 \\ & & 2 & 0 & -18 \\ \hline & 1 & 0 & -9 & 0 \end{array}$$

$$x^3 - 2x^2 - 9x + 18 = 0$$
$$(x-2)\left(x^2-9\right) = 0$$
$$(x-2)(x+3)(x-3) = 0$$

Solution set: $\{2, -3, 3\}$

19. Possible rational roots

$\pm 1, \ \pm \frac{1}{2}$

Descartes' Rule of Signs

# pos	# neg	# nonreal
2	1	0
0	1	2

Test $x = 1$:

$$\begin{array}{r|rrrr} 1 & 2 & -1 & -2 & 1 \\ & & 2 & 1 & -1 \\ \hline & 2 & 1 & -1 & 0 \end{array}$$

$$2x^3 - x^2 - 2x + 1 = 0$$
$$(x-1)\left(2x^2+x-1\right) = 0$$
$$(x-1)(2x-1)(x+1) = 0$$

Solution set: $\left\{1, \frac{1}{2}, -1\right\}$

21. Possible rational roots:

$\pm 1, \ \pm \frac{1}{3}$

Descartes' Rule of Signs:

# pos	# neg	# nonreal
1	2	0
1	0	2

Test $x = -1$:

$$\begin{array}{r|rrrr} -1 & 3 & 5 & 1 & -1 \\ & & -3 & -2 & 1 \\ \hline & 3 & 2 & -1 & 0 \end{array}$$

$$3x^3 + 5x^2 + x - 1 = 0$$
$$(x+1)\left(3x^2+2x-1\right) = 0$$
$$(x+1)(3x-1)(x+1) = 0$$

Solution set: $\left\{-1, \frac{1}{3}, -1\right\}$

23. Possible rational roots

$\pm 1, \ \pm 2, \ \pm 3, \ \pm 6,$
$\pm 9, \ \pm 18, \ \pm \frac{1}{2}, \ \pm \frac{3}{2},$
$\pm \frac{9}{2}, \ \pm \frac{1}{3}, \ \pm \frac{2}{3}, \ \pm \frac{1}{5},$
$\pm \frac{2}{5}, \ \pm \frac{3}{5}, \ \pm \frac{6}{5}, \ \pm \frac{9}{5},$
$\pm \frac{18}{5}, \ \pm \frac{1}{6}, \ \pm \frac{1}{10}, \ \pm \frac{3}{10},$
$\pm \frac{9}{10}, \ \pm \frac{1}{15}, \ \pm \frac{2}{15}, \ \pm \frac{1}{30}$

Descartes' Rule of Signs

# pos	# neg	# nonreal
2	1	0
0	1	2

Test $x = \frac{2}{3}$:

$$\begin{array}{r|rrrr} \frac{2}{3} & 30 & -47 & -9 & 18 \\ & & 20 & -18 & -18 \\ \hline & 30 & -27 & -27 & 0 \end{array}$$

$$30x^3 - 47x^2 - 9x + 18 = 0$$
$$\left(x - \tfrac{2}{3}\right)\left(30x^2 - 27x - 27\right) = 0$$
$$3\left(x - \tfrac{2}{3}\right)\left(10x^2 - 9x - 9\right) = 0$$
$$(3x-2)(2x-3)(5x+3) = 0$$

Solution set: $\left\{\frac{2}{3}, \frac{3}{2}, -\frac{3}{5}\right\}$

SECTION 5.3

25.

Possible rational roots

$\pm 1, \pm 2, \pm 3, \pm 4,$
$\pm 6, \pm 8, \pm 12, \pm 24,$
$\pm \frac{1}{3}, \pm \frac{2}{3}, \pm \frac{4}{3}, \pm \frac{8}{3},$
$\pm \frac{1}{5}, \pm \frac{2}{5}, \pm \frac{3}{5}, \pm \frac{4}{5},$
$\pm \frac{6}{5}, \pm \frac{8}{5}, \pm \frac{12}{5}, \pm \frac{24}{5},$
$\pm \frac{1}{15}, \pm \frac{2}{15}, \pm \frac{4}{15}, \pm \frac{8}{15}$

Descartes' Rule of Signs

# pos	# neg	# nonreal
2	1	0
0	1	2

Test $x = 4$:

$$\begin{array}{r|rrrr} 4 & 15 & -61 & -2 & 24 \\ & & 60 & -4 & -24 \\ \hline & 15 & -1 & -6 & 0 \end{array}$$

$$\begin{aligned} 15x^3 - 61x^2 - 2x + 24 &= 0 \\ (x-4)\left(15x^2 - x - 6\right) &= 0 \\ (x-4)(3x-2)(5x+3) &= 0 \end{aligned}$$

Solution set: $\left\{4, \frac{2}{3}, -\frac{3}{5}\right\}$

27.

Possible rational roots

$\pm 1, \pm 2, \pm 3, \pm 5,$
$\pm 6, \pm 10, \pm 15, \pm 30,$
$\pm \frac{1}{2}, \pm \frac{3}{2}, \pm \frac{5}{2}, \pm \frac{15}{2},$
$\pm \frac{1}{3}, \pm \frac{2}{3}, \pm \frac{5}{3}, \pm \frac{10}{3},$
$\pm \frac{1}{4}, \pm \frac{3}{4}, \pm \frac{5}{4}, \pm \frac{15}{4},$
$\pm \frac{1}{6}, \pm \frac{5}{6}, \pm \frac{1}{8}, \pm \frac{3}{8},$
$\pm \frac{5}{8}, \pm \frac{15}{8}, \pm \frac{1}{12}, \pm \frac{5}{12},$
$\pm \frac{1}{24}, \pm \frac{5}{24}$

Descartes' Rule of Signs

# pos	# neg	# nonreal
3	0	0
1	0	2

Test $x = \frac{3}{2}$:

$$\begin{array}{r|rrrr} \frac{3}{2} & 24 & -82 & 89 & -30 \\ & & 36 & -69 & 30 \\ \hline & 24 & -46 & 20 & 0 \end{array}$$

$$\begin{aligned} 24x^3 - 82x^2 + 89x - 30 &= 0 \\ \left(x - \tfrac{3}{2}\right)\left(24x^2 - 46x + 20\right) &= 0 \\ 2\left(x - \tfrac{3}{2}\right)\left(12x^2 - 23x + 10\right) &= 0 \\ (2x-3)(4x-5)(3x-2) &= 0 \end{aligned}$$

Solution set: $\left\{\frac{3}{2}, \frac{5}{4}, \frac{2}{3}\right\}$

29.

Possible rational roots

$\pm 1, \pm 2, \pm 3, \pm 4,$
$\pm 6, \pm 8, \pm 12, \pm 24$

Descartes' Rule of Signs

# pos	# neg	# nonreal
4	0	0
2	0	2
0	0	4

Test $x = 1$:

$$\begin{array}{r|rrrrr} 1 & 1 & -10 & 35 & -50 & 24 \\ & & 1 & -9 & 26 & -24 \\ \hline & 1 & -9 & 26 & -24 & 0 \end{array}$$

Test $x = 2$:

$$\begin{array}{r|rrrr} 2 & 1 & -9 & 26 & -24 \\ & & 2 & -14 & 24 \\ \hline & 1 & -7 & 12 & 0 \end{array}$$

$$\begin{aligned} x^4 - 10x^3 + 35x^2 - 50x + 24 &= 0 \\ (x-1)\left(x^3 - 9x^2 + 26x - 24\right) &= 0 \\ (x-1)(x-2)\left(x^2 - 7x + 12\right) &= 0 \\ (x-1)(x-2)(x-3)(x-4) &= 0 \end{aligned}$$

Solution set: $\{1, 2, 3, 4\}$

31.

Possible rational roots

$\pm 1, \pm 2, \pm 3, \pm 5,$
$\pm 6, \pm 10, \pm 15,$
± 30

Descartes' Rule of Signs

# pos	# neg	# nonreal
2	2	0
2	0	2
0	2	2
0	0	4

Test $x = 2$:

$$\begin{array}{r|rrrrr} 2 & 1 & 3 & -13 & -9 & 30 \\ & & 2 & 10 & -6 & -30 \\ \hline & 1 & 5 & -3 & -15 & 0 \end{array}$$

Test $x = -5$:

$$\begin{array}{r|rrrr} -5 & 1 & 5 & -3 & -15 \\ & & -5 & 0 & 15 \\ \hline & 1 & 0 & -3 & 0 \end{array}$$

$$\begin{aligned} x^4 + 3x^3 - 13x^2 - 9x + 30 &= 0 \\ (x-2)\left(x^3 + 5x^2 - 3x - 15\right) &= 0 \\ (x-2)(x+5)\left(x^2 - 3\right) &= 0 \end{aligned}$$

$x^2 - 3$ does not factor rationally.

Rational solutions: $\{2, -5\}$

SECTION 5.3

33. Possible rational roots

$\pm 1,\ \pm 3,\ \pm \frac{1}{2},\ \pm \frac{3}{2},\ \pm \frac{1}{4}, \pm \frac{3}{4}$

Descartes' Rule of Signs

# pos	# neg	# nonreal
3	1	0
1	1	2

$4x^4 - 8x^3 - x^2 + 8x - 3 = 0$

$(x+1)\left(4x^3 - 12x^2 + 11x + 3\right) = 0$

$(x+1)(x-1)\left(4x^2 - 8x + 3\right) = 0$

$(x+1)(x-1)(2x-3)(2x-1) = 0$

Solution set: $\left\{-1, 1, \frac{3}{2}, \frac{1}{2}\right\}$

Test $x = -1$:

-1	4	-8	-1	8	-3
		-4	12	-11	3
	4	-12	11	-3	0

Test $x = 1$:

1	4	-12	11	-3
		4	-8	3
	4	-8	3	0

35. Possible rational roots

$\pm 1,\ \pm 2,\ \pm 4,\ \pm 5,\ \pm 8,\ \pm 10, \pm 20,\ \pm 40, \pm \frac{1}{2}, \pm \frac{5}{2}$

Descartes' Rule of Signs

# pos	# neg	# nonreal
1	3	0
1	1	2

$2x^4 - x^3 - 2x^2 - 4x - 40 = 0$

$(x+2)\left(2x^3 - 5x^2 + 8x - 20\right) = 0$

$(x+2)\left(x - \frac{5}{2}\right)\left(2x^2 + 8\right) = 0$

$2x^2 + 8$ does not factor rationally.

Rational Solutions: $\left\{-2, \frac{5}{2}\right\}$

Test $x = -2$:

-2	2	-1	-2	-4	-40
		-4	10	-16	40
	2	-5	8	-20	0

Test $x = \frac{5}{2}$:

$\frac{5}{2}$	2	-5	8	-20
		5	0	20
	2	0	8	0

37. Possible rational roots

$\pm 1,\ \pm \frac{1}{2},\ \pm \frac{1}{3},\ \pm \frac{1}{4},\ \pm \frac{1}{6},\ \pm \frac{1}{9}, \pm \frac{1}{12}, \pm \frac{1}{18},\ \pm \frac{1}{36}$

Descartes' Rule of Signs

# pos	# neg	# nonreal
3	1	0
1	1	2

$36x^4 - x^2 + 2x - 1 = 0$

$\left(x + \frac{1}{2}\right)\left(36x^3 - 18x^2 + 8x - 2\right) = 0$

$\left(x + \frac{1}{2}\right)\left(x - \frac{1}{3}\right)\left(36x^2 - 6x + 6\right) = 0$

$36x^2 - 6x + 6$ does not factor rationally. Rational Solutions: $\left\{-\frac{1}{2}, \frac{1}{3}\right\}$

Test $x = -\frac{1}{2}$:

$-\frac{1}{2}$	36	0	-1	2	-1
		-18	9	-4	1
	36	-18	8	-2	0

Test $x = \frac{1}{3}$:

$\frac{1}{3}$	36	-18	8	-2
		12	-2	2
	36	-6	6	0

39. Possible rat. roots

$\pm 1,\ \pm 2,\ \pm 3,\ \pm 4,\ \pm 6,\ \pm 12$

Descartes' Rule of Signs

# pos	# neg	# nonreal
2	3	0
2	1	2
0	3	2
0	1	4

$x^5 + 3x^4 - 5x^3 - 15x^2 + 4x + 12 = 0$

$(x+1)\left(x^4 + 2x^3 - 7x^2 - 8x + 12\right) = 0$

$(x+1)(x-1)\left(x^3 + 3x^2 - 4x - 12\right) = 0$

$(x+1)(x-1)(x-2)\left(x^2 + 5x + 6\right) = 0$

$(x+1)(x-1)(x-2)(x+2)(x+3) = 0$

Solution set: $\{-1, 1, 2, -2, -3\}$

Test $x = -1$:

-1	1	3	-5	-15	4	12
		-1	-2	7	8	-12
	1	2	-7	-8	12	0

Test $x = 1$:

1	1	2	-7	-8	12
		1	3	-4	-12
	1	3	-4	-12	0

Test $x = 2$:

2	1	3	-4	-12
		2	10	12
	1	5	6	0

41. Possible rat. roots

$\pm 1, \pm 2, \pm 3, \pm 4, \pm 6, \pm 12, \pm \frac{1}{2}, \pm \frac{3}{2}, \pm \frac{1}{4}, \pm \frac{3}{4}$

Descartes' Rule of Signs

# pos	# neg	# nonreal
4	1	0
2	1	2
0	1	4

Test $x = 3$:

$$\begin{array}{r|rrrrrr} 3 & 4 & -12 & 15 & -45 & -4 & 12 \\ & & 12 & 0 & 45 & 0 & -12 \\ \hline & 4 & 0 & 15 & 0 & -4 & 0 \end{array}$$

Test $x = -\frac{1}{2}$:

$$\begin{array}{r|rrrrr} -\frac{1}{2} & 4 & 0 & 15 & 0 & -4 \\ & & -2 & 1 & -8 & 4 \\ \hline & 4 & -2 & 16 & -8 & 0 \end{array}$$

Test $x = \frac{1}{2}$:

$$\begin{array}{r|rrrr} \frac{1}{2} & 4 & -2 & 16 & -8 \\ & & 2 & 0 & 8 \\ \hline & 4 & 0 & 16 & 0 \end{array}$$

$$4x^5 - 12x^4 + 15x^3 - 45x^2 - 4x + 12 = 0$$
$$(x-3)(4x^4 + 15x^2 - 4) = 0$$
$$(x-3)(x+\tfrac{1}{2})(4x^3 - 2x^2 + 16x - 8) = 0$$
$$(x-3)(x+\tfrac{1}{2})(x-\tfrac{1}{2})(4x^2 + 16) = 0$$

$4x^2 + 16$ does not factor rationally. Rational Solutions: $\left\{3, -\frac{1}{2}, \frac{1}{2}\right\}$

43. First, factor out the common factor of x: $x^7 - 12x^5 + 48x^3 - 64x = x(x^6 - 12x^4 + 48x^2 - 64)$

Possible rat. roots

$\pm 1, \pm 2, \pm 4, \pm 8, \pm 16, \pm 32, \pm 64$

Descartes' Rule of Signs

# pos	# neg	# zero	# nonreal
3	3	1	0
3	1	1	2
1	3	1	2
1	1	1	4

Test $x = 2$:

$$\begin{array}{r|rrrrrrr} 2 & 1 & 0 & -12 & 0 & 48 & 0 & -64 \\ & & 2 & 4 & -16 & -32 & 32 & 64 \\ \hline & 1 & 2 & -8 & -16 & 16 & 32 & 0 \end{array}$$

Test $x = 2$:

$$\begin{array}{r|rrrrrr} 2 & 1 & 2 & -8 & -16 & 16 & 32 \\ & & 2 & 8 & 0 & -32 & -32 \\ \hline & 1 & 4 & 0 & -16 & -16 & 0 \end{array}$$

Test $x = 2$:

$$\begin{array}{r|rrrrr} 2 & 1 & 4 & 0 & -16 & -16 \\ & & 2 & 12 & 24 & 16 \\ \hline & 1 & 6 & 12 & 8 & 0 \end{array}$$

Test $x = -2$:

$$\begin{array}{r|rrrr} -2 & 1 & 6 & 12 & 8 \\ & & -2 & -8 & -8 \\ \hline & 1 & 4 & 4 & 0 \end{array}$$

$$x^7 - 12x^5 + 48x^3 - 64x = 0$$
$$x(x^6 - 12x^4 + 48x^2 - 64) = 0$$
$$x(x-2)(x^5 + 2x^4 - 8x^3 - 16x^2 + 16x + 32) = 0$$
$$x(x-2)(x-2)(x^4 + 4x^3 - 16x - 16) = 0$$
$$x(x-2)(x-2)(x-2)(x^3 + 6x^2 + 12x + 8) = 0$$
$$x(x-2)(x-2)(x-2)(x+2)(x^2 + 4x + 4) = 0$$
$$x(x-2)(x-2)(x-2)(x+2)(x+2)(x+2) = 0 \Rightarrow \text{Sol. set} = \{0, 2, 2, 2, -2, -2, -2\}$$

45.

Possible rational roots

$\pm 1, \pm 2, \pm 3, \pm 6$

Test $x = 3$:

$$\begin{array}{r|rrrr} 3 & 1 & -3 & -2 & 6 \\ & & 3 & 0 & -6 \\ \hline & 1 & 0 & -2 & 0 \end{array}$$

Descartes' Rule of Signs

# pos	# neg	# nonreal
2	1	0
0	1	2

$x^3 - 3x^2 - 2x + 6 = 0$

$(x-3)(x^2-2) = 0$

$x - 3 = 0$ **or** $x^2 - 2 = 0$

$x = 3$ $\qquad x = \pm\sqrt{2}$

Solution set: $\left\{3, -\sqrt{2}, \sqrt{2}\right\}$

47.

Possible rational roots

$\pm 1, \pm \frac{1}{2}$

Test $x = \frac{1}{2}$:

$$\begin{array}{r|rrrr} \frac{1}{2} & 2 & -1 & 2 & -1 \\ & & 1 & 0 & 1 \\ \hline & 2 & 0 & 2 & 0 \end{array}$$

Descartes' Rule of Signs

# pos	# neg	# nonreal
3	0	0
1	0	2

$2x^3 - x^2 + 2x - 1 = 0$

$\left(x - \frac{1}{2}\right)(2x^2 + 2) = 0$

$x - \frac{1}{2} = 0$ **or** $2x^2 + 2 = 0$

$x = \frac{1}{2}$ $\qquad x = \pm\sqrt{-1}$

$x = \pm i$

Solution set: $\left\{\frac{1}{2}, -i, i\right\}$

49.

Possible rational roots

$\pm 1, \pm 2, \pm 4, \pm 8$

± 16

Descartes' Rule of Signs

# pos	# neg	# nonreal
2	2	0
2	0	2
0	2	2
0	0	4

$x^4 - 2x^3 - 8x^2 + 8x + 16 = 0$

$(x-2)(x^3 - 8x - 8) = 0$

$(x-2)(x+2)(x^2 - 2x - 4) = 0$

Use the quadratic formula.

Solution set: $\left\{2, -2, 1 \pm \sqrt{5}\right\}$

Test $x = 2$:

$$\begin{array}{r|rrrrr} 2 & 1 & -2 & -8 & 8 & 16 \\ & & 2 & 0 & -16 & -16 \\ \hline & 1 & 0 & -8 & -8 & 0 \end{array}$$

Test $x = -2$:

$$\begin{array}{r|rrrr} -2 & 1 & 0 & -8 & -8 \\ & & -2 & 4 & 8 \\ \hline & 1 & -2 & -4 & 0 \end{array}$$

51.

Possible rational roots

$\pm 1, \pm 3, \pm 9, \pm \frac{1}{2}$

$\pm \frac{3}{2}, \pm \frac{9}{2}$

Descartes' Rule of Signs

# pos	# neg	# nonreal
1	3	0
1	1	2

$2x^4 + x^3 + 17x^2 + 9x - 9 = 0$

$(x+1)(2x^3 - x^2 + 18x - 9) = 0$

$(x+1)\left(x - \frac{1}{2}\right)(2x^2 + 18) = 0$

$2x^2 + 18 = 0 \Rightarrow x = \pm\sqrt{-9}$

Solution set: $\left\{-1, \frac{1}{2}, \pm 3i\right\}$

Test $x = -1$:

$$\begin{array}{r|rrrrr} -1 & 2 & 1 & 17 & 9 & -9 \\ & & -2 & 1 & -18 & 9 \\ \hline & 2 & -1 & 18 & -9 & 0 \end{array}$$

Test $x = \frac{1}{2}$:

$$\begin{array}{r|rrrr} \frac{1}{2} & 2 & -1 & 18 & -9 \\ & & 1 & 0 & 9 \\ \hline & 2 & 0 & 18 & 0 \end{array}$$

53. Possible rat. roots: $\pm 1,\ \pm 5,\ \pm 25$

Descartes' Rule of Signs

# pos	# neg	# nonreal
5	0	0
3	0	2
1	0	4

$$x^5-3x^4+28x^3-76x^2+75x-25=0$$
$$(x-1)\left(x^4-2x^3+26x^2-50x+25\right)=0$$
$$(x-1)(x-1)\left(x^3-x^2+25x-25\right)=0$$
$$(x-1)(x-1)(x-1)\left(x^2+25\right)=0$$
$$x^2+25=0\Rightarrow x=\pm\sqrt{-25}$$

Solution set: $\{1, 1, 1, \pm 5i\}$

Test $x=1$:
$$\begin{array}{r|rrrrrr} 1 & 1 & -3 & 28 & -76 & 75 & -25 \\ & & 1 & -2 & 26 & -50 & 25 \\ \hline & 1 & -2 & 26 & -50 & 25 & 0 \end{array}$$

Test $x=1$:
$$\begin{array}{r|rrrrr} 1 & 1 & -2 & 26 & -50 & 25 \\ & & 1 & -1 & 25 & -25 \\ \hline & 1 & -1 & 25 & -25 & 0 \end{array}$$

Test $x=1$:
$$\begin{array}{r|rrrr} 1 & 1 & -1 & 25 & -25 \\ & & 1 & 0 & 25 \\ \hline & 1 & 0 & 25 & 0 \end{array}$$

55. Possible rat. roots: $\pm 1,\ \pm 2,\ \pm 3,\ \pm 4,\ \pm 6,\ \pm 12,\ \pm \frac{1}{2},\ \pm \frac{3}{2}$

Descartes' Rule of Signs

# pos	# neg	# nonreal
4	1	0
2	1	2
0	1	4

$$2x^5-3x^4+6x^3-9x^2-8x+12=0$$
$$(x-1)\left(2x^4-x^3+5x^2-4x-12\right)=0$$
$$(x-1)(x+1)\left(2x^3-3x^2+8x-12\right)=0$$
$$(x-1)(x+1)\left(x-\tfrac{3}{2}\right)\left(2x^2+8\right)=0$$
$$2x^2+8=0\Rightarrow x=\pm\sqrt{-4}$$

Solution set: $\left\{1, -1, \frac{3}{2}, \pm 2i\right\}$

Test $x=1$:
$$\begin{array}{r|rrrrrr} 1 & 2 & -3 & 6 & -9 & -8 & 12 \\ & & 2 & -1 & 5 & -4 & -12 \\ \hline & 2 & -1 & 5 & -4 & -12 & 0 \end{array}$$

Test $x=-1$:
$$\begin{array}{r|rrrrr} -1 & 2 & -1 & 5 & -4 & -12 \\ & & -2 & 3 & -8 & 12 \\ \hline & 2 & -3 & 8 & -12 & 0 \end{array}$$

Test $x=\frac{3}{2}$:
$$\begin{array}{r|rrrr} \frac{3}{2} & 2 & -3 & 8 & -12 \\ & & 3 & 0 & 12 \\ \hline & 2 & 0 & 8 & 0 \end{array}$$

57. If $(1+i)$ is a root, then so is $(1-i)$, and $x-(1+i)$ and $x-(1-i)$ are factors. Then $[x-(1+i)][x-(1-i)]=x^2-2x+2$ is a factor. Divide it out:

$$\begin{array}{r|l} & x-3 \\ x^2-2x+2 & x^3-5x^2+8x-6 \\ & \underline{x^3-2x^2+2x} \\ & -3x^2+6x-6 \\ & \underline{-3x^2+6x-6} \\ & 0 \end{array}$$

$$x^3-5x^2+8x-6=0$$
$$\left(x^2-2x+2\right)(x-3)=0$$

Solution set: $\{1+i, 1-i, 3\}$

59. If $(1+i)$ is a root, then so is $(1-i)$, and $x-(1+i)$ and $x-(1-i)$ are factors. Then $[x-(1+i)][x-(1-i)] = x^2-2x+2$ is a factor. Divide it out:

$$\begin{array}{r|l} & x^2 - 9 \\ x^2-2x+2 & x^4-2x^3-7x^2+18x-18 \\ & x^4-2x^3+2x^2 \\ & -9x^2+18x-18 \\ & -9x^2+18x-18 \\ & 0 \end{array}$$

$$x^4-2x^3-7x^2+18x-18=0$$
$$(x^2-2x+2)(x^2-9)=0$$
$$(x^2-2x+2)(x+3)(x-3)=0$$

Solution set: $\{1+i, 1-i, -3, 3\}$

61. Possible rational roots

$\pm 1, \pm 2, \pm 3, \pm 6, \pm\frac{1}{3}, \pm\frac{2}{3}$

Descartes' Rule of Signs

# pos	# neg	# nonreal
1	2	0
1	0	2

Test $x=-1$:

$$\begin{array}{r|rrrr} -1 & 3 & -4 & -13 & -6 \\ & & -3 & 7 & 6 \\ \hline & 3 & -7 & -6 & 0 \end{array}$$

$$x^3-\frac{4}{3}x^2-\frac{13}{3}x-2=0$$
$$3x^3-4x^2-13x-6=0$$
$$(x+1)(3x^2-7x-6)=0$$
$$(x+1)(3x+2)(x-3)=0$$

Solution set: $\{-1, -\frac{2}{3}, 3\}$

63.
$$x^{-5}-8x^{-4}+25x^{-3}-38x^{-2}+28x^{-1}-8=0$$
$$x^5(x^{-5}-8x^{-4}+25x^{-3}-38x^{-2}+28x^{-1}-8)=x^5(0)$$
$$1-8x+25x^2-38x^3+28x^4-8x^5=0$$

Possible rat. roots

$\pm 1, \pm\frac{1}{2}, \pm\frac{1}{4}, \pm\frac{1}{8}$

Descartes' Rule of Signs

# pos	# neg	# nonreal
5	0	0
3	0	2
1	0	4

$$8x^5-28x^4+38x^3-25x^2+8x-1=0$$
$$(x-1)(8x^4-20x^3+18x^2-7x+1)=0$$
$$(x-1)(x-1)(8x^3-12x^2+6x-1)=0$$
$$(x-1)(x-1)(x-\tfrac{1}{2})(8x^2-8x+2)=0$$
$$(x-1)(x-1)(x-\tfrac{1}{2})(4x-2)(2x-1)=0$$

Solution set: $\{1, 1, \frac{1}{2}, \frac{1}{2}, \frac{1}{2}\}$

Test $x=1$:

$$\begin{array}{r|rrrrrr} 1 & 8 & -28 & 38 & -25 & 8 & -1 \\ & & 8 & -20 & 18 & -7 & 1 \\ \hline & 8 & -20 & 18 & -7 & 1 & 0 \end{array}$$

Test $x=1$:

$$\begin{array}{r|rrrrr} 1 & 8 & -20 & 18 & -7 & 1 \\ & & 8 & -12 & 6 & -1 \\ \hline & 8 & -12 & 6 & -1 & 0 \end{array}$$

Test $x=\frac{1}{2}$:

$$\begin{array}{r|rrrr} \frac{1}{2} & 8 & -12 & 6 & -1 \\ & & 4 & -4 & 1 \\ \hline & 8 & -8 & 2 & 0 \end{array}$$

65. Let $x = R_1$. Then $x + 10 = R_2$ and $x + 50 = R_3$.

$$\frac{1}{R} = \frac{1}{R_1} + \frac{1}{R_2} + \frac{1}{R_3}$$
$$\frac{1}{6} = \frac{1}{x} + \frac{1}{x+10} + \frac{1}{x+50}$$
$$6x(x+10)(x+50) \cdot \frac{1}{6} = 6x(x+10)(x+50)\left(\frac{1}{x} + \frac{1}{x+10} + \frac{1}{x+50}\right)$$
$$x(x+10)(x+50) = 6(x+10)(x+50) + 6x(x+50) + 6x(x+10)$$
$$x^3 + 60x^2 + 500x = 6x^2 + 360x + 3000 + 6x^2 + 300x + 6x^2 + 60x$$
$$x^3 + 42x^2 - 220x - 3000 = 0$$
$$(x-10)\left(x^2 + 52x + 300\right) = 0$$

10	1	42	−220	−3000
		10	520	3000
	1	52	300	0

Use the quadratic formula on the second factor. The two solutions from that factor are negative. The only solution that makes sense is $x = 10$. The resistances are 10, 20 and 60 ohms.

67. Let $x =$ the height. Then $x + 7 =$ the length, and $x + 4 =$ the width.
The volume is $x(x+7)(x+4) = x^3 + 11x^2 + 28x$.

$$x^3 + 11x^2 + 28x = 4420$$
$$x^3 + 11x^2 + 28x - 4420 = 0$$
$$(x-13)\left(x^2 + 24x + 340\right) = 0$$

Test $x = 13$:

13	1	11	28	−4420
		13	312	4420
	1	24	340	0

The only real solution is $x = 13$.
The height is 13 inches.

69. Possible rat'l roots

$\pm 1, \pm 2, \pm 3,$
$\pm 5, \pm 6, \pm 9,$
$\pm 10, \pm 15, \pm 18,$
$\pm 25, \pm 30, \pm 45,$
$\pm 50, \pm 75, \pm 90,$
$\pm 150, \pm 225,$
± 450

Descartes' Rule of Signs

# pos	# neg	# nonreal
3	1	0
1	1	2

$$-x^4 + 5x^3 + 91x^2 - 545x + 550 = 100$$
$$x^4 - 5x^3 - 91x^2 + 545x - 450 = 0$$
$$(x-1)\left(x^3 - 4x^2 - 95x + 450\right) = 0$$
$$(x-1)(x-5)\left(x^2 + x - 90\right) = 0$$
$$(x-1)(x-5)(x-9)(x+10) = 0$$

Solution set: $\{1, 5, 9, -10\}$

Test $x = 1$:

1	1	−5	−91	545	−450
		1	−4	−95	450
	1	−4	−95	450	0

Test $x = 5$:

5	1	−4	−95	450
		5	5	−450
	1	1	−90	0

The only solutions between 0 and 9 are 1, 5, and 9 miles.

71. Answers may vary.

SECTION 5.3

73. A coordinate of a point on the parabola has coordinates $(x, 16 - x^2)$.

Thus, the area of the rectangle is $A = 2x(16 - x^2) = 32x - 2x^3$.

$$32x - 2x^3 = 42$$
$$-2x^3 + 32x - 42 = 0$$
$$x^3 - 16x + 21 = 0$$
$$(x - 3)(x^2 + 3x - 7) = 0$$

Test $x = 3$:

3	1	0	−16	21
		3	9	−21
	1	3	−7	0

Using the quadratic formula on the second factor yields the solutions $x = \dfrac{-3 \pm \sqrt{37}}{2} \approx 1.54$ or -4.54.

The only solutions that make sense are $x = 3$ or $x = 1.54$.

Points: (3, 7) or (1.54, 13.63)

75. $\sqrt{72a^3b^5c} = \sqrt{36a^2b^4}\sqrt{2abc} = 6ab^2\sqrt{2abc}$

77. $\sqrt{18a^2b} + a\sqrt{50b} = \sqrt{9a^2}\sqrt{2b} + a\sqrt{25}\sqrt{2b} = 3a\sqrt{2b} + 5a\sqrt{2b} = 8a\sqrt{2b}$

79. $\dfrac{2}{\sqrt{3}-1} = \dfrac{2\left(\sqrt{3}+1\right)}{\left(\sqrt{3}-1\right)\left(\sqrt{3}+1\right)} = \dfrac{2\left(\sqrt{3}+1\right)}{\left(\sqrt{3}\right)^2 - 1^2} = \dfrac{2\left(\sqrt{3}+1\right)}{3-1} = \dfrac{2\left(\sqrt{3}+1\right)}{2} = \sqrt{3}+1$

Exercises 5.4 (page 535)

1. $P(a)$ and $P(b)$

3. x_l and x_r

5. $x^4 - 4 = \left(\sqrt{2}\right)^4 - 4 = 4 - 4 = 0;$

The only possible rational roots are ± 1, ± 2, ± 4. None of these satisfy the equation, so $\sqrt{2}$ is irrational.

7. $x^4 - 49 = \left(\sqrt{7}\right)^4 - 49 = 49 - 49 = 0;$

The only possible rational roots are ± 1, ± 7, ± 49. None of these satisfy the equation, so $\sqrt{7}$ is irrational.

9. $P(x) = 2x^2 + x - 3$

$P(-2) = 3;\ P(-1) = -2$

Thus, there is a root between -2 and -1.

11. $P(x) = 3x^3 - 11x^2 - 14x$

$P(4) = -40;\ P(5) = 30$

Thus, there is a root between 4 and 5.

13. $P(x) = x^4 - 8x^2 + 15$

$P(1) = 8;\ P(2) = -1$

Thus, there is a root between 1 and 2.

15.
$$30x^3 + 10 = 61x^2 + 39x$$
$$30x^3 - 61x^2 - 39x + 10 = 0$$

$P(x) = 30x^3 - 61x^2 - 39x + 10$

$P(2) = -72;\ P(3) = 154$

Thus, there is a root between 2 and 3.

17.
$$30x^3 + 10 = 61x^2 + 39x$$
$$30x^3 - 61x^2 - 39x + 10 = 0$$

$P(x) = 30x^3 - 61x^2 - 39x + 10;\ P(0) = 10$

$P(1) = -60;$ There is a root between 0 and 1.

19. $P(x) = x^2 - 3$. Note that $P(1) = -2$ and $P(2) = 1$. There is a root between 1 and 2.

STEP	x_l	c	x_r	$P(x_l)$	$P(x_c)$	$P(x_r)$
0	1	1.5	2	−2.0	−0.75	1.0
1	1.5	1.75	2	−0.75	0.0625	1.0
2	1.5	1.625	1.75	−0.75	−0.3594	0.0625
3	1.625	1.6875	1.75	−0.3594	−0.1523	0.0625
4	1.6875	1.71875	1.75	−0.1523	−0.0459	0.0625
5	1.71875	1.734375	1.75			

To the nearest tenth, the solution is $x = 1.7$.

21. $P(x) = x^2 - 5$. Note that $P(-3) = 4$ and $P(-2) = -1$. There is a root between -3 and -2.

STEP	x_l	c	x_r	$P(x_l)$	$P(x_c)$	$P(x_r)$
0	−3	−2.5	−2	4.0	1.25	−1.0
1	−2.5	−2.25	−2	1.25	0.0625	−1.0
2	−2.25	−2.125	−2	0.0625	−0.4844	−1.0
3	−2.25	−2.1875	−2.125	0.0625	−0.2148	−0.4844
4	−2.25	−2.21875	−2.1875	0.0625	−0.0771	−0.2148
5	−2.25	−2.234375	−2.21875			

To the nearest tenth, the solution is $x = -2.2$.

23. $P(x) = x^3 - x^2 - 2$. Note that $P(1) = -2$ and $P(2) = 2$. There is a root between 1 and 2.

STEP	x_l	c	x_r	$P(x_l)$	$P(x_c)$	$P(x_r)$
0	1	1.5	2	−2.0	−0.875	2.0
1	1.5	1.75	2	−0.875	0.2969	2.0
2	1.5	1.625	1.75	−0.875	−0.3496	0.2969
3	1.625	1.6875	1.75	−0.3496	−0.0422	0.2969
4	1.6875	1.71875	1.75	−0.0422	0.1233	0.2969
5	1.6875	1.703125	1.71875			

To the nearest tenth, the solution is $x = 1.7$.

25. $P(x) = 3x^4 + 3x^3 - x^2 - 4x - 4$. Note that $P(-2) = 24$ and $P(-1) = -1$.
There is a root between -2 and -1.

STEP	x_l	c	x_r	$P(x_l)$	$P(x_c)$	$P(x_r)$
0	−2	−1.5	−1	24.0	4.8125	−1.0
1	−1.5	−1.25	−1	4.8125	0.9023	−1.0
2	−1.25	−1.125	−1	0.9023	−0.2317	−1.0
3	−1.25	−1.1875	−1.125	0.9023	0.2818	−0.2317
4	−1.1875	−1.15625	−1.125	0.2818	0.0127	−0.2317
5	−1.15625	−1.140625	−1.125	0.0127	−0.1125	−0.2317
6	−1.15625	−1.1484375	−1.140625	0.0127	−0.0507	−0.1125
7	−1.15625	−1.15234375	−1.1484375	0.0127	−0.0192	−0.0507
8	−1.15625	−1.154296875	−1.15234375			

To the nearest tenth, the solution is $x = -1.2$.

27. Find the x-coordinates of any x-intercepts:

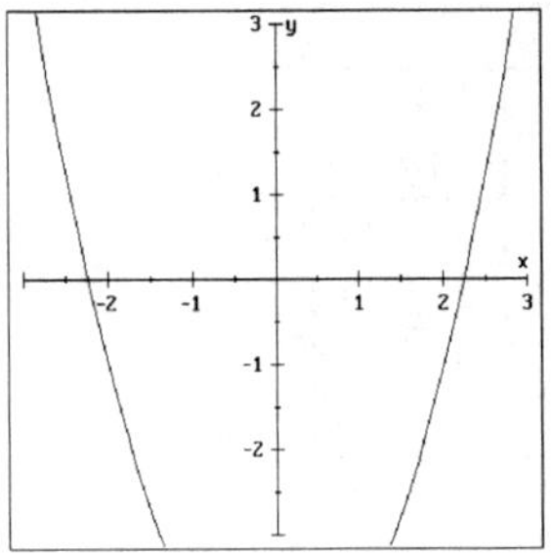

$\{-2.2, 2.2\}$

29. Find the x-coordinates of any x-intercepts:

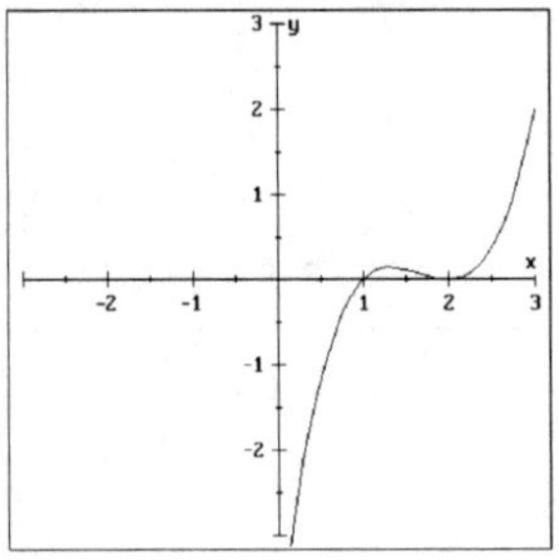

$\{1, 2\}$: Bisection fails for $x = 2$ since the solution cannot be bracketed by a negative y-coordinate and a positive y-coordinate.

31. Let $h =$ the height of the box. Then the volume of the box is $16(10)h$, or$160h$. $160h = 1000 \Rightarrow 160h - 1000 = 0$. Let $f(h) = 160h - 1000$. $f(4) = -360$ while $f(8) = 280$, so there is a solution between $h = 4$ and $h = 8$.

33. A point on the graph has coordinates (x, x^3). Then the distance to the origin is found by the distance formula:

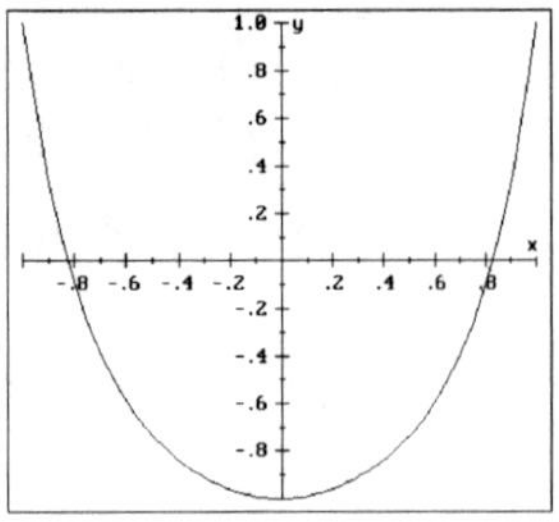

$$d = \sqrt{(x-0)^2 + (x^3-0)^2} = \sqrt{x^2 + x^6} = 1$$
$$\sqrt{x^2 + x^6} = 1$$
$$x^2 + x^6 = 1$$

$x^2 + x^6 - 1 = 0$: Graph and find intercepts. The x-coordinates are about 0.83 and -0.83. Find the y-coordinates using $y = x^3$. The points are $(0.83, 0.56)$ and $(-0.83, -0.56)$.

35. **Answers may vary.**

37. $y = \dfrac{5x-3}{x^2-4} = \dfrac{5x-3}{(x+2)(x-2)}$

vertical: $x = -2$, $x = 2$
horizontal: $y = 0$
slant: none

39. $y = \dfrac{x^2}{x-2} = x + 2 + \dfrac{4}{x-2}$

vertical: $x = 2$
horizontal: none
slant: $y = x + 2$

Chapter 5 Review (page 539)

1. $P(1) = 4(1)^4 + 2(1)^3 - 3(1)^2 - 2 = 1$
The remainder is 1.

2. $P(2) = 4(2)^4 + 2(2)^3 - 3(2)^2 - 2 = 66$
The remainder is 66.

3. $P(-3) = 4(-3)^4 + 2(-3)^3 - 3(-3)^2 - 2 = 241 \Rightarrow$ The remainder is 241.

CHAPTER 5 REVIEW

4. $P(-2) = 4(-2)^4 + 2(-2)^3 - 3(-2)^2 - 2 = 34 \Rightarrow$ The remainder is 34.

5. $P(2) = (2)^3 + 4(2)^2 - 2(2) + 4 = 24 \Rightarrow$ The remainder is 24. $\Rightarrow$ not a factor

6. $P(-3) = 2(-3)^4 + 10(-3)^3 + 4(-3)^2 + 7(-3) + 21 = -72 \Rightarrow$ The remainder is -72.
$\Rightarrow$ not a factor

7. $P(5) = (5)^5 - 3125 = 0$
The remainder is 0. $\Rightarrow$ factor

8. $P(6) = (6)^5 - 6(6)^4 - 4(6) + 24 = 0$
The remainder is 0. $\Rightarrow$ factor

9.
$$\begin{array}{r|rrrrr} 3 & 3 & 0 & 2 & 3 & 7 \\ & & 9 & 27 & 87 & 270 \\ \hline & 3 & 9 & 29 & 90 & 277 \end{array}$$
$$3x^3 + 9x^2 + 29x + 90 + \frac{277}{x-3}$$

10.
$$\begin{array}{r|rrrrr} 2 & 2 & 0 & -3 & 3 & -1 \\ & & 4 & 8 & 10 & 26 \\ \hline & 2 & 4 & 5 & 13 & 25 \end{array}$$
$$2x^3 + 4x^2 + 5x + 13 + \frac{25}{x-2}$$

11.
$$\begin{array}{r|rrrrrr} -2 & 5 & -4 & 3 & -2 & 1 & -1 \\ & & -10 & 28 & -62 & 128 & -258 \\ \hline & 5 & -14 & 31 & -64 & 129 & -259 \end{array}$$
$$5x^4 - 14x^3 + 31x^2 - 64x + 129 + \frac{-259}{x+2}$$

12.
$$\begin{array}{r|rrrrrr} -1 & 4 & 2 & -1 & 3 & 2 & 1 \\ & & -4 & 2 & -1 & -2 & 0 \\ \hline & 4 & -2 & 1 & 2 & 0 & 1 \end{array}$$
$$4x^4 - 2x^3 + x^2 + 2x + \frac{1}{x+1}$$

13.
$$\begin{array}{r|rrrr} 3 & 5 & 2 & -1 & 1 \\ & & 15 & 51 & 150 \\ \hline & 5 & 17 & 50 & 151 \end{array} \quad P(3) = 151$$

14.
$$\begin{array}{r|rrrr} -3 & 5 & 2 & -1 & 1 \\ & & -15 & 39 & -114 \\ \hline & 5 & -13 & 38 & -113 \end{array}$$
$$P(-3) = -113$$

15.
$$\begin{array}{r|rrrr} \frac{1}{2} & 5 & 2 & -1 & 1 \\ & & \frac{5}{2} & \frac{9}{4} & \frac{5}{8} \\ \hline & 5 & \frac{9}{2} & \frac{5}{4} & \frac{13}{8} \end{array} \quad P\left(\tfrac{1}{2}\right) = \tfrac{13}{8}$$

16.
$$\begin{array}{r|rrrr} i & 5 & 2 & -1 & 1 \\ & & 5i & -5+2i & -2-6i \\ \hline & 5 & 2+5i & -6+2i & -1-6i \end{array}$$
$$P(i) = -1 - 6i$$

17. $x = 3$ is a solution, so $(x-3)$ is a factor.
Use synthetic division to divide by $(x-3)$.
$$\begin{array}{r|rrrr} 3 & 2 & -3 & -11 & 6 \\ & & 6 & 9 & -6 \\ \hline & 2 & 3 & -2 & 0 \end{array}$$

$2x^3 - 3x^2 - 11x + 6 = 0$
$(x-3)(2x^2 + 3x - 2) = 0$
$(x-3)(2x-1)(x+2) = 0$
Solution set: $\left\{3, \tfrac{1}{2}, -2\right\}$

18. $x = -2$ is a solution, so $(x+2)$ is a factor. Use synthetic division to divide by $(x+2)$.

$$\begin{array}{r|rrrrr} -2 & 1 & 4 & -1 & -20 & -20 \\ & & -2 & -4 & 10 & 20 \\ \hline & 1 & 2 & -5 & -10 & 0 \end{array}$$

$$x^4 + 4x^3 - x^2 - 20x - 20 = 0$$
$$(x+2)\left(x^3 + 2x^2 - 5x - 10\right) = 0$$

Use the fact that $x = -2$ is a double root and divide the depressed polynomial by $(x+2)$:

$$\begin{array}{r|rrrr} -2 & 1 & 2 & -5 & -10 \\ & & -2 & 0 & 10 \\ \hline & 1 & 0 & -5 & 0 \end{array}$$

$$(x+2)\left(x^3 + 2x^2 - 5x - 10\right) = 0$$
$$(x+2)(x+2)\left(x^2 - 5\right) = 0$$

Solution set: $\left\{-2, -2, \sqrt{5}, -\sqrt{5}\right\}$

19. $2(x+1)(x-2)\left(x-\frac{3}{2}\right) = 2(x^2 - x - 2)\left(x - \frac{3}{2}\right) = 2\left(x^3 - \frac{5}{2}x^2 - \frac{1}{2}x + 3\right)$
$= 2x^3 - 5x^2 - x + 6$

20. $2(x-1)(x+3)\left(x-\frac{1}{2}\right) = 2(x^2 + 2x - 3)\left(x - \frac{1}{2}\right) = 2\left(x^3 + \frac{3}{2}x^2 - 4x + \frac{3}{2}\right)$
$= 2x^3 + 3x^2 - 8x + 3$

21. $(x-2)(x+5)(x-i)(x+i) = \left(x^2 + 3x - 10\right)\left(x^2 - i^2\right) = \left(x^2 + 3x - 10\right)\left(x^2 + 1\right)$
$= x^4 + 3x^3 - 9x^2 + 3x - 10$

22. $(x+3)(x-2)(x-i)(x+i) = \left(x^2 + x - 6\right)\left(x^2 - i^2\right) = \left(x^2 + x - 6\right)\left(x^2 + 1\right)$
$= x^4 + x^3 - 5x^2 + x - 6$

23. $3x^6 - 4x^5 + 3x + 2 = 0$
6 roots

24. $2x^6 - 5x^4 + 5x^3 - 4x^2 + x - 12 = 0$
6 roots

25. $3x^{65} - 4x^{50} + 3x^{17} + 2x = 0$
65 roots

26. $x^{1984} - 12 = 0$
1984 roots

27. $P(x) = x^4 - 16$
4 linear factors, 4 zeros

28. $P(x) = x^{40} + x^{30}$
40 linear factors, 40 zeros

29. $P(x) = 4x^5 + 2x^3$
5 linear factors, 5 zeros

30. $P(x) = x^3 - 64x$
3 linear factors, 3 zeros

31. $2 - i$ is also a root.

32. $-i = 0 - i$, so $0 + i = i$ is also a root.

33. If $-i$ is a root, then i is a root also:
$$(x-4)(x+i)(x-i) = 0$$
$$(x-4)\left(x^2 - i^2\right) = 0$$
$$(x-4)\left(x^2 + 1\right) = 0$$
$$x^3 - 4x^2 + x - 4 = 0$$

34. If i is a root, then $-i$ is a root also:
$$(x+5)(x-i)(x+i) = 0$$
$$(x+5)\left(x^2 - i^2\right) = 0$$
$$(x+5)\left(x^2 + 1\right) = 0$$
$$x^3 + 5x^2 + x + 5 = 0$$

35. $P(x) = 3x^4 + 2x^3 - 4x + 2$: 2 sign variations $\Rightarrow$ 2 or 0 positive roots
$P(-x) = 3(-x)^4 + 2(-x)^3 - 4(-x) + 2$
$= 3x^4 - 2x^3 + 4x + 2$: 2 sign variations $\Rightarrow$ 2 or 0 negative roots

# pos	# neg	# nonreal
2	2	0
2	0	2
0	2	2
0	0	4

36. $P(x) = 2x^4 - 3x^3 + 5x^2 + x - 5$: 3 sign variations $\Rightarrow$ 3 or 1 positive roots
$P(-x) = 2(-x)^4 - 3(-x)^3 + 5(-x)^2 + (-x) - 5$
$= 2x^4 + 3x^3 + 5x^2 - x - 5$: 1 sign variation $\Rightarrow$ 1 negative root

# pos	# neg	# nonreal
3	1	0
1	1	2

37. $4x^5 + 3x^4 + 2x^3 + x^2 + x = 7$
$4x^5 + 3x^4 + 2x^3 + x^2 + x - 7 = 0$
$P(x) = 4x^5 + 3x^4 + 2x^3 + x^2 + x - 7$: 1 sign variation $\Rightarrow$ 1 positive root
$P(-x) = 4(-x)^5 + 3(-x)^4 + 2(-x)^3 + (-x)^2 + (-x) - 7$
$= -4x^5 + 3x^4 - 2x^3 + x^2 - x - 7$: 4 sign variations $\Rightarrow$ 4 or 2 or 0 negative roots

# pos	# neg	# nonreal
1	4	0
1	2	2
1	0	4

38. $P(x) = 3x^7 - 4x^5 + 3x^3 + x - 4$: 3 sign variations $\Rightarrow$ 3 or 1 positive roots
$P(-x) = 3(-x)^7 - 4(-x)^5 + 3(-x)^3 + (-x) - 4$
$= -3x^7 + 4x^5 - 3x^3 - x - 4$: 2 sign variations $\Rightarrow$ 2 or 0 negative roots

# pos	# neg	# nonreal
3	2	2
3	0	4
1	2	4
1	0	6

39. $P(x) = x^4 + x^2 + 24{,}567$: 0 sign variations $\Rightarrow$ 0 positive roots
$P(-x) = (-x)^4 + (-x)^2 + 24{,}567$
$= x^4 + x^2 + 24{,}567$: 0 sign variations $\Rightarrow$ 0 negative roots

# pos	# neg	# nonreal
0	0	4

40. $P(x) = -x^7 - 5$: 0 sign variations $\Rightarrow$ 0 positive roots
$P(-x) = -(-x)^7 - 5$
$= x^7 - 5$: 1 sign variation $\Rightarrow$ 1 negative root

# pos	# neg	# nonreal
0	1	6

41. $P(x) = 5x^3 - 4x^2 - 2x + 4$

2	5	−4	−2	4
		10	12	20
	5	6	10	24

Upper bound: 2

−1	5	−4	−2	4
		−5	9	−7
	5	−9	7	−3

Lower bound: -1

42. $P(x) = x^4 + 3x^3 - 5x^2 - 9x + 1$

2	1	3	−5	−9	1
		2	10	10	2
	1	5	5	1	3

Upper bound: 2

−5	1	3	−5	−9	1
		−5	10	−25	170
	1	−2	5	−34	171

Lower bound: -5

43. num: $\pm 1, \pm 2, \pm 3, \pm 6$; den: $\pm 1, \pm 2$
possible roots: $\pm 1, \pm 2, \pm 3, \pm 6, \pm \frac{1}{2}, \pm \frac{3}{2}$

44. num: $\pm 1, \pm 2, \pm 5, \pm 10$; den: $\pm 1, \pm 2, \pm 4$
possible roots: $\pm 1, \pm 2, \pm 5, \pm 10, \pm \frac{1}{2}, \pm \frac{5}{2}, \pm \frac{1}{4}, \pm \frac{5}{4}$

45. $x^3 - 10x^2 + 29x - 20 = 0$
Possible rational roots
$\pm 1, \pm 2, \pm 4, \pm 5, \pm 10, \pm 20$
Descartes' Rule of Signs

# pos	# neg	# nonreal
3	0	0
1	0	2

Test $x = 1$:

1	1	−10	29	−20
		1	−9	20
	1	−9	20	0

$x^3 - 10x^2 + 29x - 20 = 0$
$(x-1)(x^2 - 9x + 20) = 0$
$(x-1)(x-5)(x-4) = 0$
Solution set: $\{1, 5, 4\}$

46. $x^3 - 8x^2 - x + 8 = 0$
Possible rational roots
$\pm 1, \pm 2, \pm 4, \pm 8$
Descartes' Rule of Signs

# pos	# neg	# nonreal
2	1	0
0	1	2

Test $x = 1$:

1	1	−8	−1	8
		1	−7	−8
	1	−7	−8	0

$x^3 - 8x^2 - x + 8 = 0$
$(x-1)(x^2 - 7x - 8) = 0$
$(x-1)(x-8)(x+1) = 0$
Solution set: $\{1, 8, -1\}$

47. Possible rational roots

$\pm 1, \pm 2, \pm 3, \pm 5,$
$\pm 6, \pm 10, \pm 15, \pm 30,$
$\pm \frac{1}{2}, \pm \frac{3}{2}, \pm \frac{5}{2}, \pm \frac{15}{2}$

Test $x = -2$:

$$\begin{array}{r|rrrr} -2 & 2 & 17 & 41 & 30 \\ & & -4 & -26 & -30 \\ \hline & 2 & 13 & 15 & 0 \end{array}$$

Descartes' Rule of Signs

# pos	# neg	# nonreal
0	3	0
0	1	2

$$2x^3 + 17x^2 + 41x + 30 = 0$$
$$(x+2)\left(2x^2 + 13x + 15\right) = 0$$
$$(x+2)(2x+3)(x+5) = 0$$

Solution set: $\left\{-2, -\frac{3}{2}, -5\right\}$

48. Possible rational roots

$\pm 1, \pm \frac{1}{3}$

Test $x = \frac{1}{3}$:

$$\begin{array}{r|rrrr} \frac{1}{3} & 3 & 2 & 2 & -1 \\ & & 1 & 1 & 1 \\ \hline & 3 & 3 & 3 & 0 \end{array}$$

Descartes' Rule of Signs

# pos	# neg	# nonreal
1	2	0
1	0	2

$$3x^3 + 2x^2 + 2x - 1 = 0$$
$$\left(x - \frac{1}{3}\right)\left(3x^2 + 3x + 3\right) = 0$$

$3x^2 + 3x + 3$ does not factor rationally.

Rational solutions: $\left\{\frac{1}{3}\right\}$

49. Possible rat. roots

$\pm 1, \pm 2, \pm 3, \pm 4,$
$\pm 6, \pm 9, \pm 12,$
$\pm 18, \pm 36, \pm \frac{1}{2},$
$\pm \frac{3}{2}, \pm \frac{9}{2}, \pm \frac{1}{4},$
$\pm \frac{3}{4}, \pm \frac{9}{4}$

Descartes' Rule of Signs

# pos	# neg	# nonreal
2	2	0
2	0	2
0	2	2
0	0	4

$$4x^4 - 25x^2 + 36 = 0$$
$$(x-2)\left(4x^3 + 8x^2 - 9x - 18\right) = 0$$
$$(x-2)(x+2)\left(4x^2 - 9\right) = 0$$
$$(x-2)(x+2)(2x+3)(2x-3) = 0$$

Solution set: $\left\{2, -2, -\frac{3}{2}, \frac{3}{2}\right\}$

Test $x = 2$:

$$\begin{array}{r|rrrrr} 2 & 4 & 0 & -25 & 0 & 36 \\ & & 8 & 16 & -18 & -36 \\ \hline & 4 & 8 & -9 & -18 & 0 \end{array}$$

Test $x = -2$:

$$\begin{array}{r|rrrr} -2 & 4 & 8 & -9 & -18 \\ & & -8 & 0 & 18 \\ \hline & 4 & 0 & -9 & 0 \end{array}$$

50. Possible rat. roots

$\pm 1, \pm 2, \pm 4,$
$\pm 8, \pm 16, \pm 32,$
$\pm \frac{1}{2}$

Descartes' Rule of Signs

# pos	# neg	# nonreal
2	2	0
2	0	2
0	2	2
0	0	4

$$2x^4 - 11x^3 - 6x^2 + 64x + 32 = 0$$
$$(x-4)\left(2x^3 - 3x^2 - 18x - 8\right) = 0$$
$$(x-4)(x-4)\left(2x^2 + 5x + 2\right) = 0$$
$$(x-4)(x-4)(2x+1)(x+2) = 0$$

Solution set: $\left\{4, 4, -\frac{1}{2}, -2\right\}$

Test $x = 4$:

$$\begin{array}{r|rrrrr} 4 & 2 & -11 & -6 & 64 & 32 \\ & & 8 & -12 & -72 & -32 \\ \hline & 2 & -3 & -18 & -8 & 0 \end{array}$$

Test $x = 4$:

$$\begin{array}{r|rrrr} 4 & 2 & -3 & -18 & -8 \\ & & 8 & 20 & 8 \\ \hline & 2 & 5 & 2 & 0 \end{array}$$

51. Possible rational roots

$\pm 1,\ \pm 2,\ \pm 4,\ \pm 8$
$\pm 16,\ \pm \frac{1}{3},\ \pm \frac{2}{3},\ \pm \frac{4}{3}$
$\pm \frac{8}{3},\ \pm \frac{16}{3}$

Test $x = \frac{1}{3}$:

$$\begin{array}{r|rrrr} \frac{1}{3} & 3 & -1 & 48 & -16 \\ & & 1 & 0 & 16 \\ \hline & 3 & 0 & 48 & 0 \end{array}$$

Descartes' Rule of Signs

# pos	# neg	# nonreal
3	0	0
1	0	2

$3x^3 - x^2 + 48x - 16 = 0$
$\left(x - \frac{1}{3}\right)\left(3x^2 + 48\right) = 0$
$x = \frac{1}{3}$ **or** $x = \pm\sqrt{-16}$
$x = \pm 4i$

Solution set: $\left\{\frac{1}{3}, -4i, 4i\right\}$

52. Possible rational roots

$\pm 1,\ \pm 2,\ \pm 4,\ \pm 5$
$\pm 10,\ \pm 20$

Descartes' Rule of Signs

# pos	# neg	# nonreal
2	2	0
2	0	2
0	2	2
0	0	4

$x^4 - 2x^3 - 9x^2 + 8x + 20 = 0$
$(x-2)\left(x^3 - 9x - 10\right) = 0$
$(x-2)(x-2)\left(x^2 - 2x - 5\right) = 0$
Use the quadratic formula.

Solution set: $\left\{2, -2, 1 \pm \sqrt{6}\right\}$

Test $x = 2$:

$$\begin{array}{r|rrrrr} 2 & 1 & -2 & -9 & 8 & 20 \\ & & 2 & 0 & -18 & -20 \\ \hline & 1 & 0 & -9 & -10 & 0 \end{array}$$

Test $x = -2$:

$$\begin{array}{r|rrrr} -2 & 1 & 0 & -9 & -10 \\ & & -2 & 4 & 10 \\ \hline & 1 & -2 & -5 & 0 \end{array}$$

53. $P(x) = 5x^3 + 37x^2 + 59x + 18$
$P(-1) = -9;\ P(0) = 18$

54. $P(x) = 6x^3 - x^2 - 10x - 3$
$P(1) = -8;\ P(2) = 21$

55. $P(x) = x^3 - 2x^2 - 9x - 2$. Note that $P(4) = -6$ and $P(5) = 28$.
There is a root between 4 and 5.

STEP	x_l	c	x_r	$P(x_l)$	$P(x_c)$	$P(x_r)$
0	4	4.5	5	−8.0	8.125	21.0
1	4	4.25	4.5	−8.0	0.3906	8.125
2	4	4.125	4.25	−8.0	−2.9668	0.3906
3	4.125	4.1875	4.25	−2.9668	−1.3293	0.3906
4	4.1875	4.21875	4.25	−1.3293	−0.4798	0.3906
5	4.21875	4.234375	4.25	−0.4798	−0.0472	0.3906
6	4.234375	4.2421875	2.25	−0.0472	0.1711	0.3906
7	4.234275	4.23823125	4.2421875			

To the nearest tenth, the solution is $x = 4.2$.

56. $P(x) = 6x^2 - 13x - 5$. Note that $P(2) = -7$ and $P(3) = 10$.
There is a root between 2 and 3.

STEP	x_l	c	x_r	$P(x_l)$	$P(x_c)$	$P(x_r)$
0	2	2.5	3	−7.0	0	10.0

The solution is exactly $x = 2.5$.

57. Find the x-coordinates of any positive x-intercepts:

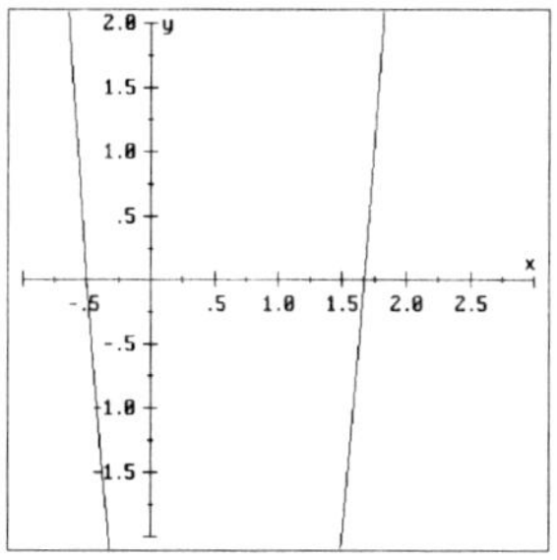

$x = 1.67$

58. Find the x-coordinates of any positive x-intercepts:

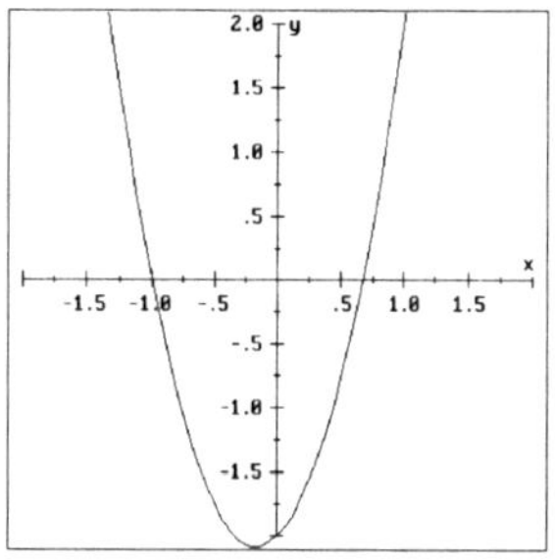

$x = 0.67$

59. Let $x =$ the length of the shortest panel. Then $x + 2$ and $x + 5$ are the lengths of the other panels. Since each panel has an area of 60, the widths are $\frac{60}{x}$, $\frac{60}{x+2}$ and $\frac{60}{x+5}$, respectively.

$$15 = \frac{60}{x} + \frac{60}{x+2} + \frac{60}{x+5}$$

$$x(x+2)(x+5)(15) = x(x+2)(x+5)\left(\frac{60}{x} + \frac{60}{x+2} + \frac{60}{x+5}\right)$$

$$15x(x+2)(x+5) = 60(x+2)(x+5) + 60x(x+5) + 60x(x+2)$$

$$15x^3 + 105x^2 + 150x = 60x^2 + 420x + 600 + 60x^2 + 300x + 60x^2 + 120x$$

$$15x^3 - 75x^2 - 690x - 600 = 0$$

$$15\left(x^3 - 5x^2 - 46x - 40\right) = 0$$

$$x^3 - 5x^2 - 46x - 40 = 0$$

$$(x - 10)\left(x^2 + 5x + 4\right) = 0$$

$$(x - 10)(x + 4)(x + 1) = 0$$

Test $x = 10$:

10	1	−5	−46	−40
		10	50	40
	1	5	4	0

The positive solution $x = 10$ is the only one that makes sense.
The dimensions are 10 m by 6 m, 12 m by 5 m, and 15 m by 4 m.

60. Let $x =$ the radius. Then $x + 3$ is the height.

$$\pi r^2 h = V$$

$$r^2 h = \frac{V}{\pi}$$

$$x^2(x+3) = \frac{19000}{\pi}$$

$$x^3 + 3x^2 = 6047.89$$

$$x^3 + 3x^2 - 6047.89 = 0$$

Graph $y = x^3 + 3x^2 - 6047.89$ and find the x-coordinates of any x-intercepts:

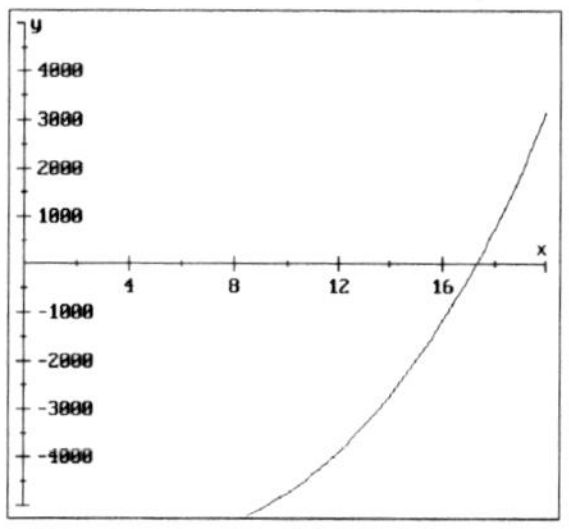

The radius should be about 17.27 feet.

Chapter 5 Test (page 546)

1. $P(-2) = (-2)^2 + 5(-2) + 6 = 4 - 10 + 6 = 0$; -2 is a zero of $P(x)$.

2. $P(2) = 3(2)^3 - 9(2) - 5$
$= 3(8) - 18 - 5 = \boxed{1}$

$$\begin{array}{r|l} & 3x^2 + 6x + 3 \\ \hline x-2 & 3x^3 + 0x^2 - 9x - 5 \\ & \underline{3x^3 - 6x^2} \\ & \quad 6x^2 - 9x \\ & \quad \underline{6x^2 - 12x} \\ & \qquad 3x - 5 \\ & \qquad \underline{3x - 6} \\ & \qquad\qquad \boxed{1} \end{array}$$

3. $P(-2) = (-2)^5 + 2$
$= -32 + 2 = \boxed{-30}$

$$\begin{array}{r|l} & x^4 - 2x^3 + 4x^2 - 8x + 16 \\ \hline x+2 & x^5 + 0x^4 + 0x^3 + 0x^2 + 0x + 2 \\ & \underline{x^5 + 2x^4} \\ & \; -2x^4 + 0x^3 \\ & \; \underline{-2x^4 - 4x^3} \\ & \qquad 4x^3 + 0x^2 \\ & \qquad \underline{4x^3 + 8x^2} \\ & \qquad\; -8x^2 + 0x \\ & \qquad\; \underline{-8x^2 - 16x} \\ & \qquad\qquad 16x + 2 \\ & \qquad\qquad \underline{16x + 32} \\ & \qquad\qquad\qquad \boxed{-30} \end{array}$$

4. $P(3) = 2(3)^4 - 10(3)^3 + 4(3^2) + 7(3) + 21 = -30 \Rightarrow$ remainder $= -30$
$x - 3$ is not a factor of the polynomial.

5.
$$\begin{array}{r|rrrr} 2 & 2 & -3 & -4 & -1 \\ & & 4 & 2 & -4 \\ \hline & 2 & 1 & -2 & -5 \end{array}$$
$(x-2)(2x^2 + x - 2) - 5$

6.
$$\begin{array}{r|rrrr} -1 & 2 & -3 & -4 & -1 \\ & & -2 & 5 & -1 \\ \hline & 2 & -5 & 1 & -2 \end{array}$$
$(x+1)(2x^2 - 5x + 1) - 2$

7.
$$\begin{array}{r|rrr} 5 & 2 & -7 & -15 \\ & & 10 & 15 \\ \hline & 2 & 3 & 0 \end{array} \quad \boxed{2x+3}$$

8.
$$\begin{array}{r|rrrr} -2 & 3 & 7 & 2 & 0 \\ & & -6 & -2 & 0 \\ \hline & 3 & 1 & 0 & 0 \end{array} \quad \boxed{3x^2 + x}$$

9.
$$\begin{array}{r|rrrr} 1 & 3 & -2 & 0 & 4 \\ & & 3 & 1 & 1 \\ \hline & 3 & 1 & 1 & 5 \end{array} \quad P(1) = 5$$

10.
$$\begin{array}{r|rrrr} -2 & 3 & -2 & 0 & 4 \\ & & -6 & 16 & -32 \\ \hline & 3 & -8 & 16 & -28 \end{array} \quad P(-2) = -28$$

11.
$$\begin{array}{r|rrrr} -\frac{1}{3} & 3 & -2 & 0 & 4 \\ & & -1 & 1 & -\frac{1}{3} \\ \hline & 3 & -3 & 1 & \frac{11}{3} \end{array} \quad P\left(-\tfrac{1}{3}\right) = \tfrac{11}{3}$$

12.
$$\begin{array}{r|rrrr} i & 3 & -2 & 0 & 4 \\ & & 3i & -3-2i & 2-3i \\ \hline & 3 & -2+3i & -3-2i & 6-3i \end{array}$$
$P(i) = 6 - 3i$

13. $(x-5)(x+1)(x-0) = (x^2 - 4x - 5)x = x^3 - 4x^2 - 5x$

14. $(x-i)(x+i)\left(x - \sqrt{3}\right)\left(x + \sqrt{3}\right) = (x^2 - i^2)(x^2 - 3) = (x^2+1)(x^2-3) = x^4 - 2x^2 - 3$

CHAPTER 5 TEST

15. If i is a root, then $-i$ is a root also:

$$\begin{aligned}(x-2)(x-i)(x+i)&=0\\(x-2)\left(x^2-i^2\right)&=0\\(x-2)\left(x^2+1\right)&=0\\x^3-2x^2+x-2&=0\end{aligned}$$

16. If $2+i$ is a root, then $2-i$ is a root also:

$$\begin{aligned}(x-1)[x-(2+i)][x-(2-i)]&=0\\(x-1)\left[x^2-(2-i)x-(2+i)x+(2+i)(2-i)\right]&=0\\(x-1)\left[x^2-2x+ix-2x-ix+4-i^2\right]&=0\\(x-1)\left[x^2-4x+5\right]&=0\\x^3-5x^2+9x-5&=0\end{aligned}$$

17. $P(x)=3x^3+2x^2-4x+1$
3 linear factors, 3 zeros

18. If $3-2i$ is a root, then $3+2i$ is also a root.

19. $P(x)=3x^5-2x^4+2x^2-x-3$: 3 sign variations $\Rightarrow$ 3 or 1 positive roots
$P(-x)=3(-x)^5-2(-x)^4+2(-x)^2-(-x)-3$
$=-3x^5-2x^4+2x^2+x-4$: 2 sign variations $\Rightarrow$ 2 or 0 negative roots

# pos	# neg	# nonreal
3	2	0
3	0	2
1	2	2
1	0	4

20. $P(x)=2x^3-5x^2-2x-1$: 1 sign variation $\Rightarrow$ 1 positive root
$P(-x)=2(-x)^3-5(-x)^2-2(-x)-1$
$=-2x^3-5x^2+2x-1$: 2 sign variations $\Rightarrow$ 2 or 0 negative roots

# pos	# neg	# nonreal
1	2	0
1	0	2

21. $P(x)=x^5-x^4-5x^3+5x^2+4x-5$

$$\begin{array}{r|rrrrrr}3 & 1 & -1 & -5 & 5 & 4 & -5\\ & & 3 & 6 & 3 & 24 & 84\\\hline & 1 & 2 & 1 & 8 & 28 & 79\end{array}$$

Upper bound: 3

$$\begin{array}{r|rrrrrr}-3 & 1 & -1 & -5 & 5 & 4 & -5\\ & & -3 & 12 & -21 & 48 & -156\\\hline & 1 & -4 & 7 & -16 & 52 & -161\end{array}$$

Lower bound: -3

22. $P(x)=2x^3-11x^2+10x+3$

$$\begin{array}{r|rrrr}6 & 2 & -11 & 10 & 3\\ & & 12 & 6 & 96\\\hline & 2 & 1 & 16 & 99\end{array}$$

Upper bound: 6

$$\begin{array}{r|rrrr}-1 & 2 & -11 & 10 & 3\\ & & -2 & 13 & -23\\\hline & 2 & -13 & 23 & -20\end{array}$$

Lower bound: -1

CHAPTER 5 TEST

23. num: $\pm 1, \pm 2, \pm 4, \pm 8$; den: ± 1; possible roots: $\pm 1, \pm 2, \pm 4, \pm 8$

24. num: $\pm 1, \pm 2$; den: $\pm 1, \pm 5$; possible roots: $\pm 1, \pm 2, \pm \frac{1}{5}, \pm \frac{2}{5}$

25. Possible rational roots

$\pm 1, \pm 2, \pm 3, \pm 6, \pm \frac{1}{2}, \pm \frac{3}{2}$

Test $x = 2$:

$$\begin{array}{r|rrrr} 2 & 2 & 3 & -11 & -6 \\ & & 4 & 14 & 6 \\ \hline & 2 & 7 & 3 & 0 \end{array}$$

Descartes' Rule of Signs

# pos	# neg	# nonreal
1	2	0
1	0	2

$2x^3 + 3x^2 - 11x - 6 = 0$

$(x-2)(2x^2 + 7x + 3) = 0$

$(x-2)(2x+1)(x+3) = 0$

Solution set: $\{2, -\frac{1}{2}, -3\}$

26. Possible rational roots

$\pm 1, \pm 2$

Test $x = 2$:

$$\begin{array}{r|rrrr} 2 & 1 & -2 & 1 & -2 \\ & & 2 & 0 & 2 \\ \hline & 1 & 0 & 1 & 0 \end{array}$$

Descartes' Rule of Signs

# pos	# neg	# nonreal
3	0	0
1	0	2

$x^3 - 2x^2 + x - 2 = 0$

$(x-2)(x^2+1) = 0$

$x - 2 = 0$ **or** $x^2 + 1 = 0$

$x = 2$ $\quad x = \pm\sqrt{-1}$

$x = \pm i$

Solution set: $\{2, -i, i\}$

27. Possible rational roots

$\pm 1, \pm 2, \pm 3, \pm 6, \pm 9, \pm 18, \pm 27, \pm 54$

Descartes' Rule of Signs

# pos	# neg	# nonreal
1	3	0
1	1	2

$x^4 + x^3 + 3x^2 + 9x - 54 = 0$

$(x-2)(x^3 + 3x^2 + 9x + 27) = 0$

$(x-2)(x+3)(x^2+9) = 0$

$x^2 + 9 = 0 \Rightarrow x^2 = -9 \Rightarrow x = \pm 3i$

Solution set: $\{2, -3, \pm 3i\}$

Test $x = 2$:

$$\begin{array}{r|rrrrr} 2 & 1 & 1 & 3 & 9 & -54 \\ & & 2 & 6 & 18 & 54 \\ \hline & 1 & 3 & 9 & 27 & 0 \end{array}$$

Test $x = -3$:

$$\begin{array}{r|rrrr} -3 & 1 & 3 & 9 & 27 \\ & & -3 & 0 & -27 \\ \hline & 1 & 0 & 9 & 0 \end{array}$$

28. Possible rat. roots

$\pm 1, \pm 2, \pm 4, \pm 8$

Descartes' Rule of Signs

# pos	# neg	# nonreal
1	4	0
1	2	2
1	0	4

$x^5 + 2x^4 + 3x^3 + 6x^2 - 4x - 8 = 0$

$(x-1)(x^4 + 3x^3 + 6x^2 + 12x + 8) = 0$

$(x-1)(x+1)(x^3 + 2x^2 + 4x + 8) = 0$

$(x-1)(x+1)(x+2)(x^2+4) = 0$

$x^2 + 4 = 0 \Rightarrow x = \pm\sqrt{-4}$

Solution set: $\{1, -1, -2, \pm 2i\}$

Test $x = 1$:

$$\begin{array}{r|rrrrrr} 1 & 1 & 2 & 3 & 6 & -4 & -8 \\ & & 1 & 3 & 6 & 12 & 8 \\ \hline & 1 & 3 & 6 & 12 & 8 & 0 \end{array}$$

Test $x = -1$:

$$\begin{array}{r|rrrrr} -1 & 1 & 3 & 6 & 12 & 8 \\ & & -1 & -2 & -4 & -8 \\ \hline & 1 & 2 & 4 & 8 & 0 \end{array}$$

Test $x = -2$:

$$\begin{array}{r|rrrr} -2 & 1 & 2 & 4 & 8 \\ & & -2 & 0 & -8 \\ \hline & 1 & 0 & 4 & 0 \end{array}$$

29. $P(1) = 5$, $P(2) = 28$; The Intermediate Value Theorem does not guarantee a zero between 1 and 2.

CHAPTER 5 TEST

30. $P(x) = x^2 - 11$. Note that $P(3) = -2$ and $P(4) = 5$. There is a root between 3 and 4.

STEP	x_l	c	x_r	$P(x_l)$	$P(x_c)$	$P(x_r)$
0	3	3.5	4	-2.0	1.25	5.0
1	3	3.25	3.5	-2.0	-0.4375	1.25
2	3.25	3.375	3.5	-0.4375	0.3906	1.25
3	3.25	3.3125	3.375	-0.4375	-0.0273	0.3906
4	3.3125	3.34375	3.375			

To the nearest tenth, the solution is $x = 3.3$.

Cumulative Review Exercises (page 547)

1. $f(x) = 3^x - 2$; Shift $y = 3^x$ D2.

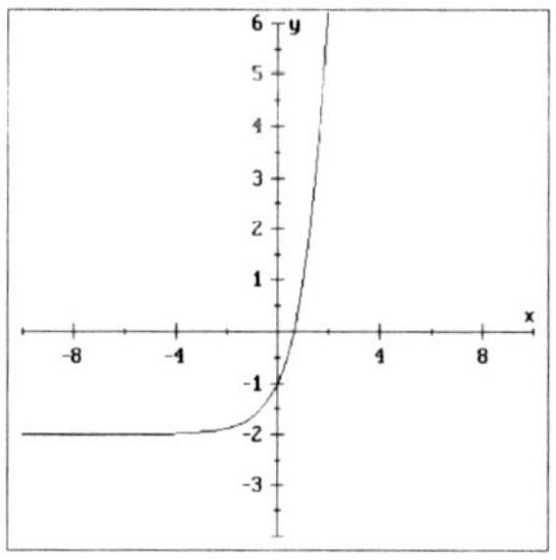

2. $f(x) = 2e^x$; Stretch $y = e^x$ vertically by a factor of 2.

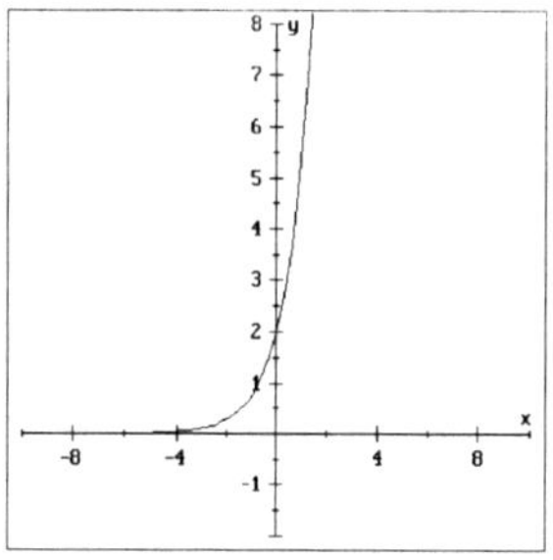

3. $f(x) = \log_3 x$

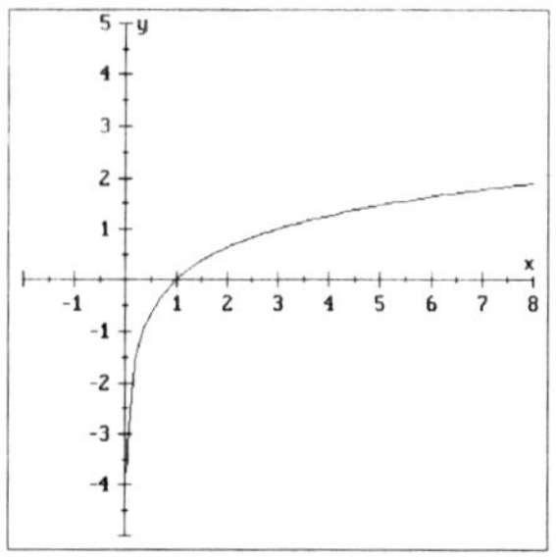

4. $f(x) = \ln(x - 2)$; Shift $y = \ln x$ R2.

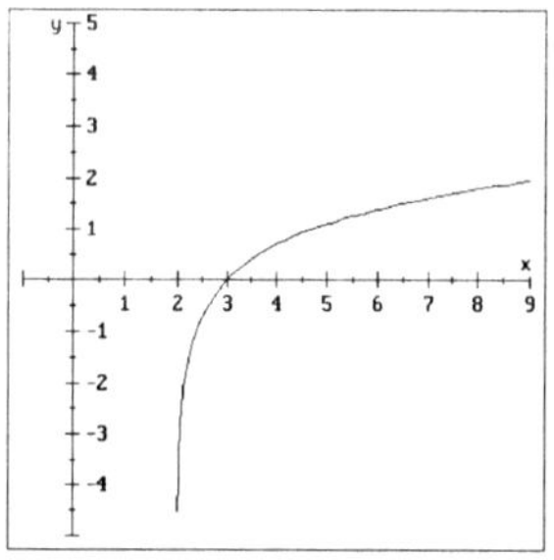

5. $\log_2 64 = 6$ (because $2^6 = 64$)

6. $\log_{1/2} 8 = -3$ $\left(\text{because } \left(\frac{1}{2}\right)^{-3} = 8\right)$

7. $\ln e^3 = 3 \ln e = 3$

8. $2^{\log_2 2} = 2$

9. $\log abc = \log a + \log b + \log c$

10. $\log \frac{a^2 b}{c} = \log a^2 b - \log c$
$= \log a^2 + \log b - \log c$
$= 2 \log a + \log b - \log c$

CUMULATIVE REVIEW EXERCISES

11. $\log\sqrt{\dfrac{ab}{c^3}} = \log\left(\dfrac{ab}{c^3}\right)^{1/2}$
$= \frac{1}{2}\log\left(\dfrac{ab}{c^3}\right)$
$= \frac{1}{2}(\log ab - \log c^3)$
$= \frac{1}{2}(\log a + \log b - 3\log c)$

12. $\ln\dfrac{\sqrt{ab^2}}{c} = \ln\dfrac{(ab^2)^{1/2}}{c}$
$= \ln\left(ab^2\right)^{1/2} - \ln c$
$= \frac{1}{2}\ln\left(ab^2\right) - \ln c$
$= \frac{1}{2}(\ln a + \ln b^2) - \ln c$
$= \frac{1}{2}(\ln a + 2\ln b) - \ln c$
$= \frac{1}{2}\ln a + \ln b - \ln c$

13. $3\ln a - 3\ln b = \ln a^3 - \ln b^3 = \ln\dfrac{a^3}{b^3}$

14. $\dfrac{1}{2}\log a + 3\log b - \dfrac{2}{3}\log c = \log a^{1/2} + \log b^3 - \log c^{2/3} = \log\dfrac{a^{1/2}b^3}{c^{2/3}} = \log\dfrac{\sqrt{a}\,b^3}{\sqrt[3]{c^2}}$

15.
$$\begin{aligned} 3^{x+1} &= 8 \\ \log 3^{x+1} &= \log 8 \\ (x+1)\log 3 &= \log 8 \\ x+1 &= \frac{\log 8}{\log 3} \\ x &= \frac{\log 8}{\log 3} - 1 \end{aligned}$$

16.
$$\begin{aligned} 3^{x-1} &= 3^{2x} \\ x-1 &= 2x \\ -1 &= x \end{aligned}$$

17.
$$\begin{aligned} \log x + \log 2 &= 3 \\ \log 2x &= 3 \\ 10^3 &= 2x \\ 1000 &= 2x \\ 500 &= x \end{aligned}$$

18.
$$\begin{aligned} \log(x+1) + \log(x-1) &= 1 \\ \log(x+1)(x-1) &= 1 \\ \log\left(x^2-1\right) &= 1 \\ 10^1 &= x^2 - 1 \\ 10 &= x^2 - 1 \\ 11 &= x^2 \\ \pm\sqrt{11} &= x \end{aligned}$$

Only the positive answer, $x = \sqrt{11}$, checks.

19. $P(1) = 9$

1	4	0	3	2
		4	4	7
	4	4	7	9

20. $P(-2) = -36$

−2	4	0	3	2
		−8	16	−38
	4	−8	19	−36

21. $P\left(\frac{1}{2}\right) = 4$

$\frac{1}{2}$	4	0	3	2
		2	1	2
	4	2	4	4

22. $P(i) = 2 - i$

i	4	0	3	2
		$4i$	-4	$-i$
	4	$4i$	-1	$2-i$

23. factor

−1	1	2	−1	−2
		−1	−1	2
	1	1	−2	0

24. not a factor

2	1	2	−1	−2
		2	8	14
	1	4	7	12

CUMULATIVE REVIEW EXERCISES

25.
$$\begin{array}{r|rrrr} 1 & 1 & 2 & -1 & -2 \\ & & 1 & 3 & 2 \\ \hline & 1 & 3 & 2 & 0 \end{array} \quad \text{factor}$$

26.
$$\begin{array}{r|rrrr} -2 & 1 & 2 & -1 & -2 \\ & & -2 & 0 & 2 \\ \hline & 1 & 0 & -1 & 0 \end{array} \quad \text{factor}$$

27. $x^{12} - 4x^8 + 2x^4 + 12 = 0 \Rightarrow$ 12 roots

28. $x^{2000} - 1 = 0 \Rightarrow$ 2000 roots

29. $P(x) = x^4 + 2x^3 - 3x^2 + x + 2$: 2 sign variations $\Rightarrow$ 2 or 0 positive roots

$P(-x) = (-x)^4 + 2(-x)^3 - 3(-x)^2 + (-x) + 2$

$= x^4 - 2x^3 - 3x^2 - x + 2$: 2 sign variations $\Rightarrow$ 2 or 0 negative roots

# pos	# neg	# nonreal
2	2	0
2	0	2
0	2	2
0	0	4

30. $P(x) = x^4 - 3x^3 - 2x^2 - 3x - 5$

1 sign variation $\Rightarrow$ 1 positive root

$P(-x) = (-x)^4 - 3(-x)^3 - 2(-x)^2 - 3(-x) - 5$

$= x^4 + 3x^3 - 2x^2 + 3x - 3$

3 sign variations $\Rightarrow$ 3 or 1 negative roots

# pos	# neg	# nonreal
1	3	0
1	1	2

31. Possible rational roots

$\pm 1, \pm 3, \pm 9$

Descartes' Rule of Signs

# pos	# neg	# nonreal
1	2	0
1	0	2

Test $x = -1$:
$$\begin{array}{r|rrrr} -1 & 1 & 1 & -9 & -9 \\ & & -1 & 0 & 9 \\ \hline & 1 & 0 & -9 & 0 \end{array}$$

$x^3 + x^2 - 9x - 9 = 0$

$(x+1)(x^2 - 9) = 0$

$(x+1)(x+3)(x-3) = 0$

Solution set: $\{-1, -3, 3\}$

32. Possible rational roots

$\pm 1, \pm 2$

Descartes' Rule of Signs

# pos	# neg	# nonreal
2	1	0
0	1	2

Test $x = -1$:
$$\begin{array}{r|rrrr} -1 & 1 & -2 & -1 & 2 \\ & & -1 & 3 & -2 \\ \hline & 1 & -3 & 2 & 0 \end{array}$$

$x^3 - 2x^2 - x + 2 = 0$

$(x+1)(x^2 - 3x + 2) = 0$

$(x+1)(x-1)(x-2) = 0$

Solution set: $\{-1, 1, 2\}$

Exercises 6.1 (page 560)

1. system

3. consistent

5. independent

7. consistent

9. dependent

11. is

13. $\begin{cases} y = -3x + 5 \\ x - 2y = -3 \end{cases}$

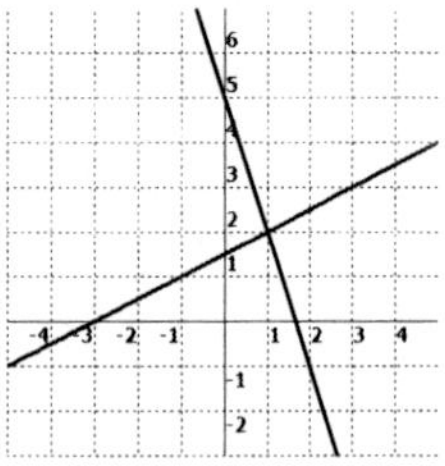

solution: $(1, 2)$

15. $\begin{cases} 3x + 2y = 2 \\ -2x + 3y = 16 \end{cases}$

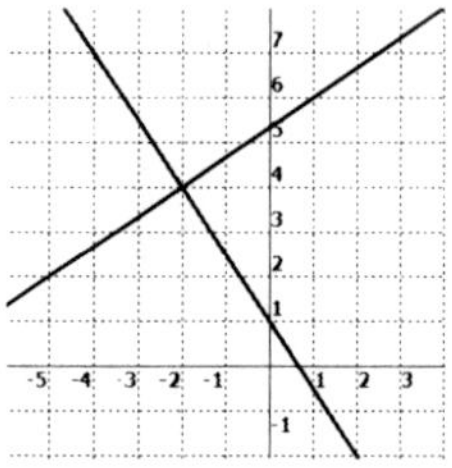

solution: $(-2, 4)$

17. $\begin{cases} y = -x + 5 \\ 3x + 3y = 27 \end{cases}$

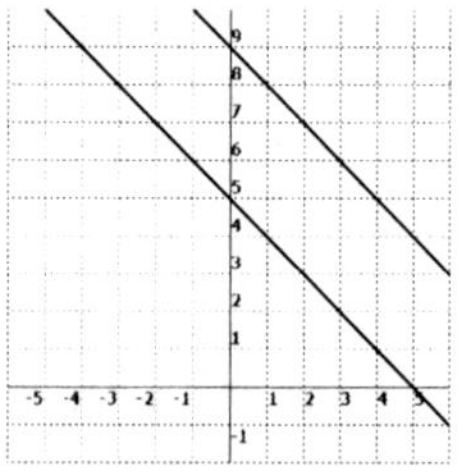

no solution
inconsistent system

19. $\begin{cases} y = -x + 6 \\ 5x + 5y = 30 \end{cases}$

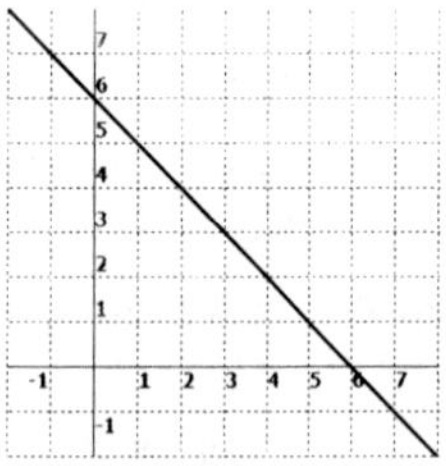

dependent equations
infinitely many solutions

21. $\begin{cases} y = -5.7x + 7.8 \\ y = 37.2 - 19.1x \end{cases}$

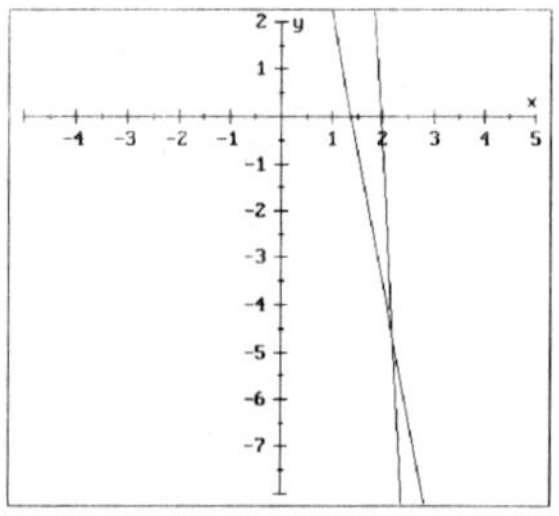

solution: $(2.2, -4.7)$

23. $\begin{cases} y = \dfrac{5.5 - 2.7x}{3.5} \\ 5.3x - 9.2y = 6.0 \end{cases}$

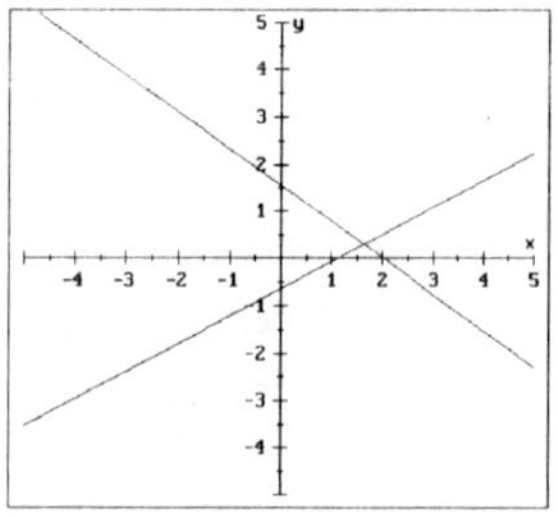

solution: $(1.7, 0.3)$

25. $\begin{cases} (1) \quad y = x - 1 \\ (2) \quad y = 2x \end{cases}$

Substitute $y = x - 1$ from (1) into (2):

$$y = 2x$$
$$x - 1 = 2x$$
$$-1 = x$$

Substitute and solve for y:

$y = 2x$

$y = 2(-1) = -2$

$(-1, -2)$

27. $\begin{cases} (1) \quad 2x + 3y = 0 \\ (2) \quad y = 3x - 11 \end{cases}$

Substitute $y = 3x - 11$ from (2) into (1):

$$2x + 3y = 0$$
$$2x + 3(3x - 11) = 0$$
$$2x + 9x - 33 = 0$$
$$11x = 33$$
$$x = 3$$

Substitute and solve for y:

$y = 3x - 11$

$y = 3(3) - 11 = -2$

$(3, -2)$

29. $\begin{cases} (1) & 4x+3y=3 \\ (2) & 2x-6y=-1 \end{cases}$

Substitute $y=\dfrac{3-4x}{3}$ from (1) into (2):

$$\begin{aligned} 2x-6y&=-1 \\ 2x-6\cdot\frac{3-4x}{3}&=-1 \\ 2x-2(3-4x)&=-1 \\ 2x-6+8x&=-1 \\ 10x&=5 \\ x&=\tfrac{1}{2} \end{aligned}$$

Substitute and solve for y:

$$\begin{aligned} 4x+3y&=3 \\ 4\left(\tfrac{1}{2}\right)+3y&=3 \\ 2+3y&=3 \\ 3y&=1 \\ y&=\tfrac{1}{3} \end{aligned}$$

Solution: $\left(\frac{1}{2},\frac{1}{3}\right)$

31. $\begin{cases} (1) & x+3y=1 \\ (2) & 2x+6y=3 \end{cases}$

Substitute $x=1-3y$ from (1) into (2):

$$\begin{aligned} 2x+6y&=3 \\ 2(1-3y)+6y&=3 \\ 2-6y+6y&=3 \\ 2&\neq 3 \end{aligned}$$

Inconsistent system $\Rightarrow$ No solution

33. $\begin{cases} (1) & y=3x-6 \\ (2) & x=\frac{1}{3}y+2 \end{cases}$

Substitute $x=\frac{1}{3}y+2$ from (2) into (1):

$y=3x-6$

$y=3\left(\frac{1}{3}y+2\right)-6$

$y=y+6-6$

$0=0$

Dependent equations

General solution: $(x,3x-6)$

35. $\begin{aligned} 5x-3y&=12 \Rightarrow \times(-1) \\ \underline{2x-3y}&\underline{=\ 3} \end{aligned}$ $\quad \begin{aligned} -5x+3y&=-12 \\ \underline{2x-3y}&\underline{=\ \ 3} \\ -3x&=-9 \\ x&=3 \end{aligned}$ $\quad \begin{aligned} 2x-3y&=3 \\ 2(3)-3y&=3 \\ 6-3y&=3 \\ -3y&=-3 \\ y&=1 \end{aligned}$

Solution: $(3,1)$

37. $\begin{aligned} x-7y&=-11 \Rightarrow \times(-8) \\ \underline{8x+2y}&\underline{=\ 28} \end{aligned}$ $\quad \begin{aligned} -8x+56y&=88 \\ \underline{8x+2y}&\underline{=\ 28} \\ 58y&=116 \\ y&=2 \end{aligned}$ $\quad \begin{aligned} x-7y&=-11 \\ x-7(2)&=-11 \\ x-14&=-11 \\ x&=3 \end{aligned}$

Solution: $(3,2)$

39. $\begin{aligned} 3(x-y)=y-9 &\Rightarrow 3x-3y=y-9 \Rightarrow 3x-4y=-9 \Rightarrow \times(5) \\ \underline{5(x+y)=-15} &\Rightarrow \underline{5x+5y=-15} \Rightarrow \underline{5x+5y=-15} \Rightarrow \times(4) \end{aligned}$ $\quad \begin{aligned} 15x-20y&=-45 \\ \underline{20x+20y}&\underline{=-60} \\ 35x&=-105 \\ x&=-3 \end{aligned}$

$$\begin{aligned} 5x+5y&=-15 \\ 5(-3)+5y&=-15 \\ -15+5y&=-15 \\ 5y&=0 \\ y&=0 \end{aligned}$$

Solution: $(-3,0)$

41.
$$2=\frac{1}{x+y}\Rightarrow 2(x+y)=1\Rightarrow 2x+2y=1$$
$$2=\frac{3}{x-y}\Rightarrow 2(x-y)=3\Rightarrow 2x-2y=3$$
$$4x=4$$
$$x=1$$

$$2x+2y=1$$
$$2(1)+2y=1$$
$$2+2y=1$$
$$2y=-1$$
$$y=-\tfrac{1}{2}$$

Solution: $(1,-\tfrac{1}{2})$

43.
$$y+2x=5\Rightarrow 2x+y=5$$
$$0.5y=2.5-x\Rightarrow x+0.5y=2.5\Rightarrow\times(-2)$$

$$2x+y=5$$
$$-2x-y=-5$$
$$0=0$$

Dependent Equations

$$2x+y=5$$
$$y=-2x+5$$

Gen. sol.: $(x,-2x+5)$

45.
$$x+2(x-y)=2\Rightarrow x+2x-2y=2\Rightarrow 3x-2y=2$$
$$3(y-x)-y=5\Rightarrow 3y-3x-y=5\Rightarrow -3x+2y=5$$
$$0\neq 7$$

No Solution

Inconsistent system

47.
$$x+\frac{y}{3}=\frac{5}{3}\Rightarrow\times(3)\quad 3x+y=5\Rightarrow 3x+y=5$$
$$\frac{x+y}{3}=3-x\Rightarrow\times(3)\quad x+y=9-3x\Rightarrow 4x+y=9\Rightarrow\times(-1)$$

$$3x+y=5$$
$$-4x-y=-9$$
$$-x=-4$$
$$x=4$$

$$3x+y=5$$
$$3(4)+y=5$$
$$12+y=5$$
$$y=-7$$

Solution: $(4,-7)$

49.
$$\frac{3}{2}x+\frac{1}{3}y=2\Rightarrow\times(6)\quad 9x+2y=12\Rightarrow$$
$$\frac{2}{3}x+\frac{1}{9}y=1\Rightarrow\times(9)\quad 6x+y=9\Rightarrow\times(-2)$$

$$9x+2y=12$$
$$-12x-2y=-18$$
$$-3x=-6$$
$$x=2$$

$$6x+y=9$$
$$6(2)+y=9$$
$$12+y=9$$
$$y=-3$$

Solution: $(2,-3)$

51.
$$\frac{x-y}{5}+\frac{x+y}{2}=6\Rightarrow\times(10)\quad 2(x-y)+5(x+y)=60\Rightarrow 2x-2y+5x+5y=60$$
$$\frac{x-y}{2}-\frac{x+y}{4}=3\Rightarrow\times(8)\quad 4(x-y)-2(x+y)=24\Rightarrow 4x-4y-2x-2y=24$$

$$7x+3y=60\Rightarrow\times(2)$$
$$2x-6y=24$$

$$14x+6y=120$$
$$2x-6y=24$$
$$16x=144$$
$$x=9$$

$$7x+3y=60$$
$$7(9)+3y=60$$
$$63+3y=60$$
$$3y=-3$$
$$y=-1$$

Solution: $(9,-1)$

53.

$$\begin{array}{ll} (1) & x+y+z=3 \\ (2) & 2x+y+z=4 \\ (3) & 3x+y-z=5 \end{array} \qquad \text{Add (1) and (3):} \; \begin{array}{ll} (1) & x+\;y+z=3 \\ (3) & 3x+\;y-z=5 \\ \hline (4) & 4x+2y \quad\;\; =8 \end{array} \qquad \text{Add equations (2) and (3):} \; \begin{array}{ll} (2) & 2x+\;y+z=4 \\ (3) & 3x+\;y-z=5 \\ \hline (5) & 5x+2y \quad\;\; =9 \end{array}$$

Solve the system of two equations and two unknowns formed by equations (4) and (5):

$$\begin{array}{l} 4x+2y=8 \Rightarrow \times(-1) \\ \underline{5x+2y=9 \Rightarrow} \end{array} \quad \begin{array}{r} -4x-2y=-8 \\ 5x+2y=\;\;\,9 \\ \hline x \quad\;\; =\;\;\,1 \end{array} \quad \begin{array}{r} 4x+2y=8 \\ 4(1)+2y=8 \\ 2y=4 \\ y=2 \end{array} \quad \begin{array}{r} x+y+z=3 \\ 1+2+z=3 \\ 3+z=3 \\ z=0 \end{array} \quad \begin{array}{l} \text{Solution:} \\ (1,2,0) \end{array}$$

55.

$$\begin{array}{ll} (1) & x-y+z=0 \\ (2) & x+y+2z=-1 \\ (3) & -x-y+z=0 \end{array} \qquad \text{Add (1) and (2):} \; \begin{array}{ll} (1) & x-y+\;\;z=\;\;\,0 \\ (2) & x+y+2z=-1 \\ \hline (4) & 2x \quad\;\; +3z=-1 \end{array} \qquad \text{Add equations (2) and (3):} \; \begin{array}{ll} (2) & x+y+2z=-1 \\ (3) & -x-y+\;\;z=\;\;\,0 \\ \hline (5) & 3z=-1 \\ & z=-\frac{1}{3} \end{array}$$

Solve the system of two equations and two unknowns formed by equations (4) and (5):

$$\begin{array}{r} 2x+3z=-1 \\ 2x+3\left(-\frac{1}{3}\right)=-1 \\ 2x=0 \\ x=0 \end{array} \quad \begin{array}{r} x-y+z=0 \\ 0-y+\left(-\frac{1}{3}\right)=0 \\ -y=\frac{1}{3} \\ y=-\frac{1}{3} \end{array} \quad \begin{array}{l} \text{Solution:} \\ \left(0,-\frac{1}{3},-\frac{1}{3}\right) \end{array}$$

57.

$$\begin{array}{ll} (1) & 2x+y=4 \\ (2) & x-z=2 \\ (3) & y+z=1 \end{array} \qquad \text{Add (2) and (3):} \; \begin{array}{ll} (2) & x \quad\;\; -z=2 \\ (3) & \quad\;\; y+z=1 \\ \hline (4) & x+y \quad\;\; =3 \end{array}$$

Solve the system of two equations and two unknowns formed by equations (1) and (4):

$$\begin{array}{l} 2x+y=4 \Rightarrow \times(-1) \\ \underline{x+y=3 \Rightarrow} \end{array} \quad \begin{array}{r} -2x-y=-4 \\ x+y=\;\;\,3 \\ \hline -x \quad\;\; =-1 \\ x \quad\;\; =\;\;\,1 \end{array} \quad \begin{array}{r} x+y=3 \\ 1+y=3 \\ y=2 \end{array} \quad \begin{array}{r} y+z=1 \\ 2+z=1 \\ z=-1 \end{array} \quad \begin{array}{l} \text{Solution:} \\ (1,2,-1) \end{array}$$

59.

$$\begin{array}{ll} (1) & x+y+z=6 \\ (2) & 2x+y+3z=17 \\ (3) & x+y+2z=11 \end{array} \qquad \text{Add (1) and } -(2)\text{:} \; \begin{array}{ll} (1) & x+y+\;\;z=\;\;\,6 \\ -(2) & -2x-y-3z=-17 \\ \hline (4) & -x \quad\;\; -2z=-11 \end{array} \qquad \text{Add equations (1) and } -(3)\text{:} \; \begin{array}{ll} (1) & x+y+\;\;z=\;\;\,6 \\ -(3) & -x-y-2z=-11 \\ \hline (5) & -z=-5 \\ & z=\;\;\,5 \end{array}$$

Solve the system of two equations and two unknowns formed by equations (4) and (5):

$$\begin{array}{r} -x-2z=-11 \\ -x-2(5)=-11 \\ -x=-1 \\ x=1 \end{array} \quad \begin{array}{r} x+y+z=6 \\ 1+y+5=6 \\ y=0 \end{array} \quad \text{Solution: } (1,0,5)$$

61. $\begin{array}{ll} (1) & x+y+z=3 \\ (2) & x+z=2 \\ (3) & 2x+2y+2z=3 \end{array}$ Add $-2\cdot(1)$ and (3): $\begin{array}{lr} -2\cdot(1) & -2x-2y-2z=-6 \\ (3) & 2x+2y+2z=\ \ 3 \\ \hline (4) & 0\neq -3 \end{array}$

No solution; inconsistent system

63. $\begin{array}{ll} (1) & x+2y-z=2 \\ (2) & 2x-y=-1 \\ (3) & 3x+y+z=1 \end{array}$ Add (1) and (3): $\begin{array}{ll} (1) & x+2y-z=2 \\ (3) & 3x+\ \ y+z=1 \\ \hline (4) & 4x+3y\ \ \ \ \ =3 \end{array}$

Solve the system of two equations and two unknowns formed by equations (2) and (4):

$\begin{array}{l} 2x-\ \ y=-1 \Rightarrow \times(3) \\ \underline{4x+3y=\ \ 3 \Rightarrow} \end{array}$ $\begin{array}{rcl} 6x-3y &=& -3 \\ 4x+3y &=& 3 \\ \hline 10x &=& 0 \\ x &=& 0 \end{array}$ $\begin{array}{r} 2x-y=-1 \\ 2(0)-y=-1 \\ -y=-1 \\ y=1 \end{array}$ $\begin{array}{r} 3x+y+z=1 \\ 3(0)+1+z=1 \\ z=0 \end{array}$ Solution: $(0,1,0)$

65. $\begin{array}{ll} (1) & 3x+4y+2z=4 \\ (2) & 6x-2y+z=4 \\ (3) & 3x-8y-6z=-3 \end{array}$ Add (1) and $2\cdot(2)$: $\begin{array}{ll} (1) & 3x+4y+2z=\ \ 4 \\ 2\cdot(2) & 12x-4y+2z=\ \ 8 \\ \hline (4) & 15x\ \ \ \ \ \ \ +4z=12 \end{array}$ Add equations $2\cdot(1)$ and (3): $\begin{array}{ll} 2\cdot(1) & 6x+8y+4z=\ \ 8 \\ (3) & 3x-8y-6z=-3 \\ \hline (5) & 9x\ \ \ \ \ \ \ -2z=\ \ 5 \end{array}$

Solve the system of two equations and two unknowns formed by equations (4) and (5):

$\begin{array}{l} 15x+4z=12 \Rightarrow \\ \underline{\ \ 9x-2z=\ \ 5 \Rightarrow} \times(2) \end{array}$ $\begin{array}{rcl} 15x+4z &=& 12 \\ 18x-4z &=& 10 \\ \hline 33x &=& 22 \\ x &=& \frac{2}{3} \end{array}$ $\begin{array}{r} 15x+4z=12 \\ 15\left(\frac{2}{3}\right)+4z=12 \\ 4z=2 \\ z=\frac{1}{2} \end{array}$ $\begin{array}{r} 3x+4y+2z=4 \\ 3\left(\frac{2}{3}\right)+4y+2\left(\frac{1}{2}\right)=4 \\ 4y=1 \\ y=\frac{1}{4} \end{array}$

Solution: $\left(\frac{2}{3},\frac{1}{4},\frac{1}{2}\right)$

67. $\begin{array}{ll} (1) & 2x-y-z=0 \\ (2) & x-2y-z=-1 \\ (3) & x-y-2z=-1 \end{array}$ Add (1) and $-2\cdot(2)$: $\begin{array}{ll} (1) & 2x-\ \ y-\ \ z=0 \\ -2\cdot(2) & -2x+4y+2z=2 \\ \hline (4) & 3y+\ \ z=2 \end{array}$ Add equations (1) and $-2\cdot(3)$: $\begin{array}{ll} (1) & 2x-\ \ y-\ \ z=0 \\ (3) & -2x+2y+4z=2 \\ \hline (5) & y+3z=2 \end{array}$

Solve the system of two equations and two unknowns formed by equations (4) and (5):

$\begin{array}{l} 3y+\ \ z=2 \Rightarrow \\ \underline{\ \ y+3z=2 \Rightarrow} \times(-3) \end{array}$ $\begin{array}{rcl} 3y+\ \ z &=& 2 \\ -3y-9z &=& -6 \\ \hline -8z &=& -4 \\ z &=& \frac{1}{2} \end{array}$ $\begin{array}{r} y+3z=2 \\ y+3\left(\frac{1}{2}\right)=2 \\ y=\frac{1}{2} \end{array}$ $\begin{array}{r} x-y-2z=-1 \\ x-\frac{1}{2}-2\left(\frac{1}{2}\right)=-1 \\ x=\frac{1}{2} \end{array}$

Solution: $\left(\frac{1}{2},\frac{1}{2},\frac{1}{2}\right)$

69.
$$\begin{array}{llr}
(1) & (x+y)+(y+z)+(z+x)=6 \Rightarrow & 2x+2y+2z=6 \\
(2) & (x-y)+(y-z)+(z-x)=0 \Rightarrow & 0=0 \\
(3) & x+y+2z=4 \Rightarrow & x+y+2z=4
\end{array}$$

Add (1) and $-2\cdot(3)$:
$$\begin{array}{lrcr}
(1) & 2x+2y+2z &=& 6 \\
-2\cdot(3) & -2x-2y-4z &=& -8 \\ \hline
(4) & -2z &=& -2 \\
 & z &=& 1
\end{array}$$

Since (2) is always true, the equations are dependent. z must equal 1. Then, from (3), $x+y=2$. Let $x=$ any real #. Then $y=2-x$. Solution: $\boxed{(x, 2-x, 1)}$

71. Let $x=$ cost of a hamburger and let $y=$ cost of the fries. Then $\begin{cases} (1) & 2x+4y=8 \\ (2) & 3x+2y=8 \end{cases}$

$$\begin{array}{ll}
2x+4y=8 & \\
3x+2y=8 \Rightarrow \times(-2) &
\end{array}
\qquad
\begin{array}{rcr}
2x+4y &=& 8 \\
-6x-4y &=& -16 \\ \hline
-4x &=& -8 \\
x &=& 2
\end{array}
\qquad
\begin{array}{r}
2x+4y=8 \\
2(2)+4y=8 \\
4y=4 \\
y=1
\end{array}$$

A hamburger costs \$2, while an order of fries costs \$1.

73. Let $x=$ acres of corn and let $y=$ acres of soybeans. Then $\begin{cases} (1) & x+y=350 \\ (2) & x=y+100 \end{cases}$

Substitute $x=y+100$ from (2) into (1):
$$\begin{aligned}
x+y&=350 \\
y+100+y&=350 \\
2y&=250 \\
y&=125
\end{aligned}$$

Substitute and solve for x:
$$\begin{aligned}
x&=y+100 \\
&=125+100=225
\end{aligned}$$

The farmer should plant 225 acres of corn and 125 acres of soybeans.

75. Let $b=$ speed in still water. Let $c=$ the speed of the current. Then the following system applies:

$$\begin{array}{ll}
3(b+c)=30 & 3b+3c=30 \Rightarrow \times\ (5) \\
5(b-c)=30 & 5b-5c=30 \Rightarrow \times\ (3)
\end{array}
\qquad
\begin{array}{rcr}
15b+15c &=& 150 \\
15b-15c &=& 90 \\ \hline
30b &=& 240 \\
b &=& 8
\end{array}$$

The boat has a speed of 8 kilometers per hour in still water.

77. Let $x=$ grams of 9% alloy. Let $y=$ grams of 84% alloy. Then the following system applies: (note: 34% of 60 is $0.34\times 60=20.4$)

$$\begin{array}{l}
x+y=60 \Rightarrow \times\ (-9) \\
0.09x+0.84y=20.4 \Rightarrow \times\ (100)
\end{array}
\qquad
\begin{array}{rcr}
-9x-9y &=& -540 \\
9x+84y &=& 2040 \\ \hline
75y &=& 1500 \\
y &=& 20
\end{array}$$

She must use 40 grams of the 9% and 20 grams of the 84% alloy.

79.
$$\begin{array}{rl}
448x=112y & \Rightarrow 448x-112y=0 \Rightarrow \times(-1) \\
448(x+1)=192(y-1) & \Rightarrow 448x-192y=-640 \Rightarrow
\end{array}
\qquad
\begin{array}{rcr}
-448x+112y &=& 0 \\
448x-192y &=& -640 \\ \hline
-80y &=& -640 \\
y &=& 8
\end{array}$$

$$\begin{aligned}
448x&=112y \\
448x&=112(8) \\
x&=2
\end{aligned}$$

The lever has a length of 10 feet.

81.

$$E(x) = 43.53x + 742.72 \qquad R(x) = 89.95x$$

$$\begin{aligned} E(x) &= R(x) \\ 43.53x + 742.72 &= 89.95x \\ 742.72 &= 46.42x \\ 16 &= x \Rightarrow \boxed{\text{Daily production should be 16 pairs.}} \end{aligned}$$

83. Let x = hours at fast food restaurant, y = hours at gas station and z = janitorial hours.

(1) $x = 15$
(2) $x + y + z = 30$
(3) $5.7x + 6.3y + 10z = 198.50$

Substitute $x = 15$ into (2) and (3):

(2) $15 + y + z = 30$
(3) $5.7(15) + 6.3y + 10z = 198.5$

Solve the system of two equations and two unknowns formed by equations (2) and (3):

$$\begin{aligned} y + z &= 15 \Rightarrow \times(-10) & -10y - 10z &= -150 \\ 6.3y + 10z &= 113 \Rightarrow \times(-3) & 6.3y + 10z &= 113 \\ & & -3.7y &= -37 \\ & & y &= 10 \end{aligned}$$

$$\begin{aligned} y + z &= 15 \\ 10 + z &= 15 \\ z &= 5 \end{aligned}$$

He spends 15 hours cooking, 10 hours at the gas station and 5 hours doing janitorial work.

85. Let x = # between 0-14, y = # between 15-49 and z = # 50 or over.

(1) $x + y + z = 3$
(2) $x + y = 2.61$
(3) $y + z = 1.95$

Add (1) and $-(2)$:

$$\begin{aligned} (1) \quad x + y + z &= 3 \\ -(2) \quad -x - y \quad &= -2.61 \\ z &= 0.39 \end{aligned}$$

Substitute $z = 0.39$ into (3):

$$\begin{aligned} y + z &= 1.95 \\ y + 0.39 &= 1.95 \\ y &= 1.56 \end{aligned}$$

Substitute $y = 1.56$ into (2):

$$\begin{aligned} x + y &= 2.61 \\ x + 1.56 &= 2.61 \\ x &= 1.05 \end{aligned}$$

There are 1.05 million between 0-14, 1.56 million between 15-49 and 0.39 million over 50.

87. Let x = the smallest angle, y = the middle angle and z = the largest angle.

(1) $x + y + z = 180$
(2) $z = x + y + 20$
(3) $z = 3x + 10$

Substitute (3) into (1):

$$\begin{aligned} x + y + 3x + 10 &= 180 \\ 4x + y &= 170 \quad (4) \end{aligned}$$

Substitute (3) and (2):

$$\begin{aligned} 3x + 10 &= x + y + 20 \\ 2x - y &= 10 \quad (5) \end{aligned}$$

Solve the system of two equations and two unknowns formed by equations (4) and (5):

$$\begin{aligned} 4x + y &= 170 \\ 2x - y &= 10 \\ 6x &= 180 \\ x &= 30 \end{aligned}$$

$$\begin{aligned} 4x + y &= 170 \\ 4(30) + y &= 170 \\ 120 + y &= 170 \\ y &= 50 \end{aligned}$$

$$\begin{aligned} z &= 3x + 10 \\ z &= 3(30) + 10 \\ z &= 90 + 10 \\ z &= 100 \end{aligned}$$

Solution: The angles have measures of 30°, 50° and 100°.

89-93. Answers may vary.

95. $y = 3^x$; points: $(0, 1)$, $(1, 3)$

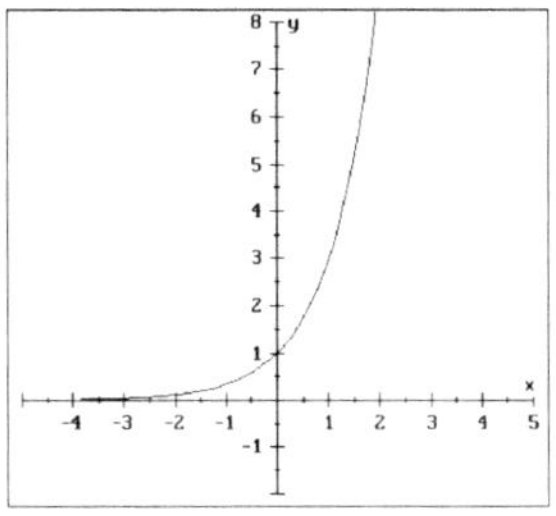

97.
$$\begin{aligned} 3 &= \log_2 x \\ 2^3 &= x \\ 8 &= x \end{aligned}$$

99. $\log \dfrac{x}{y^2 z} = \log x - \log\left(y^2 z\right) = \log x - \left(\log y^2 + \log z\right) = \log x - 2\log y - \log z$

101. $\log x + 3\log y - \dfrac{1}{2}\log z = \log x + \log y^3 + \log z^{-1/2} = \log x + \log y^3 + \log \dfrac{1}{\sqrt{z}} = \log \dfrac{xy^3}{\sqrt{z}}$

Exercises 6.2 (page 574)

1. matrix

3. coefficient

5. equation

7. row equivalent

9. interchanged

11. adding, multiple

13. $\begin{cases} (1) & x + y = 5 \\ (2) & x - 2y = -4 \end{cases} \Rightarrow \begin{cases} (1) & x + y = 5 \\ (2) & -3y = -9 \end{cases} \Rightarrow \begin{cases} (1) & x + y = 5 \\ (2) & y = 3 \end{cases}$

$-(1) + (2) \Rightarrow (2)$ $\quad -\frac{1}{3}(2) \Rightarrow (2)$

From (2): $y = 3$ From (1): $x + y = 5$, $x + 3 = 5$, $x = 2$ Solution: $(2, 3)$

15. $\begin{cases} (1) & x - y = 1 \\ (2) & 2x - y = 8 \end{cases} \Rightarrow \begin{cases} (1) & x - y = 1 \\ (2) & y = 6 \end{cases}$

$-2(1) + (2) \Rightarrow (2)$

From (2): $y = 6$ From (1): $x - y = 1$, $x - 6 = 1$, $x = 7$ Solution: $(7, 6)$

17. $\begin{cases} (1) & x + 2y - z = 2 \\ (2) & x - 3y + 2z = 1 \\ (3) & x + y - 3z = -6 \end{cases} \Rightarrow \begin{cases} (1) & x + 2y - z = 2 \\ (2) & -5y + 3z = -1 \\ (3) & -y - 2z = -8 \end{cases} \Rightarrow \begin{cases} (1) & x + 2y - z = 2 \\ (2) & -5y + 3z = -1 \\ (3) & 13z = 39 \end{cases} \Rightarrow$

$-(1) + (2) \Rightarrow (2)$, $-(1) + (3) \Rightarrow (3)$ $\quad -5(3) + (2) \Rightarrow (3)$

continued on next page

17. continued

$$\begin{cases}(1) & x+2y-z=2\\(2) & y-\frac{3}{5}z=\frac{1}{5}\\(3) & z=3\end{cases}$$
$-\frac{1}{5}(2)\Rightarrow(2)$
$\frac{1}{13}(3)\Rightarrow(3)$

From (3): $z=3$

From (2): $y-\dfrac{3}{5}z=\dfrac{1}{5}$
$$y-\frac{3}{5}(3)=\frac{1}{5}$$
$$y=2$$

From (1): $x+2y-z=2$
$$x+2(2)-(3)=2$$
$$x=1$$

Solution: $(1,2,3)$

19. $$\begin{cases}(1) & x-y-z=-3\\(2) & 5x+y=6\\(3) & y+z=4\end{cases}\Rightarrow\begin{cases}(1) & x-y-z=-3\\(2) & 6y+5z=21\\(3) & y+z=4\end{cases}\Rightarrow\begin{cases}(1) & x-y-\phantom{\frac{5}{6}}z=-3\\(2) & y+\frac{5}{6}z=\frac{7}{2}\\(3) & -\phantom{\frac{5}{6}}z=-3\end{cases}\Rightarrow$$
$-5(1)+(2)\Rightarrow(2)$

$\frac{1}{6}(2)\Rightarrow(2)$
$-6(3)+(2)\Rightarrow(3)$

$$\begin{cases}(1) & x-y-\phantom{\frac{5}{6}}z=-3\\(2) & y+\frac{5}{6}z=\frac{7}{2}\\(3) & z=3\end{cases}\Rightarrow$$
$-(3)\Rightarrow(3)$

From (3): $z=3$

From (2): $y+\frac{5}{6}z=\frac{7}{2}$
$$y+\tfrac{5}{6}(3)=\tfrac{7}{2}$$
$$y=1$$

From (1): $x-y-z=-3$
$$x-(1)-(3)=-3$$
$$x=1$$

Solution: $(1,1,3)$

21. row echelon form

23. reduced row echelon form

25. $$\left[\begin{array}{cc|c}2 & 1 & 3\\1 & -3 & 5\end{array}\right]\Rightarrow\left[\begin{array}{cc|c}1 & -3 & 5\\2 & 1 & 3\end{array}\right]\Rightarrow\left[\begin{array}{cc|c}1 & -3 & 5\\0 & 7 & -7\end{array}\right]\Rightarrow\left[\begin{array}{cc|c}1 & -3 & 5\\0 & 1 & -1\end{array}\right]$$
$R_1\Leftrightarrow R_2$ $\quad -2R_1+R_2\Rightarrow R_2$ $\quad \frac{1}{7}R_2\Rightarrow R_2$

From R_2: $y=-1$

From R_1: $x-3y=5$
$$x-3(-1)=5$$
$$x+3=5$$
$$x=2$$

Solution: $(2,-1)$

27. $$\left[\begin{array}{cc|c}1 & -7 & -2\\5 & -2 & -10\end{array}\right]\Rightarrow\left[\begin{array}{cc|c}1 & -7 & -2\\0 & 33 & 0\end{array}\right]\Rightarrow\left[\begin{array}{cc|c}1 & -7 & -2\\0 & 1 & 0\end{array}\right]$$
$-5R_1+R_2\Rightarrow R_2$ $\quad \frac{1}{33}R_2\Rightarrow R_2$

From R_2: $y=0$

From R_1: $x-7y=-2$
$$x-7(0)=-2$$
$$x=-2$$

Solution: $(-2,0)$

29. $$\left[\begin{array}{cc|c}2 & -1 & 5\\1 & 3 & 6\end{array}\right]\Rightarrow\left[\begin{array}{cc|c}1 & 3 & 6\\2 & -1 & 5\end{array}\right]\Rightarrow\left[\begin{array}{cc|c}1 & 3 & 6\\0 & -7 & -7\end{array}\right]\Rightarrow\left[\begin{array}{cc|c}1 & 3 & 6\\0 & 1 & 1\end{array}\right]$$
$R_1\Leftrightarrow R_2$ $\quad -2R_1+R_2\Rightarrow R_2$ $\quad -\frac{1}{7}R_2\Rightarrow R_2$

From R_2: $y=1$

From R_1: $x+3y=6$
$$x+3(1)=6$$
$$x=3$$

Solution: $(3,1)$

31. $\left[\begin{array}{cc|c} 1 & -2 & 3 \\ -2 & 4 & 6 \end{array}\right] \Rightarrow \left[\begin{array}{cc|c} 1 & -2 & 3 \\ 0 & 0 & 12 \end{array}\right] \Rightarrow$ From R_2, $0x + 0y = 12$. This is impossible.

$2R_1 + R_2 \Rightarrow R_2$

No solution $\Rightarrow$ inconsistent system

33. $\left[\begin{array}{cc|c} 2 & -1 & 7 \\ -1 & \frac{1}{3} & -\frac{7}{3} \end{array}\right] \Rightarrow \left[\begin{array}{cc|c} 1 & -\frac{1}{3} & \frac{7}{3} \\ 2 & -1 & 7 \end{array}\right] \Rightarrow \left[\begin{array}{cc|c} 1 & -\frac{1}{3} & \frac{7}{3} \\ 0 & -\frac{1}{3} & \frac{7}{3} \end{array}\right] \Rightarrow \left[\begin{array}{cc|c} 1 & -\frac{1}{3} & \frac{7}{3} \\ 0 & 1 & -7 \end{array}\right]$

$R_1 \Leftrightarrow -R_2$ $\quad -2R_1 + R_2 \Rightarrow R_2$ $\quad -3R_2 \Rightarrow R_2$

From R_2: $y = -7$ From R_1: $x - \frac{1}{3}y = \frac{7}{3}$ Solution: $(0, -7)$

$x - \frac{1}{3}(-7) = \frac{7}{3}$

$x + \frac{7}{3} = \frac{7}{3}$

$x = 0$

35. $\left[\begin{array}{ccc|c} 1 & -1 & 1 & 3 \\ 2 & -1 & 1 & 4 \\ 1 & 2 & -1 & -1 \end{array}\right] \Rightarrow \left[\begin{array}{ccc|c} 1 & -1 & 1 & 3 \\ 0 & 1 & -1 & -2 \\ 0 & 3 & -2 & -4 \end{array}\right] \Rightarrow \left[\begin{array}{ccc|c} 1 & -1 & 1 & 3 \\ 0 & 1 & -1 & -2 \\ 0 & 0 & 1 & 2 \end{array}\right] \Rightarrow$

$-2R_1 + R_2 \Rightarrow R_2$ $\quad -3R_2 + R_3 \Rightarrow R_3$

$-R_1 + R_3 \Rightarrow R_3$

From R_3: $z = 2$ From R_1: $x - y + z = 3$

From R_2: $y - z = -2$ $\quad x - (0) + (2) = 3$

$y - (2) = -2$ $\quad x = 1$

$y = 0$ Solution: $(1, 0, 2)$

37. $\left[\begin{array}{ccc|c} 1 & 1 & -1 & -1 \\ 3 & 1 & 0 & 4 \\ 0 & 1 & -2 & -4 \end{array}\right] \Rightarrow \left[\begin{array}{ccc|c} 1 & 1 & -1 & -1 \\ 0 & -2 & 3 & 7 \\ 0 & 1 & -2 & -4 \end{array}\right] \Rightarrow \left[\begin{array}{ccc|c} 1 & 1 & -1 & -1 \\ 0 & 1 & -\frac{3}{2} & -\frac{7}{2} \\ 0 & 0 & -1 & -1 \end{array}\right] \Rightarrow \left[\begin{array}{ccc|c} 1 & 1 & -1 & -1 \\ 0 & 1 & -\frac{3}{2} & -\frac{7}{2} \\ 0 & 0 & 1 & 1 \end{array}\right]$

$-3R_1 + R_2 \Rightarrow R_2$ $\quad -\frac{1}{2}R_2 \Rightarrow R_2$ $\quad -R_3 \Rightarrow R_3$

$R_2 + 2R_3 \Rightarrow R_3$

From (3): $z = 1$ From (1): $x + y - z = -1$ Solution:

From (2): $y - \frac{3}{2}z = -\frac{7}{2}$ $\quad x + (-2) - (1) = -1$ $\quad (2, -2, 1)$

$y - \frac{3}{2}(1) = -\frac{7}{2}$ $\quad x = 2$

$y = -2$

39. $\left[\begin{array}{ccc|c} 1 & -1 & 1 & 2 \\ 2 & 1 & 1 & 5 \\ 3 & 0 & -4 & -5 \end{array}\right] \Rightarrow \left[\begin{array}{ccc|c} 1 & -1 & 1 & 2 \\ 0 & 3 & -1 & 1 \\ 0 & 3 & -7 & -11 \end{array}\right] \Rightarrow \left[\begin{array}{ccc|c} 1 & -1 & 1 & 2 \\ 0 & 1 & -\frac{1}{3} & \frac{1}{3} \\ 0 & 0 & 6 & 12 \end{array}\right] \Rightarrow$

$-2R_1 + R_2 \Rightarrow R_2$ $\quad \frac{1}{3}R_2 \Rightarrow R_2$

$-3R_1 + R_3 \Rightarrow R_3$ $\quad -R_3 + R_2 \Rightarrow R_3$

continued on next page

39. continued

$$\begin{bmatrix} 1 & -1 & 1 & | & 2 \\ 0 & 1 & -\frac{1}{3} & | & \frac{1}{3} \\ 0 & 0 & 1 & | & 2 \end{bmatrix}$$
$\frac{1}{6}R_3 \Rightarrow R_3$

From (3): $z = 2$

From (2): $y - \frac{1}{3}z = \frac{1}{3}$

$y - \frac{1}{3}(2) = \frac{1}{3}$

$y = 1$

From (1): $x - y + z = 2$

$x - (1) + (2) = 2$

$x = 1$

Solution: $(1, 1, 2)$

41. $$\begin{bmatrix} 1 & 1 & -1 & | & 5 \\ 1 & 1 & 1 & | & 2 \\ 3 & 3 & -1 & | & 12 \end{bmatrix} \Rightarrow \begin{bmatrix} 1 & 1 & -1 & | & 5 \\ 0 & 0 & 2 & | & -3 \\ 0 & 0 & 2 & | & -3 \end{bmatrix} \Rightarrow \begin{bmatrix} 1 & 1 & -1 & | & 5 \\ 0 & 0 & 1 & | & -\frac{3}{2} \\ 0 & 0 & 1 & | & -\frac{3}{2} \end{bmatrix}$$

$-R_1 + R_2 \Rightarrow R_2$, $-3R_1 + R_3 \Rightarrow R_3$; $\frac{1}{2}R_2 \Rightarrow R_2$, $\frac{1}{2}R_3 \Rightarrow R_3$

From R_2: $z = -\frac{3}{2}$ From R_1: $x + y - z = 5$

$x + y - \left(-\frac{3}{2}\right) = 5$

$y = -x + \frac{7}{2}$

Solution: $\left(x, -x + \frac{7}{2}, -\frac{3}{2}\right)$

43. $$\begin{bmatrix} 2 & -1 & 1 & | & 6 \\ 3 & 1 & -1 & | & 2 \\ -1 & 3 & -3 & | & 8 \end{bmatrix} \Rightarrow \begin{bmatrix} -1 & 3 & -3 & | & 8 \\ 3 & 1 & -1 & | & 2 \\ 2 & -1 & 1 & | & 6 \end{bmatrix} \Rightarrow \begin{bmatrix} -1 & 3 & -3 & | & 8 \\ 0 & 10 & -10 & | & 26 \\ 0 & 5 & -5 & | & 22 \end{bmatrix} \Rightarrow$$

$R_1 \Leftrightarrow R_3$; $3R_1 + R_2 \Rightarrow R_2$, $2R_1 + R_3 \Rightarrow R_3$

$$\begin{bmatrix} -1 & 3 & -3 & | & 8 \\ 0 & 10 & -10 & | & 26 \\ 0 & 0 & 0 & | & -18 \end{bmatrix}$$
$-2R_3 + R_2 \Rightarrow R_3$

R_3 indicates $0x + 0y + 0z = -18$. This is impossible.

No solution $\Rightarrow$ inconsistent system

45. $$\begin{bmatrix} 1 & -2 & | & 7 \\ 0 & 1 & | & 3 \end{bmatrix} \Rightarrow \begin{bmatrix} 1 & 0 & | & 13 \\ 0 & 1 & | & 3 \end{bmatrix} \Rightarrow \text{Solution: } (13, 3)$$
$2R_2 + R_1 \Rightarrow R_1$

47. $$\begin{bmatrix} 1 & 2 & -1 & | & 3 \\ 0 & 1 & 3 & | & 1 \\ 0 & 0 & 1 & | & -2 \end{bmatrix} \Rightarrow \begin{bmatrix} 1 & 0 & -7 & | & 1 \\ 0 & 1 & 3 & | & 1 \\ 0 & 0 & 1 & | & -2 \end{bmatrix} \Rightarrow \begin{bmatrix} 1 & 0 & 0 & | & -13 \\ 0 & 1 & 0 & | & 7 \\ 0 & 0 & 1 & | & -2 \end{bmatrix}$$

$-2R_2 + R_1 \Rightarrow R_1$; $7R_3 + R_1 \Rightarrow R_1$, $-3R_3 + R_2 \Rightarrow R_2$

Solution: $(-13, 7, -2)$

49. $$\begin{bmatrix} 1 & -1 & | & 7 \\ 1 & 1 & | & 13 \end{bmatrix} \Rightarrow \begin{bmatrix} 1 & -1 & | & 7 \\ 0 & 2 & | & 6 \end{bmatrix} \Rightarrow \begin{bmatrix} 1 & -1 & | & 7 \\ 0 & 1 & | & 3 \end{bmatrix} \Rightarrow \begin{bmatrix} 1 & 0 & | & 10 \\ 0 & 1 & | & 3 \end{bmatrix}$$

$-R_1 + R_2 \Rightarrow R_2$; $\frac{1}{2}R_2 \Rightarrow R_2$; $R_2 + R_1 \Rightarrow R_1$

Solution: $(10, 3)$

51. $$\begin{bmatrix} 1 & -\frac{1}{2} & | & 0 \\ 1 & 2 & | & 0 \end{bmatrix} \Rightarrow \begin{bmatrix} 1 & -\frac{1}{2} & | & 0 \\ 0 & \frac{5}{2} & | & 0 \end{bmatrix} \Rightarrow \begin{bmatrix} 1 & -\frac{1}{2} & | & 0 \\ 0 & 1 & | & 0 \end{bmatrix} \Rightarrow \begin{bmatrix} 1 & 0 & | & 0 \\ 0 & 1 & | & 0 \end{bmatrix}$$

$-R_1 + R_2 \Rightarrow R_2$; $\frac{2}{5}R_2 \Rightarrow R_2$; $\frac{1}{2}R_2 + R_1 \Rightarrow R_1$

Solution: $(0, 0)$

SECTION 6.2

53. $\left[\begin{array}{ccc|c} 1 & 1 & 2 & 0 \\ 1 & 1 & 1 & 2 \\ 1 & 0 & 1 & 1 \end{array}\right] \Rightarrow \underset{\substack{-R_1+R_2 \Rightarrow R_2 \\ -R_1+R_3 \Rightarrow R_3}}{\left[\begin{array}{ccc|c} 1 & 1 & 2 & 0 \\ 0 & 0 & -1 & 2 \\ 0 & -1 & -1 & 1 \end{array}\right]} \Rightarrow \underset{R_2 \Leftrightarrow R_3}{\left[\begin{array}{ccc|c} 1 & 1 & 2 & 0 \\ 0 & -1 & -1 & 1 \\ 0 & 0 & -1 & 2 \end{array}\right]} \Rightarrow \underset{\substack{-R_2 \Rightarrow R_2 \\ -R_3 \Rightarrow R_3}}{\left[\begin{array}{ccc|c} 1 & 1 & 2 & 0 \\ 0 & 1 & 1 & -1 \\ 0 & 0 & 1 & -2 \end{array}\right]}$

$\underset{-R_2+R_1 \Rightarrow R_1}{\left[\begin{array}{ccc|c} 1 & 0 & 1 & 1 \\ 0 & 1 & 1 & -1 \\ 0 & 0 & 1 & -2 \end{array}\right]} \Rightarrow \underset{\substack{-R_3+R_1 \Rightarrow R_1 \\ -R_3+R_2 \Rightarrow R_2}}{\left[\begin{array}{ccc|c} 1 & 0 & 0 & 3 \\ 0 & 1 & 0 & 1 \\ 0 & 0 & 1 & -2 \end{array}\right]}$ Solution: $(3, 1, -2)$

55. $\left[\begin{array}{ccc|c} 2 & 1 & -2 & 1 \\ -1 & 1 & -3 & 0 \\ 4 & 3 & 0 & 4 \end{array}\right] \Rightarrow \underset{\substack{R_1+2R_2 \Rightarrow R_2 \\ -2R_1+R_3 \Rightarrow R_3}}{\left[\begin{array}{ccc|c} 2 & 1 & -2 & 1 \\ 0 & 3 & -8 & 1 \\ 0 & 1 & 4 & 2 \end{array}\right]} \Rightarrow \underset{\substack{-3R_1+R_2 \Rightarrow R_1 \\ -3R_3+R_2 \Rightarrow R_3}}{\left[\begin{array}{ccc|c} -6 & 0 & -2 & -2 \\ 0 & 3 & -8 & 1 \\ 0 & 0 & -20 & -5 \end{array}\right]} \Rightarrow \underset{-\frac{1}{20}R_3 \Rightarrow R_3}{\left[\begin{array}{ccc|c} -6 & 0 & -2 & -2 \\ 0 & 3 & -8 & 1 \\ 0 & 0 & 1 & \frac{1}{4} \end{array}\right]}$

$\underset{\substack{2R_3+R_1 \Rightarrow R_1 \\ 8R_3+R_2 \Rightarrow R_2}}{\left[\begin{array}{ccc|c} -6 & 0 & 0 & -\frac{3}{2} \\ 0 & 3 & 0 & 3 \\ 0 & 0 & 1 & \frac{1}{4} \end{array}\right]} \Rightarrow \underset{\substack{-\frac{1}{6}R_1 \Rightarrow R_1 \\ \frac{1}{3}R_2 \Rightarrow R_2}}{\left[\begin{array}{ccc|c} 1 & 0 & 0 & \frac{1}{4} \\ 0 & 1 & 0 & 1 \\ 0 & 0 & 1 & \frac{1}{4} \end{array}\right]}$ Solution: $\left(\frac{1}{4}, 1, \frac{1}{4}\right)$

57. $\left[\begin{array}{cccc|c} 2 & -2 & 3 & 1 & 2 \\ 1 & 1 & 1 & 1 & 5 \\ -1 & 2 & -3 & 2 & 2 \\ 1 & 1 & 2 & -1 & 4 \end{array}\right] \Rightarrow \underset{\substack{-2R_2+R_1 \Rightarrow R_2 \\ 2R_3+R_1 \Rightarrow R_3 \\ -2R_4+R_1 \Rightarrow R_4}}{\left[\begin{array}{cccc|c} 2 & -2 & 3 & 1 & 2 \\ 0 & -4 & 1 & -1 & -8 \\ 0 & 2 & -3 & 5 & 6 \\ 0 & -4 & -1 & 3 & -6 \end{array}\right]} \Rightarrow \underset{\substack{-2R_1+R_2 \Rightarrow R_1 \\ 2R_3+R_2 \Rightarrow R_3 \\ -R_4+R_2 \Rightarrow R_4}}{\left[\begin{array}{cccc|c} -4 & 0 & -5 & -3 & -12 \\ 0 & -4 & 1 & -1 & -8 \\ 0 & 0 & -5 & 9 & 4 \\ 0 & 0 & 2 & -4 & -2 \end{array}\right]}$

$\underset{\frac{1}{2}R_4 \Rightarrow R_4}{\left[\begin{array}{cccc|c} -4 & 0 & -5 & -3 & -12 \\ 0 & -4 & 1 & -1 & -8 \\ 0 & 0 & -5 & 9 & 4 \\ 0 & 0 & 1 & -2 & -1 \end{array}\right]} \Rightarrow \underset{\substack{-R_1+R_3 \Rightarrow R_1 \\ 5R_2+R_3 \Rightarrow R_2 \\ 5R_4+R_3 \Rightarrow R_4}}{\left[\begin{array}{cccc|c} 4 & 0 & 0 & 12 & 16 \\ 0 & -20 & 0 & 4 & -36 \\ 0 & 0 & -5 & 9 & 4 \\ 0 & 0 & 0 & -1 & -1 \end{array}\right]} \Rightarrow \underset{\substack{\frac{1}{4}R_1 \Rightarrow R_1 \\ \frac{1}{4}R_2 \Rightarrow R_2 \\ -R_4 \Rightarrow R_4}}{\left[\begin{array}{cccc|c} 1 & 0 & 0 & 3 & 4 \\ 0 & -5 & 0 & 1 & -9 \\ 0 & 0 & -5 & 9 & 4 \\ 0 & 0 & 0 & 1 & 1 \end{array}\right]}$

$\Rightarrow \underset{\substack{-3R_4+R_1 \Rightarrow R_1 \\ -R_4+R_2 \Rightarrow R_2 \\ -9R_4+R_3 \Rightarrow R_3}}{\left[\begin{array}{cccc|c} 1 & 0 & 0 & 0 & 1 \\ 0 & -5 & 0 & 0 & -10 \\ 0 & 0 & -5 & 0 & -5 \\ 0 & 0 & 0 & 1 & 1 \end{array}\right]} \Rightarrow \underset{\substack{-\frac{1}{5}R_2 \Rightarrow R_2 \\ -\frac{1}{5}R_3 \Rightarrow R_3}}{\left[\begin{array}{cccc|c} 1 & 0 & 0 & 0 & 1 \\ 0 & 1 & 0 & 0 & 2 \\ 0 & 0 & 1 & 0 & 1 \\ 0 & 0 & 0 & 1 & 1 \end{array}\right]}$ Solution: $(1, 2, 1, 1)$

59. $\left[\begin{array}{cccc|c} 1 & 1 & 0 & 1 & 4 \\ 1 & 0 & 1 & 1 & 2 \\ 2 & 2 & 1 & 2 & 8 \\ 1 & -1 & 1 & -1 & -2 \end{array}\right] \Rightarrow \left[\begin{array}{cccc|c} 1 & 1 & 0 & 1 & 4 \\ 0 & 1 & -1 & 0 & 2 \\ 0 & 0 & 1 & 0 & 0 \\ 0 & 2 & -1 & 2 & 6 \end{array}\right] \Rightarrow \left[\begin{array}{cccc|c} 1 & 0 & 1 & 1 & 2 \\ 0 & 1 & -1 & 0 & 2 \\ 0 & 0 & 1 & 0 & 0 \\ 0 & 0 & 1 & 2 & 2 \end{array}\right] \Rightarrow$

$-R_2 + R_1 \Rightarrow R_2$, $-2R_1 + R_3 \Rightarrow R_3$, $-R_4 + R_1 \Rightarrow R_4$

$-R_2 + R_1 \Rightarrow R_1$, $-2R_2 + R_4 \Rightarrow R_4$

$\left[\begin{array}{cccc|c} 1 & 0 & 0 & 1 & 2 \\ 0 & 1 & 0 & 0 & 2 \\ 0 & 0 & 1 & 0 & 0 \\ 0 & 0 & 0 & 2 & 2 \end{array}\right] \Rightarrow \left[\begin{array}{cccc|c} 1 & 0 & 0 & 0 & 1 \\ 0 & 1 & 0 & 0 & 2 \\ 0 & 0 & 1 & 0 & 0 \\ 0 & 0 & 0 & 1 & 1 \end{array}\right]$ Solution: $(1, 2, 0, 1)$

$-R_3 + R_1 \Rightarrow R_1$, $R_3 + R_2 \Rightarrow R_2$, $-R_3 + R_4 \Rightarrow R_4$

$-\frac{1}{2}R_4 + R_1 \Rightarrow R_1$, $\frac{1}{2}R_4 \Rightarrow R_4$

61. $\left[\begin{array}{ccc|c} \frac{1}{3} & \frac{3}{4} & -\frac{2}{3} & -2 \\ 1 & \frac{1}{2} & \frac{1}{3} & 1 \\ \frac{1}{6} & -\frac{1}{8} & -1 & 0 \end{array}\right] \Rightarrow \left[\begin{array}{ccc|c} 1 & \frac{9}{4} & -2 & -6 \\ 6 & 3 & 2 & 6 \\ 4 & -3 & -24 & 0 \end{array}\right] \Rightarrow \left[\begin{array}{ccc|c} 1 & \frac{9}{4} & -2 & -6 \\ 0 & -\frac{21}{2} & 14 & 42 \\ 0 & -12 & -16 & 24 \end{array}\right] \Rightarrow$

$3R_1 \Rightarrow R_1$, $6R_2 \Rightarrow R_2$, $24R_3 \Rightarrow R_3$

$-6R_1 + R_2 \Rightarrow R_2$, $-4R_1 + R_3 \Rightarrow R_3$

$\left[\begin{array}{ccc|c} 1 & \frac{9}{4} & -2 & -6 \\ 0 & 1 & -\frac{4}{3} & -4 \\ 0 & 3 & 4 & -6 \end{array}\right] \Rightarrow \left[\begin{array}{ccc|c} 1 & 0 & 1 & 3 \\ 0 & 1 & -\frac{4}{3} & -4 \\ 0 & 0 & 8 & 6 \end{array}\right] \Rightarrow \left[\begin{array}{ccc|c} 1 & 0 & 0 & \frac{9}{4} \\ 0 & 1 & 0 & -3 \\ 0 & 0 & 1 & \frac{3}{4} \end{array}\right]$ Solution: $\left(\frac{9}{4}, -3, \frac{3}{4}\right)$

$-\frac{2}{21}R_2 \Rightarrow R_2$, $-\frac{1}{4}R_3 \Rightarrow R_3$

$-\frac{9}{4}R_2 + R_1 \Rightarrow R_1$, $-3R_2 + R_3 \Rightarrow R_3$

$-\frac{1}{8}R_3 + R_1 \Rightarrow R_1$, $\frac{1}{6}R_3 + R_2 \Rightarrow R_2$, $\frac{1}{8}R_3 \Rightarrow R_3$

63. $\left[\begin{array}{ccc|c} \frac{1}{2} & \frac{1}{4} & -1 & 2 \\ \frac{2}{3} & \frac{1}{4} & \frac{1}{2} & \frac{3}{2} \\ \frac{2}{3} & 0 & 1 & -\frac{1}{3} \end{array}\right] \Rightarrow \left[\begin{array}{ccc|c} 1 & \frac{1}{2} & -2 & 4 \\ 8 & 3 & 6 & 18 \\ 2 & 0 & 3 & -1 \end{array}\right] \Rightarrow \left[\begin{array}{ccc|c} 1 & \frac{1}{2} & -2 & 4 \\ 0 & -1 & 22 & -14 \\ 0 & -1 & 7 & -9 \end{array}\right] \Rightarrow$

$2R_1 \Rightarrow R_1, 12R_2 \Rightarrow R_2$, $3R_3 \Rightarrow R_3$

$-8R_1 + R_2 \Rightarrow R_2$, $-2R_1 + R_3 \Rightarrow R_3$

$\left[\begin{array}{ccc|c} 1 & \frac{1}{2} & -2 & 4 \\ 0 & 1 & -22 & 14 \\ 0 & -1 & 7 & -9 \end{array}\right] \Rightarrow \left[\begin{array}{ccc|c} 1 & 0 & 9 & -3 \\ 0 & 1 & -22 & 14 \\ 0 & 0 & -15 & 5 \end{array}\right] \Rightarrow \left[\begin{array}{ccc|c} 1 & 0 & 9 & -3 \\ 0 & 1 & -22 & 14 \\ 0 & 0 & 1 & -\frac{1}{3} \end{array}\right] \Rightarrow$

$-R_2 \Rightarrow R_2$

$-\frac{1}{2}R_2 + R_1 \Rightarrow R_1$, $R_2 + R_3 \Rightarrow R_3$

$-\frac{1}{15}R_3 \Rightarrow R_3$

$\left[\begin{array}{ccc|c} 1 & 0 & 0 & 0 \\ 0 & 1 & 0 & \frac{20}{3} \\ 0 & 0 & 1 & -\frac{1}{3} \end{array}\right]$ Solution: $\left(0, \frac{20}{3}, -\frac{1}{3}\right)$

$-9R_3 + R_1 \Rightarrow R_1$, $22R_3 + R_2 \Rightarrow R_2$

SECTION 6.2

65. $\left[\begin{array}{ccc|c} 3 & -6 & 9 & 18 \\ 2 & -4 & 3 & 12 \\ 1 & -2 & 3 & 6 \end{array}\right] \Rightarrow \left[\begin{array}{ccc|c} 1 & -2 & 3 & 6 \\ 2 & -4 & 3 & 12 \\ 3 & -6 & 9 & 18 \end{array}\right] \Rightarrow \left[\begin{array}{ccc|c} 1 & -2 & 3 & 6 \\ 0 & 0 & -3 & 0 \\ 0 & 0 & 0 & 0 \end{array}\right] \Rightarrow \left[\begin{array}{ccc|c} 1 & -2 & 3 & 6 \\ 0 & 0 & 1 & 0 \\ 0 & 0 & 0 & 0 \end{array}\right]$

$R_1 \Leftrightarrow R_3$ $\quad$ $-2R_1 + R_2 \Rightarrow R_2$, $-3R_1 + R_3 \Rightarrow R_3$ $\quad$ $-\frac{1}{3}R_2 \Rightarrow R_2$

From (2): $z = 0$

From (1):
$$x - 2y + 3z = 6$$
$$x - 2y + 3(0) = 6$$
$$2y = -x + 6$$
$$y = -\tfrac{1}{2}x - 3$$

Solution: $\left(x, -\frac{1}{2}x - 3, 0\right)$

67. $\left[\begin{array}{cc|c} 1 & 1 & -2 \\ 3 & -1 & 6 \\ 2 & 2 & -4 \\ 1 & -1 & 4 \end{array}\right] \Rightarrow \left[\begin{array}{cc|c} 1 & 1 & -2 \\ 0 & -4 & 12 \\ 0 & 0 & 0 \\ 0 & -2 & 6 \end{array}\right] \Rightarrow \left[\begin{array}{cc|c} 1 & 1 & -2 \\ 0 & 1 & -3 \\ 0 & -2 & 6 \\ 0 & 0 & 0 \end{array}\right] \Rightarrow \left[\begin{array}{cc|c} 1 & 0 & 1 \\ 0 & 1 & -3 \\ 0 & 0 & 0 \\ 0 & 0 & 0 \end{array}\right]$

$-3R_1 + R_2 \Rightarrow R_2$, $-2R_1 + R_3 \Rightarrow R_3$, $-R_1 + R_4 \Rightarrow R_4$ $\quad$ $-\frac{1}{4}R_2 \Rightarrow R_2$, $R_3 \Leftrightarrow R_4$ $\quad$ $-R_2 + R_1 \Rightarrow R_1$, $2R_2 + R_3 \Rightarrow R_3$

Solution: $(1, -3)$

69. $\left[\begin{array}{ccc|c} 1 & 2 & 1 & 4 \\ 3 & -1 & -1 & 2 \end{array}\right] \Rightarrow \left[\begin{array}{ccc|c} 1 & 2 & 1 & 4 \\ 0 & -7 & -4 & -10 \end{array}\right] \Rightarrow \left[\begin{array}{ccc|c} 1 & 2 & 1 & 4 \\ 0 & 1 & \frac{4}{7} & \frac{10}{7} \end{array}\right] \Rightarrow \left[\begin{array}{ccc|c} 1 & 0 & -\frac{1}{7} & \frac{8}{7} \\ 0 & 1 & \frac{4}{7} & \frac{10}{7} \end{array}\right]$

$-3R_1 + R_2 \Rightarrow R_2$ $\quad$ $-\frac{1}{7}R_2 \Rightarrow R_2$ $\quad$ $-2R_2 + R_1 \Rightarrow R_1$

From R_1: $x - \frac{1}{7}z = \frac{8}{7}$
$$x = \tfrac{8}{7} + \tfrac{1}{7}z$$

From R_2: $y + \frac{4}{7}z = \frac{10}{7}$
$$y = \tfrac{10}{7} - \tfrac{4}{7}z$$

Solution:
$x = \frac{8}{7} + \frac{1}{7}z$, $y = \frac{10}{7} - \frac{4}{7}z$
$z =$ any real number

71. $\left[\begin{array}{cccc|c} 1 & 1 & 0 & 0 & 1 \\ 1 & 0 & 1 & 0 & 0 \\ 0 & 1 & 0 & 1 & 0 \end{array}\right] \Rightarrow \left[\begin{array}{cccc|c} 1 & 1 & 0 & 0 & 1 \\ 0 & 1 & -1 & 0 & 1 \\ 0 & 1 & 0 & 1 & 0 \end{array}\right] \Rightarrow \left[\begin{array}{cccc|c} 1 & 0 & 1 & 0 & 0 \\ 0 & 1 & -1 & 0 & 1 \\ 0 & 0 & 1 & 1 & -1 \end{array}\right] \Rightarrow$

$-R_2 + R_1 \Rightarrow R_2$ $\quad$ $-R_2 + R_1 \Rightarrow R_1$, $-R_2 + R_3 \Rightarrow R_3$

$\left[\begin{array}{cccc|c} 1 & 0 & 0 & -1 & 1 \\ 0 & 1 & 0 & 1 & 0 \\ 0 & 0 & 1 & 1 & -1 \end{array}\right] \Rightarrow$

$-R_3 + R_1 \Rightarrow R_1$, $R_3 + R_2 \Rightarrow R_2$

From R_1: $w - z = 1$, $w = 1 + z$

From R_2: $x + z = 0$, $x = -z$

From R_3: $y + z = -1$, $y = -1 - z$

Solution:
$w = 1 + z$,
$x = -z$,
$y = -1 - z$,
$z =$ any real #

73. $\left[\begin{array}{cc|c} 1 & 1 & 3 \\ 2 & 1 & 1 \\ 3 & 2 & 2 \end{array}\right] \Rightarrow \left[\begin{array}{cc|c} 1 & 1 & 3 \\ 0 & -1 & -5 \\ 0 & -1 & -7 \end{array}\right] \Rightarrow \left[\begin{array}{cc|c} 1 & 1 & 3 \\ 0 & -1 & -5 \\ 0 & 0 & -2 \end{array}\right]$

$-2R_1 + R_2 \Rightarrow R_2$, $-3R_1 + R_3 \Rightarrow R_3$ $\quad$ $-R_2 + R_3 \Rightarrow R_3$

R_3 indicates that $0x + 0y = -2$. This is impossible. The system is inconsistent. $\Rightarrow$ no solution

SECTION 6.2

75. Let $p =$ speed with no wind. Let $w =$ the speed of the wind. Then the following system applies:

$$\begin{cases} p + w = 300 \\ p - w = 220 \end{cases} \quad \left[\begin{array}{cc|c} 1 & 1 & 300 \\ 1 & -1 & 220 \end{array}\right] \Rightarrow \underset{-R_2 + R_1 \Rightarrow R_2}{\left[\begin{array}{cc|c} 1 & 1 & 300 \\ 0 & 2 & 80 \end{array}\right]} \Rightarrow \underset{\frac{1}{2}R_2 \Rightarrow R_2}{\left[\begin{array}{cc|c} 1 & 1 & 300 \\ 0 & 1 & 40 \end{array}\right]} \Rightarrow \underset{-R_2 + R_1 \Rightarrow R_1}{\left[\begin{array}{cc|c} 1 & 0 & 260 \\ 0 & 1 & 40 \end{array}\right]}$$

The plane has a speed of 260 miles per hour with no wind, so it could travel 1300 miles in 5 hours.

77. Let $d =$ width of a dictionary, $a =$ width of an atlas and $t =$ width of a thesaurus.

$$\begin{cases} 3d + 5a + t = 35 \\ 6d + 2t = 35 \\ 2d + 4a + 3t = 35 \end{cases} \quad \left[\begin{array}{ccc|c} 3 & 5 & 1 & 35 \\ 6 & 0 & 2 & 35 \\ 2 & 4 & 3 & 35 \end{array}\right] \Rightarrow \underset{\substack{-2R_1 + R_2 \Rightarrow R_2 \\ -\frac{2}{3}R_1 + R_3 \Rightarrow R_3}}{\left[\begin{array}{ccc|c} 3 & 5 & 1 & 35 \\ 0 & -10 & 0 & -35 \\ 0 & \frac{2}{3} & \frac{7}{3} & \frac{35}{3} \end{array}\right]} \Rightarrow \underset{\substack{-\frac{1}{10}R_2 \Rightarrow R_2 \\ 3R_3 \Rightarrow R_3}}{\left[\begin{array}{ccc|c} 3 & 5 & 1 & 35 \\ 0 & 1 & 0 & 3.5 \\ 0 & 2 & 7 & 35 \end{array}\right]}$$

$$\Rightarrow \underset{\substack{-5R_2 + R_1 \Rightarrow R_1 \\ -2R_2 + R_3 \Rightarrow R_3}}{\left[\begin{array}{ccc|c} 3 & 0 & 1 & 17.5 \\ 0 & 1 & 0 & 3.5 \\ 0 & 0 & 7 & 28 \end{array}\right]} \Rightarrow \underset{\substack{-\frac{1}{7}R_3 + R_1 \Rightarrow R_1 \\ \frac{1}{7}R_3 \Rightarrow R_3}}{\left[\begin{array}{ccc|c} 3 & 0 & 0 & 13.5 \\ 0 & 1 & 0 & 3.5 \\ 0 & 0 & 1 & 4 \end{array}\right]} \Rightarrow \underset{\frac{1}{3}R_1 \Rightarrow R_1}{\left[\begin{array}{ccc|c} 1 & 0 & 0 & 4.5 \\ 0 & 1 & 0 & 3.5 \\ 0 & 0 & 1 & 4 \end{array}\right]}$$

Dictionaries are 4.5 in. wide. Atlases are 3.5 in. wide. Thesauruses are 4 in. wide.

79. Let A, B and C represent the number of ounces of each food.

$$\begin{cases} A + 2B + 2C = 22 \\ A + B + C = 12 \\ 2A + B + 2C = 20 \end{cases} \quad \left[\begin{array}{ccc|c} 1 & 2 & 2 & 22 \\ 1 & 1 & 1 & 12 \\ 2 & 1 & 2 & 20 \end{array}\right] \Rightarrow \underset{\substack{-R_1 + R_2 \Rightarrow R_2 \\ -2R_1 + R_3 \Rightarrow R_3}}{\left[\begin{array}{ccc|c} 1 & 2 & 2 & 22 \\ 0 & -1 & -1 & -10 \\ 0 & -3 & -2 & -24 \end{array}\right]} \Rightarrow$$

$$\underset{\substack{2R_2 + R_1 \Rightarrow R_1 \\ -3R_2 + R_3 \Rightarrow R_3 \\ -R_2 \Rightarrow R_2}}{\left[\begin{array}{ccc|c} 1 & 0 & 0 & 2 \\ 0 & 1 & 1 & 10 \\ 0 & 0 & 1 & 6 \end{array}\right]} \Rightarrow \underset{-R_3 + R_2 \Rightarrow R_2}{\left[\begin{array}{ccc|c} 1 & 0 & 0 & 2 \\ 0 & 1 & 0 & 4 \\ 0 & 0 & 1 & 6 \end{array}\right]}$$

2 ounces of Food A, 4 ounces of Food B, and 6 ounces of Food C should be used.

81. Answers may vary.

83. Answers may vary.

85.

$$\left[\begin{array}{ccc|c} 1 & 1 & 1 & 14 \\ 2 & 3 & -2 & -7 \\ 1 & -5 & 1 & 8 \end{array}\right] \Rightarrow \underset{\substack{-2R_1 + R_2 \Rightarrow R_2 \\ -R_1 + R_3 \Rightarrow R_3}}{\left[\begin{array}{ccc|c} 1 & 1 & 1 & 14 \\ 0 & 1 & -4 & -35 \\ 0 & -6 & 0 & -6 \end{array}\right]} \Rightarrow \underset{R_2 \Leftrightarrow R_3}{\left[\begin{array}{ccc|c} 1 & 1 & 1 & 14 \\ 0 & -6 & 0 & -6 \\ 0 & 1 & -4 & -35 \end{array}\right]} \Rightarrow$$

$$\underset{\substack{\frac{1}{6}R_2 + R_1 \Rightarrow R_1 \\ \frac{1}{6}R_2 + R_3 \Rightarrow R_3}}{\left[\begin{array}{ccc|c} 1 & 0 & 1 & 13 \\ 0 & -6 & 0 & -6 \\ 0 & 0 & -4 & -36 \end{array}\right]} \Rightarrow \underset{\substack{\frac{1}{4}R_3 + R_1 \Rightarrow R_1 \\ -\frac{1}{6}R_2 \Rightarrow R_2 \\ -\frac{1}{4}R_3 \Rightarrow R_3}}{\left[\begin{array}{ccc|c} 1 & 0 & 0 & 4 \\ 0 & 1 & 0 & 1 \\ 0 & 0 & 1 & 9 \end{array}\right]}$$

$x^2 = 4 \Rightarrow \boxed{x = \pm 2}$

$y^2 = 1 \Rightarrow \boxed{y = \pm 1}$

$z^2 = 9 \Rightarrow \boxed{z = \pm 3}$

87. $y = mx + b$

89. equal

91. $y = mx + b$
$y = 2x + 7$

93. $x = 2$

Exercises 6.3 (page 588)

1. i, j

3. corresponding

5. columns, rows

7. additive identity

9. $x = 2, y = 5$

11. $x + y = 3$
$3 + x = 4 \Rightarrow x = 1$
$5y = 10 \Rightarrow y = 2$

13. $A + B = \begin{bmatrix} 2 & 1 & -1 \\ -3 & 2 & 5 \end{bmatrix} + \begin{bmatrix} -3 & 1 & 2 \\ -3 & -2 & -5 \end{bmatrix} = \begin{bmatrix} -1 & 2 & 1 \\ -6 & 0 & 0 \end{bmatrix}$

15. additive inverse of $A = \begin{bmatrix} -5 & 2 & -7 \\ 5 & 0 & -3 \\ 2 & -3 & 5 \end{bmatrix}$

17. $A - B = \begin{bmatrix} -3 & 2 & -2 \\ -1 & 4 & -5 \end{bmatrix} - \begin{bmatrix} 3 & -3 & -2 \\ -2 & 5 & -5 \end{bmatrix} = \begin{bmatrix} -6 & 5 & 0 \\ 1 & -1 & 0 \end{bmatrix}$

19. $5A = 5\begin{bmatrix} 3 & -3 \\ 0 & -2 \end{bmatrix} = \begin{bmatrix} 15 & -15 \\ 0 & -10 \end{bmatrix}$

21. $5A = 5\begin{bmatrix} 5 & 15 & -2 \\ -2 & -5 & 1 \end{bmatrix}$
$= \begin{bmatrix} 25 & 75 & -10 \\ -10 & -25 & 5 \end{bmatrix}$

23. $5A + 3B = 5\begin{bmatrix} 3 & 1 & -2 \\ -4 & 3 & -2 \end{bmatrix} + 3\begin{bmatrix} 1 & -2 & 2 \\ -5 & -5 & 3 \end{bmatrix} = \begin{bmatrix} 15 & 5 & -10 \\ -20 & 15 & -10 \end{bmatrix} + \begin{bmatrix} 3 & -6 & 6 \\ -15 & -15 & 9 \end{bmatrix}$
$= \begin{bmatrix} 18 & -1 & -4 \\ -35 & 0 & -1 \end{bmatrix}$

25. $\begin{bmatrix} 2 & 3 \\ 3 & -2 \end{bmatrix}_{2\times 2} \begin{bmatrix} 1 & 2 \\ 0 & -2 \end{bmatrix}_{2\times 2} = \begin{bmatrix} (2)(1) + (3)(0) & (2)(2) + (3)(-2) \\ (3)(1) + (-2)(0) & (3)(2) + (-2)(-2) \end{bmatrix}_{2\times 2} = \begin{bmatrix} 2 & -2 \\ 3 & 10 \end{bmatrix}$

27. $\begin{bmatrix} -4 & -2 \\ 21 & 0 \end{bmatrix}_{2\times 2} \begin{bmatrix} -5 & 6 \\ 21 & -1 \end{bmatrix}_{2\times 2} = \begin{bmatrix} (-4)(-5) + (-2)(21) & (-4)(6) + (-2)(-1) \\ (21)(-5) + (0)(21) & (21)(6) + (0)(-1) \end{bmatrix}_{2\times 2}$
$= \begin{bmatrix} -22 & -22 \\ -105 & 126 \end{bmatrix}$

29. $\begin{bmatrix} 2 & 1 & 3 \\ 1 & 2 & -1 \\ 0 & 1 & 0 \end{bmatrix}_{3\times 3} \begin{bmatrix} 1 & 2 & 3 \\ 2 & -2 & 1 \\ 0 & 0 & 1 \end{bmatrix}_{3\times 3}$

$$= \begin{bmatrix} (2)(1)+(1)(2)+(3)(0) & (2)(2)+(1)(-2)+(3)(0) & (2)(3)+(1)(1)+(3)(1) \\ (1)(1)+(2)(2)+(-1)(0) & (1)(2)+(2)(-2)+(-1)(0) & (1)(3)+(2)(1)+(-1)(1) \\ (0)(1)+(1)(2)+(0)(0) & (0)(2)+(1)(-2)+(0)(0) & (0)(3)+(1)(1)+(0)(1) \end{bmatrix}_{3\times 3}$$

$$= \begin{bmatrix} 4 & 2 & 10 \\ 5 & -2 & 4 \\ 2 & -2 & 1 \end{bmatrix}$$

31. $\begin{bmatrix} 1 \\ -2 \\ -3 \end{bmatrix}_{3\times 1} [4 \quad -5 \quad -6]_{1\times 3} = \begin{bmatrix} (1)(4) & (1)(-5) & (1)(-6) \\ (-2)(4) & (-2)(-5) & (-2)(-6) \\ (-3)(4) & (-3)(-5) & (-3)(-6) \end{bmatrix}_{3\times 3} = \begin{bmatrix} 4 & -5 & -6 \\ -8 & 10 & 12 \\ -12 & 15 & 18 \end{bmatrix}$

33. $[1 \quad 2 \quad 3]_{1\times 3} \begin{bmatrix} 4 & 5 & 6 \\ 7 & 8 & 9 \end{bmatrix}_{2\times 3}$

Not possible

35. $\begin{bmatrix} 2 & 3 & 4 \\ 1 & 2 & 3 \\ -2 & 2 & 2 \end{bmatrix}_{3\times 3} \begin{bmatrix} -1 \\ 2 \\ 3 \end{bmatrix}_{3\times 1} = \begin{bmatrix} (2)(-1)+(3)(2)+(4)(3) \\ (1)(-1)+(2)(2)+(3)(3) \\ (-2)(-1)+(2)(2)+(2)(3) \end{bmatrix}_{3\times 1} = \begin{bmatrix} 16 \\ 12 \\ 12 \end{bmatrix}$

37. $\begin{bmatrix} 1 & 2 & 3 \\ 4 & 5 & 6 \\ 7 & 8 & 9 \end{bmatrix}_{3\times 3} \begin{bmatrix} 1 & 2 \\ 3 & 4 \end{bmatrix}_{2\times 2}$

Not possible

39. $AB = \begin{bmatrix} 2.3 & -1.7 & 3.1 \\ -2 & 3.5 & 1 \\ -8 & 4.7 & 9.1 \end{bmatrix} \begin{bmatrix} -2.5 \\ 5.2 \\ -7 \end{bmatrix} = \begin{bmatrix} -36.29 \\ 16.2 \\ -19.26 \end{bmatrix}$

41. $A^2 = \begin{bmatrix} 2.3 & -1.7 & 3.1 \\ -2 & 3.5 & 1 \\ -8 & 4.7 & 9.1 \end{bmatrix} \begin{bmatrix} 2.3 & -1.7 & 3.1 \\ -2 & 3.5 & 1 \\ -8 & 4.7 & 9.1 \end{bmatrix} = \begin{bmatrix} -16.11 & 4.71 & 33.64 \\ -19.6 & 20.35 & 6.4 \\ -100.6 & 72.82 & 62.71 \end{bmatrix}$

43. $A(B+C) = \begin{bmatrix} 2 & 3 \\ 1 & 3 \end{bmatrix} \left(\begin{bmatrix} 2 & 1 & -5 \\ 1 & 1 & 2 \end{bmatrix} + \begin{bmatrix} -2 & -1 & 6 \\ 0 & -1 & -1 \end{bmatrix} \right) = \begin{bmatrix} 2 & 3 \\ 1 & 3 \end{bmatrix} \begin{bmatrix} 0 & 0 & 1 \\ 1 & 0 & 1 \end{bmatrix} = \begin{bmatrix} 3 & 0 & 5 \\ 3 & 0 & 4 \end{bmatrix}$

$$AB + AC = \begin{bmatrix} 2 & 3 \\ 1 & 3 \end{bmatrix} \begin{bmatrix} 2 & 1 & -5 \\ 1 & 1 & 2 \end{bmatrix} + \begin{bmatrix} 2 & 3 \\ 1 & 3 \end{bmatrix} \begin{bmatrix} -2 & -1 & 6 \\ 0 & -1 & -1 \end{bmatrix}$$

$$= \begin{bmatrix} 7 & 5 & -4 \\ 5 & 4 & 1 \end{bmatrix} + \begin{bmatrix} -4 & -5 & 9 \\ -2 & -4 & 3 \end{bmatrix} = \begin{bmatrix} 3 & 0 & 5 \\ 3 & 0 & 4 \end{bmatrix}$$

SECTION 6.3

45. $3(AB) = 3\left(\begin{bmatrix} 2 & 3 \\ 1 & 3 \end{bmatrix}\begin{bmatrix} 2 & 1 & -5 \\ 1 & 1 & 2 \end{bmatrix}\right) = 3\begin{bmatrix} 7 & 5 & -4 \\ 5 & 4 & 1 \end{bmatrix} = \begin{bmatrix} 21 & 15 & -12 \\ 15 & 12 & 3 \end{bmatrix}$

$(3A)B = \left(3\begin{bmatrix} 2 & 3 \\ 1 & 3 \end{bmatrix}\right)\begin{bmatrix} 2 & 1 & -5 \\ 1 & 1 & 2 \end{bmatrix} = \begin{bmatrix} 6 & 9 \\ 3 & 9 \end{bmatrix}\begin{bmatrix} 2 & 1 & -5 \\ 1 & 1 & 2 \end{bmatrix} = \begin{bmatrix} 21 & 15 & -12 \\ 15 & 12 & 3 \end{bmatrix}$

47. $A - BC = \begin{bmatrix} 1 & 3 \\ 2 & 5 \end{bmatrix} - \begin{bmatrix} -1 \\ 3 \end{bmatrix}[3 \quad 2] = \begin{bmatrix} 1 & 3 \\ 2 & 5 \end{bmatrix} - \begin{bmatrix} -3 & -2 \\ 9 & 6 \end{bmatrix} = \begin{bmatrix} 4 & 5 \\ -7 & -1 \end{bmatrix}$

49. $CB - AB = [3 \quad 2]\begin{bmatrix} -1 \\ 3 \end{bmatrix} - \begin{bmatrix} 1 & 3 \\ 2 & 5 \end{bmatrix}\begin{bmatrix} -1 \\ 3 \end{bmatrix} = [3] - \begin{bmatrix} 8 \\ 13 \end{bmatrix} \Rightarrow$ not possible

51. $ABC = \begin{bmatrix} 1 & 3 \\ 2 & 5 \end{bmatrix}\begin{bmatrix} -1 \\ 3 \end{bmatrix}[3 \quad 2] = \begin{bmatrix} 8 \\ 13 \end{bmatrix}[3 \quad 2] = \begin{bmatrix} 24 & 16 \\ 39 & 26 \end{bmatrix}$

53. $A^2B = \begin{bmatrix} 1 & 3 \\ 2 & 5 \end{bmatrix}\begin{bmatrix} 1 & 3 \\ 2 & 5 \end{bmatrix}\begin{bmatrix} -1 \\ 3 \end{bmatrix} = \begin{bmatrix} 7 & 18 \\ 12 & 31 \end{bmatrix}\begin{bmatrix} -1 \\ 3 \end{bmatrix} = \begin{bmatrix} 47 \\ 81 \end{bmatrix}$

55. $Q = \begin{bmatrix} 200 & 300 & 100 \\ 100 & 200 & 200 \end{bmatrix}, C = \begin{bmatrix} 5 \\ 2 \\ 4 \end{bmatrix}$

$QC = \begin{bmatrix} 200 & 300 & 100 \\ 100 & 200 & 200 \end{bmatrix}\begin{bmatrix} 5 \\ 2 \\ 4 \end{bmatrix} = \begin{bmatrix} 2000 \\ 1700 \end{bmatrix}$ Cost of balls from Supplier 1; Cost of balls from Supplier 2

57. $Q = \begin{bmatrix} 217 & 23 & 319 \\ 347 & 24 & 340 \\ 3 & 97 & 750 \end{bmatrix}, P = \begin{bmatrix} 0.75 \\ 1.00 \\ 1.25 \end{bmatrix}$

$QP = \begin{bmatrix} 217 & 23 & 319 \\ 347 & 24 & 340 \\ 3 & 97 & 750 \end{bmatrix}\begin{bmatrix} 0.75 \\ 1.00 \\ 1.25 \end{bmatrix} = \begin{bmatrix} 584.50 \\ 709.25 \\ 1036.75 \end{bmatrix}$ \$ spent by adult males; \$ spent by adult females; \$ spent by children

59. $A^2 = \begin{bmatrix} 0 & 1 & 1 \\ 1 & 0 & 0 \\ 0 & 1 & 0 \end{bmatrix}^2 = \begin{bmatrix} 0 & 1 & 1 \\ 1 & 0 & 0 \\ 0 & 1 & 0 \end{bmatrix}\begin{bmatrix} 0 & 1 & 1 \\ 1 & 0 & 0 \\ 0 & 1 & 0 \end{bmatrix} = \begin{bmatrix} 1 & 1 & 0 \\ 0 & 1 & 1 \\ 1 & 0 & 0 \end{bmatrix}$

61. $A^2 = \begin{bmatrix} 0 & 2 & 1 & 0 \\ 2 & 0 & 1 & 0 \\ 1 & 1 & 0 & 2 \\ 0 & 0 & 2 & 0 \end{bmatrix}^2 = \begin{bmatrix} 0 & 2 & 1 & 0 \\ 2 & 0 & 1 & 0 \\ 1 & 1 & 0 & 2 \\ 0 & 0 & 2 & 0 \end{bmatrix}\begin{bmatrix} 0 & 2 & 1 & 0 \\ 2 & 0 & 1 & 0 \\ 1 & 1 & 0 & 2 \\ 0 & 0 & 2 & 0 \end{bmatrix} = \begin{bmatrix} 5 & 1 & 2 & 2 \\ 1 & 5 & 2 & 2 \\ 2 & 2 & 6 & 0 \\ 2 & 2 & 0 & 4 \end{bmatrix}$

A^2 represents the number of ways two cities can be linked through one intermediary.

63. Let $A = \begin{bmatrix} 1 & 1 \\ 1 & 1 \end{bmatrix}$ and $B = \begin{bmatrix} 1 & 0 \\ 0 & 0 \end{bmatrix}$.

$$(AB)^2 = \left(\begin{bmatrix} 1 & 1 \\ 1 & 1 \end{bmatrix}\begin{bmatrix} 1 & 0 \\ 0 & 0 \end{bmatrix}\right)^2 = \begin{bmatrix} 1 & 0 \\ 1 & 0 \end{bmatrix}^2 = \begin{bmatrix} 1 & 0 \\ 1 & 0 \end{bmatrix}\begin{bmatrix} 1 & 0 \\ 1 & 0 \end{bmatrix} = \begin{bmatrix} 1 & 0 \\ 1 & 0 \end{bmatrix}$$

$$A^2B^2 = \begin{bmatrix} 1 & 1 \\ 1 & 1 \end{bmatrix}^2\begin{bmatrix} 1 & 0 \\ 0 & 0 \end{bmatrix}^2 = \begin{bmatrix} 2 & 2 \\ 2 & 2 \end{bmatrix}\begin{bmatrix} 1 & 0 \\ 0 & 0 \end{bmatrix} = \begin{bmatrix} 2 & 0 \\ 2 & 0 \end{bmatrix}. \ (AB)^2 \neq A^2B^2$$

65. Let $A = \begin{bmatrix} 1 & 2 \\ 1 & 2 \end{bmatrix}$ and $B = \begin{bmatrix} 2 & 2 \\ -1 & -1 \end{bmatrix}$. $AB = \begin{bmatrix} 1 & 2 \\ 1 & 2 \end{bmatrix}\begin{bmatrix} 2 & 2 \\ -1 & -1 \end{bmatrix} = \begin{bmatrix} 0 & 0 \\ 0 & 0 \end{bmatrix}$

67. $(3x+2)(2x-3) - (2-x) = 6x^2 - 9x + 4x - 6 - 2 + x = 6x^2 - 4x - 8$

69. $\dfrac{1+\frac{1}{x}}{1-\frac{1}{x}} = \dfrac{x\left(1+\frac{1}{x}\right)}{x\left(1-\frac{1}{x}\right)} = \dfrac{x+1}{x-1}$

71.
$$\begin{aligned} s &= \frac{n(a+l)}{2} \\ 2s &= n(a+l) \\ \frac{2s}{n} &= a+l \\ \frac{2s}{n} - l &= a \end{aligned}$$

Exercises 6.4 (page 598)

1. $AB = BA = I$

3. $[I \mid A^{-1}]$

5. $\left[\begin{array}{cc|cc} 3 & -4 & 1 & 0 \\ -2 & 3 & 0 & 1 \end{array}\right] \Rightarrow \left[\begin{array}{cc|cc} 1 & -\frac{4}{3} & \frac{1}{3} & 0 \\ -2 & 3 & 0 & 1 \end{array}\right] \Rightarrow \left[\begin{array}{cc|cc} 1 & -\frac{4}{3} & \frac{1}{3} & 0 \\ 0 & \frac{1}{3} & \frac{2}{3} & 1 \end{array}\right] \Rightarrow$

$\frac{1}{3}R_1 \Rightarrow R_1$; $2R_1 + R_2 \Rightarrow R_2$

$\left[\begin{array}{cc|cc} 1 & 0 & 3 & 4 \\ 0 & \frac{1}{3} & \frac{2}{3} & 1 \end{array}\right] \Rightarrow \left[\begin{array}{cc|cc} 1 & 0 & 3 & 4 \\ 0 & 1 & 2 & 3 \end{array}\right] \Rightarrow$ Inverse: $\begin{bmatrix} 3 & 4 \\ 2 & 3 \end{bmatrix}$

$4R_2 + R_1 \Rightarrow R_1$; $3R_2 \Rightarrow R_2$

7. $\left[\begin{array}{cc|cc} 3 & 7 & 1 & 0 \\ 2 & 5 & 0 & 1 \end{array}\right] \Rightarrow \left[\begin{array}{cc|cc} 1 & \frac{7}{3} & \frac{1}{3} & 0 \\ 2 & 5 & 0 & 1 \end{array}\right] \Rightarrow \left[\begin{array}{cc|cc} 1 & \frac{7}{3} & \frac{1}{3} & 0 \\ 0 & \frac{1}{3} & -\frac{2}{3} & 1 \end{array}\right] \Rightarrow$

$\frac{1}{3}R_1 \Rightarrow R_1$; $-2R_1 + R_2 \Rightarrow R_2$

$\left[\begin{array}{cc|cc} 1 & 0 & 5 & -7 \\ 0 & \frac{1}{3} & -\frac{2}{3} & 1 \end{array}\right] \Rightarrow \left[\begin{array}{cc|cc} 1 & 0 & 5 & -7 \\ 0 & 1 & -2 & 3 \end{array}\right] \Rightarrow$ Inverse: $\begin{bmatrix} 5 & -7 \\ -2 & 3 \end{bmatrix}$

$-7R_2 + R_1 \Rightarrow R_1$; $3R_2 \Rightarrow R_2$

9. $\left[\begin{array}{ccc|ccc} 1 & 0 & 3 & 1 & 0 & 0 \\ -1 & 1 & 3 & 0 & 1 & 0 \\ -2 & 1 & 1 & 0 & 0 & 1 \end{array}\right] \Rightarrow \left[\begin{array}{ccc|ccc} 1 & 0 & 3 & 1 & 0 & 0 \\ 0 & 1 & 6 & 1 & 1 & 0 \\ 0 & 1 & 7 & 2 & 0 & 1 \end{array}\right] \Rightarrow \left[\begin{array}{ccc|ccc} 1 & 0 & 3 & 1 & 0 & 0 \\ 0 & 1 & 6 & 1 & 1 & 0 \\ 0 & 0 & 1 & 1 & -1 & 1 \end{array}\right] \Rightarrow$

$R_1 + R_2 \Rightarrow R_2$, $2R_1 + R_3 \Rightarrow R_3$; $-R_2 + R_3 \Rightarrow R_3$

continued on next page

9. continued

$$\left[\begin{array}{ccc|ccc} 1 & 0 & 0 & -2 & 3 & -3 \\ 0 & 1 & 0 & -5 & 7 & -6 \\ 0 & 0 & 1 & 1 & -1 & 1 \end{array}\right] \Rightarrow \text{Inverse: } \begin{bmatrix} -2 & 3 & -3 \\ -5 & 7 & -6 \\ 1 & -1 & 1 \end{bmatrix}$$

$-3R_3 + R_1 \Rightarrow R_1$
$-6R_3 + R_2 \Rightarrow R_2$

11.

$$\left[\begin{array}{ccc|ccc} 3 & 2 & 1 & 1 & 0 & 0 \\ 1 & 1 & -1 & 0 & 1 & 0 \\ 4 & 3 & 1 & 0 & 0 & 1 \end{array}\right] \Rightarrow \left[\begin{array}{ccc|ccc} 1 & 1 & -1 & 0 & 1 & 0 \\ 3 & 2 & 1 & 1 & 0 & 0 \\ 4 & 3 & 1 & 0 & 0 & 1 \end{array}\right] \Rightarrow \left[\begin{array}{ccc|ccc} 1 & 1 & -1 & 0 & 1 & 0 \\ 0 & -1 & 4 & 1 & -3 & 0 \\ 0 & -1 & 5 & 0 & -4 & 1 \end{array}\right] \Rightarrow$$

$R_1 \Leftrightarrow R_2$ $\qquad$ $-3R_1 + R_2 \Rightarrow R_2$, $-4R_1 + R_3 \Rightarrow R_3$

$$\left[\begin{array}{ccc|ccc} 1 & 1 & -1 & 0 & 1 & 0 \\ 0 & 1 & -4 & -1 & 3 & 0 \\ 0 & -1 & 5 & 0 & -4 & 1 \end{array}\right] \Rightarrow \left[\begin{array}{ccc|ccc} 1 & 0 & 3 & 1 & -2 & 0 \\ 0 & 1 & -4 & -1 & 3 & 0 \\ 0 & 0 & 1 & -1 & -1 & 1 \end{array}\right] \Rightarrow$$

$-R_2 \Rightarrow R_2$ $\qquad$ $-R_2 + R_1 \Rightarrow R_1$, $R_2 + R_3 \Rightarrow R_3$

$$\left[\begin{array}{ccc|ccc} 1 & 0 & 0 & 4 & 1 & -3 \\ 0 & 1 & 0 & -5 & -1 & 4 \\ 0 & 0 & 1 & -1 & -1 & 1 \end{array}\right] \Rightarrow \text{Inverse} = \begin{bmatrix} 4 & 1 & -3 \\ -5 & -1 & 4 \\ -1 & -1 & 1 \end{bmatrix}$$

$-3R_3 + R_1 \Rightarrow R_1$
$4R_3 + R_2 \Rightarrow R_2$

13.

$$\left[\begin{array}{ccc|ccc} 1 & 3 & 5 & 1 & 0 & 0 \\ 0 & 1 & 6 & 0 & 1 & 0 \\ 1 & 4 & 11 & 0 & 0 & 1 \end{array}\right] \Rightarrow \left[\begin{array}{ccc|ccc} 1 & 3 & 5 & 1 & 0 & 0 \\ 0 & 1 & 6 & 0 & 1 & 0 \\ 0 & 1 & 6 & -1 & 0 & 1 \end{array}\right] \Rightarrow \left[\begin{array}{ccc|ccc} 1 & 3 & 5 & 1 & 0 & 0 \\ 0 & 1 & 6 & 0 & 1 & 0 \\ 0 & 0 & 0 & -1 & -1 & 1 \end{array}\right]$$

$-R_1 + R_3 \Rightarrow R_3$ $\qquad$ $-R_2 + R_3 \Rightarrow R_3$

Since the original matrix cannot be changed into the identity, there is no inverse matrix.

15.

$$\left[\begin{array}{ccc|ccc} 1 & 2 & 3 & 1 & 0 & 0 \\ 0 & 1 & 2 & 0 & 1 & 0 \\ 0 & 0 & 1 & 0 & 0 & 1 \end{array}\right] \Rightarrow \left[\begin{array}{ccc|ccc} 1 & 0 & -1 & 1 & -2 & 0 \\ 0 & 1 & 2 & 0 & 1 & 0 \\ 0 & 0 & 1 & 0 & 0 & 1 \end{array}\right] \Rightarrow \left[\begin{array}{ccc|ccc} 1 & 0 & 0 & 1 & -2 & 1 \\ 0 & 1 & 0 & 0 & 1 & -2 \\ 0 & 0 & 1 & 0 & 0 & 2 \end{array}\right] \Rightarrow$$

$-2R_2 + R_1 \Rightarrow R_1$ $\qquad$ $R_3 + R_1 \Rightarrow R_1$, $-2R_3 + R_2 \Rightarrow R_2$

$$\text{Inverse} = \begin{bmatrix} 1 & -2 & 1 \\ 0 & 1 & -2 \\ 0 & 0 & 1 \end{bmatrix}$$

17.

$$\left[\begin{array}{ccc|ccc} 1 & 6 & 4 & 1 & 0 & 0 \\ 1 & -2 & -5 & 0 & 1 & 0 \\ 2 & 4 & -1 & 0 & 0 & 1 \end{array}\right] \Rightarrow \left[\begin{array}{ccc|ccc} 1 & 6 & 4 & 1 & 0 & 0 \\ 0 & -8 & -9 & -1 & 1 & 0 \\ 0 & -8 & -9 & -2 & 0 & 1 \end{array}\right] \Rightarrow \left[\begin{array}{ccc|ccc} 1 & 6 & 4 & 1 & 0 & 0 \\ 0 & -8 & -9 & -1 & 1 & 0 \\ 0 & 0 & 0 & -1 & -1 & 1 \end{array}\right]$$

$-R_1 + R_2 \Rightarrow R_2$, $-2R_1 + R_3 \Rightarrow R_3$ $\qquad$ $-R_2 + R_3 \Rightarrow R_3$

Since the original matrix cannot be changed into the identity, there is no inverse matrix.

SECTION 6.4

19. Inverse $= \begin{bmatrix} 1 & -2 & 1 & 0 \\ 0 & 1 & -2 & 1 \\ 0 & 0 & 1 & -2 \\ 0 & 0 & 0 & 1 \end{bmatrix}$

21. Inverse $= \begin{bmatrix} 8 & -2 & -6 \\ -5 & 2 & 4 \\ 2 & 0 & -2 \end{bmatrix}$

23. Inverse $= \begin{bmatrix} -2.5 & 5 & 3 & 5.5 \\ 5.5 & -8 & -6 & -9.5 \\ -1 & 3 & 1 & 3 \\ -5.5 & 9 & 6 & 10.5 \end{bmatrix}$

25. $\begin{bmatrix} 3 & -4 \\ -2 & 3 \end{bmatrix}\begin{bmatrix} x \\ y \end{bmatrix} = \begin{bmatrix} 1 \\ 5 \end{bmatrix}$

$$\begin{bmatrix} x \\ y \end{bmatrix} = \begin{bmatrix} 3 & -4 \\ -2 & 3 \end{bmatrix}^{-1}\begin{bmatrix} 1 \\ 5 \end{bmatrix}$$

$$\begin{bmatrix} x \\ y \end{bmatrix} = \begin{bmatrix} 3 & 4 \\ 2 & 3 \end{bmatrix}\begin{bmatrix} 1 \\ 5 \end{bmatrix}$$

$$\begin{bmatrix} x \\ y \end{bmatrix} = \begin{bmatrix} 23 \\ 17 \end{bmatrix}$$

27. $\begin{bmatrix} 3 & -4 \\ -2 & 3 \end{bmatrix}\begin{bmatrix} x \\ y \end{bmatrix} = \begin{bmatrix} 0 \\ 0 \end{bmatrix}$

$$\begin{bmatrix} x \\ y \end{bmatrix} = \begin{bmatrix} 3 & -4 \\ -2 & 3 \end{bmatrix}^{-1}\begin{bmatrix} 0 \\ 0 \end{bmatrix}$$

$$\begin{bmatrix} x \\ y \end{bmatrix} = \begin{bmatrix} 3 & 4 \\ 2 & 3 \end{bmatrix}\begin{bmatrix} 0 \\ 0 \end{bmatrix}$$

$$\begin{bmatrix} x \\ y \end{bmatrix} = \begin{bmatrix} 0 \\ 0 \end{bmatrix}$$

29. $\begin{bmatrix} 2 & 1 & -1 \\ 2 & 2 & -1 \\ -1 & -1 & 1 \end{bmatrix}\begin{bmatrix} x \\ y \\ z \end{bmatrix} = \begin{bmatrix} 2 \\ 4 \\ -1 \end{bmatrix}$

$$\begin{bmatrix} x \\ y \\ z \end{bmatrix} = \begin{bmatrix} 2 & 1 & -1 \\ 2 & 2 & -1 \\ -1 & -1 & 1 \end{bmatrix}^{-1}\begin{bmatrix} 2 \\ 4 \\ -1 \end{bmatrix}$$

$$\begin{bmatrix} x \\ y \\ z \end{bmatrix} = \begin{bmatrix} 1 & 0 & 1 \\ -1 & 1 & 0 \\ 0 & 1 & 2 \end{bmatrix}\begin{bmatrix} 2 \\ 4 \\ -1 \end{bmatrix} = \begin{bmatrix} 1 \\ 2 \\ 2 \end{bmatrix}$$

31. $\begin{bmatrix} -2 & 1 & -3 \\ 2 & 3 & 0 \\ 1 & 0 & 1 \end{bmatrix}\begin{bmatrix} x \\ y \\ z \end{bmatrix} = \begin{bmatrix} 2 \\ -3 \\ 5 \end{bmatrix}$

$$\begin{bmatrix} x \\ y \\ z \end{bmatrix} = \begin{bmatrix} -2 & 1 & -3 \\ 2 & 3 & 0 \\ 1 & 0 & 1 \end{bmatrix}^{-1}\begin{bmatrix} 2 \\ -3 \\ 5 \end{bmatrix}$$

$$\begin{bmatrix} x \\ y \\ z \end{bmatrix} = \begin{bmatrix} 3 & -1 & 9 \\ -2 & 1 & -6 \\ -3 & 1 & -8 \end{bmatrix}\begin{bmatrix} 2 \\ -3 \\ 5 \end{bmatrix} = \begin{bmatrix} 54 \\ -37 \\ -49 \end{bmatrix}$$

33. $\begin{bmatrix} 5 & 3 \\ -7 & 5 \end{bmatrix}\begin{bmatrix} x \\ y \end{bmatrix} = \begin{bmatrix} 13 \\ -9 \end{bmatrix}$

$\begin{bmatrix} x \\ y \end{bmatrix} = \begin{bmatrix} 5 & 3 \\ -7 & 5 \end{bmatrix}^{-1}\begin{bmatrix} 13 \\ -9 \end{bmatrix}$

$\begin{bmatrix} x \\ y \end{bmatrix} = \begin{bmatrix} 2 \\ 1 \end{bmatrix}$

35. $\begin{bmatrix} 5 & 2 & 3 \\ 2 & 0 & 5 \\ 3 & 0 & 1 \end{bmatrix}\begin{bmatrix} x \\ y \\ z \end{bmatrix} = \begin{bmatrix} 12 \\ 7 \\ 4 \end{bmatrix}$

$\begin{bmatrix} x \\ y \\ z \end{bmatrix} = \begin{bmatrix} 5 & 2 & 3 \\ 2 & 0 & 5 \\ 3 & 0 & 1 \end{bmatrix}^{-1}\begin{bmatrix} 12 \\ 7 \\ 4 \end{bmatrix}$

$= \begin{bmatrix} 1 \\ 2 \\ 1 \end{bmatrix}$

37. $\begin{bmatrix} 23 & 27 \\ 21 & 22 \end{bmatrix}\begin{bmatrix} x \\ y \end{bmatrix} = \begin{bmatrix} 127 \\ 108 \end{bmatrix}$

$\begin{bmatrix} x \\ y \end{bmatrix} = \begin{bmatrix} 23 & 27 \\ 21 & 22 \end{bmatrix}^{-1}\begin{bmatrix} 127 \\ 108 \end{bmatrix}$

$\begin{bmatrix} x \\ y \end{bmatrix} = \begin{bmatrix} 2 \\ 3 \end{bmatrix}$

2 of model A and 3 of model B can be made.

39. $AB = \begin{bmatrix} 17 \\ 43 \end{bmatrix}$

$A^{-1}AB = IB = B$

$B = A^{-1}AB = \begin{bmatrix} 1 & 1 \\ 2 & 3 \end{bmatrix}^{-1}\begin{bmatrix} 17 \\ 43 \end{bmatrix}$

$= \begin{bmatrix} 8 \\ 9 \end{bmatrix} \Rightarrow$ "HI"

41. No

43. $A^2 = \begin{bmatrix} -1 & -1 \\ 1 & 1 \end{bmatrix}\begin{bmatrix} -1 & -1 \\ 1 & 1 \end{bmatrix} = \begin{bmatrix} 0 & 0 \\ 0 & 0 \end{bmatrix}$

45. $\begin{bmatrix} 1 & 0 & 0 \\ -2 & -3 & -2 \\ 3 & 6 & 1 \end{bmatrix}\begin{bmatrix} x \\ y \\ z \end{bmatrix} = \begin{bmatrix} 0 \\ 0 \\ 0 \end{bmatrix}$

$\begin{bmatrix} x \\ y \\ z \end{bmatrix} = \begin{bmatrix} 1 & 0 & 0 \\ -2 & -3 & -2 \\ 3 & 6 & 1 \end{bmatrix}^{-1}\begin{bmatrix} 0 \\ 0 \\ 0 \end{bmatrix} = \begin{bmatrix} 0 \\ 0 \\ 0 \end{bmatrix}$

47.
$$\begin{aligned} AB &= AC \\ A^{-1}AB &= A^{-1}AC \\ IB &= IC \\ B &= C \end{aligned}$$

49.
$$\begin{aligned} (I-B)(I+B) &= I^2 + IB - BI - B^2 \\ &= I + B - B - B^2 \\ &= I - B^2 \\ &= I - \mathbf{0} = I \end{aligned}$$

Thus, $I - B$ and $I + B$ are inverses.

51. $y = \dfrac{3x-5}{x^2-4} = \dfrac{3x-5}{(x+2)(x-2)}$

domain: $(-\infty, -2) \cup (-2, 2) \cup (2, \infty)$

53. $y = \dfrac{3x-5}{\sqrt{x^2+4}}$

domain: $(-\infty, \infty)$

55. $y = x^2$; range: $[0, \infty)$

57. $y = \log x$; range: all $(-\infty, \infty)$

Exercises 6.5 (page 610)

1. $|A|$, det A

3. 0

5. 0

SECTION 6.5

7. $\begin{vmatrix} 2 & 1 \\ -2 & 3 \end{vmatrix} = (2)(3) - (1)(-2)$
$= 6 - (-2) = 8$

9. $\begin{vmatrix} 2 & -3 \\ -3 & 5 \end{vmatrix} = (2)(5) - (-3)(-3)$
$= 10 - 9 = 1$

11. $M_{21} = \begin{vmatrix} -2 & 3 \\ 8 & 9 \end{vmatrix} = (-2)(9) - (3)(8)$
$= -18 - 24 = -42$

13. $M_{33} = \begin{vmatrix} 1 & -2 \\ 4 & 5 \end{vmatrix} = (1)(5) - (-2)(4)$
$= 5 + 8 = 13$

15. $C_{21} = -\begin{vmatrix} -2 & 3 \\ 8 & 9 \end{vmatrix} = -[(-2)(9) - (3)(8)]$
$= -[-18 - 24] = 42$

17. $C_{33} = \begin{vmatrix} 1 & -2 \\ 4 & 5 \end{vmatrix} = (1)(5) - (-2)(4)$
$= 5 + 8 = 13$

19. $$\begin{vmatrix} 2 & -3 & 5 \\ -2 & 1 & 3 \\ 1 & 3 & -2 \end{vmatrix} = 2\begin{vmatrix} 1 & 3 \\ 3 & -2 \end{vmatrix} - (-3)\begin{vmatrix} -2 & 3 \\ 1 & -2 \end{vmatrix} + 5\begin{vmatrix} -2 & 1 \\ 1 & 3 \end{vmatrix}$$
$$= 2(-11) + 3(1) + 5(-7) = -22 + 3 - 35 = -54$$

21. $$\begin{vmatrix} 1 & -1 & 2 \\ 2 & 1 & 3 \\ 1 & 1 & -1 \end{vmatrix} = 1\begin{vmatrix} 1 & 3 \\ 1 & -1 \end{vmatrix} - (-1)\begin{vmatrix} 2 & 3 \\ 1 & -1 \end{vmatrix} + 2\begin{vmatrix} 2 & 1 \\ 1 & 1 \end{vmatrix}$$
$$= 1(-4) + 1(-5) + 2(1) = -4 - 5 + 2 = -7$$

23. $$\begin{vmatrix} 2 & 1 & -1 \\ 1 & 3 & 5 \\ 2 & -5 & 3 \end{vmatrix} = 2\begin{vmatrix} 3 & 5 \\ -5 & 3 \end{vmatrix} - 1\begin{vmatrix} 1 & 5 \\ 2 & 3 \end{vmatrix} + (-1)\begin{vmatrix} 1 & 3 \\ 2 & -5 \end{vmatrix}$$
$$= 2(34) - 1(-7) - 1(-11) = 68 + 7 + 11 = 86$$

25. $$\begin{vmatrix} 0 & 1 & -3 \\ -3 & 5 & 2 \\ 2 & -5 & 3 \end{vmatrix} = 0\begin{vmatrix} 5 & 2 \\ -5 & 3 \end{vmatrix} - 1\begin{vmatrix} -3 & 2 \\ 2 & 3 \end{vmatrix} + (-3)\begin{vmatrix} -3 & 5 \\ 2 & -5 \end{vmatrix}$$
$$= 0 - 1(-13) - 3(5) = 0 + 13 - 15 = -2$$

27. $$\begin{vmatrix} 0 & 0 & 1 & 0 \\ -2 & 1 & 0 & 1 \\ 1 & 0 & 1 & 2 \\ 2 & 0 & 1 & 2 \end{vmatrix} = 0(***) - 0(***) + 1\begin{vmatrix} -2 & 1 & 1 \\ 1 & 0 & 2 \\ 2 & 0 & 2 \end{vmatrix} - 0(***)$$
$$= 1\left(-2\begin{vmatrix} 0 & 2 \\ 0 & 2 \end{vmatrix} - 1\begin{vmatrix} 1 & 2 \\ 2 & 2 \end{vmatrix} + 1\begin{vmatrix} 1 & 0 \\ 2 & 0 \end{vmatrix}\right) = -2(0) - 1(-2) + 1(0) = 2$$

29. $$\begin{vmatrix} 10 & 20 & 10 & 30 \\ -2 & 1 & -3 & 1 \\ -1 & 0 & 1 & -2 \\ 2 & -1 & -1 & 3 \end{vmatrix} = \begin{vmatrix} 10 & 20 & 10 & 30 \\ 0 & 0 & -4 & 4 \\ 0 & 2 & 2 & 1 \\ 0 & -5 & -3 & -3 \end{vmatrix} \quad \begin{matrix} \\ R_2 + R_4 \\ R_1 + R_3 \\ -2R_1 + R_4 \end{matrix}$$

$$= 10\begin{vmatrix} 0 & -4 & 4 \\ 2 & 2 & 1 \\ -5 & -3 & -3 \end{vmatrix} \quad \text{(expand along first column)}$$

$$= 10\left(0\begin{vmatrix} 2 & 1 \\ -3 & -3 \end{vmatrix} - (-4)\begin{vmatrix} 2 & 1 \\ -5 & -3 \end{vmatrix} + 4\begin{vmatrix} 2 & 2 \\ -5 & -3 \end{vmatrix}\right)$$

$$= 10[0 + 4(-1) + 4(4)] = 10(12) = 120$$

31. R_1 and R_2 have been switched. This multiplies the determinant by -1. TRUE

33. R_1 and R_2 have both been multiplied by -1. This multiplies the determinant by -1 twice. FALSE

35. R_1 and R_2 have been switched. This multiplies the determinant by -1. However, R_3 has been multiplied by -1, which also multiplies the determinant by -1. Thus, the determinant remains equal to 3.

37. R_1 has been added to R_3. This leaves the determinant equal to 3.

39. $$x = \frac{\begin{vmatrix} 7 & 2 \\ -4 & -3 \end{vmatrix}}{\begin{vmatrix} 3 & 2 \\ 2 & -3 \end{vmatrix}} = \frac{-13}{-13} = 1 \quad y = \frac{\begin{vmatrix} 3 & 7 \\ 2 & -4 \end{vmatrix}}{\begin{vmatrix} 3 & 2 \\ 2 & -3 \end{vmatrix}} = \frac{-26}{-13} = 2$$

41. $$x = \frac{\begin{vmatrix} 3 & -1 \\ 9 & -7 \end{vmatrix}}{\begin{vmatrix} 1 & -1 \\ 3 & -7 \end{vmatrix}} = \frac{-12}{-4} = 3 \quad y = \frac{\begin{vmatrix} 1 & 3 \\ 3 & 9 \end{vmatrix}}{\begin{vmatrix} 1 & -1 \\ 3 & -7 \end{vmatrix}} = \frac{0}{-4} = 0$$

43. $$x = \frac{\begin{vmatrix} 2 & 2 & 1 \\ 2 & -1 & 1 \\ 4 & 1 & 3 \end{vmatrix}}{\begin{vmatrix} 1 & 2 & 1 \\ 1 & -1 & 1 \\ 1 & 1 & 3 \end{vmatrix}} = \frac{-6}{-6} = 1 \quad y = \frac{\begin{vmatrix} 1 & 2 & 1 \\ 1 & 2 & 1 \\ 1 & 4 & 3 \end{vmatrix}}{\begin{vmatrix} 1 & 2 & 1 \\ 1 & -1 & 1 \\ 1 & 1 & 3 \end{vmatrix}} = \frac{0}{-6} = 0 \quad z = \frac{\begin{vmatrix} 1 & 2 & 2 \\ 1 & -1 & 2 \\ 1 & 1 & 4 \end{vmatrix}}{\begin{vmatrix} 1 & 2 & 1 \\ 1 & -1 & 1 \\ 1 & 1 & 3 \end{vmatrix}} = \frac{-6}{-6} = 1$$

SECTION 6.5

45. $x = \dfrac{\begin{vmatrix} 5 & -1 & 1 \\ 10 & -3 & 2 \\ 0 & 3 & 1 \end{vmatrix}}{\begin{vmatrix} 2 & -1 & 1 \\ 3 & -3 & 2 \\ 1 & 3 & 1 \end{vmatrix}} = \frac{-5}{-5} = 1 \quad y = \dfrac{\begin{vmatrix} 2 & 5 & 1 \\ 3 & 10 & 2 \\ 1 & 0 & 1 \end{vmatrix}}{\begin{vmatrix} 2 & -1 & 1 \\ 3 & -3 & 2 \\ 1 & 3 & 1 \end{vmatrix}} = \frac{5}{-5} = -1 \quad z = \dfrac{\begin{vmatrix} 2 & -1 & 5 \\ 3 & -3 & 10 \\ 1 & 3 & 0 \end{vmatrix}}{\begin{vmatrix} 2 & -1 & 1 \\ 3 & -3 & 2 \\ 1 & 3 & 1 \end{vmatrix}} = \frac{-10}{-5} = 2$

47. Rewrite system:
$3x + 2y + 3z = 66$
$2x + 6y - z = 36$
$3x + 1y + 6z = 96$

$x = \dfrac{\begin{vmatrix} 66 & 2 & 3 \\ 36 & 6 & -1 \\ 96 & 1 & 6 \end{vmatrix}}{\begin{vmatrix} 3 & 2 & 3 \\ 2 & 6 & -1 \\ 3 & 1 & 6 \end{vmatrix}} = \dfrac{198}{33} = 6$

$y = \dfrac{\begin{vmatrix} 3 & 66 & 3 \\ 2 & 36 & -1 \\ 3 & 96 & 6 \end{vmatrix}}{\begin{vmatrix} 3 & 2 & 3 \\ 2 & 6 & -1 \\ 3 & 1 & 6 \end{vmatrix}} = \dfrac{198}{33} = 6 \quad z = \dfrac{\begin{vmatrix} 3 & 2 & 66 \\ 2 & 6 & 36 \\ 3 & 1 & 96 \end{vmatrix}}{\begin{vmatrix} 3 & 2 & 3 \\ 2 & 6 & -1 \\ 3 & 1 & 6 \end{vmatrix}} = \dfrac{396}{33} = 12$

49. $p = \dfrac{\begin{vmatrix} 0 & -1 & 3 & -1 \\ -1 & 1 & 0 & -1 \\ 2 & 0 & -1 & 0 \\ 7 & -2 & 0 & 3 \end{vmatrix}}{\begin{vmatrix} 2 & -1 & 3 & -1 \\ 1 & 1 & 0 & -1 \\ 3 & 0 & -1 & 0 \\ 1 & -2 & 0 & 3 \end{vmatrix}} = \dfrac{-15}{-18} = \dfrac{5}{6} \quad q = \dfrac{\begin{vmatrix} 2 & 0 & 3 & -1 \\ 1 & -1 & 0 & -1 \\ 3 & 2 & -1 & 0 \\ 1 & 7 & 0 & 3 \end{vmatrix}}{\begin{vmatrix} 2 & -1 & 3 & -1 \\ 1 & 1 & 0 & -1 \\ 3 & 0 & -1 & 0 \\ 1 & -2 & 0 & 3 \end{vmatrix}} = \dfrac{-12}{-18} = \dfrac{2}{3}$

$r = \dfrac{\begin{vmatrix} 2 & -1 & 0 & -1 \\ 1 & 1 & -1 & -1 \\ 3 & 0 & 2 & 0 \\ 1 & -2 & 7 & 3 \end{vmatrix}}{\begin{vmatrix} 2 & -1 & 3 & -1 \\ 1 & 1 & 0 & -1 \\ 3 & 0 & -1 & 0 \\ 1 & -2 & 0 & 3 \end{vmatrix}} = \dfrac{-9}{-18} = \dfrac{1}{2} \quad s = \dfrac{\begin{vmatrix} 2 & -1 & 3 & 0 \\ 1 & 1 & 0 & -1 \\ 3 & 0 & -1 & 2 \\ 1 & -2 & 0 & 7 \end{vmatrix}}{\begin{vmatrix} 2 & -1 & 3 & -1 \\ 1 & 1 & 0 & -1 \\ 3 & 0 & -1 & 0 \\ 1 & -2 & 0 & 3 \end{vmatrix}} = \dfrac{-45}{-18} = \dfrac{5}{2}$

51. $\begin{vmatrix} x & y & 1 \\ 0 & 0 & 1 \\ 4 & 6 & 1 \end{vmatrix} = 0$

$$x\begin{vmatrix} 0 & 1 \\ 6 & 1 \end{vmatrix} - y\begin{vmatrix} 0 & 1 \\ 4 & 1 \end{vmatrix} + 1\begin{vmatrix} 0 & 0 \\ 4 & 6 \end{vmatrix} = 0$$
$$x(-6) - y(-4) + 1(0) = 0$$
$$-6x + 4y = 0$$
$$3x - 2y = 0$$

53. $\begin{vmatrix} x & y & 1 \\ -2 & 3 & 1 \\ 5 & -3 & 1 \end{vmatrix} = 0$

$$x\begin{vmatrix} 3 & 1 \\ -3 & 1 \end{vmatrix} - y\begin{vmatrix} -2 & 1 \\ 5 & 1 \end{vmatrix} + 1\begin{vmatrix} -2 & 3 \\ 5 & -3 \end{vmatrix} = 0$$
$$x(6) - y(-7) + 1(-9) = 0$$
$$6x + 7y - 9 = 0$$
$$6x + 7y = 9$$

55. $\pm\frac{1}{2}\begin{vmatrix} 0 & 0 & 1 \\ 12 & 0 & 1 \\ 12 & 5 & 1 \end{vmatrix} = \pm\frac{1}{2}(60)$

$= 30$ square units

57. $\pm\frac{1}{2}\begin{vmatrix} 2 & 3 & 1 \\ 10 & 8 & 1 \\ 0 & 20 & 1 \end{vmatrix} = \pm\frac{1}{2}(146)$

$= 73$ square units

59. $\begin{vmatrix} a & b \\ c & d \end{vmatrix} = ad - bc; \begin{vmatrix} b & a \\ d & c \end{vmatrix} = bc - ad = -(ad - bc)$

61. $\begin{vmatrix} a & b \\ c & d \end{vmatrix} = ad - bc$

$$\begin{vmatrix} a & b + ka \\ c & d + kc \end{vmatrix} = a(d + kc) - (b + ka)c = ad + akc - bc - akc = ad - bc$$

63. $\begin{vmatrix} 3 & x \\ 1 & 2 \end{vmatrix} = \begin{vmatrix} 2 & -1 \\ x & -5 \end{vmatrix}$

$$6 - x = -10 + x$$
$$-2x = -16$$
$$x = 8$$

65. $\begin{vmatrix} 3 & x & 1 \\ x & 0 & -2 \\ 4 & 0 & 1 \end{vmatrix} = \begin{vmatrix} 2 & x \\ x & 4 \end{vmatrix}$

$$-x\begin{vmatrix} x & -2 \\ 4 & 1 \end{vmatrix} = 8 - x^2$$
$$-x(x + 8) = 8 - x^2$$
$$-x^2 - 8x = -x^2 + 8$$
$$x = -1$$

67. Let $x =$ \$ invested in HiTech, $y =$ \$ invested in SaveTel, and $z =$ \$ invested in OilCo.

$$\begin{cases} x + y + z = 20{,}000 \\ y + z = 3x \\ 0.10x + 0.05y + 0.06z = 0.066(20{,}000) \end{cases} \qquad \begin{aligned} x + y + z &= 20000 \\ -3x + y + z &= 0 \\ 10x + 5y + 6z &= 132000 \end{aligned}$$

$$x = \frac{\begin{vmatrix} 20000 & 1 & 1 \\ 0 & 1 & 1 \\ 132000 & 5 & 6 \end{vmatrix}}{\begin{vmatrix} 1 & 1 & 1 \\ -3 & 1 & 1 \\ 10 & 5 & 6 \end{vmatrix}} = \frac{20000}{4} = 5000, \quad y = \frac{\begin{vmatrix} 1 & 20000 & 1 \\ -3 & 0 & 1 \\ 10 & 132000 & 6 \end{vmatrix}}{\begin{vmatrix} 1 & 1 & 1 \\ -3 & 1 & 1 \\ 10 & 5 & 6 \end{vmatrix}} = \frac{32000}{4} = 8000$$

continued on next page

67. continued

$$z = \frac{\begin{vmatrix} 1 & 1 & 20000 \\ -3 & 1 & 0 \\ 10 & 5 & 132000 \end{vmatrix}}{\begin{vmatrix} 1 & 1 & 1 \\ -3 & 1 & 1 \\ 10 & 5 & 6 \end{vmatrix}} = \frac{28000}{4} = 7000 \Rightarrow$$ He should invest \$5000 in HiTech, \$8000 in SaveTel, and \$7000 in OilCo.

69. $\begin{vmatrix} 1 & 3 & 4 \\ 0 & 5 & 2 \\ 0 & 0 & 2 \end{vmatrix} = 10$

$1 \cdot 5 \cdot 2 = 10$

71. $\begin{vmatrix} 1 & 2 & 4 & 3 \\ 0 & 2 & 2 & 1 \\ 0 & 0 & 3 & 2 \\ 0 & 0 & 0 & 4 \end{vmatrix} = 24$

$1 \cdot 2 \cdot 3 \cdot 4 = 24$

73. $3(1)(1) + 2(-2)(1) + (-1)(2)(3) = -7$

$1(1)(-1) + 3(-2)(3) + 1(2)(2) = -15$

$-7 - (-15) = 8$

75-77. Answers may vary.

79. yes

81. $\begin{vmatrix} 2.3 & 5.7 & 6.1 \\ 3.4 & 6.2 & 8.3 \\ 5.8 & 8.2 & 9.2 \end{vmatrix} = 21.468$

83. $x^2 + 3x - 4 = (x + 4)(x - 1)$

85. $9x^3 - x = x(9x^2 - 1)$

$= x(3x + 1)(3x - 1)$

87. $\frac{1}{x-2} + \frac{2}{2x-1} = \frac{1(2x-1)}{(x-2)(2x-1)} + \frac{2(x-2)}{(2x-1)(x-2)} = \frac{2x-1+2x-4}{(x-2)(2x-1)} = \frac{4x-5}{(x-2)(2x-1)}$

89. $\frac{2}{x^2+1} + \frac{1}{x} = \frac{2x}{x(x^2+1)} + \frac{x^2+1}{x(x^2+1)} = \frac{x^2+2x+1}{x(x^2+1)}$

Exercises 6.6 (page 620)

1. first-degree, second-degree

3. $\frac{3x-1}{x(x-1)} = \frac{A}{x} + \frac{B}{x-1}$

$\frac{3x-1}{x(x-1)} = \frac{A(x-1)}{x(x-1)} + \frac{Bx}{x(x-1)}$

$\frac{3x-1}{x(x-1)} = \frac{Ax - A + Bx}{x(x-1)}$

$\frac{3x-1}{x(x-1)} = \frac{(A+B)x - A}{x(x-1)}$

$\begin{cases} A + B = 3 \\ -A = -1 \end{cases} \Rightarrow A = 1, B = 2$

$\frac{3x-1}{x(x-1)} = \frac{1}{x} + \frac{2}{x-1}$

SECTION 6.6

5. $\dfrac{2x-15}{x(x-3)} = \dfrac{A}{x} + \dfrac{B}{x-3}$

$\dfrac{2x-15}{x(x-3)} = \dfrac{A(x-3)}{x(x-3)} + \dfrac{Bx}{x(x-3)}$

$\dfrac{2x-15}{x(x-3)} = \dfrac{Ax-3A+Bx}{x(x-3)}$

$\dfrac{2x-15}{x(x-3)} = \dfrac{(A+B)x-3A}{x(x-3)}$

$\begin{cases} A+B = 2 \\ -3A = -15 \end{cases} \Rightarrow A=5,\ B=-3$

$\dfrac{2x-15}{x(x-3)} = \dfrac{5}{x} - \dfrac{3}{x-3}$

7. $\dfrac{3x+1}{(x+1)(x-1)} = \dfrac{A}{x+1} + \dfrac{B}{x-1}$

$\dfrac{3x+1}{(x+1)(x-1)} = \dfrac{A(x-1)}{(x+1)(x-1)} + \dfrac{B(x+1)}{(x+1)(x-1)}$

$\dfrac{3x+1}{(x+1)(x-1)} = \dfrac{Ax-A+Bx+B}{(x+1)(x-1)}$

$\dfrac{3x+1}{(x+1)(x-1)} = \dfrac{(A+B)x+(-A+B)}{(x+1)(x-1)}$

$\begin{cases} A+B=3 \\ -A+B=1 \end{cases} \Rightarrow A=1,\ B=2$

$\dfrac{3x+1}{(x+1)(x-1)} = \dfrac{1}{x+1} + \dfrac{2}{x-1}$

9. $\dfrac{-4}{x^2-2x} =$

$\dfrac{-4}{x(x-2)} = \dfrac{A}{x} + \dfrac{B}{x-2}$

$\dfrac{-4}{x(x-2)} = \dfrac{A(x-2)}{x(x-2)} + \dfrac{Bx}{x(x-2)}$

$\dfrac{-4}{x(x-2)} = \dfrac{Ax-2A+Bx}{x(x-2)}$

$\dfrac{-4}{x(x-2)} = \dfrac{(A+B)x-2A}{x(x-2)}$

$\begin{cases} A+B = 0 \\ -2A = -4 \end{cases} \Rightarrow A=2,\ B=-2$

$\dfrac{4}{x^2-2x} = \dfrac{2}{x} - \dfrac{2}{x-2}$

11. $\dfrac{-2x+11}{(x+2)(x-3)} = \dfrac{A}{x+2} + \dfrac{B}{x-3}$

$\dfrac{-2x+11}{(x+2)(x-3)} = \dfrac{A(x-3)}{(x+2)(x-3)} + \dfrac{B(x+2)}{(x+2)(x-3)}$

$\dfrac{-2x+11}{(x+2)(x-3)} = \dfrac{Ax-3A+Bx+2B}{(x+2)(x-3)}$

$\dfrac{-2x+11}{(x+2)(x-3)} = \dfrac{(A+B)x+(-3A+2B)}{(x+2)(x-3)}$

$\dfrac{-2x+11}{x^2-x-6} = \dfrac{-2x+11}{(x+2)(x-3)}$

$\begin{cases} A+B=-2 \\ -3A+2B=11 \end{cases} \Rightarrow A=-3,\ B=1$

$\dfrac{-2x+11}{(x+2)(x-3)} = \dfrac{-3}{x+2} + \dfrac{1}{x-3}$

13. $\dfrac{3x-23}{(x+3)(x-1)} = \dfrac{A}{x+3} + \dfrac{B}{x-1}$

$\dfrac{3x-23}{(x+3)(x-1)} = \dfrac{A(x-1)}{(x+3)(x-1)} + \dfrac{B(x+3)}{(x+3)(x-1)}$

$\dfrac{3x-23}{(x+3)(x-1)} = \dfrac{Ax-A+Bx+3B}{(x+3)(x-1)}$

$\dfrac{3x-23}{(x+3)(x-1)} = \dfrac{(A+B)x+(-A+3B)}{(x+3)(x-1)}$

$\dfrac{3x-23}{x^2+2x-3} = \dfrac{3x-23}{(x+3)(x-1)}$

$\begin{cases} A+B=3 \\ -A+3B=-23 \end{cases} \Rightarrow A=8,\ B=-5$

$\dfrac{3x-23}{(x+3)(x-1)} = \dfrac{8}{x+3} - \dfrac{5}{x-1}$

15. $\dfrac{9x-31}{2x^2-13x+15}=\dfrac{9x-31}{(2x-3)(x-5)}=\dfrac{A}{2x-3}+\dfrac{B}{x-5}$

$$\dfrac{9x-31}{(2x-3)(x-5)}=\dfrac{A(x-5)}{(2x-3)(x-5)}+\dfrac{B(2x-3)}{(2x-3)(x-5)}$$

$$\dfrac{9x-31}{(2x-3)(x-5)}=\dfrac{Ax-5A+2Bx-3B}{(2x-3)(x-5)}$$

$$\dfrac{9x-31}{(2x-3)(x-5)}=\dfrac{(A+2B)x+(-5A-3B)}{(2x-3)(x-5)}$$

$\begin{cases} A+2B= & 9 \\ -5A-3B= & -31 \end{cases} \Rightarrow A=5,\ B=2$ $\quad \dfrac{9x-31}{(2x-3)(x-5)}=\dfrac{5}{2x-3}+\dfrac{2}{x-5}$

17. $\dfrac{4x^2+4x-2}{x(x^2-1)}=\dfrac{4x^2+4x-2}{x(x+1)(x-1)}=\dfrac{A}{x}+\dfrac{B}{x+1}+\dfrac{C}{x-1}$

$$\dfrac{4x^2+4x-2}{x(x+1)(x-1)}=\dfrac{A(x+1)(x-1)}{x(x+1)(x-1)}+\dfrac{Bx(x-1)}{x(x+1)(x-1)}+\dfrac{Cx(x+1)}{x(x+1)(x-1)}$$

$$\dfrac{4x^2+4x-2}{x(x+1)(x-1)}=\dfrac{Ax^2-A+Bx^2-Bx+Cx^2+Cx}{x(x+1)(x-1)}$$

$$\dfrac{4x^2+4x-2}{x(x+1)(x-1)}=\dfrac{(A+B+C)x^2+(-B+C)x+(-A)}{x(x+1)(x-1)}$$

$\begin{cases} A+B+C= & 4 \\ \quad -B+C= & 4 \\ -A \quad\quad = & -2 \end{cases} \Rightarrow \begin{matrix} A=2 \\ B=-1 \\ C=3 \end{matrix}$ $\quad \dfrac{4x^2+4x-2}{x(x+1)(x-1)}=\dfrac{2}{x}-\dfrac{1}{x+1}+\dfrac{3}{x-1}$

19. $\dfrac{x^2+x+3}{x(x^2+3)}=\dfrac{A}{x}+\dfrac{Bx+C}{x^2+3}$

$$=\dfrac{A(x^2+3)}{x(x^2+3)}+\dfrac{(Bx+C)x}{x(x^2+3)}$$

$$=\dfrac{Ax^2+3A+Bx^2+Cx}{x(x^2+3)}$$

$$=\dfrac{(A+B)x^2+Cx+3A}{x(x^2+3)}$$

$\begin{cases} A+B \quad = 1 \\ \quad\quad C = 1 \\ 3A \quad\quad = 3 \end{cases} \Rightarrow \begin{matrix} A=1 \\ B=0 \\ C=1 \end{matrix}$

$$\dfrac{x^2+x+3}{x(x^2+3)}=\dfrac{1}{x}+\dfrac{1}{x^2+3}$$

21. $\dfrac{3x^2+8x+11}{(x+1)(x^2+2x+3)}=\dfrac{A}{x+1}+\dfrac{Bx+C}{x^2+2x+3}$

$$\dfrac{3x^2+8x+11}{(x+1)(x^2+2x+3)}=\dfrac{A(x^2+2x+3)}{(x+1)(x^2+2x+3)}+\dfrac{(Bx+C)(x+1)}{(x+1)(x^2+2x+3)}$$

$$\dfrac{3x^2+8x+11}{(x+1)(x^2+2x+3)}=\dfrac{Ax^2+2Ax+3A+Bx^2+Bx+Cx+C}{(x+1)(x^2+2x+3)}$$

$$\dfrac{3x^2+8x+11}{(x+1)(x^2+2x+3)}=\dfrac{(A+B)x^2+(2A+B+C)x+(3A+C)}{(x+1)(x^2+2x+3)}$$

continued on next page

21. continued

$$\begin{cases} A+B \quad\;\; = 3 \\ 2A+B+C = 8 \\ 3A \quad\;\; +C = 11 \end{cases} \Rightarrow \begin{matrix} A=3 \\ B=0 \\ C=2 \end{matrix} \qquad \frac{3x^2+8x+11}{(x+1)(x^2+2x+3)} = \frac{3}{x+1} + \frac{2}{x^2+2x+3}$$

23. $$\frac{5x^2+9x+3}{x(x+1)^2} = \frac{A}{x} + \frac{B}{x+1} + \frac{C}{(x+1)^2}$$

$$= \frac{A(x+1)^2}{x(x+1)^2} + \frac{Bx(x+1)}{x(x+1)^2} + \frac{Cx}{x(x+1)^2}$$

$$= \frac{Ax^2+2Ax+A+Bx^2+Bx+Cx}{x(x+1)^2}$$

$$= \frac{(A+B)x^2+(2A+B+C)x+A}{x(x+1)^2}$$

$$\begin{cases} A+B \quad\;\; = 5 \\ 2A+B+C = 9 \\ A \quad\;\; = 3 \end{cases} \Rightarrow \begin{matrix} A=3 \\ B=2 \\ C=1 \end{matrix} \qquad \frac{5x^2+9x+3}{x(x+1)^2} = \frac{3}{x} + \frac{2}{x+1} + \frac{1}{(x+1)^2}$$

25. $$\frac{-2x^2+x-2}{x^2(x-1)} = \frac{A}{x} + \frac{B}{x^2} + \frac{C}{x-1}$$

$$= \frac{Ax(x-1)}{x^2(x-1)} + \frac{B(x-1)}{x^2(x-1)} + \frac{Cx^2}{x^2(x-1)}$$

$$= \frac{Ax^2-Ax+Bx-B+Cx^2}{x^2(x-1)}$$

$$= \frac{(A+C)x^2+(-A+B)x+(-B)}{x^2(x-1)}$$

$$\begin{cases} A \quad\;\; +C = -2 \\ -A+B \quad\;\; = 1 \\ \quad -B \quad\;\; = -2 \end{cases} \Rightarrow \begin{matrix} A=1 \\ B=2 \\ C=-3 \end{matrix} \qquad \frac{-2x^2+x-2}{x^2(x-1)} = \frac{1}{x} + \frac{2}{x^2} - \frac{3}{x-1}$$

27. $$\frac{3x^2-13x+18}{x^3-6x^2+9x} = \frac{3x^2-13x+18}{x(x-3)^2} = \frac{A}{x} + \frac{B}{x-3} + \frac{C}{(x-3)^2}$$

$$= \frac{A(x-3)^2}{x(x-3)^2} + \frac{Bx(x-3)}{x(x-3)^2} + \frac{Cx}{x(x-3)^2}$$

$$= \frac{Ax^2-6Ax+9A+Bx^2-3Bx+Cx}{x(x-3)^2}$$

$$= \frac{(A+B)x^2+(-6A-3B+C)x+9A}{x(x-3)^2}$$

$$\begin{cases} A+B \quad\;\; = 3 \\ -6A-3B+C = -13 \\ 9A \quad\;\; = 18 \end{cases} \Rightarrow \begin{matrix} A=2 \\ B=1 \\ C=2 \end{matrix} \qquad \frac{3x^2-13x+18}{x(x-3)^2} = \frac{2}{x} + \frac{1}{x-3} + \frac{2}{(x-3)^2}$$

29. $$\frac{x^2-2x-3}{(x-1)^3} = \frac{A}{x-1}+\frac{B}{(x-1)^2}+\frac{C}{(x-1)^3} = \frac{A(x-1)^2}{(x-1)^3}+\frac{B(x-1)}{(x-1)^3}+\frac{C}{(x-1)^3} = \frac{Ax^2-2Ax+A+Bx-B+C}{(x-1)^3} = \frac{Ax^2+(-2A+B)x+(A-B+C)}{(x-1)^3}$$

$$\begin{cases} A = 1 \\ -2A+B = -2 \\ A-B+C = -3 \end{cases} \Rightarrow \begin{matrix} A=1 \\ B=0 \\ C=-4 \end{matrix} \quad \frac{x^2-2x-3}{(x-1)^3} = \frac{1}{x-1} - \frac{4}{(x-1)^3}$$

31. $$\frac{x^3+4x^2+2x+1}{x^4+x^3+x^2} =$$

$$\frac{x^3+4x^2+2x+1}{x^2(x^2+x+1)} = \frac{A}{x}+\frac{B}{x^2}+\frac{Cx+D}{x^2+x+1} = \frac{Ax(x^2+x+1)}{x^2(x^2+x+1)}+\frac{B(x^2+x+1)}{x^2(x^2+x+1)}+\frac{(Cx+D)x^2}{x^2(x^2+x+1)} = \frac{Ax^3+Ax^2+Ax+Bx^2+Bx+B+Cx^3+Dx^2}{x^2(x^2+x+1)} = \frac{(A+C)x^3+(A+B+D)x^2+(A+B)x+B}{x^2(x^2+x+1)}$$

$$\begin{cases} A + C = 1 \\ A+B+D = 4 \\ A+B = 2 \\ B = 1 \end{cases} \Rightarrow \begin{matrix} A=1 \\ B=1 \\ C=0 \\ D=2 \end{matrix} \quad \frac{x^3+4x^2+2x+1}{x^2(x^2+x+1)} = \frac{1}{x}+\frac{1}{x^2}+\frac{2}{x^2+x+1}$$

33. $$\frac{4x^3+5x^2+3x+4}{x^2(x^2+1)} = \frac{A}{x}+\frac{B}{x^2}+\frac{Cx+D}{x^2+1} = \frac{Ax(x^2+1)}{x^2(x^2+1)}+\frac{B(x^2+1)}{x^2(x^2+1)}+\frac{(Cx+D)x^2}{x^2(x^2+1)} = \frac{Ax^3+Ax+Bx^2+B+Cx^3+Dx^2}{x^2(x^2+1)} = \frac{(A+C)x^3+(B+D)x^2+Ax+B}{x^2(x^2+1)}$$

$$\begin{cases} A + C = 4 \\ B + D = 5 \\ A = 3 \\ B = 4 \end{cases} \Rightarrow \begin{matrix} A=3 \\ B=4 \\ C=1 \\ D=1 \end{matrix} \quad \frac{4x^3+5x^2+3x+4}{x^2(x^2+1)} = \frac{3}{x}+\frac{4}{x^2}+\frac{x+1}{x^2+1}$$

35. $$\frac{-x^2-3x-5}{x^3+x^2+2x+2}=\frac{-x^2-3x-5}{x^2(x+1)+2(x+1)}=\frac{-x^2-3x-5}{(x+1)(x^2+2)}$$

$$\begin{aligned}\frac{-x^2-3x-5}{(x+1)(x^2+2)}&=\frac{A}{x+1}+\frac{Bx+C}{x^2+2}\\&=\frac{A(x^2+2)}{(x+1)(x^2+2)}+\frac{(Bx+C)(x+1)}{(x+1)(x^2+2)}\\&=\frac{Ax^2+2A+Bx^2+Bx+Cx+C}{(x+1)(x^2+2)}\\&=\frac{(A+B)x^2+(B+C)x+(2A+C)}{(x+1)(x^2+2)}\end{aligned}$$

$$\begin{cases}A+B=-1\\B+C=-3\\2A+C=-5\end{cases}\Rightarrow\begin{array}{l}A=-1\\B=0\\C=-3\end{array}\qquad\frac{-x^2-3x-5}{(x+1)(x^2+2)}=\frac{-1}{x+1}-\frac{3}{x^2+2}$$

37. $$\begin{aligned}\frac{x^3+4x^2+3x+6}{(x^2+2)(x^2+x+2)}&=\frac{Ax+B}{x^2+2}+\frac{Cx+D}{x^2+x+2}\\&=\frac{(Ax+B)(x^2+x+2)}{(x^2+2)(x^2+x+2)}+\frac{(Cx+D)(x^2+2)}{(x^2+2)(x^2+x+2)}\\&=\frac{Ax^3+Ax^2+2Ax+Bx^2+Bx+2B+Cx^3+2Cx+Dx^2+2D}{(x^2+2)(x^2+x+2)}\\&=\frac{(A+C)x^3+(A+B+D)x^2+(2A+B+2C)x+(2B+2D)}{(x^2+2)(x^2+x+2)}\end{aligned}$$

$$\begin{cases}A+C=1\\A+B+D=4\\2A+B+2C=3\\2B+2D=6\end{cases}\Rightarrow\begin{array}{l}A=1\\B=1\\C=0\\D=2\end{array}\qquad\frac{x^3+4x^2+3x+6}{(x^2+2)(x^2+x+2)}=\frac{x+1}{x^2+2}+\frac{2}{x^2+x+2}$$

39. $$\begin{aligned}&\frac{2x^4+6x^3+20x^2+22x+25}{x(x^2+2x+5)^2}\\&=\frac{A}{x}+\frac{Bx+C}{x^2+2x+5}+\frac{Dx+E}{(x^2+2x+5)^2}\\&=\frac{A(x^2+2x+5)^2}{x(x^2+2x+5)^2}+\frac{(Bx+C)(x)(x^2+2x+5)}{x(x^2+2x+5)^2}+\frac{(Dx+E)x}{x(x^2+2x+5)^2}\\&=\frac{(A+B)x^4+(4A+2B+C)x^3+(14A+5B+2C+D)x^2+(20A+5C+E)x+(25A)}{x(x^2+2x+5)^2}\end{aligned}$$

$$\begin{cases}A+B=2\\4A+2B+C=6\\14A+5B+2C+D=20\\20A+5C+E=22\\25A=25\end{cases}\Rightarrow\begin{array}{l}A=1\\B=1\\C=0\\D=1\\E=2\end{array}$$

continued on next page

39. continued

$$\frac{2x^4+6x^3+20x^2+22x+25}{x(x^2+2x+5)^2}=\frac{1}{x}+\frac{x}{x^2+2x+5}+\frac{x+2}{(x^2+2x+5)^2}$$

41. Use long division first: $\frac{x^3}{x^2+3x+2}=x-3+\frac{7x+6}{x^2+3x+2}=x-3+\frac{7x+6}{(x+1)(x+2)}$

$$\begin{aligned}\frac{7x+6}{(x+1)(x+2)}&=\frac{A}{x+1}+\frac{B}{x+2}\\&=\frac{A(x+2)}{(x+1)(x+2)}+\frac{B(x+1)}{(x+1)(x+2)}\\&=\frac{Ax+2A+Bx+B}{(x+1)(x+2)}\\&=\frac{(A+B)x+(2A+B)}{(x+1)(x+2)}\end{aligned}$$

$\begin{cases}A+B=7\\2A+B=6\end{cases}\Rightarrow A=-1,\ B=8\quad x-3+\frac{7x+6}{(x+1)(x+2)}=x-3-\frac{1}{x+1}+\frac{8}{x+2}$

43. Use long division first: $\frac{3x^3+3x^2+6x+4}{3x^3+x^2+3x+1}=1+\frac{2x^2+3x+3}{3x^3+x^2+3x+1}=1+\frac{2x^2+3x+3}{(3x+1)(x^2+1)}$

$$\begin{aligned}\frac{2x^2+3x+3}{(3x+1)(x^2+1)}&=\frac{A}{3x+1}+\frac{Bx+C}{x^2+1}\\&=\frac{A(x^2+1)}{(3x+1)(x^2+1)}+\frac{(Bx+C)(3x+1)}{(3x+1)(x^2+1)}\\&=\frac{Ax^2+A+3Bx^2+Bx+3Cx+C}{(3x+1)(x^2+1)}\\&=\frac{(A+3B)x^2+(B+3C)x+(A+C)}{(3x+1)(x^2+1)}\end{aligned}$$

$\begin{cases}A+3B=2\\B+3C=3\\A+C=3\end{cases}\Rightarrow\begin{matrix}A=2\\B=0\\C=1\end{matrix}\quad 1+\frac{2x^2+3x+3}{(3x+1)(x^2+1)}=1+\frac{2}{3x+1}+\frac{1}{x^2+1}$

45. Use long division first: $\frac{x^3+3x^2+2x+1}{x^3+x^2+x}=1+\frac{2x^2+x+1}{x^3+x^2+x}=1+\frac{2x^2+x+1}{x(x^2+x+1)}$

$$\begin{aligned}\frac{2x^2+x+1}{x(x^2+x+1)}&=\frac{A}{x}+\frac{Bx+C}{x^2+x+1}\\&=\frac{A(x^2+x+1)}{x(x^2+x+1)}+\frac{(Bx+C)x}{x(x^2+x+1)}\\&=\frac{Ax^2+Ax+A+Bx^2+Cx}{x(x^2+x+1)}\\&=\frac{(A+B)x^2+(A+C)x+(A)}{x(x^2+x+1)}\end{aligned}$$

continued on next page...

45. continued

$$\frac{2x^2+x+1}{x(x^2+x+1)}=\frac{(A+B)x^2+(A+C)x+(A)}{x(x^2+x+1)}$$

$$\begin{cases} A+B & =2 \\ A \quad +C & =1 \\ A & =1 \end{cases} \Rightarrow \begin{array}{l} A=1 \\ B=1 \\ C=0 \end{array} \quad 1+\frac{2x^2+x+1}{x(x^2+x+1)}=1+\frac{1}{x}+\frac{x}{x^2+x+1}$$

47. Use long division first: $\dfrac{2x^4+2x^3+3x^2-1}{(x^2-x)(x^2+1)}=2+\dfrac{4x^3+x^2+2x-1}{x(x-1)(x^2+1)}$

$$\begin{aligned}\frac{4x^3+x^2+2x-1}{x(x-1)(x^2+1)}&=\frac{A}{x}+\frac{B}{x-1}+\frac{Cx+D}{x^2+1}\\&=\frac{A(x-1)(x^2+1)}{x(x-1)(x^2+1)}+\frac{Bx(x^2+1)}{x(x-1)(x^2+1)}+\frac{(Cx+D)(x)(x-1)}{x(x-1)(x^2+1)}\\&=\frac{Ax^3-Ax^2+Ax-A+Bx^3+Bx+Cx^3-Cx^2+Dx^2-Dx}{x(x-1)(x^2+1)}\\&=\frac{(A+B+C)x^3+(-A-C+D)x^2+(A+B-D)x+(-A)}{x(x-1)(x^2+1)}\end{aligned}$$

$$\begin{cases} A+B+C & =4 \\ -A \quad -C+D & =1 \\ A+B \quad -D & =2 \\ -A & =-1 \end{cases} \Rightarrow \begin{array}{l} A=1 \\ B=3 \\ C=0 \\ D=2 \end{array} \quad 2+\frac{4x^3+x^2+2x-1}{x(x-1)(x^2+1)}=2+\frac{1}{x}+\frac{3}{x-1}+\frac{2}{x^2+1}$$

49. $x^3+1=(x+1)(x^2-x+1)\Rightarrow$ not prime

51. $\sqrt{8a^3b}=\sqrt{4a^2}\sqrt{2ab}=2|a|\sqrt{2ab}$

53.
$$\begin{aligned}\sqrt{18x^5}+x^2\sqrt{50x}-5x^2\sqrt{2x}&=\sqrt{9x^4}\sqrt{2x}+x^2\sqrt{25}\sqrt{2x}-5x^2\sqrt{2x}\\&=3x^2\sqrt{2x}+5x^2\sqrt{2x}-5x^2\sqrt{2x}=3x^2\sqrt{2x}\end{aligned}$$

55.
$$\begin{aligned}\sqrt{x-5}&=x-7\\\left(\sqrt{x-5}\right)^2&=(x-7)^2\\x-5&=x^2-14x+49\\0&=x^2-15x+54\\0&=(x-9)(x-6)\end{aligned}$$

$x=9$ or $x=6$; $x=6$ does not check and is extraneous.

Exercises 6.7 (page 629)

1. half-plane, boundary

3. is not

SECTION 6.7

5. $2x + 3y < 12$

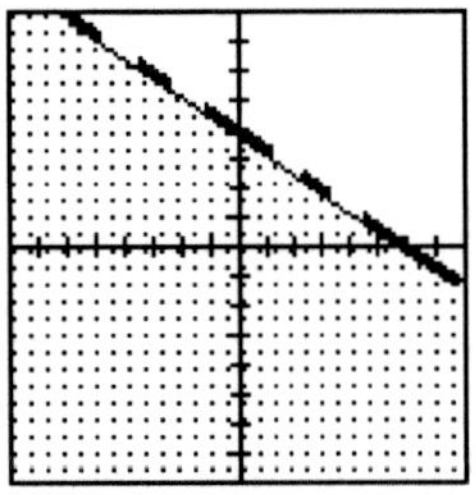

7. $x < 3$

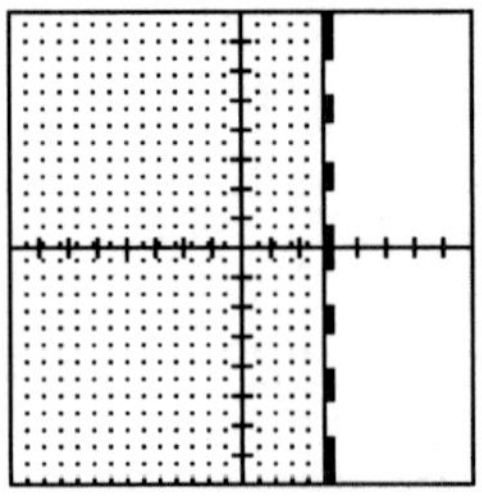

9. $4x - y > 4$

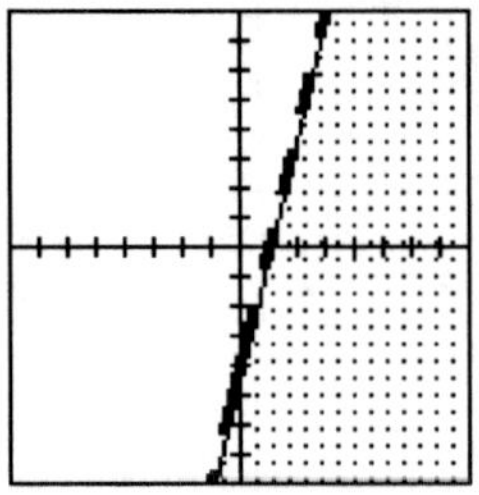

11. $y > 2x$

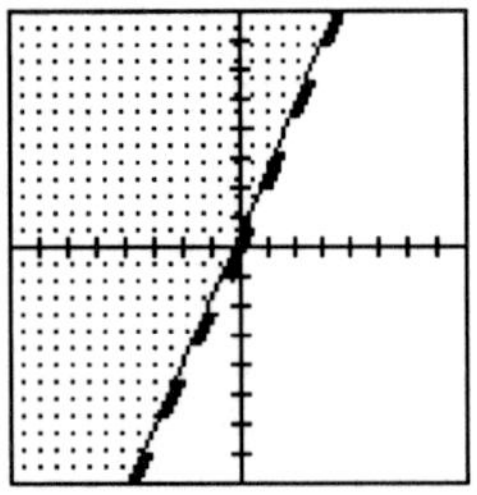

13. $y \le \frac{1}{2}x + 1$

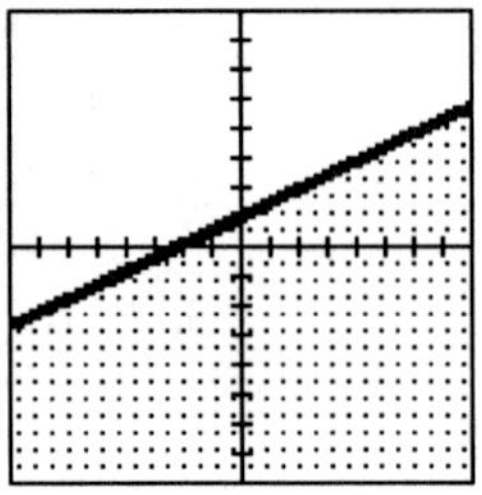

15. $2y \ge 3x - 2$

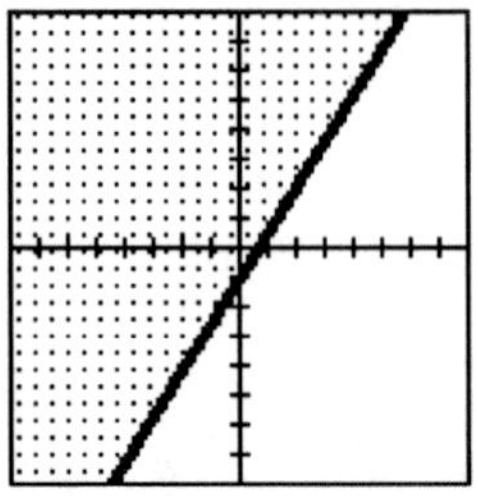

17. $y < x^2$

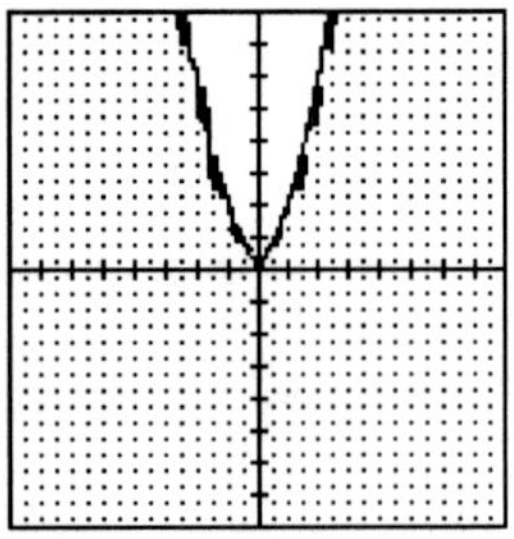

19. $x^2 + y^2 \le 4$

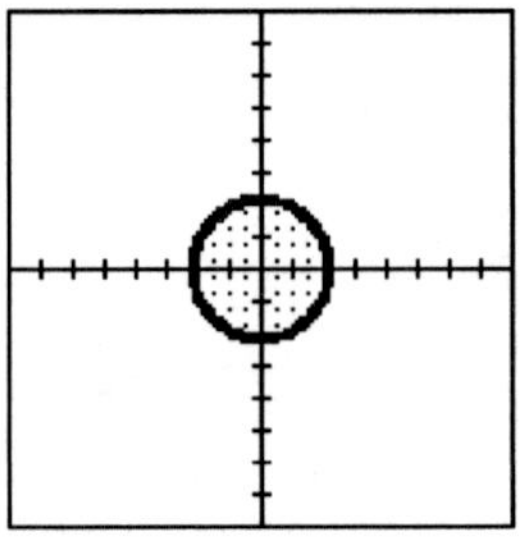

21. $\begin{cases} y < 3 \\ x \ge 2 \end{cases}$

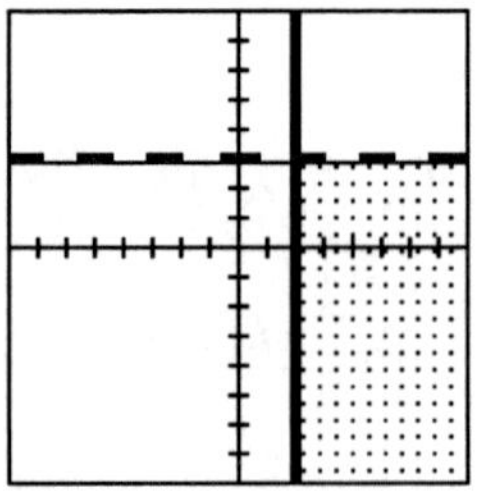

23. $\begin{cases} y \ge 1 \\ x < 2 \end{cases}$

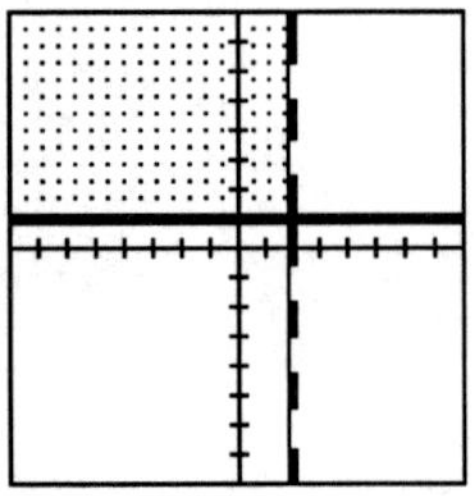

25. $\begin{cases} y \le x - 2 \\ y \ge 2x + 1 \end{cases}$

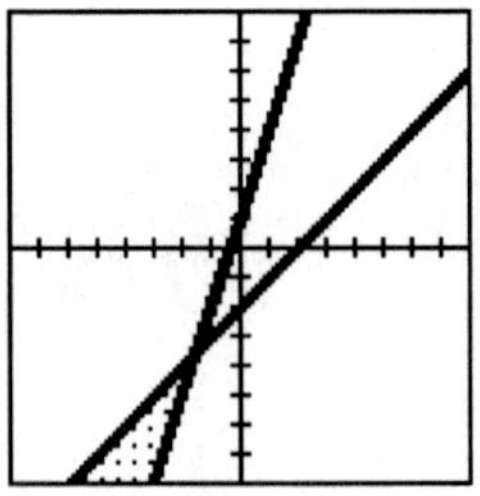

27. $\begin{cases} x + y < 2 \\ x + y \le 1 \end{cases}$

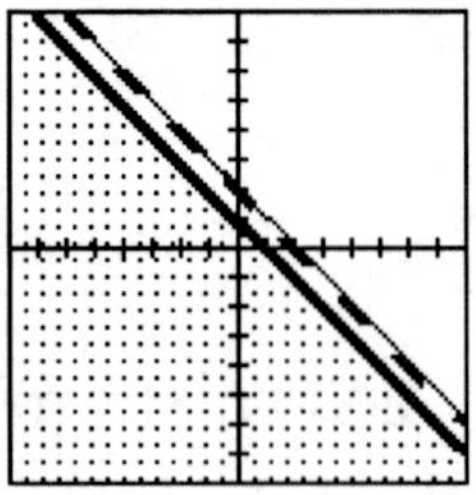

29. $\begin{cases} x + 2y < 3 \\ 2x - 4y < 8 \end{cases}$

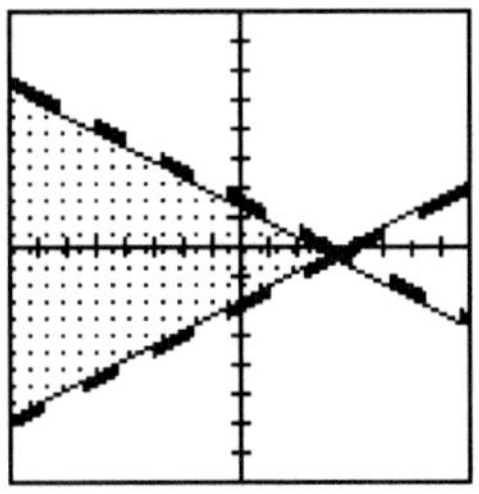

31. $\begin{cases} 2x - 3y \geq 6 \\ 3x + 2y < 6 \end{cases}$

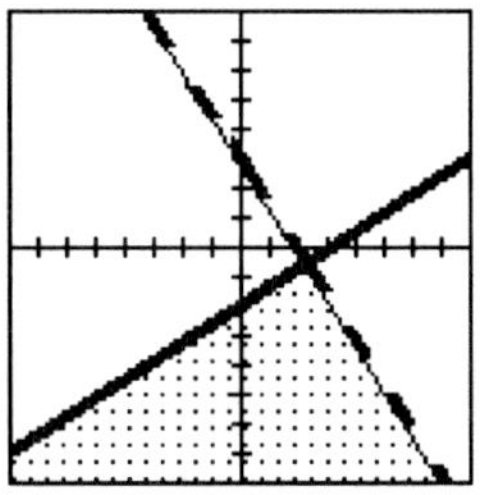

33. $\begin{cases} y \geq x^2 - 4 \\ y \leq \frac{1}{2}x \end{cases}$

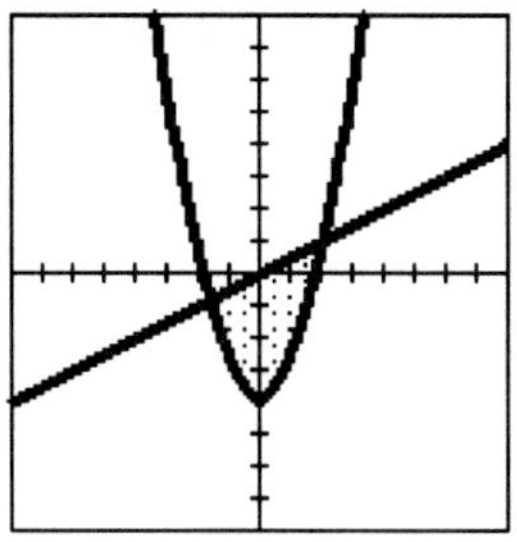

35. $\begin{cases} y \geq x^2 \\ y < 4 - x^2 \end{cases}$

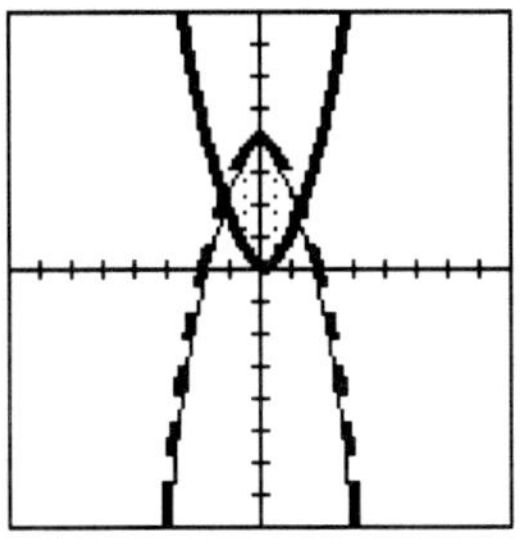

37. $\begin{cases} 2x - y \leq 0 \\ x + 2y \leq 10 \\ y \geq 0 \end{cases}$

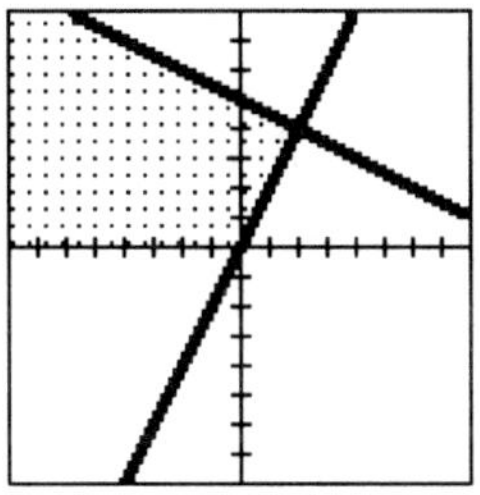

39. $\begin{cases} x - 2y \geq 0 \\ x - y \leq 2 \\ x \geq 0 \end{cases}$

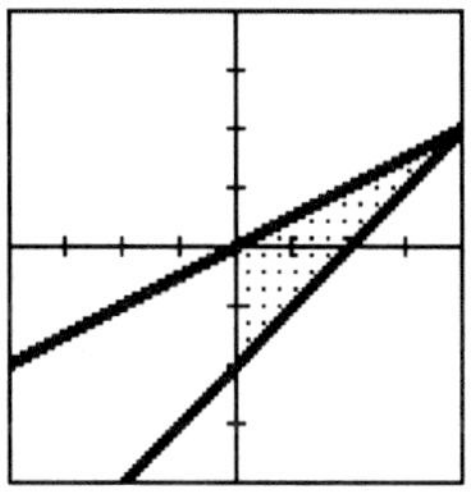

41. $\begin{cases} x + y \leq 4 \\ x - y \leq 4 \\ x \geq 0, y \geq 0 \end{cases}$

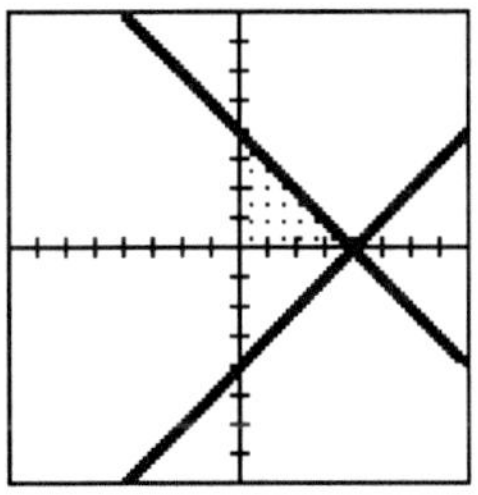

43. $\begin{cases} 3x - 2y \leq 6 \\ x + 2y \leq 10 \\ x \geq 0, y \geq 0 \end{cases}$

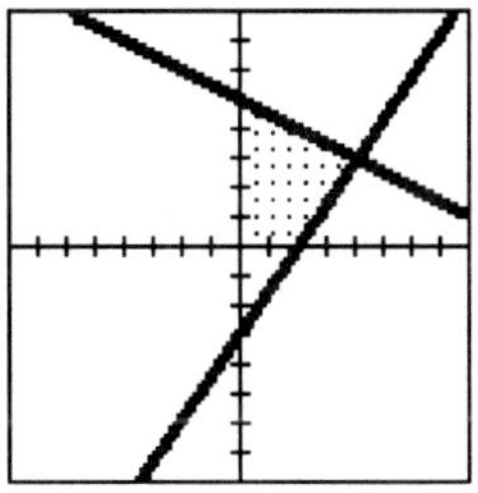

45. $\begin{cases} 6s + 4l \leq 60 \\ s \geq 0, l \geq 0 \end{cases}$

47-51. Answers may vary.

53. one, one

55. 0

Exercises 6.8 (page 639)

1. constraints

3. objective

5.

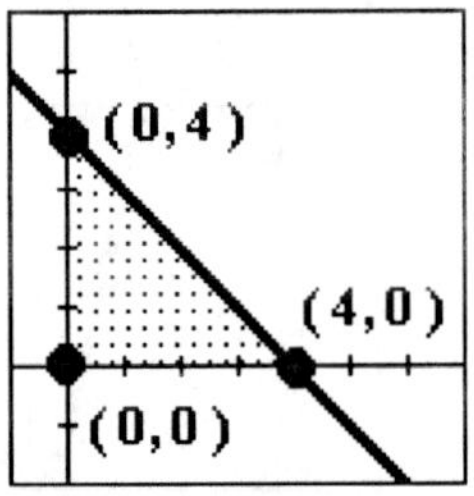

Point	$P = 2x + 3y$
(0, 0)	$= 2(0) + 3(0) = 0$
(0, 4)	$= 2(0) + 3(4) = 12$
(4, 0)	$= 2(4) + 3(0) = 8$

Max: $P = 12$ at $(0, 4)$

7.

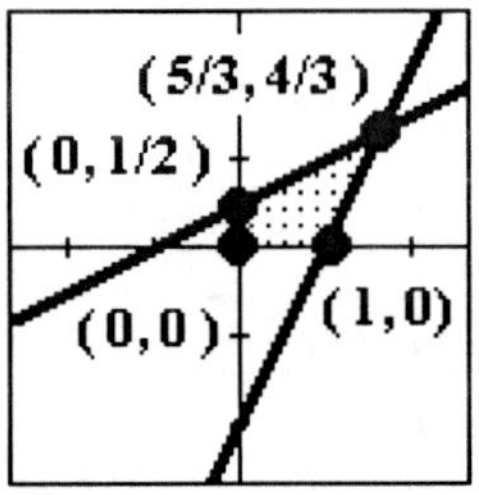

Point	$P = y + \frac{1}{2}x$
(0, 0)	$= 0 + \frac{1}{2}(0) = 0$
$\left(0, \frac{1}{2}\right)$	$= \frac{1}{2} + \frac{1}{2}(0) = \frac{1}{2}$
$\left(\frac{5}{3}, \frac{4}{3}\right)$	$= \frac{4}{3} + \frac{1}{2}\left(\frac{5}{3}\right) = \frac{13}{6}$
(1, 0)	$= 0 + \frac{1}{2}(1) = \frac{1}{2}$

Max: $P = \frac{13}{6}$ at $\left(\frac{5}{3}, \frac{4}{3}\right)$

9.

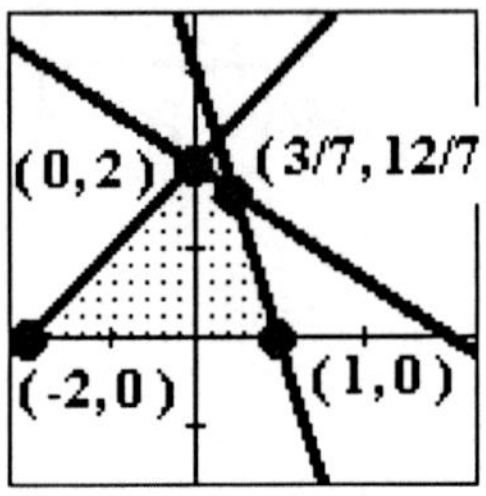

Point	$P = 2x + y$
(−2, 0)	$= 2(-2) + 0 = -4$
(1, 0)	$= 2(1) + 0 = 2$
$\left(\frac{3}{7}, \frac{12}{7}\right)$	$= 2\left(\frac{3}{7}\right) + \frac{12}{7} = \frac{18}{7}$
(0, 2)	$= 2(0) + 2 = 2$

Max: $P = \frac{18}{7}$ at $\left(\frac{3}{7}, \frac{12}{7}\right)$

11.

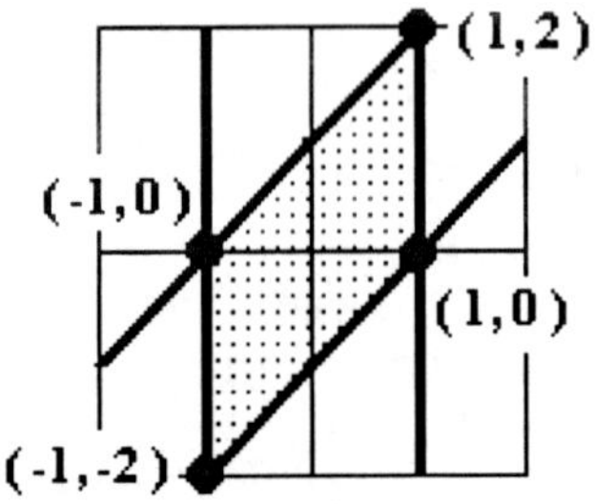

Point	$P = 3x - 2y$
(1, 0)	$= 3(1) - 2(0) = 3$
(1, 2)	$= 3(1) - 2(2) = -1$
(−1, 0)	$= 3(-1) - 2(0) = -3$
(−1, −2)	$= 3(-1) - 2(-2) = 1$

Max: $P = 3$ at $(1, 0)$

13.

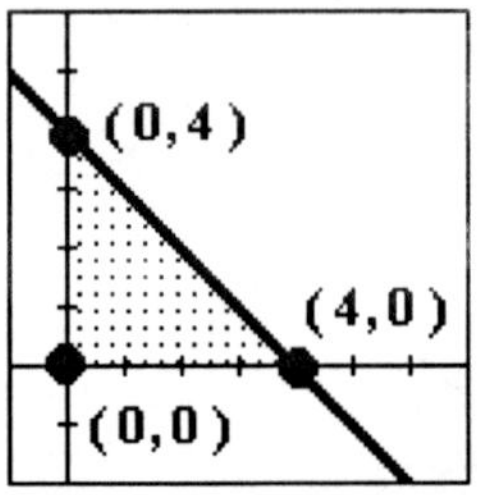

Point	$P = 5x + 12y$
(0, 0)	$= 5(0) + 12(0) = 0$
(0, 4)	$= 5(0) + 12(4) = 48$
(4, 0)	$= 5(4) + 12(0) = 20$

Min: $P = 0$ at $(0, 0)$

15.

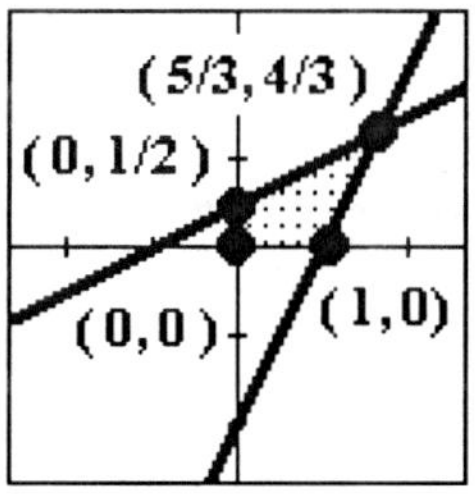

Point	$P = 3y + x$
$(0, 0)$	$= 3(0) + 0 = 0$
$(0, \frac{1}{2})$	$= 3(\frac{1}{2}) + 0 = \frac{3}{2}$
$(\frac{5}{3}, \frac{4}{3})$	$= 3(\frac{4}{3}) + \frac{5}{3} = \frac{17}{3}$
$(1, 0)$	$= 3(0) + 1 = 1$

Min: $P = 0$ at $(0, 0)$

17.

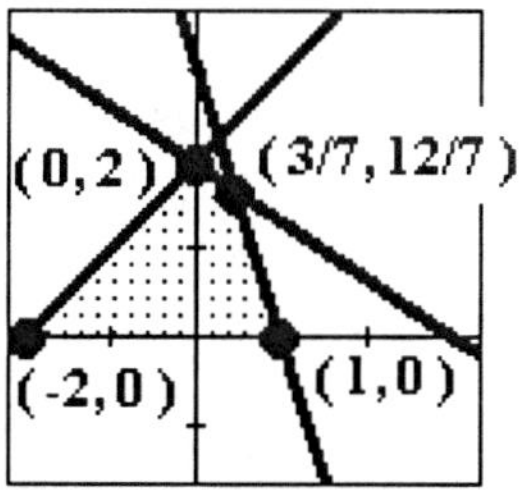

Point	$P = 6x + 2y$
$(-2, 0)$	$= 6(-2) + 2(0) = -12$
$(1, 0)$	$= 6(1) + 2(0) = 6$
$(\frac{3}{7}, \frac{12}{7})$	$= 6(\frac{3}{7}) + 2(\frac{12}{7}) = 6$
$(0, 2)$	$= 6(0) + 2(2) = 4$

Min: $P = -12$ at $(-2, 0)$

19.

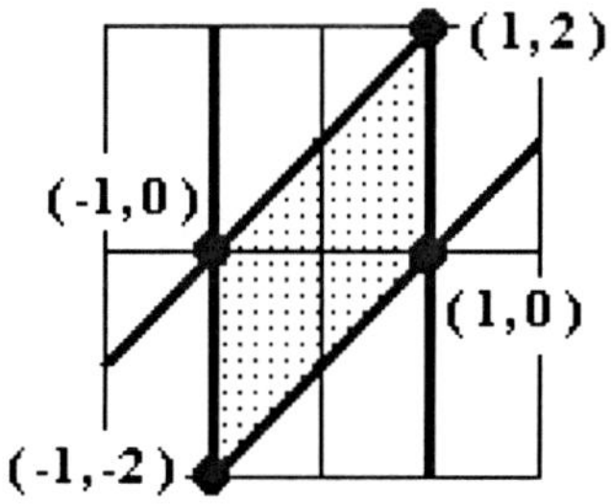

Point	$P = 2x - 2y$
$(1, 0)$	$= 2(1) - 2(0) = 2$
$(1, 2)$	$= 2(1) - 2(2) = -2$
$(-1, 0)$	$= 2(-1) - 2(0) = -2$
$(-1, -2)$	$= 2(-1) - 2(-2) = 2$

Min: $P = -2$ on the edge joining $(1, 2)$ and $(-1, 0)$

21. Let $x =$ # tables and $y =$ # chairs.

Maximize $P = 100x + 80y$

subject to $\begin{cases} 2x + 3y \leq 42 \\ 6x + 2y \leq 42 \\ x \geq 0, y \geq 0 \end{cases}$

Point	$P = 100x + 80y$
$(7, 0)$	$= 100(7) + 80(0) = 700$
$(3, 12)$	$= 100(3) + 80(12) = 1260$
$(0, 14)$	$= 100(0) + 80(14) = 1120$

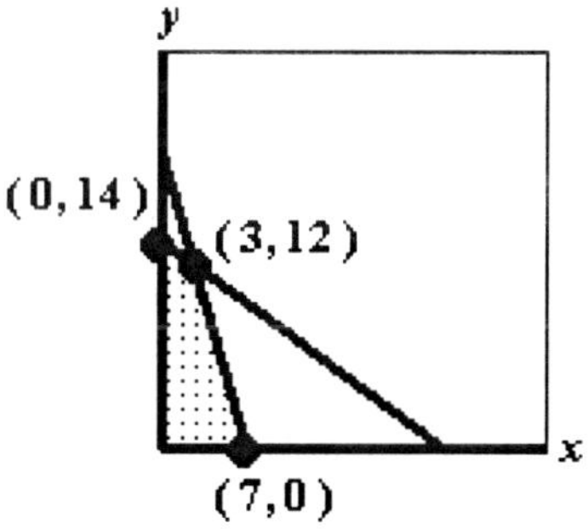

They should make 3 tables and 12 chairs, for a maximum profit of \$1260.

23. Let $x =$ # IBM and $y =$ # Apple.
Maximize $P = 50x + 40y$

subject to $\begin{cases} x + y \leq 60 \\ 20 \leq x \leq 30 \\ 30 \leq y \leq 50 \end{cases}$

Point	$P = 50x + 40y$
(20, 30)	$= 50(20) + 40(30) = 2200$
(30, 30)	$= 50(30) + 40(30) = 2700$
(20, 40)	$= 50(20) + 40(40) = 2600$

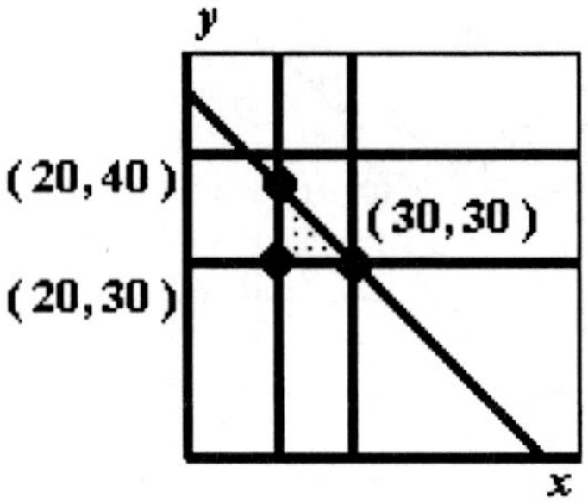

She should stock 30 IBM and 30 Apple computers, for a maximum commission of \$2700.

25. Let $x =$ # DVRs and $y =$ # TVs.
Maximize $P = 40x + 32y$

subject to $\begin{cases} 3x + 4y \leq 180 \\ 2x + 3y \leq 120 \\ 2x + y \leq 60 \\ x \geq 0, y \geq 0 \end{cases}$

Point	$P = 40x + 32y$
(0, 0)	$= 40(0) + 32(0) = 0$
(0, 40)	$= 40(0) + 32(40) = 1280$
(15, 30)	$= 40(15) + 32(30) = 1560$
(30, 0)	$= 40(30) + 32(0) = 1200$

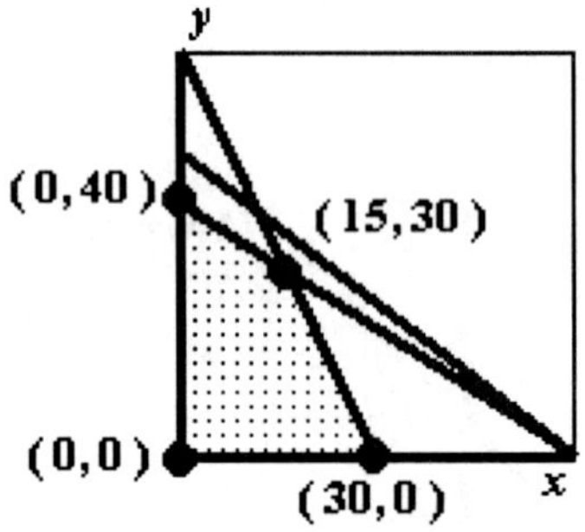

15 DVRs and 30 TVs should be made, for a maximum profit of \$1560.

27. Let $x =$ \$ in stocks and $y =$ \$ in bonds.
Maximize $P = 0.09x + 0.07y$

subject to $\begin{cases} x + y \leq 200000 \\ x \geq 100000 \\ y \geq 50000 \end{cases}$

Point	$P = 0.09x + 0.07y$
(100000, 50000)	$= 12500$
(150000, 50000)	$= 17000$
(100000, 100000)	$= 16000$

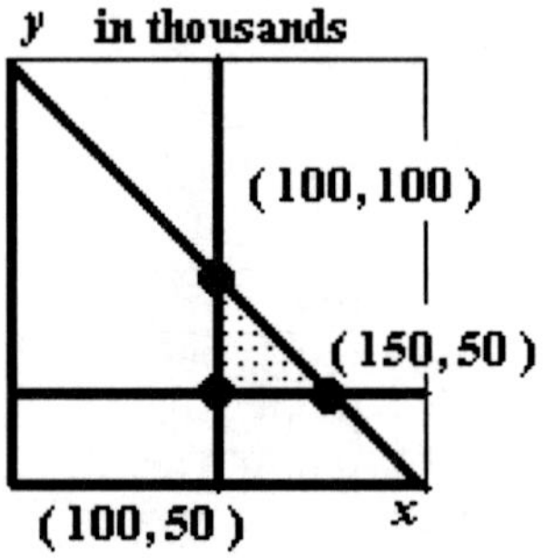

She should invest \$150,000 in stocks and \$50,000 in bonds, for a maximum return of \$17,000.

29. Let x = # buses and y = # trucks.
Minimize $P = 350x + 200y$

subject to $\begin{cases} 40x + 10y \geq 100 \\ 3x + 6y \geq 18 \\ x \geq 0, y \geq 0 \end{cases}$

Point	$P = 350x + 200y$
(0, 10)	$= 350(0) + 200(10) = 2000$
(2, 2)	$= 350(2) + 200(2) = 1100$
(6, 0)	$= 350(6) + 200(0) = 2100$

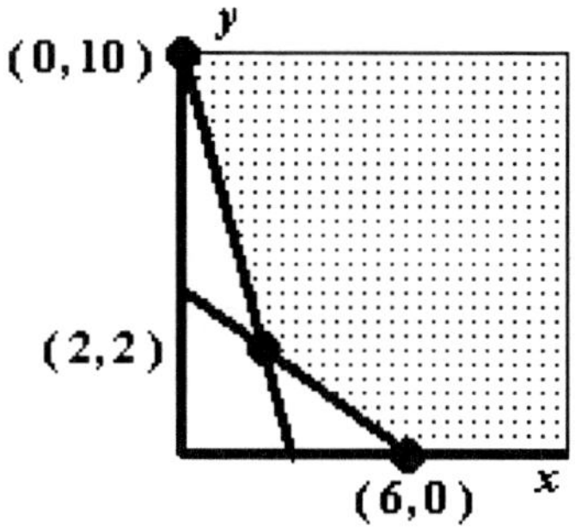

2 buses and 2 trucks should be rented, for a minimum cost of $1100.

31. **Answers may vary.**

33. $\begin{bmatrix} 1 & 2 & 3 \\ 1 & -2 & 3 \\ 0 & 2 & -3 \\ 2 & 0 & 6 \end{bmatrix} \Rightarrow \begin{bmatrix} 1 & 2 & 3 \\ 0 & -4 & 0 \\ 0 & 2 & -3 \\ 0 & -4 & 0 \end{bmatrix} \Rightarrow \begin{bmatrix} 1 & 2 & 3 \\ 0 & 1 & 0 \\ 0 & 2 & -3 \\ 0 & 0 & 0 \end{bmatrix} \Rightarrow \begin{bmatrix} 1 & 0 & 3 \\ 0 & 1 & 0 \\ 0 & 0 & -3 \\ 0 & 0 & 0 \end{bmatrix} \Rightarrow \begin{bmatrix} 1 & 0 & 0 \\ 0 & 1 & 0 \\ 0 & 0 & 1 \\ 0 & 0 & 0 \end{bmatrix}$

$-R_1 + R_2 \Rightarrow R_2$, $-2R_1 + R_4 \Rightarrow R_4$; $-\frac{1}{4}R_2 \Rightarrow R_2$, $-R_4 + R_2 \Rightarrow R_4$; $-2R_2 + R_1 \Rightarrow R_1$, $-2R_2 + R_3 \Rightarrow R_3$; $R_3 + R_1 \Rightarrow R_1$, $-\frac{1}{3}R_3 \Rightarrow R_3$

35. $\left(\frac{1}{3} - 3y, y, -\frac{1}{3}\right)$

Chapter 6 Review (page 642)

1. $\begin{cases} 2x - y = -1 \\ x + y = 7 \end{cases}$

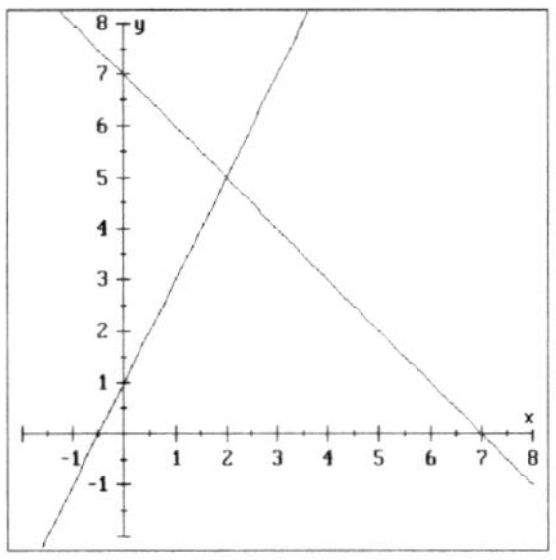

solution: $(2, 5)$

2. $\begin{cases} 5x + 2y = 1 \\ 2x - y = -5 \end{cases}$

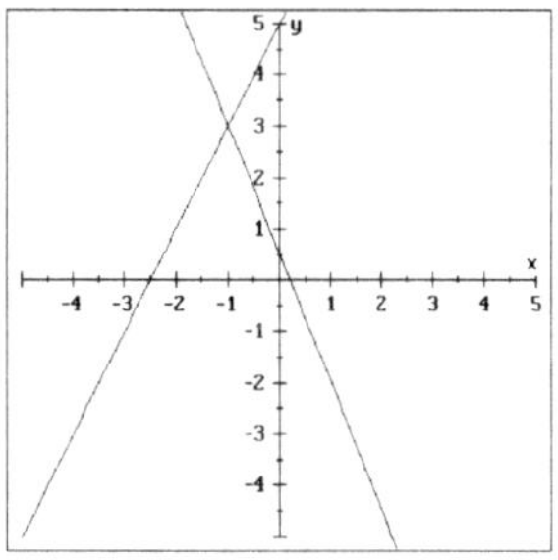

solution: $(-1, 3)$

3. $\begin{cases} y = 5x + 7 \\ x = y - 7 \end{cases}$

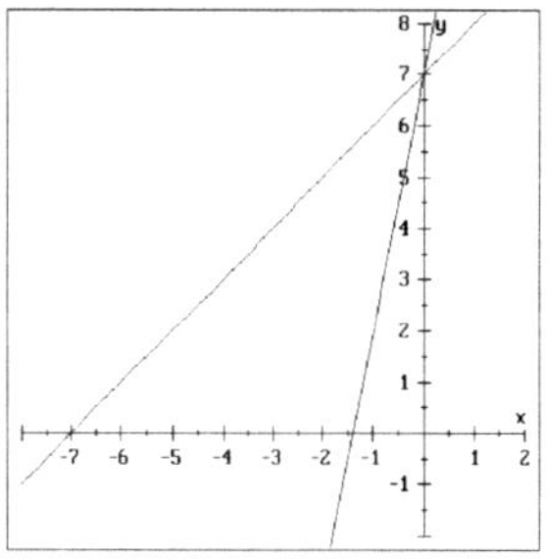

solution: $(0, 7)$

4. $\begin{cases} 3x + 2y = 6 \\ y = -\frac{3}{2}x + 3 \end{cases}$

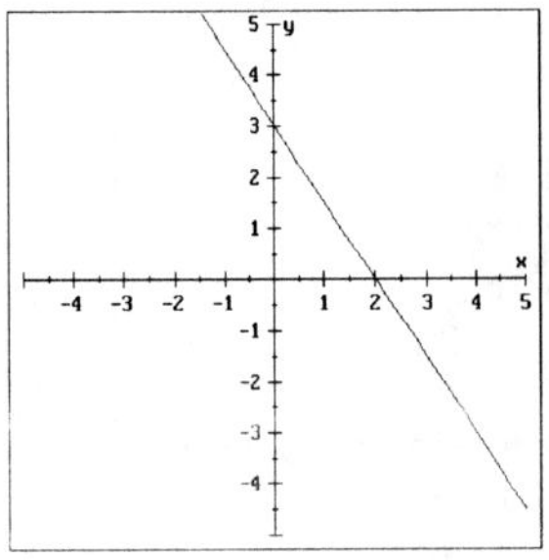

infinitely many solutions
dependent equations

5. $\begin{cases} 4x - y = 4 \\ y = 4(x - 2) \end{cases}$

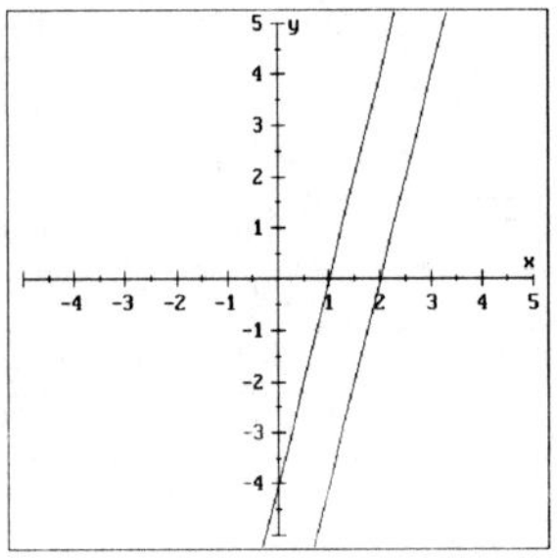

no solutions
inconsistent system

6. $\begin{cases} (1) \quad 2y + x = 0 \\ (2) \quad x = y + 3 \end{cases}$

Substitute $x = y + 3$ into (1):

$$\begin{aligned} 2y + x &= 0 \\ 2y + (y + 3) &= 0 \\ 3y + 3 &= 0 \\ 3y &= -3 \\ y &= -1 \end{aligned}$$

Substitute and solve for x:

$x = y + 3$

$x = -1 + 3 = 2$

$\boxed{x = 2, y = -1}$

7. $\begin{cases} (1) \quad 2x + y = -3 \\ (2) \quad x - y = 3 \end{cases}$

Substitute $x = y + 3$ into (1):

$$\begin{aligned} 2x + y &= -3 \\ 2(y + 3) + y &= -3 \\ 2y + 6 + y &= -3 \\ 3y &= -9 \\ y &= -3 \end{aligned}$$

Substitute and solve for x:

$x = y + 3$

$x = -3 + 3 = 0$

$\boxed{x = 0, y = -3}$

8. $\begin{cases} (1) \quad \dfrac{x + y}{2} + \dfrac{x - y}{3} = 1 \\ (2) \quad y = 3x - 2 \end{cases}$

Substitute $y = 3x - 2$ into (1):

$$\begin{aligned} \frac{x + 3x - 2}{2} + \frac{x - (3x - 2)}{3} &= 1 \\ \frac{4x - 2}{2} + \frac{-2x + 2}{3} &= 1 \end{aligned}$$

$$\begin{aligned} 3(4x - 2) + 2(-2x + 2) &= 6 \\ 12x - 6 - 4x + 4 &= 6 \\ 8x &= 8 \\ x &= 1 \end{aligned}$$

Substitute and solve for y:

$y = 3x - 2$

$y = 3(1) - 2 = 1$

$\boxed{x = 1, y = 1}$

9. $\begin{cases} (1) & y = 3x - 4 \\ (2) & 9x - 3y = 12 \end{cases}$

Substitute $y = 3x - 4$ into (2):

$$\begin{aligned} 9x - 3y &= 12 \\ 9x - 3(3x - 4) &= 12 \\ 9x - 9x + 12 &= 12 \\ 0 &= 0 \end{aligned}$$

Dependent equations

General solution: $(x, 3x - 4)$

10. $\begin{cases} (1) & x = -\frac{3}{2}y + 3 \\ (2) & 2x + 3y = 4 \end{cases}$

Substitute $x = -\frac{3}{2}y + 3$ into (2):

$$\begin{aligned} 2x + 3y &= 4 \\ 2\left(-\tfrac{3}{2}y + 3\right) + 3y &= 4 \\ -3y + 6 + 3y &= 4 \\ 6 &\neq 4 \end{aligned}$$

Inconsistent system $\Rightarrow$ No solution

11.

$$\begin{array}{l} x + 5y = 7 \\ 3x + y = -7 \Rightarrow \times (-5) \end{array} \qquad \begin{array}{rl} x + 5y &= 7 \\ -15x - 5y &= 35 \\ \hline -14x &= 42 \\ x &= -3 \end{array} \qquad \begin{array}{rl} x + 5y &= 7 \\ -3 + 5y &= 7 \\ 5y &= 10 \\ y &= 2 \end{array}$$

Solution: $\boxed{x = -3,\ y = 2}$

12.

$$\begin{array}{l} 2x + 3y = 11 \Rightarrow \times (3) \\ 3x - 7y = -41 \Rightarrow \times (-2) \end{array} \qquad \begin{array}{rl} 6x + 9y &= 33 \\ -6x + 14y &= 82 \\ \hline 23y &= 115 \\ y &= 5 \end{array} \qquad \begin{array}{rl} 2x + 3y &= 11 \\ 2x + 3(5) &= 11 \\ 2x &= -4 \\ x &= -2 \end{array}$$

Solution: $\boxed{x = -2,\ y = 5}$

13.

$$\begin{array}{l} 2(x + y) - x = 0 \Rightarrow x + 2y = 0 \Rightarrow \\ 3(x + y) + 2y = 1 \Rightarrow 3x + 5y = 1 \Rightarrow \end{array} \begin{array}{rl} -3x - 6y &= 0 \\ 3x + 5y &= 1 \\ \hline -y &= 1 \\ y &= -1 \end{array} \qquad \begin{array}{rl} x + 2y &= 0 \\ x + 2(-1) &= 0 \\ x &= 2 \end{array}$$

Solution: $\boxed{\begin{array}{l} x = 2 \\ y = -1 \end{array}}$

14.

$$\begin{array}{l} 8x + 12y = 24 \\ 2x + 3y = 4 \Rightarrow \times (-4) \end{array} \qquad \begin{array}{rl} 8x + 12y &= 24 \\ -8x - 12y &= -16 \\ \hline 0 &\neq 8 \end{array}$$

$\Rightarrow$ Inconsistent system: No solution

15.

$$\begin{array}{l} 3x - y = 4 \Rightarrow \times (-3) \\ 9x - 3y = 12 \end{array} \qquad \begin{array}{rl} -9x + 3y &= -12 \\ 9x - 3y &= 12 \\ \hline 0 &= 0 \end{array} \qquad \begin{array}{l} 3x - y = 4 \\ 3x - 4 = y \end{array}$$

Dependent equations, General Solution: $(x, 3x - 4)$

16.

$$\begin{array}{ll} (1) & 3x + 2y - z = 2 \\ (2) & x + y - z = 0 \\ (3) & 2x + 3y - z = 1 \end{array}$$

Add (1) and $-(2)$:

$$\begin{array}{lrl} (1) & 3x + 2y - z &= 2 \\ -(2) & -x - y + z &= 0 \\ \hline (4) & 2x + y &= 2 \end{array}$$

Add equations (1) and $-(3)$:

$$\begin{array}{lrl} (1) & 3x + 2y - z &= 2 \\ -(3) & -2x - 3y + z &= -1 \\ \hline (5) & x - y &= 1 \end{array}$$

Solve the system of two equations and two unknowns formed by equations (4) and (5):

$$\begin{array}{rl} 2x + y &= 2 \\ x - y &= 1 \\ \hline 3x &= 3 \\ x &= 1 \end{array} \qquad \begin{array}{rl} 2x + y &= 2 \\ 2(1) + y &= 2 \\ y &= 0 \end{array} \qquad \begin{array}{rl} x + y - z &= 0 \\ 1 + 0 - z &= 0 \\ -z &= -1 \\ z &= 1 \end{array}$$

Solution: $\boxed{x = 1,\ y = 0,\ z = 1}$

17. (1) $5x - y + z = 3$ (2) $3x + y + 2z = 2$ (3) $x + y = 2$

Add (1) and (2):

$$\begin{array}{ll} (1) & 5x - y + z = 3 \\ (2) & 3x + y + 2z = 2 \\ \hline (4) & 8x + 3z = 5 \end{array}$$

Add equations (1) and (3):

$$\begin{array}{ll} (1) & 5x - y + z = 3 \\ (3) & x + y = 2 \\ \hline (5) & 6x + z = 5 \end{array}$$

Solve the system of two equations and two unknowns formed by equations (4) and (5):

$$\begin{array}{l} 8x + 3z = 5 \\ 6x + z = 5 \Rightarrow \times\ (-3) \end{array} \qquad \begin{array}{rl} 8x + 3z = & 5 \\ -18x - 3z = & -15 \\ \hline -10x = & -10 \\ x = & 1 \end{array} \qquad \begin{array}{r} 6x + z = 5 \\ 6(1) + z = 5 \\ z = -1 \end{array} \qquad \begin{array}{r} x + y = 2 \\ 1 + y = 2 \\ y = 1 \end{array}$$

Solution: $\boxed{x = 1, y = 1, z = -1}$

18. (1) $2x - y + z = 1$ (2) $x - y + 2z = 3$ (3) $x - y + z = 1$

Add (1) and $-$(2):

$$\begin{array}{ll} (1) & 2x - y + z = 1 \\ -(2) & -x + y - 2z = -3 \\ \hline (4) & x - z = -2 \end{array}$$

Add equations (1) and $-$(3):

$$\begin{array}{ll} (1) & 2x - y + z = 1 \\ -(3) & -x + y - z = -1 \\ \hline (5) & x = 0 \end{array}$$

Solve the system of two equations and two unknowns formed by equations (4) and (5):

$$\begin{array}{r} x - z = -2 \\ 0 - z = -2 \\ z = 2 \end{array} \qquad \begin{array}{r} x - y + z = 1 \\ 0 - y + 2 = 1 \\ -y = -1 \\ y = 1 \end{array}$$

Solution: $\boxed{x = 0, y = 1, z = 2}$

19. Let x = cost of fake fur and let y = cost of leather. Then $\begin{cases} (1) & 25x + 15y = 9300 \\ (2) & 10x + 30y = 12600 \end{cases}$

$$\begin{array}{l} 25x + 15y = 9300 \Rightarrow \times\ (-2) \\ 10x + 30y = 12600 \end{array} \qquad \begin{array}{rl} -50x + 30y = & -18600 \\ 10x + 30y = & 12600 \\ \hline -40x = & -6000 \\ x = & 150 \end{array} \qquad \begin{array}{r} 25x + 15y = 9300 \\ 25(150) + 15y = 9300 \\ 15y = 5550 \\ y = 370 \end{array}$$

The fake fur coats cost \$150 while the leather coats cost \$370. The cost will be \$10,400.

20. Let x = # adult tickets, y = # senior tickets and z = # children tickets.

(1) $x + y + z = 1800$ (2) $5x + 4y + 2.5z = 7425$ (3) $y + z = 900$

Add -4(1) and (2):

$$\begin{array}{ll} -4(1) & -4x - 4y - 4z = -7200 \\ (2) & 5x + 4y + 2.5z = 7425 \\ \hline (4) & x - 1.5z = 225 \end{array}$$

Add equations (1) and $-$(3):

$$\begin{array}{ll} (1) & x + y + z = 1800 \\ -(3) & -y - z = -900 \\ \hline (5) & x = 900 \end{array}$$

Solve the system of two equations and two unknowns formed by equations (4) and (5):

$$\begin{array}{r} x - 1.5z = 225 \\ 900 - 1.5z = 225 \\ -1.5z = -675 \\ z = 450 \end{array} \qquad \begin{array}{r} y + z = 900 \\ y + 450 = 900 \\ y = 450 \end{array}$$

There were 900 adult tickets, 450 senior tickets, and 450 children's tickets sold.

CHAPTER 6 REVIEW

21. $\begin{bmatrix} 2 & 5 & | & 7 \\ 3 & -1 & | & 2 \end{bmatrix} \Rightarrow \begin{bmatrix} 1 & -6 & | & -5 \\ 3 & -1 & | & 2 \end{bmatrix} \Rightarrow \begin{bmatrix} 1 & -6 & | & -5 \\ 0 & 17 & | & 17 \end{bmatrix} \Rightarrow \begin{bmatrix} 1 & -6 & | & -5 \\ 0 & 1 & | & 1 \end{bmatrix} \Rightarrow$

$-R_1 + R_2 \Rightarrow R_1$ $\quad -3R_1 + R_2 \Rightarrow R_2$ $\quad \frac{1}{17}R_2 \Rightarrow R_2$

$\begin{bmatrix} 1 & 0 & | & 1 \\ 0 & 1 & | & 1 \end{bmatrix}$ Solution: $\boxed{x = 1, y = 1}$

$6R_2 + R_1 \Rightarrow R_1$

22. $\begin{bmatrix} 3 & -1 & | & -4 \\ -6 & 2 & | & 8 \end{bmatrix} \Rightarrow \begin{bmatrix} 3 & -1 & | & -4 \\ 0 & 0 & | & 0 \end{bmatrix} \Rightarrow \begin{matrix} 3x - y = -4 \\ 3x + 4 = y \end{matrix}$ Dependent equations; General Solution: $(x, 3x + 4)$

$2R_1 + R_2 \Rightarrow R_2$

23. $\begin{bmatrix} 1 & 3 & -1 & | & 8 \\ 2 & 1 & -2 & | & 11 \\ 1 & -1 & 5 & | & -8 \end{bmatrix} \Rightarrow \begin{bmatrix} 1 & 3 & -1 & | & 8 \\ 0 & -5 & 0 & | & -5 \\ 0 & -4 & 6 & | & -16 \end{bmatrix} \Rightarrow \begin{bmatrix} 1 & 3 & -1 & | & 8 \\ 0 & 1 & 0 & | & 1 \\ 0 & -4 & 6 & | & -16 \end{bmatrix} \Rightarrow$

$-2R_1 + R_2 \Rightarrow R_2$, $-R_1 + R_3 \Rightarrow R_3$ $\quad -\frac{1}{5}R_2 \Rightarrow R_2$

$\begin{bmatrix} 1 & 0 & -1 & | & 5 \\ 0 & 1 & 0 & | & 1 \\ 0 & 0 & 6 & | & -12 \end{bmatrix} \Rightarrow \begin{bmatrix} 1 & 0 & 0 & | & 3 \\ 0 & 1 & 0 & | & 1 \\ 0 & 0 & 1 & | & -2 \end{bmatrix}$ Solution: $\boxed{x = 3, y = 1, z = -2}$

$-3R_2 + R_1 \Rightarrow R_1$, $4R_2 + R_3 \Rightarrow R_3$ $\quad \frac{1}{6}R_3 + R_1 \Rightarrow R_1$, $\frac{1}{6}R_3 \Rightarrow R_3$

24. $\begin{bmatrix} 1 & 3 & 1 & | & 3 \\ 2 & -1 & 1 & | & -11 \\ 3 & 2 & 3 & | & 2 \end{bmatrix} \Rightarrow \begin{bmatrix} 1 & 3 & 1 & | & 3 \\ 0 & -7 & -1 & | & -17 \\ 0 & -7 & 0 & | & -7 \end{bmatrix} \Rightarrow \begin{bmatrix} 1 & 3 & 1 & | & 3 \\ 0 & 1 & 0 & | & 1 \\ 0 & -7 & -1 & | & -17 \end{bmatrix} \Rightarrow$

$-2R_1 + R_2 \Rightarrow R_2$, $-3R_1 + R_3 \Rightarrow R_3$ $\quad R_2 \Leftrightarrow -\frac{1}{7}R_3$

$\begin{bmatrix} 1 & 0 & 1 & | & 0 \\ 0 & 1 & 0 & | & 1 \\ 0 & 0 & -1 & | & -10 \end{bmatrix} \Rightarrow \begin{bmatrix} 1 & 0 & 0 & | & -10 \\ 0 & 1 & 0 & | & 1 \\ 0 & 0 & 1 & | & 10 \end{bmatrix}$ Solution: $\boxed{x = -10, y = 1, z = 10}$

$-3R_2 + R_1 \Rightarrow R_1$, $7R_2 + R_3 \Rightarrow R_3$ $\quad R_3 + R_1 \Rightarrow R_1$, $-R_3 \Rightarrow R_3$

25. $\begin{bmatrix} 1 & 1 & 1 & | & 4 \\ 3 & -2 & -2 & | & -3 \\ 4 & -1 & -1 & | & 0 \end{bmatrix} \Rightarrow \begin{bmatrix} 1 & 1 & 1 & | & 4 \\ 0 & -5 & -5 & | & -15 \\ 0 & -5 & -5 & | & -16 \end{bmatrix} \Rightarrow \begin{bmatrix} 1 & 1 & 1 & | & 4 \\ 0 & -5 & -5 & | & -15 \\ 0 & 0 & 0 & | & -1 \end{bmatrix}$

$-3R_1 + R_2 \Rightarrow R_2$, $-4R_1 + R_3 \Rightarrow R_3$ $\quad -R_2 + R_3 \Rightarrow R_3$

The last row indicates $0x + 0y + 0z = -1$. This is impossible. $\Rightarrow$ no solution

26. $-4 = x, x = -4, 0 = x + 4, x + 7 = y \Rightarrow \boxed{x = -4, y = 3}$

27. $\begin{bmatrix} 3 & 2 & 1 \\ 3 & 2 & 1 \end{bmatrix} + \begin{bmatrix} -2 & 1 & 3 \\ 1 & -2 & 1 \end{bmatrix} = \begin{bmatrix} 1 & 3 & 4 \\ 4 & 0 & 2 \end{bmatrix}$

28. $\begin{bmatrix} 2 & 3 & 5 \\ 1 & -2 & 4 \\ 2 & 1 & -2 \end{bmatrix} - \begin{bmatrix} 0 & -2 & 1 \\ 3 & 4 & -2 \\ 6 & -4 & 1 \end{bmatrix} = \begin{bmatrix} 2 & 5 & 4 \\ -2 & -6 & 6 \\ -4 & 5 & -3 \end{bmatrix}$

29. $\begin{bmatrix} 1 & -2 \\ -3 & 1 \end{bmatrix}\begin{bmatrix} 2 & 3 \\ -1 & 2 \end{bmatrix} = \begin{bmatrix} 4 & -1 \\ -7 & -7 \end{bmatrix}$

30. $\begin{bmatrix} -2 & 3 & 5 \\ 1 & -2 & -3 \end{bmatrix}\begin{bmatrix} 2 & 1 \\ -1 & 2 \\ -2 & 3 \end{bmatrix} = \begin{bmatrix} -17 & 19 \\ 10 & -12 \end{bmatrix}$

31. $\begin{bmatrix} 1 & -3 & 2 \end{bmatrix}\begin{bmatrix} 2 \\ 1 \\ 3 \end{bmatrix} = \begin{bmatrix} 5 \end{bmatrix}$

32. $\begin{bmatrix} 1 \\ 2 \\ 1 \\ 5 \end{bmatrix}\begin{bmatrix} 2 & -1 & 1 & 3 \end{bmatrix} = \begin{bmatrix} 2 & -1 & 1 & 3 \\ 4 & -2 & 2 & 6 \\ 2 & -1 & 1 & 3 \\ 10 & -5 & 5 & 15 \end{bmatrix}$

33. $\begin{bmatrix} 1 & -5 & 3 \\ 2 & 1 & -1 \end{bmatrix}\begin{bmatrix} 2 \\ -2 \\ 3 \end{bmatrix}\begin{bmatrix} 1 & -1 \\ -1 & 3 \end{bmatrix}\begin{bmatrix} 1 \\ -2 \end{bmatrix} = \begin{bmatrix} 21 \\ -1 \end{bmatrix}\begin{bmatrix} 1 & -1 \\ -1 & 3 \end{bmatrix}\begin{bmatrix} 1 \\ -2 \end{bmatrix} \Rightarrow$ not possible

34. $\begin{bmatrix} 1 & -3 & 2 \end{bmatrix}\begin{bmatrix} 2 \\ 1 \\ -5 \end{bmatrix} + \begin{bmatrix} 1 & -3 \end{bmatrix}\begin{bmatrix} 2 \\ 5 \end{bmatrix} = \begin{bmatrix} -11 \end{bmatrix} + \begin{bmatrix} -13 \end{bmatrix} = \begin{bmatrix} -24 \end{bmatrix}$

35. $\left(\begin{bmatrix} 1 & -3 \\ 3 & 1 \end{bmatrix} + \begin{bmatrix} -1 & 3 \\ 1 & 1 \end{bmatrix}\right)\begin{bmatrix} 1 \\ -5 \end{bmatrix} = \begin{bmatrix} 0 & 0 \\ 4 & 2 \end{bmatrix}\begin{bmatrix} 1 \\ -5 \end{bmatrix} = \begin{bmatrix} 0 \\ -6 \end{bmatrix}$

36. $\left[\begin{array}{cc|cc} 2 & 3 & 1 & 0 \\ 3 & 5 & 0 & 1 \end{array}\right] \Rightarrow \underset{\frac{1}{2}R_1 \Rightarrow R_1}{\left[\begin{array}{cc|cc} 1 & \frac{3}{2} & \frac{1}{2} & 0 \\ 3 & 5 & 0 & 1 \end{array}\right]} \Rightarrow \underset{-3R_1 + R_2 \Rightarrow R_2}{\left[\begin{array}{cc|cc} 1 & \frac{3}{2} & \frac{1}{2} & 0 \\ 0 & \frac{1}{2} & -\frac{3}{2} & 1 \end{array}\right]} \Rightarrow$

$\underset{-3R_2 + R_1 \Rightarrow R_1}{\left[\begin{array}{cc|cc} 1 & 0 & 5 & -3 \\ 0 & \frac{1}{2} & -\frac{3}{2} & 1 \end{array}\right]} \Rightarrow \underset{2R_2 \Rightarrow R_2}{\left[\begin{array}{cc|cc} 1 & 0 & 5 & -3 \\ 0 & 1 & -3 & 2 \end{array}\right]} \Rightarrow$ Inverse: $\begin{bmatrix} 5 & -3 \\ -3 & 2 \end{bmatrix}$

37. $\left[\begin{array}{ccc|ccc} 1 & 0 & 0 & 1 & 0 & 0 \\ 2 & 0 & -2 & 0 & 1 & 0 \\ 1 & 2 & 2 & 0 & 0 & 1 \end{array}\right] \Rightarrow \underset{R_2 \Leftrightarrow R_3}{\left[\begin{array}{ccc|ccc} 1 & 0 & 0 & 1 & 0 & 0 \\ 1 & 2 & 2 & 0 & 0 & 1 \\ 2 & 0 & -2 & 0 & 1 & 0 \end{array}\right]} \Rightarrow \underset{\substack{-R_1 + R_2 \Rightarrow R_2 \\ -2R_1 + R_3 \Rightarrow R_3}}{\left[\begin{array}{ccc|ccc} 1 & 0 & 0 & 1 & 0 & 0 \\ 0 & 2 & 2 & -1 & 0 & 1 \\ 0 & 0 & -2 & -2 & 1 & 0 \end{array}\right]} \Rightarrow$

$\underset{R_2 + R_3 \Rightarrow R_3}{\left[\begin{array}{ccc|ccc} 1 & 0 & 0 & 1 & 0 & 0 \\ 0 & 2 & 0 & -3 & 1 & 1 \\ 0 & 0 & -2 & -2 & 1 & 0 \end{array}\right]} \Rightarrow \underset{\substack{\frac{1}{2}R_2 \Rightarrow R_2 \\ -\frac{1}{2}R_3 \Rightarrow R_3}}{\left[\begin{array}{ccc|ccc} 1 & 0 & 0 & 1 & 0 & 0 \\ 0 & 1 & 0 & -\frac{3}{2} & \frac{1}{2} & \frac{1}{2} \\ 0 & 0 & 1 & 1 & -\frac{1}{2} & 0 \end{array}\right]}$: Inverse $= \begin{bmatrix} 1 & 0 & 0 \\ -\frac{3}{2} & \frac{1}{2} & \frac{1}{2} \\ 1 & -\frac{1}{2} & 0 \end{bmatrix}$

38. $$\left[\begin{array}{ccc|ccc}1&0&8&1&0&0\\3&7&6&0&1&0\\1&2&3&0&0&1\end{array}\right] \Rightarrow \left[\begin{array}{ccc|ccc}1&0&8&1&0&0\\0&7&-18&-3&1&0\\0&2&-5&-1&0&1\end{array}\right] \Rightarrow \left[\begin{array}{ccc|ccc}1&0&8&1&0&0\\0&1&-\frac{18}{7}&-\frac{3}{7}&\frac{1}{7}&0\\0&2&-5&-1&0&1\end{array}\right] \Rightarrow$$

$-3R_1 + R_2 \Rightarrow R_2$, $-R_1 + R_3 \Rightarrow R_3$; $\frac{1}{7}R_2 \Rightarrow R_2$

$$\left[\begin{array}{ccc|ccc}1&0&8&1&0&0\\0&1&-\frac{18}{7}&-\frac{3}{7}&\frac{1}{7}&0\\0&0&\frac{1}{7}&-\frac{1}{7}&-\frac{2}{7}&1\end{array}\right] \Rightarrow \left[\begin{array}{ccc|ccc}1&0&0&9&16&-56\\0&1&0&-3&-5&18\\0&0&1&-1&-2&7\end{array}\right] \Rightarrow$$

$-2R_2 + R_3 \Rightarrow R_3$; $-56R_3 + R_1 \Rightarrow R_1$, $18R_3 + R_2 \Rightarrow R_2$, $7R_3 \Rightarrow R_3$

Inverse: $$\begin{bmatrix}9&16&-56\\-3&-5&18\\-1&-2&7\end{bmatrix}$$

39. $$\left[\begin{array}{cc|cc}-6&4&1&0\\-3&2&0&1\end{array}\right] \Rightarrow \left[\begin{array}{cc|cc}1&-\frac{2}{3}&-\frac{1}{6}&0\\-3&2&0&1\end{array}\right] \Rightarrow \left[\begin{array}{cc|cc}1&-\frac{2}{3}&-\frac{1}{6}&0\\0&0&-\frac{1}{2}&1\end{array}\right] \Rightarrow \text{No inverse exists.}$$

$-\frac{1}{6}R_1 \Rightarrow R_1$; $3R_1 + R_2 \Rightarrow R_2$

40. $$\left[\begin{array}{ccc|ccc}4&4&1&1&0&0\\1&1&1&0&1&0\\-1&-1&0&0&0&1\end{array}\right] \Rightarrow \left[\begin{array}{ccc|ccc}1&1&1&0&1&0\\4&4&1&1&0&0\\-1&-1&0&0&0&1\end{array}\right] \Rightarrow$$

$R_1 \Leftrightarrow R_2$

$$\left[\begin{array}{ccc|ccc}1&1&1&0&1&0\\0&0&-3&1&-4&0\\0&0&1&0&1&1\end{array}\right] \Rightarrow \left[\begin{array}{ccc|ccc}1&1&1&0&1&0\\0&0&0&1&-1&3\\0&0&1&0&1&1\end{array}\right] : \text{No inverse exists.}$$

$-4R_1 + R_2 \Rightarrow R_2$, $R_1 + R_3 \Rightarrow R_3$; $3R_3 + R_2 \Rightarrow R_2$

41. $$\begin{bmatrix}4&-1&2\\1&1&2\\1&0&1\end{bmatrix}\begin{bmatrix}x\\y\\z\end{bmatrix} = \begin{bmatrix}0\\1\\0\end{bmatrix}$$

$$\begin{bmatrix}x\\y\\z\end{bmatrix} = \begin{bmatrix}4&-1&2\\1&1&2\\1&0&1\end{bmatrix}^{-1}\begin{bmatrix}0\\1\\0\end{bmatrix} = \begin{bmatrix}1&1&-4\\1&2&-6\\-1&-1&5\end{bmatrix}\begin{bmatrix}0\\1\\0\end{bmatrix} = \begin{bmatrix}1\\2\\-1\end{bmatrix}$$

42. $$\begin{bmatrix}1&3&1&3\\1&4&1&3\\0&1&1&0\\1&2&-1&2\end{bmatrix}\begin{bmatrix}w\\x\\y\\z\end{bmatrix} = \begin{bmatrix}1\\2\\1\\1\end{bmatrix}$$

$$\begin{bmatrix}w\\x\\y\\z\end{bmatrix} = \begin{bmatrix}1&3&1&3\\1&4&1&3\\0&1&1&0\\1&2&-1&2\end{bmatrix}^{-1}\begin{bmatrix}1\\2\\1\\1\end{bmatrix} = \begin{bmatrix}3&-5&5&3\\-1&1&0&0\\1&-1&1&0\\0&1&-2&-1\end{bmatrix}\begin{bmatrix}1\\2\\1\\1\end{bmatrix} = \begin{bmatrix}1\\1\\0\\-1\end{bmatrix}$$

43. $$\begin{vmatrix}3&-2\\1&-3\end{vmatrix} = (3)(-3) - (-2)(1) = -9 + 2 = -7$$

44. $\begin{vmatrix} 1 & -2 & 3 \\ 2 & -1 & 3 \\ 1 & -1 & 0 \end{vmatrix} = 1\begin{vmatrix} -1 & 3 \\ -1 & 0 \end{vmatrix} - (-2)\begin{vmatrix} 2 & 3 \\ 1 & 0 \end{vmatrix} + 3\begin{vmatrix} 2 & -1 \\ 1 & -1 \end{vmatrix}$

$= 1(3) + 2(-3) + 3(-1) = 3 - 6 - 3 = -6$

45. $\begin{vmatrix} 1 & 3 & -1 \\ 1 & 2 & 1 \\ 1 & 0 & 2 \end{vmatrix} = 1\begin{vmatrix} 2 & 1 \\ 0 & 2 \end{vmatrix} - 3\begin{vmatrix} 1 & 1 \\ 1 & 2 \end{vmatrix} + (-1)\begin{vmatrix} 1 & 2 \\ 1 & 0 \end{vmatrix}$

$= 1(4) - 3(1) - 1(-2) = 4 - 3 + 2 = 3$

46. Expand along 3rd row... $\begin{vmatrix} 1 & 2 & 3 & 4 \\ -1 & 3 & -3 & 2 \\ 0 & 0 & 0 & -1 \\ 3 & 3 & 4 & 3 \end{vmatrix} = 0(*) - 0(*) + 0(*) - (-1)\begin{vmatrix} 1 & 2 & 3 \\ -1 & 3 & -3 \\ 3 & 3 & 4 \end{vmatrix}$

$= 1\left(1\begin{vmatrix} 3 & -3 \\ 3 & 4 \end{vmatrix} - 2\begin{vmatrix} -1 & -3 \\ 3 & 4 \end{vmatrix} + 3\begin{vmatrix} -1 & 3 \\ 3 & 3 \end{vmatrix}\right)$

$= 1(21) - 2(5) + 3(-12) = -25$

47. $x = \dfrac{\begin{vmatrix} -5 & 3 \\ -4 & 1 \end{vmatrix}}{\begin{vmatrix} 1 & 3 \\ -2 & 1 \end{vmatrix}} = \dfrac{7}{7} = 1 \quad y = \dfrac{\begin{vmatrix} 1 & -5 \\ -2 & -4 \end{vmatrix}}{\begin{vmatrix} 1 & 3 \\ -2 & 1 \end{vmatrix}} = \dfrac{-14}{7} = -2$

48. $x = \dfrac{\begin{vmatrix} -1 & -1 & 1 \\ -4 & -1 & 3 \\ -1 & -3 & 1 \end{vmatrix}}{\begin{vmatrix} 1 & -1 & 1 \\ 2 & -1 & 3 \\ 1 & -3 & 1 \end{vmatrix}} = \dfrac{2}{2} = 1 \quad y = \dfrac{\begin{vmatrix} 1 & -1 & 1 \\ 2 & -4 & 3 \\ 1 & -1 & 1 \end{vmatrix}}{\begin{vmatrix} 1 & -1 & 1 \\ 2 & -1 & 3 \\ 1 & -3 & 1 \end{vmatrix}} = \dfrac{0}{2} = 0 \quad z = \dfrac{\begin{vmatrix} 1 & -1 & -1 \\ 2 & -1 & -4 \\ 1 & -3 & -1 \end{vmatrix}}{\begin{vmatrix} 1 & -1 & 1 \\ 2 & -1 & 3 \\ 1 & -3 & 1 \end{vmatrix}} = \dfrac{-4}{2} = -2$

49. $x = \dfrac{\begin{vmatrix} 7 & -3 & 1 \\ -9 & 1 & -3 \\ 3 & 1 & 1 \end{vmatrix}}{\begin{vmatrix} 1 & -3 & 1 \\ 1 & 1 & -3 \\ 1 & 1 & 1 \end{vmatrix}} = \dfrac{16}{16} = 1 \quad y = \dfrac{\begin{vmatrix} 1 & 7 & 1 \\ 1 & -9 & -3 \\ 1 & 3 & 1 \end{vmatrix}}{\begin{vmatrix} 1 & -3 & 1 \\ 1 & 1 & -3 \\ 1 & 1 & 1 \end{vmatrix}} = \dfrac{-16}{16} = -1$

$z = \dfrac{\begin{vmatrix} 1 & -3 & 7 \\ 1 & 1 & -9 \\ 1 & 1 & 3 \end{vmatrix}}{\begin{vmatrix} 1 & -3 & 1 \\ 1 & 1 & -3 \\ 1 & 1 & 1 \end{vmatrix}} = \dfrac{48}{16} = 3$

50. $w = \dfrac{\begin{vmatrix} 4 & 1 & -1 & 1 \\ 4 & 1 & 0 & 1 \\ 0 & 1 & 2 & 1 \\ 2 & 0 & 1 & 1 \end{vmatrix}}{\begin{vmatrix} 1 & 1 & -1 & 1 \\ 2 & 1 & 0 & 1 \\ 0 & 1 & 2 & 1 \\ 1 & 0 & 1 & 1 \end{vmatrix}} = \dfrac{-4}{-4} = 1$ $\quad x = \dfrac{\begin{vmatrix} 1 & 4 & -1 & 1 \\ 2 & 4 & 0 & 1 \\ 0 & 0 & 2 & 1 \\ 1 & 2 & 1 & 1 \end{vmatrix}}{\begin{vmatrix} 1 & 1 & -1 & 1 \\ 2 & 1 & 0 & 1 \\ 0 & 1 & 2 & 1 \\ 1 & 0 & 1 & 1 \end{vmatrix}} = \dfrac{0}{-4} = 0$

$y = \dfrac{\begin{vmatrix} 1 & 1 & 4 & 1 \\ 2 & 1 & 4 & 1 \\ 0 & 1 & 0 & 1 \\ 1 & 0 & 2 & 1 \end{vmatrix}}{\begin{vmatrix} 1 & 1 & -1 & 1 \\ 2 & 1 & 0 & 1 \\ 0 & 1 & 2 & 1 \\ 1 & 0 & 1 & 1 \end{vmatrix}} = \dfrac{4}{-4} = -1$ $\quad z = \dfrac{\begin{vmatrix} 1 & 1 & -1 & 4 \\ 2 & 1 & 0 & 4 \\ 0 & 1 & 2 & 0 \\ 1 & 0 & 1 & 2 \end{vmatrix}}{\begin{vmatrix} 1 & 1 & -1 & 1 \\ 2 & 1 & 0 & 1 \\ 0 & 1 & 2 & 1 \\ 1 & 0 & 1 & 1 \end{vmatrix}} = \dfrac{-8}{-4} = 2$

51. $\begin{vmatrix} 3a & 3b & 3c \\ d & e & f \\ g & h & i \end{vmatrix} = 3\begin{vmatrix} a & b & c \\ d & e & f \\ g & h & i \end{vmatrix} = 21$

52. $\begin{vmatrix} a & b & c \\ d+g & e+h & f+i \\ g & h & i \end{vmatrix} = \begin{vmatrix} a & b & c \\ d & e & f \\ g & h & i \end{vmatrix} = 7$

53.
$$\begin{aligned} \frac{7x+3}{x^2+x} = \frac{7x+3}{x(x+1)} &= \frac{A}{x} + \frac{B}{x+1} \\ &= \frac{A(x+1)}{x(x+1)} + \frac{Bx}{x(x+1)} \\ &= \frac{Ax + A + Bx}{x(x+1)} \\ &= \frac{(A+B)x + A}{x(x+1)} \end{aligned}$$

$\begin{cases} A + B = 7 \\ A \quad\;\; = 3 \end{cases} \Rightarrow \begin{matrix} A = 3 \\ B = 4 \end{matrix} \quad \dfrac{7x+3}{x(x+1)} = \dfrac{3}{x} + \dfrac{4}{x+1}$

54.
$$\begin{aligned} \frac{4x^3+3x+x^2+2}{x^4+x^2} = \frac{4x^3+x^2+3x+2}{x^2(x^2+1)} &= \frac{A}{x} + \frac{B}{x^2} + \frac{Cx+D}{x^2+1} \\ &= \frac{Ax(x^2+1)}{x^2(x^2+1)} + \frac{B(x^2+1)}{x^2(x^2+1)} + \frac{(Cx+D)x^2}{x^2(x^2+1)} \\ &= \frac{Ax^3 + Ax + Bx^2 + B + Cx^3 + Dx^2}{x^2(x^2+1)} \\ &= \frac{(A+C)x^3 + (B+D)x^2 + Ax + B}{x^2(x^2+1)} \end{aligned}$$

continued on next page

54. continued

$$\begin{cases} A \quad + C \quad = 4 \\ \quad B \quad + D = 1 \\ A \quad\quad\quad = 3 \\ \quad B \quad\quad = 2 \end{cases} \Rightarrow \begin{matrix} A = 3 \\ B = 2 \\ C = 1 \\ D = -1 \end{matrix} \qquad \frac{4x^3 + x^2 + 3x + 2}{x^2(x^2+1)} = \frac{3}{x} + \frac{2}{x^2} + \frac{x-1}{x^2+1}$$

55.

$$\frac{x^2+5}{x^3+x^2+5x} = \frac{x^2+5}{x(x^2+x+5)} = \frac{A}{x} + \frac{Bx+C}{x^2+x+5}$$
$$= \frac{A(x^2+x+5)}{x(x^2+x+5)} + \frac{(Bx+C)x}{x(x^2+x+5)}$$
$$= \frac{Ax^2 + Ax + 5A + Bx^2 + Cx}{x(x^2+x+5)}$$
$$= \frac{(A+B)x^2 + (A+C)x + (5A)}{x(x^2+x+5)}$$

$$\begin{cases} A + B \quad = 1 \\ A \quad + C = 0 \\ 5A \quad\quad = 5 \end{cases} \Rightarrow \begin{matrix} A = 1 \\ B = 0 \\ C = -1 \end{matrix} \qquad \frac{x^2+5}{x(x^2+x+5)} = \frac{1}{x} - \frac{1}{x^2+x+5}$$

56.

$$\frac{x^2+1}{(x+1)^3} = \frac{A}{x+1} + \frac{B}{(x+1)^2} + \frac{C}{(x+1)^3}$$
$$= \frac{A(x+1)^2}{(x+1)^3} + \frac{B(x+1)}{(x+1)^3} + \frac{C}{(x+1)^3}$$
$$= \frac{Ax^2 + 2Ax + A + Bx + B + C}{(x+1)^3}$$
$$= \frac{Ax^2 + (2A+B)x + (A+B+C)}{(x+1)^3}$$

$$\begin{cases} A \quad\quad = 1 \\ 2A + B \quad = 0 \\ A + B + C = 1 \end{cases} \Rightarrow \begin{matrix} A = 1 \\ B = -2 \\ C = 2 \end{matrix} \qquad \frac{x^2+1}{(x+1)^3} = \frac{1}{x+1} - \frac{2}{(x+1)^2} + \frac{2}{(x+1)^3}$$

57. $y \geq -2x - 1$

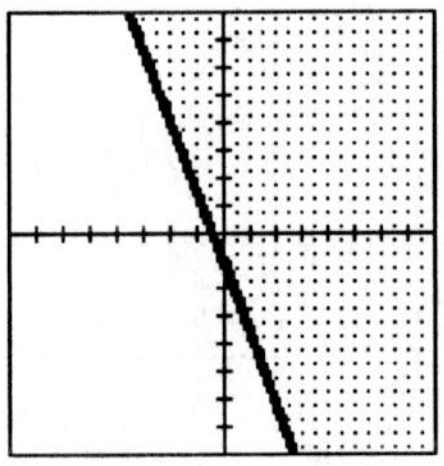

58. $x^2 + y^2 > 4$

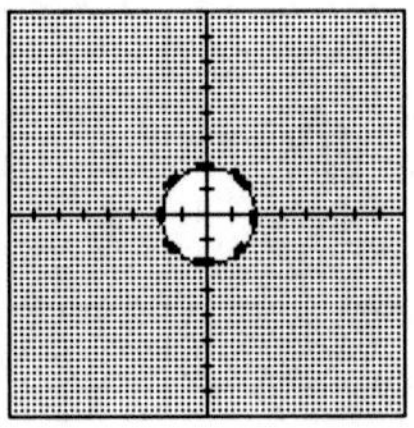

59. $\begin{cases} 3x + 2y \le 6 \\ x - y > 3 \end{cases}$

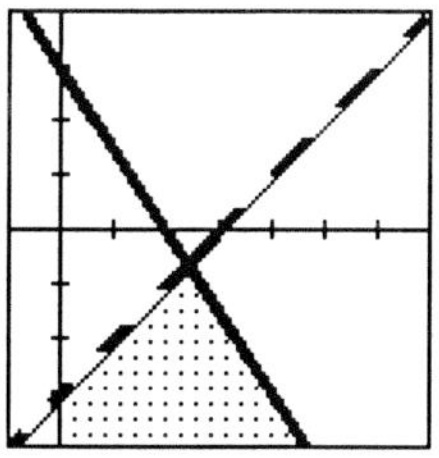

60. $\begin{cases} y \le x^2 + 1 \\ y \ge x^2 - 1 \end{cases}$

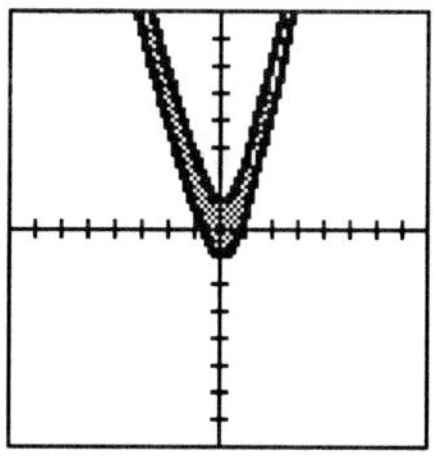

61.

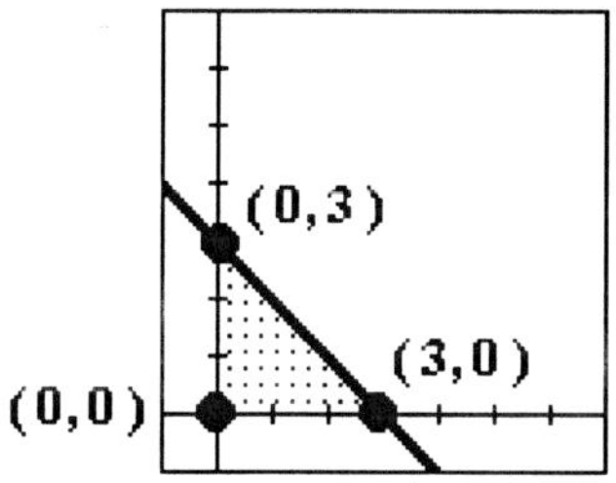

Point	$P = 2x + y$
(0, 0)	$= 2(0) + 0 = 0$
(0, 3)	$= 2(0) + 3 = 3$
(3, 0)	$= 2(3) + 0 = 6$

Max: $P = 6$ at (3, 0)

62.

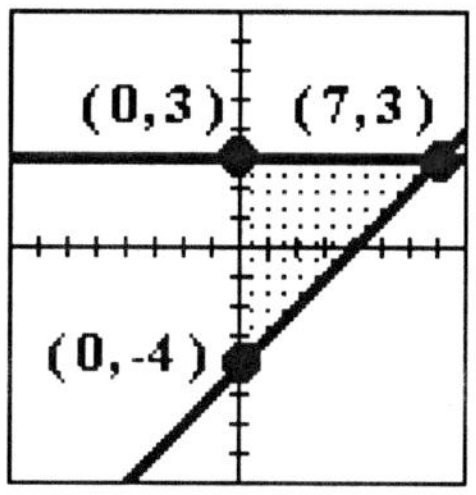

Point	$P = 2x - 3y$
(0, 3)	$= 2(0) - 3(3) = -9$
(7, 3)	$= 2(7) - 3(3) = 5$
(0, −4)	$= 2(0) - 3(-4) = 12$

Max: $P = 12$ at (0, −4)

63.

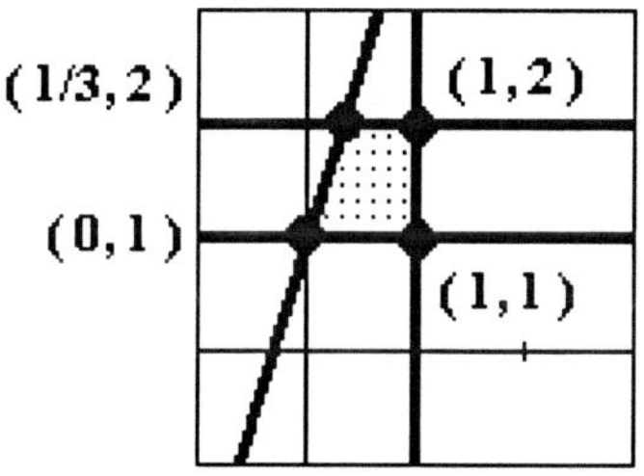

Point	$P = 3x - y$
(0, 1)	$= 3(0) - 1 = -1$
(1, 1)	$= 3(1) - 1 = 2$
(1, 2)	$= 3(1) - 2 = 1$
$\left(\frac{1}{3}, 2\right)$	$= 3\left(\frac{1}{3}\right) - 2 = -1$

Max: $P = 2$ at (1, 1)

64.

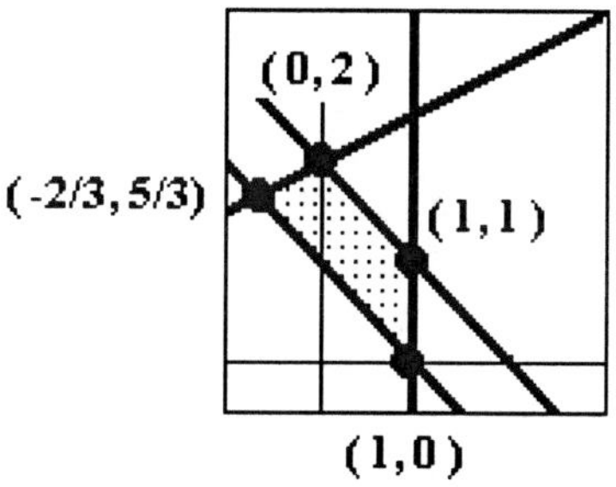

Point	$P = y - 2x$
(0, 2)	$= 2 - 2(0) = 2$
(1, 1)	$= 1 - 2(1) = -1$
(1, 0)	$= 0 - 2(1) = -2$
$\left(-\frac{2}{3}, \frac{5}{3}\right)$	$= \frac{5}{3} - 2\left(-\frac{2}{3}\right) = 3$

Max: $P = 3$ at $\left(-\frac{2}{3}, \frac{5}{3}\right)$

65. Let $x =$ bags of Fertilizer x and $y =$ bags of Fertilizer y.
Maximize $P = 6x + 5y$

subject to $\begin{cases} 6x + 10y \le 20000 \\ 8x + 6y \le 16400 \\ 6x + 4y \le 12000 \\ x \ge 0, y \ge 0 \end{cases}$

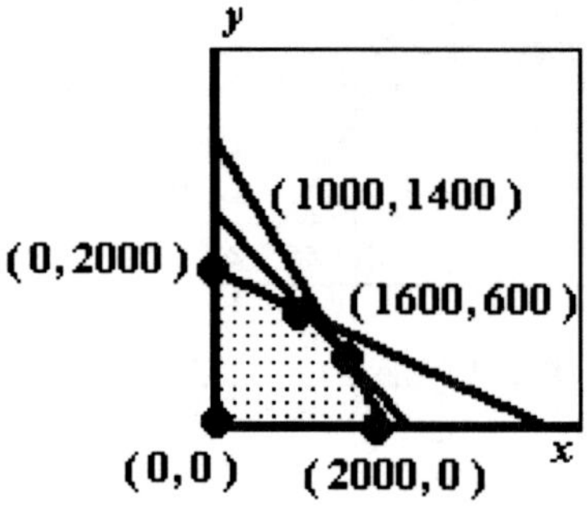

Point	$P = 6x + 5y$
(0, 0)	$= 0$
(0, 2000)	$= 10000$
(1000, 1400)	$= 13000$
(1600, 600)	$= 6(1600) + 5(600) = 12600$
(2000, 0)	$= 6(2000) + 5(0) = 12000$

1000 bags of x and 1400 bags of y should be made, for a maximum profit of \$12,600.

Chapter 6 Test (page 650)

1. $\begin{cases} x - 3y = -5 \\ 2x - y = 0 \end{cases}$

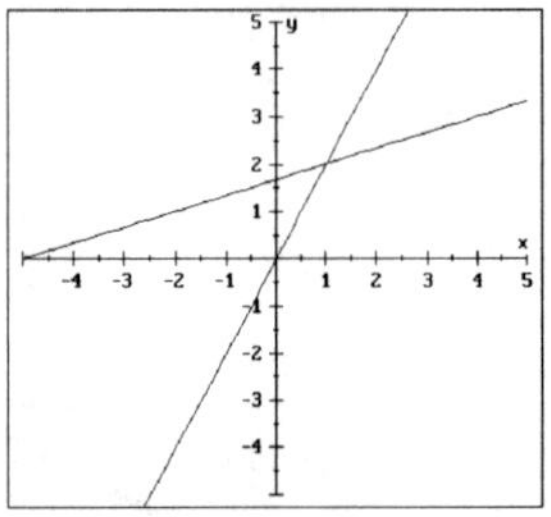

solution: $(1, 2)$

2. $\begin{cases} x = 2y + 5 \\ y = 2x - 4 \end{cases}$

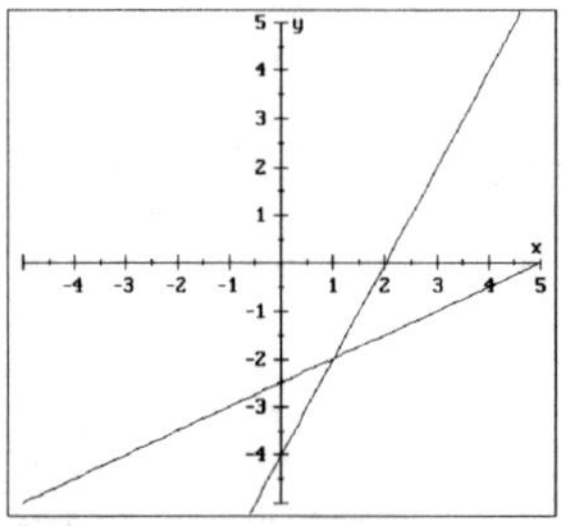

solution: $(1, -2)$

3. $\begin{aligned} 3x + y &= 0 \Rightarrow \times\ (5) \\ 2x - 5y &= 17 \end{aligned}$ $\quad \begin{aligned} 15x + 5y &= 0 \\ 2x - 5y &= 17 \\ \hline 17x &= 17 \\ x &= 1 \end{aligned}$ $\quad \begin{aligned} 3x + y &= 0 \\ 3(1) + y &= 0 \\ y &= -3 \end{aligned}$

Solution: $\boxed{x = 1, y = -3}$

4. $\begin{aligned} \frac{x+y}{2} + x &= 7 \Rightarrow 3x + y = 14 \Rightarrow & 3x + y &= 14 \\ \frac{x-y}{2} - y &= -6 \Rightarrow x - 3y = -12 \Rightarrow & -3x + 9y &= 36 \\ & & 10y &= 50 \\ & & y &= 5 \end{aligned}$ $\quad \begin{aligned} x - 3y &= -12 \\ x - 3(5) &= -12 \\ x &= 3 \end{aligned}$

Solution: $\boxed{\begin{aligned} x &= 3 \\ y &= 5 \end{aligned}}$

5. Let $x =$ liters of 20% solution and $y =$ liters of 45% solution. The following system applies:

$\begin{aligned} x + y &= 10 \Rightarrow \times\ (-2) \\ 0.2x + 0.45y &= 3 \Rightarrow \times\ (10) \end{aligned}$ $\quad \begin{aligned} -2x - 2y &= -20 \\ 2x + 4.5y &= 30 \\ \hline 2.5y &= 10 \\ y &= 4 \end{aligned}$

She must use 4 liters of the 45% and 6 liters of the 20% solution.

CHAPTER 6 TEST

6. Let $x = \#$ from Ace, $y = \#$ from Hi-Fi and $z = \#$ from CD World.

$$\begin{array}{ll} (1) & x+y+z=175 \\ (2) & -x-y+z=25 \\ (3) & 170x+165y+160z=28500 \end{array}$$

Add (1) and (2):

$$\begin{array}{lr} (1) & x+y+\ z=175 \\ (2) & -x-y+\ z=\ \ 25 \\ \hline (4) & 2z=200 \\ & z=100 \end{array}$$

Add equations 170(2) and (3):

$$\begin{array}{lr} 170(2) & -170x-170y+170z=\ \ 4250 \\ (3) & 170x+165y+160z=28500 \\ \hline (5) & -\ \ 5y+330z=32750 \end{array}$$

Solve the system of two equations and two unknowns formed by equations (4) and (5):

$$\begin{aligned} -5y+330z&=32750 \\ -5y+330(100)&=32750 \\ -5y&=-250 \\ y&=50 \end{aligned} \qquad \begin{aligned} x+y+z&=175 \\ x+50+100&=175 \\ y&=25 \end{aligned}$$

Ace buys 25 units per month. Hi-Fi buys 50 units per month. CD World buys 100 units per month.

7. $\left[\begin{array}{cc|c} 3 & -2 & 4 \\ 2 & 3 & 7 \end{array}\right] \Rightarrow \left[\begin{array}{cc|c} 1 & -5 & -3 \\ 2 & 3 & 7 \end{array}\right] \Rightarrow \left[\begin{array}{cc|c} 1 & -5 & -3 \\ 0 & 13 & 13 \end{array}\right] \Rightarrow \left[\begin{array}{cc|c} 1 & -5 & -3 \\ 0 & 1 & 1 \end{array}\right]$

$-R_2+R_1 \Rightarrow R_1 \qquad -2R_1+R_2 \Rightarrow R_2 \qquad \frac{1}{13}R_2 \Rightarrow R_2$

From R_2: $y=1$ From R_1: $x-5y=-3$, $x-5(1)=-3$, $x=2$ Solution: $\boxed{x=2,\ y=1}$

8. $\left[\begin{array}{ccc|c} 1 & 3 & -1 & 6 \\ 2 & -1 & -2 & -2 \\ 1 & 2 & 1 & 6 \end{array}\right] \Rightarrow \left[\begin{array}{ccc|c} 1 & 3 & -1 & 6 \\ 0 & -7 & 0 & -14 \\ 0 & 1 & -2 & 0 \end{array}\right] \Rightarrow \left[\begin{array}{ccc|c} 1 & 3 & -1 & 6 \\ 0 & 1 & 0 & 2 \\ 0 & 0 & -14 & -14 \end{array}\right] \Rightarrow$

$-2R_1+R_2 \Rightarrow R_2$, $-R_3+R_1 \Rightarrow R_3 \qquad -\frac{1}{7}R_2 \Rightarrow R_2$, $7R_3+R_2 \Rightarrow R_3$

$\left[\begin{array}{ccc|c} 1 & 3 & -1 & 6 \\ 0 & 1 & 0 & 2 \\ 0 & 0 & 1 & 1 \end{array}\right]$

$-\frac{1}{14}R_3 \Rightarrow R_3$

From (3): $z=1$ From (2): $y=2$

From (1): $x+3y-z=6$, $x+3(2)-(1)=6$, $x=1$

Solution: $\boxed{x=1,\ y=2,\ z=1}$

9. $\left[\begin{array}{ccc|c} 1 & 2 & 3 & -5 \\ 3 & 1 & -2 & 7 \\ 0 & 1 & -1 & 2 \end{array}\right] \Rightarrow \left[\begin{array}{ccc|c} 1 & 2 & 3 & -5 \\ 0 & -5 & -11 & 22 \\ 0 & 1 & -1 & 2 \end{array}\right] \Rightarrow \left[\begin{array}{ccc|c} 1 & 2 & 3 & -5 \\ 0 & 1 & -1 & 2 \\ 0 & -5 & -11 & 22 \end{array}\right] \Rightarrow$

$-3R_1+R_2 \Rightarrow R_2 \qquad R_2 \Leftrightarrow R_3$

$\left[\begin{array}{ccc|c} 1 & 0 & 5 & -9 \\ 0 & 1 & -1 & 2 \\ 0 & 0 & -16 & 32 \end{array}\right] \Rightarrow \left[\begin{array}{ccc|c} 1 & 0 & 5 & -9 \\ 0 & 1 & -1 & 2 \\ 0 & 0 & 1 & -2 \end{array}\right] \Rightarrow \left[\begin{array}{ccc|c} 1 & 0 & 0 & 1 \\ 0 & 1 & 0 & 0 \\ 0 & 0 & 1 & -2 \end{array}\right]$ Solution: $\boxed{\begin{array}{l} x=1 \\ y=0 \\ z=-2 \end{array}}$

$-2R_2+R_1 \Rightarrow R_1$, $5R_2+R_3 \Rightarrow R_3 \qquad -\frac{1}{16}R_3 \Rightarrow R_3 \qquad -5R_3+R_1 \Rightarrow R_1$, $R_2+R_3 \Rightarrow R_2$

10. $\left[\begin{array}{ccc|c} 1 & 2 & 1 & 0 \\ 3 & -2 & -2 & 7 \\ 4 & 0 & -1 & 7 \end{array}\right] \Rightarrow \left[\begin{array}{ccc|c} 1 & 2 & 1 & 0 \\ 0 & -8 & -5 & 7 \\ 0 & -8 & -5 & 7 \end{array}\right] \Rightarrow \left[\begin{array}{ccc|c} 1 & 2 & 1 & 0 \\ 0 & 1 & \frac{5}{8} & -\frac{7}{8} \\ 0 & 0 & 0 & 0 \end{array}\right] \Rightarrow$

$-3R_1 + R_2 \Rightarrow R_2$ $\quad$ $-R_2 + R_3 \Rightarrow R_3$

$-4R_1 + R_3 \Rightarrow R_3$ $\quad$ $-\frac{1}{8}R_2 \Rightarrow R_2$

$\left[\begin{array}{ccc|c} 1 & 0 & -\frac{1}{4} & \frac{7}{4} \\ 0 & 1 & \frac{5}{8} & -\frac{7}{8} \\ 0 & 0 & 0 & 0 \end{array}\right] \Rightarrow$ Solution: $x = \frac{7}{4} + \frac{1}{4}z$, $y = -\frac{7}{8} - \frac{5}{8}z$, $z =$ any real number

$-2R_2 + R_1 \Rightarrow R_1$

Note: This answer is equivalent to the answer provided in the textbook.

11. $$3\begin{bmatrix} 2 & -3 & 5 \\ 0 & 3 & -1 \end{bmatrix} - 5\begin{bmatrix} -2 & 1 & -1 \\ 0 & 3 & 2 \end{bmatrix} = \begin{bmatrix} 6 & -9 & 15 \\ 0 & 9 & -3 \end{bmatrix} - \begin{bmatrix} -10 & 5 & -5 \\ 0 & 15 & 10 \end{bmatrix} = \begin{bmatrix} 16 & -14 & 20 \\ 0 & -6 & -13 \end{bmatrix}$$

12. $$\begin{bmatrix} 1 & 2 & 3 \end{bmatrix}\begin{bmatrix} 2 & -2 \\ -2 & 2 \\ 1 & 0 \end{bmatrix}\begin{bmatrix} 3 \\ -2 \end{bmatrix} = \begin{bmatrix} 1 & 2 \end{bmatrix}\begin{bmatrix} 3 \\ -2 \end{bmatrix} = \begin{bmatrix} -1 \end{bmatrix}$$

13. $\left[\begin{array}{cc|cc} 5 & 19 & 1 & 0 \\ 2 & 7 & 0 & 1 \end{array}\right] \Rightarrow \left[\begin{array}{cc|cc} 1 & \frac{19}{5} & \frac{1}{5} & 0 \\ 2 & 7 & 0 & 1 \end{array}\right] \Rightarrow \left[\begin{array}{cc|cc} 1 & \frac{19}{5} & \frac{1}{5} & 0 \\ 0 & -\frac{3}{5} & -\frac{2}{5} & 1 \end{array}\right] \Rightarrow$

$\frac{1}{5}R_1 \Rightarrow R_1$ $\quad$ $-2R_1 + R_2 \Rightarrow R_2$

$\left[\begin{array}{cc|cc} 1 & \frac{19}{5} & \frac{1}{5} & 0 \\ 0 & 1 & \frac{2}{3} & -\frac{5}{3} \end{array}\right] \Rightarrow \left[\begin{array}{cc|cc} 1 & 0 & -\frac{7}{3} & \frac{19}{3} \\ 0 & 1 & \frac{2}{3} & -\frac{5}{3} \end{array}\right] \Rightarrow$ Inverse: $\begin{bmatrix} -\frac{7}{3} & \frac{19}{3} \\ \frac{2}{3} & -\frac{5}{3} \end{bmatrix}$

$-\frac{5}{3}R_2 \Rightarrow R_1$ $\quad$ $-\frac{19}{5}R_2 + R_1 \Rightarrow R_1$

14. $\left[\begin{array}{ccc|ccc} -1 & 3 & -2 & 1 & 0 & 0 \\ 4 & 1 & 4 & 0 & 1 & 0 \\ 0 & 3 & -1 & 0 & 0 & 1 \end{array}\right] \Rightarrow \left[\begin{array}{ccc|ccc} 1 & -3 & 2 & -1 & 0 & 0 \\ 0 & 13 & -4 & 4 & 1 & 0 \\ 0 & 3 & -1 & 0 & 0 & 1 \end{array}\right] \Rightarrow$

$4R_1 + R_2 \Rightarrow R_2$

$-R_1 \Rightarrow R_1$

$\left[\begin{array}{ccc|ccc} 1 & -3 & 2 & -1 & 0 & 0 \\ 0 & 1 & 0 & 4 & 1 & -4 \\ 0 & 3 & -1 & 0 & 0 & 1 \end{array}\right] \Rightarrow \left[\begin{array}{ccc|ccc} 1 & 0 & 2 & 11 & 3 & -12 \\ 0 & 1 & 0 & 4 & 1 & -4 \\ 0 & 0 & -1 & -12 & -3 & 13 \end{array}\right] \Rightarrow$

$-4R_3 + R_2 \Rightarrow R_2$ $\quad$ $3R_2 + R_1 \Rightarrow R_1$

$-3R_2 + R_3 \Rightarrow R_3$

$\left[\begin{array}{ccc|ccc} 1 & 0 & 0 & -13 & -3 & 14 \\ 0 & 1 & 0 & 4 & 1 & -4 \\ 0 & 0 & 1 & 12 & 3 & -13 \end{array}\right]$ Inverse: $\begin{bmatrix} -13 & -3 & 14 \\ 4 & 1 & -4 \\ 12 & 3 & -13 \end{bmatrix}$

$2R_3 + R_1 \Rightarrow R_1$

$-R_3 \Rightarrow R_3$

CHAPTER 6 TEST

15. $\begin{bmatrix} 5 & 19 \\ 2 & 7 \end{bmatrix} \begin{bmatrix} x \\ y \end{bmatrix} = \begin{bmatrix} 3 \\ 2 \end{bmatrix}$

$$\begin{bmatrix} x \\ y \end{bmatrix} = \begin{bmatrix} 5 & 19 \\ 2 & 7 \end{bmatrix}^{-1} \begin{bmatrix} 3 \\ 2 \end{bmatrix} = \begin{bmatrix} -\frac{7}{3} & \frac{19}{3} \\ \frac{2}{3} & -\frac{5}{3} \end{bmatrix} \begin{bmatrix} 3 \\ 2 \end{bmatrix} = \begin{bmatrix} \frac{17}{3} \\ -\frac{4}{3} \end{bmatrix}$$

16. $\begin{bmatrix} -1 & 3 & -2 \\ 4 & 1 & 4 \\ 0 & 3 & -1 \end{bmatrix} \begin{bmatrix} x \\ y \\ z \end{bmatrix} = \begin{bmatrix} 1 \\ 3 \\ -1 \end{bmatrix}$

$$\begin{bmatrix} x \\ y \\ z \end{bmatrix} = \begin{bmatrix} -1 & 3 & -2 \\ 4 & 1 & 4 \\ 0 & 3 & -1 \end{bmatrix}^{-1} \begin{bmatrix} 1 \\ 3 \\ -1 \end{bmatrix}$$

$$\begin{bmatrix} x \\ y \\ z \end{bmatrix} = \begin{bmatrix} -13 & -3 & 14 \\ 4 & 1 & -4 \\ 12 & 3 & -13 \end{bmatrix} \begin{bmatrix} 1 \\ 3 \\ -1 \end{bmatrix} = \begin{bmatrix} -36 \\ 11 \\ 34 \end{bmatrix}$$

17. $\begin{vmatrix} 3 & -5 \\ -3 & 1 \end{vmatrix} = (3)(1) - (-5)(-3) = 3 - 15 = -12$

18. $\begin{vmatrix} 3 & 5 & -1 \\ -2 & 3 & -2 \\ 1 & 5 & -3 \end{vmatrix} = 3\begin{vmatrix} 3 & -2 \\ 5 & -3 \end{vmatrix} - 5\begin{vmatrix} -2 & -2 \\ 1 & -3 \end{vmatrix} + (-1)\begin{vmatrix} -2 & 3 \\ 1 & 5 \end{vmatrix}$

$$= 3(1) - 5(8) - 1(-13) = 3 - 40 + 13 = -24$$

19. $y = \dfrac{\begin{vmatrix} 3 & 3 \\ -3 & 2 \end{vmatrix}}{\begin{vmatrix} 3 & -5 \\ -3 & 1 \end{vmatrix}} = \dfrac{15}{-12} = -\dfrac{5}{4}$

20. $y = \dfrac{\begin{vmatrix} 3 & 2 & -1 \\ -2 & 1 & -2 \\ 1 & 0 & -3 \end{vmatrix}}{\begin{vmatrix} 3 & 5 & -1 \\ -2 & 3 & -2 \\ 1 & 5 & -3 \end{vmatrix}} = \dfrac{-24}{-24} = 1$

21. $\dfrac{5x}{2x^2 - x - 3} = \dfrac{5x}{(2x-3)(x+1)} = \dfrac{A}{2x-3} + \dfrac{B}{x+1}$

$$= \frac{A(x+1)}{(2x-3)(x+1)} + \frac{B(2x-3)}{(2x-3)(x+1)}$$

$$= \frac{Ax + A + 2Bx - 3B}{(2x-3)(x+1)}$$

$$= \frac{(A+2B)x + (A-3B)}{(2x-3)(x+1)}$$

$\begin{cases} A + 2B = 5 \\ A - 3B = 0 \end{cases} \rightarrow \begin{matrix} A = 3 \\ B = 1 \end{matrix} \quad \dfrac{5x}{(2x-3)(x+1)} = \dfrac{3}{2x-3} + \dfrac{1}{x+1}$

CHAPTER 6 TEST

22. $$\frac{3x^2+x+2}{x^3+2x} = \frac{3x^2+x+2}{x(x^2+2)} = \frac{A}{x} + \frac{Bx+C}{x^2+2}$$

$$= \frac{A(x^2+2)}{x(x^2+2)} + \frac{(Bx+C)x}{x(x^2+2)}$$

$$= \frac{Ax^2+2A+Bx^2+Cx}{x(x^2+2)}$$

$$= \frac{(A+B)x^2+Cx+2A}{x(x^2+2)}$$

$$\begin{cases} A+B & =3 \\ \quad C & =1 \\ 2A & =2 \end{cases} \Rightarrow \begin{array}{l} A=1 \\ B=2 \\ C=1 \end{array} \qquad \frac{3x^2+x+2}{x(x^2+2)} = \frac{1}{x} + \frac{2x+1}{x^2+2}$$

23. $\begin{cases} x-3y \geq 3 \\ x+3y \leq 3 \end{cases}$

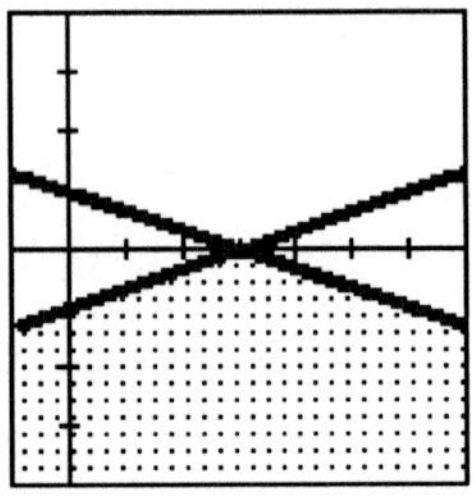

24. $\begin{cases} 3x+4y \leq 12 \\ 3x+4y \geq 6 \\ x \geq 0,\ y \geq 0 \end{cases}$

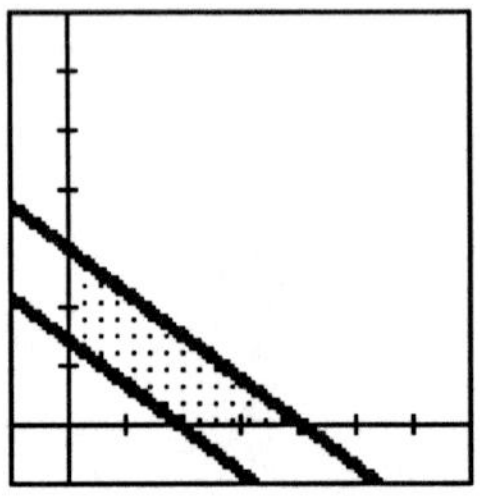

25.

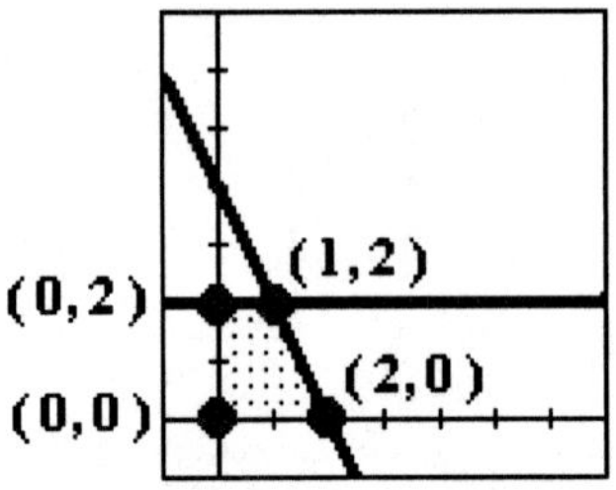

Point	$P=3x+2y$
(0, 2)	$=3(0)+2(2)=4$
(1, 2)	$=3(1)+2(2)=7$
(2, 0)	$=3(2)+2(0)=6$
(0, 0)	$=3(0)+2(0)=0$

Max: $P=7$ at $(1, 2)$

26.

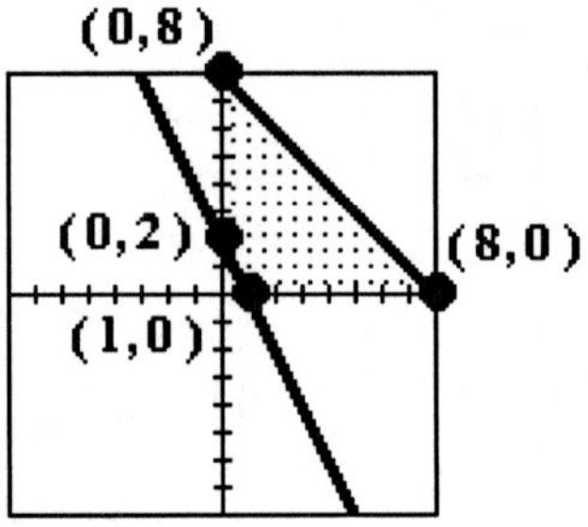

Point	$P=y-x$
(0, 2)	$=2-0=2$
(0, 8)	$=8-0=8$
(8, 0)	$=0-8=-8$
(1, 0)	$=0-1=-1$

Min: $P=-8$ at $(8, 0)$

Exercises 7.1 (page 666)

1. $(2, -5), 3$

3. $(0, 0), \sqrt{5}$

5. to the left

7. downward

9. directrix, focus

11. Two squared variables: circle

13. One squared variable: parabola

15.
$$(x-h)^2+(y-k)^2=r^2$$
$$(x-0)^2+(y-0)^2=7^2$$
$$\boxed{x^2+y^2=49}$$
$$\boxed{x^2+y^2-49=0}$$

17.
$$r=\sqrt{(3-2)^2+(2-(-2))^2}=\sqrt{17}$$
$$(x-h)^2+(y-k)^2=r^2$$
$$(x-2)^2+(y-(-2))^2=\left(\sqrt{17}\right)^2$$
$$\boxed{(x-2)^2+(y+2)^2=17}$$
$$x^2-4x+4+y^2+4y+4=17$$
$$\boxed{x^2+y^2-4x+4y-9=0}$$

19.
$$\begin{aligned} 3x+\ y&=1 \Rightarrow \times\ (3) \\ -2x-3y&=4 \end{aligned} \qquad \begin{aligned} 9x+3y&=3 \\ -2x-3y&=4 \\ \hline 7x\qquad&=7 \\ x\qquad&=1 \end{aligned}$$
$$\begin{aligned} 3x+y&=1 \\ 3(1)+y&=1 \\ y&=-2 \end{aligned}$$
Center: $(1, -2)$
$$(x-h)^2+(y-k)^2=r^2$$
$$(x-1)^2+(y-(-2))^2=6^2$$
$$\boxed{(x-1)^2+(y+2)^2=36}$$
$$x^2-2x+1+y^2+4y+4=36$$
$$\boxed{x^2+y^2-2x+4y-31=0}$$

21.
$$x^2+y^2=4$$
$$(x-0)^2+(y-0)^2=2^2$$
$$C(0,0),\ r=2$$

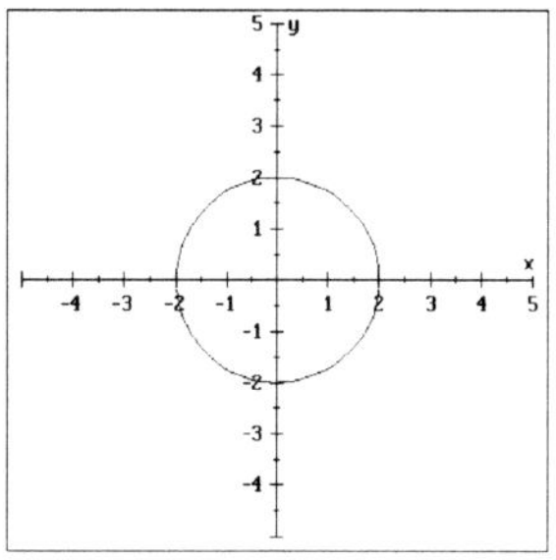

23.
$$3x^2+3y^2-12x-6y=12$$
$$x^2-4x+y^2-2y=4$$
$$x^2-4x+4+y^2-2y+1=4+4+1$$
$$(x-2)^2+(y-1)^2=3^2$$
$$C(2,1),\ r=3$$

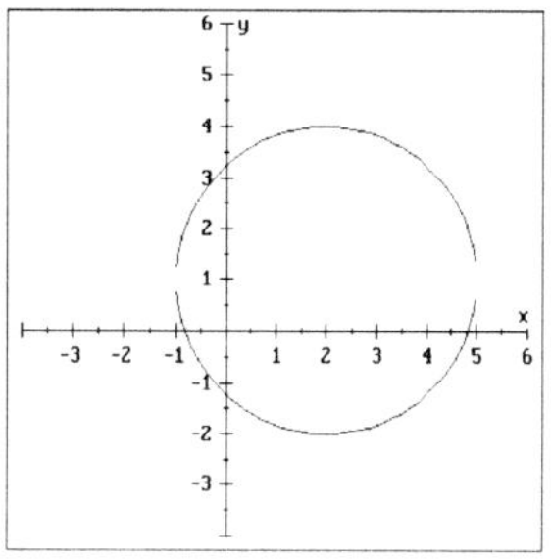

25.
$$x^2=12y$$
$$(x-0)^2=4\cdot 3(y-0)$$
$p=3$, opens up

V $(0,0)$, $F(0,3)$, D: $y=-3$

27.
$$(y-3)^2=20x$$
$$(y-3)^2=4\cdot 5(x-0)$$
$p=5$, opens right

V $(0,3)$, $F(5,3)$, D: $x=-5$

29. $(x+2)^2 = -24(y-1)$
$(x-(-2))^2 = 4\cdot(-6)(y-1)$
$p=-6$, opens down
V $(-2,1)$, $F(-2,-5)$, D: $y=7$

31. Vertical (up), $p=3$
$(x-h)^2 = 4p(y-k)$
$(x-0)^2 = 4(3)(y-0)$
$x^2 = 12y$

33. Horizontal (left), $p=-3$
$(y-k)^2 = 4p(x-h)$
$(y-0)^2 = 4(-3)(x-0)$
$y^2 = -12x$

35. Vertical (down), $p=-3$
$(x-h)^2 = 4p(y-k)$
$(x-3)^2 = 4(-3)(y-5)$
$(x-3)^2 = -12(y-5)$

37. Vertical (down), $p=-7$
$(x-h)^2 = 4p(y-k)$
$(x-3)^2 = 4(-7)(y-5)$
$(x-3)^2 = -28(y-5)$

39. Vertical (down), $p=-1$
$(x-h)^2 = 4p(y-k)$
$(x-0)^2 = 4(-1)(y-2)$
$x^2 = -4(y-2)$

41. Horizontal (right), $p=2$
$(y-k)^2 = 4p(x-h)$
$(y-(-5))^2 = 4(2)(x-1)$
$(y+5)^2 = 8(x-1)$

43. $(x-2)^2 = 4p(y-2)$
$(0-2)^2 = 4p(0-2)$
$4 = -8p$
$-\frac{1}{2} = p$
$-2 = 4p$
$(x-2)^2 = -2(y-2)$

OR $(y-2)^2 = 4p(x-2)$
$(0-2)^2 = 4p(0-2)$
$4 = -8p$
$-\frac{1}{2} = p$
$-2 = 4p$
$(y-2)^2 = -2(x-2)$

45. $(x-(-4))^2 = 4p(y-6)$
$(0+4)^2 = 4p(3-6)$
$16 = -12p$
$-\frac{4}{3} = p$
$-\frac{16}{3} = 4p$
$(x+4)^2 = -\frac{16}{3}(y-6)$

OR $(y-6)^2 = 4p(x-(-4))$
$(3-6)^2 = 4p(0+4)$
$9 = 16p$
$\frac{9}{16} = p$
$\frac{9}{4} = 4p$
$(y-6)^2 = \frac{9}{4}(x+4)$

47. $(x-6)^2 = 4p(y-8)$
$(5-6)^2 = 4p(10-8)$
$1 = 8p$
$\frac{1}{8} = p$
$\frac{1}{2} = 4p$
$(x-6)^2 = \frac{1}{2}(y-8)$

OR $(y-8)^2 = 4p(x-6)$
$(10-8)^2 = 4p(5-6)$
$4 = -4p$
$-1 = p$
$-4 = 4p$
$(y-8)^2 = -4(x-6)$

Check to see which equation is satisfied by $(5,6)$ as well. Answer: $(y-8)^2 = -4(x-6)$

SECTION 7.1

49.
$$(x-3)^2 = 4p(y-1) \quad \textbf{OR} \quad (y-1)^2 = 4p(x-3)$$
$$(4-3)^2 = 4p(3-1) \qquad (3-1)^2 = 4p(4-3)$$
$$1 = 8p \qquad 4 = 4p$$
$$\frac{1}{8} = p \qquad (y-1)^2 = 4(x-3)$$
$$\frac{1}{2} = 4p$$
$$(x-3)^2 = \tfrac{1}{2}(y-1)$$

Check to see which equation is satisfied by $(2, 3)$ as well. Answer: $(x-3)^2 = \frac{1}{2}(y-1)$

51.
$$y = x^2 + 4x + 5$$
$$y - 5 = x^2 + 4x$$
$$y - 5 + 4 = x^2 + 4x + 4$$
$$y - 1 = (x+2)^2$$

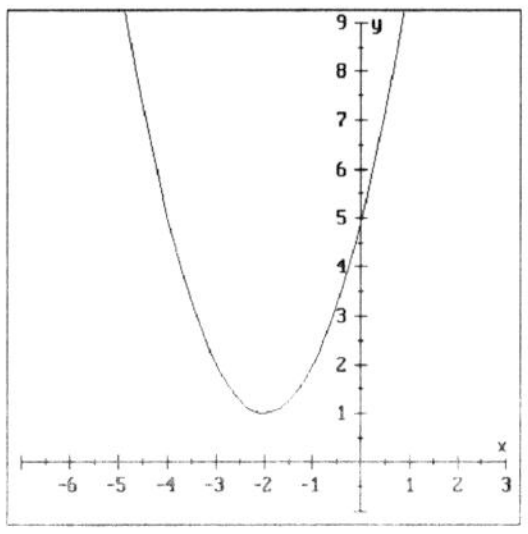

53.
$$y^2 + 4x - 6y = -1$$
$$y^2 - 6y = -4x - 1$$
$$y^2 - 6y + 9 = -4x - 1 + 9$$
$$(y-3)^2 = -4x + 8$$
$$(y-3)^2 = -4(x-2)$$

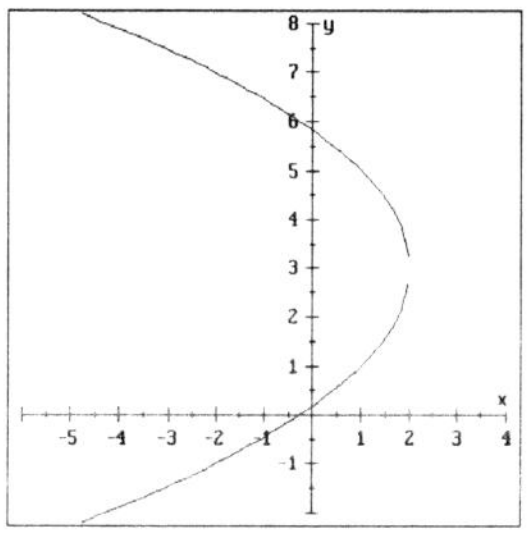

55.
$$y^2 - 4y = 4x - 8$$
$$y^2 - 4y + 4 = 4x - 8 + 4$$
$$(y-2)^2 = 4x - 4$$
$$(y-2)^2 = 4(x-1)$$

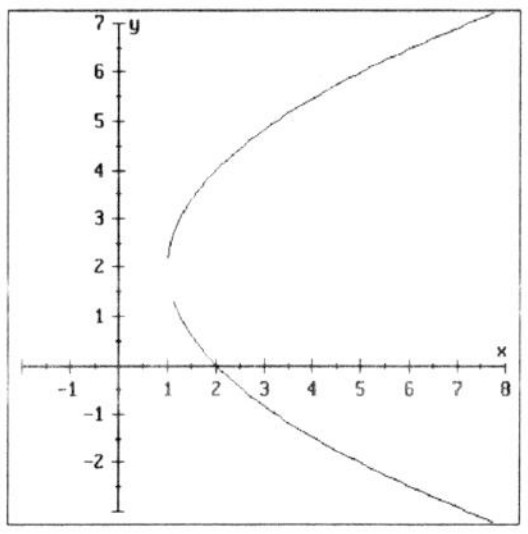

57.
$$y^2 - 4y = -8x + 20$$
$$y^2 - 4y + 4 = -8x + 20 + 4$$
$$(y-2)^2 = -8x + 24$$
$$(y-2)^2 = -8(x-3)$$

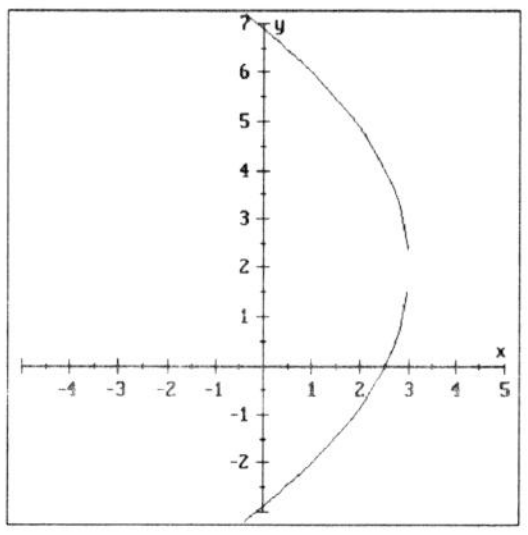

59. $x^2 - 6y + 22 = -4x$

$$x^2 + 4x = 6y - 22$$
$$x^2 + 4x + 4 = 6y - 22 + 4$$
$$(x + 2)^2 = 6y - 18$$
$$(x + 2)^2 = 6(y - 3)$$

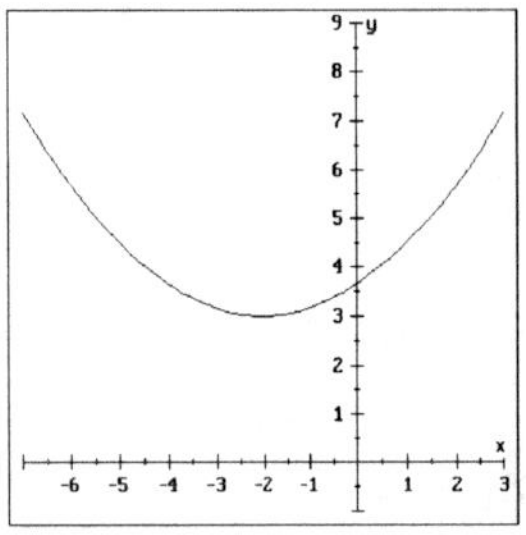

61. $4x^2 - 4x + 32y = 47$

$$x^2 - x = -8y + \tfrac{47}{4}$$
$$x^2 - x + \tfrac{1}{4} = -8y + \tfrac{47}{4} + \tfrac{1}{4}$$
$$\left(x - \tfrac{1}{2}\right)^2 = -8y + 12$$
$$\left(x - \tfrac{1}{2}\right)^2 = -8\left(y - \tfrac{3}{2}\right)$$

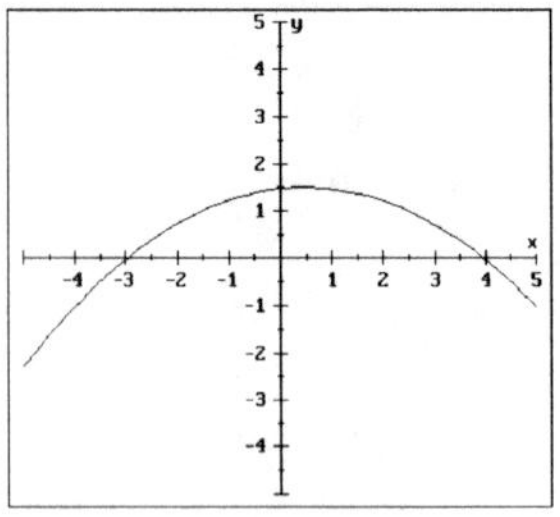

63. $y^2 = 4x - 12$

$$y = \pm\sqrt{4x - 12}$$

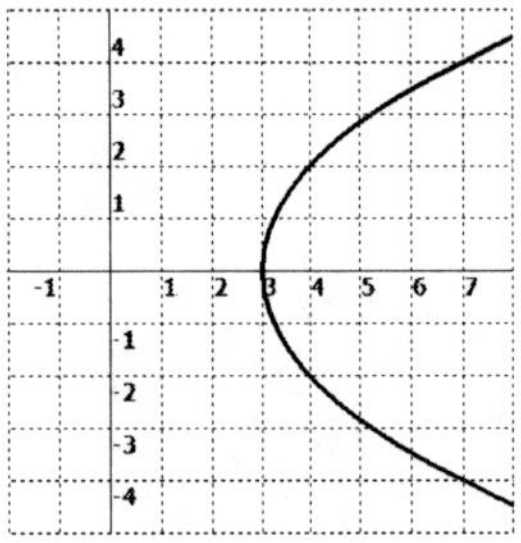

65. Check the coordinates:

$$x^2 + y^2 = 50^2 + 70^2$$
$$= 2500 + 4900 = 7400$$

$7400 < 8100$, so the city can receive.

67. Graph both circles: $\begin{cases} x^2 + y^2 = 1600 \\ x^2 + (y - 35)^2 = 625 \end{cases}$

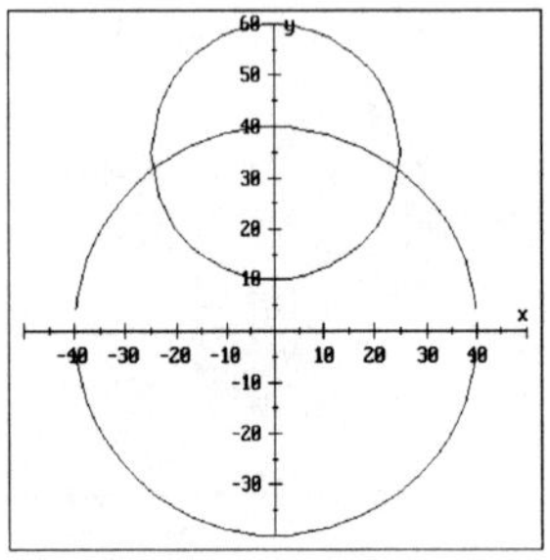

The point farthest from the transmitter $(0, 0)$ is the point $(0, 60)$. The greatest distance is 60 miles.

69. $C(4, 0)$, $r = 4$

$$(x - h)^2 + (y - k)^2 = r^2$$
$$(x - 4)^2 + (y - 0)^2 = 4^2$$
$$(x - 4)^2 + y^2 = 16$$

71. $x^2 + y^2 = 16$: $C(0,0)$, $r = 4$
Small gear: $C(7,0)$, $r = 3$
$$(x-h)^2 + (y-k)^2 = r^2$$
$$(x-7)^2 + (y-0)^2 = 3^2$$
$$(x-7)^2 + y^2 = 9$$

73. Find the distance to the focus:
$4p = 8 \Rightarrow p = 2$
It will be hottest 2 feet from the vertex.

75. The vertex is $(0, 0)$, while $(15, -10)$ is on the curve (vertical parabola):
$$(x-h)^2 = 4p(y-k)$$
$$(15-0)^2 = 4p(-10-0)$$
$$225 = -40p$$
$$-\tfrac{45}{2} = 4p \Rightarrow x^2 = -\tfrac{45}{2}y$$

77. The depth is the y-coordinate when $x = 4$.
[4 feet on each side of the vertex]
$y = \frac{1}{16}x^2$
$y = \frac{1}{16}(4)^2 = 1$
The depth is 1 foot.

79.
$$s = -16t^2 + 80\sqrt{3}t$$
$$-\tfrac{1}{16}s = t^2 - 5\sqrt{3}t$$
$$-\tfrac{1}{16}s + \tfrac{75}{4} = t^2 - 5\sqrt{3}t + \tfrac{75}{4}$$
$$-\tfrac{1}{16}(s - 300) = \left(t - \tfrac{5\sqrt{3}}{2}\right)^2$$
The maximum height is 300 feet.

81. Place the vertex at $(0, 0)$, with the focus at $(1, 0) \Rightarrow p = 1, 4p = 4$.
$$(y-k)^2 = 4p(x-h)$$
$$y^2 = 4x$$
Let $x = 10$: $y^2 = 4x$
$$y^2 = 4(10) \Rightarrow y = \pm\sqrt{40}$$
The width $= 2\sqrt{40} \approx 12.6$ cm.

83. The vertex is $(0, 0)$, while $(315, -630)$ is on the curve (vertical parabola):
$$(x-h)^2 = 4p(y-k)$$
$$(315-0)^2 = 4p(-630-0)$$
$$99{,}225 = -2520p$$
$$-\tfrac{19845}{504} = p \Rightarrow 4p = -\tfrac{19845}{126}$$
Let $y = -430$:
$$x^2 = -\tfrac{19845}{126}y$$
$$x^2 = -\tfrac{19845}{126}(-430)$$
$$x = \pm\sqrt{-\tfrac{19845}{126}(-430)}$$
$$x \approx \pm 260$$
The width is about 520 feet.

85.
$$(y-2)^2 = 8(x-1)$$
$$y^2 - 4y + 4 = 8x - 8$$
$$y^2 - 8x - 4y + 12 = 0$$
$$0x^2 + 0xy + y^2 - 8x - 4y + 12 = 0$$

87.

$(x-h)^2+(y-k)^2=r^2$	$(x-h)^2+(y-k)^2=r^2$
$(0-h)^2+(8-k)^2=r^2$	$(5-h)^2+(3-k)^2=r^2$
$h^2+64-16k+k^2=r^2$	$25-10h+h^2+9-6k+k^2=r^2$
$h^2+k^2-r^2=16k-64$	$h^2+k^2-r^2=10h+6k-34$

$$\begin{aligned}(x-h)^2+(y-k)^2&=r^2\\(4-h)^2+(6-k)^2&=r^2\\16-8h+h^2+36-12k+k^2&=r^2\\h^2+k^2-r^2&=8h+12k-52\end{aligned}$$

$$\begin{cases}16k-64=10h+6k-34 & \Rightarrow \quad 10k-10h=30\\16k-64=8h+12k-52 & \Rightarrow \quad 4k-8h=12\end{cases} \Rightarrow k=3,\ h=0$$

Substitute into one of the above equations to get $r=5$. Circle: $x^2+(y-3)^2=25$

89.
$y=ax^2+bx+c$ $\quad$ $y=ax^2+bx+c$ $\quad$ $y=ax^2+bx+c$

$8=a(1)^2+b(1)+c$ $\quad$ $-1=a(-2)^2+b(-2)+c$ $\quad$ $15=a(2)^2+b(2)+c$

$8=a+b+c$ $\quad$ $-1=4a-2b+c$ $\quad$ $15=4a+2b+c$

$$\begin{cases}a+b+c=8\\4a-2b+c=-1\\4a+2b+c=15\end{cases} \Rightarrow a=1,\ b=4,\ c=3 \Rightarrow y=x^2+4x+3$$

91. The stone hits the ground when $s=0$:

$0=-16t^2+128t$

$0=-16t(t-8)$

It hits the ground after 8 seconds.

Find s when $t=x$:

$s=-16x^2+128x$

Find s when $t=8-x$:

$$\begin{aligned}s&=-16(8-x)^2+128(8-x)\\&=-16\left(64-16x+x^2\right)+1024-128x\\&=-1024+256x-16x^2+1024-128x\\&=-16x^2+128x\end{aligned}$$

93. $x^2+4x+\boxed{4}$

95. $x^2-7x+\boxed{\frac{49}{4}}$

97.
$$\begin{aligned}x^2+4x&=5\\x^2+4x-5&=0\\(x+5)(x-1)&=0\\x=-5 \text{ or } x&=1\end{aligned}$$

99.
$$\begin{aligned}x^2-7x-18&=0\\(x+2)(x-9)&=0\\x=-2 \text{ or } x&=9\end{aligned}$$

Exercises 7.2 (page 684)

1. sum, constant

3. vertices

5. $(a,0)$, $(-a,0)^2$

SECTION 7.2

7. $2a = 26 \Rightarrow$ String: 26 inches long
$2b = 10 \Rightarrow b = 5$
$b^2 = a^2 - c^2$
$5^2 = 13^2 - c^2 \Rightarrow c = 12$
Thumbtacks: $2c = 24$ inches apart

9. Both variables squared with equal coefficients: circle

11. One variable squared: parabola

13. Both variables squared with unequal coefficients: ellipse

15. $a = 4, b = 3$; horizontal
$C(0,0)$
$$\frac{x^2}{16} + \frac{y^2}{9} = 1$$

17. $c = 3, a = 5$; horizontal
$$b^2 = a^2 - c^2 = 25 - 9 = 16$$
$$\frac{x^2}{25} + \frac{y^2}{16} = 1$$

19. $c = 1, b = \frac{4}{3}$; vertical
$$a^2 = b^2 + c^2 = \tfrac{16}{9} + 1 = \tfrac{25}{9}$$
$$\frac{x^2}{16/9} + \frac{y^2}{25/9} = 1$$
$$\frac{9x^2}{16} + \frac{9y^2}{25} = 1$$

21. $c = 3, a = 4$; vertical
$$b^2 = a^2 - c^2 = 16 - 9 = 7$$
$$\frac{x^2}{7} + \frac{y^2}{16} = 1$$

23. vertical
$$\frac{(x-3)^2}{4} + \frac{(y-4)^2}{9} = 1$$

25. horizontal
$$\frac{(x-3)^2}{9} + \frac{(y-4)^2}{4} = 1$$

27. Center: $(3, 4)$, $b = 4$, $c = 5$, horizontal
$$a^2 = b^2 + c^2 = 16 + 25 = 41$$
$$\frac{(x-3)^2}{41} + \frac{(y-4)^2}{16} = 1$$

29. Center: $(0, 4)$, $c = 4$, $a = 6$, horizontal
$$b^2 = a^2 - c^2 = 36 - 16 = 20$$
$$\frac{x^2}{36} + \frac{(y-4)^2}{20} = 1$$

31. Center: $(0, 0)$, $c = 6$, $a = 10$, horizontal
$$b^2 = a^2 - c^2 = 100 - 36 = 64$$
$$\frac{x^2}{100} + \frac{y^2}{64} = 1$$

33. $$\frac{x^2}{25} + \frac{y^2}{9} = 1$$
Center: $(0, 0)$, $a = 5$, $b = 3$, horizontal

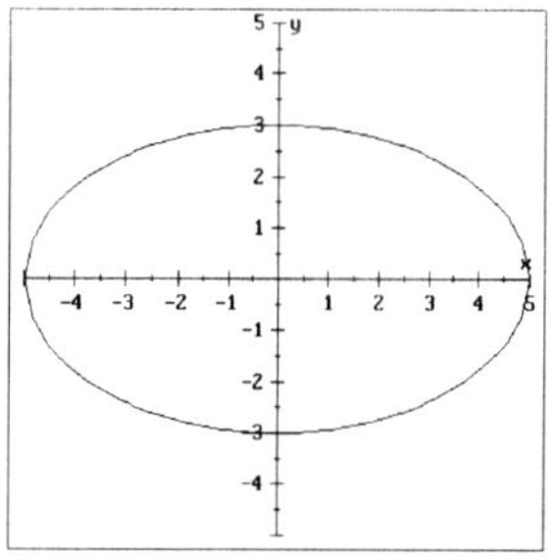

35. $\dfrac{x^2}{25}+\dfrac{y^2}{49}=1$

Center: $(0, 0)$, $a=7$, $b=5$, vertical

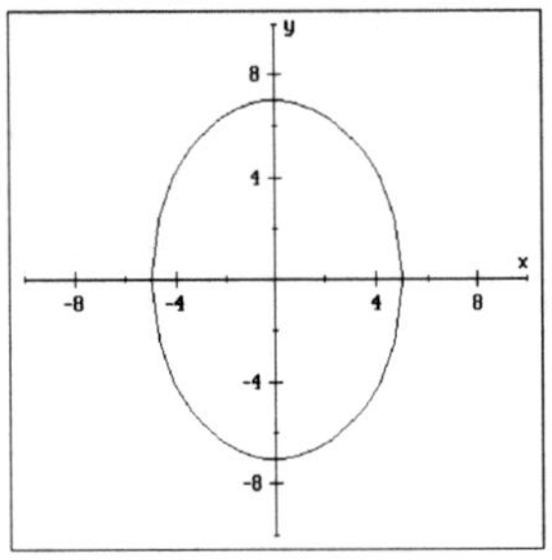

37. $\dfrac{x^2}{16}+\dfrac{(y+2)^2}{36}=1$

Center: $(0, -2)$, $a=6$, $b=4$, vertical

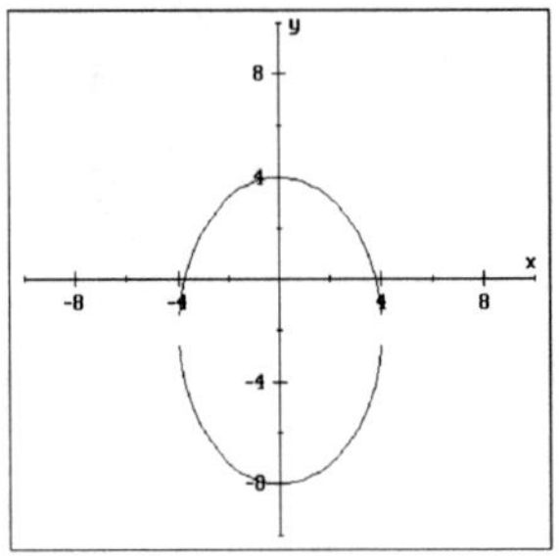

39. $\dfrac{(x-4)^2}{49}+\dfrac{(y-2)^2}{9}=1$

Center: $(4, 2)$, $a=7$, $b=3$, horizontal

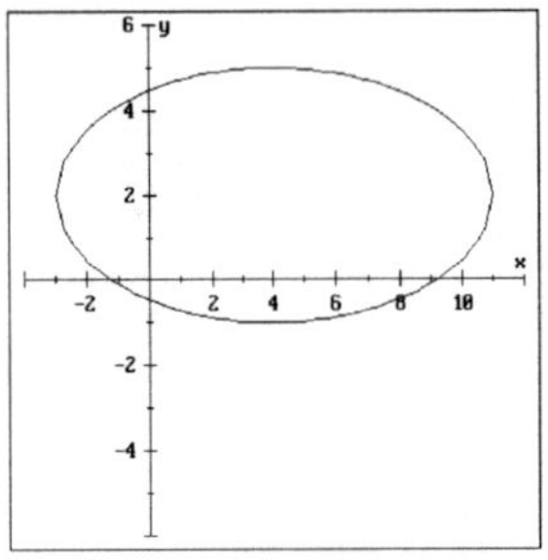

41.

$$\begin{aligned} 4x^2+y^2-2y&=15\\ 4x^2+y^2-2y+1&=15+1\\ 4x^2+(y-1)^2&=16\\ \frac{4x^2}{16}+\frac{(y-1)^2}{16}&=\frac{16}{16}\\ \frac{x^2}{4}+\frac{(y-1)^2}{16}&=1 \end{aligned}$$

43.

$$\begin{aligned} 9x^2+4y^2+18x+16y-11&=0\\ 9(x^2+2x)+4(y^2+4y)&=11\\ 9(x^2+2x+1)+4(y^2+4y+4)&=11+9+16\\ 9(x+1)^2+4(y+2)^2&=36\\ \frac{9(x+1)^2}{36}+\frac{4(y+2)^2}{36}&=\frac{36}{36}\\ \frac{(x+1)^2}{4}+\frac{(y+2)^2}{9}&=1 \end{aligned}$$

45.
$$x^2 + 4y^2 - 4x + 8y + 4 = 0$$
$$x^2 - 4x + 4(y^2 + 2y) = -4$$
$$x^2 - 4x + 4 + 4(y^2 + 2y + 1) = -4 + 4 + 4$$
$$(x-2)^2 + 4(y+1)^2 = 4$$
$$\frac{(x-2)^2}{4} + \frac{(y+1)^2}{1} = 1$$
Center: $(2, -1)$, $a = 2$, $b = 1$, horizontal

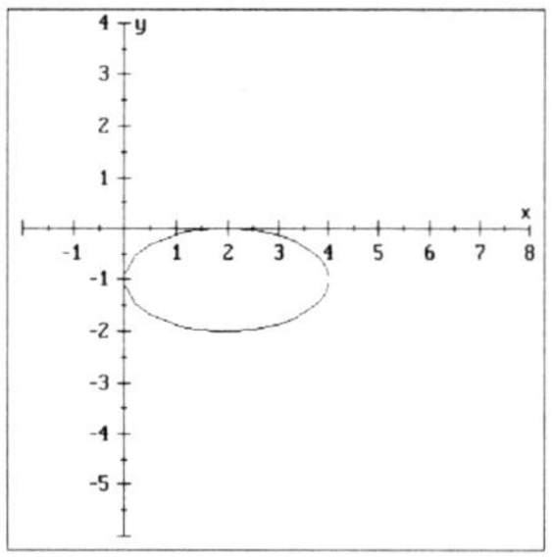

47.
$$16x^2 + 25y^2 - 160x - 200y + 400 = 0$$
$$16(x^2 - 10x) + 25(y^2 - 8y) = -400$$
$$16(x^2 - 10x + 25) + 25(y^2 - 8y + 16) = -400 + 400 + 400$$
$$16(x-5)^2 + 25(y-4)^2 = 400$$
$$\frac{(x-5)^2}{25} + \frac{(y-4)^2}{16} = 1$$
Center: $(5, 4)$, $a = 5$, $b = 4$, horizontal

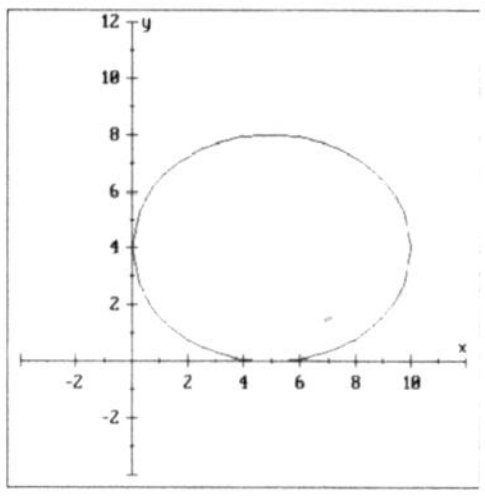

49.
$$\frac{x^2}{4} + \frac{y^2}{36} = 1$$
$$9x^2 + 4y^2 = 36$$
$$4y^2 = 36 - 9x^2$$
$$y^2 = \frac{36 - 9x^2}{4}$$
$$y = \pm\sqrt{\frac{36 - 9x^2}{4}}$$

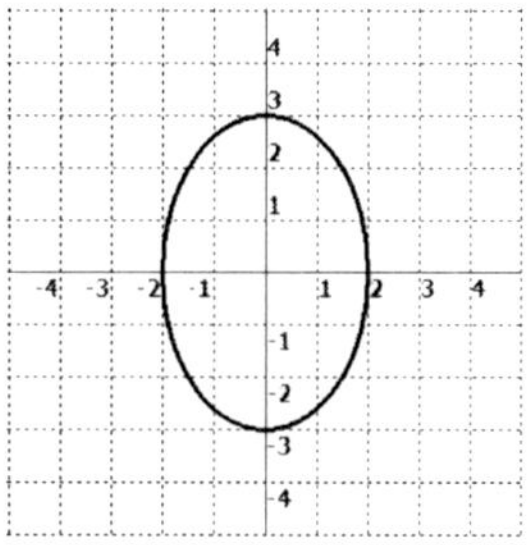

51. $C(0,0)$, $a = 30$, $b = 20$, horizontal
$$\frac{(x-0)^2}{30^2} + \frac{(y-0)^2}{20^2} = 1$$
$$\frac{x^2}{900} + \frac{y^2}{400} = 1$$

53. $a = 50$, $b = 30$

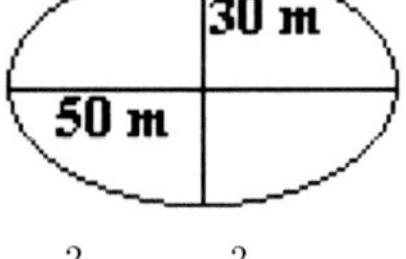

$$\frac{x^2}{2500} + \frac{y^2}{900} = 1$$

$$c^2 = a^2 - b^2$$
$$= 2500 - 900$$
$$= 1600$$
$$c = 40$$

$$\frac{40^2}{2500} + \frac{y^2}{900} = 1$$
$$\frac{y^2}{900} = \frac{900}{2500}$$
$$\frac{y}{30} = \pm\frac{30}{50}$$
$$y = \pm 18$$
The focal width is 36 meters.

55. $a = 24, b = 12$

$$\frac{x^2}{144} + \frac{y^2}{576} = 1$$
$$\frac{x^2}{144} + \frac{(-12)^2}{576} = 1$$
$$\frac{x^2}{144} = \pm\frac{432}{576}$$
$$x \approx \pm 10.4$$

The width is about 20.8 inches.

57. The farthest distance $= a + c$:

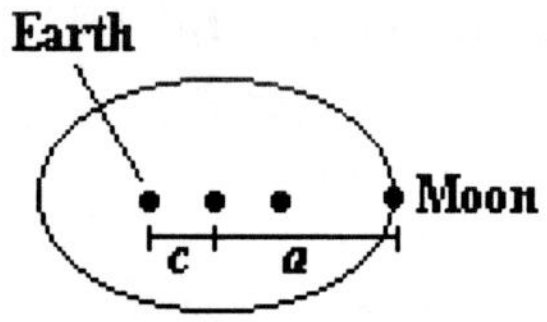

$$a = \frac{378000}{2} = 189000$$
$$\frac{c}{a} = \frac{11}{200}$$
$$c = \frac{11a}{200} = 10395$$

distance $= a + c = 199{,}395$ miles

59. $FB = \sqrt{(c-0)^2 + (0-b)^2} = \sqrt{c^2 + b^2} = \sqrt{a^2} = |a| = a\ (a \geq 0)$

61. The equation of the ellipse is $\frac{x^2}{a^2} + \frac{y^2}{b^2} = 1$.

Let $x = c$ and solve for y^2:

$$\frac{x^2}{a^2} + \frac{y^2}{b^2} = 1$$
$$\frac{c^2}{a^2} + \frac{y^2}{b^2} = 1$$
$$y^2 = b^2\left(1 - \frac{c^2}{a^2}\right)$$
$$= b^2\left(1 - \frac{a^2 - b^2}{a^2}\right)$$
$$= b^2\left(1 - 1 + \frac{b^2}{a^2}\right) = \frac{b^4}{a^2}$$

Thus, $y = \pm\sqrt{\frac{b^4}{a^2}} = \pm\frac{b^2}{a}$.

The coordinates of A' and A are $\left(c, \frac{b^2}{a}\right)$ and $\left(c, -\frac{b^2}{a}\right)$. Therefore, the focal width is $\frac{2b^2}{a}$.

63. Let the origin be at the midpoint of the line segment between the two thumbtacks and let the x-axis be parallel to that segment. Then $2a = 6$, so $a = 3$. Also, $2c = 2$, so $c = 1$. Find b:

$b^2 = a^2 - c^2 = 3^2 - 1^2 = 8$. The equation is $\frac{x^2}{a^2} + \frac{y^2}{b^2} = 1$, or $\frac{x^2}{9} + \frac{y^2}{8} = 1$.

65. Answers may vary.

67. **Answers may vary.**

69. $$B + C = \begin{bmatrix} 1 & 2 \\ 2 & 0 \\ -1 & 1 \end{bmatrix} + \begin{bmatrix} 0 & 1 \\ -1 & 1 \\ -2 & 0 \end{bmatrix} = \begin{bmatrix} 1 & 3 \\ 1 & 1 \\ -3 & 1 \end{bmatrix}$$

71. $$AC + B = \begin{bmatrix} 3 & -1 & 2 \\ 0 & 2 & -1 \\ 3 & 1 & 1 \end{bmatrix}\begin{bmatrix} 0 & 1 \\ -1 & 1 \\ -2 & 0 \end{bmatrix} + \begin{bmatrix} 1 & 2 \\ 2 & 0 \\ -1 & 1 \end{bmatrix} = \begin{bmatrix} -3 & 2 \\ 0 & 2 \\ -3 & 4 \end{bmatrix} + \begin{bmatrix} 1 & 2 \\ 2 & 0 \\ -1 & 1 \end{bmatrix} = \begin{bmatrix} -2 & 4 \\ 2 & 2 \\ -4 & 5 \end{bmatrix}$$

Exercises 7.3 (page 698)

1. difference, constant

3. $(a, 0), (-a, 0)$

5. transverse axis

7. Both variables squared with equal coefficients and same sign: circle

9. One variable squared: parabola

11. Both variables squared with unequal coefficients and same sign: ellipse

13. Both variables squared with opposite signs: hyperbola

15. $a = 5, c = 7$; horizontal

$$\begin{aligned} b^2 &= c^2 - a^2 \\ &= 49 - 25 = 24 \end{aligned}$$

$$\frac{x^2}{25} - \frac{y^2}{24} = 1$$

17. $a = 2, b = 3$; horizontal

$$\frac{(x-2)^2}{4} - \frac{(y-4)^2}{9} = 1$$

19. $a = 3$; vertical

$$\frac{(y-3)^2}{9} - \frac{(x-5)^2}{b^2} = 1$$

$$\frac{(8-3)^2}{9} - \frac{(1-5)^2}{b^2} = 1$$

$$\frac{25}{9} - \frac{16}{b^2} = 1$$

$$\frac{25}{9} - \frac{16}{b^2} = 1$$

$$-\frac{16}{b^2} = -\frac{16}{9}$$

$$b^2 = 9$$

$$\frac{(y-3)^2}{9} - \frac{(x-5)^2}{9} = 1$$

21. Center: $(0, 0)$, $a = 3$, $c = 5$; vertical

$$\begin{aligned} b^2 &= c^2 - a^2 \\ &= 25 - 9 = 16 \end{aligned}$$

$$\frac{y^2}{9} - \frac{x^2}{16} = 1$$

23. $c = 6, a = 2$; horizontal

$$\begin{aligned} b^2 &= c^2 - a^2 \\ &= 36 - 4 = 32 \end{aligned}$$

$$\frac{(x-1)^2}{4} - \frac{(y-4)^2}{32} = 1$$

25.

$$\frac{x^2}{a^2} - \frac{y^2}{b^2} = 1$$

$$\frac{4^2}{a^2} - \frac{2^2}{b^2} = 1$$

$$\frac{16}{a^2} - \frac{4}{b^2} = 1$$

$$\frac{16}{a^2} = 1 + \frac{4}{b^2}$$

$$\frac{64}{a^2} = 4 + \frac{16}{b^2}$$

$$\frac{x^2}{a^2} - \frac{y^2}{b^2} = 1$$

$$\frac{8^2}{a^2} - \frac{(-6)^2}{b^2} = 1$$

$$\frac{64}{a^2} - \frac{36}{b^2} = 1$$

$$\frac{64}{a^2} - \frac{36}{b^2} = 1$$

$$4 + \frac{16}{b^2} - \frac{36}{b^2} = 1$$

$$3 = \frac{20}{b^2}$$

$$b^2 = \frac{20}{3}$$

$$a^2 = 10$$

$$\frac{x^2}{10} - \frac{3y^2}{20} = 1$$

SECTION 7.3

27. $$4(x-1)^2 - 9(y+2)^2 = 36$$
$$\frac{4(x-1)^2}{36} - \frac{9(y+2)^2}{36} = \frac{36}{36}$$
$$\frac{(x-1)^2}{9} - \frac{(y+2)^2}{4} = 1$$
$a = 3, b = 2$

Area $= (2a)(2b) = (6)(4) = 24$ sq. units

29. $$x^2 + 6x - y^2 + 2y = -11$$
$$x^2 + 6x - \left(y^2 - 2y\right) = -11$$
$$x^2 + 6x + 9 - \left(y^2 - 2y + 1\right) = -11 + 9 - 1$$
$$(x+3)^2 - (y-1)^2 = -3$$
$$\frac{(x+3)^2}{-3} - \frac{(y-1)^2}{-3} = 1$$
$$\frac{(y-1)^2}{3} - \frac{(x+3)^2}{3} = 1 \Rightarrow a = \sqrt{3}, b = \sqrt{3};$$
Area $= (2a)(2b) = \left(2\sqrt{3}\right)\left(2\sqrt{3}\right) = 12$ sq. units

31. $$(2a)(2b) = 36$$
$$4(2b) = 36$$
$$b = \tfrac{9}{2}$$
$$\frac{(x+2)^2}{4} - \frac{4(y+4)^2}{81} = 1$$
OR
$$\frac{(y+4)^2}{4} - \frac{4(x+2)^2}{81} = 1$$

33. Center: $(0,0)$, $a = 6$, $b = \frac{5}{4}$
$$\frac{x^2}{6^2} - \frac{y^2}{\left(\frac{5}{4}\right)^2} = 1$$
$$\frac{x^2}{36} - \frac{16y^2}{25} = 1$$

35. $$\frac{x^2}{9} - \frac{y^2}{4} = 1$$
Center: $(0,0)$, $a = 3$, $b = 2$, horizontal

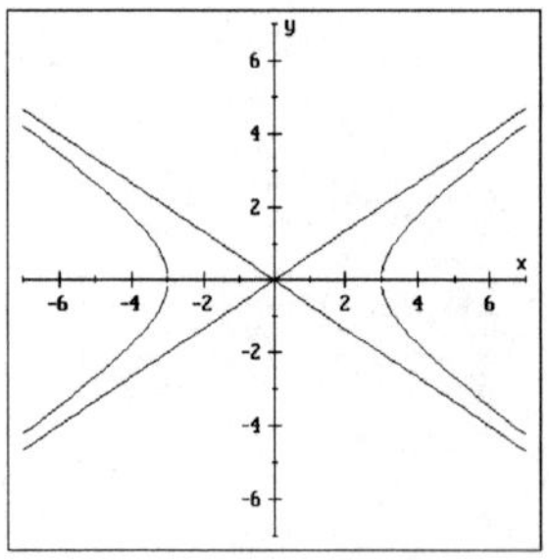

37. $$4x^2 - 3y^2 = 36$$
$$\frac{4x^2}{36} - \frac{3y^2}{36} = \frac{36}{36}$$
$$\frac{x^2}{9} - \frac{y^2}{12} = 1$$
Center: $(0,0)$, $a = 3$, $b = 2\sqrt{3}$, horizontal

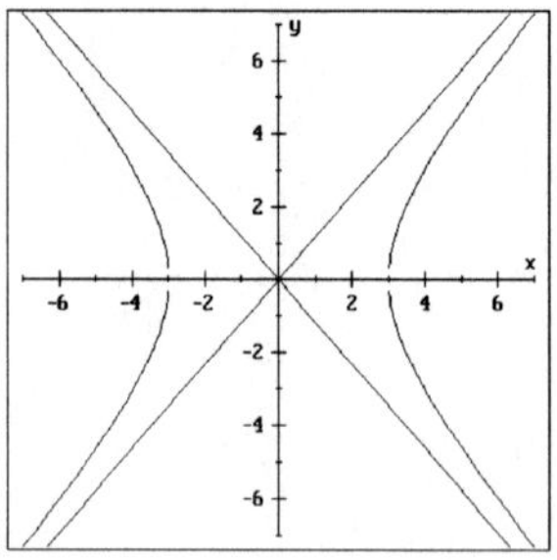

39. $y^2 - x^2 = 1$

$$\frac{y^2}{1} - \frac{x^2}{1} = 1$$

Center: $(0, 0)$, $a = 1$, $b = 1$, vertical

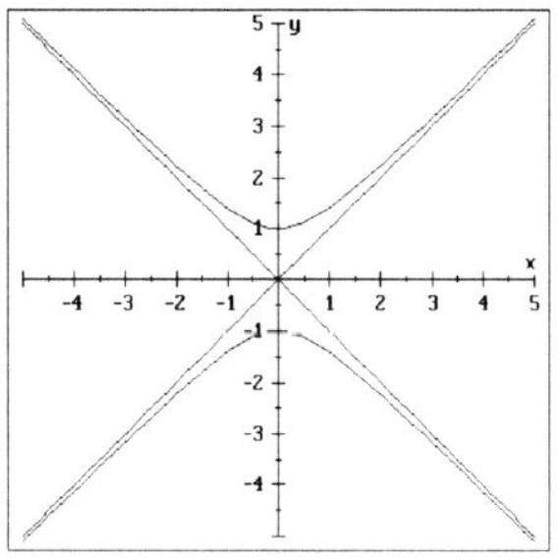

41. $$\frac{(x+2)^2}{9} - \frac{y^2}{4} = 1$$

Center: $(-2, 0)$, $a = 3$, $b = 2$, horizontal

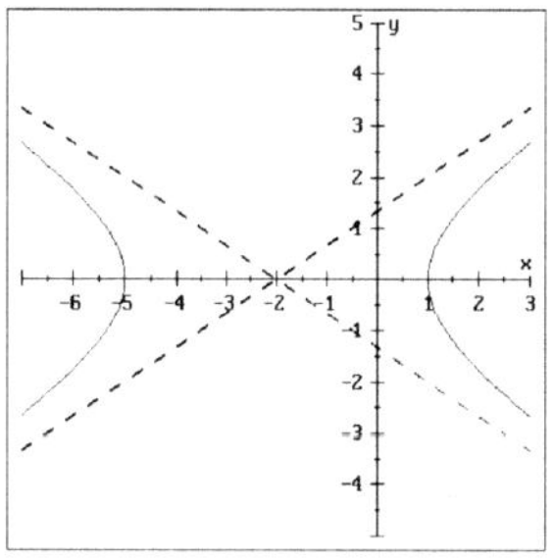

43. $4(y-2)^2 - 9(x+1)^2 = 36$

$$\frac{4(y-2)^2}{36} - \frac{9(x+1)^2}{36} = \frac{36}{36}$$

$$\frac{(y-2)^2}{9} - \frac{(x+1)^2}{4} = 1$$

Center: $(-1, 2)$, $a = 3$, $b = 2$, vertical

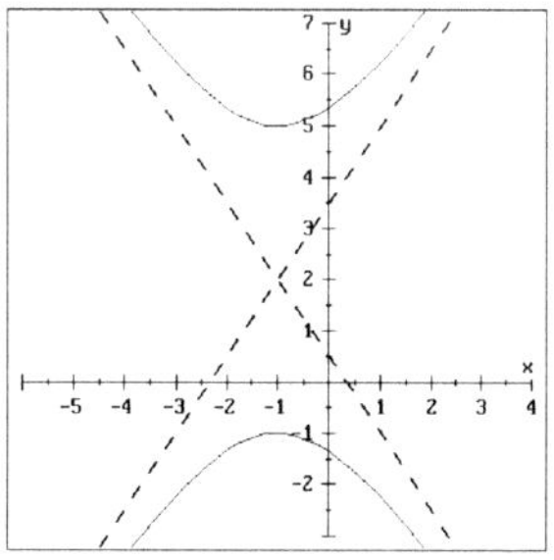

45.

$$4x^2 - 2y^2 + 8x - 8y = 8$$
$$4(x^2 + 2x) - 2(y^2 + 4y) = 8$$
$$4(x^2 + 2x + 1) - 2(y^2 + 4y + 4) = 8 + 4 - 8$$
$$4(x+1)^2 - 2(y+2)^2 = 4$$
$$\frac{4(x+1)^2}{4} - \frac{2(y+2)^2}{4} = \frac{4}{4}$$
$$\frac{(x+1)^2}{1} - \frac{(y+2)^2}{2} = 1$$

Center: $(-1, -2)$, $a = 1$, $b = \sqrt{2}$, horizontal

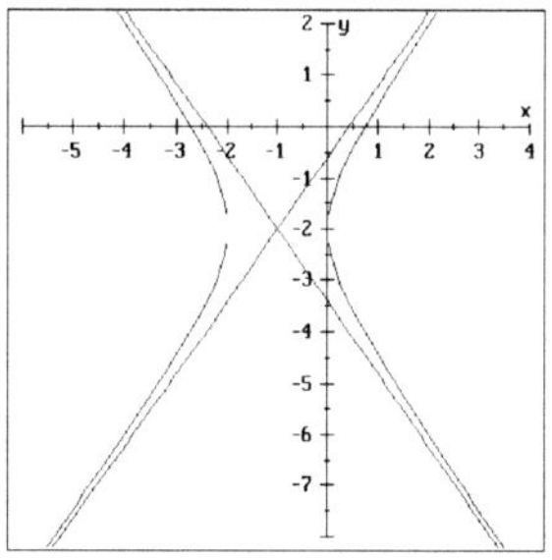

47.

$$y^2 - 4x^2 + 6y + 32x = 59$$
$$y^2 + 6y - 4(x^2 - 8x) = 59$$
$$y^2 + 6y + 9 - 4(x^2 - 8x + 16) = 59 + 9 - 64$$
$$(y+3)^2 - 4(x-4)^2 = 4$$
$$\frac{(y+3)^2}{4} - \frac{(x-4)^2}{1} = 1$$

Center: $(4, -3)$, $a = 2$, $b = 1$, vertical

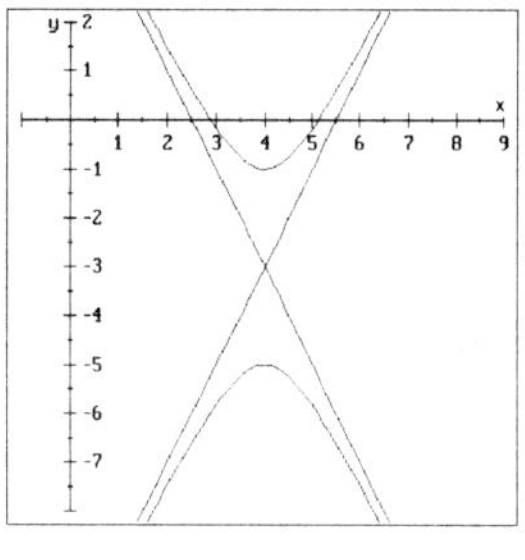

SECTION 7.3

49. $-xy = 6$

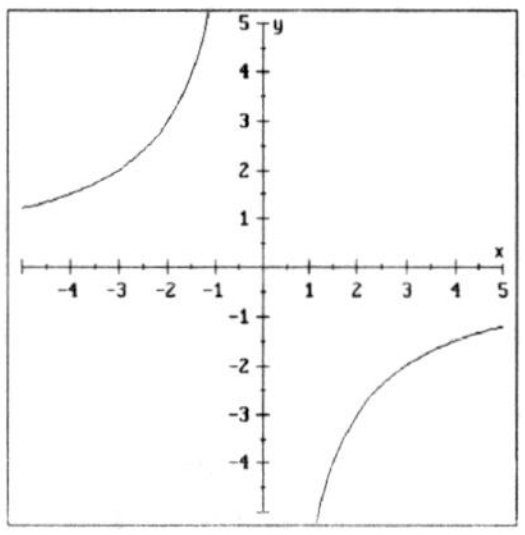

51. Foci: $(-2, 1), (8, 1)$

Center: $(3, 1), c = 5$

$2a = 6 \Rightarrow a = 3, b = 4$

$$\frac{(x-3)^2}{9} - \frac{(y-1)^2}{16} = 1$$

53. The distance between the point (x, y) and the line $x = -2$ is the difference between the y-coordinates, or $y - (-2) = y + 2$.

$$\sqrt{(x-0)^2 + (y-3)^2} = \frac{3}{2}(y+2)$$

$$x^2 + (y-3)^2 = \frac{9}{4}(y+2)^2$$

$$4x^2 + 4(y-3)^2 = 9(y+2)^2$$

$$4x^2 + 4(y^2 - 6y + 9) = 9(y^2 + 4y + 4)$$

$$4x^2 - 5y^2 - 60y = 0$$

55. $x^2 - \frac{y^2}{4} = 1$

$$-\frac{y^2}{4} = -x^2 + 1$$

$$y^2 = 4x^2 - 4$$

$$y = \pm\sqrt{4x^2 - 4}$$

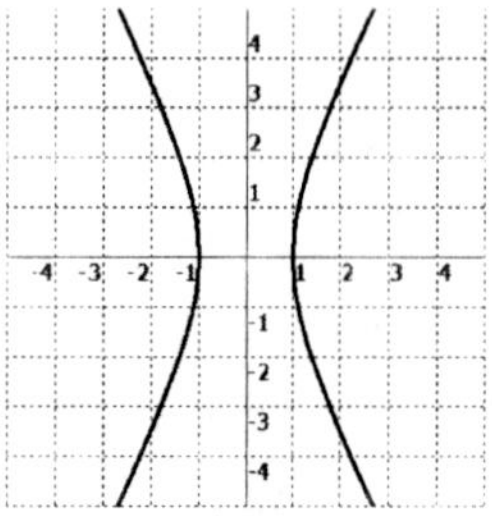

57.

$$xy = k$$

$$(12)(2) = k$$

$$24 = k$$

59. $9y^2 - x^2 = 81$

$$\frac{9y^2}{81} - \frac{x^2}{81} = \frac{81}{81}$$

$$\frac{y^2}{9} - \frac{x^2}{81} = 1$$

$a = 3 \Rightarrow 3$ units

61. $2a = 24 \Rightarrow a = 12$

$c = 13 \Rightarrow b = 5$

$$\frac{x^2}{144} - \frac{y^2}{25} = 1$$

63. $2a = 12 \Rightarrow a = 6$

$c = 10 \Rightarrow b = 8$

$$\frac{x^2}{36} - \frac{y^2}{64} = 1$$

65. Answers may vary.

67. Answers may vary.

69.
$$\begin{aligned} f(x) &= 3x - 2 \\ y &= 3x - 2 \\ x &= 3y - 2 \\ x + 2 &= 3y \\ \frac{x+2}{3} &= y \Rightarrow f^{-1}(x) = \frac{x+2}{3} \end{aligned}$$

71.
$$\begin{aligned} f(x) &= \frac{5x}{x+2} \\ y &= \frac{5x}{x+2} \\ x &= \frac{5y}{y+2} \\ x(y+2) &= 5y \\ xy + 2x &= 5y \\ 2x &= 5y - xy \\ 2x &= y(5 - x) \\ \frac{2x}{5-x} &= y \Rightarrow f^{-1}(x) = \frac{2x}{5-x} \end{aligned}$$

73. $f(g(x)) = f\left((x+1)^2\right) = \left((x+1)^2\right)^2 + 1 = (x+1)^4 + 1$

75. $f(f(x)) = f(x^2 + 1) = (x^2 + 1)^2 + 1$

Exercises 7.4 (page 707)

1. graphs

3. $\begin{cases} 8x^2 + 32y^2 = 256 \\ x = 2y \end{cases}$

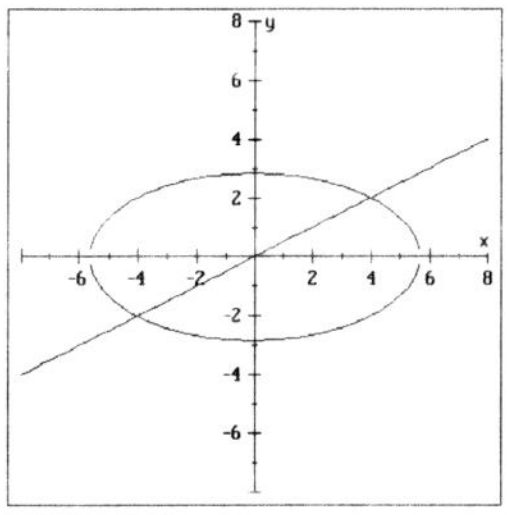

$(4, 2), (-4, -2)$

5. $\begin{cases} x^2 + y^2 = 90 \\ y = x^2 \end{cases}$

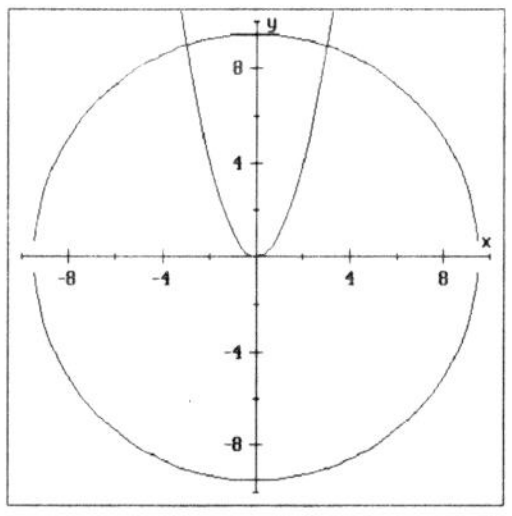

$(3, 9), (-3, 9)$

7. $\begin{cases} x^2 + y^2 = 25 \\ 12x^2 + 64y^2 = 768 \end{cases}$

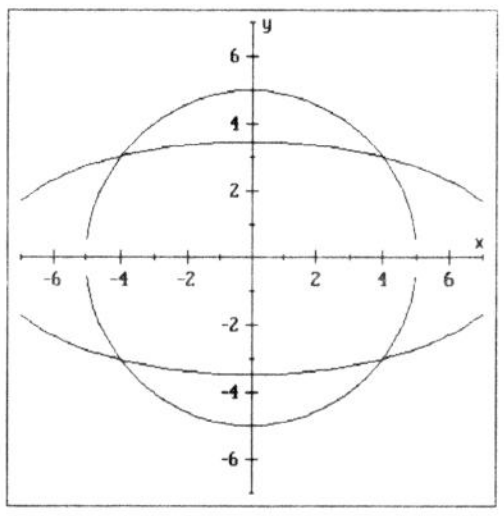

$(-4, 3), (4, 3)$
$(-4, -3), (4, -3)$

9. $\begin{cases} x^2 - 13 = -y^2 \\ y = 2x - 4 \end{cases}$

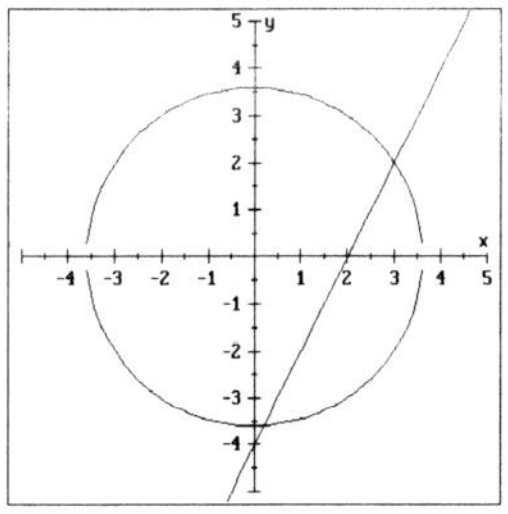

$\left(\frac{1}{5}, -\frac{18}{5}\right), (3, 2)$

11. $\begin{cases} x^2 - 6x - y = -5 \\ x^2 - 6x + y = -5 \end{cases}$

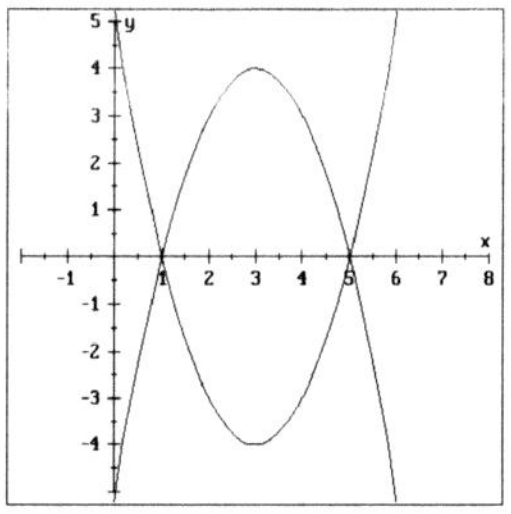

$(1, 0), (5, 0)$

13. $\begin{cases} y = x + 1 \\ y = x^2 + x \end{cases}$

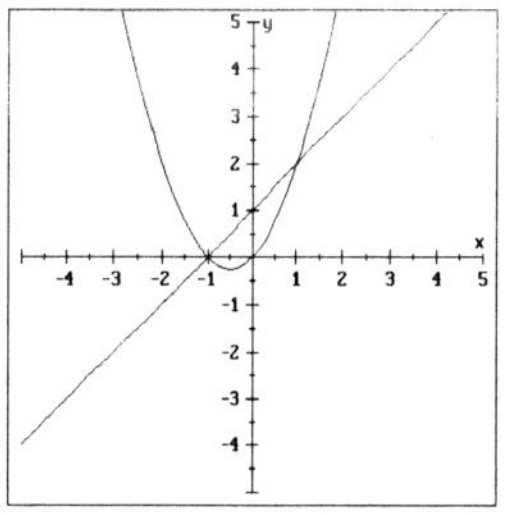

$(1, 2), (-1, 0)$

SECTION 7.4

15. $\begin{cases} 6x^2 + 9y^2 = 10 \Rightarrow y = \pm\sqrt{\frac{10-6x^2}{9}} \\ 3y - 2x = 0 \quad\Rightarrow y = \frac{2}{3}x \end{cases}$

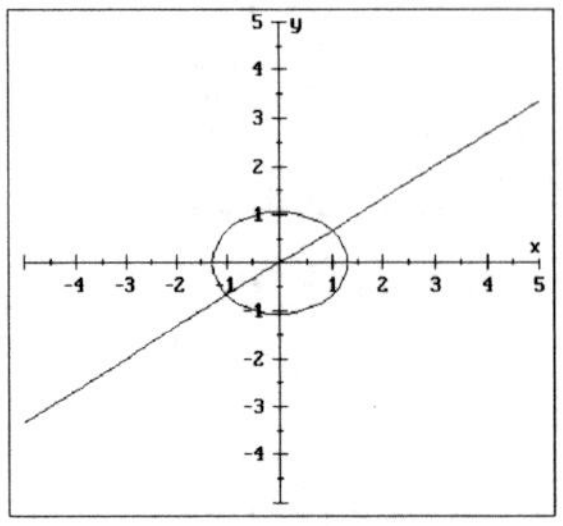

(1, 0.67), (−1, −0.67)

17.

$$5x + 3y = 15 \Rightarrow y = \frac{15 - 5x}{3}$$
$$25x^2 + 9y^2 = 225$$
$$25x^2 + 9\left(\frac{15 - 5x}{3}\right)^2 = 225$$
$$25x^2 + (15 - 5x)^2 = 225$$
$$25x^2 + 225 - 150x + 25x^2 = 225$$
$$50x^2 - 150x = 0$$
$$50x(x - 3) = 0$$

$x = 0$	$x = 3$
$y = \frac{15 - 5(0)}{3} = 5$	$y = \frac{15 - 5(3)}{3} = 0$
(0, 5)	(3, 0)

19.

$$x + y = 2 \Rightarrow y = 2 - x$$
$$x^2 + y^2 = 2$$
$$x^2 + (2 - x)^2 = 2$$
$$x^2 + 4 - 4x + x^2 = 2$$
$$2x^2 - 4x + 2 = 0$$
$$2(x - 1)(x - 1) = 0$$

$x = 1$
$y = 2 - 1 = 1$
(1, 1)

21.

$$x + y = 3 \Rightarrow y = 3 - x$$
$$x^2 + y^2 = 5$$
$$x^2 + (3 - x)^2 = 5$$
$$x^2 + 9 - 6x + x^2 = 5$$
$$2x^2 - 6x + 4 = 0$$
$$2(x - 1)(x - 2) = 0$$

$x = 1$	$x = 2$
$y = 3 - 1 = 2$	$y = 3 - 2 = 1$
(1, 2)	(2, 1)

23.

$$y = x^2 - 1 \Rightarrow x^2 = y + 1$$
$$x^2 + y^2 = 13$$
$$y + 1 + y^2 = 13$$
$$y^2 + y - 12 = 0$$
$$(y - 3)(y + 4) = 0$$

$y = 3$	$y = -4$
$3 + 1 = x^2 \Rightarrow x = \pm 2$	$-4 + 1 = x^2$
(2, 3), (−2, 3)	no real solutions

25.

$$y = x^2 \Rightarrow x^2 = y$$
$$x^2 + y^2 = 30$$
$$y + y^2 = 30$$
$$y^2 + y - 30 = 0$$
$$(y - 5)(y + 6) = 0$$

$y = 5$	$y = -6$
$5 = x^2 \Rightarrow x = \pm\sqrt{5}$	$-6 = x^2$
$\left(\sqrt{5}, 5\right), \left(-\sqrt{5}, 5\right)$	no real solutions

27.

$$\begin{aligned} 2x^2 + y^2 &= 6 \\ x^2 - y^2 &= 3 \\ \hline 3x^2 &= 9 \\ x^2 &= 3 \\ x &= \pm\sqrt{3} \end{aligned} \qquad \begin{aligned} 2x^2 + y^2 &= 6 \\ 2(3) + y^2 &= 6 \\ y^2 &= 0 \\ y &= 0 \end{aligned}$$

$\left(\sqrt{3}, 0\right), \left(-\sqrt{3}, 0\right)$

SECTION 7.4

29.
$$\begin{array}{rcr} x^2+y^2 &=& 20 \\ x^2-y^2 &=& -12 \\ \hline 2x^2 &=& 8 \\ x^2 &=& 4 \\ x &=& \pm 2 \end{array} \qquad \begin{aligned} x^2+y^2 &= 20 \\ 4+y^2 &= 20 \\ y^2 &= 16 \\ y &= \pm 4 \end{aligned}$$

$(2, 4), (-2, 4), (2, -4), (-2, -4)$

31.
$$\begin{aligned} y = x^2 - 10 \Rightarrow x^2 &= y + 10 \\ y^2 &= 40 - x^2 \\ y^2 &= 40 - y - 10 \\ y^2 + y - 30 &= 0 \\ (y-5)(y+6) &= 0 \end{aligned}$$

$y = 5$: $x^2 = 5 + 10 \Rightarrow x = \pm\sqrt{15}$; $\left(\sqrt{15}, 5\right), \left(-\sqrt{15}, 5\right)$

$y = -6$: $x^2 = -6 + 10 \Rightarrow x = \pm 2$; $(2, -6), (-2, -6)$

33.
$$\begin{aligned} y = x^2 - 4 \Rightarrow x^2 &= y + 4 \\ x^2 - y^2 &= -16 \\ y + 4 - y^2 &= -16 \\ y^2 - y - 20 &= 0 \\ (y-5)(y+4) &= 0 \end{aligned}$$

$y = 5$: $x^2 = 5 + 4 \Rightarrow x = \pm 3$; $(3, 5), (-3, 5)$

$y = -4$: $x^2 = -4 + 4 \Rightarrow x = 0$; $(0, -4)$

35.
$$\begin{array}{rcr} x^2 - y^2 &=& -5 \\ 3x^2 + 2y^2 &=& 30 \\ \hline 2x^2 - 2y^2 &=& -10 \\ 3x^2 + 2y^2 &=& 30 \\ \hline 5x^2 &=& 20 \\ x^2 &=& 4 \\ x &=& \pm 2 \end{array} \qquad \begin{aligned} 3x^2 + 2y^2 &= 30 \\ 3(4) + 2y^2 &= 30 \\ 2y^2 &= 18 \\ y^2 &= 9 \\ y &= \pm 3 \end{aligned}$$

$(2, 3), (-2, 3), (2, -3), (-2, -3)$

37.
$$\begin{array}{l} \frac{1}{x} + \frac{2}{y} = 1 \\ \frac{2}{x} - \frac{1}{y} = \frac{1}{3} \\ \hline \end{array} \qquad \begin{array}{rcl} \frac{1}{x} + \frac{2}{y} &=& 1 \\ \frac{4}{x} - \frac{2}{y} &=& \frac{2}{3} \\ \hline \frac{5}{x} &=& \frac{5}{3} \\ x &=& 3 \end{array} \qquad \begin{aligned} \frac{1}{x} + \frac{2}{y} &= 1 \\ \frac{1}{3} + \frac{2}{y} &= 1 \\ \frac{2}{y} &= \frac{2}{3} \\ y &= 3 \end{aligned}$$

$(3, 3)$

39.
$$\begin{aligned} 3y^2 &= xy \\ 3y^2 - xy &= 0 \\ y(3y - x) &= 0 \\ y = 0 &\text{ or } x = 3y \end{aligned}$$

$y = 0$:
$$\begin{aligned} 2x^2 + xy - 84 &= 0 \\ 2x^2 + x(0) - 84 &= 0 \\ 2x^2 &= 84 \\ x^2 &= 42 \\ x &= \pm\sqrt{42} \end{aligned}$$
$\left(\sqrt{42}, 0\right), \left(-\sqrt{42}, 0\right)$

$x = 3y$:
$$\begin{aligned} 2x^2 + xy - 84 &= 0 \\ 2(3y)^2 + (3y)y - 84 &= 0 \\ 18y^2 + 3y^2 - 84 &= 0 \\ 21y^2 &= 84 \\ y^2 = 4 \Rightarrow y &= \pm 2 \end{aligned}$$
$(6, 2), (-6, -2)$

41.
$$xy = \tfrac{1}{6} \Rightarrow y = \tfrac{1}{6x}$$
$$y + x = 5xy$$
$$\frac{1}{6x} + x = \frac{5x}{6x}$$
$$1 + 6x^2 = 5x$$
$$6x^2 - 5x + 1 = 0$$
$$6x^2 - 5x + 1 = 0$$
$$(2x-1)(3x-1) = 0$$

$x = \frac{1}{2}$	$x = \frac{1}{3}$
$y = \frac{1}{6(1/2)} = \frac{1}{3}$	$y = \frac{1}{6(1/3)} = \frac{1}{2}$
$\left(\frac{1}{2}, \frac{1}{3}\right)$	$\left(\frac{1}{3}, \frac{1}{2}\right)$

43. Let $x =$ width and $y =$ length.
$$\begin{cases} xy = 63 \\ 2x + 2y = 32 \end{cases}$$
$$xy = 63 \Rightarrow y = \tfrac{63}{x}$$
$$2x + 2y = 32$$
$$2x + 2\left(\frac{63}{x}\right) = 32$$
$$2x^2 + 126 = 32x$$
$$2x^2 - 32x + 126 = 0$$
$$2(x-9)(x-7) = 0$$

$x = 9$	$x = 7$
$y = \frac{63}{9} = 7$	$y = \frac{63}{7} = 9$

The dimensions are 9 cm by 7 cm.

45. Let $x =$ width and $y =$ length.
$$\begin{cases} xy = 8000 \\ 2x + y = 260 \end{cases}$$
$$xy = 8000 \Rightarrow y = \tfrac{8000}{x}$$
$$2x + y = 260$$
$$2x + \left(\frac{8000}{x}\right) = 260$$
$$2x^2 + 8000 = 260x$$
$$2x^2 - 260x + 8000 = 0$$
$$2(x-50)(x-80) = 0$$

$x = 50$	$x = 80$
$y = \frac{8000}{50} = 160$	$y = \frac{8000}{80} = 100$

The dimensions are 50 ft by 160 ft or 80 ft by 100 ft.

47. Let $x =$ Carol's principal.

	I	P	r
Carol	67.50	x	$\frac{67.50}{x}$
John	94.50	$x + 150$	$\frac{94.50}{x+150}$

$$\boxed{\text{John's rate}} = \boxed{\text{Carol's rate}} + 0.015$$
$$\frac{94.50}{x+150} = \frac{67.50}{x} + 0.015$$
$$94.5x = 67.5(x + 150) + 0.015x(x + 150)$$
$$0.015x^2 - 24.75x + 10{,}125 = 0$$
$$x^2 - 1650x + 675{,}000 = 0$$
$$(x - 750)(x - 900) = 0$$
$$x - 750 = 0 \quad \textbf{or} \quad x - 900 = 0$$
$$x = 750 \qquad\qquad x = 900 \Rightarrow$$

Carol invested either \$750 at 9% or she invested \$900 at 7.5% interest.

49.
$$\begin{cases} y = \frac{1}{10}x \\ y = -\frac{1}{300}x^2 + \frac{1}{5}x \end{cases}$$
$$\tfrac{1}{10}x = -\tfrac{1}{300}x^2 + \tfrac{1}{5}x$$
$$30x = -x^2 + 60x$$
$$x^2 - 30x = 0$$
$$x(x - 30) = 0$$
$$x = 0 \quad \textbf{or} \quad x - 30 = 0$$
$$x = 30$$
$$x = 30 \Rightarrow y = \tfrac{1}{10}(30) = 3 \quad \boxed{(30, 3)}$$

51. $\begin{cases} y = x^2 \\ x + y = 2 \end{cases}$

$$x + y = 2$$
$$x + x^2 = 2$$
$$x^2 + x - 2 = 0$$
$$(x + 2)(x - 1) = 0$$
$$x + 2 = 0 \quad \textbf{or} \quad x - 1 = 0$$
$$x = -2 \qquad\qquad x = 1$$

$x = -2 \Rightarrow y = (-2)^2 = 4$

$x = 1 \Rightarrow y = 1^2 = 1$

There are potential collision points at $(-2, 4)$ and $(1, 1)$.

53. $\begin{cases} x = 2y \\ (x - 120)^2 + y^2 = 100^2 \end{cases}$

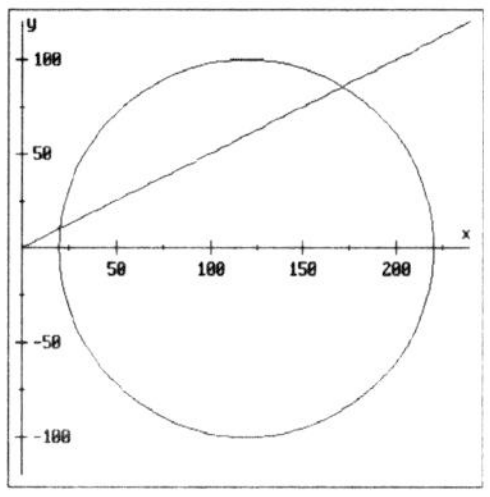

The x-coordinate of the point where the line crosses the circle closest to the origin has the approximate coordinates (20.5, 10.25). The distance to the origin is about 23 miles.

55-59. Answers may vary.

61. $y = \dfrac{3x + 1}{x - 1}$

Vertical: $x = 1$; Horizontal: $y = \frac{3}{1} = 3$

Slant: none

63. $y = \dfrac{3x + 1}{x^2 - 1} = \dfrac{3x + 1}{(x - 1)(x + 1)}$

Vertical: $x = 1, x = -1$

Horizontal: $y = 0$; Slant: none

65. $f(x) = \dfrac{3x^2 + 5}{5x^2 + 3}$; Replace x with $-x$:

Identical $\Rightarrow$ symmetric about y-axis

67. $f(x) = \dfrac{3x^3}{5x^2 + 3}$; Replace x with $-x$:

Opposite $\Rightarrow$ symmetric about origin

Chapter 7 Review (page 711)

1.
$$(x - h)^2 + (y - k)^2 = r^2$$
$$(x - 0)^2 + (y - 0)^2 = 4^2$$
$$x^2 + y^2 = 16$$

2.
$$r = \sqrt{(6 - 0)^2 + (8 - 0)^2} = 10$$
$$(x - h)^2 + (y - k)^2 = r^2$$
$$(x - 0)^2 + (y - 0)^2 = 10^2$$
$$x^2 + y^2 = 100$$

3.
$$(x - h)^2 + (y - k)^2 = r^2$$
$$(x - 3)^2 + (y - (-2))^2 = 5^2$$
$$(x - 3)^2 + (y + 2)^2 = 25$$

4.
$$r = \sqrt{(-2 - 1)^2 + (4 - 0)^2} = 5$$
$$(x - h)^2 + (y - k)^2 = r^2$$
$$(x - (-2))^2 + (y - 4)^2 = 5^2$$
$$(x + 2)^2 + (y - 4)^2 = 25$$

CHAPTER 7 REVIEW

5.

$$C\left(\frac{-2+12}{2}, \frac{4+16}{2}\right) = C(5, 10)$$

$$r = \sqrt{(12-5)^2 + (16-10)^2} = \sqrt{85}$$

$$(x-h)^2 + (y-k)^2 = r^2$$

$$(x-5)^2 + (y-10)^2 = \left(\sqrt{85}\right)^2$$

$$(x-5)^2 + (y-10)^2 = 85$$

6.

$$C\left(\frac{-3+7}{2}, \frac{-6+10}{2}\right) = C(2, 2)$$

$$r = \sqrt{(7-2)^2 + (10-2)^2} = \sqrt{89}$$

$$(x-h)^2 + (y-k)^2 = r^2$$

$$(x-2)^2 + (y-2)^2 = \left(\sqrt{89}\right)^2$$

$$(x-2)^2 + (y-2)^2 = 89$$

7.

$$x^2 + y^2 - 6x + 4y = 3$$

$$x^2 - 6x + y^2 + 4y = 3$$

$$x^2 - 6x + 9 + y^2 + 4y + 4 = 3 + 9 + 4$$

$$(x-3)^2 + (y+2)^2 = 16$$

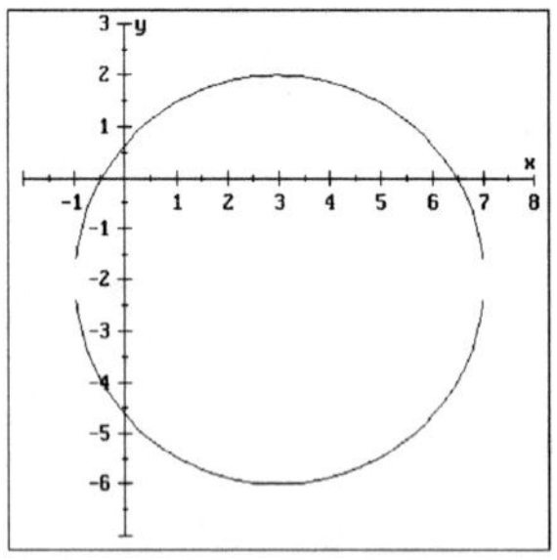

8.

$$x^2 + 4x + y^2 - 10y = -13$$

$$x^2 + 4x + 4 + y^2 - 10y + 25 = -13 + 4 + 25$$

$$(x+2)^2 + (y-5)^2 = 16$$

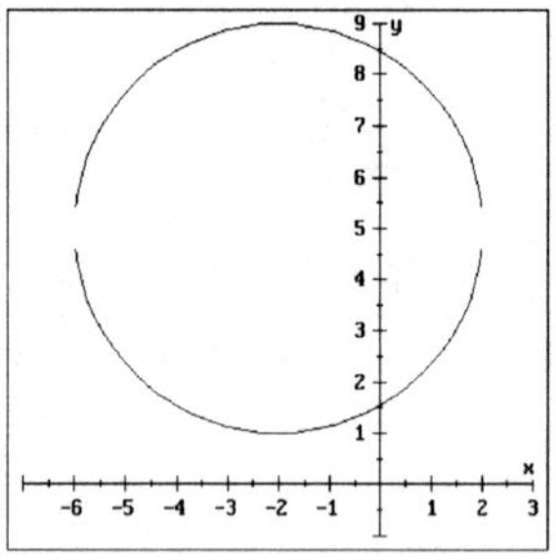

9. Horizontal

$$(y-0)^2 = 4p(x-0)$$

$$(4-0)^2 = 4p(-8-0)$$

$$16 = -32p$$

$$-2 = 4p$$

$$y^2 = -2x$$

10. Vertical

$$(x-0)^2 = 4p(y-0)$$

$$(-8-0)^2 = 4p(4-0)$$

$$64 = 16p$$

$$16 = 4p$$

$$x^2 = 16y$$

11. Vertical

$$(x+2)^2 = 4p(y-3)$$

$$(-4+2)^2 = 4p(-8-3)$$

$$4 = 4p(-11)$$

$$-\frac{4}{11} = 4p$$

$$(x+2)^2 = -\frac{4}{11}(y-3)$$

12.
$$x^2 - 4y - 2x + 9 = 0$$
$$x^2 - 2x = 4y - 9$$
$$x^2 - 2x + 1 = 4y - 9 + 1$$
$$(x-1)^2 = 4(y-2)$$

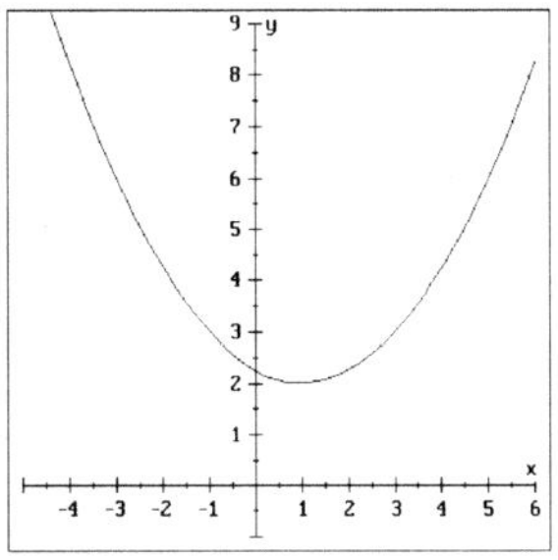

13.
$$y^2 - 6y = 4x - 13$$
$$y^2 - 6y + 9 = 4x - 13 + 9$$
$$(y-3)^2 = 4(x-1)$$

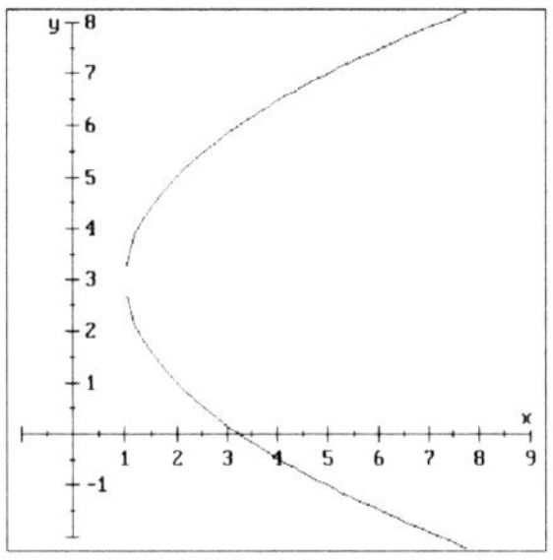

14. $a = 6$, $b = 4$, horizontal
$$\frac{x^2}{36} + \frac{y^2}{16} = 1$$

15. $a = 5$, $b = 2$, vertical
$$\frac{x^2}{4} + \frac{y^2}{25} = 1$$

16. $a = 4$, $b = 3$, horizontal $\Rightarrow \dfrac{(x+2)^2}{16} + \dfrac{(y-3)^2}{9} = 1$

17.
$$4x^2 + y^2 - 16x + 2y = -13$$
$$4(x^2 - 4x) + y^2 + 2y = -13$$
$$4(x^2 - 4x + 4) + y^2 + 2y + 1 = -13 + 16 + 1$$
$$4(x-2)^2 + (y+1)^2 = 4$$
$$\frac{(x-2)^2}{1} + \frac{(y+1)^2}{4} = 1$$
Center: $(2, -1)$, $a = 2$, $b = 1$, vertical

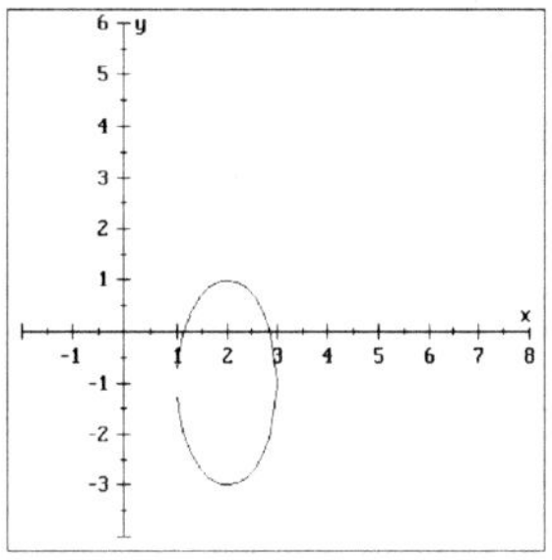

18. $a = 2$, $c = 4$; horizontal
$$b^2 = c^2 - a^2 = 16 - 4 = 12$$
$$\frac{x^2}{4} - \frac{y^2}{12} = 1$$

19. $a = 3$, $c = 5$; vertical
$$b^2 = c^2 - a^2 = 25 - 9 = 16$$
$$\frac{y^2}{9} - \frac{x^2}{16} = 1$$

20. $C(0, 3)$, $a = 3$, $c = 5$; horizontal
$$b^2 = c^2 - a^2 = 25 - 9 = 16$$
$$\frac{x^2}{9} - \frac{(y-3)^2}{16} = 1$$

21. $C(3, 0)$, $a = 3$, $c = 5$; vertical
$$b^2 = c^2 - a^2 = 25 - 9 = 16$$
$$\frac{y^2}{9} - \frac{(x-3)^2}{16} = 1$$

22. $y = \pm \dfrac{b}{a}x \Rightarrow y = \pm \dfrac{4}{5}x$

23.
$$\begin{aligned} 9x^2 - 4y^2 - 16y - 18x &= 43 \\ 9\left(x^2 - 2x\right) - 4\left(y^2 + 4y\right) &= 43 \\ 9\left(x^2 - 2x + 1\right) - 4\left(y^2 + 4y + 4\right) &= 43 + 9 - 16 \\ 9(x-1)^2 - 4(y+2)^2 &= 36 \\ \frac{(x-1)^2}{4} - \frac{(y+2)^2}{9} &= 1 \end{aligned}$$

Center: $(1, -2)$, $a = 2$, $b = 3$, horizontal

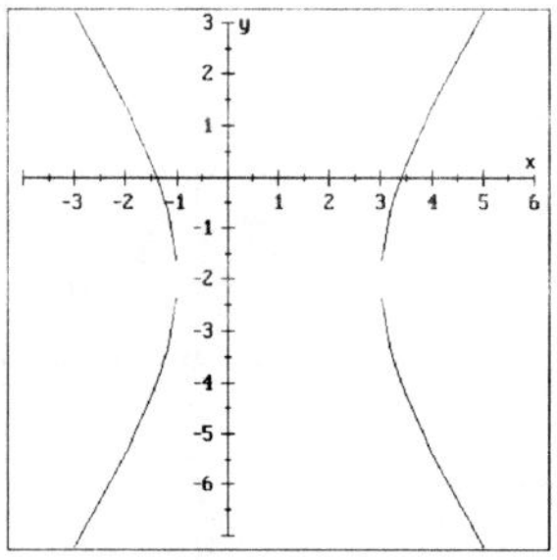

24. $4xy = 1$

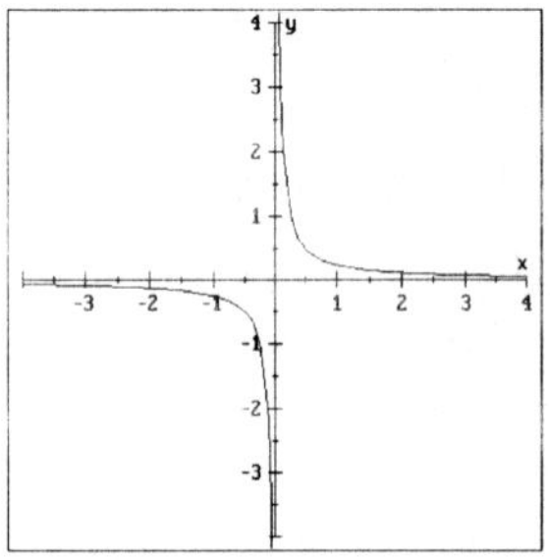

25. $\begin{cases} x^2 + y^2 = 16 \\ y = x + 4 \end{cases}$

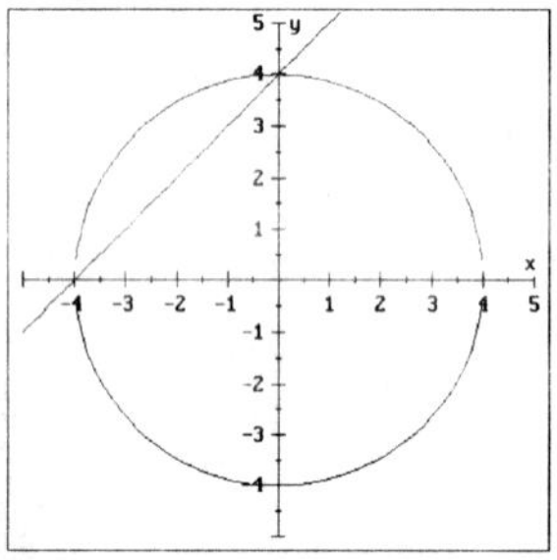

$(-4, 0), (0, 4)$

26. $\begin{cases} 3x^2 + y^2 = 52 \\ x^2 - y^2 = 12 \end{cases}$

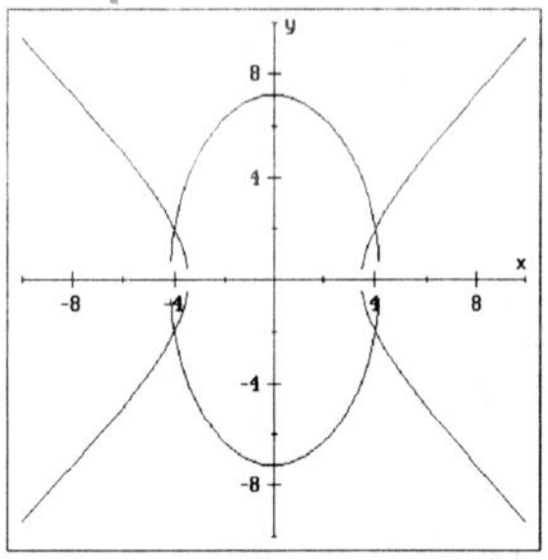

$(-4, 2), (4, 2), (-4, -2), (4, -2)$

27. $\begin{cases} \dfrac{x^2}{16} + \dfrac{y^2}{12} = 1 \\ x^2 - \dfrac{y^2}{3} = 1 \end{cases}$

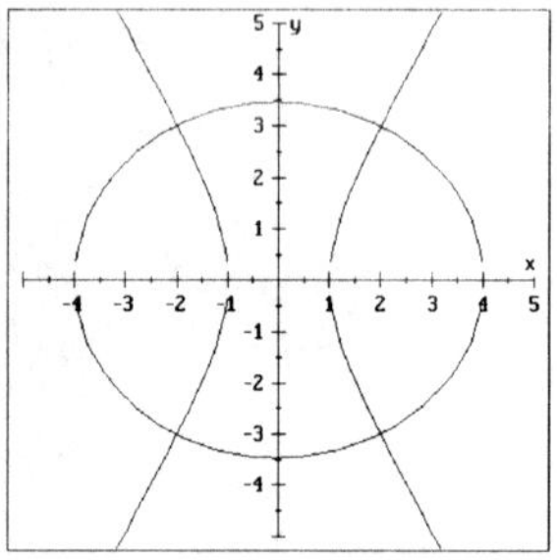

$(-2, 3), (2, 3), (-2, -3), (2, -3)$

28.
$$\begin{aligned} 3x^2 + y^2 &= 52 \\ x^2 - y^2 &= 12 \\ \hline 4x^2 &= 64 \\ x^2 &= 16 \\ x &= \pm 4 \end{aligned} \qquad \begin{aligned} 3x^2 + y^2 &= 52 \\ 3(16) + y^2 &= 52 \\ y^2 &= 4 \\ y &= \pm 2 \end{aligned}$$

$(4, 2), (-4, 2), (4, -2), (-4, -2)$

29. $-\sqrt{3}y+4\sqrt{3}=3x \Rightarrow x=\dfrac{-\sqrt{3}y+4\sqrt{3}}{3}$

$$x^2+y^2=16$$
$$\left(\frac{-\sqrt{3}y+4\sqrt{3}}{3}\right)^2+y^2=16$$
$$3y^2-24y+48+9y^2=144$$
$$12y^2-24y-96=0$$
$$12(y-4)(y+2)=0$$

$y=4$: $x=\dfrac{-\sqrt{3}(4)+4\sqrt{3}}{3}=0 \Rightarrow (0,4)$

$y=-2$: $x=\dfrac{-\sqrt{3}(-2)+4\sqrt{3}}{3}=2\sqrt{3} \Rightarrow \left(2\sqrt{3},-2\right)$

30. $\frac{x^2}{16}+\frac{y^2}{12}=1$, $x^2-\frac{y^2}{3}=1$

$$3x^2+4y^2=48$$
$$3x^2-y^2=3$$
$$5y^2=45$$

$5y^2=45$, $y^2=9$, $y=\pm 3$

$3x^2-y^2=3$, $3x^2-9=3$, $3x^2=12$, $x^2=4$, $x=\pm 2 \Rightarrow (2,3), (-2,3), (2,-3), (-2,-3)$

Chapter 7 Test (page 719)

1.
$$(x-h)^2+(y-k)^2=r^2$$
$$(x-2)^2+(y-3)^2=3^2$$
$$(x-2)^2+(y-3)^2=9$$

2.
$$C\left(\frac{-2+6}{2},\frac{-2+8}{2}\right)=C(2,3)$$
$$r=\sqrt{(6-2)^2+(8-3)^2}=\sqrt{41}$$
$$(x-h)^2+(y-k)^2=r^2$$
$$(x-2)^2+(y-3)^2=\left(\sqrt{41}\right)^2$$
$$(x-2)^2+(y-3)^2=41$$

3.
$$r=\sqrt{(7-2)^2+(7-(-5))^2}=13$$
$$(x-h)^2+(y-k)^2=r^2$$
$$(x-2)^2+(y-(-5))^2=13^2$$
$$(x-2)^2+(y+5)^2=169$$

4.
$$x^2+y^2-4x+6y+4=0$$
$$x^2-4x+y^2+6y=-4$$
$$x^2-4x+4+y^2+6y+9=-4+4+9$$
$$(x-2)^2+(y+3)^2=9$$
$$C(2,-3),\ r=3$$

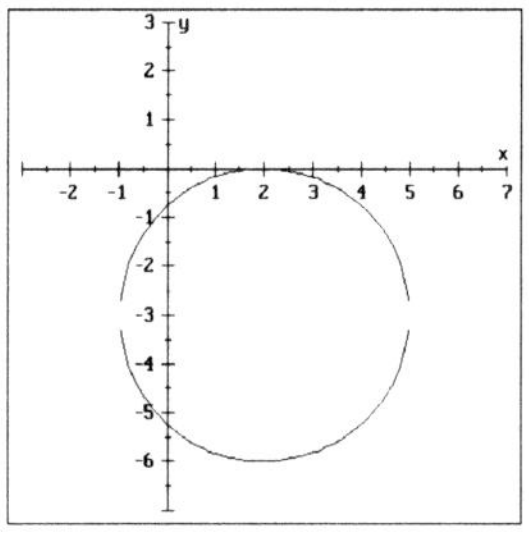

CHAPTER 7 TEST

5. Vertical (up), $p = 4$

$$(x-h)^2 = 4p(y-k)$$
$$(x-3)^2 = 4(4)(y-2)$$
$$(x-3)^2 = 16(y-2)$$

6. Horizontal

$$(y+6)^2 = 4p(x-4)$$
$$(-4+6)^2 = 4p(3-4)$$
$$4 = -4p$$
$$-4 = 4p$$
$$(y+6)^2 = -4(x-4)$$

7. $(x-2)^2 = 4p(y+3)$ **OR** $(y+3)^2 = 4p(x-2)$

$$(0-2)^2 = 4p(0+3) \qquad (0+3)^2 = 4p(0-2)$$
$$4 = 4p(3) \qquad 9 = 4p(-2)$$
$$\frac{4}{3} = 4p \qquad -\frac{9}{2} = 4p$$
$$(x-2)^2 = \tfrac{4}{3}(y+3) \qquad (y+3)^2 = -\tfrac{9}{2}(x-2)$$

8.

$$x^2 - 6x - 8y = 7$$
$$x^2 - 6x = 8y + 7$$
$$x^2 - 6x + 9 = 8y + 7 + 9$$
$$(x-3)^2 = 8(y+2)$$

Vertex: $(3, -2)$, vertical

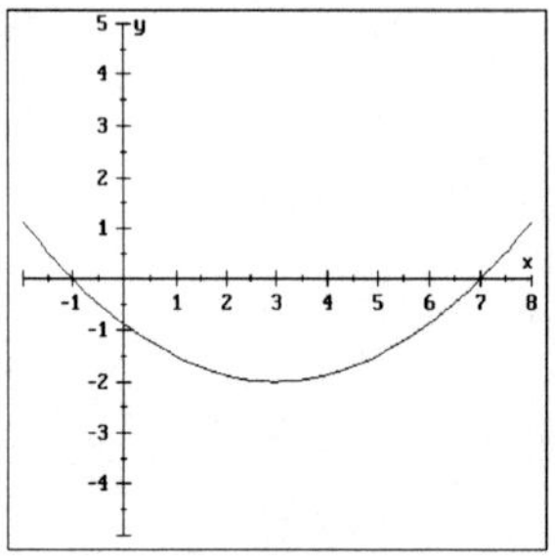

9. $a = 10$, $c = 6$, horizontal

$$b^2 = a^2 - c^2$$
$$= 100 - 36 = 64$$
$$\frac{x^2}{100} + \frac{y^2}{64} = 1$$

10. $b = 12$, $c = 5$, horizontal

$$a^2 = b^2 + c^2$$
$$= 144 + 25 = 169$$
$$\frac{x^2}{169} + \frac{y^2}{144} = 1$$

11. $a = 6$, $b = 2$, vertical

$$\frac{(x-2)^2}{4} + \frac{(y-3)^2}{36} = 1$$

12.

$$9x^2 + 4y^2 - 18x - 16y - 11 = 0$$
$$9(x^2 - 2x) + 4(y^2 - 4y) = 11$$
$$9(x^2 - 2x + 1) + 4(y^2 - 4y + 4) = 11 + 9 + 16$$
$$9(x-1)^2 + 4(y-2)^2 = 36$$
$$\frac{(x-1)^2}{4} + \frac{(y-2)^2}{9} = 1$$

Center: $(1, 2)$, $a = 3$, $b = 2$, vertical

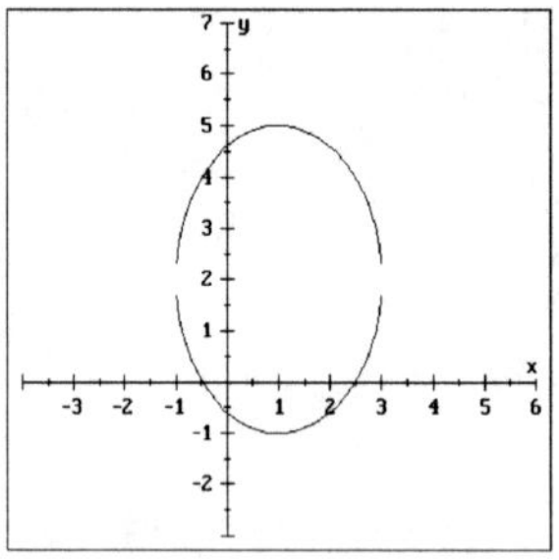

CHAPTER 7 TEST

13. $a = 5, c = 13$; horizontal

$b^2 = c^2 - a^2$

$= 169 - 25 = 144$

$$\frac{x^2}{25} - \frac{y^2}{144} = 1$$

14. $C(0,0), a = 6, c = \frac{13}{2}$

horizontal

$b^2 = c^2 - a^2$

$= \frac{169}{4} - 36 = \frac{25}{4}$

$$\frac{x^2}{36} - \frac{4y^2}{25} = 1$$

15. $a = 8, c = 10$; horizontal

$b^2 = c^2 - a^2$

$= 100 - 64 = 36$

$$\frac{(x-2)^2}{64} - \frac{(y+1)^2}{36} = 1$$

16.

$$x^2 - 4y^2 + 16y = 8$$
$$x^2 - 4(y^2 - 4y) = 8$$
$$x^2 - 4(y^2 - 4y + 4) = 8 - 16$$
$$x^2 - 4(y-2)^2 = -8$$
$$\frac{(y-2)^2}{2} - \frac{x^2}{8} = 1$$

Center: $(0, 2), a = \sqrt{2}, b = \sqrt{8}$, vertical

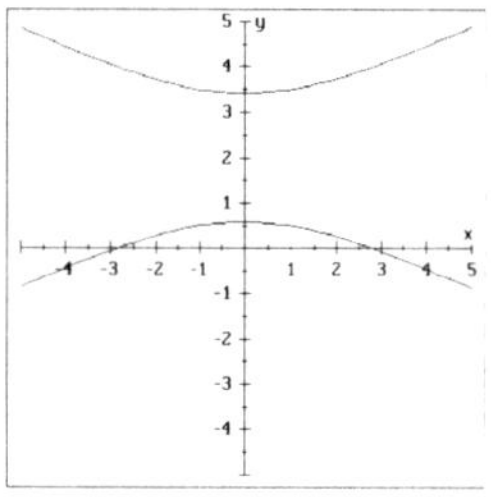

17. $y = x^2 - 3 \Rightarrow x^2 = y + 3$

$$x^2 + y^2 = 23$$
$$y + 3 + y^2 = 23$$
$$y^2 + y - 20 = 0$$
$$(y-4)(y+5) = 0$$

$y = 4$

$4 + 3 = x^2 \Rightarrow x = \pm\sqrt{7}$

$\left(\sqrt{7}, 4\right), \left(-\sqrt{7}, 4\right)$

$y = -5$

$-5 + 3 = x^2$

no real solutions

18. $x^2 + y^2 = 27 \Rightarrow x^2 = 27 - y^2$

$$2x^2 - 3y^2 = 9$$
$$2(27 - y^2) - 3y^2 = 9$$
$$54 - 2y^2 - 3y^2 = 9$$
$$45 = 5y^2$$
$$9 = y^2$$

$y = 3$

$x^2 = 27 - 3^2 \Rightarrow x = \pm 3\sqrt{2}$

$\left(3\sqrt{2}, 3\right), \left(-3\sqrt{2}, 3\right)$

$y = -3$

$x^2 = 27 - (-3)^2 \Rightarrow x = \pm 3\sqrt{2}$

$\left(3\sqrt{2}, -3\right), \left(-3\sqrt{2}, -3\right)$

19. $y^2 - 4y - 6x - 14 = 0$

$$y^2 - 4y = 6x + 14$$
$$y^2 - 4x + 4 = 6x + 14 + 4$$
$$(y-2)^2 = 6(x+3) \Rightarrow \text{Parabola}$$

20.

$$2x^2 + 3y^2 - 4x + 12y + 8 = 0$$
$$2(x^2 - 2x) + 3(y^2 + 4y) = -8$$
$$2(x^2 - 2x + 1) + 3(y^2 + 4y + 4) = -8 + 2 + 12$$
$$2(x-1)^2 + 3(y+2)^2 = 6$$
$$\frac{(x-1)^2}{3} + \frac{(y+2)^2}{2} = 1 \Rightarrow \text{ellipse}$$

Cumulative Review Exercises (page 720)

1. $64^{2/3} = \left(64^{1/3}\right)^2 = 4^2 = 16$

2. $8^{-1/3} = \dfrac{1}{8^{1/3}} = \dfrac{1}{2}$

3. $\dfrac{y^{2/3}y^{5/3}}{y^{1/3}} = \dfrac{y^{7/3}}{y^{1/3}} = y^{6/3} = y^2$

4. $\dfrac{\left(x^{5/3}\right)\left(x^{1/2}\right)}{x^{3/4}} = \dfrac{x^{13/6}}{x^{3/4}} = x^{34/24} = x^{17/12}$

5. $\left(x^{2/3} - x^{1/3}\right)\left(x^{2/3} + x^{1/3}\right) = x^{4/3} + x^{3/3} - x^{3/3} - x^{2/3} = x^{4/3} - x^{2/3}$

6. $\left(x^{-1/2} + x^{1/2}\right)^2 = \left(x^{-1/2} + x^{1/2}\right)\left(x^{-1/2} + x^{1/2}\right) = x^{-2/2} + x^0 + x^0 + x^{2/2} = \dfrac{1}{x} + 2 + x$

7. $\sqrt[3]{-27x^3} = \sqrt[3]{(-3x)^3} = -3x$

8. $\sqrt{48t^3} = \sqrt{16t^2}\sqrt{3t} = 4t\sqrt{3t}$

9. $\sqrt[3]{\dfrac{128x^4}{2x}} = \sqrt[3]{64x^3} = 4x$

10. $\sqrt{x^2 + 6x + 9} = \sqrt{(x+3)^2} = x + 3$

11. $\sqrt{50} - \sqrt{8} + \sqrt{32} = 5\sqrt{2} - 2\sqrt{2} + 4\sqrt{2} = 7\sqrt{2}$

12. $-3\sqrt[4]{32} - 2\sqrt[4]{162} + 5\sqrt[4]{48} = -3 \cdot 2\sqrt[4]{2} - 2 \cdot 3\sqrt[4]{2} + 5 \cdot 2\sqrt[4]{3} = -12\sqrt[4]{2} + 10\sqrt[4]{3}$

13. $3\sqrt{2}\left(2\sqrt{3} - 4\sqrt{12}\right) = 6\sqrt{6} - 12\sqrt{24} = 6\sqrt{6} - 12 \cdot 2\sqrt{6} = -18\sqrt{6}$

14. $\dfrac{5}{\sqrt[3]{x}} = \dfrac{5\sqrt[3]{x^2}}{\sqrt[3]{x}\sqrt[3]{x^2}} = \dfrac{5\sqrt[3]{x^2}}{x}$

15.
$$\begin{aligned} \frac{\sqrt{x}+2}{\sqrt{x}-1} &= \frac{(\sqrt{x}+2)(\sqrt{x}+1)}{(\sqrt{x}-1)(\sqrt{x}+1)} \\ &= \frac{x + 3\sqrt{x} + 2}{x - 1} \end{aligned}$$

16. $\sqrt[6]{x^3y^3} = \left(x^3y^3\right)^{1/6} = x^{3/6}y^{3/6} = x^{1/2}y^{1/2} = (xy)^{1/2} = \sqrt{xy}$

17.
$$\begin{aligned} 5\sqrt{x+2} &= x + 8 \\ \left(5\sqrt{x+2}\right)^2 &= (x+8)^2 \\ 25(x+2) &= x^2 + 16x + 64 \\ 25x + 50 &= x^2 + 16x + 64 \\ 0 &= x^2 - 9x + 14 \\ 0 &= (x-2)(x-7) \\ x = 2 \text{ or } x &= 7 \text{ (both check)} \end{aligned}$$

18.
$$\begin{aligned} \sqrt{x} + \sqrt{x+2} &= 2 \\ \sqrt{x+2} &= 2 - \sqrt{x} \\ \left(\sqrt{x+2}\right)^2 &= \left(2 - \sqrt{x}\right)^2 \\ x + 2 &= 4 - 4\sqrt{x} + x \\ 4\sqrt{x} &= 2 \\ \left(4\sqrt{x}\right)^2 &= 2^2 \\ 16x &= 4 \\ x &= \frac{1}{4} \end{aligned}$$

CUMULATIVE REVIEW EXERCISES

19.
$$\begin{aligned} 2x^2 + x - 3 &= 0 \\ x^2 + \frac{1}{2}x &= \frac{3}{2} \\ x^2 + \frac{1}{2}x + \frac{1}{16} &= \frac{24}{16} + \frac{1}{16} \\ \left(x + \tfrac{1}{4}\right)^2 &= \tfrac{25}{16} \\ x + \tfrac{1}{4} &= \pm \tfrac{5}{4} \\ x &= -\tfrac{1}{4} \pm \tfrac{5}{4} \\ x = 1 \text{ or } x &= -\tfrac{3}{2} \end{aligned}$$

20. $3x^2 + 4x - 1 = 0 \Rightarrow a = 3, b = 4, c = -1$
$$\begin{aligned} x &= \frac{-b \pm \sqrt{b^2 - 4ac}}{2a} \\ &= \frac{-4 \pm \sqrt{4^2 - 4(3)(-1)}}{2(3)} \\ &= \frac{-4 \pm \sqrt{16 + 12}}{6} \\ &= \frac{-4 \pm \sqrt{28}}{6} = \frac{-2 \pm \sqrt{7}}{3} \end{aligned}$$

21. $(3 + 5i) + (4 - 3i) = 3 + 5i + 4 - 3i = 7 + 2i$

22. $(7 - 4i) - (12 + 3i) = 7 - 4i - 12 - 3i = -5 - 7i$

23. $(2 - 3i)(2 + 3i) = 4 + 6i - 6i - 9i^2 = 4 - 9(-1) = 4 + 9 = 13 + 0i$

24. $(3 + i)(3 - 3i) = 9 - 9i + 3i - 3i^2 = 9 - 6i - 3(-1) = 9 - 6i + 3 = 12 - 6i$

25.
$$\begin{aligned} (3 - 2i) - (4 + i)^2 = 3 - 2i - \left(16 + 8i + i^2\right) = 3 - 2i - (15 + 8i) &= 3 - 2i - 15 - 8i \\ &= -12 - 10i \end{aligned}$$

26. $\dfrac{5}{3 - i} = \dfrac{5(3 + i)}{(3 - i)(3 + i)} = \dfrac{5(3 + i)}{9 - i^2} = \dfrac{5(3 + i)}{10} = \dfrac{3 + i}{2} = \dfrac{3}{2} + \dfrac{1}{2}i$

27. $|3 + 2i| = \sqrt{3^2 + 2^2} = \sqrt{13}$

28. $|5 - 6i| = \sqrt{5^2 + (-6)^2} = \sqrt{61}$

29. $2x^2 + 4x = k \Rightarrow 2x^2 + 4x - k = 0$

$a = 2, b = 4, c = -k$: Set $b^2 - 4ac = 0$.
$$\begin{aligned} b^2 - 4ac &= 0 \\ 4^2 - 4(2)(-k) &= 0 \\ 16 + 8k &= 0 \\ k &= -2 \end{aligned}$$

30. $y = \dfrac{1}{2}x^2 - x + 1$: $a = \dfrac{1}{2}, b = -1, c = 1$
$$x = -\frac{b}{2a} = -\frac{-1}{2\left(\frac{1}{2}\right)} = 1$$
$$y = \frac{1}{2}(1)^2 - 1 + 1 = \frac{1}{2}$$

31.
$$\begin{aligned} x^2 - x - 6 &> 0 \\ (x + 2)(x - 3) &> 0 \end{aligned}$$
factors $= 0$: $x = -2, x = 3$

intervals: $(-\infty, -2), (-2, 3), (3, \infty)$

interval	test number	value of $x^2 - x - 6$
$(-\infty, -2)$	-3	$+6$
$(-2, 3)$	0	-6
$(3, \infty)$	4	$+6$

Solution: $(-\infty, -2) \cup (3, \infty)$

32.
$$\begin{aligned} x^2 - x - 6 &\le 0 \\ (x + 2)(x - 3) &\le 0 \end{aligned}$$
factors $= 0$: $x = -2, x = 3$

intervals: $(-\infty, -2), (-2, 3), (3, \infty)$

interval	test number	value of $x^2 - x - 6$
$(-\infty, -2)$	-3	$+6$
$(-2, 3)$	0	-6
$(3, \infty)$	4	$+6$

Solution: $[-2, 3]$

CUMULATIVE REVIEW EXERCISES

33. $f(-1) = 3(-1)^2 + 2 = 3 + 2 = 5$

34. $(g \circ f)(2) = g(f(2)) = g(3(2)^2 + 2)$
$= g(14)$
$= 2(14) - 1 = 27$

35. $(f \circ g)(x) = f(g(x))$
$= f(2x - 1)$
$= 3(2x - 1)^2 + 2$
$= 3(4x^2 - 4x + 1) + 2$
$= 12x^2 - 12x + 3 + 2$
$= 12x^2 - 12x + 5$

36. $(g \circ f)(x) = g(f(x))$
$= g(3x^2 + 2)$
$= 2(3x^2 + 2) - 1$
$= 6x^2 + 4 - 1$
$= 6x^2 + 3$

37. $y = \log_2 x \Rightarrow 2^y = x$

38. $3^b = a \Rightarrow \log_3 a = b$

39. $\log_x 25 = 2 \Rightarrow x^2 = 25 \Rightarrow x = 5$

40. $\log_5 125 = x \Rightarrow 5^x = 125 \Rightarrow x = 3$

41. $\log_3 x = -3 \Rightarrow 3^{-3} = x \Rightarrow x = \frac{1}{27}$

42. $\log_5 x = 0 \Rightarrow 5^0 = x \Rightarrow x = 1$

43. $y = \log_2 x$; inverse: $y = 2^x$

44. $\log_{10} 10^x = x$, so $y = x$.

45. $\log 98 = \log (14 \cdot 7) = \log 14 + \log 7 = 1.1461 + 0.8451 = 1.9912$

46. $\log 2 = \log \frac{14}{7} = \log 14 - \log 7 = 1.1461 - 0.8451 = 0.3010$

47. $\log 49 = \log 7^2 = 2 \log 7 = 2(0.8451) = 1.6902$

48. $\log \frac{7}{5} = \log \frac{7}{10/2} = \log 7 - \log \frac{10}{2} = \log 7 - (\log 10 - \log 2) = \log 7 - \log 10 + \log 2$
$= 0.8451 - 1 + 0.3010 = 0.1461$

49.
$$\begin{aligned}
2^{x+2} &= 3^x \\
\log 2^{x+2} &= \log 3^x \\
(x+2) \log 2 &= x \log 3 \\
x \log 2 + 2 \log 2 &= x \log 3 \\
2 \log 2 &= x \log 3 - x \log 2 \\
2 \log 2 &= x (\log 3 - \log 2) \\
\frac{2 \log 2}{\log 3 - \log 2} &= x
\end{aligned}$$

50.
$$\begin{aligned}
2 \log 5 + \log x - \log 4 &= 2 \\
\log 5^2 + \log x - \log 4 &= 2 \\
\log \frac{25x}{4} &= 2 \\
10^2 &= \frac{25x}{4} \\
400 &= 25x \\
16 &= x
\end{aligned}$$

51.
$$\begin{aligned}
A &= A_0\left(1 + \frac{r}{k}\right)^{kt} \\
&= 9000\left(1 + \frac{-0.12}{1}\right)^{1(9)} \\
&\approx \$2848.31
\end{aligned}$$

52. $\log_6 8 = \frac{\log 8}{\log 6} \approx 1.16056$

CUMULATIVE REVIEW EXERCISES

53. $\begin{cases} 2x + y = 5 \\ x - 2y = 0 \end{cases}$

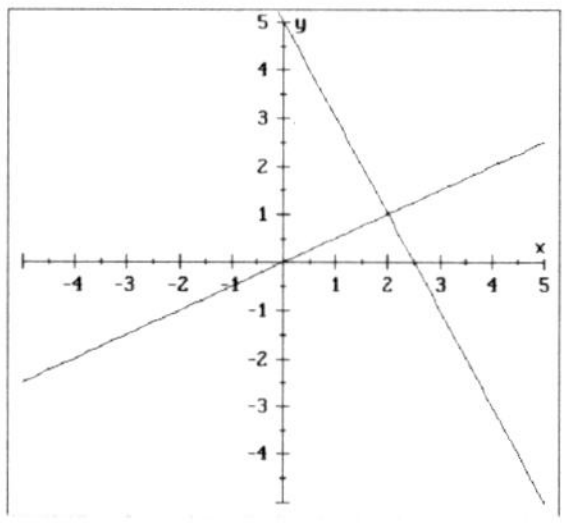

solution: $(2, 1)$

54. $\begin{cases} (1) \quad 3x + y = 4 \\ (2) \quad 2x - 3y = -1 \end{cases}$

Substitute $y = -3x + 4$ from (1) into (2):

$$\begin{aligned} 2x - 3y &= -1 \\ 2x - 3(-3x + 4) &= -1 \\ 2x + 9x - 12 &= -1 \\ 11x &= 11 \\ x &= 1 \end{aligned}$$

Substitute and solve for y:

$y = -3x + 4 = -3(1) + 4 = 1$

$\boxed{x = 1, y = 1}$

55.

$$\begin{aligned} x + 2y &= -2 \\ 2x - \ y &= \ 6 \Rightarrow \times 2 \end{aligned} \qquad \begin{aligned} x + 2y &= -2 \\ 4x - 2y &= 12 \\ \hline 5x \qquad &= 10 \\ x \qquad &= 2 \end{aligned} \qquad \begin{aligned} 2x - y &= 6 \\ 2(2) - y &= 6 \\ -y &= 2 \\ y &= -2 \end{aligned}$$

Solution: $\boxed{x = 2, y = -2}$

56.

$$\begin{aligned} \frac{x}{10} + \frac{y}{5} &= \frac{1}{2} \Rightarrow \times 10 \\ \frac{x}{2} - \frac{y}{5} &= \frac{13}{10} \Rightarrow \times 10 \end{aligned} \qquad \begin{aligned} x + 2y &= 5 \\ 5x - 2y &= 13 \\ \hline 6x \qquad &= 18 \\ x \qquad &= 3 \end{aligned} \qquad \begin{aligned} x + 2y &= 5 \\ 3 + 2y &= 5 \\ 2y &= 2 \\ y &= 1 \end{aligned}$$

Solution: $\boxed{x = 3, y = 1}$

57. $\begin{vmatrix} 3 & -2 \\ 1 & -1 \end{vmatrix} = 3(-1) - (-2)1$

$= -3 + 2 = -1$

58. $y = \dfrac{\begin{vmatrix} 4 & -1 \\ 3 & -7 \end{vmatrix}}{\begin{vmatrix} 4 & -3 \\ 3 & 4 \end{vmatrix}} = \dfrac{-25}{25} = -1$

59. $\begin{bmatrix} x \\ y \\ z \end{bmatrix} = \begin{bmatrix} 1 & 1 & 1 \\ 2 & -1 & -1 \\ 1 & -2 & 1 \end{bmatrix}^{-1} \begin{bmatrix} 1 \\ -4 \\ 4 \end{bmatrix} = \begin{bmatrix} \frac{1}{3} & \frac{1}{3} & 0 \\ \frac{1}{3} & 0 & -\frac{1}{3} \\ \frac{1}{3} & -\frac{1}{3} & \frac{1}{3} \end{bmatrix} \begin{bmatrix} 1 \\ -4 \\ 4 \end{bmatrix} = \begin{bmatrix} -1 \\ -1 \\ 3 \end{bmatrix}$

60. $\begin{bmatrix} x \\ y \\ z \end{bmatrix} = \begin{bmatrix} 1 & 2 & 3 \\ 3 & 2 & 1 \\ 2 & 3 & 1 \end{bmatrix}^{-1} \begin{bmatrix} 6 \\ 6 \\ 6 \end{bmatrix} = \begin{bmatrix} -\frac{1}{12} & \frac{7}{12} & -\frac{1}{3} \\ -\frac{1}{12} & -\frac{5}{12} & \frac{2}{3} \\ \frac{5}{12} & \frac{1}{12} & -\frac{1}{3} \end{bmatrix} \begin{bmatrix} 6 \\ 6 \\ 6 \end{bmatrix} = \begin{bmatrix} 1 \\ 1 \\ 1 \end{bmatrix}$

SECTION 8.1

Exercises 8.1 (page 729)

1. power

3. first

5. $7 \cdot 6 \cdot 5 \cdot 4 \cdot 3 \cdot 2 \cdot 1$

7. $(n-1)!$

9. $5! = 5 \cdot 4 \cdot 3 \cdot 2 \cdot 1 = 120$

11. $3! \cdot 6! = 6 \cdot 720 = 4320$

13. $6! + 6! = 720 + 720 = 1440$

15. $\dfrac{9!}{12!} = \dfrac{9!}{12 \cdot 11 \cdot 10 \cdot 9!} = \dfrac{1}{1320}$

17. $\dfrac{5! \cdot 7!}{9!} = \dfrac{5! \cdot 7!}{9 \cdot 8 \cdot 7!} = \dfrac{120}{72} = \dfrac{5}{3}$

19. $\dfrac{18!}{6!(18-6)!} = \dfrac{18!}{6! \cdot 12!} = \dfrac{18 \cdot 17 \cdot 16 \cdot 15 \cdot 14 \cdot 13 \cdot 12!}{6! \cdot 12!} = \dfrac{13{,}366{,}080}{720} = 18{,}564$

21. Row 5 of Pascal's triangle: 1 5 10 10 5 1
$(a+b)^5 = a^5 + 5a^4b + 10a^3b^2 + 10a^2b^3 + 5ab^4 + b^5$

23. Row 3 of Pascal's triangle: 1 3 3 1
$(x-y)^3 = x^3 + 3x^2(-y) + 3x(-y)^2 + (-y)^3 = x^3 - 3x^2y + 3xy^2 - y^3$

25. $(a+b)^3 = a^3 + \dfrac{3!}{1!2!}a^2b + \dfrac{3!}{2!1!}ab^2 + b^3 = a^3 + 3a^2b + 3ab^2 + b^3$

27. $(a-b)^5 = a^5 + \dfrac{5!}{1!4!}a^4(-b) + \dfrac{5!}{2!3!}a^3(-b)^2 + \dfrac{5!}{3!2!}a^2(-b)^3 + \dfrac{5!}{4!1!}a(-b)^4 + (-b)^5$
$= a^5 - 5a^4b + 10a^3b^2 - 10a^2b^3 + 5ab^4 - b^5$

29. $(2x+y)^3 = (2x)^3 + \dfrac{3!}{1!2!}(2x)^2y + \dfrac{3!}{2!1!}(2x)y^2 + y^3 = 8x^3 + 12x^2y + 6xy^2 + y^3$

31. $(x-2y)^3 = x^3 + \dfrac{3!}{1!2!}x^2(-2y) + \dfrac{3!}{2!1!}x(-2y)^2 + (-2y)^3 = x^3 - 6x^2y + 12xy^2 - 8y^3$

33. $(2x+3y)^4 = (2x)^4 + \dfrac{4!}{1!3!}(2x)^3(3y) + \dfrac{4!}{2!2!}(2x)^2(3y)^2 + \dfrac{4!}{3!1!}(2x)(3y)^3 + (3y)^4$
$= 16x^4 + 96x^3y + 216x^2y^2 + 216xy^3 + 81y^4$

35. $(x-2y)^4 = x^4 + \dfrac{4!}{1!3!}x^3(-2y) + \dfrac{4!}{2!2!}x^2(-2y)^2 + \dfrac{4!}{3!1!}x(-2y)^3 + (-2y)^4$
$= x^4 - 8x^3y + 24x^2y^2 - 32xy^3 + 16y^4$

37. $(x-3y)^5 = x^5 + \dfrac{5!}{1!4!}x^4(-3y) + \dfrac{5!}{2!3!}x^3(-3y)^2 + \dfrac{5!}{3!2!}x^2(-3y)^3 + \dfrac{5!}{4!1!}x(-3y)^4 + (-3y)^5$
$= x^5 - 15x^4y + 90x^3y^2 - 270x^2y^3 + 405xy^4 - 243y^5$

SECTION 8.1

39. $\left(\frac{x}{2}+y\right)^4 = \left(\frac{x}{2}\right)^4 + \frac{4!}{1!3!}\left(\frac{x}{2}\right)^3 y + \frac{4!}{2!2!}\left(\frac{x}{2}\right)^2 y^2 + \frac{4!}{3!1!}\left(\frac{x}{2}\right) y^3 + y^4$
$= \frac{1}{16}x^4 + \frac{1}{2}x^3y + \frac{3}{2}x^2y^2 + 2xy^3 + y^4$

41. The 3rd term will involve b^2.
$\frac{4!}{2!2!}a^2b^2 = 6a^2b^2$

43. The 5th term will involve b^4.
$\frac{7!}{3!4!}a^3b^4 = 35a^3b^4$

45. The 6th term will involve $(-b)^5$.
$\frac{5!}{0!5!}a^0(-b)^5 = -b^5$

47. The 5th term will involve b^4.
$\frac{17!}{13!4!}a^{13}b^4 = 2380a^{13}b^4$

49. The 2nd term will involve $\left(-\sqrt{2}\right)^1$.
$\frac{4!}{3!1!}a^3\left(-\sqrt{2}\right)^1 = -4\sqrt{2}\,a^3$

51. The 5th term will involve $\left(\sqrt{3}b\right)^4$.
$\frac{9!}{5!4!}a^5\left(\sqrt{3}b\right)^4 = 1134a^5b^4$

53. The 3rd term will involve y^2.
$\frac{4!}{2!2!}\left(\frac{x}{2}\right)^2 y^2 = \frac{3}{2}x^2y^2$

55. The 10th term will involve $\left(-\frac{s}{2}\right)^9$.
$\frac{11!}{2!9!}\left(\frac{r}{2}\right)^2\left(-\frac{s}{2}\right)^9 = -\frac{55}{2048}r^2s^9$

57. The 4th term will involve b^3.
$\frac{n!}{(n-3)!3!}a^{n-3}b^3$

59. The rth term will involve b^{r-1}.
$\frac{n!}{(n-r+1)!(r-1)!}a^{n-r+1}b^{r-1}$

61. Answers may vary.

63. Each term will contain $a^n\left(-\frac{1}{a}\right)^{10-n}$, or $a^n\left(-a^{n-10}\right)$. The constant term will occur when this product is equal to 1, or when the sum of the exponents equals 0.
$n+n-10=0 \Rightarrow n=5$
$\frac{10!}{5!5!}a^5\left(-\frac{1}{a}\right)^5 = -252$

65. $\frac{n!}{0!(n-0)!} = \frac{n!}{0!n!} = \frac{n!}{1\cdot n!} = 1$

67. Answers may vary.

69. Answers may vary.

71. $3x^3y^2z^4 - 6xyz^5 + 15x^2yz^2 = 3xyz^2(x^2yz^2 - 2z^3 + 5x)$

73. $a^4 - b^4 = (a^2+b^2)(a^2-b^2) = (a^2+b^2)(a+b)(a-b)$

75. $\frac{\frac{1}{x}+\frac{1}{3}}{\frac{1}{x}-\frac{1}{3}} = \frac{3x\left(\frac{1}{x}+\frac{1}{3}\right)}{3x\left(\frac{1}{x}-\frac{1}{3}\right)} = \frac{3+x}{3-x}$

Exercises 8.2 (page 738)

1. domain **3.** series **5.** infinite

7. Summation notation **9.** 6 **11.** $5c$

13. $f(1) = 5(1)(1-1) = 0$ $f(2) = 5(2)(2-1) = 10$ $f(3) = 5(3)(3-1) = 30$
$f(4) = 5(4)(4-1) = 60$ $f(5) = 5(5)(5-1) = 100$ $f(6) = 5(6)(6-1) = 150$

15. 1, 6, 11, 16, ... Add 5 to get the next term. The next term is 21.

17. $a, a+d, a+2d, a+3d, \ldots$ Add d to get the next term. The next term is $a+4d$.

19. 1, 3, 6, 10, ... The difference between terms increases by 1 each time. The next term is $10+5=15$.

21. $1+2+3+4+5=15$ **23.** $3+3+3+3+3=15$

25. $2\left(\frac{1}{3}\right)^1 + 2\left(\frac{1}{3}\right)^2 + 2\left(\frac{1}{3}\right)^3 + 2\left(\frac{1}{3}\right)^4 + 2\left(\frac{1}{3}\right)^5 = \frac{2}{3} + \frac{2}{9} + \frac{2}{27} + \frac{2}{81} + \frac{2}{243} = \frac{242}{243}$

27. $[3(1)-2]+[3(2)-2]+[3(3)-2]+[3(4)-2]+[3(5)-2] = 1+4+7+10+13 = 35$

29. $a_1 = 3$
$a_2 = 2a_1 + 1 = 2(3) + 1 = 7$
$a_3 = 2a_2 + 1 = 2(7) + 1 = 15$
$a_4 = 2a_3 + 1 = 2(15) + 1 = 31$

31. $a_1 = -4$
$a_2 = \frac{a_1}{2} = \frac{-4}{2} = -2$
$a_3 = \frac{a_2}{2} = \frac{-2}{2} = -1$
$a_4 = \frac{a_3}{2} = \frac{-1}{2} = -\frac{1}{2}$

33. $a_1 = k$
$a_2 = a_1^2 = k^2$
$a_3 = a_2^2 = \left(k^2\right)^2 = k^4$
$a_4 = a_3^2 = \left(k^4\right)^2 = k^8$

35. $a_1 = 8$
$a_2 = \frac{2a_1}{k} = \frac{2(8)}{k} = \frac{16}{k}$
$a_3 = \frac{2a_2}{k} = \frac{2\left(\frac{16}{k}\right)}{k} = \frac{32}{k^2}$
$a_4 = \frac{2a_3}{k} = \frac{2\left(\frac{32}{k^2}\right)}{k} = \frac{64}{k^3}$

37. alternating **39.** not alternating

41. $\sum_{k=1}^{5} 2k = 2\sum_{k=1}^{5} k = 2(1+2+3+4+5) = 2(15) = 30$

43. $\sum_{k=3}^{4} \left(-2k^2\right) = -2\sum_{k=3}^{4} k^2 = -2\left(3^2+4^2\right) = -2(25) = -50$

45. $\sum_{k=1}^{5} (3k-1) = 3\sum_{k=1}^{5} k - \sum_{k=1}^{5} 1 = 3(1+2+3+4+5) - 5(1) = 3(15) - 5 = 40$

47. $\sum_{k=1}^{1000} \frac{1}{2} = 1000\left(\frac{1}{2}\right) = 500$

49. $\sum_{x=3}^{4} \frac{1}{x} = \frac{1}{3} + \frac{1}{4} = \frac{4}{12} + \frac{3}{12} = \frac{7}{12}$

51. $$\sum_{x=1}^{4}(4x+1)^2 - \sum_{x=1}^{4}(4x-1)^2 = \sum_{x=1}^{4}\left(16x^2+8x+1\right) - \sum_{x=1}^{4}\left(16x^2-8x+1\right)$$
$$= \sum_{x=1}^{4}16x = 16\sum_{x=1}^{4}x = 16(1+2+3+4) = 16(10) = 160$$

53. $$\sum_{x=6}^{8}(5x-1)^2 + \sum_{x=6}^{8}(10x-1) = \sum_{x=6}^{8}\left(25x^2-10x+1\right) + \sum_{x=6}^{8}(10x-1)$$
$$= \sum_{x=6}^{8}\left(25x^2\right) = 25\sum_{x=6}^{8}x^2 = 25\left(6^2+7^2+8^2\right) = 3725$$

55. Answers may vary.

57. Answers may vary.

59. $$\frac{8}{12} = \frac{4}{x}$$
$$8x = 48$$
$$x = 6 \text{ cm}$$

61. $$AB^2 + BC^2 = AC^2$$
$$24^2 + 10^2 = AC^2$$
$$676 = AC^2$$
$$26 \text{ ft} = AC$$

Exercises 8.3 (page 744)

1. $(n-1)d$

3. infinite

5. $a_n = a + (n-1)d$

7. Arithmetic means

9. $1, 3, 5, 7, 9, 11$

11. $$a_n = a + (n-1)d$$
$$a_3 = 5 + (3-1)d$$
$$2 = 5 + 2d$$
$$-3 = 2d$$
$$-\frac{3}{2} = d \Rightarrow 5, \frac{7}{2}, 2, \frac{1}{2}, -1, -\frac{5}{2}$$

13. $$a_n = a + (n-1)d$$
$$a_7 = a + (7-1)\frac{5}{2}$$
$$24 = a + 6\left(\frac{5}{2}\right)$$
$$24 = a + 15$$
$$9 = a \Rightarrow 9, \frac{23}{2}, 14, \frac{33}{2}, 19, \frac{43}{2}$$

15. $$a_n = a + (n-1)d$$
$$a_{40} = 6 + (40-1)8$$
$$= 6 + 39(8) = 6 + 312 = 318$$

17. $$a_n = a + (n-1)d$$
$$a_6 = -2 + (6-1)d$$
$$28 = -2 + 5d$$
$$30 = 5d \Rightarrow d = 6$$

SECTION 8.3

19. $a_n = a + (n-1)d$
$a_{55} = -8 + (55-1)7$
$= -8 + 54(7) = -8 + 378 = 370$

21. 5th term: $14 = a + 4d$; 2nd term: $5 = a + d$

Solve the system: $\begin{cases} a + 4d = 14 \\ a + d = 5 \end{cases}$

$a = 2, d = 3$
$a_{15} = 2 + 14(3) = 44$

23. $a = 10, a_5 = 20$
$20 = 10 + 4d$
$10 = 4d$
$\frac{5}{2} = d \Rightarrow 10, \boxed{\frac{25}{2}, 15, \frac{35}{2}}, 20$

25. $a = -7, a_6 = \frac{2}{3}$
$\frac{2}{3} = -7 + 5d$
$\frac{23}{3} = 5d$
$\frac{23}{15} = d$
$-7, \boxed{-\frac{82}{15}, -\frac{59}{15}, -\frac{12}{5}, -\frac{13}{15}}, \frac{2}{3}$

27. $a = 5, d = 2$
$a_{15} = a + (n-1)d = 5 + 14(2) = 33$
$S_{15} = \frac{n(a + a_{15})}{2} = \frac{15(5+33)}{2} = 285$

29. $a = \frac{27}{2}, d = \frac{3}{2}$
$a_{20} = a + (n-1)d = \frac{27}{2} + 19\left(\frac{3}{2}\right) = 42$
$S_{20} = \frac{n(a + a_{20})}{2} = \frac{20\left(\frac{27}{2} + 42\right)}{2} = 555$

31. $d = \frac{1}{2}, a_{25} = 10$
$10 = a + 24\left(\frac{1}{2}\right)$
$10 = a + 12 \Rightarrow a = -2$
$a_{30} = a + (n-1)d = -2 + 29\left(\frac{1}{2}\right) = \frac{25}{2}$
$S_{30} = \frac{n(a + a_{30})}{2} = \frac{30\left(-2 + \frac{25}{2}\right)}{2} = 157\frac{1}{2}$

33. $a = 1, d = 1, n = a_{200} = 200; S_{200} = \frac{n(a + a_{200})}{2} = \frac{200(1+200)}{2} = 20{,}100$

35. $a = 180, d = 180, n = 8 - 2 = 6$
$a_6 = a + (n-1)d = 180 + 5(180)$
$= 1080°$

$a = 180, d = 180, n = 12 - 2 = 10$
$a_{10} = a + (n-1)d = 180 + 9(180)$
$= 1800°$

37. $a = 5500, d = -105, n = 49$
Note: $n = 49$ occurs at the <u>beginning</u> of the 49th month, right after the 48th payment has been made
49th term $= a + (n-1)d$
$= 5500 + 48(-105)$
$= \$460$

39. $a = 237{,}500; d = 150{,}000;$
$a_{10} = a + (n-1)d$
$= 237{,}500 + 9(150{,}000)$
$= \$1{,}587{,}500$

41. $a_3 = a + (n-1)d$
$= 16 + 2(32) = 80$ feet

43. $a = 1, d = 1, n = 20, a_{20} = 20$
$S = \dfrac{n(a + a_{20})}{2} = \dfrac{20(1+20)}{2} = 210$ logs

45. Answers may vary.

47. Answers may vary.

49.
$$\begin{aligned} x + \sqrt{x+3} &= 9 \\ \sqrt{x+3} &= 9 - x \\ \left(\sqrt{x+3}\right)^2 &= (9-x)^2 \\ x + 3 &= 81 - 18x + x^2 \\ 0 &= x^2 - 19x + 78 \\ 0 &= (x-6)(x-13) \\ x &= 6 \text{ or } x = 13 \end{aligned}$$
$x = 13$ does not check and is extraneous.

51.
$$\begin{aligned} \frac{1}{x} + \frac{2}{x^2} + \frac{1}{x^3} &= 0 \\ x^3\left(\frac{1}{x} + \frac{2}{x^2} + \frac{1}{x^3}\right) &= x^3(0) \\ x^2 + 2x + 1 &= 0 \\ (x+1)(x+1) &= 0 \\ x &= -1 \end{aligned}$$

53.
$$\begin{aligned} x^4 - 1 &= 0 \\ \left(x^2+1\right)\left(x^2-1\right) &= 0 \end{aligned}$$
$x^2 + 1 = 0$ **OR** $x^2 - 1 = 0$
$x^2 = -1$ $\quad x^2 = 1$
$x = \pm\sqrt{-1}$ $\quad x = \pm\sqrt{1}$
$x = \pm i$ $\quad x = \pm 1$

Exercises 8.4 (page 753)

1. r^{n-1}

3. ar^{n-1}

5. infinite

7. Geometric means

9. 10, 20, 40, 80

11. $-2, -6, -18, -54$

13. $3, 3\sqrt{2}, 6, 6\sqrt{2}$

15.
$$\begin{aligned} a_4 &= ar^{4-1} \\ 54 &= 2r^3 \\ 27 &= r^3 \\ 3 &= r \Rightarrow 2, 6, 18, 54 \end{aligned}$$

17. $a = \frac{1}{4}, r = 4;\ a_6 = ar^{6-1} = \left(\frac{1}{4}\right)4^5 = 256$

19.
$$\begin{aligned} a_2 &= ar^1 \\ 6 &= ar^1 \\ a_3 &= ar^2 \\ -18 &= ar^2 \end{aligned} \qquad \begin{aligned} ar^2 &= -18 \\ ar \cdot r &= -18 \\ 6r &= -18 \\ r &= -3 \end{aligned}$$
$r = -3$: $a = \dfrac{6}{r} = \dfrac{6}{-3} = -2$
$a_5 = ar^4 = -2(-3)^4 = -162$

SECTION 8.4

21.
$$a_5 = ar^4$$
$$20 = 10r^4$$
$$2 = r^4$$
$\sqrt[4]{2} = r$ (problem specifies positive)

$10, \boxed{10\sqrt[4]{2}, 10\sqrt{2}, 10\sqrt[4]{8}}, 20$

23.
$$a_6 = ar^5$$
$$2048 = 2r^5$$
$$1024 = r^5$$
$$4 = r$$
$2, \boxed{8, 32, 128, 512}, 2048$

25. $a = 4, r = 2, n = 5$
$$S_5 = \frac{a - ar^n}{1 - r} = \frac{4 - 4(2)^5}{1 - 2} = \frac{-124}{-1} = 124$$

27. $a = 2, r = -3, n = 10$
$$S_{10} = \frac{a - ar^n}{1 - r} = \frac{2 - 2(-3)^{10}}{1 - (-3)} = \frac{-118{,}096}{4} = -29{,}524$$

29. $a = 3, r = \frac{3}{2}, n = 6$
$$S_6 = \frac{a - ar^n}{1 - r} = \frac{3 - 3\left(\frac{3}{2}\right)^6}{1 - \frac{3}{2}} = \frac{-\frac{1995}{64}}{-\frac{1}{2}} = \frac{1995}{32}$$

31. $a = 6, r = \frac{2}{3}$
$$S_\infty = \frac{a}{1 - r} = \frac{6}{1 - \frac{2}{3}} = \frac{6}{\frac{1}{3}} = 18$$

33. $a = 12, r = -\frac{1}{2}$
$$S_\infty = \frac{a}{1 - r} = \frac{12}{1 - \left(-\frac{1}{2}\right)} = \frac{12}{\frac{3}{2}} = 8$$

35. $a = \frac{5}{10} = \frac{1}{2}, r = \frac{1}{10}$
$$S_\infty = \frac{a}{1 - r} = \frac{\frac{1}{2}}{1 - \frac{1}{10}} = \frac{\frac{1}{2}}{\frac{9}{10}} = \frac{5}{9}$$

37. $a = \frac{25}{100} = \frac{1}{4}, r = \frac{1}{100}$
$$S_\infty = \frac{a}{1 - r} = \frac{\frac{1}{4}}{1 - \frac{1}{100}} = \frac{\frac{1}{4}}{\frac{99}{100}} = \frac{25}{99}$$

39. $a = 623, r = 1.10$
$$a_9 = ar^8 = 623(1.1)^8 \approx 1335 \text{ students}$$
$$\#\text{ professors} = \frac{1335}{60} \approx 22.25$$
23 professors will be needed.

41. $a = 100, r = 0.6, a_5 = 100(0.6)^4 = 12.96$ ft

Down	Up
$a = 100, r = 0.6$	$a = 60, r = 0.6$
$S_\infty = \frac{a}{1 - r}$	$S_\infty = \frac{a}{1 - r}$
$= \frac{100}{1 - 0.6}$	$= \frac{60}{1 - 0.6}$
$= 250$	$= 150$

Total vertical distance: 400 ft

43. $a = 10, r = 0.95$
$$a_{14} = ar^{13} = 10(0.95)^{13} \approx 5.13 \text{ meters}$$

45. $a = 1000,\ r = 1 + \dfrac{0.0675}{365},\ n = 365$

$$ar^{365} = 1000\left(1 + \frac{0.0675}{365}\right)^{365} \approx \$1069.82$$

The interest will be \$69.82.

47. $a = c,\ r = 0.80,\ n = 5$

$ar^5 = c(0.80)^5 = 0.32768c$, or about $\frac{1}{3}c$

49. $a = 5 \times 10^9,\ r = 2$

$n = (3010 - 1990)/30 = 34$

$ar^{34} = 5 \times 10^9 (2)^{34} \approx 8.6 \times 10^{19}$

51. $a = 50{,}000;\ r = 1.06,\ n = 22$

$$ar^{22} = 50{,}000(1.06)^{22} \approx \$180{,}176.87$$

53. $a = 1000,\ r = 1 + \dfrac{0.07}{4},\ n = 40$

$$ar^{40} = 1000\left(1 + \frac{0.07}{4}\right)^{40} \approx \$2001.60$$

55. $a = 1000,\ r = 1 + \dfrac{0.07}{365},\ n = 3650$

$$ar^{3650} = 1000\left(1 + \frac{0.07}{365}\right)^{3650} \approx \$2013.62$$

57. $a = 2000,\ r = 1 + \dfrac{0.11}{4},\ n = 180$

$$ar^{180} = 2000\left(1 + \frac{0.11}{4}\right)^{180} \approx \$264{,}094.58$$

59. $a = 1000,\ r = 0.8$

$$S_\infty = \frac{a}{1 - r} = \frac{1000}{1 - 0.8} = 5000$$

61. $a = 1,\ r = 2,\ n = 64$

$$S = \frac{a - ar^n}{1 - r} = \frac{1 - 1(2)^{64}}{1 - 2} \approx 1.8447 \times 10^{19} \text{ grains}$$

63. no

65. $(3 + 2i) + (2 - 5i) = 3 + 2i + 2 - 5i = 5 - 3i$

67. $(3 + i)(3 - i) = 9 - 3i + 3i - i^2 = 9 - (-1) = 10 + 0i$

69. $\dfrac{2 + 3i}{2 - i} = \dfrac{(2 + 3i)(2 + i)}{(2 - i)(2 + i)} = \dfrac{4 + 2i + 6i + 3i^2}{4 - i^2} = \dfrac{4 + 8i + 3(-1)}{4 - (-1)} = \dfrac{1 + 8i}{5} = \dfrac{1}{5} + \dfrac{8}{5}i$

71. $i^{127} = i^{124} i^3 = (i^4)^{31} i^3 = 1^{31} i^3 = i^3 = -i$

Exercises 8.5 (page 759)

1. two

3. $n = k + 1$

5.

$n=1$	$n=2$
$5(1) \stackrel{?}{=} \frac{5(1)(1+1)}{2}$	$5+5(2) \stackrel{?}{=} \frac{5(2)(2+1)}{2}$
$5=5$	$15 \stackrel{?}{=} \frac{10(3)}{2}$
	$15=15$

$n=3$	$n=4$
$5+10+5(3) \stackrel{?}{=} \frac{5(3)(3+1)}{2}$	$5+10+15+5(4) \stackrel{?}{=} \frac{5(4)(4+1)}{2}$
$30 \stackrel{?}{=} \frac{15(4)}{2}$	$50 \stackrel{?}{=} \frac{20(5)}{2}$
$30=30$	$50=50$

7.

$n=1$	$n=2$
$3(1)+4 \stackrel{?}{=} \frac{1(3(1)+11)}{2}$	$7+3(2)+4 \stackrel{?}{=} \frac{2(3(2)+11)}{2}$
$7 \stackrel{?}{=} \frac{1(14)}{2}$	$17 \stackrel{?}{=} \frac{2(17)}{2}$
$7=7$	$17=17$

$n=3$	$n=4$
$7+10+3(3)+4 \stackrel{?}{=} \frac{3(3(3)+11)}{2}$	$7+10+13+3(4)+4 \stackrel{?}{=} \frac{4(3(4)+11)}{2}$
$30 \stackrel{?}{=} \frac{3(20)}{2}$	$46 \stackrel{?}{=} \frac{4(23)}{2}$
$30=30$	$46=46$

9.

Check $n=1$: $2 \stackrel{?}{=} 1(1+1)$ True for $n=1$

$2=2$

Assume for $n=k$: $2+4+6+\cdots+2k = k(k+1)$

Show for $n=k+1$: $\boxed{2+4+6+\cdots+2k} + 2(k+1) = \boxed{k(k+1)} + 2(k+1)$

$$2+4+6+\cdots+2(k+1) = k^2+k+2k+2$$
$$2+4+6+\cdots+2(k+1) = k^2+3k+2$$
$$2+4+6+\cdots+2(k+1) = (k+1)(k+2)$$

Since this is what results when $n=k+1$ is in the formula, we have shown that the formula works for $n=k+1$ if it works for $n=k$.

11.

Check $n = 1$: $4(1) - 1 \stackrel{?}{=} 1(2(1) + 1)$ True for $n = 1$

$$3 = 3$$

Assume for $n = k$ and show for $n = k + 1$:

$$3 + 7 + 11 + \cdots + (4k - 1) = k(2k + 1)$$

$$\boxed{3 + 7 + 11 + \cdots + (4k - 1)} + 4(k + 1) - 1 = \boxed{k(2k + 1)} + 4(k + 1) - 1$$

$$3 + 7 + 11 + \cdots + [4(k + 1) - 1] = 2k^2 + k + 4k + 4 - 1$$

$$3 + 7 + 11 + \cdots + [4(k + 1) - 1] = 2k^2 + 5k + 3$$

$$3 + 7 + 11 + \cdots + [4(k + 1) - 1] = (k + 1)(2k + 3)$$

$$3 + 7 + 11 + \cdots + [4(k + 1) - 1] = (k + 1)(2(k + 1) + 1)$$

Since this is what results when $n = k + 1$ is in the formula, we have shown that the formula works for $n = k + 1$ if it works for $n = k$.

13.

Check $n = 1$: $14 - 4(1) \stackrel{?}{=} 12(1) - 2(1)^2$ True for $n = 1$

$$10 = 10$$

Assume for $n = k$ and show for $n = k + 1$:

$$10 + 6 + 2 + \cdots + (14 - 4k) = 12k - 2k^2$$

$$\boxed{10 + 6 + 2 + \cdots + (14 - 4k)} + 14 - 4(k + 1) = \boxed{12k - 2k^2} + 14 - 4(k + 1)$$

$$10 + 6 + 2 + \cdots + (14 - 4(k + 1)) = 12k - 2k^2 + 14 - 4k - 4$$

$$10 + 6 + 2 + \cdots + (14 - 4(k + 1)) = 12k + 12 - 2k^2 - 4k - 2$$

$$10 + 6 + 2 + \cdots + (14 - 4(k + 1)) = 12(k + 1) - 2(k^2 + 2k + 1)$$

$$10 + 6 + 2 + \cdots + (14 - 4(k + 1)) = 12(k + 1) - 2(k + 1)^2$$

Since this is what results when $n = k + 1$ is in the formula, we have shown that the formula works for $n = k + 1$ if it works for $n = k$.

15.

Check $n = 1$: $3(1) - 1 \stackrel{?}{=} \dfrac{1(3(1)+1)}{2}$ True for $n = 1$

$2 = 2$

Assume for $n = k$ and show for $n = k + 1$:

$$2 + 5 + 8 + \cdots + (3k - 1) = \frac{k(3k+1)}{2}$$

$$\boxed{2 + 5 + 8 + \cdots + (3k - 1)} + 3(k+1) - 1 = \boxed{\frac{k(3k+1)}{2}} + 3(k+1) - 1$$

$$2 + 5 + 8 + \cdots + (3(k+1) - 1) = \frac{k(3k+1)}{2} + \frac{2 \cdot (3(k+1) - 1)}{2}$$

$$2 + 5 + 8 + \cdots + (3(k+1) - 1) = \frac{3k^2 + k + 6k + 6 - 2}{2}$$

$$2 + 5 + 8 + \cdots + (3(k+1) - 1) = \frac{3k^2 + 7k + 4}{2}$$

$$2 + 5 + 8 + \cdots + (3(k+1) - 1) = \frac{(k+1)(3k+4)}{2}$$

$$2 + 5 + 8 + \cdots + (3(k+1) - 1) = \frac{(k+1)(3(k+1)+1)}{2}$$

Since this is what results when $n = k + 1$ is in the formula, we have shown that the formula works for $n = k + 1$ if it works for $n = k$.

17.

Check $n = 1$: $1^2 \stackrel{?}{=} \dfrac{1(1+1)(2(1)+1)}{6}$ True for $n = 1$

$1 = 1$

Assume for $n = k$ and show for $n = k + 1$:

$$1^2 + 2^2 + 3^2 + \cdots + k^2 = \frac{k(k+1)(2k+1)}{6}$$

$$\boxed{1^2 + 2^2 + 3^2 + \cdots + k^2} + (k+1)^2 = \boxed{\frac{k(k+1)(2k+1)}{6}} + (k+1)^2$$

$$1^2 + 2^2 + 3^2 + \cdots + (k+1)^2 = \frac{(2k^2 + k)(k+1)}{6} + \frac{6(k+1)(k+1)}{6}$$

$$1^2 + 2^2 + 3^2 + \cdots + (k+1)^2 = \frac{(2k^2 + k + 6(k+1))(k+1)}{6}$$

$$1^2 + 2^2 + 3^2 + \cdots + (k+1)^2 = \frac{(2k^2 + 7k + 6)(k+1)}{6}$$

$$1^2 + 2^2 + 3^2 + \cdots + (k+1)^2 = \frac{(2k+3)(k+2)(k+1)}{6}$$

$$1^2 + 2^2 + 3^2 + \cdots + (k+1)^2 = \frac{(k+1)(k+2)(2k+3)}{6}$$

Since this is what results when $n = k + 1$ is in the formula, we have shown that the formula works for $n = k + 1$ if it works for $n = k$.

19.

Check $n = 1$: $\frac{5}{3}(1) - \frac{4}{3} \stackrel{?}{=} 1\left(\frac{5}{6}(1) - \frac{1}{2}\right)$ True for $n = 1$

$\frac{1}{3} = \frac{1}{3}$

Assume for $n = k$ and show for $n = k + 1$:

$$\frac{1}{3} + 2 + \frac{11}{3} + \cdots + \left(\frac{5}{3}k - \frac{4}{3}\right) = k\left(\frac{5}{6}k - \frac{1}{2}\right)$$

$$\boxed{\frac{1}{3} + 2 + \frac{11}{3} + \cdots + \left(\frac{5}{3}k - \frac{4}{3}\right)} + \left(\frac{5}{3}(k+1) - \frac{4}{3}\right) = \boxed{k\left(\frac{5}{6}k - \frac{1}{2}\right)} + \left(\frac{5}{3}(k+1) - \frac{4}{3}\right)$$

$$\frac{1}{3} + 2 + \frac{11}{3} + \cdots + \left(\frac{5}{3}(k+1) - \frac{4}{3}\right) = \frac{5}{6}k^2 - \frac{1}{2}k + \frac{5}{3}k + \frac{5}{3} - \frac{4}{3}$$

$$\frac{1}{3} + 2 + \frac{11}{3} + \cdots + \left(\frac{5}{3}(k+1) - \frac{4}{3}\right) = \frac{5}{6}k^2 + \frac{7}{6}k + \frac{1}{3}$$

$$\frac{1}{3} + 2 + \frac{11}{3} + \cdots + \left(\frac{5}{3}(k+1) - \frac{4}{3}\right) = (k+1)\left(\frac{5}{6}k + \frac{1}{3}\right)$$

$$\frac{1}{3} + 2 + \frac{11}{3} + \cdots + \left(\frac{5}{3}(k+1) - \frac{4}{3}\right) = (k+1)\left(\frac{5}{6}k + \frac{5}{6} - \frac{1}{2}\right)$$

$$\frac{1}{3} + 2 + \frac{11}{3} + \cdots + \left(\frac{5}{3}(k+1) - \frac{4}{3}\right) = (k+1)\left(\frac{5}{6}(k+1) - \frac{1}{2}\right)$$

Since this is what results when $n = k + 1$ is in the formula, we have shown that the formula works for $n = k + 1$ if it works for $n = k$.

21.

Check $n = 1$: $\left(\frac{1}{2}\right)^1 \stackrel{?}{=} 1 - \left(\frac{1}{2}\right)^1$ True for $n = 1$

$\frac{1}{2} = \frac{1}{2}$

Assume for $n = k$ and show for $n = k + 1$:

$$\frac{1}{2} + \frac{1}{4} + \frac{1}{8} + \cdots + \left(\frac{1}{2}\right)^k = 1 - \left(\frac{1}{2}\right)^k$$

$$\boxed{\frac{1}{2} + \frac{1}{4} + \frac{1}{8} + \cdots + \left(\frac{1}{2}\right)^k} + \left(\frac{1}{2}\right)^{k+1} = \boxed{1 - \left(\frac{1}{2}\right)^k} + \left(\frac{1}{2}\right)^{k+1}$$

$$\frac{1}{2} + \frac{1}{4} + \frac{1}{8} + \cdots + \left(\frac{1}{2}\right)^{k+1} = 1 - 2\left(\frac{1}{2}\right)\left(\frac{1}{2}\right)^k + \left(\frac{1}{2}\right)^{k+1}$$

$$\frac{1}{2} + \frac{1}{4} + \frac{1}{8} + \cdots + \left(\frac{1}{2}\right)^{k+1} = 1 - 2\left(\frac{1}{2}\right)^{k+1} + \left(\frac{1}{2}\right)^{k+1}$$

$$\frac{1}{2} + \frac{1}{4} + \frac{1}{8} + \cdots + \left(\frac{1}{2}\right)^{k+1} = 1 - \left(\frac{1}{2}\right)^{k+1}$$

Since this is what results when $n = k + 1$ is in the formula, we have shown that the formula works for $n = k + 1$ if it works for $n = k$.

23.

Check $n = 1$: $2^{1-1} \stackrel{?}{=} 2^1 - 1$ True for $n = 1$ $1 = 1$
Assume for $n = k$ and show for $n = k + 1$: $2^0 + 2^1 + 2^2 + \cdots + 2^{k-1} = 2^k - 1$ $\boxed{2^0 + 2^1 + 2^2 + \cdots + 2^{k-1}} + 2^k = \boxed{2^k - 1} + 2^k$ $2^0 + 2^1 + 2^2 + \cdots + 2^k = 2 \cdot 2^k - 1$ $2^0 + 2^1 + 2^2 + \cdots + 2^k = 2^{k+1} - 1$
Since this is what results when $n = k + 1$ is in the formula, we have shown that the formula works for $n = k + 1$ if it works for $n = k$.

25.

Check $n = 1$: $x - y$ is a factor of $x^1 - y^1$. True for $n = 1$
Assume for $n = k$ and show for $n = k + 1$: Thus, we assume that $x^k - y^k = (x - y)(\text{SOMETHING})$. $x^{k+1} - y^{k+1} = x^{k+1} - xy^k + xy^k - y^{k+1}$ $= x(x^k - y^k) + y^k(x - y)$ $= x(x - y)(\text{SOMETHING}) + y^k(x - y)$ $= (x - y)[x(\text{SOMETHING}) + y^k]$
We have shown that $x - y$ is a factor of $x^{k+1} - y^{k+1}$ if it is a factor of $x^k - y^k$.

27. The formula is true for $n = 3$, since a triangle has $180° = (3 - 2) \cdot 180°$. Next, assume that a polygon with k sides has an angle sum of $(k - 2) \cdot 180°$. Take a polygon with $k + 1$ sides. Consider two adjacent sides with a common endpoint. Connect the endpoints which are NOT common to both sides. The figure is now a polygon with k sides with a triangle adjacent to it.

$$\boxed{\begin{array}{c}\text{Sum of angles of}\\(k+1)\text{-sided polygon}\end{array}} = \boxed{\begin{array}{c}\text{Sum of angles of}\\k\text{-sided polygon}\end{array}} + \boxed{\begin{array}{c}\text{Sum of angles of}\\\text{triangle}\end{array}}$$

$$= (k - 2) \cdot 180° + 180° = (k - 1) \cdot 180° = [(k + 1) - 2] \cdot 180°$$

Thus, the formula works for $n = k + 1$ if it works for $n = k$.

29. Assume for $n = k$ and show for $n = k + 1$:

$$1 + 2 + 3 + \cdots + k = \tfrac{k}{2}(k + 1) + 1$$

$$\boxed{1 + 2 + 3 + \cdots + k} + (k + 1) = \boxed{\tfrac{k}{2}(k + 1) + 1} + (k + 1)$$

$$1 + 2 + 3 + \cdots + (k + 1) = \tfrac{1}{2}k(k + 1) + (k + 1) + 2$$

$$1 + 2 + 3 + \cdots + (k + 1) = \tfrac{1}{2}(k + 1)(k + 2)$$

$$1 + 2 + 3 + \cdots + (k + 1) = \tfrac{(k+1)}{2}(k + 2)$$

The formula works for $n = k + 1$ if it works for $n = k$. However, the formula does not work for $n = 1$. Thus, the formula does not work for all natural numbers.

SECTION 8.5

31.

Check $n=1$: $7^1-1=7-1=6$ True for $n=1$ Thus, 7^1-1 is divisible by 6.	
Assume for $n=k$ and show for $n=k+1$: 7^k-1 is divisible by 6, so $7^k-1=6\cdot x$, where x is some natural number. Then $7^{k+1}-1=7^k\cdot 7-1=(6x+1)\cdot 7-1=42x+6=6(7x+1)$ Thus, $7^{k+1}-1$ is divisible by 6.	
We have shown that $7^{k+1}-1$ is divisible by 6 if 7^k-1 is divisible by 6.	

33.

Check $n=1$: $1+r^1 \stackrel{?}{=} \frac{1-r^2}{1-r}$ True for $n=1$

$$1+r \stackrel{?}{=} \frac{(1+r)(1-r)}{1-r}$$

$$1+r=1+r$$

Assume for $n=k$ and show for $n=k+1$:

$$1+r+r^2+r^3+\cdots+r^k=\frac{1-r^{k+1}}{1-r}$$

$$\boxed{1+r+r^2+r^3+\cdots+r^k}+r^{k+1}=\boxed{\frac{1-r^{k+1}}{1-r}}+r^{k+1}$$

$$1+r+r^2+r^3+\cdots+r^{k+1}=\frac{1-r^{k+1}+r^{k+1}(1-r)}{1-r}$$

$$1+r+r^2+r^3+\cdots+r^{k+1}=\frac{1-r^{k+1}+r^{k+1}-r^{k+2}}{1-r}$$

$$1+r+r^2+r^3+\cdots+r^{k+1}=\frac{1-r^{(k+1)+1}}{1-r}$$

Since this is what results when $n=k+1$ is in the formula, we have shown that the formula works for $n=k+1$ if it works for $n=k$.

35.

Check $n=1$: $a^m a^1=a^m\cdot a=a^{m+1}$, by definition   True for $n=1$

Assume for $n=k$ and show for $n=k+1$:

$$a^m a^k=a^{m+k}$$

$$a^m a^k a=a^{m+k}a$$

$$a^m a^{k+1}=a^{m+k+1}$$

$$a^m a^{k+1}=a^{m+(k+1)}$$

We have shown that the formula works for $n=k+1$ if it works for $n=k$.

37. **a.** 1 **b.** 3 **c.** 7 **d.** 15

39. $3x + 4y \leq 12$

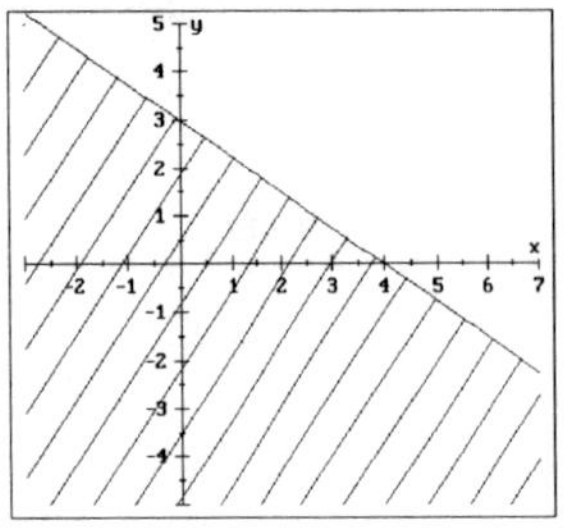

41. $2x + y > 5$

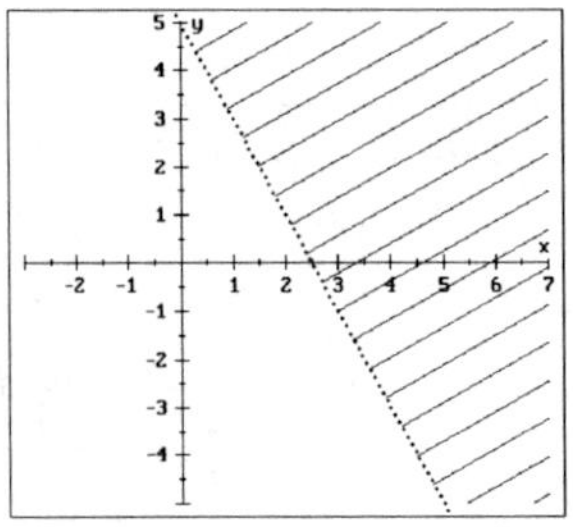

Exercises 8.6 (page 769)

1. $6 \cdot 4 = 24$

3. $\dfrac{n!}{(n-r)!}$

5. 1

7. $\dfrac{n!}{r!(n-r)!}$

9. 1

11. $\dfrac{n!}{a! \cdot b! \cdots}$

13. $P(7,4) = \dfrac{7!}{(7-4)!} = \dfrac{7!}{3!} = \dfrac{7 \cdot 6 \cdot 5 \cdot 4 \cdot 3!}{3!} = 7 \cdot 6 \cdot 5 \cdot 4 = 840$

15. $C(7,4) = \dfrac{7!}{4!(7-4)!} = \dfrac{7!}{4!3!} = \dfrac{7 \cdot 6 \cdot 5 \cdot 4!}{4!3!} = \dfrac{7 \cdot 6 \cdot 5}{3 \cdot 2 \cdot 1} = 35$

17. $P(5,5) = \dfrac{5!}{(5-5)!} = \dfrac{5!}{0!} = \dfrac{5!}{1} = 5 \cdot 4 \cdot 3 \cdot 2 \cdot 1 = 120$

19. $\dbinom{5}{4} = \dfrac{5!}{4!(5-4)!} = \dfrac{5 \cdot 4!}{4!1!} = \dfrac{5}{1} = 5$

21. $\dbinom{5}{0} = \dfrac{5!}{0!(5-0)!} = \dfrac{5!}{0!5!} = \dfrac{1}{1} = 1$

23. $P(5,4) \cdot C(5,3) = \dfrac{5!}{(5-4)!} \cdot \dfrac{5!}{3!(5-3)!} = \dfrac{5!}{1!} \cdot \dfrac{5!}{3!2!} = 120 \cdot 10 = 1200$

25. $\dbinom{5}{3}\dbinom{4}{3}\dbinom{3}{3} = \dfrac{5!}{3!(5-3)!} \cdot \dfrac{4!}{3!(4-3)!} \cdot \dfrac{3!}{3!(3-3)!} = 10 \cdot 4 \cdot 1 = 40$

27. $\dbinom{68}{66} = \dfrac{68!}{66!(68-66)!} = \dfrac{68 \cdot 67 \cdot 66!}{66!2!} = \dfrac{68 \cdot 67}{2 \cdot 1} = 2278$

29. $8 \cdot 6 \cdot 3 = 144$

31. $8 \cdot 10 \cdot 10 \cdot 10 \cdot 10 \cdot 10 \cdot 10 = 8{,}000{,}000$

33. Consider the e and the r to be a block that cannot be divided, say x. Then the problem becomes finding the number of ways to rearrange the letters in the word $numbx$. This can be done in $5!$, or 120 ways. For each of these possibilities, the e and the r could be reversed, doubling the number of possibilities. The answer is $2 \cdot 120$, or 240 ways.

SECTION 8.6

35. The word must appear as $\square\, L\, U\, \square\, \square$, where one of the Fs must appear in each box. This can be done in 3!, or 6 ways.

37. $8! = 40{,}320$

39. The line will look like this:

$$\underset{\text{M}}{\underline{5}}\ \underset{\text{M}}{\underline{4}}\ \underset{\text{M}}{\underline{3}}\ \underset{\text{M}}{\underline{2}}\ \underset{\text{M}}{\underline{1}}\ \underset{\text{W}}{\underline{5}}\ \underset{\text{W}}{\underline{4}}\ \underset{\text{W}}{\underline{3}}\ \underset{\text{W}}{\underline{2}}\ \underset{\text{W}}{\underline{1}}$$

Then there are $5 \cdot 4 \cdot 3 \cdot 2 \cdot 1 \cdot 5 \cdot 4 \cdot 3 \cdot 2 \cdot 1 = 14{,}400$ arrangements.

41. $P(30, 3) = \dfrac{30!}{27!} = 24{,}360$

43. $(8 - 1)! = 7! = 5040$

45. Consider the two people who must sit together as a single person, so that there are 5 "people" who must be arranged in a circle. This can be done in $(5 - 1)! = 4! = 24$ ways. However, the two people who have been seated next to each other could be switched, so that the number of arrangements is doubled. There are $2 \cdot 24 = 48$ possible arrangements.

47. Consider Sally and John as a single person and Martha and Peter as a single person, so that there are 5 "people" who must be arranged in a circle. This can be done in $(5 - 1)! = 4! = 24$ ways. However, each group of 2 people could be switched, so that the number of arrangements will equal $2 \cdot 2 \cdot 24$, or 96.

49. $\dbinom{10}{4} = \dfrac{10!}{4!6!} = 210$

51. $4! = 24$

53. $7! = 5040$

55. $\dfrac{6!}{3!2!1!} = 60$

57. $25 \cdot 24 \cdot 9 \cdot 9 \cdot 8 \cdot 7 = 2{,}721{,}600$

59. $\dbinom{6}{3}\dbinom{8}{3} = 20 \cdot 56 = 1120$

61. $\dbinom{12}{4}\dbinom{10}{3} = 495 \cdot 120 = 59{,}400$

63. $\underset{\text{H}}{\underline{17}} \cdot \underset{\text{W}}{\underline{16}} = 272$

65. $\dbinom{8}{2} = 28$

67. $\dbinom{10}{5} = 252$

69. $\dbinom{30}{5} = 142{,}506$

71. $\dbinom{12}{10} = 66$

73. For each topping, you have two choices, select or do not select:

$$\underset{T_1}{\underline{2}}\ \underset{T_2}{\underline{2}}\ \underset{T_3}{\underline{2}}\ \underset{T_4}{\underline{2}}\ \underset{T_5}{\underline{2}}\ \underset{T_6}{\underline{2}}\ \underset{T_7}{\underline{2}}\ \underset{T_8}{\underline{2}}$$

$2 \cdot 2 \cdot 2 \cdot 2 \cdot 2 \cdot 2 \cdot 2 \cdot 2 = 256$

75. 9th row of triangle: 1 8 28 56 70 56 28 8 1; 4th number in row: $\boxed{56}$

77. $C(n,n) = \dfrac{n!}{n!(n-n)!} = \dfrac{n!}{n!0!} = \dfrac{n!}{n!} = 1$

79. $\dbinom{n}{n-r} = \dfrac{n!}{(n-r)!(n-(n-r))!} = \dfrac{n!}{(n-r)!r!} = \dfrac{n!}{r!(n-r)!} = \dbinom{n}{r}$

81. **Answers may vary.**

83. $\log_x 16 = 4$
$x^4 = 16 \Rightarrow x = 2$

85. $\log_{\sqrt{7}} 49 = x$
$\left(\sqrt{7}\right)^x = 49 \Rightarrow x = 4$

87. true

89. true

Exercises 8.7 (page 776)

1. experiment

3. $\frac{n(E)}{n(S)}$

5. $\{(1,H), (2,H), (3,H), (4,H), (5,H), (6,H), (1,T), (2,T), (3,T), (4,T), (5,T), (6,T)\}$

7. $\{A, B, C, D, E, F, G, H, I, J, K, L, M, N, O, P, Q, R, S, T, U, V, W, X, Y, Z\}$

9. $\frac{1}{6}$

11. $\frac{4}{6} = \frac{2}{3}$

13. $\frac{19}{42}$

15. $\frac{13}{42}$

17. $\frac{3}{8}$

19. $\frac{0}{8} = 0$

21. rolls of 4: $\{(1,3), (2,2), (3,1)\}$
Probability $= \dfrac{3}{36} = \dfrac{1}{12}$

23. $\dfrac{\text{\# aces}}{\text{\# cards}} \cdot \dfrac{\text{\# aces}}{\text{\# cards}} = \dfrac{4}{52} \cdot \dfrac{4}{52} = \dfrac{1}{169}$

25. $\dfrac{\text{\# red}}{\text{\# eggs}} = \dfrac{5}{12}$

27. $\dfrac{\text{\# ways to get 13 cards of the same suit}}{\text{\# ways to get 13 cards from the deck of 52}} = \dfrac{4 \cdot \binom{13}{13}}{\binom{52}{13}} = \dfrac{4}{6.350136 \times 10^{11}} \approx 6.3 \times 10^{-12}$

29. impossible $\Rightarrow 0$

31. $\dfrac{\text{\# face cards}}{\text{\# cards in deck}} = \dfrac{12}{52} = \dfrac{3}{13}$

33. $\dfrac{\text{\# ways to get 5 orange}}{\text{\# ways to get 5 cubes}} = \dfrac{\binom{5}{5}}{\binom{6}{5}} = \dfrac{1}{6}$

35. rolls of 11: $(1,4,6), (1,5,5), (1,6,4), (2,3,6), (2,4,5), (2,5,4), (2,6,3), (3,2,6), (3,3,5)$
$(3,4,4), (3,5,3), (3,6,2), (4,1,6), (4,2,5), (4,3,4), (4,4,3), (4,5,2), (4,6,1), (5,1,5),$
$(5,2,4), (5,3,3), (5,4,2), (5,5,1), (6,1,4), (6,2,3), (6,3,2), (6,4,1)$
Probability $= \frac{27}{216} = \frac{1}{8}$

37. $\frac{\binom{5}{3}}{2^5} = \frac{10}{32} = \frac{5}{16}$

39. $SSSS, SSSF, SSFS, SSFF, SFSS, SFSF, SFFS, SFFF,$
$FSSS, FSSF, FSFS, FSFF, FFSS, FFSF, FFFS, FFFF$

41. $\frac{4}{16} = \frac{1}{4}$

43. $\frac{4}{16} = \frac{1}{4}$

45. $\frac{16}{16} = 1$

47. $\frac{32}{119}$

49. $\frac{\binom{8}{4}}{\binom{10}{4}} = \frac{70}{210} = \frac{1}{3}$

51. $P(A \cap B) = P(A) \cdot P(B|A)$
$= 0.3(0.6) = 0.18$

53. $P(A \cap B) = P(A) \cdot P(B|A)$
$= 0.2(0.7) = 0.14$

55. $P(A \cap B) = P(A) \cdot P(B|A)$
$0.25 = 0.75P(B|A)$
$0.33 \approx P(B|A)$
$33\% \approx P(B|A)$

57. no

59. $|x + 3| = 7$
$x + 3 = 7$ **OR** $x + 3 = -7$
$x = 4$ $\quad$ $x = -10$

61. $|x - 3| < 7$
$-7 < x - 3 < 7$
$-4 < x < 10$
$(-4, 10)$

Chapter 8 Review (page 779)

1. $6! = 6 \cdot 5 \cdot 4 \cdot 3 \cdot 2 \cdot 1 = 720$

2. $7! \cdot 0! \cdot 1! \cdot 3! = 5040 \cdot 1 \cdot 1 \cdot 6 = 30{,}240$

3. $\frac{8!}{7!} = \frac{8 \cdot 7!}{7!} = 8$

4. $\frac{5! \cdot 7! \cdot 8!}{6! \cdot 9!} = \frac{5! \cdot 7 \cdot 6! \cdot 8!}{6! \cdot 9 \cdot 8!} = \frac{5! \cdot 7}{9} = \frac{280}{3}$

5. $(x + y)^3 = x^3 + \frac{3!}{1!2!}x^2y + \frac{3!}{2!1!}xy^2 + y^3 = x^3 + 3x^2y + 3xy^2 + y^3$

6. $(p + q)^4 = p^4 + \frac{4!}{1!3!}p^3q + \frac{4!}{2!2!}p^2q^2 + \frac{4!}{3!1!}pq^3 + q^4 = p^4 + 4p^3q + 6p^2q^2 + 4pq^3 + q^4$

7. $(a - b)^5 = a^5 + \frac{5!}{1!4!}a^4(-b) + \frac{5!}{2!3!}a^3(-b)^2 + \frac{5!}{3!2!}a^2(-b)^3 + \frac{5!}{4!1!}a(-b)^4 + (-b)^5$
$= a^5 - 5a^4b + 10a^3b^2 - 10a^2b^3 + 5ab^4 - b^5$

8. $(2a - b)^3 = (2a)^3 + \frac{3!}{1!2!}(2a)^2(-b) + \frac{3!}{2!1!}(2a)(-b)^2 + (-b)^3 = 8a^3 - 12a^2b + 6ab^2 - b^3$

9. The 4th term will involve b^3.
$\frac{8!}{5!3!}a^5b^3 = 56a^5b^3$

10. The 3rd term will involve $(-y)^2$.
$\frac{5!}{3!2!}(2x)^3(-y)^2 = 80x^3y^2$

11. The 7th term will involve $(-y)^6$.

$$\frac{9!}{3!6!}x^3(-y)^6 = 84x^3y^6$$

12. The 4th term will involve 7^3.

$$\frac{6!}{3!3!}(4x)^3 7^3 = 439{,}040x^3$$

13. $4^3 - 1 = 63$

14. $\dfrac{4^2+2}{2} = \dfrac{18}{2} = 9$

15.
$$\begin{aligned} a_1 &= 5 \\ a_2 &= 3a_1 + 2 = 3(5) + 2 = 17 \\ a_3 &= 3a_2 + 2 = 3(17) + 2 = 53 \\ a_4 &= 3a_3 + 2 = 3(53) + 2 = 161 \end{aligned}$$

16.
$$\begin{aligned} a_1 &= -2 \\ a_2 &= 2a_1^2 = 2(-2)^2 = 8 \\ a_3 &= 2a_2^2 = 2(8)^2 = 128 \\ a_4 &= 2a_3^2 = 2(128)^2 = 32{,}768 \end{aligned}$$

17. $\displaystyle\sum_{k=1}^{4} 3k^2 = 3\sum_{k=1}^{4} k^2 = 3\left(1^2 + 2^2 + 3^2 + 4^2\right) = 3(30) = 90$

18. $\displaystyle\sum_{k=1}^{10} 6 = 10(6) = 60$

19.
$$\begin{aligned} \sum_{k=5}^{8}\left(k^3 + 3k^2\right) = \sum_{k=5}^{8} k^3 + 3\sum_{k=5}^{8} k^2 &= \left(5^3 + 6^3 + 7^3 + 8^3\right) + 3\left(5^2 + 6^2 + 7^2 + 8^2\right) \\ &= 1718 \end{aligned}$$

20. $\displaystyle\sum_{k=1}^{30}\left(\frac{3}{2}k - 12\right) - \frac{3}{2}\sum_{k=1}^{30} k = \frac{3}{2}\sum_{k=1}^{30} k - \sum_{k=1}^{30} 12 - \frac{3}{2}\sum_{k=1}^{30} k = -\sum_{k=1}^{30} 12 = -360$

21.
$$\begin{aligned} &a = 5,\ d = 4 \\ a_{29} &= a + (n-1)d \\ &= 5 + (29-1)4 = 117 \end{aligned}$$

22.
$$\begin{aligned} &a = 8,\ d = 7 \\ a_{40} &= a + (n-1)d \\ &= 8 + (40-1)7 = 281 \end{aligned}$$

23.
$$\begin{aligned} &a = 6,\ d = -7 \\ a_{15} &= a + (n-1)d \\ &= 6 + (15-1)(-7) = -92 \end{aligned}$$

24.
$$\begin{aligned} &a = \tfrac{1}{2},\ d = -2 \\ a_{35} &= a + (n-1)d \\ &= \tfrac{1}{2} + (35-1)(-2) = -\tfrac{135}{2} \end{aligned}$$

25.
$$\begin{aligned} &a = 2,\ a_5 = 8 \\ &8 = 2 + 4d \\ &6 = 4d \\ &\frac{3}{2} = d \Rightarrow 2, \boxed{\frac{7}{2}, 5, \frac{13}{2}}, 8 \end{aligned}$$

26.
$$\begin{aligned} &a = 10,\ a_7 = 100 \\ &100 = 10 + 6d \\ &90 = 6d \\ &15 = d \Rightarrow 10, \boxed{25, 40, 55, 70, 85}, 100 \end{aligned}$$

27.
$$\begin{aligned} &a = 5,\ d = 4 \\ a_{40} &= a + (n-1)d = 5 + 39(4) = 161 \\ S_{40} &= \frac{n(a + a_{40})}{2} = \frac{40(5 + 161)}{2} = 3320 \end{aligned}$$

28.
$$\begin{aligned} &a = 8,\ d = 7 \\ a_{40} &= a + (n-1)d = 8 + 39(7) = 281 \\ S_{40} &= \frac{n(a + a_{40})}{2} = \frac{40(8 + 281)}{2} = 5780 \end{aligned}$$

CHAPTER 8 REVIEW

29. $a = 6, d = -7$

$$a_{40} = a + (n-1)d = 6 + 39(-7) = -267;\ S_{40} = \frac{n(a + a_{40})}{2} = \frac{40(6 - 267)}{2} = -5220$$

30. $a = \frac{1}{2}, d = -2$

$$a_{40} = a + (n-1)d = \frac{1}{2} + 39(-2) = -\frac{155}{2};\ S_{40} = \frac{n(a + a_{40})}{2} = \frac{40\left(\frac{1}{2} - \frac{155}{2}\right)}{2} = -1540$$

31. $a = 81, r = \frac{1}{3}$

$$a_{11} = ar^{n-1} = 81\left(\tfrac{1}{3}\right)^{10} = \tfrac{1}{729}$$

32. $a = 2, r = 3$

$$a_9 = ar^{n-1} = 2(3)^8 = 13{,}122$$

33. $a = 9, r = \frac{1}{2}$

$$a_{15} = ar^{n-1} = 9\left(\frac{1}{2}\right)^{14} = \frac{9}{16{,}384}$$

34. $a = 8, r = -\frac{1}{5}$

$$a_7 = ar^{n-1} = 8\left(-\frac{1}{5}\right)^6 = \frac{8}{15{,}625}$$

35.

$$a_5 = ar^4$$
$$8 = 2r^4$$
$$4 = r^4$$
$$\pm\sqrt[4]{4} = r \Rightarrow r = \pm\sqrt{2}$$

Use $r = +\sqrt{2}$:

$2, \boxed{2\sqrt{2}, 4, 4\sqrt{2}}, 8$

36.

$$a_6 = ar^5$$
$$64 = -2r^5$$
$$-32 = r^5$$
$$-2 = r$$

$-2, \boxed{4, -8, 16, -32}, 64$

37. $a_3 = ar^2$

$64 = 4r^2$

$16 = r^2 \Rightarrow r = 4$ (problem specifies positive) $\Rightarrow 4, \boxed{16}, 64$

38. $a = 81, r = \frac{1}{3}, n = 8$

$$S_8 = \frac{a - ar^n}{1 - r} = \frac{81 - 81\left(\frac{1}{3}\right)^8}{1 - \frac{1}{3}} = \frac{81 - \frac{1}{81}}{\frac{2}{3}} = \frac{3280}{27}$$

39. $a = 2, r = 3, n = 8$

$$S_8 = \frac{a - ar^n}{1 - r} = \frac{2 - 2(3)^8}{1 - 3} = \frac{-13{,}120}{-2} = 6560$$

40. $a = 9, r = \frac{1}{2}, n = 8$

$$S_8 = \frac{a - ar^n}{1 - r} = \frac{9 - 9\left(\frac{1}{2}\right)^8}{1 - \frac{1}{2}} = \frac{\frac{2295}{256}}{\frac{1}{2}} = \frac{2295}{128}$$

41. $a = 8, r = -\frac{1}{5}, n = 8$

$$S_8 = \frac{a - ar^n}{1 - r} = \frac{8 - 8\left(-\frac{1}{5}\right)^8}{1 - \left(-\frac{1}{5}\right)} = \frac{\frac{3{,}124{,}992}{390{,}625}}{\frac{6}{5}} = \frac{520{,}832}{78{,}125}$$

42. $a = \frac{1}{3}, r = 3, n = 8$

$$S_8 = \frac{a - ar^n}{1 - r} = \frac{\frac{1}{3} - \frac{1}{3}(3)^8}{1 - 3} = \frac{-\frac{6560}{3}}{-2} = \frac{3280}{3}$$

43. $a_7 = ar^6$

$$= 2\sqrt{2}\left(\sqrt{2}\right)^6 = 16\sqrt{2}$$

44. $a = \dfrac{1}{3}, r = \dfrac{1}{2}$

$$S_\infty = \frac{a}{1 - r} = \frac{\frac{1}{3}}{1 - \frac{1}{2}} = \frac{\frac{1}{3}}{\frac{1}{2}} = \frac{2}{3}$$

45. $a = \dfrac{1}{5}, r = -\dfrac{2}{3}$

$$S_\infty = \frac{a}{1 - r} = \frac{\frac{1}{5}}{1 - \left(-\frac{2}{3}\right)} = \frac{\frac{1}{5}}{\frac{5}{3}} = \frac{3}{25}$$

46. $a = \dfrac{1}{3}, r = \dfrac{3}{2} > 1 \Rightarrow$ no sum

47. $a = \dfrac{1}{2}, r = \dfrac{1}{2}$

$$S_\infty = \frac{a}{1 - r} = \frac{\frac{1}{2}}{1 - \frac{1}{2}} = \frac{\frac{1}{2}}{\frac{1}{2}} = 1$$

48. $a = \dfrac{3}{10}, r = \dfrac{1}{10}$

$$S_\infty = \frac{a}{1 - r} = \frac{\frac{3}{10}}{1 - \frac{1}{10}} = \frac{\frac{3}{10}}{\frac{9}{10}} = \frac{1}{3}$$

49. $a = \dfrac{9}{10}, r = \dfrac{1}{10}$

$$S_\infty = \frac{a}{1 - r} = \frac{\frac{9}{10}}{1 - \frac{1}{10}} = \frac{\frac{9}{10}}{\frac{9}{10}} = 1$$

50. $a = \dfrac{17}{100}, r = \dfrac{1}{100}$

$$S_\infty = \frac{a}{1 - r} = \frac{\frac{17}{100}}{1 - \frac{1}{100}} = \frac{\frac{17}{100}}{\frac{99}{100}} = \frac{17}{99}$$

51. $a = \dfrac{45}{100}, r = \dfrac{1}{100}$

$$S_\infty = \frac{a}{1 - r} = \frac{\frac{45}{100}}{1 - \frac{1}{100}} = \frac{\frac{45}{100}}{\frac{99}{100}} = \frac{5}{11}$$

52. $a = 3000, r = 1 + \dfrac{0.0775}{365}, n = 2190$

$$ar^{365} = 3000\left(1 + \frac{0.0775}{365}\right)^{2190} \approx \$4775.81$$

The amount will be \$4775.81.

53. $a = 4000, r = 1.05, n = 10$

$ar^{10} = 4000(1.05)^{10} \approx 6516$ in 10 years

$ar^{-5} = 4000(1.05)^{-5} \approx 3134$ 5 years ago

54. $a = 10{,}000;\ r = 0.90;\ ar^{10} = 10{,}000(0.90)^{10} \approx \$3486.78$ in 10 years

CHAPTER 8 REVIEW

55.

$n = 1$

$$1^3 \stackrel{?}{=} \frac{1^2(1+1)^2}{4}$$
$$1 = 1$$

$n = 2$

$$1^3 + 2^3 \stackrel{?}{=} \frac{2^2(2+1)^2}{4}$$
$$9 \stackrel{?}{=} \frac{4(9)}{4}$$
$$9 = 9$$

$n = 3$

$$1^3 + 2^3 + 3^3 \stackrel{?}{=} \frac{3^2(3+1)^2}{4}$$
$$36 \stackrel{?}{=} \frac{9(16)}{4}$$
$$36 = 36$$

$n = 4$

$$1^3 + 2^3 + 3^3 + 4^3 \stackrel{?}{=} \frac{4^2(4+1)^2}{4}$$
$$100 \stackrel{?}{=} \frac{16(25)}{4}$$
$$100 = 100$$

56. Check $n = 1$: $1^3 = \frac{1^2(1+1)^2}{4}$ True for $n = 1$

$$1 = 1$$

Assume for $n = k$ and show for $n = k + 1$:

$$1^3 + 2^3 + 3^3 + \cdots + k^3 = \frac{k^2(k+1)^2}{4}$$

$$\boxed{1^3 + 2^3 + 3^3 + \cdots + k^3} + (k+1)^3 = \boxed{\frac{k^2(k+1)^2}{4}} + (k+1)^3$$

$$1^3 + 2^3 + 3^3 + \cdots + (k+1)^3 = \frac{k^2(k+1)^2 + 4(k+1)^3}{4}$$

$$1^2 + 2^2 + 3^2 + \cdots + (k+1)^2 = \frac{(k+1)^2[k^2 + 4(k+1)]}{4}$$

$$1^2 + 2^2 + 3^2 + \cdots + (k+1)^2 = \frac{(k+1)^2(k+2)^2}{4}$$

Since this is what results when $n = k + 1$ is in the formula, we have shown that the formula works for $n = k + 1$ if it works for $n = k$.

57. $P(8, 5) = \frac{8!}{(8-5)!} = \frac{8!}{3!} = 8 \cdot 7 \cdot 6 \cdot 5 \cdot 4 = 6720$

58. $C(7, 4) = \frac{7!}{4!(7-4)!} = \frac{7!}{4!3!} = \frac{7 \cdot 6 \cdot 5 \cdot 4!}{4! \cdot 3 \cdot 2 \cdot 1} = 35$

59. $0! \cdot 1! = 1 \cdot 1 = 1$

60. $P(10, 2) \cdot C(10, 2) = \frac{10!}{8!} \cdot \frac{10!}{2!8!} = 90 \cdot 45 = 4050$

61. $P(8, 6) \cdot C(8, 6) = \frac{8!}{2!} \cdot \frac{8!}{6!2!} = 20{,}160 \cdot 28 = 564{,}480$

62. $C(8,5)\cdot C(6,2)=\dfrac{8!}{5!3!}\cdot\dfrac{6!}{2!4!}=56\cdot 15=840$

63. $C(7,5)\cdot P(4,0)=\dfrac{7!}{5!2!}\cdot\dfrac{4!}{4!}=21\cdot 1=21$

64. $C(12,10)\cdot C(11,0)=\dfrac{12!}{10!2!}\cdot\dfrac{11!}{0!11!}=66\cdot 1=66$

65. $\dfrac{P(8,5)}{C(8,5)}=\dfrac{6720}{56}=120$

66. $\dfrac{C(8,5)}{C(13,5)}=\dfrac{56}{1287}$

67. $\dfrac{C(6,3)}{C(10,3)}=\dfrac{20}{120}=\dfrac{1}{6}$

68. $\dfrac{C(13,5)}{C(52,5)}=\dfrac{1287}{2{,}598{,}960}=\dfrac{33}{66{,}640}$

69. Consider each set of two people who must sit together as a single person, so that there are 8 "people" who must be arranged in a circle. This can be done in $(8-1)!=7!=5040$ ways. However, each pair seated next to each other could be switched, so that the number of arrangements is multiplied by 4. There are $4\cdot 5040=20{,}160$ possible arrangements.

70. $\dfrac{9!}{2!2!}=90{,}720$

71.

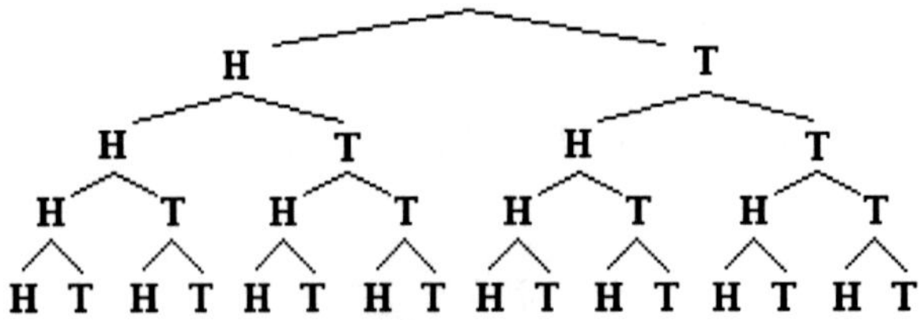

72. $\dbinom{4}{3}\dbinom{4}{2}=4\cdot 6=24$

73. $\dfrac{\binom{4}{3}\binom{4}{2}}{\binom{52}{5}}=\dfrac{24}{2{,}598{,}960}=\dfrac{1}{108{,}290}$

74. $1-\dfrac{1}{108{,}290}=\dfrac{108{,}289}{108{,}290}$

75. $\dfrac{\binom{4}{4}\binom{4}{4}\binom{4}{4}\binom{4}{1}}{\binom{52}{13}}\approx\dfrac{4}{6.35\times 10^{11}}\approx 6.3\times 10^{-12}$

76. $\dfrac{\binom{8}{3}\binom{6}{2}}{\binom{14}{5}}=\dfrac{840}{2002}=\dfrac{60}{143}$

77. $\dfrac{13+13}{52}=\dfrac{1}{2}$

78. $\dfrac{26+4-2}{52}=\dfrac{7}{13}$

79. $\dfrac{1}{\binom{52}{5}}=\dfrac{1}{2{,}598{,}960}$

80. $\dfrac{4\binom{13}{5}}{\binom{52}{5}}=\dfrac{5148}{2{,}598{,}960}=\dfrac{33}{16{,}660}$

81. Use the tree diagram in #71. $\Rightarrow \dfrac{15}{16}$

Chapter 8 Test (page 786)

1. $3! \cdot 0! \cdot 4! \cdot 1! = 6 \cdot 1 \cdot 24 \cdot 1 = 144$

2. $\dfrac{2! \cdot 4! \cdot 6! \cdot 8!}{3! \cdot 5! \cdot 7!} = 2! \cdot \dfrac{4!}{3!} \cdot \dfrac{6!}{5!} \cdot \dfrac{8!}{7!}$
$= 2 \cdot 4 \cdot 6 \cdot 8 = 384$

3. The 2nd term will involve $(2y)^1$.
$\dfrac{5!}{4!1!}x^4(2y)^1 = 10x^4y$

4. The 7th term will involve $(-b)^6$.
$\dfrac{8!}{2!6!}(2a)^2(-b)^6 = 112a^2b^6$

5. $\displaystyle\sum_{k=1}^{3}(4k+1) = 4\sum_{k=1}^{3}k + \sum_{k=1}^{3}1 = 4(1+2+3) + 3(1) = 24 + 3 = 27$

6. $\displaystyle\sum_{k=2}^{4}(3k-21) = 3\sum_{k=2}^{4}k - \sum_{k=2}^{4}21 = 3(2+3+4) - 3(21) = 27 - 63 = -36$

7. $a = 2, d = 3$
$a_{10} = a + (n-1)d = 2 + 9(3) = 29$
$S_{10} = \dfrac{n(a + a_{10})}{2} = \dfrac{10(2+29)}{2} = 155$

8. $a = 5, d = -4$
$a_{10} = a + (n-1)d = 5 + 9(-4) = -31$
$S_{10} = \dfrac{n(a + a_{10})}{2} = \dfrac{10(5-31)}{2} = -130$

9. $a = 4, a_5 = 24$
$24 = 4 + 4d$
$20 = 4d$
$5 = d \Rightarrow 4, \boxed{9, 14, 19}, 24$

10. $a_4 = ar^3$
$-54 = -2r^3$
$27 = r^3$
$3 = r \Rightarrow -2, \boxed{-6, -18}, -54$

11. $a = \frac{1}{4}, r = 2, n = 10$
$S_{10} = \dfrac{a - ar^n}{1-r} = \dfrac{\frac{1}{4} - \frac{1}{4}(2)^{10}}{1-2}$
$= \dfrac{-\frac{1023}{4}}{-1}$
$= \dfrac{1023}{4} = 255.75$

12. $a = 6, r = \frac{1}{3}, n = 10$
$S_{10} = \dfrac{a - ar^n}{1-r} = \dfrac{6 - 6\left(\frac{1}{3}\right)^{10}}{1 - \frac{1}{3}}$
$= \dfrac{\frac{354{,}288}{59{,}049}}{\frac{2}{3}}$
$= \dfrac{177{,}144}{19{,}683} \approx 9$

13. $a = c;\ r = 0.75$
$ar^3 = c(0.75)^3 \approx \$0.42c$ in 3 years

14. $a = c;\ r = 1.10$
$ar^4 = c(1.10)^4 \approx \$1.46c$ in 4 years

CHAPTER 8 TEST

15.

Check $n = 1$: $3 = \frac{1}{2}(1)(1+5)$ True for $n = 1$ $3 = 3$
Assume for $n = k$ and show for $n = k+1$: $$3+4+5+\cdots+(k+2) = \frac{1}{2}k(k+5)$$ $$\boxed{3+4+5+\cdots+(k+2)} + ((k+1)+2) = \boxed{\frac{1}{2}k(k+5)} + ((k+1)+2)$$ $$3+4+5+\cdots+((k+1)+2) = \frac{1}{2}k^2 + \frac{7}{2}k + 3$$ $$3+4+5+\cdots+((k+1)+2) = \frac{1}{2}\left(k^2+7k+6\right)$$ $$3+4+5+\cdots+((k+1)+2) = \frac{1}{2}(k+1)(k+6) = \frac{1}{2}(k+1)((k+1)+5)$$
Since this is what results when $n = k+1$ is in the formula, we have shown that the formula works for $n = k+1$ if it works for $n = k$.

16. $8 \cdot 10 \cdot 10 \cdot 10 \cdot 10 \cdot 10 = 800{,}000$

17. $P(7, 2) = \frac{7!}{5!} = 42$

18. $P(4, 4) = \frac{4!}{0!} = 24$

19. $C(8, 2) = \frac{8!}{2!6!} = 28$

20. $C(12, 0) = \frac{12!}{0!12!} = 1$

21. $4!4! = 576$

22. $(6-1)! = 5! = 120$

23. $\frac{5!}{2!} = 60$

24. $\{(H, H, H), (H, H, T), (H, T, H), (H, T, T), (T, H, H), (T, H, T), (T, T, H), (T, T, T)\}$

25. $\frac{1}{6}$

26. $\frac{4+4}{52} = \frac{2}{13}$

27. $\frac{\binom{13}{5}}{\binom{52}{5}} = \frac{33}{66{,}640}$

28. rolls of 9: $\{(3, 6), (4, 5), (5, 4), (6, 3)\}$
Probability $= \frac{4}{36} = \frac{1}{9}$

29. $\frac{\#\text{ blue}}{\#\text{ all}} \cdot \frac{\#\text{ blue}}{\#\text{ all}} = \frac{30}{50} \cdot \frac{29}{49} = \frac{87}{245}$

30. $\frac{\binom{18}{4}}{\binom{20}{4}} = \frac{3060}{4845} = \frac{12}{19}$

Exercise 9.1

1. interest

3. annual

5. future

7. interest

9. periodic rate

11. principal, periodic rate, frequency of compounding, number of years

13. effective

15. $I = Prt = 4500(0.0325)(4) = \$585$

17.
$$I = Prt$$
$$814 = P(0.055)(4)$$
$$\frac{814}{(0.055)(4)} = P$$
$$\$3700 = P$$

19.
$$I = Prt$$
$$7500 = (50{,}000)(r)(2.5)$$
$$\frac{7500}{(50{,}000)(2.5)} = r$$
$$0.06 = r \Rightarrow 6\%$$

21. $A_0 = 1200,\ i = 0.08,\ k = 1,\ n = 1$
$$A_n = A_0(1+i)^{kn}$$
$$A_1 = 1200(1+0.08)^{1\cdot 1} = \$1296$$

23. $A_0 = 1200,\ i = 0.08,\ k = 1,\ n = 5$
$$A_n = A_0(1+i)^{kn}$$
$$A_5 = 1200(1+0.08)^{1\cdot 5} \approx \$1763.19$$

25. $A_0 = 1200,\ i = 0.03,\ k = 1,\ n = 10$
$$A_n = A_0(1+i)^{kn}$$
$$A_{10} = 1200(1+0.03)^{1\cdot 10} \approx \$1612.70$$

27. $A_0 = 1200,\ i = 0.09,\ k = 1,\ n = 10$
$$A_n = A_0(1+i)^{kn}$$
$$A_{10} = 1200(1+0.09)^{1\cdot 10} \approx \$2840.84$$

29. $A_0 = 1200,\ i = \frac{0.06}{2},\ k = 2,\ n = 15$
$$A_n = A_0(1+i)^{kn}$$
$$A_{15} = 1200\left(1+\tfrac{0.06}{2}\right)^{2\cdot 15} \approx \$2912.71$$

31. $A_0 = 1200,\ i = \frac{0.06}{12},\ k = 12,\ n = 15$
$$A_n = A_0(1+i)^{kn}$$
$$A_{15} = 1200\left(1+\tfrac{0.06}{12}\right)^{12\cdot 15} \approx \$2944.91$$

33. $k = 4,\ i = \frac{0.06}{4}$
$$R = (1+i)^k - 1$$
$$= \left(1+\tfrac{0.06}{4}\right)^4 - 1$$
$$\approx 0.0614 = 6.14\%$$

35. $k = 2,\ i = \frac{0.095}{2}$
$$R = (1+i)^k - 1$$
$$= \left(1+\tfrac{0.095}{2}\right)^2 - 1$$
$$\approx 0.0973 = 9.73\%$$

37. $A_6 = 20{,}000;\ i = \frac{0.06}{2},\ k = 2,\ n = 6$
$$A_0 = A_n(1+i)^{-kn}$$
$$= 20{,}000\left(1+\tfrac{0.06}{2}\right)^{-2\cdot 6}$$
$$\approx \$14{,}027.60$$

39. $A_6 = 20{,}000;\ i = \frac{0.09}{12},\ k = 12,\ n = 6$
$$A_0 = A_n(1+i)^{-kn}$$
$$= 20{,}000\left(1+\tfrac{0.09}{12}\right)^{-12\cdot 6}$$
$$\approx \$11{,}678.47$$

41.
$$I = Prt$$
$$175 = (2500)(r)(2)$$
$$\frac{175}{(2500)(2)} = r$$
$$0.035 = r \Rightarrow 3.5\%, \text{ or } 3\tfrac{1}{2}\%$$

43. $A_0 = 7000,\ i = \frac{0.06}{4},\ k = 4,\ n = 18$
$$A_n = A_0(1+i)^{kn}$$
$$A_{18} = 7000\left(1+\tfrac{0.06}{4}\right)^{4\cdot 18} \approx \$20{,}448.11$$

45. $A_0 = 147{,}500;\ i = \frac{0.075}{2},\ k = 2,\ n = 12$

$A_n = A_0(1+i)^{kn}$

$A_{12} = 147500\left(1 + \frac{0.075}{2}\right)^{2\cdot 12} \approx \$356{,}867.13$

47. $A_0 = 137{,}000;\ i = \frac{0.11}{1},\ k = 1,\ n = 4$

$A_n = A_0(1+i)^{kn}$

$A_4 = 137{,}000\left(1 + \frac{0.11}{1}\right)^{1\cdot 4} \approx \$207{,}976$

49. $A_0 = 4.3,\ i = \frac{0.072}{1},\ k = 1,\ n = 10$

$A_n = A_0(1+i)^{kn}$

$A_{10} = 4.3\left(1 + \frac{0.072}{1}\right)^{1\cdot 10} \approx \$8.62$ million ft^3

51.

NOW account

$k = 4,\ i = \frac{0.072}{4}$

$R = (1+i)^k - 1$

$= \left(1 + \frac{0.072}{4}\right)^4 - 1$

$\approx 0.0740 = 7.40\%$

Money Market

$k = 12,\ i = \frac{0.069}{12}$

$R = (1+i)^k - 1$

$= \left(1 + \frac{0.069}{12}\right)^{12} - 1$

$\approx 0.0712 = 7.12\%$

53. $A_0 = 1230,\ i = \frac{0.12}{360},\ k = 360,\ n = 0.25$

$A_n = A_0(1+i)^{kn}$

$A_{0.25} = 1230\left(1 + \frac{0.12}{360}\right)^{360(0.25)} \approx \$1267.45$

55. $A_{1.5} = 4200;\ i = \frac{0.0575}{12},\ k = 12,\ n = 1.5$

$A_0 = A_n(1+i)^{-kn} = 4200\left(1 + \frac{0.0575}{12}\right)^{-12(1.5)}$

$\approx \$3853.73$

57.

First ten years

$A_0 = 12{,}000;\ i = \frac{0.075}{12},\ k = 12,\ n = 10$

$A_n = A_0(1+i)^{kn} = 12{,}000\left(1 + \frac{0.075}{12}\right)^{12(10)}$

$\approx \$25{,}344.78$

Next four years

$A_0 = 37{,}344.78;\ i = \frac{0.075}{12},\ k = 12,\ n = 4$

$A_n = A_0(1+i)^{kn} = 37{,}344.78\left(1 + \frac{0.075}{12}\right)^{12(4)}$

$\approx \$50{,}363.14$

59. Answers may vary.

61. Answers may vary.

63. $\dfrac{x^2 - 2x - 15}{2x^2 - 9x - 5} = \dfrac{(x-5)(x+3)}{(2x+1)(x-5)} = \dfrac{x+3}{2x+1}$

65. $\dfrac{(3-x)(x+3)}{-x^2+9} = \dfrac{(3-x)(x+3)}{9-x^2} = \dfrac{(3-x)(x+3)}{(3+x)(3-x)} = 1$

Exercise 9.2

1. annuities

3. payments, interest

5. regular deposit, periodic rate, frequency of compounding, number of years

7. $P = 100,\ i = \frac{0.06}{1},\ k = 1,\ n = 10$

$$A_n = \frac{P\left[(1+i)^{kn} - 1\right]}{i} = \frac{100\left[\left(1 + \frac{0.06}{1}\right)^{1\cdot 10} - 1\right]}{\frac{0.06}{1}} \approx \$1318.08$$

SECTION 9.2

9. $P = 100,\ i = \frac{0.06}{1},\ k = 1,\ n = 3$

$$A_n = \frac{P\left[(1+i)^{kn} - 1\right]}{i} = \frac{100\left[\left(1+\frac{0.06}{1}\right)^{1\cdot 3} - 1\right]}{\frac{0.06}{1}} \approx \$318.36$$

11. $P = 100,\ i = \frac{0.04}{1},\ k = 1,\ n = 10$

$$A_n = \frac{P\left[(1+i)^{kn} - 1\right]}{i} = \frac{100\left[\left(1+\frac{0.04}{1}\right)^{1\cdot 10} - 1\right]}{\frac{0.04}{1}} \approx \$1200.61$$

13. $P = 100,\ i = \frac{0.095}{1},\ k = 1,\ n = 10$

$$A_n = \frac{P\left[(1+i)^{kn} - 1\right]}{i} = \frac{100\left[\left(1+\frac{0.095}{1}\right)^{1\cdot 10} - 1\right]}{\frac{0.095}{1}} \approx \$1556.03$$

15. $P = 100,\ i = \frac{0.08}{2},\ k = 2,\ n = 15$

$$A_n = \frac{P\left[(1+i)^{kn} - 1\right]}{i} = \frac{100\left[\left(1+\frac{0.08}{2}\right)^{2\cdot 15} - 1\right]}{\frac{0.08}{2}} \approx \$5608.49$$

17. $P = 100,\ i = \frac{0.08}{12},\ k = 12,\ n = 15$

$$A_n = \frac{P\left[(1+i)^{kn} - 1\right]}{i} = \frac{100\left[\left(1+\frac{0.08}{12}\right)^{12\cdot 15} - 1\right]}{\frac{0.08}{12}} \approx \$34{,}603.82$$

19. $A_n = 20{,}000;\ i = \frac{0.04}{1},\ k = 1,\ n = 10$

$$P = \frac{A_n i}{(1+i)^{kn} - 1} = \frac{20{,}000\left(\frac{0.04}{1}\right)}{\left(1+\frac{0.04}{1}\right)^{1\cdot 10} - 1} \approx \$1665.81$$

21. $A_n = 20{,}000;\ i = \frac{0.09}{2},\ k = 2,\ n = 10$

$$P = \frac{A_n i}{(1+i)^{kn} - 1} = \frac{20{,}000\left(\frac{0.09}{2}\right)}{\left(1+\frac{0.09}{2}\right)^{2\cdot 10} - 1} \approx \$637.52$$

23. $P = 200,\ i = \frac{0.06}{12},\ k = 12,\ n = 1$

$$A_n = \frac{P\left[(1+i)^{kn} - 1\right]}{i} = \frac{200\left[\left(1+\frac{0.06}{12}\right)^{12\cdot 1} - 1\right]}{\frac{0.06}{12}} \approx \$2467.11$$

25. $P = 135{,}000;\ i = \frac{0.087}{12},\ k = 12,\ n = 20$

$$A_n = \frac{P\left[(1+i)^{kn} - 1\right]}{i} = \frac{135{,}000\left[\left(1+\frac{0.087}{12}\right)^{12\cdot 20} - 1\right]}{\frac{0.087}{12}} \approx \$86{,}803{,}923.58$$

27. $A_n = 750{,}000;\ i = \frac{0.0975}{12},\ k = 12,\ n = 2.5$

$$P = \frac{A_n i}{(1+i)^{kn} - 1} = \frac{750{,}000\left(\frac{0.0975}{12}\right)}{\left(1+\frac{0.0975}{12}\right)^{12\cdot 2.5} - 1} \approx \$22{,}177.71$$

29.

Bank A	Bank B
$A_n = 10{,}000;\ i = \frac{0.055}{1},\ k = 1,\ n = 20$	$A_n = 10{,}000;\ i = \frac{0.0535}{12},\ k = 12,\ n = 20$

Bank A:

$$P = \frac{A_n i}{(1+i)^{kn} - 1} = \frac{10{,}000\left(\frac{0.055}{1}\right)}{\left(1+\frac{0.055}{1}\right)^{1\cdot 20} - 1} \approx \$286.79 \text{ per year}$$

Bank B:

$$P = \frac{A_n i}{(1+i)^{kn} - 1} = \frac{10{,}000\left(\frac{0.0535}{12}\right)}{\left(1+\frac{0.0535}{12}\right)^{12\cdot 20} - 1} \approx \$23.36 \text{ per month, or } \$280.32 \text{ per year}$$

Bank B requires the lower annual contributions.

31.

First 15 years	Next 15 years
$P = 100,\ i = \frac{0.08}{12},\ k = 12,\ n = 15$	$P = 34{,}603.82,\ i = \frac{0.08}{12},\ k = 12,\ n = 15$

First 15 years:

$$A_n = \frac{P\left[(1+i)^{kn} - 1\right]}{i} = \frac{100\left[\left(1+\frac{0.08}{12}\right)^{12\cdot 15} - 1\right]}{\frac{0.08}{12}} \approx \$34.603.82$$

Next 15 years:

$$A_n = P(1+i)^{kn} = 34{,}603.82\left(1+\frac{0.08}{12}\right)^{12\cdot 15} \approx \$114{,}432.12$$

33.

Last 5 years	First 5 years
$A_n = 13{,}500;\ i = \frac{0.09}{1},\ k = 1,\ n = 5$	$A_n = 8774.07;\ i = \frac{0.09}{1},\ k = 1,\ n = 5$

Last 5 years:

$$PV = A_n(1+i)^{-kn} = 13{,}500\left(1+\frac{0.09}{1}\right)^{-1(5)} \approx \$8774.07$$

First 5 years:

$$P = \frac{A_n i}{(1+i)^{kn} - 1} = \frac{8774.07\left(\frac{0.09}{1}\right)}{\left(1+\frac{0.09}{1}\right)^{1\cdot 5} - 1} \approx \$1466.08 \text{ per year}$$

35.

$$\begin{aligned} \frac{2(5x-12)}{x} &= 8 \\ 2(5x-12) &= 8x \\ 10x - 24 &= 8x \\ 2x &= 24 \\ x &= 12 \end{aligned}$$

37.

$$\begin{aligned} \sqrt{2x+3} &= 3 \\ \left(\sqrt{2x+3}\right)^2 &= 3^2 \\ 2x+3 &= 9 \\ 2x &= 6 \\ x &= 3 \end{aligned}$$

Exercise 9.3

1. present value **3.** promissory note **5.** amortizing

7. $P = 3500,\ k = 1,\ i = \dfrac{0.0525}{1},\ n = 25$

$$A_0 = \frac{P\left[1-(1+i)^{-kn}\right]}{i} = \frac{3500\left[1-\left(1+\frac{0.0525}{1}\right)^{-1\cdot 25}\right]}{\frac{0.0525}{1}} \approx \$48{,}116.14$$

SECTION 9.3

9. $A = 25{,}000;\ k = 12, i = \dfrac{0.12}{12}, n = 15$

$$P = \frac{Ai}{1-(1+i)^{-kn}} = \frac{25{,}000\left(\frac{0.12}{12}\right)}{1-\left(1+\frac{0.12}{12}\right)^{-12\cdot 15}} \approx \$300.04$$

11. $P = 700, k = 4, i = \dfrac{0.0625}{4}, n = 15$

$$A_0 = \frac{P\left[1-(1+i)^{-kn}\right]}{i} = \frac{700\left[1-\left(1+\frac{0.0625}{4}\right)^{-4\cdot 15}\right]}{\frac{0.0625}{4}} \approx \$27{,}128.43$$

13. $P = 12{,}000;\ k = 1, i = \dfrac{0.085}{1}, n = 15$

$$A_0 = \frac{P\left[1-(1+i)^{-kn}\right]}{i} = \frac{12{,}000\left[1-\left(1+\frac{0.085}{1}\right)^{-1\cdot 15}\right]}{\frac{0.085}{1}} \approx \$99{,}650.84$$

15. $A = 21{,}700;\ k = 12, i = \dfrac{0.084}{12}, n - 4$

$$P = \frac{Ai}{1-(1+i)^{-kn}} = \frac{21{,}700\left(\frac{0.084}{12}\right)}{1-\left(1+\frac{0.084}{12}\right)^{-12\cdot 4}} \approx \$533.84$$

17.

15-year

$A = 130{,}000;\ k = 12, i = \dfrac{0.12}{12}, n = 15$

$$P = \frac{Ai}{1-(1+i)^{-kn}} = \frac{130{,}000\left(\frac{0.12}{12}\right)}{1-\left(1+\frac{0.12}{12}\right)^{-12\cdot 15}} \approx \$1560.22$$

20-year

$A = 130{,}000;\ k = 12, i = \dfrac{0.11}{12}, n = 20$

$$P = \frac{Ai}{1-(1+i)^{-kn}} = \frac{130{,}000\left(\frac{0.11}{12}\right)}{1-\left(1+\frac{0.11}{12}\right)^{-12\cdot 20}} \approx \$1341.84$$

19.

From 55 to 80

$P = 5000;\ k = 12, i = \frac{0.0875}{12}, n = 25$

$$A_0 = \frac{P\left[1-(1+i)^{-kn}\right]}{i} = \frac{5000\left[1-\left(1+\frac{0.0875}{12}\right)^{-12\cdot 25}\right]}{\frac{0.0875}{12}} \approx \$608{,}166.24$$

From 20 to 55

$A_n = 608{,}166.24;\ i = \frac{0.0875}{12}, k = 12, n = 35$

$$P = \frac{A_n i}{(1+i)^{kn}-1} = \frac{608{,}166.24\left(\frac{0.0875}{12}\right)}{\left(1+\frac{0.0875}{12}\right)^{12\cdot 35}-1} \approx \$220.13$$

21.

First ten years

$$A_n = \frac{500\left[\left(1+\frac{0.0725}{4}\right)^{4\cdot 10}-1\right]}{\frac{0.0725}{4}} \approx \$29{,}003.32$$

Next ten years (2 parts)

$$A_n = \frac{1500\left[\left(1+\frac{0.0725}{4}\right)^{4\cdot 10}-1\right]}{\frac{0.0725}{4}} \approx \$87{,}009.96$$

$$A_n = 29{,}003.32\left(1+\frac{0.0725}{4}\right)^{4(10)} \approx \$59{,}496.55$$

Total $= \$87{,}009.96 + \$59{,}496.55 = \$146{,}506.51$

23. $\dfrac{6\sqrt{30}}{3\sqrt{5}} = \dfrac{6}{3} \cdot \dfrac{\sqrt{30}}{\sqrt{5}} = 2\sqrt{\dfrac{30}{5}} = 2\sqrt{6}$

25. $3\sqrt{5x} + 5\sqrt{20x} = 3\sqrt{5x} + 5\sqrt{4}\sqrt{5x} = 3\sqrt{5x} + 10\sqrt{5x} = 13\sqrt{5x}$

Chapter 9 Review

1. $I = Prt = 2000(0.09)(5) = 900$
$2000 + 900 = \$2900$

2. $A_n = 2000\left(1 + \dfrac{0.09}{1}\right)^{1 \cdot 5} \approx \$3077.25$

3. $A_n = 2350\left(1 + \frac{0.14}{360}\right)^{360(1/6)} \approx \$2405.47$

4. $A_n = 2000\left(1 + \dfrac{0.076}{4}\right)^{4 \cdot 16} \approx \$6670.80$

5. BigBank: $R = \left(1 + \dfrac{0.063}{4}\right)^4 - 1 = 0.0645$ BestBank: $R = \left(1 + \dfrac{0.0621}{365}\right)^{365} - 1 = 0.0641$
BigBank has the better return.

6. $A_6 = 7900;\ i = \frac{0.0575}{2},\ k = 2,\ n = 6$
$A_0 = A_n(1+i)^{-kn}$
$= 7900\left(1 + \frac{0.0575}{2}\right)^{-2 \cdot 6} \approx \$5622.23$

7. $A_n = \dfrac{500\left[\left(1 + \frac{0.05}{1}\right)^{1 \cdot 13} - 1\right]}{\frac{0.05}{1}}$
$\approx \$8856.49$

8. $A_n = \dfrac{150\left[\left(1 + \frac{0.08}{12}\right)^{12 \cdot 20} - 1\right]}{\frac{0.08}{12}}$
$\approx \$88{,}353.06$

9. $P = \dfrac{40{,}700\left(\frac{0.075}{12}\right)}{\left(1 + \frac{0.075}{12}\right)^{12 \cdot 7} - 1} \approx \$369.89$

10. $A_0 = \dfrac{250\left[1 - \left(1 + \frac{0.065}{2}\right)^{-2 \cdot 20}\right]}{\frac{0.065}{2}}$
$\approx \$5552.11$

11. $A_0 = \dfrac{50{,}000\left[1 - \left(1 + \frac{0.096}{1}\right)^{-1 \cdot 20}\right]}{\frac{0.096}{1}}$
$\approx \$437{,}563.50$

12. $P = \dfrac{150{,}500\left(\frac{0.1075}{12}\right)}{1 - \left(1 + \frac{0.1075}{12}\right)^{-12 \cdot 15}} \approx \$1687.03$; Total paid $= 1687.03 \cdot 180 = \$303{,}665.40$

13. $P = \dfrac{150{,}500\left(\frac{0.1075}{12}\right)}{1 - \left(1 + \frac{0.1075}{12}\right)^{-12 \cdot 30}} \approx \$1404.89$; Total paid $= 1404.89 \cdot 360 = \$505{,}760.40$

Chapter 9 Test

1. compound **2.** periodic **3.** effective **4.** annual

5. annuities **6.** sinking fund **7.** present value **8.** amortizing

9. $I = Prt = 1300(0.05)(10) = 650$
$1300 + 650 = \$1950$

10. $A_n = 1300\left(1 + \frac{0.05}{1}\right)^{1\cdot 10} \approx \$2117.56$

11. $A_n = 1300\left(1 + \frac{0.05}{12}\right)^{12\cdot 10} \approx \$2141.11$

12. $R = \left(1 + \frac{0.05}{12}\right)^{12} - 1 = 0.05116$
$= 5.116\%$

13. $A_0 = 5000\left(1 + \frac{0.07}{4}\right)^{-4\cdot 10} \approx \$2498.00$

14. $$A_n = \frac{700\left[\left(1 + \frac{0.073}{12}\right)^{12\cdot 5} - 1\right]}{\frac{0.073}{12}}$$
$\approx \$50{,}506.10$

15. $$P = \frac{8000\left(\frac{0.065}{12}\right)}{\left(1 + \frac{0.065}{12}\right)^{12\cdot 5} - 1} \approx \$113.20$$

16. $$A_0 = \frac{1000\left[1 - \left(1 + \frac{0.068}{12}\right)^{-12\cdot 15}\right]}{\frac{0.068}{12}}$$
$\approx \$112{,}652.71$

17. $$P = \frac{90{,}000\left(\frac{0.0895}{12}\right)}{1 - \left(1 + \frac{0.0895}{12}\right)^{-12\cdot 15}} \approx \$910.16$$

Cumulative Review Exercises

1. $$\begin{cases} 2x + y = 8 \\ x - 2y = -1 \end{cases}$$

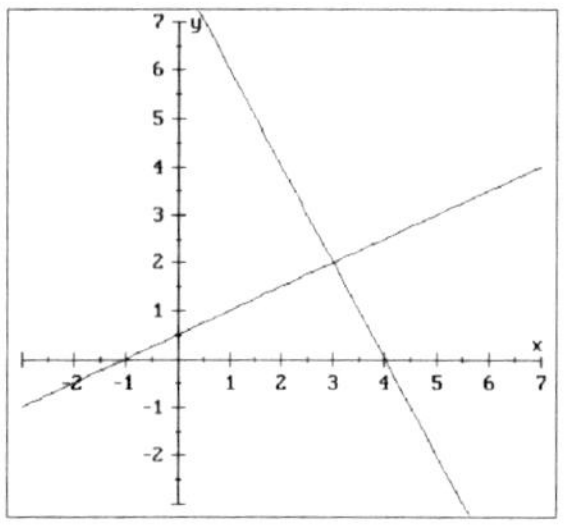

(3, 2)

2. $$\begin{cases} 3x = -y + 2 \\ y + x - 4 = -2x \end{cases}$$

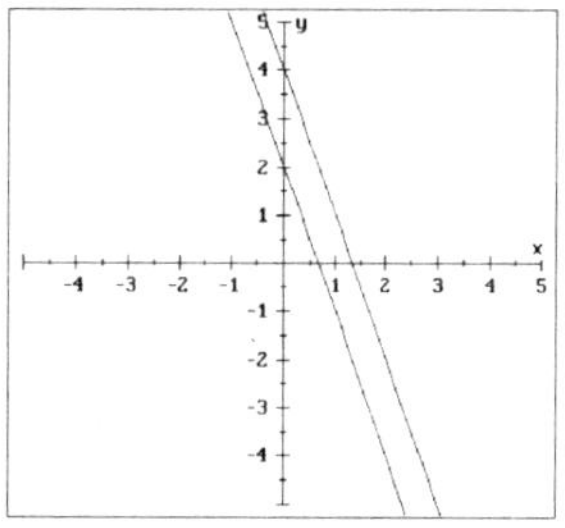

no solution, inconsistent system

3.
$$\begin{array}{lll} 5x = 3y + 12 \Rightarrow 5x - 3y = 12 \Rightarrow & 5x - 3y = 12 & 2x - 3y = 3 \\ \underline{2x - 3y = 3} \quad \Rightarrow \underline{2x - 3y = 3} \Rightarrow & \underline{-2x + 3y = -3} & 2(3) - 3y = 3 \\ & 3x = 9 & -3y = -3 \\ & x = 3 & y = 1 \end{array}$$

Solution: $\boxed{x = 3, y = 1}$

CUMULATIVE REVIEW EXERCISES

4. $$\begin{bmatrix} x \\ y \\ z \end{bmatrix} = \begin{bmatrix} 2 & 1 & -1 \\ 1 & -1 & 1 \\ 1 & 1 & -3 \end{bmatrix}^{-1} \begin{bmatrix} 7 \\ 2 \\ 2 \end{bmatrix} = \begin{bmatrix} 3 \\ 2 \\ 1 \end{bmatrix}$$

5. $$\left[\begin{array}{ccc|c} 2 & 1 & -1 & 0 \\ 1 & -1 & 1 & 3 \\ 1 & 1 & -3 & -5 \end{array}\right] \Rightarrow \left[\begin{array}{ccc|c} 1 & 1 & -3 & -5 \\ 1 & -1 & 1 & 3 \\ 2 & 1 & -1 & 0 \end{array}\right] \Rightarrow \left[\begin{array}{ccc|c} 1 & 1 & -3 & -5 \\ 0 & -2 & 4 & 8 \\ 0 & -1 & 5 & 10 \end{array}\right] \Rightarrow$$

$R_1 \Leftrightarrow R_3$

$-R_1 + R_2 \Rightarrow R_2$
$-2R_1 + R_3 \Rightarrow R_3$

$$\left[\begin{array}{ccc|c} 1 & 0 & -1 & -1 \\ 0 & 1 & -2 & -4 \\ 0 & 0 & 3 & 6 \end{array}\right] \Rightarrow \left[\begin{array}{ccc|c} 1 & 0 & 0 & 1 \\ 0 & 1 & 0 & 0 \\ 0 & 0 & 1 & 2 \end{array}\right]$$

Solution: $\boxed{x = 1, y = 0, z = 2}$

$\frac{1}{2}R_2 + R_1 \Rightarrow R_1$
$-\frac{1}{2}R_2 + R_3 \Rightarrow R_3$
$-\frac{1}{2}R_2 \Rightarrow R_2$

$\frac{1}{3}R_3 + R_1 \Rightarrow R_1$
$\frac{2}{3}R_3 + R_2 \Rightarrow R_2$
$\frac{1}{3}R_3 \Rightarrow R_3$

6. $$\left[\begin{array}{cccc|c} 2 & -2 & 3 & 1 & 2 \\ 1 & 1 & 1 & 1 & 5 \\ -1 & 2 & -3 & 2 & 2 \\ 1 & 1 & 2 & -1 & 4 \end{array}\right] \Rightarrow \left[\begin{array}{cccc|c} 2 & -2 & 3 & 1 & 2 \\ 0 & -4 & 1 & -1 & -8 \\ 0 & 2 & -3 & 5 & 6 \\ 0 & -4 & -1 & 3 & -6 \end{array}\right] \Rightarrow \left[\begin{array}{cccc|c} -4 & 0 & -5 & -3 & -12 \\ 0 & -4 & 1 & -1 & -8 \\ 0 & 0 & -5 & 9 & 4 \\ 0 & 0 & 2 & -4 & -2 \end{array}\right]$$

$-2R_2 + R_1 \Rightarrow R_2$
$2R_3 + R_1 \Rightarrow R_3$
$-2R_4 + R_1 \Rightarrow R_4$

$-2R_1 + R_2 \Rightarrow R_1$
$2R_3 + R_2 \Rightarrow R_3$
$-R_4 + R_2 \Rightarrow R_4$

$$\left[\begin{array}{cccc|c} -4 & 0 & -5 & -3 & -12 \\ 0 & -4 & 1 & -1 & -8 \\ 0 & 0 & -5 & 9 & 4 \\ 0 & 0 & 1 & -2 & -1 \end{array}\right] \Rightarrow \left[\begin{array}{cccc|c} 4 & 0 & 0 & 12 & 16 \\ 0 & -20 & 0 & 4 & -36 \\ 0 & 0 & -5 & 9 & 4 \\ 0 & 0 & 0 & -1 & -1 \end{array}\right] \Rightarrow \left[\begin{array}{cccc|c} 1 & 0 & 0 & 3 & 4 \\ 0 & -5 & 0 & 1 & -9 \\ 0 & 0 & -5 & 9 & 4 \\ 0 & 0 & 0 & 1 & 1 \end{array}\right]$$

$\frac{1}{2}R_4 \Rightarrow R_4$

$-R_1 + R_3 \Rightarrow R_1$
$5R_2 + R_3 \Rightarrow R_2$
$5R_4 + R_3 \Rightarrow R_4$

$\frac{1}{4}R_1 \Rightarrow R_1$
$\frac{1}{4}R_2 \Rightarrow R_2$
$-R_4 \Rightarrow R_4$

$$\Rightarrow \left[\begin{array}{cccc|c} 1 & 0 & 0 & 0 & 1 \\ 0 & -5 & 0 & 0 & -10 \\ 0 & 0 & -5 & 0 & -5 \\ 0 & 0 & 0 & 1 & 1 \end{array}\right] \Rightarrow \left[\begin{array}{cccc|c} 1 & 0 & 0 & 0 & 1 \\ 0 & 1 & 0 & 0 & 2 \\ 0 & 0 & 1 & 0 & 1 \\ 0 & 0 & 0 & 1 & 1 \end{array}\right]$$

Solution:
$\boxed{x = 1, y = 2, z = 1, t = 1}$

$-3R_4 + R_1 \Rightarrow R_1$
$-R_4 + R_2 \Rightarrow R_2$
$-9R_4 + R_3 \Rightarrow R_3$

$-\frac{1}{5}R_2 \Rightarrow R_2$
$-\frac{1}{5}R_3 \Rightarrow R_3$

7. $$\begin{bmatrix} 2 & 1 \\ 1 & 4 \end{bmatrix} + \begin{bmatrix} -1 & 2 \\ 2 & 3 \end{bmatrix} = \begin{bmatrix} 1 & 3 \\ 3 & 7 \end{bmatrix}$$

8. $$\begin{bmatrix} -1 & 2 \\ 2 & 3 \end{bmatrix} - \begin{bmatrix} 2 & 1 \\ 1 & 4 \end{bmatrix} = \begin{bmatrix} -3 & 1 \\ 1 & -1 \end{bmatrix}$$

9. $$\begin{bmatrix} 2 & 1 \\ 1 & 4 \end{bmatrix} \begin{bmatrix} 2 & 0 & -1 \\ -1 & 2 & 2 \end{bmatrix} = \begin{bmatrix} 3 & 2 & 0 \\ -2 & 8 & 7 \end{bmatrix}$$

CUMULATIVE REVIEW EXERCISES

10. $\begin{bmatrix}-1 & 2\\ 2 & 3\end{bmatrix}^2+2\begin{bmatrix}2 & 1\\ 1 & 4\end{bmatrix}=\begin{bmatrix}-1 & 2\\ 2 & 3\end{bmatrix}\begin{bmatrix}-1 & 2\\ 2 & 3\end{bmatrix}+\begin{bmatrix}4 & 2\\ 2 & 8\end{bmatrix}=\begin{bmatrix}5 & 4\\ 4 & 13\end{bmatrix}+\begin{bmatrix}4 & 2\\ 2 & 8\end{bmatrix}=\begin{bmatrix}9 & 6\\ 6 & 21\end{bmatrix}$

11. $\left[\begin{array}{cc|cc}2 & 6 & 1 & 0\\ 2 & 4 & 0 & 1\end{array}\right]\Rightarrow\left[\begin{array}{cc|cc}1 & 3 & \frac{1}{2} & 0\\ 2 & 4 & 0 & 1\end{array}\right]\Rightarrow\left[\begin{array}{cc|cc}1 & 3 & \frac{1}{2} & 0\\ 0 & -2 & -1 & 1\end{array}\right]\Rightarrow$

$\frac{1}{2}R_1\Rightarrow R_1$ $\quad -2R_1+R_2\Rightarrow R_2$

$\left[\begin{array}{cc|cc}1 & 3 & \frac{1}{2} & 0\\ 0 & 1 & \frac{1}{2} & -\frac{1}{2}\end{array}\right]\Rightarrow\left[\begin{array}{cc|cc}1 & 0 & -1 & \frac{3}{2}\\ 0 & 1 & \frac{1}{2} & -\frac{1}{2}\end{array}\right]\Rightarrow$ Inverse: $\begin{bmatrix}-1 & \frac{3}{2}\\ \frac{1}{2} & -\frac{1}{2}\end{bmatrix}$

$-\frac{1}{2}R_2\Rightarrow R_2$ $\quad -3R_2+R_1\Rightarrow R_1$

12. $\left[\begin{array}{ccc|ccc}1 & -1 & 1 & 1 & 0 & 0\\ 1 & 4 & 0 & 0 & 1 & 0\\ 2 & 4 & 1 & 0 & 0 & 1\end{array}\right]\Rightarrow\left[\begin{array}{ccc|ccc}1 & -1 & 1 & 1 & 0 & 0\\ 0 & 5 & -1 & -1 & 1 & 0\\ 0 & 6 & -1 & -2 & 0 & 1\end{array}\right]\Rightarrow\left[\begin{array}{ccc|ccc}1 & -1 & 1 & 1 & 0 & 0\\ 0 & 1 & 0 & -1 & -1 & 1\\ 0 & 6 & -1 & -2 & 0 & 1\end{array}\right]$

$-R_1+R_2\Rightarrow R_2$ $\quad -R_2+R_3\Rightarrow R_2$

$-2R_2+R_3\Rightarrow R_3$

$\Rightarrow\left[\begin{array}{ccc|ccc}1 & 0 & 1 & 0 & -1 & 1\\ 0 & 1 & 0 & -1 & -1 & 1\\ 0 & 0 & -1 & 4 & 6 & -5\end{array}\right]\Rightarrow\left[\begin{array}{ccc|ccc}1 & 0 & 0 & 4 & 5 & -4\\ 0 & 1 & 0 & -1 & -1 & 1\\ 0 & 0 & 1 & -4 & -6 & 5\end{array}\right]\Rightarrow \text{Inv}=\begin{bmatrix}4 & 5 & -4\\ -1 & -1 & 1\\ -4 & -6 & 5\end{bmatrix}$

$R_2+R_1\Rightarrow R_1$ $\quad R_3+R_1\Rightarrow R_1$

$-6R_2+R_3\Rightarrow R_3$ $\quad -R_3\Rightarrow R_3$

13. $\begin{vmatrix}-3 & 5\\ 4 & 7\end{vmatrix}=(-3)(7)-(5)(4)=-21-20=-41$

14. $\begin{vmatrix}2 & -3 & 2\\ 0 & 1 & -1\\ 1 & -2 & 1\end{vmatrix}=2\begin{vmatrix}1 & -1\\ -2 & 1\end{vmatrix}-(-3)\begin{vmatrix}0 & -1\\ 1 & 1\end{vmatrix}+2\begin{vmatrix}0 & 1\\ 1 & -2\end{vmatrix}$

$=2(-1)+3(1)+2(-1)=-2+3-2=-1$

15. $x=\dfrac{\begin{vmatrix}11 & 3\\ 24 & 5\end{vmatrix}}{\begin{vmatrix}4 & 3\\ -2 & 5\end{vmatrix}}$

16. $y=\dfrac{\begin{vmatrix}4 & 11\\ -2 & 24\end{vmatrix}}{\begin{vmatrix}4 & 3\\ -2 & 5\end{vmatrix}}$

17. $\dfrac{-x+1}{(x+1)(x+2)}=\dfrac{A}{x+1}+\dfrac{B}{x+2}$

$=\dfrac{A(x+2)}{(x+1)(x+2)}+\dfrac{B(x+1)}{(x+1)(x+2)}$

$=\dfrac{Ax+2A+Bx+B}{(x+1)(x+2)}$

$=\dfrac{(A+B)x+(2A+B)}{(x+1)(x+2)}$

$\begin{cases}A+B=-1\\ 2A+B=1\end{cases}\Rightarrow A=2,\ B=-3$

$\dfrac{-x+1}{(x+1)(x+2)}=\dfrac{2}{x+1}-\dfrac{3}{x+2}$

18.
$$\frac{x-4}{(2x-5)^2}=\frac{A}{2x-5}+\frac{B}{(2x-5)^2}$$
$$=\frac{A(2x-5)}{(2x-5)^2}+\frac{B}{(2x-5)^2}$$
$$=\frac{2Ax-5A+B}{(2x-5)^2}$$
$$=\frac{2Ax+(-5A+B)}{(2x-5)^2}$$

$$\begin{cases} 2A = 1 \\ -5A+B=-4 \end{cases} \Rightarrow \begin{matrix} A=\frac{1}{2} \\ B=-\frac{3}{2} \end{matrix}$$
$$\frac{x-4}{(2x-5)^2}=\frac{\frac{1}{2}}{2x-5}-\frac{\frac{3}{2}}{(2x-5)^2}$$

19. $y \le 2x+6$

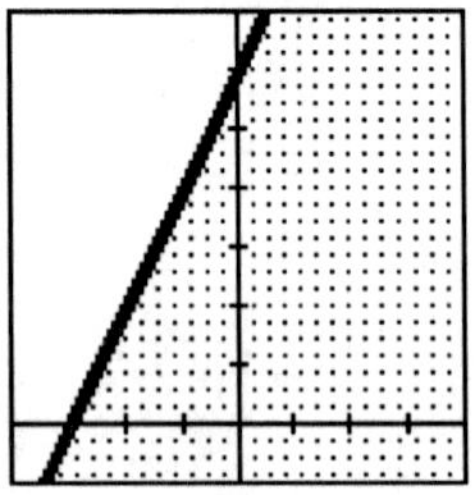

20. $\begin{cases} 2x+3y \ge 6 \\ 2x-3y \le 6 \end{cases}$

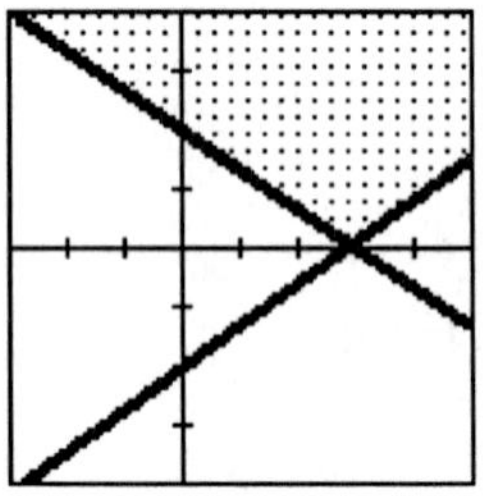

21.
$$(x-h)^2+(y-k)^2=r^2$$
$$(x-0)^2+(y-0)^2=4^2$$
$$x^2+y^2=16$$

22.
$$(x-h)^2+(y-k)^2=r^2$$
$$(x-2)^2+(y+3)^2=11^2$$
$$(x-2)^2+(y+3)^2=121$$

23.
$$x^2+y^2-4y=12$$
$$x^2+y^2-4y+4=12+4$$
$$(x-0)^2+(y-2)^2=4^2$$
$$C(0,2),\ r=4$$

24.
$$x^2-2y-2x=-7$$
$$x^2-2x+1=2y-7+1$$
$$(x-1)^2=2(y-3)$$
$$V(1,3)$$

CUMULATIVE REVIEW EXERCISES

25.
$$x^2 + 4y^2 + 2x = 3$$
$$x^2 + 2x + 1 + 4y^2 = 3 + 1$$
$$(x+1)^2 + 4y^2 = 4$$
$$\frac{(x+1)^2}{4} + \frac{y^2}{1} = 1$$

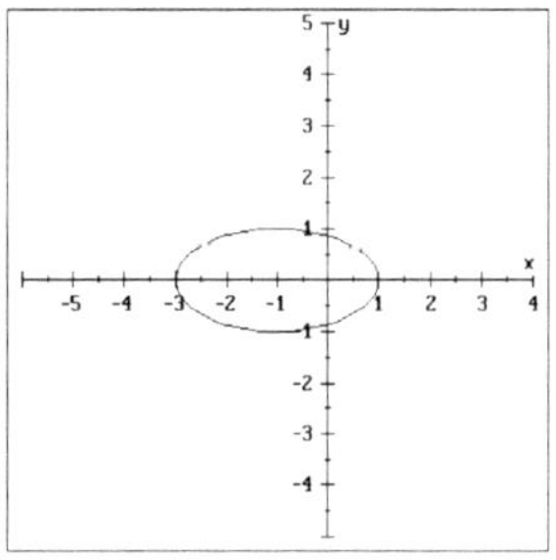

26.
$$x^2 - 9y^2 - 4x = 5$$
$$x^2 - 4x + 4 - 9y^2 = 5 + 4$$
$$(x-2)^2 - 9y^2 = 9$$
$$\frac{(x-2)^2}{9} - \frac{y^2}{1} = 1$$

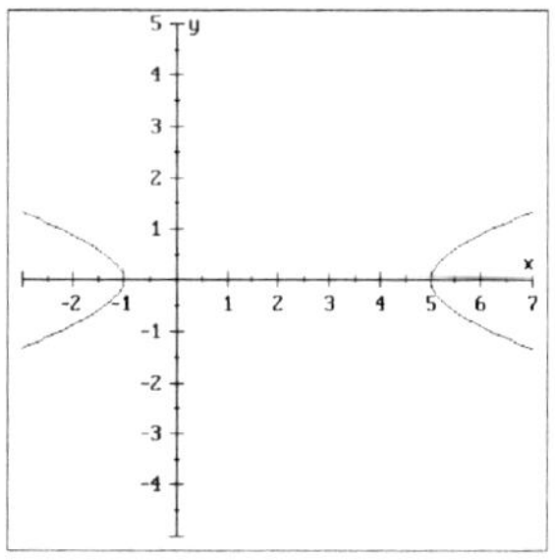

27. $a = 6, b = 4$
$$\frac{x^2}{36} + \frac{y^2}{16} = 1$$

28. $a = 5, c = 2$
$$b^2 = a^2 - c^2 = 25 - 4 = 21$$
$$\frac{(x-2)^2}{21} + \frac{(y-3)^2}{25} = 1$$

29. $a = 2, c = 3$
$$b^2 = c^2 - a^2 = 9 - 4 = 5$$
$$\frac{x^2}{4} - \frac{y^2}{5} = 1$$

30. $2a = 6, 2b = 6 \Rightarrow a = 3, b = 3$
$$\frac{(y-4)^2}{9} - \frac{(x-2)^2}{9} = 1$$

31. The 2nd term will involve $(2y)^1$.
$$\frac{8!}{7!1!}x^7(2y)^1 = 16x^7y$$

32. The 6th term will involve $(2y)^5$.
$$\frac{8!}{3!5!}x^3(2y)^5 = 1792x^3y^5$$

33. $\displaystyle\sum_{k=1}^{5} 2 = 5(2) = 10$

34. $\displaystyle\sum_{k=2}^{6}(3k+1) = 3\sum_{k=2}^{6} k + \sum_{k=2}^{6} 1 = 3(2+3+4+5+6) + 5(1) = 60 + 5 = 65$

35. $a = -2, d = 3$
$$a_6 = a + (n-1)d = -2 + 5(3) = 13$$
$$S_6 = \frac{n(a + a_6)}{2} = \frac{6(-2+13)}{2} = 33$$

36. $a = \frac{1}{9}, r = 3, n = 6$
$$S_6 = \frac{a - ar^n}{1 - r} = \frac{\frac{1}{9} - \frac{1}{9}(3)^6}{1 - 3} = \frac{-\frac{728}{9}}{-2} = \frac{364}{9}$$

37. $P(8,4) = \dfrac{8!}{(8-4)!} = \dfrac{8!}{4!} = 1680$

38. $P(24,0) = \dfrac{24!}{(24-0)!} = \dfrac{24!}{24!} = 1$

CUMULATIVE REVIEW EXERCISES

39. $C(12, 10) = \dfrac{12!}{10!2!} = 66$

40. $P(4, 4) \cdot C(6, 6) = 24 \cdot 1 = 24$

41. $4! \cdot 6! = 17{,}280$

42. $\dbinom{12}{4} = \dfrac{12!}{4!8!} = 495$

43. $\dfrac{2}{36} = \dfrac{1}{18}$

44. $\dfrac{\binom{26}{5}}{\binom{52}{5}} = \dfrac{65{,}780}{2{,}598{,}960} = \dfrac{253}{9996}$

45. $(0.6)(0.8) = 0.48$

46.

Check $n = 1$: $3(1) + 1 \stackrel{?}{=} \dfrac{1(3(1) + 5)}{2}$ True for $n = 1$

$4 = 4$

Assume for $n = k$ and show for $n = k + 1$:

$$4 + 7 + 10 + \cdots + (3k + 1) = \frac{k(3k + 5)}{2}$$

$$\boxed{4 + 7 + 10 + \cdots + (3k + 1)} + 3(k + 1) + 1 = \boxed{\frac{k(3k + 5)}{2}} + 3(k + 1) + 1$$

$$4 + 7 + 10 + \cdots + (3(k + 1) + 1) = \frac{k(3k + 5)}{2} + \frac{2 \cdot (3(k + 1) + 1)}{2}$$

$$4 + 7 + 10 + \cdots + (3(k + 1) + 1) = \frac{3k^2 + 5k + 6k + 6 + 2}{2}$$

$$4 + 7 + 10 + \cdots + (3(k + 1) + 1) = \frac{3k^2 + 11k + 8}{2}$$

$$4 + 7 + 10 + \cdots + (3(k + 1) + 1) = \frac{(k + 1)(3k + 8)}{2}$$

$$4 + 7 + 10 + \cdots + (3(k + 1) + 1) = \frac{(k + 1)(3(k + 1) + 5)}{2}$$

Since this is what results when $n = k + 1$ in the formula, we have shown that the formula works for $n = k + 1$ if it works for $n = k$.

47. $A_0 = 10{,}000\left(1 + \dfrac{0.085}{1}\right)^{-1 \cdot 12}$
$\approx \$3757.02$

48. $P = \dfrac{110{,}000\left(\frac{0.0875}{12}\right)}{1 - \left(1 + \frac{0.0875}{12}\right)^{-12 \cdot 20}} \approx \$972.08$

CPSIA information can be obtained
at www.ICGtesting.com
Printed in the USA
FFOW03n0956270816
27204FF